Birkhäuser Classics

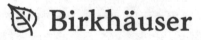

Birkhäuser

Modern Birkhäuser Classics

Many of the original research and survey monographs in pure and applied mathematics, as well as textbooks, published by Birkhäuser in recent decades have been groundbreaking and have come to be regarded as foundational to the subject. Through the MBC Series, a select number of these modern classics, entirely uncorrected, are being re-released in paperback (and as eBooks) to ensure that these treasures remain accessible to new generations of students, scholars, and researchers.

Werner O. Amrein · Anne Boutet de Monvel · Vladimir Georgescu

C_0-Groups, Commutator Methods and Spectral Theory of N-Body Hamiltonians

Reprint of the 1996 Edition

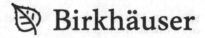 Birkhäuser

Werner O. Amrein
Ecole de Physique
Université de Genève
Genève, Switzerland

Vladimir Georgescu
Département de Mathématiques
Université de Cergy-Pontoise
Cergy-Pontoise, France

Anne Boutet de Monvel
Institut de Mathématiques de Jussieu
Université Paris Diderot
Paris, France

ISBN 978-3-0348-0732-6   ISBN 978-3-0348-0733-3   (eBook)
DOI 10.1007/978-3-0348-0733-3
Springer Basel Heidelberg New York Dordrecht London

Library of Congress Control Number: 2013953218

Mathematics Subject Classification (2010): 46L60, 47-02, 46-02, 47Dxx, 47N50, 81U10

© Springer Basel 1996
Reprint of the 1st edition 1996 by Birkhäuser Verlag, Switzerland
Originally published as volume 135 in the Progress in Mathematics series

Cover design: deblik, Berlin

Printed on acid-free paper

Springer Basel is part of Springer Science+Business Media
(www.birkhauser-science.com)

Contents

Chapter 1
Some Spaces of Functions and Distributions

Chapter 2
Real Interpolation of Banach Spaces

Chapter 3
C_0-Groups and Functional Calculi

Chapter 4
Some Examples of C_0-Groups

Chapter 5
Automorphisms Associated to C_0-Representations

Chapter 6
Unitary Representations and Regularity

Chapter 7
The Conjugate Operator Method

Chapter 8
An Algebraic Framework for the Many-Body Problem

Chapter 9
Spectral Theory of N-Body Hamiltonians

Chapter 10
Quantum-Mechanical N-Body Systems

To our daughters

Vera, Violaine, Sonia, Tiphaine

Koyu, le religieux, dit:

*seule une personne de compréhension réduite
désire arranger les choses en séries complètes.*

*C'est l'incomplétude qui est désirable.
En tout, mauvaise est la régularité.*

*Dans les palais d'autrefois, on laissait toujours
un bâtiment inachevé, obligatoirement.*

(Tsuredzure Gusa, par Yoshida No Kaneyoshi, XIVème siècle)

Preface

The relevance of commutator methods in spectral and scattering theory has been known for a long time, and numerous interesting results have been obtained by such methods. The reader may find a description and references in the books by Putnam [Pu], Reed-Simon [RS] and Baumgärtel-Wollenberg [BW] for example. A new point of view emerged around 1979 with the work of E. Mourre in which the method of locally conjugate operators was introduced. His idea proved to be remarkably fruitful in establishing detailed spectral properties of N-body Hamiltonians. A problem that was considered extremely difficult before that time, the proof of the absence of a singularly continuous spectrum for such operators, was then solved in a rather straightforward manner (by E. Mourre himself for $N = 3$ and by P. Perry, I. Sigal and B. Simon for general N). The Mourre estimate, which is the main input of the method, also has consequences concerning the behaviour of N-body systems at large times. A deeper study of such propagation properties allowed I. Sigal and A. Soffer in 1985 to prove existence and completeness of wave operators for N-body systems with short range interactions without implicit conditions on the potentials (for $N = 3$, similar results were obtained before by means of purely time-dependent methods by V. Enss and by K. Sinha, M. Krishna and P. Muthuramalingam).

Our interest in commutator methods was raised by the major achievements mentioned above. In studying these papers we arrived at the conviction that the field of applications of the method of locally conjugate operators was by no means exhausted and also that the theory itself could be improved on an abstract level such as to cover most of the known results in spectral and scattering theory and to obtain these results under sharper and more natural conditions. The present monograph is a presentation of the principal outcomes of our efforts in this direction.

It turned out that, in order to arrive at the refined version of the locally conjugate operator method we were looking for, we had to have recourse to certain non-Hilbertian techniques which are rarely used in spectral and scattering theory, such as real interpolation theory and C_0-groups of automorphisms of C*-

algebras. This explains to some extent the structure of our text. As suggested
by its title, it may be divided into three parts:

(a) a first part (Chapters 1–4) containing essentially a self-contained presentation of certain aspects of real interpolation theory and a rather advanced
study of multi-parameter C_0-groups, their functional calculi and the scales of
Besov spaces associated with them,

(b) a second part devoted to a study of commutator expansions with precise
estimates on the remainders (Chapters 5 and 6) and to a general version of the
conjugate operator method (Chapter 7), and

(c) a third part (Chapters 8–10) containing a new algebraic framework for the
N-body problem and a study of the spectral properties of N-body Hamiltonians
based on the results of Chapter 7.

As regards part (b), we mention that the deepest estimates on the remainders in the commutator expansions are not used further on in this text. These
estimates are important in scattering theory, and the proof of their most refined form involves in an essential way the tools and results of the preceding
chapters; this is our second motivation for the quite elaborate presentation of
these tools in the first part.

We pass now to a more detailed description of the contents of our text.
Chapter 1 contains rather standard material from distribution theory. Its first
purpose is to introduce some notations and to fix the terminology. However,
we also prove several estimates, the usefulness of which will appear only much
later. Chapter 2, devoted to real interpolation theory of Banach spaces, has
been included because, from our experience, the main audience for which this
text is intended is unfamiliar with this material. This chapter may be used as
a short introduction to the basic ideas of real interpolation theory. We were
obliged to omit many important subjects (for example we say nothing about the
interpolation of L^p-spaces), but our proofs are complete and relatively simple.

In Chapter 3 we consider a strongly continuous representation W of $\mathbb{R}^n$ in
a Banach space $\mathbf{F}$ and make a detailed study of the Besov scale $\{\mathbf{F}_{s,p} \mid s \in
\mathbb{R}, \, p \in [1, +\infty]\}$ and the functional calculus associated to W. As regards the
Besov scale, we go beyond the situation studied by H. Triebel who considered
the case $s > 0$. If $n = 1$ or if $\mathbf{F}$ is a Hilbert space and W is unitary, the
extension to $s \leq 0$ is straightforward. The remaining cases are considerably
more complicated, and one relevant question remains open (see the introduction
to Chapter 3). An important tool that we use is an abstract version of the
Littlewood-Paley dyadic decomposition method. This tool is also essential in
order to obtain good bounds on the norm of functions of the n-component
generator $A \equiv (A_1, \ldots, A_n)$ of W. From the technical point of view this chapter,
together with Chapter 7, forms the core of the book.

Some special situations of the general framework of Chapter 3 are considered
in Chapters 4, 5 and 6. In Chapter 4 we treat the classical case of weighted
Besov and Sobolev spaces on $\mathbb{R}^n$ and also the case when W is induced by a flow
on $\mathbb{R}^n$. In Chapter 5 we consider the group of automorphisms $\mathscr{W}$ induced by
W in the Banach algebra $\mathscr{B} \equiv B(\mathbf{F})$ of bounded linear operators on $\mathbf{F}$. If one
applies the theory of Chapter 3 to the group $\mathscr{W}$ in $\mathscr{B}$, one gets a Besov scale

$\{\mathscr{B}_{s,p}\}$; for $s > 0$, the property $T \in \mathscr{B}_{s,p}$ may be interpreted as a regularity property of the operator T relative to the group W (for integer s this can be reexpressed in terms of boundedness of multiple commutators of T with the generator A of W). By using the functional calculus associated to W, one then gets so-called (left) commutator expansions of the form

$$T\psi(A) = \sum_{|\alpha| \leq m-1} \frac{(-1)^{|\alpha|}}{\alpha!} \psi^{(\alpha)}(A) \operatorname{ad}_A^\alpha(T) + \mathscr{R}_m;$$

here m is a positive integer, $\alpha = (\alpha_1, \ldots, \alpha_n)$ with $\alpha_k \in \{0, 1, 2, \ldots\}$ is a multi-index, $\psi^{(\alpha)}(x) = (\partial/\partial x_1)^{\alpha_1} \cdot \ldots \cdot (\partial/\partial x_n)^{\alpha_n} \psi(x)$ and $\operatorname{ad}_A^\alpha(T)$ is a multiple commutator defined by the rules $\operatorname{ad}_{A_j}(T) = [A_j, T]$, $\operatorname{ad}_A^\alpha = \operatorname{ad}_{A_1}^{\alpha_1} \circ \ldots \circ \operatorname{ad}_{A_n}^{\alpha_n}$. A particular case of our results (Theorem 5.5.3) states that the remainder $\mathscr{R}_m$ is a bounded operator from $\mathbf{F}_{s,p}$ to $\mathbf{F}_{s+\mu,p}$ if ψ is a symbol of class $S^{m-\mu}(\mathbb{R}^n)$ with $\mu > 0$ (provided that T is sufficiently regular with respect to W); this is one of the deepest results of the book.

The framework of Chapter 6 is a special case of that discussed in Chapter 5, namely $\mathbf{F}$ is a Hilbert space and W is a unitary representation of $\mathbb{R}^n$. If H is a self-adjoint (unbounded) operator in $\mathbf{F}$, its regularity with respect to W is defined in terms of that of its resolvent. The purpose of this chapter is to specify regularity properties of functions $\varphi(H)$ of H given the regularity class of H and to study boundedness properties of operators of the form $\psi_1(A)[\varphi(H) - \varphi(H')]\psi_2(A)$, where H' is a second self-adjoint operator. Results on boundedness of operators of this form are especially important in scattering theory.

Chapter 7 is devoted to the conjugate operator method. This is a way of controlling the behaviour of the resolvent of a self-adjoint operator in the neighbourhood of its spectrum. A preliminary discussion of the usefulness of such an investigation may be found in Section 7.1. In the next two sections we present the essence of the conjugate operator method. Consequences and extensions are treated in Sections 7.4 and 7.5, while the final section contains the first important examples (pseudo-differential operators). The material of Chapter 7 is based on our own research; a rather detailed comparison with other approaches, as well as historical remarks and a description of the main ideas of the method in simple situations, may be found in the introduction to Chapter 7.

The last part of the book (Chapters 8–10) is devoted to some aspects of the spectral analysis of classes of Hamiltonians having a many-channel structure. Such operators describe composite physical systems, and physicists are interested in characterizing spectral and scattering properties of the Hamiltonian of the total system in terms of the properties of subsystems. This is somewhat vague, and the purpose of Chapter 8 is to discuss a framework in which such Hamiltonians can be defined and studied in a natural way. The formalism is purely algebraic; to describe the Hamiltonian, we introduce the concept of an observable affiliated to a graded C*-algebra. A more detailed motivation of this point of view is given in the introduction to Chapter 8. In this chapter we also give a meaning at an algebraic level to the so-called Mourre estimate, and one of the main steps of the verification of this estimate for N-body Hamiltonians (namely the reduction to subsystems) is achieved at this level of generality. The graded C*-algebras that are specific to the standard N-body situation are studied in Chapter 9 where we also present our results on the spectral analysis of non-relativistic N-body Hamiltonians. The connection with the quantum-mechanical formulation of the N-body problem is made in the final Chapter 10.

This monograph is not meant to be a review of the research on commutator methods in general and not even on the method of locally conjugate operators. Following the work of E. Mourre, there appeared a small number of papers concerned with the abstract side of the theory and a somewhat larger number devoted to applications. Many of these papers are cited in the bibliography, and some of them are discussed in the text. We feel that the theory is still in a state of evolution; various theoretical questions are still open, and certainly numerous interesting applications will still be found.

We hardly touch upon questions related to scattering, except that in Chapter 7 we give criteria for existence and completeness of local relative wave operators in two-body type situations which, when applied to pseudo-differential operators, give sharp results. We plan to consider the many-body scattering theory in a companion volume.

We are much indebted to our colleague and friend Jean-Jacques Sansuc for his involvement in this project. His encouragement through numerous stimulating discussions, and his criticism and suggestions had a considerable impact on the final form of our text, and his familiarity with all intricacies of TEX was of invaluable help to us. It is a great pleasure for us to express here our deep gratitude to him. We wish to extend our thanks to Louis Boutet de Monvel who, during long years, has patiently listened to and answered many of our questions and helped us with pertinent advice.

Finally we acknowledge financial support from the Centre National pour la Recherche Scientifique (C.N.R.S. France), the Swiss National Science Foundation and the University of Paris 7.

Paris and Geneva
June 1995

Comments on notations

A detailed index of symbols may be found at the end of the text. However, we point out some special conventions that we use.

(1) The Laplace operator Δ on $\mathbb{R}^n$ is defined with a minus sign, i.e.

$$\Delta = -\sum_{k=1}^{n}(\partial/\partial x_k)^2.$$

(2) The usual Lebesgue measure on $\mathbb{R}^n$ is denoted by dx, but we find it more convenient to use as basic measure what we call Fourier measure $\underline{d}x$ defined as $\underline{d}x = (2\pi)^{-n/2}dx$.

(3) An absolutely continuous measure φ on $\mathbb{R}^n$ will be identified with its density $\varphi(x)$ with respect to Fourier measure: $\int_{\mathbb{R}^n} f(x)\varphi(dx) \equiv \int_{\mathbb{R}^n} f(x)\varphi(x)\underline{d}(x)$. Furthermore, we shall use the notation on the r.h.s. also if the measure φ is not absolutely continuous and even if φ is a distribution.

(4) We say "integrable measure" instead of "bounded measure".

(5) If $\mathbf{F}$ is a topological vector space, then $\mathbf{F}^*$ is the space of *anti*-linear continuous forms on $\mathbf{F}$. In particular a distribution (hence also a measure) on $\mathbb{R}^n$ is an anti-linear form on test functions.

(6) The generator A of a one-parameter C_0-group $W(s)$ is defined such that, formally, $W(s) = \exp(iAs)$.

(7) We set $\mathbb{N} = \{0, 1, 2, \ldots\}$ and $\langle x \rangle = (1 + |x|^2)^{1/2}$ if $x \in \mathbb{R}^n$.

The reader might be bothered by the abundance of indices in symbols denoting abstract Banach spaces or function spaces. In general we adopt the following rule: if $\mathbf{F}$ is a Banach space equipped with a representation of $\mathbb{R}^n$, then $\mathbf{F}_s$ or $\mathbf{F}_{s,p}$ denotes a space of the associated Sobolev or Besov scale respectively.

However, function spaces on $\mathbb{R}^n$ are naturally equipped with two representations of $\mathbb{R}^n$, the group of multiplication operators $\{e^{ia \cdot x}\}_{a \in \mathbb{R}^n}$ and the translation group. In order to avoid an accumulation of lower indices, we use them only for the first one of these representations and write indices referring to the translation group as upper indices. Thus upper indices describe the local regularity of functions, whereas the lower indices characterize their rate of decay at infinity (see the beginning of Section 4.1 for a more detailed explanation in a special situation, and also Section 5.4).

In one case our notation is not consistent with these rules. In order to avoid confusions, we have written $\mathcal{M}_w$ (page 78) for the space which should be denoted by $\mathcal{M}^w$ according to the above conventions. However, for special choices of $\mathcal{M}_w$ (see page 85), we use the notations $\mathcal{M}^r$ and $\mathcal{M}^{(\omega)}$ which are in agreement with our rules.

CHAPTER 1

Some Spaces of Functions and Distributions

This chapter is devoted to some aspects of distribution theory. In Section 1.1 we define various spaces of smooth functions as well as some differential operators; we insist on an invariant definition of these objects, since this will be important in the study of N-body Hamiltonians. Various definitions and facts from distribution theory are reviewed in Section 1.2. The reader should be familiar with the contents of these two sections because we fix notations and terminology which are in some cases not quite standard. Sections 1.3 and 1.4 are more technical and may be skipped temporarily. We prove results relating local regularity properties of a distribution with the behaviour at infinity of its Fourier transform. We also establish an identity due to A.P. Calderón giving a representation of a distribution in terms of its derivatives of a fixed order plus a regular term. Finally, we use this representation in order to prove several facts that will be needed in later chapters.

1.1. Calculus on Euclidean Spaces

By a *euclidean space* we mean a finite-dimensional real vector space X provided with a scalar product $(\cdot, \cdot)$. We shall write $|\cdot|$ for the norm in X. Sometimes, in order to avoid confusions, we shall add a subscript specifying the space, e.g. $(\cdot, \cdot)_X$ or $|\cdot|_X$. The dimension of X will often be denoted by the letter n. If $x \in X$, we set $\langle x \rangle = (1 + |x|^2)^{1/2}$. Clearly $\langle x + y \rangle \le \sqrt{2} \langle x \rangle \cdot \langle y \rangle$ $\forall x, y \in X$.

Each subspace Y of a euclidean space X is itself a euclidean space if one considers on it the scalar product induced by that defined on X. We denote by $Y^\perp$ the orthogonal complement of a subspace Y in X and by π_Y the orthogonal projection operator of X onto Y. We use the symbol $\mathbf{O}$ for the space (or subspace) $\{0\}$ consisting only of the zero vector. The standard example of a euclidean space of dimension n is the space $\mathbb{R}^n$ consisting of n-tuples $x = (x_1, \ldots, x_n)$ of real numbers with scalar product $(x, y)_{\mathbb{R}^n} = \sum_{k=1}^n x_k y_k$. Of course, each euclidean space of dimension n can be identified with $\mathbb{R}^n$ by choosing in X an orthonormal basis $\{v_1, \ldots, v_n\}$, and we shall occasionally use such an identification. However, our emphasis here is on objects which are invariantly defined: this is not only important in the applications we have in mind but also gives a formal simplicity

W. O. Amrein et al., *C₀-Groups, Commutator Methods and Spectral Theory of N-Body Hamiltonians*, Modern Birkhäuser Classics,
DOI: 10.1007/978-3-0348-0733-3_1, © Springer Basel 1996

which is often of great help.

Let us recall the invariant definition of the derivative of a real function f defined in a neighbourhood of a point $x \in X$: one says that f *is differentiable at* x if there exists a vector $f'(x) \in X$ such that for $h \in X$ with $|h|$ sufficiently small: $f(x+h) = f(x) + (h, f'(x)) + o(|h|)$. Then $f'(x)$ is uniquely defined and is called the *derivative of* f *at* x. If $v \in X$ is a fixed vector in X, then the *derivative of* f *at* X *in the direction* v is given by the formula:

$$(1.1.1) \qquad \partial_v f(x) = (v, f'(x)) = \lim_{\varepsilon \to 0} \varepsilon^{-1}(f(x + \varepsilon v) - f(x)).$$

Higher order derivatives $f''(x)$, $f'''(x)$, etc. are obtained by requiring that a higher order development in h of the difference $f(x+h) - f(x)$ exists; we refer to Section 1.1 of [H] for details. Notice that, since there is a bijective correspondence between quadratic (or symmetric bilinear) forms on X and symmetric linear operators in X, one can define the second-order derivative (if it exists) of the real function f as the unique symmetric operator $f''(x) : X \to X$ such that for sufficiently small $h \in X$:

$$f(x + h) = f(x) + (h, f'(x)) + \frac{1}{2}(h, f''(x)h) + o(|h|^2).$$

This invariant presentation of $f''(x)$ allows a neat definition of the *Laplace-Beltrami operator* Δ associated to X:

$$(1.1.2) \qquad (\Delta f)(x) = - \operatorname{Tr} f''(x),$$

where $\operatorname{Tr} A$ is the trace of a linear operator $A : X \to X$.

Let $\{v_1, \ldots, v_n\}$ be any (algebraic) basis of X and set $\partial_j = \partial_{v_j}$. The dual basis $\{v^1, \ldots, v^n\}$ in X is defined by the conditions $(v^j, v_k) = \delta_{jk}$, and one sets $g^{jk} = (v^j, v^k)$. It is possible to give an explicit expression for Δf in the coordinate system associated to $\{v_1, \ldots, v_n\}$, namely:

$$(1.1.3) \qquad \Delta f = - \sum_{i,j=1}^n g^{ij} \partial_i \partial_j f.$$

If f is a complex-valued function on X which is k times differentiable at a point x, we denote by $\nabla^{(k)} f(x)$ its k-th order derivative at x (we avoid the notation $f^{(k)}(x)$ used in Section 1.1 of [H] in order not to be in conflict with the notation $f^{(\alpha)}$ that will be introduced below). Then $\nabla^{(k)} f(x)$ is a (complex-valued) symmetric k-linear form on X, and we define its norm $|\nabla^{(k)} f(x)|$ by the formula:

$$|\nabla^{(k)} f(x)| = \Big[\sum_{\substack{1 \leq i_1, \ldots, i_k \leq n \\ 1 \leq j_1, \ldots, j_k \leq n}} g^{i_1 j_1} \ldots g^{i_k j_k} \partial_{i_1} \ldots \partial_{i_k} f(x) \partial_{j_1} \ldots \partial_{j_k} \overline{f}(x) \Big]^{1/2}.$$

The advantage of this definition is that the right-hand side is independent of the chosen basis, so $|\nabla^{(k)} f(x)|$ is invariantly defined.

Assume that the basis $v_1, \ldots, v_n$ is *orthonormal* (so that $v_j = v^j$ and $g^{jk} = \delta_{jk}$). Recall that a multi-index α is a n-tuple $(\alpha_1, \ldots, \alpha_n)$ of integers $\alpha_j \geq 0$. The following conventions will be used:

$$|\alpha| = \alpha_1 + \cdots + \alpha_n, \quad \alpha! = \alpha_1! \ldots \alpha_n!, \quad \partial^\alpha = \partial_1^{\alpha_1} \ldots \partial_n^{\alpha_n}$$

and

$$x^\alpha = x_1^{\alpha_1} \ldots x_n^{\alpha_n} \text{ if } x = x_1 v_1 + \cdots + x_n v_n \in X \quad (x_j \in \mathbb{R}).$$

Observe that the operator ∂^α and the number x^α depend on the chosen basis. If α, β are multi-indices, we write $\beta \leq \alpha$ if $\beta_1 \leq \alpha_1, \ldots, \beta_n \leq \alpha_n$, and then we define $\alpha - \beta = (\alpha_1 - \beta_1, \ldots, \alpha_n - \beta_n)$ and

$$\binom{\alpha}{\beta} = \prod_{j=1}^{n} \binom{\alpha_j}{\beta_j}.$$

We can now express the norm $|\nabla^{(k)} f(x)|$ in a more compact form, namely

$$(1.1.4) \qquad |\nabla^{(k)} f(x)| = \Big[\sum_{|\alpha|=k} \frac{k!}{\alpha!} |\partial^\alpha f(x)|^2 \Big]^{1/2}.$$

From now on, if an orthonormal basis in X is given, we shall set $\partial^\alpha f = f^{(\alpha)}$.

Let k be an integer ≥ 0 or $k = \infty$. We shall use the following standard notations:

$C^k(X) = \{ f : X \to \mathbb{C} \mid f$ has continuous derivatives up to order $k \}$,

$C_0^k(X) = \{ f \in C^k(X) \mid f$ has compact support $\}$,

$BC^k(X) = \{ f \in C^k(X) \mid f$ and its derivatives of order $\leq k$ are bounded $\}$,

$C_\infty^k(X) = \{ f \in C^k(X) \mid f$ and its derivatives of order $\leq k$ tend to 0 at infinity $\}$.

We also set: $C(X) = C^0(X)$, $C_0(X) = C_0^0(X)$, $BC(X) = BC^0(X)$, $C_\infty(X) = C_\infty^0(X)$. Observe that, if $k < \infty$, then $BC^k(X)$ is a Banach algebra for the norm:

$$(1.1.5) \qquad \|f\|_{BC^k} = \sup_{x \in X} \sum_{j=0}^{k} \frac{1}{j!} |\nabla^{(j)} f(x)|,$$

and $C_\infty^k(X)$ is a Banach subalgebra.

We shall use the notation $C_{\text{pol}}^k(X)$ for the subalgebra of $C^k(X)$ consisting of functions whose derivatives have at most polynomial growth at infinity (i.e. for each $j \leq k$ there are constants $c = c(j, f)$, $m = m(j, f)$ such that $|\nabla^{(j)} f(x)| \leq c \langle x \rangle^m$). Observe that $C_{\text{pol}}^\infty(X)$ is an algebra containing the polynomials and invariant under differentiation.

We shall need a very restricted class of symbols for the construction of the functional calculus of generators of C_0-groups. We shall say that $f : X \to \mathbb{C}$ is a *symbol of degree m* (m any real number) if $f \in C^\infty(X)$ and for any $k \in \mathbb{N}$ there is a constant c_k such that $|\nabla^{(k)} f(x)| \leq c_k \langle x \rangle^{m-k}$, $\forall x \in X$. If an orthonormal basis is given in X, this is equivalent with the requirement that $|f^{(\alpha)}(x)| \leq c_\alpha \langle x \rangle^{m-|\alpha|}$

for all $\alpha \in \mathbb{N}^n$ and all $x \in X$. We denote by $S^m(X)$ the vector space of all symbols of degree m and observe that:

$$m_1 \leq m_2 \Longrightarrow S^{m_1}(X) \subset S^{m_2}(X),$$
$$S^{m_1}(X) \cdot S^{m_2}(X) \subset S^{m_1+m_2}(X).$$

In particular, if $m \leq 0$ then $S^m(X)$ is an algebra (for the usual multiplication of functions). Observe also that $f \in S^m(X) \Longrightarrow f^{(\alpha)} \in S^{m-|\alpha|}(X)$ for each $\alpha \in \mathbb{N}^n$. The function $\langle x \rangle^m$ clearly belongs to $S^m(X)$ (for any $m \in \mathbb{R}$). We denote by $S^\infty(X)$ the union of all the spaces $S^m(X)$. Then $S^\infty(X)$ is an algebra which contains all polynomials and is invariant under differentiation. One could also introduce the space $S^{-\infty}(X) := \bigcap_{m \in \mathbb{R}} S^m(X)$ which is however nothing else than the space $\mathscr{S}(X)$ of tempered test functions:

(1.1.6)

$$\mathscr{S}(X) = \{f \in C^\infty(X) \mid \|f\|_k := \sup_{x \in X} \langle x \rangle^k \sum_{j=0}^{k} \frac{1}{j!} |\nabla^{(j)} f(x)| < \infty \text{ for each } k \in \mathbb{N}\}.$$

$\mathscr{S}(X)$ is a Fréchet space for the topology defined by the family of norms that appear in the definition (1.1.6).

We shall now recall Taylor's formula and then derive a consequence that will be useful in the study of C_0-groups. Choose an orthonormal basis $\{v_1, \ldots, v_n\}$ in X. If m is an integer ≥ 1 and $f : X \to \mathbf{F}$ is a Banach space-valued function of class C^m, then for each multi-index α with $|\alpha| = m$ we denote by f_α the continuous function on $X \times X$ given by:

(1.1.7) $$f_\alpha(x, y) = |\alpha| \int_0^1 f^{(\alpha)}(x + \tau y)(1 - \tau)^{|\alpha|-1} d\tau.$$

Clearly $f_\alpha(x, 0) = f^{(\alpha)}(x)$ and, if $f \in C^{m+k}(X)$, then $f_\alpha \in C^k(X \times X)$. Now Taylor's formula reads as follows:

(1.1.8) $$f(x + y) = \sum_{|\alpha|<m} \frac{y^\alpha}{\alpha!} f^{(\alpha)}(x) + \sum_{|\alpha|=m} \frac{y^\alpha}{\alpha!} f_\alpha(x, y).$$

There is a converse to Taylor's theorem which is useful for proving that a function is of class C^m. We quote a special case of a result proved in §2 of [AR] :

PROPOSITION 1.1.1. *Let $U \subset X$ be a convex open set, $\mathbf{F}$ a Banach space and f and $\{g_\alpha\}_{|\alpha|\leq m}$ continuous $\mathbf{F}$-valued functions on U such that:*

$$\lim_{|y|\to 0} \frac{1}{|y|^m} \left\| f(x + y) - \sum_{|\alpha|\leq m} \frac{y^\alpha}{\alpha!} g_\alpha(x) \right\|_{\mathbf{F}} = 0,$$

uniformly in x on each compact subset of U. Then f is of class C^m on U and $f^{(\alpha)} = g_\alpha$.

Next we show how derivatives of order α with $0 < |\alpha| < m$ can be expressed in terms of translates of f and of derivatives of order exactly m. In particular, this allows one to estimate $f^{(\alpha)}$ in terms of f and of the family $\{f^{(\beta)}\}_{|\beta|=m}$.

PROPOSITION 1.1.2. *Let X be identified with $\mathbb{R}^n$ by means of an orthonormal basis and let $\ell \geq 2$ be an integer. Let $H_1, \ldots, H_n$ be finite real sets, each of them containing exactly ℓ distinct points, and set $G = H_1 \times \cdots \times H_n \subset X$. Then there is a family $\{Q_y\}_{y \in G}$ of real polynomials on X such that:*

(a) If $P(z) = \sum_{|\alpha| < \ell} a_\alpha z^\alpha$ is a polynomial of degree $\leq \ell - 1$ on X with coefficients a_α in an arbitrary vector space, then $P(z) = \sum_{y \in G} P(y) Q_y(z)$, $\forall z \in X$; in particular $\alpha! a_\alpha = \sum_{y \in G} C_y^\alpha P(y)$ where $C_y^\alpha = Q_y^{(\alpha)}(0)$ are real constants.

(b) If $\mathbf{F}$ is a Banach space and $f : X \to \mathbf{F}$ is of class C^ℓ, then for each $x \in X$ and each multi-index α with $0 < |\alpha| < \ell$ one has (with the same constants C_y^α as above):

(1.1.9)
$$f^{(\alpha)}(x) = \sum_{y \in G} C_y^\alpha f(x+y) - \sum_{y \in G} C_y^\alpha \int_0^1 \ell(1-\tau)^{\ell-1} \sum_{|\beta| = \ell} \frac{f^{(\beta)}(x + \tau y)}{\beta!} y^\beta d\tau.$$

PROOF. Part (b) of the proposition follows immediately from Part (a) by taking $P(z) = \sum_{|\beta| < \ell} \frac{z^\beta}{\beta!} f^{(\beta)}(x)$ and then by using Taylor's formula (1.1.8).

In order to prove Part (a) it is obviously sufficient to construct polynomials Q_y such that $z^\alpha = \sum_{y \in G} y^\alpha Q_y(z)$ for all $z \in X$ and all α such that $|\alpha| < \ell$. If Q_y has the property $Q_y(z) = L_{y_1}^1(z_1) \cdot \ldots \cdot L_{y_n}^n(z_n)$ then the preceding identity is fulfilled if L_s^i ($1 \leq i \leq n$, $s \in H_i$) are polynomials on $\mathbb{R}$ such that $t^k = \sum_{s \in H_i} s^k L_s^i(t)$ for all $t \in \mathbb{R}$ and $k = 0, 1, \ldots, \ell - 1$. We take

$$L_s^i(t) = \prod_{\substack{u \in H_i \\ u \neq s}} (t - u)(s - u)^{-1}.$$

Then L_s^i is a polynomial of degree $\ell - 1$, $L_s^i(s) = 1$ and $L_s^i(t) = 0$ if $t \neq s$, $t \in H_i$. The polynomials $t \mapsto t^k$ and $t \mapsto \sum_{s \in H_i} s^k L_s^i(t)$ have degree $\leq \ell - 1$ and coincide on H_i (which has ℓ points), so they are identical. $\square$

1.2. Distributions, Fourier Transforms

On each euclidean space X there is a canonical riemannian (euclidean) measure dx. However, we shall not use this measure dx in the abstract setting but only for the case where $X = \mathbb{R}^n$. It will be more convenient to work with the measure $(2\pi)^{-(\dim X)/2} dx \equiv \underline{d}x$ which we shall call the *Fourier measure* associated to X. It is the unique translation invariant Borel measure on X such that the volume of any unit cube is equal to $(2\pi)^{(\dim X)/2}$. We choose this normalisation in order to avoid having to write powers of 2π in various expressions (e.g. in the Fourier transformations and in the Hausdorff-Young inequality).

The above definition associates canonically a measure to each euclidean space. This association has the following fundamental property: if X is the direct sum of two euclidean spaces Y and Z, with Fourier measures $\underline{d}y$ and $\underline{d}z$ respectively, then the Fourier measure on X is just the product measure:

(1.2.1)
$$\underline{d}x = \underline{d}y \otimes \underline{d}z.$$

One can associate to an euclidean space other measures having the factorization property (1.2.1). In fact, if $\kappa > 0$ and $c \geq 0$ are constants, then $\kappa^{\dim X} \exp(-c|x|^2)\underline{d}x$ defines such a measure. If c is taken to be strictly positive, then much of the theory below can be extended to the case of infinite-dimensional euclidean spaces. For example, if one takes $c = 1/2$ and κ such that the above measure is a probability measure, then the space $\mathscr{H}(X)$ that we shall introduce below will become the usual Fock space associated to X.

Having introduced the Fourier measure on an euclidean space X, one can define various function spaces over X, for example $L^p(X)$, $L^p_{\text{loc}}(X)$ for $1 \leq p \leq \infty$. For $p < \infty$ the norm in $L^p(X)$ is given by

$$(1.2.2) \qquad \|f\|_{L^p(X)} = \left[\int_X |f(x)|^p \underline{d}x\right]^{1/p}.$$

Since the Hilbert space $L^2(X)$ will play a special role in our considerations, we shall denote it from now on by $\mathscr{H}(X)$, and we shall write $\|\cdot\| = \|\cdot\|_{L^2(X)}$ and $\langle f, g\rangle = \int_X \overline{f(x)}g(x)\underline{d}x$ for its norm and scalar product respectively. We shall always identify $\mathscr{H}(X)$ with its adjoint $\mathscr{H}(X)^*$ through the Riesz isomorphism (see Section 2.8). We shall write $\mathbb{B}(X)$ for the Banach space $B(\mathscr{H}(X))$ of all bounded linear operators in $\mathscr{H}(X)$ and $\mathbb{K}(X)$ for the set $K(\mathscr{H}(X))$ of all compact operators in $\mathscr{H}(X)$.

We denote by $\mathscr{S}^*(X)$ the space of tempered distributions on X, i.e. the space of all continuous, anti-linear mappings $\mathscr{S}(X) \to \mathbb{C}$ equipped with the strong topology, i.e. the topology of uniform convergence on bounded subsets of $\mathscr{S}(X)$ (see §4, Chapter VII in [Sch]). Since $\mathscr{S}(X) \subset \mathscr{H}(X)$ continuously and densely, and since we have identified $\mathscr{H}(X)$ and $\mathscr{H}(X)^*$, we obtain a scale:

$$(1.2.3) \qquad \mathscr{S}(X) \subset \mathscr{H}(X) \subset \mathscr{S}^*(X)$$

such that, if $\langle \cdot, \cdot \rangle : \mathscr{S}(X) \times \mathscr{S}^*(X) \to \mathbb{C}$ is the antiduality between $\mathscr{S}(X)$ and $\mathscr{S}^*(X)$ (antilinear in the first and linear in the second argument), then for $f \in \mathscr{S}(X)$ and $g \in \mathscr{H}(X)$, if g is considered as an element of $\mathscr{S}^*(X)$, the number $\langle f, g\rangle$ is just the scalar product in $\mathscr{H}(X)$ introduced above. For this reason we do not distinguish between the scalar product in $\mathscr{H}(X)$ and the antiduality of $\mathscr{S}(X)$ and $\mathscr{S}^*(X)$; also we shall occasionally use the following formal notation for the action of a distribution $g \in \mathscr{S}^*(X)$ on a test function $f \in \mathscr{S}(X)$:

$$\langle f, g\rangle = \int_X \overline{f(x)}g(x)\underline{d}x.$$

If g is a locally integrable function on X with $\int_X \langle x\rangle^m |g(x)|dx < \infty$ for some $m \in \mathbb{R}$, then the r.h.s. above (the integral being interpreted in the Lebesgue sense) defines a tempered distribution. The distributions defined in such a way will be called *functions*. The second inclusion in (1.2.3) is also continuous and dense, because $\mathscr{S}(X)$ is a *reflexive* locally convex space. The reflexivity of $\mathscr{S}(X)$ also allows us to identify $[\mathscr{S}^*(X)]^*$ with $\mathscr{S}(X)$ by setting $\langle f, g\rangle \equiv \overline{\langle g, f\rangle}$ for $f \in \mathscr{S}^*(X)$, $g \in \mathscr{S}(X)$.

If $\mathbf{F}$ is a Banach space such that $\mathscr{S}(X) \subset \mathbf{F} \subset \mathscr{S}^*(X)$, with dense and continuous embeddings, then its adjoint $\mathbf{F}^*$ can be realized as a subspace of $\mathscr{S}^*(X)$, more precisely $\mathscr{S}(X) \subset \mathbf{F}^* \subset \mathscr{S}^*(X)$, and the dual norm is

$$(1.2.4) \qquad \|g\|_{\mathbf{F}^*} = \sup_{\substack{f \in \mathscr{S}(X) \\ \|f\|_{\mathbf{F}} \leq 1}} |\langle f, g \rangle|.$$

This applies in particular to $\mathbf{F} = L^p(X)$ for $1 \leq p < \infty$ and gives $L^p(X)^* = L^{p'}(X)$ with $1/p' = 1 - 1/p$.

By a *measure* on a euclidean space X we mean a complex Radon measure on X. If μ is a measure on X, we denote by $|\mu|$ its variation, i.e. the smallest positive measure on X such that $|\mu(K)| \leq |\mu|(K)$ for each compact subset K of X. A measure μ is said to be *integrable* if $|\mu|(X) < \infty$ (we prefer to use the term "integrable measure" rather than "bounded measure" in order to avoid confusion with the notion of boundedness in the distributional sense as introduced in Section 1.4). The space of all integrable measures can and will be identified with a vector subspace of $\mathscr{S}^*(X)$ through the formula $\langle f, \mu \rangle = \int_X f(x)\mu(dx)$; this subspace of $\mathscr{S}^*(X)$ is just $[C_\infty(X)]^*$, and one has

$$\int_X |\mu|(dx) = \sup_{\substack{f \in \mathscr{S}(X) \\ |f| \leq 1}} \left| \int_X f(x)\mu(dx) \right| = \sup_{\substack{f:X \to \mathbb{C} \\ f \text{ Borel}, |f| \leq 1}} \left| \int_X f(x)\mu(dx) \right|.$$

In agreement with a convention made before in the context of arbitrary distributions, we shall often write $\int_X f(x)\mu(dx) = \int_X f(x)\mu(x)\underline{dx}$ even if the measure μ is not (Lebesgue) absolutely continuous; then we write $|\mu(x)|\underline{dx}$ for $|\mu|(dx)$.

If $f \in L^1(X)$ we define its Fourier transform $\mathscr{F}f \equiv \hat{f}$ and its adjoint Fourier transform $\mathscr{F}^*f \equiv \check{f}$ by:

$$(\mathscr{F}f)(x) = \hat{f}(x) = \int_X e^{-i(x,y)} f(y)\underline{dy},$$

$$(\mathscr{F}^*f)(x) = \check{f}(x) = \int_X e^{i(x,y)} f(y)\underline{dy}.$$

$\mathscr{F}$ and $\mathscr{F}^*$ are topological isomorphims of $\mathscr{S}(X)$ onto itself such that $\mathscr{F}^*\mathscr{F}f = \mathscr{F}\mathscr{F}^*f = f$ (see Theorem 7.1.5 in [H]). Hence we can extend $\mathscr{F}$ to $\mathscr{S}^*(X)$ by defining $\mathscr{F}$ in $\mathscr{S}^*(X)$ as the adjoint mapping of the restriction of $\mathscr{F}^*$ to $\mathscr{S}(X)$; similarly $\mathscr{F}^*$ can be extended to $\mathscr{S}^*(X)$ by defining it as the adjoint of the mapping $\mathscr{F} : \mathscr{S}(X) \to \mathscr{S}(X)$. Then $\mathscr{F}$ and $\mathscr{F}^*$ are isomorphisms of $\mathscr{S}^*(X)$ onto itself and such that $\mathscr{F}^* = \mathscr{F}^{-1}$.

The restriction of $\mathscr{F}$ to $\mathscr{H}(X)$ is a unitary operator in $\mathscr{H}(X)$. If $p \in [1,2]$ and p' is defined by $p^{-1} + p'^{-1} = 1$, then $\mathscr{F}$ is a contraction from $L^p(X)$ to $L^{p'}(X)$; this is just the *Hausdorff-Young inequality* (Theorem 7.1.13 in [H]) :

$$(1.2.5) \qquad \|\mathscr{F}f\|_{L^{p'}(X)} \leq \|f\|_{L^p(X)}.$$

In the case $p = 1$ we may use the Riemann-Lebesgue lemma to obtain the stronger statement $\mathscr{F} : L^1(X) \to C_\infty(X)$. We also recall the *Young inequality*

(Corollary 4.5.2 in [H]). If one defines convolution by:

$$(1.2.6) \qquad (f * g)(x) = \int_X f(x - x')g(x')\underline{dx}',$$

then if $p, q, r \geq 1$ and $r^{-1} = p^{-1} + q^{-1} - 1$:

$$(1.2.7) \qquad \|f * g\|_{L^r(X)} \leq \|f\|_{L^p(X)}\|g\|_{L^q(X)}.$$

The fact that both inequalities (1.2.5) and (1.2.7) do not contain factors 2π is due to our choice of the Fourier measure (which appears also in the definition of the convolution product).

Let φ be any tempered distribution. Then for any $f \in \mathscr{S}(X)$ the product φf is a well defined element of $\mathscr{S}^*(X)$ (set $\langle g, \varphi f \rangle \equiv \langle g\bar{f}, \varphi \rangle$), and the mapping from $\mathscr{S}(X)$ into $\mathscr{S}^*(X)$ given by $f \mapsto \varphi f$ is obviously linear and continuous. We shall denote this operator from $\mathscr{S}(X)$ into $\mathscr{S}^*(X)$ by $\varphi(Q)$. Then we set $\varphi(P) = \mathscr{F}^*\varphi(Q)\mathscr{F}$, so that $\varphi(P) : \mathscr{S}(X) \to \mathscr{S}^*(X)$ is a second linear continuous operator associated to φ. When we wish to specify the underlying vector space, we write $\varphi(Q^X)$ and $\varphi(P^X)$ for these operators. If $a \in \mathbb{R}$, $a \neq 0$, we set $\varphi^a(x) = \varphi(ax)$ and use the notations $\varphi(aQ)$ and $\varphi(aP)$ for the operators $\varphi^a(Q)$ and $\varphi^a(P)$ respectively. We remark that $\mathscr{F}^*\varphi(P)\mathscr{F} = \varphi(-Q)$.

Observe that $C^\infty_{\text{pol}}(X) \subset \mathscr{S}^*(X)$, so that for each $\varphi \in C^\infty_{\text{pol}}(X)$, we may consider the operators $\varphi(Q)$ and $\varphi(P)$. It is easy to see that these operators have much stronger continuity properties, namely they leave $\mathscr{S}(X)$ invariant, they induce continuous operators in $\mathscr{S}(X)$ and they extend to continuous operators in $\mathscr{S}^*(X)$.

Now let Y be a subspace of X and $\varphi : Y \to \mathbb{C}$ a C^∞-function with polynomially bounded derivatives. Then we define

$$(1.2.8) \qquad \varphi(Q_Y) = (\varphi \circ \pi_Y)(Q)$$

and

$$(1.2.9) \qquad \varphi(P_Y) = (\varphi \circ \pi_Y)(P).$$

Certain operators of the above form will be occuring repeatedly and we use some special notation for them. We collect these special notations in Table 1.1. In some cases the function φ in the table depends on a parameter, so we have added a column to explain the meaning of the parameter. In the last two lines we assume that an orthonormal basis $\{v_1, \ldots, v_n\}$ of X has been fixed.

If $f \in \mathscr{S}(X)$, then the notations $P_j f = -i\partial_j f$, $P^\alpha f = (-i\partial)^\alpha f$ and $-i\partial_v f$ are consistent with the definitions made in Section 1.1. The operators $\langle Q \rangle^s$ and $\langle P \rangle^s$ are isomorphisms of $\mathscr{S}^*(X)$ onto itself, with inverses $\langle Q \rangle^{-s}$ and $\langle P \rangle^{-s}$ respectively. Moreover they leave $\mathscr{S}(X)$ invariant and are isomorphisms of $\mathscr{S}(X)$ onto itself.

To explain the notation Δ_Y, let us choose in X an algebraic basis $\{v_1, \ldots, v_n\}$ such that $v_1, \ldots, v_m \in Y$ and $v_{m+1}, \ldots, v_n \in Y^\perp$; then for $f \in \mathscr{S}(X)$ (compare

$\varphi(x)$	$\varphi(Q)$	$\varphi(P)$	Parameter
x^2	Q^2	Δ	
$|\pi_Y(x)|^2$	Q_Y^2	Δ_Y	Y is a subspace of X
$\langle x \rangle^s$	$\langle Q \rangle^s$	$\langle P \rangle^s$	$s \in \mathbb{R}$
$[1 + |\pi_Y(x)|^2]^{s/2}$	$\langle Q_Y \rangle^s$	$\langle P_Y \rangle^s$	Y is a subspace of X
$\exp[-i(v,x)]$	$\exp[-i(v,Q)]$	$T(v)$	$v \in X$
(v,x)	(v,Q)	$-i\partial_v$	$v \in X$
$(v_j, x) \equiv x_j$	Q_j	$P_j \equiv -i\partial_j$	$j = 1, \ldots, n \equiv \dim X$
x^α	Q^α	$P^\alpha \equiv (-i)^{|\alpha|}\partial^\alpha$	α is a multi-index

TABLE 1.1. Notations

with (1.1.3)) :

$$(1.2.10) \qquad \Delta_Y f = - \sum_{j,k=1}^{m} g^{jk} \partial_j \partial_k f.$$

We observe that

$$(1.2.11) \qquad \Delta = \Delta_Y + \Delta_{Y^\perp}.$$

The symbol Q stands for a vector operator (multiplication by the vector-valued function $\varphi(x) = x \in X$) and will be referred to as the *position operator*; in an orthonormal basis it is given by the n operators $Q_1, \ldots, Q_n$. Similarly P stands for a vector operator called the *momentum operator*. The family $\{T(v)\}_{v \in X}$ gives the unitary representation in $\mathscr{H}(X)$ of the additive group X by translations. In fact, from the relation $T(v) = \exp[-i(v,P)]$ we get that

$$(1.2.12) \qquad (T(v)f)(x) = f(x-v).$$

If φ is a function in $C_\infty(X)$ such that $\hat{\varphi} \in L^1(X)$, then we clearly have:

$$(1.2.13) \qquad \varphi(P) = \int_X e^{i(v,P)} \hat{\varphi}(v) dv = \int_X T(v) \check{\varphi}(v) dv.$$

In particular $\varphi(P) \in \mathbb{B}(X)$, but much more is obviously true: $\varphi(P) \in \mathbb{B}(X)$ if $\varphi \in L^\infty(X)$ and $||\varphi(P)||_{\mathbb{B}(X)} = ||\varphi||_{L^\infty}$. Various generalizations of (1.2.13) will be of importance later on. The family $\{T(v)\}_{v \in X}$ also defines a representation of the additive group X by topological automorphisms of $\mathscr{S}^*(X)$ which leave $\mathscr{S}(X)$ invariant. If $f \in \mathscr{S}^*(X)$ then we have

$$(1.2.14) \qquad \lim_{\varepsilon \to 0} \varepsilon^{-1}(T(\varepsilon v) - I)f = -\partial_v f$$

for any $v \in X$ (the limit is in the strong topology of $\mathscr{S}^*(X)$ and I denotes the identity operator). The convolution product can also be expressed in terms of functions of the momentum operator P. Let $\varphi \in \mathscr{S}^*(X)$ and $f \in \mathscr{S}(X)$. Then the formal integral

$$(\varphi * f)(x) = \int_X f(x-x') \varphi(x') dx'$$

has a rigorous meaning when interpreted as the action of the distribution φ on the test function $x' \mapsto \overline{f}(x - x')$. In the following lemma we collect some properties of the function $\varphi * f$ obtained in this way:

LEMMA 1.2.1. *If* $f \in \mathscr{S}(X)$ *and* $\varphi \in \mathscr{S}^*(X)$, *then* $\varphi * f \in C^\infty_{\text{pol}}(X)$. *Furthermore the following identities hold:*

$$(1.2.15) \qquad \varphi * f = \widehat{\varphi}(P)f = \widehat{f}(P)\varphi.$$

In particular, one has $\psi(P)f \in C^\infty_{\text{pol}}(X)$ *if* $\psi \in \mathscr{S}^*(X)$ *and* $f \in \mathscr{S}(X)$ *or if* $\psi \in \mathscr{S}(X)$ *and* $f \in \mathscr{S}^*(X)$.

PROOF. We observe that

$$(1.2.16) \qquad (\varphi * f)(x) = \int_X f(x - y)\varphi(y)\underline{dy} = \langle T(x)f^+, \varphi \rangle$$

with $f^+(y) = \overline{f(-y)}$. Since $T(\cdot) : \mathscr{S}(X) \to \mathscr{S}(X)$ is of class C^∞, with $\partial_x^\alpha T(x)g = (-i)^{|\alpha|}T(x)P^\alpha g$ for $g \in \mathscr{S}(X)$, (1.2.16) implies that $(\varphi * f)(\cdot)$ is of class C^∞. Similarly, by the continuity of φ, there is $k \in \mathbb{N}$ such that $|\langle g, \varphi \rangle| \leq c|g|_k$ for all $g \in \mathscr{S}(X)$, where $|\cdot|_k$ is as in (1.1.6). The polynomial bound on $|(\varphi * f)(x)|$ then follows from the inequality $|T(x)g|_k \leq c_k \langle x \rangle^k |g|_k$. The verification of (1.2.15) is straightforward. $\square$

It is instructive to consider the usual mollifying technique for approximating a distribution by a smooth function in the context of the preceding lemma. Let $\eta \in \mathscr{S}(X)$ with $\int_X \eta(x)dx = 1$; for $\varepsilon > 0$, define $\eta_\varepsilon(x) = \varepsilon^{-\dim X}\eta(\varepsilon^{-1}x)$. Then $(\mathscr{F}\eta_\varepsilon)(x) = \widehat{\eta}(\varepsilon x)$ and the function $\eta_\varepsilon * f$ is of class C^∞ for each $f \in \mathscr{S}^*(X)$ and converges to f (in suitable spaces) as $\varepsilon \to 0$. Lemma 1.2.1 shows that $\eta_\varepsilon * f = \widehat{\eta}(\varepsilon P)f$. Thus we can express this approximation method in the following terms: choose some $\varphi \in \mathscr{S}(X)$ with $\varphi(0) = 1$ (above $\varphi = \widehat{\eta}$) and consider the approximating function $\varphi(\varepsilon P)f$. This point of view may be generalized to an abstract setting (where P is replaced by the generator of a C_0-group of polynomial growth) and will be quite useful in Chapter 3.

We introduce also a representation of the dilation group (acting on X), which will be of considerable importance in the study of N-body hamiltonians. More precisely, for each $\tau \in \mathbb{R}$ we consider the linear continuous operator $W(\tau)$ in $\mathscr{S}^*(X)$ defined by

$$(1.2.17) \qquad (W(\tau)f)(x) = e^{\tau \dim X/2}f(e^\tau x).$$

Of course, the composition has to be interpreted in the sense of distributions (see Example 6.1.4 in [H]). The factor $\exp(\tau \dim X/2)$ has been chosen such that $W(\tau)$ is unitary as an operator in $\mathscr{H}(X)$. Clearly $W(\tau)$ maps $\mathscr{S}(X)$ into itself and its restriction to $\mathscr{S}(X)$ is a topological automorphism of $\mathscr{S}(X)$. Also, if $f \in \mathscr{S}(X)$, then $\tau \mapsto W(\tau)f \in \mathscr{S}(X)$ is a C^∞ function. Since $W(\tau)$ (in $\mathscr{S}^*(X)$) is just the adjoint of $W(-\tau)|_{\mathscr{S}(X)}$, it follows that $\{W(\tau)\}_{\tau \in \mathbb{R}}$ is a group of topological automorphisms of $\mathscr{S}^*(X)$ such that $\tau \mapsto W(\tau)f \in \mathscr{S}^*(X)$ is a C^∞ function for each $f \in \mathscr{S}^*(X)$. One may easily calculate the derivative :

$$(1.2.18) \qquad \lim_{\tau \to 0} \tau^{-1}[W(\tau) - I]f = 2iDf,$$

where D is a differential operator which, in an orthonormal basis of X, is given by the following expression:

$$(1.2.19) \qquad D = \frac{1}{4} \sum_{j=1}^{n} (P_j Q_j + Q_j P_j) \qquad (n = \dim X).$$

If f is a real function of class C^1 on X, one may write Df in an invariant form:

$$(Df)(x) = \frac{1}{2i}(x, f'(x)) + \frac{1}{4i} \dim X.$$

Let us observe that:

$$(1.2.20) \qquad \mathfrak{F}^* W(\tau)\mathfrak{F} = W(-\tau), \qquad \mathfrak{F}^* D\mathfrak{F} = -D,$$

$$(1.2.21) \qquad W(\tau)\Delta W(-\tau) = e^{-2\tau}\Delta, \qquad i[\Delta, D] = \Delta.$$

If we consider the group $\{W(\tau)\}_{\tau \in \mathbb{R}}$ as acting in $\mathcal{H}(X)$, then it is a strongly continuous unitary one-parameter group. By Stone's theorem, it has a self-adjoint infinitesimal generator. To relate its infinitesimal generator to D, we observe that $C_0^\infty(X \setminus \{0\})$ is left invariant by $\{W(\tau)\}$ and is included in the domain of its infinitesimal generator. Hence, by Nelson's lemma (Theorem 3.3.4 below), $C_0^\infty(X \setminus \{0\})$ is a core for its infinitesimal generator. By virtue of (1.2.18) this means that D is essentially self-adjoint on $C_0^\infty(X \setminus \{0\})$ and that, if we denote the closure in $\mathcal{H}(X)$ of $D|_{C_0^\infty(X \setminus \{0\})}$ by the same symbol D, then $W(\tau) = \exp(2i\tau D)$ (in the sense of the usual functional calculus in Hilbert space).

If Y is a subspace of X, one can define a continuous operator D_Y from $\mathcal{S}(X)$ into itself by choosing the orthonormal basis $\{v_1, \ldots, v_n\}$ of X in such a way that $v_1, \ldots, v_m \in Y$ and $v_{m+1}, \ldots, v_n \in Y^\perp$ and by setting

$$(1.2.22) \qquad D_Y = \frac{1}{4} \sum_{j=1}^{m} (P_j Q_j + Q_j P_j).$$

This definition is independent of the chosen basis, because for real $f \in C^1(X)$:

$$(D_Y f)(x) = \frac{1}{2i}(x, \pi_Y f'(x)) + \frac{1}{4i} \dim Y.$$

Obviously one has

$$(1.2.23) \qquad D = D_Y + D_{Y^\perp}.$$

We make a final convention. Later on we shall have to work with a family of subspaces of X, and to each of these subspaces we shall have to associate the analogue of the objects introduced above for X. One of these subspaces will be $\mathbf{O} = \{0\}$. By convention, if $X = \mathbf{O}$, we take:

$$\mathcal{H}(X) = \mathbb{C}, \quad \mathbb{B}(X) = \mathbb{K}(X) = \mathbb{C}, \quad \mathcal{S}(X) = \mathcal{S}^*(X) = \mathbb{C},$$

$$\mathfrak{F} = \mathfrak{F}^* = 1, \qquad \Delta = D = P = Q = 0.$$

1.3. Estimates of Functions and their Fourier Transforms

1.3.1. It is rather difficult, although often quite useful, to estimate the L^p-norm ($p \neq 2$) of a function in terms of its Fourier transform. We shall collect in this section some results of this nature that will be helpful in our further investigations.

Let us first point out that the Young inequality (1.2.7) can be restated as follows in a form that is sometimes more convenient (use (1.2.15)):

PROPOSITION 1.3.1 (YOUNG INEQUALITY). *Let* $1 \leq p \leq q \leq \infty$ *and define* s *by* $1 - s^{-1} = p^{-1} - q^{-1}$. *Then for each* $\varphi \in \mathscr{S}^*(X)$ *such that* $\widehat{\varphi} \in L^s(X)$, *the operator* $\varphi(P)$ *extends to a bounded operator from* $L^p(X)$ *to* $L^q(X)$ *with norm :*

$$(1.3.1) \qquad \|\varphi(P)\|_{L^p(X) \to L^q(X)} \leq \|\widehat{\varphi}\|_{L^s(X)}.$$

If $p < \infty$ *and* $q = \infty$, *then* $\varphi(P)L^p(X) \subset C_\infty(X)$.

In rough terms the preceding proposition shows that the application of an operator $\varphi(P)$ has a local smoothing effect which is unfortunately accompanied by a loss of decay at infinity.

If a function f has slow decay at infinity, its Fourier transform may be very bad locally (see Theorem 7.6.6 in [H]). If, however, the derivatives of f have better decay, the local behaviour of $\widehat{f}$ improves. On the other hand, the decay at infinity of f is controlled by the local behaviour of the derivatives of $\widehat{f}$. For the case $X = \mathbb{R}$ one may find numerous results supporting this picture in [Ti] and [Z]. The next theorem, the proof of which can be found in Chapter VI, §6.12 of [Ti], shows that there is a precise relation between the behaviour of $\widehat{f}$ at zero (or at infinity) and that of f at infinity (or at zero respectively). Observe that an assumption on the derivative f' is needed.

THEOREM 1.3.2. *Let* $f : (0, \infty) \to \mathbb{C}$ *be an absolutely continuous function, convergent to zero at infinity, and such that* $|f'(x)| \leq M x^{-1-\mu}$ *for some* $0 < \mu < 1$, $M \in \mathbb{R}$ *and all* $x > 0$. *For* $y > 0$ *let* $\widetilde{f}(y) = \int_0^\infty e^{-ixy} f(x) dx$ *(improper integral at infinity). Then*

$$\lim_{y \to +0} y^{1-\mu} \widetilde{f}(y) = \Gamma(1 - \mu) e^{-i\pi(1-\mu)/2} \lim_{x \to +\infty} x^\mu f(x)$$

if the limit on the right-hand side exists, and

$$\lim_{y \to +\infty} y^{1-\mu} \widetilde{f}(y) = \Gamma(1 - \mu) e^{-i\pi(1-\mu)/2} \lim_{x \to +0} x^\mu f(x)$$

if the limit on the right-hand side exists.

The next result is a less precise, but more general, n-dimensional version of this theorem. The decomposition that we use in its proof is a natural extension of the decomposition used by Titchmarsh. We denote by $[r]$ the integer part of a real number r, i.e. the largest integer $\leq r$.

PROPOSITION 1.3.3. *Let $k \in \mathbb{N}$ and let μ be a real number such that $\mu < n = \dim X$. Assume that $f \in L^1_{\text{loc}}(X)$, that the distributional derivatives $f^{(\alpha)}$ with $|\alpha| \le [n - \mu] + k + 1$ are locally bounded functions on $X \setminus \{0\}$, and that there is a constant $M < \infty$ such that $|f^{(\alpha)}(x)| \le M|x|^{-\mu-|\alpha|}$ on this domain. Then the distribution $\widehat{f}$ is a function of class C^k on $X \setminus \{0\}$ and there is a constant C (depending only on n, μ, k) such that $|\widehat{f}^{(\alpha)}(x)| \le CM|x|^{-(n-\mu)-|\alpha|}$ for $x \ne 0$ and $|\alpha| \le k$. If $\mu > 0$ and $|\alpha| < \mu$, then $\widehat{f}^{(\alpha)}$ is a function.*

PROOF. (i) Let us assume that the proposition has been proved for $k = 0$. Let $k \ge 1$ and let β be a multi-index with $|\beta| \le k$. We observe that $P^\beta \widehat{f} = (-1)^{|\beta|} \mathcal{F} Q^\beta f$. Now, if $\gamma \in \mathbb{N}^n$ is such that $|\gamma| \le [n - \mu] + |\beta| + 1$ and if $x \ne 0$, we get from the hypotheses of the theorem that

$$|\partial^\gamma [x^\beta f](x)| = \left| \sum_{\substack{\delta \le \beta \\ \delta \le \gamma}} \binom{\gamma}{\delta} \frac{\beta!}{\delta!} x^{\beta-\delta} f^{(\gamma-\delta)}(x) \right| \le cM|x|^{-(\mu-|\beta|)-|\gamma|}.$$

So we may apply the proposition for $k = 0$ (hence $|\alpha| = 0$), with μ replaced by $\mu - |\beta|$, to the function $Q^\beta f$ to obtain that

$$|\widehat{f}^{(\beta)}(x)| \equiv |P^\beta \widehat{f}| \le C_1 M |x|^{-(n-\mu)-|\beta|}$$

and that $\widehat{f}^{(\beta)}$ is a function if $\mu - |\beta| > 0$. This shows that the proposition is true provided that it holds for $k = 0$.

(ii) Let $k = 0$. We first show that $\widehat{f}$ is continuous and satisfies $|\widehat{f}(y)| \le CM|y|^{\mu-n}$ on $X \setminus \{0\}$. For this we fix a number $a > 0$ and we estimate separately the contribution to $\widehat{f}(y)$ coming from the regions $|x| < a^{-1}$ and $|x| > a^{-1}$. At the end we shall see that we get the best estimate by choosing a of the order $|y|$. Let $\varphi \in C^\infty(X)$ with $0 \le \varphi \le 1$, $\varphi(x) = 0$ if $|x| \le 1$ and $\varphi(x) = 1$ if $|x| \ge 2$. We write $f = f_0 + f_\infty$ with $f_0 = [1 - \varphi(aQ)]f$ and $f_\infty = \varphi(aQ)f$ and thus have $\widehat{f} = \mathcal{F}f_0 + \mathcal{F}f_\infty \equiv \widehat{f}_0 + \widehat{f}_\infty$. Clearly $f_0 \in L^1(X)$, so that $\widehat{f}_0$ is continuous on X and satisfies

(1.3.2)

$$|\widehat{f}_0(y)| \le \int_X [1 - \varphi(ax)]|f(x)|dx \le \int_{|x| \le 2/a} M|x|^{-\mu}dx = C_2 M a^{\mu-n},$$

where C_2 depends on n and μ.

One cannot estimate $\widehat{f}_\infty$ so simply because f_∞ need not decay rapidly enough at infinity (it could even grow) in order to ensure the absolute convergence of the Fourier integral. To handle this term we use the relation $Q^\alpha \widehat{f}_\infty = \mathcal{F}P^\alpha f_\infty$. In a neighbourhood of infinity $P^\alpha f_\infty$ behaves like $|x|^{-\mu-|\alpha|}$ for any $|\alpha| \le [n-\mu]+1$. If one chooses α such that $|\alpha| = [n-\mu]+1$, then $\mu+|\alpha| > n$, hence $P^\alpha f_\infty \in L^1(X)$. Then $Q^\alpha \widehat{f}_\infty$ is continuous on X. Now, for any $y_0 \ne 0$ in X, there is a multi-index α with $|\alpha| = [n-\mu]+1$ such that $y^\alpha \ne 0$ for all y in some neighbourhood of y_0. This gives the continuity of $\widehat{f}_\infty$ and hence that of $\widehat{f}$, on $X \setminus \{0\}$.

In order to get a useful estimate on $\widehat{f}_\infty$, we observe that

$$(1.3.3) \quad |(P^\alpha f_\infty)(x)| = |(-i)^{|\alpha|}\partial^\alpha(\varphi(ax)f(x))|$$

$$= |(-i)^{|\alpha|}\sum_{\beta\leq\alpha}\binom{\alpha}{\beta}a^{|\alpha|-|\beta|}\varphi^{(\alpha-\beta)}(ax)f^{(\beta)}(x)|$$

$$\leq \varphi(ax)M|x|^{-\mu-|\alpha|} + \sum_{\beta<\alpha}\binom{\alpha}{\beta}a^{|\alpha|-|\beta|}|\varphi^{(\alpha-\beta)}(ax)|M|x|^{-\mu-|\beta|}.$$

We again let $|\alpha| = [n-\mu]+1$. Then the contribution of the first term on the r.h.s. of (1.3.3) to the Fourier integral for $y^\beta\widehat{f}_\infty(y)$ is bounded by $C_3Ma^{\mu-n+|\alpha|}$ (since $\mu+|\alpha| > n$); the same bound is obtained for the contribution of each term in the last sum in (1.3.3), because for these terms the Fourier integral is restricted to the domain $\{x \in X \mid a^{-1} < |x| < 2a^{-1}\}$. In each case the constant C_3 depends only on n, μ and φ. In conclusion we get

$$(1.3.4) \qquad |y^\alpha\widehat{f}_\infty(y)| \leq C_4Ma^{\mu-n+|\alpha|} \quad \text{if } |\alpha| = [n-\mu]+1,\ y \in X.$$

Now if $\ell \geq 1$ is an integer, we have $|y|^\ell \leq c_\ell\sum_{|\alpha|=\ell}|y^\alpha|$ for some constant c_ℓ and all $y \in X$. Thus (1.3.4) implies that

$$(1.3.5) \qquad\qquad |\widehat{f}_\infty(y)| \leq C_5Ma^{\mu-n+\ell}|y|^{-\ell},$$

where $\ell = [n-\mu]+1$.

From (1.3.2) and (1.3.5) one obtains that

$$|\widehat{f}(y)| \leq C_6Ma^{\mu-n}(1 + a^\ell|y|^{-\ell}) \qquad \forall y \in X \setminus \{0\},$$

with C_6 independent of a and y. Thus we may take $a = |y|$ and conclude that $|\widehat{f}(y)| \leq CM|y|^{\mu-n}$ for all $y \neq 0$.

(iii) Now we assume $k = 0$ and $\mu > 0$, and we prove that $\widehat{f}$ is a function. Take $a = 1$ above and observe that f_0 is an integrable function with compact support, so that $\widehat{f}_0 \in C^\infty(X)$. Hence $\widehat{f}$ is a function if and only if $\widehat{f}_\infty$ is a function. Since f_∞ verifies the same estimate as f, the results obtained in (ii) show that there is a function $h \in L^1_{\mathrm{loc}}(X)$ such that $\widehat{f}_\infty(y) = h(y)$ on $X \setminus \{0\}$. Thus $\widehat{f}_\infty - h$ is a distribution with support in $\{0\}$; hence $\widehat{f}_\infty = h + g$, where g is a finite linear combination of derivatives of the Dirac measure δ, and it is enough to show that $g = 0$.

Let ψ be a function in $C_0^\infty(X)$ such that $\psi(y) = 1$ in some neighbourhood of zero. Then $\psi\widehat{f}_\infty = \psi h + g$. This equation is equivalent to $\psi(P)f_\infty = \mathcal{F}^*(\psi h) + \check{g}$. Now $\mathcal{F}^*(\psi h) \in C_\infty(X)$ because $\psi h \in L^1(X)$, and $\check{g}$ is a polynomial. Thus, to conclude that $\check{g} = 0$, it suffices to show that $\psi(P)f_\infty \in L^p(X)$ for some $p < \infty$. For this we observe that $f_\infty \in L^p(X)$ for $p > n\mu^{-1}$, because f_∞ is equal to zero near $x = 0$ and satisfies $|f_\infty(x)| \leq M|x|^{-\mu}$. By applying the Young inequality (1.3.1) one obtains that $\psi(P)f_\infty \in L^p(X)$ for $p > n\mu^{-1}$. $\square$

COROLLARY 1.3.4. *Let $g \in L^1(X)$ be rapidly decreasing at infinity (i.e. $Q^\beta g \in L^1(X)$ for each multi-index β). Assume that there is a real number $m > 0$*

such that $|g^{(\beta)}(x)| \le c_\beta |x|^{-(n-m)-|\beta|}$ for all $x \ne 0$ and each $\beta \in \mathbb{N}^n$. Then $\widehat{g} \in S^{-m}(X)$.

PROOF. $\widehat{g}$ is a function of class C^∞ because $P^\beta \widehat{g} = (-1)^{|\beta|} \cdot \mathcal{F}(Q^\beta g)$ belongs to $C_\infty(X)$ for each $\beta \in \mathbb{N}^n$. The estimates $|\widehat{g}^{(\beta)}(x)| \le c'_\beta \langle x \rangle^{-m-|\beta|}$ for $|x| \ge 1$ are obtained by applying Proposition 1.3.3 with $f = g$ and $\mu = n - m < n$. $\square$

The Fourier transforms of symbols behave very nicely outside the origin. Let us say that a distribution on X is of class $\mathscr{S}$ outside zero if it is of class C^∞ on $X \setminus \{0\}$ and decays at infinity, together with all its derivatives, more rapidly than any power of $\langle x \rangle^{-1}$.

LEMMA 1.3.5. If $f \in S^\infty(X)$, then $\mathcal{F}f$ is of class $\mathscr{S}$ outside zero.

PROOF. (i) Let $h \in S^m(X)$ for some $m \in \mathbb{R}$. Then $h^{(\alpha)} \in S^{m-|\alpha|}(X)$, and since $S^k(X) \subset L^1(X)$ if $k < -\dim X$, we have $h^{(\alpha)} \in L^1(X)$ if $|\alpha| > m + \dim X$. Thus, for each multi-index α such that $|\alpha| > m + \dim X$, $\mathcal{F}h^{(\alpha)}$ is a continuous function converging to zero at infinity. Because $\mathcal{F}h^{(\alpha)} = (iQ)^\alpha \widehat{h}$, this clearly implies the following: $\widehat{h}$ is continuous on $X \setminus \{0\}$ and for each $N < \infty$ and each $\varepsilon > 0$ there is a constant c such that $|\widehat{h}(x)| \le c|x|^{-N}$ if $|x| > \varepsilon$.

(ii) Now let $f \in S^\infty(X)$. Since $Q^\alpha f \in S^\infty(X)$ for each multi-index α and $\widehat{f}^{(\alpha)}$ is the Fourier transform of $(-iQ)^\alpha f$, the result of (i) [applied to $h = Q^\alpha f$] implies that $\widehat{f}^{(\alpha)}$ is continuous on $X \setminus \{0\}$ and of rapid decay at infinity. $\square$

If $0 < m < n$, one can completely characterize symbols of class $S^{-m}(X)$ in terms of their Fourier transforms ($n \equiv \dim X$):

PROPOSITION 1.3.6. Let m be a real number such that $0 < m < n$, and let $f \in \mathscr{S}^*(X)$. Then $f \in S^{-m}(X)$ if and only if $\widehat{f} \in L^1(X)$, $\widehat{f}$ is of class $\mathscr{S}$ outside zero and for each $\alpha \in \mathbb{N}^n$ there is a constant c_α such that

$$(1.3.6) \qquad |\partial^\alpha \widehat{f}(x)| \le c_\alpha |x|^{-(n-m)-|\alpha|} \qquad \forall x \ne 0.$$

PROOF. Proposition 1.3.3 (in which we take $\mu = m$ and let $k \to \infty$) and Lemma 1.3.5 imply that $\widehat{f}$ has the stated properties if $f \in S^{-m}(X)$. The converse follows from Corollary 1.3.4 with $g(x) = \check{f}(x) = \widehat{f}(-x)$. $\square$

If $f \in S^0(X)$ then $\widehat{f}$ is not a function in general. In fact the singularity of $\widehat{f}$ at the origin may be so strong that $\widehat{f}$ is not a measure. This happens in the very simple case where $f \in C^\infty(\mathbb{R})$, $f(x) = 0$ for $x \le -1$ and $f(x) = 1$ for $x \ge 1$; then $\widehat{f}(x) - PV(c/x)$ is a function of class C^∞ for some constant $c \ne 0$ and $PV(1/x)$ is a distribution of strictly positive order. However, the order of $\widehat{f}$ for $f \in S^0(X)$ is arbitrarily small if the concept of order of a distribution is suitably generalized. We recall that the order of a distribution is the smallest integer k such that the distribution defines a bounded linear functional on $C_0^\infty(X)$ provided with the norm $\| \cdot \|_{BC^k}$; a natural generalization of this property is that used in the next proposition. We shall show that, if $f \in S^m(X)$ for some $m \ge 0$, then $\widehat{f}$ is a distribution of order μ (in a generalized sense) for any real $\mu > m$. If $\mu > 0$ is non-integer, we denote by $[\mu]$ its integer part and introduce the norm

$$(1.3.7) \quad \|g\|_{BC^\mu} = \|g\|_{BC^{[\mu]}} + \sup_{\substack{x,y \in X \\ x \neq y}} \left[\sum_{|\alpha|=[\mu]} \frac{[\mu]!}{\alpha!} \left| \frac{g^{(\alpha)}(x) - g^{(\alpha)}(y)}{|x-y|^{\mu-[\mu]}} \right|^2 \right]^{1/2}$$

with $\|g\|_{BC^k}$ for k integer given by (1.1.5). In (1.3.7) the derivatives are taken with respect to an orthonormal basis of X, but the expression is independent of the choice of this basis.

PROPOSITION 1.3.7. *Let $m \in [0, \infty)$ and $f \in S^m(X)$. Then for each $\mu > m$ there is a constant c_μ such that*

$$(1.3.8) \qquad |\langle g, \hat{f} \rangle| \leq c_\mu \|g\|_{BC^\mu} \qquad \forall g \in \mathscr{S}(X).$$

PROOF. (i) We first prove the following: if $\theta \in C_0^\infty(X \setminus \{0\})$ and $\mu > 0$, there is a constant $c = c(\theta, \mu)$ such that for all $\varepsilon > 0$ and all $g \in \mathscr{S}(X)$

$$(1.3.9) \qquad \|\theta(\varepsilon P)g\|_{L^\infty(X)} \leq c\varepsilon^\mu \|g\|_{BC^\mu}.$$

For this we use first (1.2.15), then the relation $\theta^{(\alpha)}(0) = 0$ for each multi-index α and finally Taylor's formula (1.1.8) to obtain that

$$[\theta(\varepsilon P)g](x) = \varepsilon^{-n} \int_X \check{\theta}(\varepsilon^{-1}y)g(x-y)\underline{dy}$$

$$= \varepsilon^{-n} \int_X \check{\theta}(\varepsilon^{-1}y)[g(x-y) - \sum_{|\alpha| \leq [\mu]} \frac{(-y)^\alpha}{\alpha!} g^{(\alpha)}(x)]\underline{dy}$$

$$= \varepsilon^{-n} \sum_{|\alpha|=[\mu]} \int_X \check{\theta}(\varepsilon^{-1}y) \frac{(-y)^\alpha}{\alpha!} [g_\alpha(x,-y) - g^{(\alpha)}(x)]\underline{dy},$$

with g_α defined as in (1.1.7) if $|\alpha| \neq 0$ and $g_0(x,-y) = g(x-y)$. If $\mu \in (0,1)$, we have $|\alpha| = 0$, hence $|g_0(x,-y) - g(x)| \equiv |g(x-y) - g(x)| \leq \|g\|_{BC^\mu}|y|^\mu$. If $\mu \geq 1$, we have $|\alpha| \neq 0$ and

$$|g_\alpha(x,-y) - g^{(\alpha)}(x)| = |\alpha| \left| \int_0^1 [g^{(\alpha)}(x-\tau y) - g^{(\alpha)}(x)](1-\tau)^{|\alpha|-1}d\tau \right|$$

$$\leq \|g\|_{BC^\mu}|y|^{\mu-[\mu]}.$$

So

$$|[\theta(\varepsilon P)g](x)| \leq \varepsilon^{-n} \|g\|_{BC^\mu} \sum_{|\alpha|=[\mu]} \int_X |y|^{\mu-[\mu]} \cdot |y^\alpha \check{\theta}(\varepsilon^{-1}y)|\underline{dy}$$

$$= \varepsilon^\mu \|g\|_{BC^\mu} \sum_{|\alpha|=[\mu]} \int_X |z|^{\mu-[\mu]} |z^\alpha \check{\theta}(z)|\underline{dz},$$

which gives (1.3.9).

(ii) Now let $\mu > m$ and fix two real numbers κ and ν satisfying $m < \kappa < \nu < \mu$. Then, since $\langle P \rangle^\kappa \equiv \mathcal{F}^* \langle Q \rangle^\kappa \mathcal{F} = \mathcal{F} \langle Q \rangle^\kappa \mathcal{F}^*$, we get for $g \in \mathscr{S}(X)$:

$$(1.3.10) \qquad |\langle g, \widehat{f} \rangle| = |\langle \langle P \rangle^\kappa g, \mathcal{F} \langle Q \rangle^{-\kappa} f \rangle|$$

$$\leq \|\langle P \rangle^\kappa g\|_{L^\infty(X)} \|\mathcal{F} \langle Q \rangle^{-\kappa} f\|_{L^1(X)}.$$

Since $\langle Q \rangle^{-\kappa} f \in S^{-(\kappa - m)}(X)$, we have $\mathcal{F} \langle Q \rangle^{-\kappa} f \in L^1(X)$ by Proposition 1.3.6. To estimate the L^∞-norm of $\langle P \rangle^\kappa g$, we apply the standard Littlewood-Paley method. We choose two functions ψ and φ in $C_0^\infty(X)$ such that $\psi(x) = 0$ if $|x| \geq 1$, $\varphi(x) = 0$ if $|x| \notin (1/2, 2)$ and $1 = \psi(x) + \sum_{j=0}^\infty \varphi(2^{-j} x)$ for each $x \in X$ (one may take $\varphi(x) = u(|x|/2) - u(|x|)$, where the function $u \in C_0^\infty(\mathbb{R})$ is such that $u(t) = 1$ if $|t| \leq 1/2$ and $u(t) = 0$ if $|t| \geq 1$). Furthermore we let $\eta \in C^\infty(X)$ be such that $\eta(x) = 1$ for $|x| \geq 1/2$ and $\eta(x) = 0$ in some neighbourhood of $x = 0$. We have $\eta(x) = \eta(x) \varphi(2^{-j} x)$ for each $j \in \mathbb{N}$ and

$$\langle P \rangle^\kappa g = \langle P \rangle^\kappa \psi(P) g + \sum_{j=0}^\infty \langle P \rangle^\kappa |P|^{-\nu} \eta(P) \cdot 2^{\nu j} |2^{-j} P|^\nu \varphi(2^{-j} P) g.$$

Now it follows from Proposition 1.3.1 that $\langle P \rangle^\kappa \psi(P)$ is bounded as an operator from $L^\infty(X)$ to $L^\infty(X)$. Similarly, if we set $\xi(x) = \langle x \rangle^\kappa |x|^{-\nu} \eta(x)$, then $\xi(P)$ is bounded as an operator in $L^\infty(X)$ because $\xi \in S^{-(\nu - \kappa)}(X)$, so that $\widehat{\xi} \in L^1(X)$ by Proposition 1.3.6. Thus, if we define ϕ by $\phi(x) = |x|^\nu \varphi(x)$, we get that for some constant c' and all $g \in \mathscr{S}(X)$:

$$(1.3.11) \quad \|\langle P \rangle^\kappa g\|_{L^\infty(X)} \leq c' \|g\|_{L^\infty(X)} + c' \sum_{j=0}^\infty 2^{\nu j} \|\phi(2^{-j} P) g\|_{L^\infty(X)}.$$

The estimate (1.3.8) now follows from (1.3.10) and (1.3.11) by applying (1.3.9), with $\varepsilon = 2^{-j}$, to the function $\theta = \phi$. $\square$

We make a final remark concerning the Fourier transforms of symbols f of class $S_\delta^{-m}(\mathbb{R}^n)$. We recall that this means $f \in C^\infty$ and $|f^{(\alpha)}(x)| \leq c_\alpha \langle x \rangle^{-m - \delta |\alpha|}$ for all α. Assume that m, δ are real numbers such that $0 < \delta < 1$ and $0 < m < n(1 - \delta)/2$ and let the function f be of class C^∞ and such that $f(x) = \langle x \rangle^{-m} \exp(i|x|^{1-\delta})$ for $|x| > 1$. Then $f \in S_\delta^{-m}(\mathbb{R}^n)$ and $\widehat{f}$ is not an integrable measure. Indeed, if it were, then $f(P)$ would be a bounded operator in $L^p(\mathbb{R}^n)$ for all $p \in [1, \infty]$; but this does not hold if $|2^{-1} - p^{-1}| > m[n(1 - \delta)]^{-1}$ according to §7.4, Ch. 4 of [St1].

1.3.2. In this subsection we first state and prove a formula due to A.P. Calderón which gives a representation of a distribution in terms of its derivatives of a fixed order plus an easily controllable term (see the proof of Theorem 12 in [C1]). This is quite similar to the Sobolev integral representations and is a natural extension of the fundamental formula $f(x) = f(y) + \int_y^x f'(x) dx$ of differential calculus. Then we shall indicate some applications of this formula.

We denote by S_X the unit sphere of the euclidean space X (i.e. the set of all points $x \in X$ such that $|x| = 1$) and by $\underline{d\omega}$ the rotation invariant measure on S_X induced by the Fourier measure $\underline{dx}$ on X.

THEOREM 1.3.8. *Let $\xi \in C^\infty(X \setminus \{0\})$ have the following properties:*
(i) *there is $b < \infty$ such that $\xi(x) = 0$ if $|x| > b$;*
(ii) *there is $a > 0$ and there is a function $\widetilde{\xi}$ on S_X with $\int_{S_X} \widetilde{\xi}(\omega)\underline{d}\omega = 1$ such that $\xi(x) = |x|^{-\dim X}\widetilde{\xi}\left(\frac{x}{|x|}\right)$ if $0 < |x| < a$.*

Choose an orthonormal basis of X and define, for each multi-index $\alpha \neq 0$, $\xi_\alpha(x) = \frac{|\alpha|}{\alpha!}(ix)^\alpha \xi(x)$. Then ξ_α belongs to $L^1(X)$ and has compact support, and its Fourier transform $\widehat{\xi}_\alpha$ is a symbol of class $S^{-|\alpha|}(X)$. Furthermore, if $m \geq 1$ is an integer, let η_m be the function given by

$$\eta_m(x) = -\sum_{|\alpha|=m} \frac{m}{\alpha!}\partial^\alpha(x^\alpha \xi(x)) \quad \text{if } x \neq 0, \eta_m(0) = 0.$$

Then $\eta_m \in C_0^\infty(X \setminus \{0\})$, and one has for each $f \in \mathscr{S}^*(X)$ and each integer $m \geq 1$:

$$(1.3.12) \quad f = \sum_{|\alpha|=m} \xi_\alpha * P^\alpha f + \eta_m * f = \sum_{|\alpha|=m} P^\alpha(\xi_\alpha * f) + \eta_m * f$$

$$= \sum_{|\alpha|=m} P^\alpha \widehat{\xi}_\alpha(P)f + \widehat{\eta}_m(P)f.$$

PROOF. (i) The properties of ξ_α are easy to verify. The fact that $\widehat{\xi}_\alpha \in S^{-|\alpha|}(X)$ follows from Corollary 1.3.4 (with $g = \xi_\alpha$ and $m = |\alpha| \geq 1$), since ξ_α is homogeneous of degree $-(n - |\alpha|)$ in a neighbourhood of zero.

(ii) We shall prove the following identity in part (iii):

$$(1.3.13) \quad \sum_{|\alpha|=m} \frac{m}{\alpha!}\partial^\alpha(x^\alpha \xi(x)) = \delta(x) \quad \text{for } |x| < a,$$

where the derivatives are taken in the sense of distributions and $\delta(x)$ is the Dirac measure concentrated at the origin. (1.3.13) implies that η_m is equal to zero in a neighbourhood of the origin and that for any $f \in \mathscr{S}(X)$:

$$\sum_{|\alpha|=m} P^\alpha(\xi_\alpha * f) + \eta_m * f = \left(\sum_{|\alpha|=m} P^\alpha \xi_\alpha + \eta_m\right) * f = \delta * f = f.$$

Since $\mathscr{S}(X)$ is dense in $\mathscr{S}^*(X)$, this equality can be extended by continuity to all $f \in \mathscr{S}^*(X)$. Finally the last identity in (1.3.12) holds by virtue of (1.2.15).

(iii) It remains to prove (1.3.13), namely that for each $f \in C^\infty(X)$ satisfying $f(x) = 0$ for $|x| \geq a$:

$$(1.3.14) \quad \frac{(-1)^m}{(m-1)!}\int_X \xi(x) \sum_{|\alpha|=m} \frac{m!}{\alpha!}x^\alpha \partial^\alpha f(x)\underline{d}x = f(0).$$

This is quite easy in polar coordinates $r = |x|$, $\omega = xr^{-1}$. Notice first the identity

$$\sum_{|\alpha|=m} \frac{m!}{\alpha!}x^\alpha \partial^\alpha = r^m \left(\frac{\partial}{\partial r}\right)^m$$

valid on $C^\infty(X \setminus \{0\})$. Now we may rewrite the l.h.s. of (1.3.14) as follows:

$$\frac{(-1)^m}{(m-1)!} \lim_{\varepsilon \to +0} \int_{S_X} d\omega \widetilde{\xi}(\omega) \int_\varepsilon^a r^{m-1} \left(\frac{\partial}{\partial r}\right)^m f(r\omega).$$

By integrating m times by parts one sees that this expression is equal to $f(0)$, which proves (1.3.14). $\square$

We mention that there is another important identity due to A.P. Calderón which allows one to write a distribution as a superposition of smooth functions: it is easy to construct $\varphi \in \mathscr{S}(X)$ such that $f = \int_0^\infty \varphi(\sigma P)^2 f \sigma^{-1} d\sigma$ for all $f \in \mathscr{S}^*(X)$. This identity is called Calderón's formula (homogeneous form) and should be considered as a continuous version of the Littlewood-Paley decomposition (see part (ii) of the proof of Proposition 1.3.7). One may consult [FJW] and [JT] for a detailed presentation of the preceding formula and its applications. One may write Calderón's formula in the so-called inhomogeneous form, namely $f = \int_0^1 \varphi(\sigma P)^2 f \sigma^{-1} d\sigma + \psi(P) f$, where $\psi \in \mathscr{S}(X)$ (see [JT]). An abstract version of this identity will (implicitly) play a fundamental role in our Section 3.5.

The first equality in (1.3.12) is useful for obtaining local and global estimates on a function in terms of its derivatives. In fact $\eta_m * f$ belongs to $C^\infty(X)$ for any $f \in \mathscr{S}^*(X)$, so that the local singularities of f are the same as those of $\sum_{|\alpha|=m} \xi_\alpha * P^\alpha f$. Here ξ_α has compact support and satisfies $|\xi_\alpha(x)| \le c_\alpha |x|^{-(n-|\alpha|)}$ (where $n = \dim X$). So, if $1 \le |\alpha| \le n-1$, we have $\xi_\alpha \in L^q(X)$ for any $q < n(n-|\alpha|)^{-1}$, if $|\alpha| = n$, we have $\xi_\alpha \in L^\infty(X)$, and if $|\alpha| = n+k$ with $k \ge 1$ integer, then ξ_α is of class $C_0^{k-1}(X)$ with $\partial^\beta \xi_\alpha$ Lipschitz for $|\beta| = k-1$. This can be used together with the Young inequality or with the Hardy-Littlewood-Sobolev inequality (see Theorem V.1 of [St1]) to estimate f in terms of $\{f^{(\alpha)}\}_{|\alpha|=m}$.

The remainder of this section contains several elementary estimates and regularity results that can be obtained by this technique. Let ξ be a function satisfying the conditions of Theorem 1.3.8 with $a = 1/2$ and $b = 1$. Then the function $\xi_\varepsilon(x) = \varepsilon^{-n} \xi(x\varepsilon^{-1})$ satisfies them too (with $a = \varepsilon/2$ and $b = \varepsilon$, but with the same $\widetilde{\xi}$). By writing the first identity from (1.3.12) with ξ replaced by ξ_ε and then by taking the derivatives of some order $\le m-1$ of both sides, one obtains

$$(1.3.15) \qquad f^{(\alpha)}(x) = \sum_{|\beta|=m} \frac{m}{\beta!} \varepsilon^{m-|\alpha|} \int_X f^{(\beta)}(x - \varepsilon y) \partial_y^\alpha [y^\beta \xi(y)] \underline{dy}$$

$$+ \varepsilon^{-|\alpha|} \int_X f(x - \varepsilon y) \eta_m^{(\alpha)}(y) \underline{dy}.$$

Here $f \in \mathscr{S}^*(X)$, $m \ge 1$ is an integer, $\varepsilon > 0$ is real and α is a multi-index with $|\alpha| \le m-1$; the integrals are understood in the sense of distributions (the convolution of a distribution of compact support with an arbitrary distribution is well defined; the situation is especially simple in (1.3.15) because $|\alpha| < |\beta|$, hence $\partial^\alpha[y^\beta \xi(y)]$ is an integrable function with compact support and we may use (1.2.13)). We see that (1.3.15) expresses $f^{(\alpha)}(x)$ for $|\alpha| \le m-1$ in terms of $f^{(\beta)}(x - \varepsilon y)$ with $|\beta| = m$ and $y \in \operatorname{supp} \xi$, and of $f(x - \varepsilon y)$ with $y \in \operatorname{supp} \eta_m$. In particular, by taking into account that the support of ξ is contained in the unit ball of X, it is clear that if f is an arbitrary distribution on an arbitrary open subset Ω of X, then (1.3.15) still holds on $\Omega_\varepsilon = \{x \in \Omega \mid \mathrm{dist}(x, \partial\Omega) > \varepsilon\}$. This gives the following local regularity results which will be useful later

PROPOSITION 1.3.9. *If f is a distribution on an open set $\Omega \subset X$, and if its first order derivatives are measures, then f is a function on Ω (indeed, $f \in L^p_{\mathrm{loc}}(\Omega)$ for all $p < n/(n-1)$). If $f^{(\alpha)}$ is a measure on Ω for each $\alpha \in \mathbb{N}^n$ of order $|\alpha| = n$, then $f \in L^\infty_{\mathrm{loc}}(\Omega)$. Finally, if $f^{(\alpha)} \in L^1_{\mathrm{loc}}(\Omega)$ for all α with $|\alpha| = n$, then f is a continuous function on Ω.*

PROOF. Let Ω', Ω'' be bounded open sets such that $\overline{\Omega'} \subset \Omega''$ and $\overline{\Omega''} \subset \Omega$. Consider (1.3.15) for $x \in \Omega'$ and $\varepsilon < \mathrm{dist}(\Omega', \partial\Omega'')$. Since the last term in (1.3.15) is a function of class C^∞ on Ω', we see that $f^{(\alpha)}$ is a function on Ω' if $f^{(\beta)}$ is an integrable measure on Ω'' for $|\beta| = m$. Moreover, there is a constant C independent of f, such that

$$|f^{(\alpha)}(x)| \leq C \sum_{|\beta|=m} \int_{\Omega''} |x - y|^{-n-|\alpha|+|\beta|} |f^{(\beta)}(y)| \underline{dy}$$
$$+ \varepsilon^{-|\alpha|} \left| \int_X f(x - \varepsilon y) \eta_m^{(\alpha)}(y) \underline{dy} \right|$$

for almost every $x \in \Omega'$. This implies the first two assertions of the proposition (take $\alpha = 0$ and $m = 1$ or $m = n$). For the third one, note that there is $C < \infty$ such that

$$(1.3.16) \qquad \|f\|_{L^\infty(\Omega')} \leq C \int_{\Omega''} \left(\sum_{|\beta|=n} |f^{(\beta)}(y)| + |f(y)| \right) \underline{dy}$$

if $f^{(\beta)}$ are measures on Ω'' for $|\beta| = n$. Now assume that f is a distribution on Ω such that $f^{(\beta)} \in L^1_{\mathrm{loc}}(\Omega)$ if $|\beta| = n$. Let $\varrho \in C_0^\infty(X)$ with $\varrho \geq 0$ and $\int \varrho dx = 1$, set $\varrho_\nu(x) = \nu^{-n}\varrho(x\nu^{-1})$ for $\nu > 0$, and define $g_\nu = \varrho_\nu * f|_{\Omega''}$ for ν small enough. Then $g_\nu \in C^\infty(\Omega'')$ and for $\nu \to 0$ we have $g_\nu \to f$ in $L^1(\Omega'')$ (we have seen before that $f \in L^1_{\mathrm{loc}}(\Omega)$) and $g_\nu^{(\beta)} = \varrho_\nu * f^{(\beta)} \to f^{(\beta)}$ in $L^1(\Omega'')$ for $|\beta| = n$ (because $f^{(\beta)} \in L^1_{\mathrm{loc}}(\Omega)$ by hypothesis). From (1.3.16) we get $g_\nu \to f$ in $L^\infty(\Omega')$, so f is continuous on Ω'. This finishes the proof, because Ω' is an arbitrary open set with compact closure in Ω. $\square$

We mention that the first result of Proposition 1.3.9 can be easily improved to $f \in L^p_{\mathrm{loc}}(\Omega)$ with $p = n/(n-1)$; but for this, one has to use the Gagliardo-Nirenberg estimate (see (23) in Ch. 5, §2, [St1]).

The identity (1.3.15) allows one to estimate certain norms of f on a domain Ω in terms of its derivatives on the same domain (i.e. without having to consider derivatives on a larger domain, as in (1.3.16); this is not quite trivial even in very simple cases, e.g. if Ω is the unit ball). This will be possible for domains having the cone property, because we can choose ξ with support in a narrow truncated cone with vertex at the origin. We use the following notations: if $e \in X$, $|e| = 1$ and $0 < \theta \leq \pi$, we set $\Gamma(e, \theta) = \{x \in X \mid |x| \leq 1, (x, e) \geq |x| \cos\theta\}$. If $\theta = \pi$, then $\Gamma(e, \theta)$ is just the closed unit ball, while for $\theta < \pi$ it is the intersection of the closed unit ball with a closed circular cone with vertex at the origin, axis $\mathbb{R}_+e$ and angle θ. We also set $\widetilde{\Gamma}(e,\theta) = \{x \in \Gamma(e,\theta) \mid |x| \geq 1/2\}$. An open subset Ω of X is said to have the *cone property* if there are two numbers $\delta > 0$ and $0 < \theta \leq \pi$ such that for each $x \in X$ one can find a vector $e_x \in X$ with $|e_x| = 1$ and $x + \delta\Gamma(e_x, \theta) \subset \Omega$. If this condition is satisfied, then it is easily seen that for each $x_0 \in \Omega$ there are a number $\mu > 0$ and a vector $e \in S_X$ such that $x + \delta\Gamma(e, \theta) \subset \Omega$ for all $x \in X$ with $|x - x_0| < \mu$. This property allows one to construct a Borel function $\Omega \ni x \mapsto e_x \in S_X$ such that $x + \delta\Gamma(e_x, \theta) \subset \Omega$ for all $x \in \Omega$ [1]. We set $\Gamma_x = \Gamma(e_x, \theta)$ and we denote by χ_x^ε the characteristic function of the set $\varepsilon\Gamma_x$ for $0 < \varepsilon \leq \delta$.

We shall now prove the following preliminary estimate: *there is a constant $\gamma < \infty$, depending only on θ, m and n, such that if f is a function of class C^m on a neighbourhood of the truncated cone $x + \delta\Gamma(e, \theta)$, then for each integer $k \in [0, m-1]$ and each $\varepsilon \in (0, \delta)$:*

$$(1.3.17) \qquad \sum_{|\alpha|=k} |f^{(\alpha)}(x)| \leq \gamma \sum_{|\beta|=m} \int_{x+\varepsilon\Gamma(e,\theta)} |f^{(\beta)}(y)||y-x|^{m-k-n}dy$$

$$+\gamma\varepsilon^{-n-k} \int_{x+\varepsilon\widetilde{\Gamma}(e,\theta)} |f(y)|dy.$$

We obtain this by taking in (1.3.15) a function ξ of the form $\xi(x) = u(r)v((\omega, e))$, where $r = |x|$ and $\omega = x|x|^{-1}$ are the polar coordinates of x, and the functions $u \in C^\infty((0,\infty))$, $v \in C^\infty(\mathbb{R})$ have the following properties: (1) $u(r) = r^{-n}$ if $0 < r < 1/2$; (2) $u(r) = 0$ if $r \geq 1$; (3) $v(t) = 0$ if $t \geq -\cos\theta$. Then $\operatorname{supp}\xi \subset \Gamma(-e, \theta)$ and $\operatorname{supp}\eta_m \subset \widetilde{\Gamma}(-e, \theta)$. Since we clearly have estimates on ξ and its derivatives which are independent of e, the preceding inequality is an easy consequence of (1.3.15).

Let us choose $e = e_x$ in (1.3.17) and assume $f \in C^m(\Omega)$. In terms of the notation $|\nabla^{(k)} f(x)|$ introduced in (1.1.4) we see that there is a constant $\gamma_1 < \infty$, depending only on θ, m and n, such that for each $x \in \Omega$, $0 \leq k \leq m-1$ integer and $\varepsilon \in (0, \delta)$:

$$(1.3.18) \qquad |\nabla^{(k)} f(x)| \leq \gamma_1 \int \chi_x^\varepsilon(y-x)|y-x|^{m-k-n}|\nabla^{(m)} f(y)|dy$$

$$+\gamma_1\varepsilon^{-n-k} \int \chi_x^\varepsilon(y-x)|f(y)|dy.$$

By using the Hölder inequality we obtain that:

$$(1.3.19)$$
$$\varepsilon^k|\nabla^{(k)} f(x)| \leq \gamma_p\varepsilon^{m-\frac{n}{p}}\|\nabla^{(m)} f\|_{L^p(x+\varepsilon\Gamma_x)} + \gamma_1\varepsilon^{-n}\|f\|_{L^1(x+\varepsilon\Gamma_x)}$$

where γ_p is a constant depending only on θ, m, n and p. (1.3.19) holds for $1 \leq p \leq \infty$, with $k \leq m-n$ if $p = 1$ and $k < m - \frac{n}{p}$ if $p > 1$. In particular,

---

[1]Let $\{x_j\}$ be a sequence of points in Ω which is dense in Ω. For each j, choose a number $\mu_j > 0$ and a vector $e_j \in S_X$ such that $|y - x_j| < \mu_j \Rightarrow y + \delta\Gamma(e_j, \theta) \subset \Omega$. Then define e_x as follows: (1) if $|x-x_1| < \mu_1$, set $e_x = e_1$, (2) if $|x-x_i| \geq \mu_i$ for $1 \leq i \leq j$ and $|x-x_{j+1}| < \mu_{j+1}$, set $e_x = e_{j+1}$.

under the same conditions we get the inequality

$$(1.3.20) \quad \varepsilon^k ||\nabla^{(k)} f||_{L^\infty(\Omega)} \leq \gamma_p \varepsilon^{m-\frac{n}{p}} ||\nabla^{(m)} f||_{L^p(\Omega)} + \widetilde{\gamma}_1 \varepsilon^{-n/q} ||f||_{L^q(\Omega)}$$

where $1 \leq q \leq \infty$ and $\widetilde{\gamma}_1 < \infty$ depends only on θ, m and n. This is a Sobolev type estimate. Other such inequalities can be obtained without difficulty starting from (1.3.18). Let us mention that $C^\infty(\Omega) \cap W^{m,p}(\Omega)$ is dense in $W^{m,p}(\Omega) \equiv \{f \in L^p(\Omega) \mid f^{(\alpha)} \in L^p(\Omega)$ if $|\alpha| \leq m\}$ (see Theorem 2.3.2 in [Zi]), so it is enough to prove such estimates for functions f of class $C^\infty(\Omega)$.

We shall deduce from (1.3.18) other inequalities, which allow one to estimate the norm of differential operators from $W^{m,p}(\Omega)$ to $L^p(\Omega)$. We shall denote by the same symbol C various constants which depend only on θ, m, n, p and λ. Here θ, m, n and p have the same meaning as before, but we assume $1 < p < \infty$; and $\lambda > 0$ is a new parameter. By using the Hölder inequality in order to estimate the first term in the r.h.s. of (1.3.18), we obtain:

$$\int_{x+\varepsilon\Gamma_x} \{|\nabla^{(m)} f(y)| \cdot |y-x|^{m-k-\lambda}\} \cdot |y-x|^\lambda \frac{dy}{|y-x|^n} \leq$$

$$\leq C\varepsilon^\lambda \Big[\int_X \frac{\chi_x^\varepsilon(y-x)}{|y-x|^{n-p(m-k-\lambda)}} |\nabla^{(m)} f(y)|^p dy\Big]^{1/p}.$$

Inserting this into (1.3.18) we get

$$(1.3.21) \quad |\nabla^{(k)} f(x)|^p \leq C\varepsilon^{\lambda p} \int_X \frac{\chi_x^\varepsilon(y-x)}{|y-x|^{n-p(m-k-\lambda)}} |\nabla^{(m)} f(y)|^p dy$$

$$+ C\varepsilon^{-n-kp} \int_X \chi_x^\varepsilon(y-x) |f(y)|^p dy$$

for each $x \in \Omega$, $0 \leq k \leq m-1$, $\varepsilon \in (0,\delta)$ and $f \in C^m(\Omega)$; the constant C depends only on θ, m, n, p and λ, and we recall the conditions $1 < p < \infty$, $\lambda > 0$. We set, for an arbitrary Borel function $u : X \to \mathbb{C}$, $\mu \in \mathbb{R}$, $1 < p < \infty$ and $\varepsilon > 0$:

$$M_{p,\mu}(u; \Omega, \varepsilon) = \sup_{x_0 \in \Omega} \Big[\int_{x \in \Omega, |x-x_0| < \varepsilon} |u(x)|^p \frac{dx}{|x-x_0|^{n-p\mu}}\Big]^{1/p}.$$

If $\mu = n/p$, we set $M_{p,\mu} \equiv N_p$. By multiplying (1.3.21) by $|u(x)|^p$ and then integrating over Ω, we obtain

$$(1.3.22) \quad |||u|\nabla^{(k)} f|||_{L^p(\Omega)} \leq C\varepsilon^\lambda M_{p,m-k-\lambda}(u; \Omega, \varepsilon) ||\nabla^{(m)} f||_{L^p(\Omega)}$$

$$+ C\varepsilon^{-\frac{n}{p}-k} N_p(u; \Omega, \varepsilon) ||f||_{L^p(\Omega)}.$$

Here $\Omega \subset X$ is an open set with the cone property (characterized by two constants $\theta \in (0, \pi]$ and $\delta > 0$), $n = \dim X$, k and m are integers such that $0 \leq k < m$, and p, λ are real numbers such that $1 < p < \infty$, $\lambda > 0$. C is a constant which depends only on θ, m, n, p and λ and (1.3.22) holds for each $\varepsilon \in (0, \delta)$ and $f \in W^{m,p}(\Omega)$ (recall that $C^\infty(\Omega) \cap W^{m,p}(\Omega)$ is dense in $W^{m,p}(\Omega)$).

An assumption of the form $M_{p,\mu}(u; \Omega, \varepsilon) < \infty$ is referred to as a *condition of Stummel type*. Such conditions are very convenient in the study of linear differential operators; see [JW], especially for the case of Schrödinger operators with many-body potentials, and [Sche1] for the case of arbitrary elliptic operators. We mention the following easily proven facts:

(i) If the condition $M_{p,\mu}(u; \Omega, \varepsilon) < \infty$ is satisfied for some $\varepsilon > 0$, then it is satisfied for all $\varepsilon > 0$.

(ii) If $\mu \leq 0$, then $M_{p,\mu}(u; \Omega, \varepsilon) < \infty$ if and only if $u = 0$.

(iii) If $\mu \geq n/p$, then $M_{p,\mu}(u; \Omega, \varepsilon) < \infty$ if and only if $N_p(u; \Omega, \varepsilon) < \infty$.

(iv) If $M_{p,\mu}(u; \Omega, \varepsilon) < \infty$ holds for $\mu = \mu_0$, then it holds for all $\mu \geq \mu_0$.

Let us fix now two integers $0 \leq k < m$ and consider a differential operator of the form $L_k = \sum_{|\alpha|=k} u_\alpha P^\alpha$, where $u_\alpha : \Omega \to \mathbb{C}$ has the property $M_{p,m-k-\lambda}(u_\alpha; \Omega, 1) < \infty$ for some $\lambda > 0$. Note that, according to the properties (i)–(iv), only the values $\lambda \in [m-k-\frac{n}{p}, m-k)$ are of interest and a small λ allows u_α to have stronger singularities. We set $M_\lambda(\varepsilon) = \sup_{|\alpha|=k} M_{p,m-k-\lambda}(u_\alpha; \Omega, \varepsilon)$ and $N_\lambda(\varepsilon) = \sup_{|\alpha|=k} N_p(u_\alpha; \Omega, \varepsilon)$. Then (1.3.22) shows that L_k is a continuous operator from $W^{m,p}(\Omega)$ to $L^p(\Omega)$ and gives the following estimate for its norm:

(1.3.23)
$$\|L_k f\|_{L^p(\Omega)} \leq C\varepsilon^\lambda M_\lambda(\varepsilon) \|\nabla^{(m)} f\|_{L^p(\Omega)} + C\varepsilon^{-k-\frac{n}{p}} N_\lambda(\varepsilon) \|f\|_{L^p(\Omega)}.$$

Here C is a constant depending only on θ, m, n, p and λ, and ε is arbitrary in $(0, \delta)$. Notice that $M_\lambda(\varepsilon) \to 0$ and $N_\lambda(\varepsilon) \to 0$ as $\varepsilon \to 0$. It is convenient to introduce a new parameter $s := m(k + \frac{n}{p})(\lambda + k + \frac{n}{p})^{-1}$. Then $k < s < m$, $\lambda = \frac{m-s}{s}(k + \frac{n}{p})$ and for $\tau = \varepsilon^{\lambda+k+\frac{n}{p}}$ we shall have

$$\|L_k f\|_{L^p(\Omega)} \leq C\tau^{(m-s)/m} M_\lambda(\varepsilon) \|\nabla^{(m)} f\|_{L^p(\Omega)} + C\tau^{-s/m} N_\lambda(\varepsilon) \|f\|_{L^p(\Omega)}.$$

Let us set $\|g\|_{m,p,\Omega} = \sum_{|\alpha| \leq m} \|g^{(\alpha)}\|_{L^p(\Omega)}$ and $\|g\|_{p,\Omega} = \|g\|_{0,p,\Omega}$. As a particular case of the preceding estimate we obtain the following: there is a finite constant κ such that for each $\tau \in (0, 1]$ and each $f \in W^{m,p}(\Omega)$

(1.3.24) $$\|L_k f\|_{p,\Omega} \leq \kappa \tau^{(m-s)/m} \|f\|_{m,p,\Omega} + \kappa \tau^{-s/m} \|f\|_{p,\Omega}.$$

The meaning of estimates of this form, more precisely of estimates involving the powers τ^σ and $\tau^{\sigma-1}$ of a parameter $\tau \in (0, 1]$ where $\sigma = (m - s)/m \in (0, 1)$ is a fixed number, will become clear after the study of the real interpolation spaces in Chapter 2. Namely, according to a result which may be found in Section 2.5, (1.3.24) is equivalent to the fact that L_k is a continuous operator from the space $(W^{m,p}(\Omega), L^p(\Omega))_{\sigma,1}$ (defined by real interpolation) to the space $L^p(\Omega)$. In the special case when $\Omega = X$ and $p = 2$, this means that $L_k : \mathcal{H}^{s,1}(X) \to \mathcal{H}(X)$ is continuous, where $\mathcal{H}^{s,1}(X) = B_2^{s,1}(X)$ is a Besov space described in more details in Section 4.1.

Finally, let us explain why the Stummel type conditions are so convenient in the case of the Schrödinger hamiltonians with many-body potentials. Let $\Omega = X$, Y a subspace of X, $V^Y : Y \to \mathbb{C}$ a Borel function and $V_Y : V^Y \circ \pi_Y : X \to \mathbb{C}$ (where π_Y is the orthogonal projection of X onto Y). It is quite easy to prove that if $\mu \neq \frac{\dim Y}{p}$ then $M_{p,\mu}(V_Y; X, 1) < \infty$ if and only if $M_{p,\mu}(V^Y; Y, 1) < \infty$ [2]. Let us use this remark in the context of the usual N-body Schrödinger hamiltonians. These are operators in $\mathscr{H}(X)$ of the form $H = \Delta + \sum_{Y \in \mathscr{L}} V_Y \equiv \Delta + V$, where $\mathscr{L}$ is a finite family of subspaces of X and $V_Y = V^Y \circ \pi_Y$ for some function $V^Y : Y \to \mathbb{R}$. Let us assume that $M_{2,2-\lambda}(V^Y; Y, 1) < \infty$ for some $\lambda > 0$ such that $4 - 2\lambda \neq \dim Y$, for each $Y \in \mathscr{L}$. Then, by taking $L_0 \equiv V = \sum_{Y \in \mathscr{L}} V_Y$ and $\Omega = X$, $p = 2$, $k = 0$ in (1.3.23), we obtain

$$(1.3.25) \qquad \|Vf\| \leq \kappa\varepsilon^\lambda \|\Delta f\| + \kappa\varepsilon^{-n/2}\|f\|$$

for some constant $\kappa < \infty$ and all $\varepsilon \in (0,1)$ and $f \in \mathscr{H}^2(X) \equiv W^{2,2}(X)$. Since Δ is self-adjoint in $\mathscr{H}(X)$ with domain $\mathscr{H}^2(X)$, the Rellich-Kato criterion implies the self-adjointness of the operator $H = \Delta + V$ on the domain $\mathscr{H}^2(X)$ in $\mathscr{H}(X)$.

The condition $M_{2,2-\lambda}(V^Y; Y, 1) < \infty$ for some $\lambda > 0$, $\lambda \neq \frac{4 - \dim Y}{2}$, is a consequence of the following one: there is $p \geq 2$ such that $p > \frac{\dim Y}{2}$ and

$$\sup_{z \in Y} \int_{y \in Y, |y-z|<1} |V^Y(y)|^p dy < \infty.$$

This is easily checked by using the Hölder inequality.

1.4. Rapidly Decreasing Distributions

Both the translation group $\{T(v)\}_{v \in X}$ and the dilation group $\{W(\tau)\}_{\tau \in \mathbb{R}}$ can be used to control the behaviour at infinity of a distribution or of an operator. In later chapters we shall use the dilation group. Here we follow L. Schwartz and use the translation group (see the end of §8, Chapter VI and Theorem IX of Chapter VII in [Sch]).

DEFINITION 1.4.1. Let $f \in \mathscr{S}^*(X)$. We say that f is

(a) *weakly bounded at infinity* if for each $g \in \mathscr{S}(X)$ there is a constant $c < \infty$ such that

$$(1.4.1) \qquad |\langle g, e^{i(v,P)}f\rangle| \leq c \qquad \forall v \in X,$$

(b) *weakly vanishing at infinity* if

$$(1.4.2) \qquad \lim_{|v|\to\infty} \langle g, e^{i(v,P)}f\rangle = 0 \qquad \forall g \in \mathscr{S}(X),$$

(c) *rapidly weakly vanishing at infinity* if

$$(1.4.3) \qquad \lim_{|v|\to\infty} |v|^m \langle g, e^{i(v,P)}f\rangle = 0 \qquad \forall g \in \mathscr{S}(X), \forall m \in \mathbb{R}.$$

---

[2] Write $X = Y \oplus Y^\perp$ and observe that in the definition of $M_{p,\mu}(V_Y; X, 1)$ one may replace the unit ball $|x - x_0| < 1$ in X by the product of the unit ball $|y - y_0| < 1$ in Y with the unit ball $|z - z_0| < 1$ in $Y^\perp$. Here $x_0 = y_0 + z_0$ is the orthogonal decomposition of x_0 determined by Y.

The usual terminology for rapidly weakly vanishing distributions is "rapidly decreasing". We have added the adjective "weakly" because a C^∞ function could be unbounded but rapidly decaying at infinity as a distribution. This is the case for example for $f(x) = \langle x \rangle^k \exp(ix^2)$ (where $k \in \mathbb{N}$ and $X = \mathbb{R}$; see formula (VII,5;2) in [Sch]). In order to shorten the notation, we shall write w-*bounded* and w-*vanishing* for "weakly bounded" and "weakly vanishing" respectively.

It is obvious that if a distribution is w-bounded (or w-vanishing or rapidly w-vanishing) at infinity, then all its derivatives have the same property.

There is a more convenient way of expressing the above properties. We recall from Lemma 1.2.1 that, if $\varphi \in \mathscr{S}(X)$ and $f \in \mathscr{S}^*(X)$, then $\varphi(P)f \in C_{\text{pol}}^\infty(X)$ and

$$(1.4.4) \qquad [\varphi(P)f](x) = (\check{\varphi} * f)(x) = \langle \widehat{\varphi}, e^{i\langle x, P \rangle} f \rangle.$$

Hence, by taking into account that $P^\alpha \varphi(P)f = \psi(P)f$, where $\psi(x) \equiv x^\alpha \varphi(x)$ defines a function in $\mathscr{S}(X)$, we get:

PROPOSITION 1.4.2. (a) f *is* w-*bounded at infinity if and only if* $\varphi(P)f \in BC^\infty(X)$ *for all* $\varphi \in \mathscr{S}(X)$;

(b) f *is* w-*vanishing at infinity if and only if* $\varphi(P)f \in C_\infty^\infty(X)$ *for all* $\varphi \in \mathscr{S}(X)$;

(c) f *is rapidly* w-*vanishing at infinity if and only if* $\varphi(P)f \in \mathscr{S}(X)$ *for all* $\varphi \in \mathscr{S}(X)$.

A different characterization of the classes of distributions introduced in Definition 1.4.1 can be obtained by making use of the Calderón formula (1.3.12). By Lemma 1.2.1 each $f \in \mathscr{S}^*(X)$ defines a continuous linear operator $f(P) : \mathscr{S}(X) \to \mathscr{S}^*(X)$ and one has $\widehat{f}(P)\varphi = \widehat{\varphi}(P)f$ for $\varphi \in \mathscr{S}(X)$. It frequently happens that there is a Banach space $\mathbf{E}$ continuously embedded in $\mathscr{S}^*(X)$ such that the range of the operator $\widehat{f}(P)$ is contained in $\mathbf{E}$. If we look at $\widehat{f}(P)$ as an operator from $\mathscr{S}(X)$ to $\mathbf{E}$, its graph is obviously closed. But $\mathscr{S}(X)$ is a Fréchet space and the closed graph theorem is valid for linear mappings between Fréchet spaces (see e.g. [Y]). So $\widehat{f}(P)$ is a continuous linear operator from $\mathscr{S}(X)$ to $\mathbf{E}$. Thus $\|\widehat{f}(P)\varphi\|_{\mathbf{E}}$ is a continuous semi-norm on $\mathscr{S}(X)$, hence there are an integer $k \geq 0$ and a constant $c < \infty$ such that:

$$(1.4.5) \qquad \|\widehat{f}(P)\varphi\|_{\mathbf{E}} \leq c \sup_{x \in X} \langle x \rangle^k \sum_{j=0}^{k} \frac{1}{j!} |\nabla^{(j)} \varphi(x)|$$

for all $\varphi \in \mathscr{S}(X)$ (see (1.1.6)). Consequently $\widehat{f}(P)$ extends to a continuous operator from the completion of $\mathscr{S}(X)$ under the norm defined by the r.h.s. of (1.4.5) into $\mathbf{E}$. This completion contains all functions φ of class C^k that are of class $\mathscr{S}$ outside zero. By taking into account Lemma 1.3.5, one sees that φ belongs to this set of functions if $\widehat{\varphi} \in S^{-n-k-\varepsilon}(X)$ for some $\varepsilon > 0$. Finally, by using the relation $\widehat{f}(P)\varphi = \widehat{\varphi}(P)f$ and (1.3.12), one gets the following result:

THEOREM 1.4.3. *Let* $\mathbf{E}$ *be a Banach space continuously embedded in* $\mathscr{S}^*(X)$, *and let* $f \in \mathscr{S}^*(X)$. *Assume that* $\varphi(P)f \in \mathbf{E}$ *for all* $\varphi \in \mathscr{S}(X)$. *Then there is* *an integer $m \geq 0$ such that $\psi(P)f \in \mathbf{E}$ for all $\psi \in S^{-m}(X)$. Moreover there are distributions $f_\alpha \in \mathbf{E}$ ($|\alpha| = m$) and $f_m \in \mathbf{E}$ such that*

$$(1.4.6) \qquad\qquad f = \sum_{|\alpha|=m} P^\alpha f_\alpha + f_m.$$

COROLLARY 1.4.4. *Let $f \in \mathscr{S}^*(X)$. Then*

(a) *f is w-bounded at infinity if and only if f is a finite sum of derivatives of bounded continuous functions on X,*

(b) *f is w-vanishing at infinity if and only if f is a finite sum of derivatives of continuous functions that converge to zero at infinity,*

(c) *f is rapidly w-vanishing at infinity if and only if for each $r \in \mathbb{R}$ there are a finite number of continuous functions $f_\alpha : X \to \mathbb{C}$ satisfying $|f_\alpha(x)| \leq c\langle x \rangle^{-r}$ and $f = \sum_\alpha \partial^\alpha f_\alpha$.*

PROOF. We use Proposition 1.4.2 to characterize w-bounded, w-vanishing and rapidly w-vanishing distributions at infinity. The "only if" parts are easily obtained from Theorem 1.4.3 by taking $\mathbf{E} = BC(X)$ in (a), $\mathbf{E} = C_\infty(X)$ in (b) and $\mathbf{E} = \{g \in C(X) \mid |g(x)| \leq c\langle x \rangle^{-r}\}$ (with the natural norm) in (c). The "if" part is obvious by Definition 1.4.1. $\square$

We can now characterize rapidly decreasing distributions in terms of their Fourier transforms.

THEOREM 1.4.5. *A distribution $f \in \mathscr{S}^*(X)$ is rapidly w-vanishing at infinity if and only if its Fourier transform $\hat{f}$ belongs to $C^\infty_{\mathrm{pol}}(X)$.*

PROOF. (i) Assume that f is a rapidly w-vanishing distribution. By choosing $r > n$ in Corollary 1.4.4 (c), we get the existence of a finite number of continuous integrable functions f_α such that $f = \sum_\alpha P^\alpha f_\alpha$. Then $\hat{f} = \sum_\alpha Q^\alpha \hat{f}_\alpha$, hence $\hat{f}$ is a continuous function of at most polynomial growth. The same conclusion can be obtained for the derivatives $P^\beta \hat{f}$ of $\hat{f}$ by observing that $P^\beta \hat{f} = (-1)^{|\beta|} \mathcal{F} Q^\beta f$ and that $Q^\beta f$ is also rapidly decreasing (this last fact follows easily from Corollary 1.4.4 (c) or from (1.4.3) and the identity $e^{i(v,P)} Q^\beta = (Q+v)^\beta e^{i(v,P)}$).

(ii) Assume that $g \in C^\infty_{\mathrm{pol}}(X)$. Fix $r = 2j$, with $j \in \mathbb{N}$. By Leibniz' rule we have for any $k \in \mathbb{N}$:

$$\langle P \rangle^{2j} \langle Q \rangle^{-2k} g = \sum_{|\beta| \leq 2j} \varphi_\beta(Q) g^{(\beta)},$$

with $|\varphi_\beta(x)| \leq c\langle x \rangle^{-2k}$ for each β. We choose k so large that $\varphi_\beta(Q) g^{(\beta)} \in L^1(X)$ for each $|\beta| \leq 2j$, which is possible because $g \in C^\infty_{\mathrm{pol}}(X)$. Now set $h = \langle Q \rangle^{-2k} g$. Then $\hat{g} = \langle P \rangle^{2k} \hat{h} = \sum_{|\alpha| \leq 2k} c_\alpha \partial^\alpha \hat{h}$ for some constants c_α. Furthermore $\langle Q \rangle^{2j} \hat{h} = \mathcal{F} \langle P \rangle^{2j} h$ belongs to $L^\infty(X)$, because $\langle P \rangle^{2j} h \in L^1(X)$. Thus $|\hat{h}| \leq c\langle x \rangle^{-2j} \equiv c\langle x \rangle^{-r}$. So $\hat{g}$ is rapidly w-vanishing by Corollary 1.4.4 (c). $\square$

It is a remarkable fact that the C^∞ functions which are rapidly oscillating at infinity are rapidly w-vanishing distributions. An example which is very useful in applications is described in the next proposition.

PROPOSITION 1.4.6. *Let $S : X \to X$ be a symmetric, invertible linear operator. Denote by $\operatorname{sgn} S$ the number of positive eigenvalues minus the number of negative eigenvalues of S. Then*

$$(1.4.7) \qquad \int_X e^{-i(x,y)} e^{i(y,Sy)/2} \underline{dy} = e^{\pi i (\operatorname{sgn} S)/4} |\det S|^{-1/2} e^{-i(x,S^{-1}x)/2}$$

(where the integral means a distributional Fourier transform). In particular, the C^∞-function $f(x) = e^{i(x,Sx)/2}$ is a rapidly w-vanishing distribution.

PROOF. Let $\{v_1, \ldots, v_n\}$ be an orthonormal basis of X consisting of eigenvectors of S: $Sv_k = \lambda_k v_k$. We represent the vectors x of X as $x = x_1 v_1 + \cdots + x_n v_n$. The l.h.s. of (1.4.7) is equal to the (distributional) limit as $\varepsilon \to +0$ of the function

$$\int_X e^{-i(x,y)} e^{-(y,(\varepsilon - iS)y)/2} \underline{dy} = \prod_{k=1}^{n} (2\pi)^{-1/2} \int_{\mathbb{R}} e^{-ix_k t} e^{-(\varepsilon - i\lambda_k)t^2/2} dt.$$

(1.4.7) now follows by using the well-known formula

$$\int_{\mathbb{R}} e^{-ist} e^{-at^2/2} dt = \left(\frac{2\pi}{a} \right)^{1/2} e^{-s^2/(2a)}$$

for $\Re a > 0$, where the principal branch of the square root has been chosen (i.e. $z^{1/2} > 0$ if $z > 0$; then $\lim(\varepsilon - i\lambda)^{-1/2} = |\lambda|^{-1/2} \cdot \exp(i\pi \operatorname{sgn} \lambda/4)$ as $\varepsilon \to +0$ if $\lambda \in \mathbb{R} \setminus \{0\}$). $\square$

We end this section by pointing out some continuity properties of the operators $\varphi(P)$ with $\varphi \in C^\infty_{\mathrm{pol}}(X)$ which may look unexpected at first sight. We know from Section 1.2 that such an operator is a linear continuous mapping of $\mathscr{S}^*(X)$ into itself which leaves $\mathscr{S}(X)$ invariant and induces a continuous operator in this latter space. Since $\mathfrak{F}$ is unitary in $\mathscr{H}(X)$, it is clear that $\varphi(P)$ leaves $\mathscr{H}(X)$ invariant if and only if φ is a bounded function. The next result is very useful, although its proof is quite simple; the main point in this proof is that $\widehat{\varphi}$ is a rapidly w-vanishing distribution, but in fact much less in needed for the assertion to be true.

PROPOSITION 1.4.7. *If $\varphi \in C^\infty_{\mathrm{pol}}(X)$, then there is an integer $N \geq 0$ such that $\varphi(P)BC^{k+N}(X) \subset BC^k(X)$ for each integer $k \geq 0$. Moreover, for each k there is a constant c_k such that the norm of the operator $\varphi(\tau P) : BC^{k+N}(X) \to BC^k(X)$ is less than $c_k \langle \tau \rangle^N$ for each real τ.*

PROOF. (i) $\widehat{\varphi}$ is a rapidly decreasing distribution (see Theorem 1.4.5). By applying Corollary 1.4.4 (c), with $r > n$, to $\widehat{\varphi}$ one sees that there is an integer $N \geq 0$ such that $\varphi(x) = \sum_{|\alpha| \leq N} x^\alpha \varphi_\alpha(x)$, with $\varphi_\alpha \in C_\infty(X)$ and $\widehat{\varphi_\alpha} \in L^1(X)$. Then we have (see (1.2.13)):

$$\varphi(P) = \sum_{|\alpha| \leq N} P^\alpha \varphi_\alpha(P) \quad \text{and} \quad \varphi_\alpha(P) = \int_X e^{i(y,P)} \widehat{\varphi_\alpha}(y) \underline{dy}.$$

By using the invariance of the norm (1.1.5) of BC^k under translations one sees that $\varphi_\alpha(P)BC^k(X) \subset BC^k(X)$ for each $k \in \mathbb{N}$ and that the norm of $\varphi_\alpha(P)$

as an operator in $BC^k(X)$ is majorized by $\int_X |\widehat{\varphi_\alpha}(y)| dy$. The first assertion of the proposition follows by noticing that P^α maps $BC^{k+|\alpha|}(X)$ into $BC^k(X)$ continuously for each $k \in \mathbb{N}$.

(ii) In order to get the dependence on τ of the norm of $\varphi(\tau P)$, we assume that $\tau > 0$ and set $\tau = e^t$. The definition (1.2.17) of the dilation group W implies that $W(-t)PW(t) = e^t P$, hence $W(-t)\varphi(P)W(t) = \varphi(\tau P)$. Then we have

$$\begin{aligned}
\|P^\alpha \varphi(\tau P)f\|_{BC(X)} &= \|W(-t)\varphi(P)W(t)P^\alpha f\|_{BC(X)} \\
&= e^{-nt/2}\|\varphi(P)W(t)P^\alpha f\|_{BC(X)} \\
&\leq e^{-nt/2}c(\varphi)\|W(t)P^\alpha f\|_{BC^N(X)} \\
&\leq e^{-nt/2}c(\varphi)e^{nt/2}\max(1, e^{tN})\|P^\alpha f\|_{BC^N(X)}.
\end{aligned}$$

If we sum over $|\alpha| \leq k$, we get the claimed result. $\square$

The following situation is especially interesting. Let S be as in Proposition 1.4.6 and set $S(x) = \frac{1}{2}(x, Sx)$. Clearly $e^{i\tau S(P)}$ is unitary in $\mathcal{H}(X)$, and by Proposition 1.4.7 there is an integer $N \geq 0$ such that $e^{i\tau S(P)}BC^{k+N}(X) \subset BC^k(X)$ for each $k \in \mathbb{N}$ and each $\tau \in \mathbb{R}$; moreover:

(1.4.8) $$\|e^{i\tau S(P)}\|_{BC^{k+N}(X) \to BC^k(X)} \leq C(S, k)\langle \tau \rangle^N$$

for some constant $C(S, k)$ depending on S and k. These facts are not at all obvious (see Theorem 7.6.5 and the ensuing discussion in [H]) and have important applications in the theory of pseudo-differential operators. Let us mention that one may take $N = [n/2] + 1$ in (1.4.8) (see [H], loc. cit.), but this is not relevant for our purpose.

CHAPTER 2

Real Interpolation of Banach Spaces

In this chapter we present some results of the theory of interpolation in Banach spaces. We limit our considerations to those parts of the theory that will be useful further on in this text. In particular we develop the interpolation theory for a pair of Banach spaces $\mathbf{E}$, $\mathbf{F}$ only under the assumption that $\mathbf{E}$ is continuously embedded into $\mathbf{F}$. The case of an arbitrary compatible couple does not demand any new idea; for a detailed treatment of the general theory, see [BB], [BK], [BL], [BS], [KPS], [Tr]. The results described below are essentially due to N. Aronszjan, E. Gagliardo, J.-L. Lions and J. Peetre.

We find it convenient to formulate the interpolation theory in the category of B-spaces rather than in that of Banach spaces. The notion of B-space and some other terminology will be introduced in Section 2.1. The first three pages of this section contain only very elementary remarks and are sufficient for the reading of most parts of this monograph. The remainder of Section 2.1 (starting with Proposition 2.1.1) is rather technical and may be skipped by those who are not interested in more subtle notions (like that of Gagliardo completion). In Section 2.2 we define the real interpolation spaces in terms of the K-functional. The subsequent sections contain various properties of these interpolation spaces. In Section 2.8 we consider the real interpolation of Hilbert spaces and several related topics.

2.1. Banach Spaces and Linear Operators

Throughout this text we use the term *Banach space* for a complex Banach space. If $\mathbf{F}$ is a Banach space, we denote by $\|\cdot\|_{\mathbf{F}}$ the norm on $\mathbf{F}$; in certain special cases or if there is no risk of confusion the index will be omitted. We define the adjoint space $\mathbf{F}^*$ as the vector space of all anti-linear continuous mappings $\varphi : \mathbf{F} \to \mathbb{C}$, provided with the dual norm

$$(2.1.1) \qquad \|\varphi\|_{\mathbf{F}^*} := \sup\{|\varphi(f)| \mid f \in \mathbf{F}, \|f\|_{\mathbf{F}} \leq 1\}.$$

The (anti)-duality map $\langle \cdot, \cdot \rangle : \mathbf{F} \times \mathbf{F}^* \to \mathbb{C}$ is given by

$$(2.1.2) \qquad \langle f, \varphi \rangle := \varphi(f) \qquad \text{for } f \in \mathbf{F}, \ \varphi \in \mathbf{F}^*.$$

W. O. Amrein et al., *C0-Groups, Commutator Methods and Spectral Theory of N-Body Hamiltonians*, Modern Birkhäuser Classics,
DOI: 10.1007/978-3-0348-0733-3_2, © Springer Basel 1996

There exists a canonical isometric embedding $j : \mathbf{F} \to \mathbf{F}^{**}$ given by $j(f)(\varphi) = \overline{\varphi(f)}$ (here $f \in \mathbf{F}$, $\varphi \in \mathbf{F}^*$). $\mathbf{F}$ is always identified with a subspace of $\mathbf{F}^{**}$ through j; then the quantity $\langle \varphi, f \rangle$ is defined for $\varphi \in \mathbf{F}^*$ and $f \in \mathbf{F} \subset \mathbf{F}^{**}$ and satisfies $\langle \varphi, f \rangle = \overline{\langle f, \varphi \rangle}$. The Banach space $\mathbf{F}$ is said to be *reflexive* if $\mathbf{F} = \mathbf{F}^{**}$. If $\mathbf{F}$ is reflexive, then so is its adjoint $\mathbf{F}^*$. Furthermore, if $\mathbf{E}$ is any closed subspace of $\mathbf{F}$, then $\mathbf{E}$ and the quotient space $\mathbf{F}/\mathbf{E}$ are reflexive. Finally, each bounded closed ball in a reflexive Banach space is weakly compact. We refer to Theorem II.A.14 of [Wo] for a proof of these facts.

If $\mathbf{E}$ and $\mathbf{F}$ are Banach spaces, we denote by $B(\mathbf{E}, \mathbf{F})$ the Banach space of all continuous linear operators from $\mathbf{E}$ to $\mathbf{F}$ and by $K(\mathbf{E}, \mathbf{F})$ the subspace of all compact operators from $\mathbf{E}$ to $\mathbf{F}$. We set $B(\mathbf{E}) \equiv B(\mathbf{E}, \mathbf{E})$ and $K(\mathbf{E}) \equiv K(\mathbf{E}, \mathbf{E})$, and we shall use the symbol I for the identity operator in a Banach space. The norm in $B(\mathbf{E}, \mathbf{F})$ will sometimes be denoted by $|| \cdot ||_{\mathbf{E} \to \mathbf{F}}$.

If $T \in B(\mathbf{E}, \mathbf{F})$, then its adjoint T^* belongs to $B(\mathbf{F}^*, \mathbf{E}^*)$ and is defined, as usual, by

$$(2.1.3) \qquad \langle e, T^* f \rangle = \langle Te, f \rangle \text{ for } e \in \mathbf{E}, \ f \in \mathbf{F}^*.$$

Due to the fact that $\mathbf{E}$ is identified with a subspace of $\mathbf{E}^{**}$, the Banach space $B(\mathbf{E}, \mathbf{E}^*)$ has a rather rich structure: it is provided with an involution and with a notion of positivity. Indeed, if $T \in B(\mathbf{E}, \mathbf{E}^*)$, then the operator $T^\dagger = T^*|_{\mathbf{E}}$ belongs to the same space (if $\mathbf{E}$ is reflexive, then $T^\dagger = T^*$). We say that an operator T in $B(\mathbf{E}, \mathbf{E}^*)$ is *symmetric* if $T = T^\dagger$, in other words if $\langle e, Tf \rangle = \langle Te, f \rangle \ \forall e, f \in \mathbf{E}$; by the polarization identity (see (2.8.13)) this is equivalent to $\langle e, Te \rangle \in \mathbb{R} \ \forall e \in \mathbf{E}$. We shall say that T is *positive* if $\langle e, Te \rangle \geq 0$ for all $e \in \mathbf{E}$; note that positivity implies symmetry. We shall write $S \geq T$ if $S - T$ is positive. Observe that there is a canonical identification of $B(\mathbf{E}, \mathbf{E}^*)$ with the Banach space of all continuous sesquilinear forms on $\mathbf{E}$ (for us, a sesquilinear form is antilinear in the first variable). More precisely, one identifies the operator $T : \mathbf{E} \to \mathbf{E}^*$ with the sesquilinear form $Q(e, f) = \langle e, Tf \rangle$. Then the symmetry or positivity of an operator are expressed in terms of the sesquilinear form associated to it. Since a separately continuous sesquilinear form on a Banach space is continuous, a symmetric linear operator $\mathbf{E} \to \mathbf{E}^*$ is automatically continuous.

We recall some of the terminology related to the notion of unbounded operators. Let T be a linear map defined on a vector subspace $D(T)$ of the Banach space $\mathbf{E}$ and with values in the Banach space $\mathbf{F}$. We say that $D(T)$ *is the domain of* T and we equip it with the so-called *graph norm* (and the corresponding *graph topology*) associated to T, namely

$$(2.1.4) \qquad ||e||_T = (||e||_{\mathbf{E}}^2 + ||Te||_{\mathbf{F}}^2)^{1/2}, \qquad e \in D(T).$$

T is a *closed operator* if $D(T)$ is a Banach space (for the graph norm). If T *is densely defined* (i.e. $D(T)$ is a dense subspace of $\mathbf{E}$), then the *adjoint operator* $T^* : D(T^*) \subset \mathbf{F}^* \to \mathbf{E}^*$ is defined as follows: an element $\psi \in \mathbf{F}^*$ belongs to $D(T^*)$ if and only if there is $\varphi \in \mathbf{E}^*$ such that $\langle Te, \psi \rangle = \langle e, \varphi \rangle$ for each $e \in D(T)$; such a vector φ is uniquely determined and $T^* \psi = \varphi$. The operator T^* is always closed, but it is not densely defined in general (even if T is closed). However, if $\mathbf{F}$ is reflexive, and if T is densely defined and closed, then T^* is also densely defined and closed, and $T^{**} : D(T^{**}) \subset \mathbf{E}^{**} \to \mathbf{F}$ is an extension of T; if $\mathbf{E}$ is reflexive too, then $T^{**} = T$.

One of the aims of interpolation theory is to construct new Banach spaces starting from a given pair $\mathbf{E}$, $\mathbf{F}$. It often happens that two quite different methods of construction give Banach spaces that in fact are identical as topological vector spaces, i.e. the two methods furnish the same vector space and two different but equivalent norms; moreover, the topological vector space obtained in this way depends only on the topological vector space structure of $\mathbf{E}$ and $\mathbf{F}$. This should not be considered as a defect of the theory, but on the contrary, a very efficient tool. On the other hand this suggests the developing of interpolation theory in the category of B-spaces that we are going to introduce further on.

We first recall that a topological vector space (TVS) is a vector space $\mathbf{F}$ equipped with a topology such that the addition $\mathbf{F} \times \mathbf{F} \ni (e, f) \mapsto e + f \in \mathbf{F}$ and the multiplication with scalars $\mathbb{C} \times \mathbf{F} \ni (\lambda, f) \mapsto \lambda f \in \mathbf{F}$ are continuous functions. A sequence $\{f_n\}_{n \in \mathbb{N}}$ in a TVS is *Cauchy* if for any neighbourhood U of zero in $\mathbf{F}$ there is $N \in \mathbb{N}$ such that $f_n - f_m \in U$ if $n, m \geq N$. $\mathbf{F}$ is *sequentially complete* if each Cauchy sequence is convergent. A TVS $\mathbf{F}$ is *normable* if there is a norm on $\mathbf{F}$ such that the topology associated to it coincides with the topology of $\mathbf{F}$; such a norm will be called *admissible*. Finally we recall that two norms $||\cdot||'$, $||\cdot||''$ on a vector space $\mathbf{F}$ define the same topology on $\mathbf{F}$ if and only if they are *equivalent*, i.e. there is a constant $c \in (0, 1)$ such that $c||f||' \leq ||f||'' \leq c^{-1}||f||'$ for $f \in \mathbf{F}$; then we write $||f||' \sim ||f||''$.

By a *B-space*, or a *banachisable* TVS, we mean a normable, (sequentially) complete TVS. Alternatively, a B-space is a vector space $\mathbf{F}$ equipped with an equivalence class of norms such that $\mathbf{F}$ is a Banach space for one (and hence for all) of these norms. If there is an admissible norm on a B-space $\mathbf{F}$ that derives from a scalar product, then we say that $\mathbf{F}$ is a *H-space* or a *hilbertisable* TVS. Any Banach (resp. Hilbert) space $\mathbf{F}$ has a canonical B-space (resp. H-space) structure: the admissible norms are the norms on $\mathbf{F}$ that are equivalent to the norm that is originally given on $\mathbf{F}$. We stress the fact that two B-spaces $\mathbf{E}$, $\mathbf{F}$ are equal if and only if they are equal as topological vector spaces (i.e. they have the same underlying vector space topology). Quite often below two distinct Banach spaces are equal as B-spaces.

Numerous equations in this chapter will contain admissible norms on the B-spaces that are involved. It is understood that these equations are true for any choice of these norms; clearly, the constants that may appear will in general depend on this choice.

If $\mathbf{E}$, $\mathbf{F}$ are topological vector spaces (e.g. B-spaces), then $\mathbf{E} = \mathbf{F}$ means that $\mathbf{E}$ and $\mathbf{F}$ are equal as TVS. We write $\mathbf{E} \subset \mathbf{F}$, and say that $\mathbf{E}$ *is continuously embedded in* $\mathbf{F}$, if $\mathbf{E}$ is a vector subspace of $\mathbf{F}$ and the topology induced by $\mathbf{F}$ on $\mathbf{E}$ is weaker than the topology of $\mathbf{E}$ (equivalently, if the inclusion map $\mathbf{E} \to \mathbf{F}$ is continuous). If $\mathbf{E} \subset \mathbf{F}$ and the subset $\mathbf{E}$ of $\mathbf{F}$ is dense in $\mathbf{F}$, we write "$\mathbf{E} \subset \mathbf{F}$ *densely*". Observe that $\mathbf{E} \subset \mathbf{F}$ and $\mathbf{F} \subset \mathbf{E}$ is equivalent to $\mathbf{E} = \mathbf{F}$ (as TVS). If $\mathbf{E}$ is a B-space, $\mathbf{F}$ is a TVS and $\mathbf{E} \subset \mathbf{F}$, we say that $\mathbf{E}$ is a *B-subspace of* $\mathbf{F}$; the notion of H-*subspace* is defined similarly. Note that if $\mathbf{F}$ is a Hausdorff TVS and $\mathbf{E}$ is a vector subspace of $\mathbf{F}$, then there is at most one B-space structure on $\mathbf{E}$ such that $\mathbf{E} \subset \mathbf{F}$. More generally, if $\mathbf{E}_1$ and $\mathbf{E}_2$ *are B-subspaces of a Hausdorff*

TVS **F** *and if* $\mathbf{E}_1$ *is included in* $\mathbf{E}_2$ *as a set, then* $\mathbf{E}_1 \subset \mathbf{E}_2$ (indeed, the identity map $\mathbf{E}_1 \to \mathbf{E}_2$ will be closed and we may apply the closed graph theorem).

If **F** is a B-space, then each closed subspace **E** of **F** is a B-space (the unique B-space structure on **E** such that $\mathbf{E} \subset \mathbf{F}$ being that induced by **F**). But the class of B-subspaces of **F** is much richer if **F** is infinite dimensional: indeed, the domain of a closed operator from **F** to another B-space is always a B-subspace of **F** (the B-structure being defined by the graph topology). If **F** is a Hilbert space, then **E** is a H-subspace of **F** such that $\mathbf{E} \subset \mathbf{F}$ densely if and only if **E** is the domain of a self-adjoint operator in **F** (cf. Friedrichs' theorem, see Section 2.8). Examples of subspaces which cannot be B-subspaces are the subspaces of countable but not finite algebraic dimension (use Baire's theorem). If **E** is a B-subspace of a B-space **F** then the real and complex interpolation methods allow one to construct many other B-subspaces of **F**.

The adjoint $\mathbf{F}^*$ of an arbitrary TVS is the set of all antilinear continuous functionals $\varphi : \mathbf{F} \to \mathbb{C}$. If **F** is a B-space, then $\mathbf{F}^*$ has an obvious B-space structure. If **E**, **F** are B-spaces and $\mathbf{E} \subset \mathbf{F}$ densely, then we naturally have $\mathbf{F}^* \subset \mathbf{E}^*$ (densely if **E** is reflexive).

If **E** and **F** are B-spaces, then the set of linear continuous operators from **E** to **F** (which we denote by $B(\mathbf{E}, \mathbf{F})$) is also a B-space. If $T \in B(\mathbf{E}, \mathbf{F})$, then the notation $||T||_{\mathbf{E} \to \mathbf{F}}$ means the norm of the operator T with respect to some fixed admissible norms on **E** and **F**. As in the case of Banach spaces, the B-space $B(\mathbf{E}, \mathbf{E}^*)$ is equipped with an involution and a notion of positivity; hence it makes sense to speak of a symmetric or positive operator $T : \mathbf{E} \to \mathbf{E}^*$. If $\mathbf{E} \subset \mathbf{F}$ and $\mathbf{G} \subset \mathbf{K}$ are B-spaces and $\mathbf{E} \subset \mathbf{F}$ densely, then one can define a canonical embedding $B(\mathbf{F}, \mathbf{G}) \subset B(\mathbf{E}, \mathbf{K})$ as follows: if $T \in B(\mathbf{F}, \mathbf{G})$, we let $\widetilde{T} : \mathbf{E} \to \mathbf{K}$ be the operator given by setting $\widetilde{T}e = Te$, considered as an element of **K**. Clearly $\widetilde{T}$ is linear and continuous, so it belongs to $B(\mathbf{E}, \mathbf{K})$. Since **E** is dense in **F**, it is clear that the correspondence $T \mapsto \widetilde{T}$ is linear and injective. From now on we shall denote $\widetilde{T}$ simply by T, which gives the above-mentioned embedding of $B(\mathbf{F}, \mathbf{G})$ into $B(\mathbf{E}, \mathbf{K})$. In the preceding situation, if T is an element of $B(\mathbf{E}, \mathbf{K})$, the expression "$T \in B(\mathbf{F}, \mathbf{G})$" has an unambiguous meaning. It is equivalent to "$T(\mathbf{E}) \subset \mathbf{G}$ and $T : \mathbf{E} \to \mathbf{G}$ is continuous when **E** is provided with the topology induced by **F**". If this is the case, the operator $T : \mathbf{E} \to \mathbf{G}$ has a unique extension to a continuous mapping from **F** to **G** which we shall also denote by the letter T. Note that if $\mathbf{E}_1 \subset \mathbf{F}_1$ and $\mathbf{E}_2 \subset \mathbf{F}_2$ are B-spaces, $T \in B(\mathbf{F}_1, \mathbf{F}_2)$ and $T(\mathbf{E}_1) \subset \mathbf{E}_2$, then $T \in B(\mathbf{E}_1, \mathbf{E}_2)$ by the closed graph theorem.

In some situations it will be useful to admit the value $+\infty$ for the norm of an element of a vector space. In order to avoid confusion with the usual definition of a norm, we shall call *gauge* on a vector space **F** a mapping $|| \cdot || : \mathbf{F} \to [0, +\infty]$ such that (i) $||f|| = 0 \Rightarrow f = 0$; (ii) $||f + g|| \leq ||f|| + ||g||, \quad \forall f, g \in \mathbf{F}$; (iii) $||zf|| = |z| \cdot ||f||, \quad \forall z \in \mathbb{C}, f \in \mathbf{F}$ (with the convention $0 \cdot \infty = 0$). The equivalence of two gauges $|| \cdot ||'$, $|| \cdot ||''$ on **F** is defined exactly as in the case of norms and the same notation $||f||' \sim ||f||''$ is used in order to express it. To each gauge one may associate a normed space **E** by the following rule: **E** is the vector subspace of **F** consisting of the vectors f with $||f|| < \infty$ and the norm of **E** is the restriction of the gauge to **E**. If **E** is a Banach space, we say that the gauge is *closed*. If $\mathbf{F}$ is a TVS and we have $\mathbf{E} \subset \mathbf{F}$ (as TVS, i.e. if $\{f_n\}_{n \in \mathbb{N}}$ is a sequence in $\mathbf{E}$ such that $||f_n|| \to 0$, then $f_n \to 0$ in $\mathbf{F}$), we shall say that $|| \cdot ||$ is a *coercive gauge*. A gauge is coercive if and only if the convex set $B = \{f \in \mathbf{F} \mid ||f|| \leq 1\}$ is a bounded set in $\mathbf{F}$. A coercive gauge is called *closable* if for each Cauchy sequence $\{f_n\}_{n \in \mathbb{N}}$ in $\mathbf{E}$ with $f_n \to 0$ in $\mathbf{F}$ we have $||f_n|| \to 0$. Let $\mathbf{E}^c$ be the completion of the normed space $\mathbf{E}$. If $\mathbf{F}$ is a Hausdorff, sequentially complete TVS and $|| \cdot ||$ is a coercive gauge, then there is a natural map $\mathbf{E}^c \to \mathbf{F}$; and the gauge is closable if and only if this map is injective, so one may realize $\mathbf{E}^c$ as B-subspace of $\mathbf{F}$. Finally, we say that the gauge is *reflexive* if $\mathbf{E}^c$ is a reflexive Banach space. The next proposition describes some relations between these notions.

PROPOSITION 2.1.1. *Let $|| \cdot ||$ be a coercive gauge on a Hausdorff, sequentially complete TVS $\mathbf{F}$ and let $\mathbf{E}$ be the normed space associated to it.*

(a) *Assume that the function $|| \cdot || : \mathbf{F} \to [0, \infty]$ is lower semicontinuous (l.s.c.) or equivalently that the bounded convex set $B = \{f \in \mathbf{F} \mid ||f|| \leq 1\}$ is closed in $\mathbf{F}$. Then $|| \cdot ||$ is a closed gauge and the following Fatou property holds: if $\{f_n\}_{n \in \mathbb{N}}$ is a sequence in $\mathbf{F}$ which converges in $\mathbf{F}$ to some vector f, then $||f|| \leq \liminf_{n \to \infty} ||f_n||$. In particular, if $\{f_n\}$ is a bounded sequence in $\mathbf{E}$ and $f_n \to f$ in $\mathbf{F}$, then $f \in \mathbf{E}$. If $\mathbf{F}$ is a locally convex space, then in the preceding assertions it is sufficient to assume that $f_n \to f$ weakly in $\mathbf{F}$.*

(b) *Assume that $\mathbf{F}$ is locally convex and $|| \cdot ||$ is closed and reflexive. Then $|| \cdot || : \mathbf{F} \to [0, \infty]$ is a lower semicontinuous function.*

PROOF. (a) The Fatou property is an immediate consequence of the l.s.c. of $|| \cdot ||$. If $\mathbf{F}$ is locally convex, then the convex set B is closed if and only if it is weakly closed; hence $|| \cdot ||$ will remain l.s.c. if $\mathbf{F}$ is equipped with the weak topology. To see that $\mathbf{E}$ is complete, let $\{f_n\}$ be a Cauchy sequence in $\mathbf{E}$. Then $\{f_n\}$ is Cauchy in $\mathbf{F}$, hence there is $f \in \mathbf{F}$ such that $f_n \to f$ in $\mathbf{F}$. For each $\varepsilon > 0$ there is $N < \infty$ such that $||f_n - f_m|| \leq \varepsilon$ if $n, m \geq N$. Since $f_n - f_m \to f_n - f$ in $\mathbf{F}$ as $m \to \infty$, from the Fatou property we get $f \in \mathbf{E}$ and $||f_n - f|| \leq \varepsilon$ if $n \geq N$.

(b) The inclusion map $\mathbf{E} \to \mathbf{F}$ is continuous and remains continuous when we equip $\mathbf{E}, \mathbf{F}$ with their weak topologies. The set B is weakly compact in $\mathbf{E}$ (by the reflexivity of $\mathbf{E}$), hence its image in $\mathbf{F}$ is weakly compact too. So B is weakly closed, in particular closed, in $\mathbf{F}$. $\square$

A coercive closed gauge on a B-space need not be l.s.c. (we shall give below a simple example which clarifies this phenomenon; a more interesting situation of this type will be met in Chapter 5). But *if $|| \cdot ||$ is a coercive gauge on a Hausdorff TVS $\mathbf{F}$, one may canonically associate to it a coercive lower semicontinuous* *gauge* $|| \cdot ||_*$ *with* $|| \cdot ||_* \leq || \cdot ||$ by the following procedure [1]

(2.1.5)
$$||f||_* = \liminf_{g \to f} ||g|| = \lim_{V \in \mathscr{V}(f)} \inf_{g \in V} ||g||.$$

Here $\mathscr{V}(f)$ is the filter of neighbourhoods of f in $\mathbf{F}$. The function $|| \cdot ||_* : \mathbf{F} \to [0, \infty]$ is just the *lower semicontinuous regularization* of the convex function $|| \cdot || : \mathbf{F} \to [0, \infty]$ (see Ch. IV, §6.2 in [Bo4]). In other terms, $|| \cdot ||_*$ is the upper bound of all l.s.c. functions $\varphi : \mathbf{F} \to [-\infty, \infty]$ which are $\leq || \cdot ||$, or the largest l.s.c. function which is $\leq || \cdot ||$. If $\mathbf{E} \subset \mathbf{F}$ is the normed space associated to $|| \cdot ||$, we denote by $\widetilde{\mathbf{E}}$ the normed space associated to $|| \cdot ||_*$ and we call it the *Gagliardo completion of* $\mathbf{E}$ *with respect* to $\mathbf{F}$. $\widetilde{\mathbf{E}}$ is always a B-subspace of $\mathbf{F}$ such that $\mathbf{E} \subset \widetilde{\mathbf{E}}$. One may describe the closed unit ball $\widetilde{B} = \{f \in \mathbf{F} \mid ||f||_* \leq 1\}$ of $\widetilde{\mathbf{E}}$ as follows: $\widetilde{B} = \overline{B} \equiv$ closure in $\mathbf{F}$ of the closed unit ball $B = \{f \in \mathbf{F} \mid ||f|| \leq 1\}$ of $\mathbf{E}$ (*Proof*: Assume $||f||_* = r < \infty$. For each $V \in \mathscr{V}(f)$ let $g_V \in V$ such that the net $\{||g_V||\}$ converges to r. Then the net $f_V := r||g_V||^{-1} g_V$ has the properties $||f_V|| = r$ and $\lim f_V = f$ in $\mathbf{F}$). In particular, one has $||f||_* = ||f||$ for all $f \in \mathbf{E}$ if and only if B is closed in $\mathbf{E}$ for the topology induced by $\mathbf{F}$. If $f \in \mathbf{F}$ and $\mathbf{F}$ is metrizable, then it is clear that $f \in \widetilde{\mathbf{E}}$ if and only if there is a bounded sequence $\{f_n\}$ in $\mathbf{E}$ such that $f_n \to f$ in $\mathbf{F}$ (this is the usual definition of $\widetilde{\mathbf{E}}$, see §5.1 in [BS]; see Chapter I in [KPS] for a more geometric approach).

A rather typical example of a Gagliardo completion is the following. Let $J = [0, 1]$, $\mathbf{E} = C^1(J)$ and $\mathbf{F}$ one of the spaces $L^p(J)$, $1 \leq p \leq \infty$, or $C(J)$. Then $\mathbf{E}$, $\mathbf{F}$ are Banach spaces, $\mathbf{E} \subset \mathbf{F}$ and $\widetilde{\mathbf{E}}$ is the space of Lipschitz functions on J. The proof is an easy exercise.

If $\mathbf{F}$ is a Hausdorff locally convex space, then any convex l.s.c. function φ on $\mathbf{F}$ is the upper bound of all continuous affine functions smaller than φ (Ch. 2, §5, no. 2 [Bo2]). For gauges this description can be made more precise. First, to each coercive gauge $|| \cdot ||$ on $\mathbf{F}$ one may associate a strongly continuous seminorm on $\mathbf{F}^*$ by the formula $||\varphi||^* = \sup_{f \in B} |\langle f, \varphi \rangle|$, where $B = \{f \in \mathbf{F} \mid ||f|| \leq 1\}$ (recall that B is a bounded set in $\mathbf{F}$). $|| \cdot ||^*$ is a norm if and only if $\mathbf{E}$ is dense in $\mathbf{F}$; in this case $\mathbf{F}^* \subset \mathbf{E}^*$ and $|| \cdot ||^*$ coincides with the restriction to $\mathbf{F}^*$ of the norm on $\mathbf{E}^*$ dual to that of $\mathbf{E}$. The set $\{\varphi \in \mathbf{F}^* \mid ||\varphi||^* \leq 1\}$ is just the polar B° of B in $\mathbf{F}^*$ and $B^{\circ\circ} = \{f \in \mathbf{F} \mid |\langle f, \varphi \rangle| \leq 1 \, \forall \varphi \in B^\circ\}$ is the closure of B in $\mathbf{F}$ (B being convex; cf. Proposition 2, Ch. 4, §1 and Theorem 1, Ch. 2, §6 in [Bo2]). Hence $\widetilde{B} = B^{\circ\circ}$ or, equivalently:

(2.1.6)
$$||f||_* = \sup\{|\langle f, \varphi \rangle| \mid \varphi \in \mathbf{F}^*, \, ||\varphi||^* \leq 1\}.$$

If $\mathbf{E}$ is B-subspace of a Hausdorff TVS $\mathbf{F}$ and $|| \cdot ||_{\mathbf{E}}$ is an admissible norm on $\mathbf{E}$, we shall always extend it to a gauge on $\mathbf{F}$ by setting $||f||_{\mathbf{E}} = +\infty$ if

---

[1]It is easy to show that $||f_1 + f_2||_* \leq ||f_1||_* + ||f_2||_*$ and $||zf||_* = |z| \cdot ||f||_*$. If $||f||_* = 0$, then for each $V \in \mathscr{V}(f)$ there is $f_V \in V$ such that $\lim_V ||f_V|| = 0$. Since $\mathbf{E}$ is continuously embedded in $\mathbf{F}$, we have $\lim_V ||f_V|| = 0$ in $\mathbf{F}$ too. But $f_V \to f$ in $\mathbf{F}$ and $\mathbf{F}$ is Hausdorff, hence $f = 0$. This proves that $||f||_*$ is a l.s.c. gauge on $\mathbf{F}$. It remains to be shown that $||f||_*$ is a coercive gauge. If it is not then we can find a sequence $\{f_n\}_{n \in \mathbb{N}}$ and a closed neighbourhood U of zero in $\mathbf{F}$ such that $||f_n||_* \to 0$ and $f_n \notin U$. Since $\mathbf{F} \setminus U$ is a neighbourhood of f_n, we may find $g_n \in \mathbf{F} \setminus U$ with $||g_n|| \leq ||f_n||_* + n^{-1}$. Hence $||g_n|| \to 0$, which implies $g_n \to 0$ in $\mathbf{F}$, but this contradicts the fact that $g_n \notin U$.

$f \in \mathbf{F} \setminus \mathbf{E}$. It is clear that the l.s.c. regularizations of two admissible norms on $\mathbf{E}$ are equivalent gauges on $\mathbf{F}$, hence the normed spaces associated to them are identical as TVS. This TVS is a B-subspace $\widetilde{\mathbf{E}}$ of $\mathbf{F}$ and will be called *Gagliardo completion of the B-space $\mathbf{E}$ with respect to $\mathbf{F}$*; so we have $\mathbf{E} \subset \widetilde{\mathbf{E}} \subset \mathbf{F}$ with strict inclusions in general (if $\mathbf{F}$ is a B-space and $\mathbf{E} \neq \mathbf{F}$, then $\widetilde{\mathbf{E}}$ is of the first category in $\mathbf{F}$, cf. Lemma 1.2, Ch. I in [KPS]). If $\mathbf{E} = \widetilde{\mathbf{E}}$, we shall say that $\mathbf{E}$ is *relatively complete with respect to* $\mathbf{F}$. For example, if $\mathbf{F}$ is a B-space and $\mathbf{E}$ is a reflexive B-space, then $\mathbf{E} = \widetilde{\mathbf{E}}$. So a reflexive B-space is relatively complete with respect to any B-space in which it is embedded.

Let $\mathbf{H}$, $\mathbf{F}$ be Hausdorff locally convex TVS such that $\mathbf{H} \subset \mathbf{F}$ and let $\mathbf{E}$ be a B-subspace of $\mathbf{H}$ (hence of $\mathbf{F}$). *If $\mathbf{H}$ is reflexive, then the Gagliardo completion of $\mathbf{E}$ with respect to $\mathbf{H}$ is equal to the Gagliardo completion of $\mathbf{E}$ with respect to $\mathbf{F}$.* To prove this, let $\| \cdot \|$ be an admissible norm on $\mathbf{E}$, set $B = \{f \in \mathbf{E} \mid \|f\| \leq 1\}$ and let B' (resp. B'') be the closure of B in $\mathbf{H}$ (resp. $\mathbf{F}$). We have to show that $B' = B''$. Since we always have $B' \subset B''$, it is sufficient to prove that B' is closed in $\mathbf{F}$. Since B is bounded in $\mathbf{E}$, it is bounded in $\mathbf{H}$ too. Hence B' is a bounded closed convex set in $\mathbf{H}$. But then B' will be also weakly closed in $\mathbf{H}$ (Ch. 4, §1, no. 2 in [Bo2]) and, by the reflexivity of $\mathbf{H}$, weakly compact (Ch. 4, §2, no. 3 in [Bo2]). The canonical inclusion $\mathbf{H} \subset \mathbf{F}$ is weakly continuous (Ch. 4, §1, no. 3 in [Bo2]), hence B' is weakly compact in $\mathbf{F}$, so it is weakly closed, in particular closed, in $\mathbf{F}$. This finishes the proof.

Let $\mathbf{E}$ be a B-subspace of a reflexive Hausdorff locally convex space $\mathbf{H}$ and assume that $\mathbf{E}$ is dense in $\mathbf{H}$. Then we have a canonical embedding $\mathbf{H}^* \subset \mathbf{E}^*$ and we may consider the closure $\mathbf{G}$ of $\mathbf{H}^*$ in $\mathbf{E}^*$, equipped with the topology induced by $\mathbf{E}^*$. Since $\mathbf{H}^* \subset \mathbf{G}$ densely and $\mathbf{H}^{**} = \mathbf{H}$, we have $\mathbf{G}^* \subset \mathbf{H}$. Using (2.1.6) one may prove without any difficulty that $\mathbf{G}^*$ *is just the Gagliardo completion of $\mathbf{E}$ with respect $\mathbf{H}$.*

In the next lemma we give a more convenient expression for the l.s.c. regularization of a gauge on a Banach space.

LEMMA 2.1.2. *Let $\|\cdot\|$ be a gauge on the Banach space $\mathbf{F}$. Then $\|\cdot\|$ is coercive if and only if there is a constant $c < \infty$ such that $\|f\|_{\mathbf{F}} \leq c\|f\|$, $\forall f \in \mathbf{F}$. In this case one has*

$$(2.1.7) \qquad \|f\|_* \equiv \lim_{r \to 0} \inf_{\|e-f\|_{\mathbf{F}} < r} \|e\| = \lim_{n \to \infty} \inf_{e \in \mathbf{F}} (\|e\| + n\|f - e\|_{\mathbf{F}}).$$

PROOF. The first assertion of the lemma is obvious and the first equality in (2.1.7) is just the definition of $\|\cdot\|_*$. Denote by $|f|$ the last member of (2.1.7). If $|f| < M < \infty$, then for each $n \in \mathbb{N}$ there is $e_n \in \mathbf{E}$ with $\|e_n\| + n\|f - e_n\|_{\mathbf{F}} \leq M$. So $\|e_n\| \leq M$ and $\|f - e_n\|_{\mathbf{F}} \to 0$, which clearly implies $\|f\|_* \leq M$. This shows that $\|f\|_* \leq |f|$. Reciprocally, assume $\|f\|_* < M < \infty$ and let $\varepsilon > 0$. Then, for each $n \in \mathbb{N}$ there is e_n with $\|e_n - f\|_{\mathbf{F}} < \varepsilon/n$ and $\|e_n\| \leq M$. In particular $\|e_n\| + n\|e_n - f\|_{\mathbf{F}} \leq M + \varepsilon$, so $|f| \leq M + \varepsilon$. This proves that $|f| \leq \|f\|_*$. $\square$

Interpolation spaces are constructed starting from a pair of compatible B-spaces. Two B-spaces $\mathbf{E}_0$, $\mathbf{E}_1$ are *compatible* if the following two conditions are satisfied: (i) $\mathbf{E}_0$ and $\mathbf{E}_1$ induce the same linear structure on the set $\mathbf{E}_0 \cap \mathbf{E}_1$ (which contains at least 0); (ii) if $\{e_n\}_{n \in \mathbb{N}}$ is a sequence in $\mathbf{E}_0 \cap \mathbf{E}_1$ such that $e_n \to u$ in

$\mathbf{E}_0$ and $e_n \to v$ in $\mathbf{E}_1$, then $u = v$. Assume that condition (i) is satisfied and let $||\cdot||_k$ be an admissible norm on $\mathbf{E}_k$ for $k = 0, 1$. Then we equip $\mathbf{E}_0 \cap \mathbf{E}_1$ with the *intersection topology* defined by the norm $\max(||e||_0, ||e||_1)$ and get a normable TVS. Clearly $(\mathbf{E}_0, \mathbf{E}_1)$ *is a compatible couple if and only if* $\mathbf{E}_0 \cap \mathbf{E}_1$ *is a B-space.* Another useful criterion is: $(\mathbf{E}_0, \mathbf{E}_1)$ *is a compatible couple if and only if there is a Hausdorff TVS* $\mathbf{F}$ *such that* $\mathbf{E}_k \subset \mathbf{F}$ *for* $k = 0, 1$; *in this case, there is a least B-space* $\mathbf{G}$, *denoted* $\mathbf{E}_0 + \mathbf{E}_1$, *such that* $\mathbf{E}_k \subset \mathbf{G}$ *for* $k = 0, 1$ (*Proof*: Take $\mathbf{G} = (\mathbf{E}_0 \oplus \mathbf{E}_1)/\mathbf{K}$ where $\mathbf{E}_0 \oplus \mathbf{E}_1$ means direct sum of Banach spaces and $\mathbf{K}$ is the closed subspace consisting of elements of the form $(e, -e)$ with $e \in \mathbf{E}_0 \cap \mathbf{E}_1$). The *B*-space $\mathbf{E}_0 + \mathbf{E}_1$ is equipped with the *sum topology* defined by the norm $||e|| = \inf\{||e_0||_0 + ||e_1||_1 \mid e_k \in \mathbf{E}_k \text{ and } e = e_0 + e_1\}$. Finally, we mention that quite often compatible couples are constructed in the following way. On a vector space $\mathbf{E}$ two norms $||\cdot||_0$ and $||\cdot||_1$ are given and $\mathbf{E}_k$ is the completion of $\mathbf{E}$ for the norm $||\cdot||_k$. *Then the B-spaces* $\mathbf{E}_0$ *and* $\mathbf{E}_1$ *are compatible if and only if the norms are compatible in the following sense: if* $\{e_n\}_{n \in \mathbb{N}}$ *is a sequence in* $\mathbf{E}$ *which is Cauchy for both norms and if* $e_n \to 0$ *for one of the norms, then* $e_n \to 0$ *for the other norm too.*

Let $(\mathbf{E}_0, \mathbf{E}_1)$ be a compatible pair of *B*-spaces. Then two other *B*-spaces $\mathbf{E}_0 \cap \mathbf{E}_1$ and $\mathbf{E}_0 + \mathbf{E}_1$ have been constructed above and we have $\mathbf{E}_0 \cap \mathbf{E}_1 \subset \mathbf{E}_k \subset \mathbf{E}_0 + \mathbf{E}_1$. A *B*-space $\mathbf{E}$ such that $\mathbf{E}_0 \cap \mathbf{E}_1 \subset \mathbf{E} \subset \mathbf{E}_0 + \mathbf{E}_1$ is called an *intermediate space* for the pair $(\mathbf{E}_0, \mathbf{E}_1)$. Let $(\mathbf{F}_0, \mathbf{F}_1)$ be another compatible pair of *B*-spaces and $\mathbf{F}$ an intermediate space for it. One says that the pair $(\mathbf{E}, \mathbf{F})$ has the *interpolation property* (with respect to the couples $(\mathbf{E}_0, \mathbf{E}_1)$, $(\mathbf{F}_0, \mathbf{F}_1)$) if for each continuous linear map $T : \mathbf{E}_0 + \mathbf{E}_1 \to \mathbf{F}_0 + \mathbf{F}_1$ such that $T\mathbf{E}_k \subset \mathbf{F}_k$ for $k = 0, 1$ we have also $T\mathbf{E} \subset \mathbf{F}$. From the closed graph theorem it follows then easily that $||T||_{\mathbf{E} \to \mathbf{F}} \le c \max(||T||_{\mathbf{E}_0 \to \mathbf{F}_0}, ||T||_{\mathbf{E}_1 \to \mathbf{F}_1})$, where c is a constant depending only on the choice of the admissible norms on the six *B*-spaces which appear in the preceding relation. If $(\mathbf{E}, \mathbf{E})$ has the interpolation property with respect to the couples $(\mathbf{E}_0, \mathbf{E}_1)$, $(\mathbf{E}_0, \mathbf{E}_1)$ one says that $\mathbf{E}$ is an *interpolation space* for the pair $(\mathbf{E}_0, \mathbf{E}_1)$ (or between $\mathbf{E}_0$ and $\mathbf{E}_1$). Note that if $\mathbf{E}$ and $\mathbf{F}$ are interpolation spaces for the pair $(\mathbf{E}_0, \mathbf{E}_1)$, then $(\mathbf{E}, \mathbf{F})$ is a compatible pair ($\mathbf{E}$ and $\mathbf{F}$ being embedded in $\mathbf{E}_0 + \mathbf{E}_1$) and each interpolation space for $(\mathbf{E}, \mathbf{F})$ is an interpolation space for $(\mathbf{E}_0, \mathbf{E}_1)$ too. Being an interpolation space is a rather subtle property: for example $C^1(J)$ is clearly an intermediate space for the pair $(C^0(J), C^2(J))$ but is not an interpolation space (here $J = [0, 1]$; see Exercise 35, page 169 in [BL]; see also page 20 in [KPS] for a more elementary example).

There are many methods for constructing interpolation spaces for an arbitrary compatible pair $(\mathbf{E}_0, \mathbf{E}_1)$. For a detailed presentation of the deeper aspects of the theory one may consult [KPS], [BS], [BK]. In the remainder of this chapter we shall consider only the case $\mathbf{E}_0 \subset \mathbf{E}_1$ and we shall describe a rather particular class of interpolation spaces constructed with the help of the real interpolation method.

2.2. The K-Functional

Let $\mathbf{E}$ and $\mathbf{F}$ be *B*-spaces such that $\mathbf{E} \subset \mathbf{F}$, and let $||\cdot||_{\mathbf{E}}$ and $||\cdot||_{\mathbf{F}}$ be admissible norms on $\mathbf{E}$ and $\mathbf{F}$ respectively. We recall that $||\cdot||_{\mathbf{E}}$ is extended to a gauge on $\mathbf{F}$. For each $\tau > 0$ we introduce the so-called K-*functional* $K(\tau, \cdot)$ on $\mathbf{F}$ by the formula

$$(2.2.1) \qquad K(\tau, f) = \inf_{e \in \mathbf{E}} \left[\tau ||e||_{\mathbf{E}} + ||f - e||_{\mathbf{F}} \right] \qquad (f \in \mathbf{F}).$$

This definition is due to J. Peetre and a more precise notation is $K(\tau, f) \equiv K(\tau, f; \mathbf{F}, \mathbf{E})$. K depends on the chosen norms $|| \cdot ||_{\mathbf{E}}$ and $|| \cdot ||_{\mathbf{F}}$, and, due to the fact that $\mathbf{E} \subset \mathbf{F}$, only its behaviour near $\tau = 0$ will be important. The following relation follows immediately from the definition (2.2.1):

$$(2.2.2) \qquad K(\tau, f) \le K(\sigma, f) \le \frac{\sigma}{\tau} K(\tau, f) \qquad \text{if } 0 < \tau < \sigma.$$

Some further simple properties of the K-functional are given in the following proposition:

PROPOSITION 2.2.1. (a) *For each $\tau > 0$, $K(\tau, \cdot)$ is an admissible norm on $\mathbf{F}$. Furthermore, there is a constant $c > 0$ such that*

$$(2.2.3) \qquad c\tau ||f||_{\mathbf{F}} \le K(\tau, f) \le \min(||f||_{\mathbf{F}}, \tau ||f||_{\mathbf{E}}) \qquad \forall f \in \mathbf{F}.$$

The first inequality holds for $0 < \tau \le 1$ while the second one holds for all $\tau > 0$.

(b) *A vector $f \in \mathbf{F}$ belongs to the closure $\overline{\mathbf{E}}$ of $\mathbf{E}$ in $\mathbf{F}$ if and only if $K(\tau, f) \to 0$ as $\tau \to 0$.*

(c) *Denote by $|| \cdot ||_{\widetilde{\mathbf{E}}}$ the lower semicontinuous regularization of the gauge $|| \cdot ||_{\mathbf{E}}$ on $\mathbf{F}$ and let $\widetilde{\mathbf{E}}$ be the Gagliardo completion of $\mathbf{E}$ with respect to $\mathbf{F}$ (so $|| \cdot ||_{\widetilde{\mathbf{E}}}$ restricted to $\widetilde{\mathbf{E}}$ is an admissible norm on $\widetilde{\mathbf{E}}$). Then*

$$||f||_{\widetilde{\mathbf{E}}} = \sup_{\tau} K(\tau, f)\tau^{-1} = \lim_{\tau \to 0} K(\tau, f)\tau^{-1} \text{ and } K(\tau, f) = K(\tau, f; \mathbf{F}, \widetilde{\mathbf{E}})$$

for each $f \in \mathbf{F}$. If $f \in \overline{\mathbf{E}}$ then one also has $K(\tau, f) = K(\tau, f; \overline{\mathbf{E}}, \widetilde{\mathbf{E}})$.

PROOF. (a) It is easy to check that $K(\tau, \cdot)$ is a semi-norm on $\mathbf{F}$. The second inequality in (2.2.3) is obvious. In view of (2.2.2), the equivalence of $K(\tau, \cdot)$ and $|| \cdot ||_{\mathbf{F}}$ and the first inequality in (2.2.3) hold provided that we can prove the existence of a constant c_τ such that $||f||_{\mathbf{F}} \le c_\tau K(\tau, f)$ for all $f \in \mathbf{F}$, which we do by *reductio ad absurdum*. If one assumes that no such constant c_τ exists, then one can find a sequence $\{f_n\}$ in $\mathbf{F}$ such that $||f_n||_{\mathbf{F}} = 1$ and $K(\tau, f_n) \to 0$ as $n \to \infty$. Then there exists for each $n = 1, 2, \ldots$ a vector $e_n \in \mathbf{E}$ such that $\tau ||e_n||_{\mathbf{E}} + ||f_n - e_n||_{\mathbf{F}} \to 0$ as $n \to \infty$. Thus $e_n \to 0$ in $\mathbf{E}$ and $f_n - e_n \to 0$ in $\mathbf{F}$, hence $f_n \to 0$ in $\mathbf{F}$. This contradicts the assumption that $||f_n||_{\mathbf{F}} = 1$.

(b) Let $f \in \mathbf{F}$ and $\delta > 0$. If $K(\tau, f) \to 0$ as $\tau \to 0$, there is a number $\tau > 0$ such that $K(\tau, f) < \delta$; hence there is a vector $e_\delta \in \mathbf{E}$ such that $\tau ||e_\delta||_{\mathbf{E}} + ||f - e_\delta||_{\mathbf{F}} < \delta$. Thus $e_\delta \to f$ on $\mathbf{F}$ as $\delta \to 0$, so that $f \in \overline{\mathbf{E}}$. Reciprocally, let $f \in \overline{\mathbf{E}}$. Then, for each $\delta > 0$, there is a vector $e \in \mathbf{E}$ such that $||f - e||_{\mathbf{F}} < \delta/2$. Hence, if $0 < \tau < \delta(2||e||_{\mathbf{E}})^{-1}$, then $K(\tau, f) \le \tau ||e||_{\mathbf{E}} + ||f - e||_{\mathbf{F}} < \delta$.

(c) The equalities $||f||_{\widetilde{\mathbf{E}}} = \sup_\tau K(\tau, f)\tau^{-1} = \lim_{\tau \to 0} K(\tau, f)\tau^{-1}$ follow from (2.2.2) and Lemma 2.1.2. Set $\widetilde{K}(\tau, f) = K(\tau, f; \mathbf{F}, \widetilde{\mathbf{E}})$. We have to prove that $K(\tau, f) = \widetilde{K}(\tau, f)$ for $f \in \mathbf{F}$ (the last assertion of part (c) will then be an easy consequence). Since $||e||_{\widetilde{\mathbf{E}}} \le ||e||_{\mathbf{E}}$, the inequality $\widetilde{K} \le K$ is obvious.

Reciprocally, it is sufficient to show that $K(\tau, f) \leq \widetilde{K}(\tau, f) + \varepsilon$ for each $\varepsilon > 0$. For each $e \in \mathbf{E}$ we may find a sequence $\{e_n\}$ with $e_n \to e$ in $\mathbf{F}$ and $||e_n||_{\mathbf{E}} \leq ||e||_{\widetilde{\mathbf{E}}} + \varepsilon/(2\tau)$. Then $||e_n - e||_{\mathbf{F}} < \varepsilon/2$ for large n, hence:

$$K(\tau, f) \leq \tau ||e_n||_{\mathbf{E}} + ||f - e_n||_{\mathbf{F}} \leq \tau ||e||_{\widetilde{\mathbf{E}}} + \frac{\varepsilon}{2} + ||f - e||_{\mathbf{F}} + ||e - e_n||_{\mathbf{F}}$$

$$\leq \tau ||e||_{\widetilde{\mathbf{E}}} + ||f - e||_{\mathbf{F}} + \varepsilon.$$

Taking the lower bound over e we obtain $K(\tau, f) \leq \widetilde{K}(\tau, f) + \varepsilon$. $\square$

Now we can define for each $\theta \in (0, 1)$ and $p \in [1, +\infty]$ a gauge $|| \cdot ||_{\theta, p}$ on $\mathbf{F}$ by setting

$$(2.2.4) \qquad ||f||_{\theta, p} = \left\{ \int_0^1 \left[\frac{K(\tau, f)}{\tau^{1-\theta}} \right]^p \frac{d\tau}{\tau} \right\}^{1/p}.$$

If $p = +\infty$, we shall always interpret this equation as

$$(2.2.5) \qquad ||f||_{\theta, \infty} = \sup_{0 < \tau < 1} \tau^{\theta - 1} K(\tau, f).$$

We point out the following easy consequence of (2.2.2) and Proposition 2.2.1 (a):

LEMMA 2.2.2. *Let* $\theta \in (0, 1)$ *and* $p \in [1, +\infty]$. *There is a constant* $c = c(\theta, p) \in (0, \infty)$ *such that*

$$(2.2.6) \qquad c||f||_{\mathbf{F}} \leq ||f||_{\theta, p} \leq c^{-1}||f||_{\mathbf{E}} \qquad \forall f \in \mathbf{F},$$

$$(2.2.7) \qquad K(\tau, f) \leq c||f||_{\theta, p} \tau^{1-\theta} \qquad \forall \tau \in (0, 1), \forall f \in \mathbf{F}.$$

We now define, for $\theta \in (0, 1)$ and $p \in [1, +\infty]$:

$$(2.2.8) \qquad (\mathbf{E}, \mathbf{F})_{\theta, p} = \{f \in \mathbf{F} \mid ||f||_{\theta, p} < \infty\}.$$

This set $(\mathbf{E}, \mathbf{F})_{\theta, p}$ depends only on the B-spaces $\mathbf{E}$ and $\mathbf{F}$ (i.e. it is independent of the norms $||\cdot||_{\mathbf{E}}, ||\cdot||_{\mathbf{F}}$ used to define $||f||_{\theta, p}$, since a different choice of these norms entails a change of $|| \cdot ||_{\theta, p}$ into an equivalent gauge on $\mathbf{F}$). The restriction of the gauge $|| \cdot ||_{\theta, p}$ to the subspace $(\mathbf{E}, \mathbf{F})_{\theta, p}$ of $\mathbf{F}$ defines a norm on this subspace, and $(\mathbf{E}, \mathbf{F})_{\theta, p}$ is a Banach space for this norm. Indeed, if $f_n \in (\mathbf{E}, \mathbf{F})_{\theta, p}$ is Cauchy, i.e. $||f_n - f_m||_{\theta, p} \to 0$ as $m, n \to \infty$, then $\{f_n\}$ is also Cauchy in $\mathbf{F}$ by (2.2.6). If we denote its limit in $\mathbf{F}$ by f, then $||f_n - f||_{\theta, p} \to 0$ by Fatou's lemma.

DEFINITION 2.2.3. *Let* $\mathbf{E}, \mathbf{F}$ *be* B-spaces with $\mathbf{E} \subset \mathbf{F}$, *let* $0 < \theta < 1$ *and* $1 \leq p \leq +\infty$. *Then* $(\mathbf{E}, \mathbf{F})_{\theta, p}$, *provided with the* B-space structure associated to the norms of the form (2.2.4), is called the *real interpolation space of order* (θ, p) *associated to* $\mathbf{E}$ *and* $\mathbf{F}$.

Let $\widetilde{\mathbf{E}}$ be the Gagliardo completion of $\mathbf{E}$ with respect to $\mathbf{F}$ and $\overline{\mathbf{E}}$ the closure of $\mathbf{E}$ in $\mathbf{F}$. Then $\widetilde{\mathbf{E}}$ and $\overline{\mathbf{E}}$ are B-spaces such that $\mathbf{E} \subset \widetilde{\mathbf{E}} \subset (\mathbf{E}, \mathbf{F})_{\theta,p} \subset \overline{\mathbf{E}} \subset \mathbf{F}$. From Proposition 2.2.1 and Lemma 2.2.2 it follows easily that:

$$(2.2.9) \qquad\qquad (\mathbf{E}, \mathbf{F})_{\theta,p} = (\widetilde{\mathbf{E}}, \overline{\mathbf{E}})_{\theta,p}.$$

On the other hand, by using Proposition 2.2.1 (a) and Fatou's lemma in (2.2.4), it is easily shown that $\| \cdot \|_{\theta,p}$ *is a lower semicontinuous gauge on* $\mathbf{F}$. Hence $(\mathbf{E}, \mathbf{F})_{\theta,p}$ *is relatively complete with respect to* $\mathbf{F}$.

In the following sections we shall give various properties and characterizations of the real interpolation spaces associated to a couple $(\mathbf{E}, \mathbf{F})$ of B-spaces. The following lemma will be useful to obtain these results. The relation (2.2.10) is especially important: it makes the connection between the K-functional and approximation theory. Indeed, (2.2.10) shows that the behaviour of $K(\tau, f)$ as $\tau \to 0$ expresses the degree of approximability of $f \in \mathbf{F}$ by elements of the B-subspace $\mathbf{E}$ (see Chapters 4 and 5 in [Sh1] and also [BuS] and [Be]).

LEMMA 2.2.4. *For each* $f \in \mathbf{F}$ *there is a function* $u : (0, \infty) \to \mathbf{E}$ *of class* C^∞, *with* $u(\sigma) = 0$ *for* $\sigma \geq 1$ *such that for all* $\tau > 0$:

$$(2.2.10) \qquad K(\tau, f) \leq \tau \|u(\tau)\|_{\mathbf{E}} + \|f - u(\tau)\|_{\mathbf{F}} \leq cK(\tau, f),$$

where $c \in (0, \infty)$ *is a constant that is independent of* τ *and* f. *Furthermore, if* $v : (0, \infty) \to \mathbf{E}$ *is defined as* $v(\sigma) = -\sigma \frac{du(\sigma)}{d\sigma}$, *then* v *is of class* C^∞, $v(\sigma) = 0$ *if* $\sigma \geq 1$ *and*

$$(2.2.11) \qquad \tau \|v(\tau)\|_{\mathbf{E}} + \|v(\tau)\|_{\mathbf{F}} \leq 4cK(\tau, f) \qquad \forall \tau > 0.$$

PROOF. (i) If $f = 0$, one may take $u \equiv 0$. So let us assume that $f \in \mathbf{F} \setminus \{0\}$. Then $0 < K(\tau, f) < 2K(\tau, f)$ for each $\tau > 0$, because $K(\tau, \cdot)$ is an admissible norm on $\mathbf{F}$. Hence we may choose for each $n = 1, 2, 3, \ldots$ a vector $e_n \in \mathbf{E}$ such that

$$(2.2.12) \qquad K(2^{-n}, f) \leq 2^{-n} \|e_n\|_{\mathbf{E}} + \|f - e_n\|_{\mathbf{F}} \leq 2K(2^{-n}, f).$$

We define

$$e_\tau = \begin{cases} e_n & \text{if } 2^{-n-1} < \tau \leq 2^{-n} \text{ for some } n = 1, 2, 3, \ldots \\ 0 & \text{if } \tau > 1/2. \end{cases}$$

We notice that, if $\tau \in (2^{-n-1}, 2^{-n}]$ for some $n = 1, 2, 3, \ldots$, then

$$\tau \|e_\tau\|_{\mathbf{F}} + \|f - e_\tau\|_{\mathbf{F}} \leq 2^{-n} \|e_n\|_{\mathbf{E}} + \|f - e_n\|_{\mathbf{F}} \leq 2K(2^{-n}, f) \leq 4K(\tau, f),$$

where the last inequality follows from the second inequality in (2.2.2) (observe that $2^{-n} \tau^{-1} \leq 2$). A similar inequality holds also for $\tau > 1/2$; in this case one obtains from Proposition 2.2.1 (a) and the first inequality in (2.2.2) that :

$$(2.2.13) \qquad \tau \|e_\tau\|_{\mathbf{E}} + \|f - e_\tau\|_{\mathbf{F}} = \|f\|_{\mathbf{F}} \leq bK(1/2, f) \leq bK(\tau, f)$$

for some constant $b < \infty$ which is independent of f and τ. If we assume $b \geq 4$, then (2.2.13) is valid for all $\tau > 0$.

Now suppose that $\varphi : (0, \infty) \to \mathbb{R}$ is an integrable function with support in the interval $[1, 2]$. Then $\varphi(\tau/\sigma) = 0$ unless $\tau/2 \leq \sigma \leq \tau$. Hence, by using (2.2.13) and the first inequality in (2.2.2), one finds that for any $\tau > 0$:

$$(2.2.14) \qquad \int_0^\infty \left| \varphi \left(\frac{\tau}{\sigma} \right) \right| [\tau \|e_\sigma\|_{\mathbf{E}} + \|f - e_\sigma\|_{\mathbf{F}}] \frac{d\sigma}{\sigma} \leq$$

$$\leq 2 \int_0^\infty \left| \varphi \left(\frac{\tau}{\sigma} \right) \right| [\sigma \|e_\sigma\|_{\mathbf{E}} + \|f - e_\sigma\|_{\mathbf{F}}] \frac{d\sigma}{\sigma}$$

$$\leq 2b \int_0^\infty \left| \varphi \left(\frac{\tau}{\sigma} \right) \right| K(\sigma, f) \frac{d\sigma}{\sigma}$$

$$\leq 2b \cdot \sup_{\tau/2 \leq \varrho \leq \tau} K(\varrho, f) \cdot \int_0^\infty \left| \varphi \left(\frac{\tau}{\sigma} \right) \right| \frac{d\sigma}{\sigma}$$

$$\leq 2bK(\tau, f) \int_0^\infty |\varphi(s)| \frac{ds}{s}.$$

(ii) Now let $\eta \in C_0^\infty((0, \infty))$ be such that $0 \leq \eta \leq 2$, supp $\eta \subset [1, 2]$, $\int_0^\infty \eta(\tau) \frac{d\tau}{\tau} = 1$, $\int_0^\infty |\eta'(\tau)| d\tau \leq 4$. We define, for $0 < \tau < \infty$:

$$u(\tau) = \int_0^\infty \eta \left(\frac{\tau}{\sigma} \right) e_\sigma \frac{d\sigma}{\sigma}.$$

Clearly $u : (0, \infty) \to \mathbf{E}$ is of class C^∞ and $u(\tau) = 0$ if $\tau \geq 1$. The first inequality in (2.2.10) is evident. For the second one we write :

$$f - u(\tau) = \int_0^\infty \eta \left(\frac{\tau}{\sigma} \right) [f - e_\sigma] \frac{d\sigma}{\sigma}.$$

Then

$$\tau \|u_\tau\|_{\mathbf{E}} + \|f - u_\tau\|_{\mathbf{F}} \leq \int_0^\infty \eta \left(\frac{\tau}{\sigma} \right) [\tau \|e_\sigma\|_{\mathbf{E}} + \|f - e_\sigma\|_{\mathbf{F}}] \frac{d\sigma}{\sigma}.$$

By (2.2.14) with $\varphi = \eta$, the r.h.s. of this inequality is majorized by $cK(\tau, f)$ with $c = 2b$.

(iii) Let η and u be as in (ii) and define $\xi \in C_0^\infty((0, \infty))$ by $\xi(t) = -t\eta'(t)$. Then

$$v(\tau) \equiv -\tau \frac{d}{d\tau} u(\tau) = \int_0^\infty \xi \left(\frac{\tau}{\sigma} \right) e_\sigma \frac{d\sigma}{\sigma}.$$

Since $\int_0^\infty \xi(\sigma) \frac{d\sigma}{\sigma} = 0$, we also have

$$v(\tau) = \int_0^\infty \xi \left(\frac{\tau}{\sigma} \right) [e_\sigma - f] \frac{d\sigma}{\sigma}.$$

Then, by using (2.2.14) with $\varphi = \xi$, one gets that

$$\tau||v(\tau)||_{\mathbf{E}} + ||v(\tau)||_{\mathbf{F}} \leq$$

$$\leq \int_0^\infty \left|\xi\left(\frac{\tau}{\sigma}\right)\right| [\tau||e_\sigma||_{\mathbf{E}} + ||f - e_\sigma||_{\mathbf{F}}] \frac{d\sigma}{\sigma}$$

$$\leq cK(\tau, f) \int_0^\infty |\xi(\tau)| \frac{d\tau}{\tau} \leq 4cK(\tau, f). \quad \square$$

COROLLARY 2.2.5. *Let $f \in \mathbf{F}$ and let u and v be as in Lemma 2.2.4. Then*
(a) $f \in \overline{\mathbf{E}}$ *if and only if $\tau u(\tau) \to 0$ in $\mathbf{E}$ and $u(\tau) \to f$ in $\mathbf{F}$ as $\tau \to 0$.*
(b) *If $f \in \overline{\mathbf{E}}$, then*

(2.2.15) $$f = \lim_{\varepsilon \to +0} \int_\varepsilon^\infty v(\tau) \frac{d\tau}{\tau} \text{ in the norm of } \mathbf{F}.$$

(c) $f \in (\mathbf{E}, \mathbf{F})_{\theta, p}$ *if and only if*

$$\left\{\int_0^1 \left[\tau^\theta ||u(\tau)||_{\mathbf{E}} + \tau^\theta \left\|\frac{f - u(\tau)}{\tau}\right\|_{\mathbf{F}}\right]^p \frac{d\tau}{\tau}\right\}^{1/p} < \infty.$$

PROOF. (a) follows from Proposition 2.2.1(b) and (2.2.10). For (b) we observe that the integral on the r.h.s. of (2.2.15) is equal to $u(\varepsilon)$ and then use the result of (a). (c) is an immediate consequence of the definition (2.2.4) and of (2.2.10). $\square$

REMARK 2.2.6. The inequality (2.2.11) implies that

$$\int_0^1 [\tau||v(\tau)||_{\mathbf{E}} + ||v(\tau)||_{\mathbf{F}}] \frac{d\tau}{\tau} \leq 4c \int_0^1 K(\tau, f) \frac{d\tau}{\tau}.$$

Hence, if the last integral is finite (for example if $f \in (\mathbf{E}, \mathbf{F})_{\theta, p}$ for some $\theta \in (0, 1)$ and some $p \in [1, \infty]$), then $\int_0^\infty ||v(\tau)||_{\mathbf{F}} \frac{d\tau}{\tau} < \infty$, so that $f = \int_0^\infty v(\tau) \frac{d\tau}{\tau}$ (as a Bochner integral in $\mathbf{F}$) by Corollary 2.2.5 (b). Assume now that a B-subspace $\mathbf{G} \subset \mathbf{F}$ is given and that some $f \in \mathbf{F}$ is represented as a (Bochner) integral in $\mathbf{F}$ of the preceding form with a locally integrable function $v : (0, \infty) \to \mathbf{G}$. If we have $\int_0^\infty ||v(\tau)||_{\mathbf{G}} \tau^{-1} d\tau < \infty$ for some admissible norm on $\mathbf{G}$, then $f \in \mathbf{G}$ and $||f||_{\mathbf{G}} \leq \int_0^\infty ||v(\tau)||_{\mathbf{G}} \tau^{-1} d\tau$. This easily proven observation will be often used in what follows.

2.3. The Mean and the Trace Method

In several theoretical and practical questions, the description of the spaces $(\mathbf{E}, \mathbf{F})_{\theta, p}$ contained in Definition 2.2.3 is not convenient. For this reason, many other ways of constructing these interpolation spaces appear in the literature. We shall describe below the mean method (Proposition 2.3.2) and the trace method (Proposition 2.3.3). As before, we consider two B-spaces $\mathbf{E}$ and $\mathbf{F}$ with $\mathbf{E} \subset \mathbf{F}$ and let $|| \cdot ||_{\mathbf{E}}, || \cdot ||_{\mathbf{F}}$ be admissible norms on these spaces. We begin with some preliminary inequalities. If $\mathbf{G}$ is an arbitrary B-space and $1 \leq p \leq \infty$, we denote by $L_*^p(\mathbf{G})$ the B-space of $\mathbf{G}$-valued (equivalence classes of) functions on

$(0, \infty)$ that are p-integrable with respect to the measure $d\tau/\tau$. An admissible norm on $L_*^p(\mathbf{G})$ is given by

$$\|f\|_{L_*^p(\mathbf{G})} = \left[\int_0^\infty \|f(\tau)\|_{\mathbf{G}}^p \frac{d\tau}{\tau} \right]^{1/p} \quad \text{if } 1 \leq p < \infty,$$

$$\|f\|_{L_*^p(\mathbf{G})} = \operatorname*{ess\,sup}_{0 < \tau < \infty} \|f(\tau)\|_{\mathbf{G}} \quad \text{if } p = \infty.$$

For $0 \leq a < b \leq \infty$ the space $L_*^p((a,b); \mathbf{G})$ is defined by replacing $(0, \infty)$ by (a, b) above.

LEMMA 2.3.1. *Let* $w : (0, \infty) \to \mathbf{E}$ *be a locally integrable function such that* $w \in L_*^1(\mathbf{F})$, *and set* $g = \int_0^\infty w(\sigma) \frac{d\sigma}{\sigma} \in \mathbf{F}$. *Then, for each* $\tau > 0$:

$$(2.3.1) \qquad K(\tau, g) \leq \int_1^\infty \|\tau \sigma w(\tau\sigma)\|_{\mathbf{E}} \frac{d\sigma}{\sigma^2} + \int_0^1 \|w(\tau\sigma)\|_{\mathbf{F}} \frac{d\sigma}{\sigma}.$$

Moreover one has for each $\theta \in (0, 1)$ *and each* $p \in [1, \infty]$:

$$(2.3.2) \qquad \|g\|_{\theta, p} \leq \frac{1}{\theta} \|\tau^\theta w\|_{L_*^p(\mathbf{E})} + \frac{1}{1-\theta} \|\tau^{\theta-1} w\|_{L_*^p(\mathbf{F})}.$$

PROOF. (i) We first prove (2.3.1). For each $\tau > 0$ we set $u(\tau) = \int_\tau^\infty w(s) s^{-1} ds$, so that $g - u(\tau) = \int_0^\tau w(s) s^{-1} ds$. Then, as gauges on $\mathbf{F}$:

$$K(\tau, g) \leq \tau \|u(\tau)\|_{\mathbf{E}} + \|g - u(\tau)\|_{\mathbf{F}} \leq \int_\tau^\infty \frac{\tau}{s} \|sw(s)\|_{\mathbf{E}} \frac{ds}{s} + \int_0^\tau \|w(s)\|_{\mathbf{F}} \frac{ds}{s}.$$

The last expression is identical with the r.h.s. of (2.3.1).

(ii) The inequality (2.3.2) is easily obtained by using (2.3.1):

$$\|g\|_{\theta, p} \leq \|(\cdot)^{\theta-1} K(\cdot, g)\|_{L_*^p}$$

$$\leq \int_1^\infty \|(\sigma\cdot)^\theta w(\sigma\cdot)\|_{L_*^p(\mathbf{E})} \frac{d\sigma}{\sigma^{1+\theta}} + \int_0^1 \|(\sigma\cdot)^{\theta-1} w(\sigma\cdot)\|_{L_*^p(\mathbf{F})} \frac{d\sigma}{\sigma^\theta}$$

$$= \frac{1}{\theta} \|(\cdot)^\theta w\|_{L_*^p(\mathbf{E})} + \frac{1}{1-\theta} \|(\cdot)^{\theta-1} w\|_{L_*^p(\mathbf{F})}. \qquad \square$$

PROPOSITION 2.3.2. *Let* $\theta \in (0, 1)$ *and* $p \in [1, +\infty]$.

(a) *Let* $v : (0, \infty) \to \mathbf{E}$ *be such that* $\tau^\theta v \in L_*^p(\mathbf{E})$ *and* $\tau^{\theta-1} v \in L_*^p(\mathbf{F})$. *Then* $v \in L_*^1(\mathbf{F})$, *the vector* $f := \int_0^\infty v(\tau) \tau^{-1} d\tau$ *belongs to* $(\mathbf{E}, \mathbf{F})_{\theta, p}$ *and satisfies*

$$(2.3.3) \qquad \|f\|_{\theta, p} \leq \frac{1}{\theta} \|\tau^\theta v\|_{L_*^p(\mathbf{E})} + \frac{1}{1-\theta} \|\tau^{\theta-1} v\|_{L_*^p(\mathbf{F})}.$$

(b) *Conversely, if* $f \in (\mathbf{E}, \mathbf{F})_{\theta, p}$, *then there is a function* $v : (0, \infty) \to \mathbf{E}$ *of class* C^∞ *with* $v(\tau) = 0$ *for* $\tau \geq 1$, $v(\cdot) \in L_*^1(\mathbf{F})$, $f = \int_0^\infty v(\tau) \tau^{-1} d\tau$ *and such that (2.3.3) and the following inequality are satisfied:*

$$(2.3.4) \qquad \|\tau^\theta v\|_{L_*^p(\mathbf{E})} + \|\tau^{\theta-1} v\|_{L_*^p(\mathbf{F})} \leq c \|f\|_{\theta, p}$$

where $c < \infty$ *is some constant independent of* f.

PROOF. (a) In view of Lemma 2.3.1 it suffices to show that $v \in L^1_*(\mathbf{F})$. This can be done by using the Hölder inequality, with $p' = (1 - 1/p)^{-1}$:

$$\int_0^1 ||v(\tau)||_{\mathbf{F}} \frac{d\tau}{\tau} = \int_0^1 ||\tau^{\theta-1} v(\tau)||_{\mathbf{F}} \frac{d\tau}{\tau} \tau^{1-\theta}$$

$$\leq ||\tau^{\theta-1} v||_{L^p_*(\mathbf{F})} \cdot \left[\int_0^1 \tau^{(1-\theta)p'} \frac{d\tau}{\tau} \right]^{1/p'} < \infty,$$

and similarly

$$\int_1^\infty ||v(\tau)||_{\mathbf{E}} \frac{d\tau}{\tau} \leq ||\tau^\theta v||_{L^p_*(\mathbf{E})} \cdot \left[\int_1^\infty \tau^{-\theta p'} \frac{d\tau}{\tau} \right]^{1/p'} < \infty.$$

(b) We let v be the function given by Lemma 2.2.4. In view of (a), it is sufficient to prove the inequality (2.3.4). This is easily done by using (2.2.11) :

$$||\tau^\theta v||_{L^p_*(\mathbf{E})} + ||\tau^{\theta-1} v||_{L^p_*(\mathbf{F})} \leq 2 \cdot 4c \left[\int_0^1 \left[\frac{K(\tau, f)}{\tau^{1-\theta}} \right]^p \frac{d\tau}{\tau} \right]^{1/p} = 8c ||f||_{\theta,p}. \quad \square$$

Proposition 2.3.2 may be restated in the following terms: $(\mathbf{E}, \mathbf{F})_{\theta,p}$ is the space of vectors $f \in \mathbf{F}$ that can be represented as a "mean" $f = \int_0^\infty v(\tau) \tau^{-1} d\tau$ (with respect to the Haar measure $\tau^{-1} d\tau$ on $(0, \infty)$) of a (Bochner) integrable function $v : (0, \infty) \to \mathbf{F}$ satisfying $|||v||| := ||\tau^\theta v||_{L^p_*(\mathbf{E})} + ||\tau^{\theta-1} v||_{L^p_*(\mathbf{F})} < \infty$. Moreover, the infimum of $|||v|||$ over all such functions v is an admissible norm on $(\mathbf{E}, \mathbf{F})_{\theta,p}$.

There is a second version of the mean method (called "method of constants" in [KPS]) which is often useful in applications. Consider an arbitrary Borel function $u : (0, \infty) \to \mathbf{E}$; then $K(\tau, f) \leq \tau ||u(\tau)||_{\mathbf{E}} + ||f - u(\tau)||_{\mathbf{F}}$ for all $\tau > 0$. By using (2.2.4) we get

$$||f||_{\theta,p} \leq ||\tau^{\theta-1} K(\cdot, f)||_{L^p_*} \leq ||\tau^\theta u(\cdot)||_{L^p_*(\mathbf{E})} + ||\tau^{\theta-1} (f - u(\cdot))||_{L^p_*(\mathbf{F})}.$$

Lemma 2.2.4 shows that $(\mathbf{E}, \mathbf{F})_{\theta,p}$ is exactly the set of $f \in \mathbf{F}$ such that there is a function u for which the r.h.s. of the preceding relation is finite. Moreover, the infimum of the r.h.s. over all u is a gauge on $\mathbf{F}$ equivalent to $|| \cdot ||_{\theta,p}$.

PROPOSITION 2.3.3. Let $\theta \in (0, 1)$, $p \in [1, +\infty]$ and $p' = (1 - 1/p)^{-1}$.

(a) Let $u : (0, \infty) \to \mathbf{E}$ be such that $\tau^\theta u \in L^p_*(\mathbf{E})$ and $\tau^\theta \frac{d}{d\tau} u \in L^p_*(\mathbf{F})$ (where the derivative is interpreted in the sense of distributions). Then $\lim_{\tau \to 0} u(\tau) \equiv f$ exists in $\mathbf{F}$ and belongs to $(\mathbf{E}, \mathbf{F})_{\theta,p}$. Moreover one has:

$$(2.3.5) \qquad ||f||_{\theta,p} \leq ||\tau^\theta u||_{L^p_*(\mathbf{E})} + \frac{1}{1-\theta} \left\| \tau^\theta \frac{d}{d\tau} u \right\|_{L^p_*(\mathbf{F})}$$

and, for each $\sigma > 0$:

$$(2.3.6) \qquad ||u(\sigma) - f||_{\mathbf{F}} \leq [(1-\theta)p']^{-1/p'} \sigma^{1-\theta} \left\| \tau^\theta \frac{d}{d\tau} u \right\|_{L^p_*(\mathbf{F})}.$$

(b) *Conversely, let $f \in (\mathbf{E}, \mathbf{F})_{\theta,p}$. Then there exists a function $u : (0, \infty) \to \mathbf{E}$ of class C^∞ satisfying (2.2.10), with $u(\tau) = 0$ if $\tau \geq 1$ and such that for some constant $c < \infty$ independent of f:*

$$(2.3.7) \qquad ||\tau^\theta u||_{L^p_*(\mathbf{E})} + \left\|\tau^\theta \frac{d}{d\tau} u\right\|_{L^p_*(\mathbf{F})} \leq c||f||_{\theta,p},$$

$$(2.3.8) \qquad ||u(\sigma) - f||_{\mathbf{F}} \leq c||f||_{\theta,p}\sigma^{1-\theta} \qquad \forall \sigma > 0.$$

PROOF. (a) By writing $u'(\tau) = [\tau^{\theta-1/p}u'(\tau)] \cdot [\tau^{1-\theta-1/p'}]$ and by using the Hölder inequality, one gets that for $t > 0$:

$$\int_0^t ||u'(\tau)||_{\mathbf{F}} d\tau \leq ||\tau^\theta u'||_{L^p_*(\mathbf{F})} \cdot [(1-\theta)p']^{-1/p'} t^{1-\theta}.$$

This implies that $\{u(\sigma)\}_{\sigma>0}$ is Cauchy in $\mathbf{F}$ as $\sigma \to 0$, so that $f \equiv \lim_{\sigma \to 0} u(\sigma)$ exists. Then (2.3.6) follows from the preceding estimate and

$$||u(\sigma) - f||_{\mathbf{F}} \leq \int_0^\sigma ||u'(\tau)||_{\mathbf{F}} d\tau.$$

Next we observe that

$$K(\tau, f) \leq \tau ||u(\tau)||_{\mathbf{E}} + ||f - u(\tau)||_{\mathbf{F}} \leq \tau ||u(\tau)||_{\mathbf{E}} + \tau \int_0^1 ||u'(\tau\sigma)||_{\mathbf{F}} d\sigma.$$

Hence

$$||f||_{\theta,p} \leq ||(\cdot)^{\theta-1} K(\cdot, f)||_{L^p_*} \leq ||(\cdot)^\theta u||_{L^p_*(\mathbf{E})} + \int_0^1 ||(\sigma\cdot)^\theta u'(\sigma\cdot)||_{L^p_*(\mathbf{F})} \frac{d\sigma}{\sigma^\theta},$$

and the last expression is identical with the r.h.s. of (2.3.5).

(b) Let u and v be as in Lemma 2.2.4. The second inequality in (2.2.10) implies that $||\tau^\theta u||_{L^p_*(\mathbf{E})} = ||\tau^\theta u||_{L^p_*((0,1);\mathbf{E})} \leq c||f||_{\theta,p}$. Also, by (2.3.4):

$$\left\|\tau^\theta \frac{du}{d\tau}\right\|_{L^p_*(\mathbf{F})} = ||\tau^{\theta-1} v||_{L^p_*(\mathbf{F})} \leq c||f||_{\theta,p}.$$

Thus (2.3.7) is proven.

To prove (2.3.8) we remark that $u(\sigma) \to f$ as $\sigma \to 0$ by Corollary 2.2.5 (a), because $f \in \overline{\mathbf{E}}$ by (2.2.7) and Proposition 2.2.1 (b). Now (2.3.8) follows from (2.3.6) and (2.3.7). $\square$

We end the section with the description of a discrete version of the mean method which is sometimes easier to use. Let us first point out a discrete version of the gauge (2.2.4).

LEMMA 2.3.4. *Let $0 < \theta < 1$ and $1 \leq p \leq +\infty$. For each $f \in \mathbf{F}$ one has*

$$(2.3.9) \qquad 2^{\theta-1}||f||_{\theta,p} \leq \left\{\sum_{n=0}^\infty [2^{n(1-\theta)} K(2^{-n}, f)]^p\right\}^{1/p} \leq 4||f||_{\theta,p}.$$

PROOF. We consider the case where $p < \infty$. The argument for $p = +\infty$ is similar. If $\tau \in [2^{-n-1}, 2^{-n}]$ for some $n = 0, 1, 2, \ldots$, then by (2.2.2):

$$2^{\theta-1}\frac{K(\tau, f)}{\tau^{1-\theta}} \le 2^{n(1-\theta)}K(2^{-n}, f) \le 2\frac{K(\tau, f)}{\tau^{1-\theta}}.$$

The first inequality implies that:

$$2^{p(\theta-1)}\|f\|_{\theta,p}^p = 2^{p(\theta-1)}\sum_{n=0}^{\infty}\int_{2^{-n-1}}^{2^{-n}}\left[\frac{K(\tau, f)}{\tau^{1-\theta}}\right]^p \frac{d\tau}{\tau}$$

$$\le \sum_{n=0}^{\infty}\left[2^{n(1-\theta)}K(2^{-n}, f)\right]^p \cdot \int_{2^{-n-1}}^{2^{-n}}\frac{d\tau}{\tau},$$

and the first inequality in (2.3.9) follows because the last integral is equal to $\ln 2$ for each n. The proof of the second inequality in (2.3.9) is similar. $\square$

If $p \in [1, +\infty]$, $\mathbf{F}$ is a B-space and $\|\cdot\|_{\mathbf{F}}$ an admissible norm on $\mathbf{F}$, we denote by $\ell^p(\mathbf{F})$ the Banach space of sequences $\{f_n\}_{n=0}^{\infty}$ of vectors of $\mathbf{F}$, with

$$\|\{f_n\}\|_{\ell^p(\mathbf{F})} = \left[\sum_{n=0}^{\infty}\|f_n\|_{\mathbf{F}}^p\right]^{1/p} \quad \text{if } p < \infty,$$

$$\|\{f_n\}\|_{\ell^\infty(\mathbf{F})} = \sup_{n \in \mathbb{N}}\|f_n\|_{\mathbf{F}}.$$

PROPOSITION 2.3.5. *Let $0 < \theta < 1$ and $1 \le p \le \infty$. A vector $f \in \mathbf{F}$ belongs to $(\mathbf{E}, \mathbf{F})_{\theta,p}$ if and only if there is a sequence $\{f_n\}_{n=0}^{\infty}$ in $\mathbf{E}$ with*

(2.3.10)

$$\||\{f_n\}\||_{\theta,p} := \frac{1}{\theta}\|\{2^{-n\theta}f_n\}\|_{\ell^p(\mathbf{E})} + \frac{2}{1-\theta}\|\{2^{n(1-\theta)}f_n\}\|_{\ell^p(\mathbf{F})} < \infty$$

and such that $f = \sum_{n=0}^{\infty}f_n$. We have $\|f\|_{\theta,p} \le \||\{f_n\}\||_{\theta,p}$ and the infimum of $\||\{f_n\}\||_{\theta,p}$ over all such representations of f is an admissible norm on $(\mathbf{E}, \mathbf{F})_{\theta,p}$.

PROOF. To prove that $\|f\|_{\theta,p} \le \||\{f_n\}\||_{\theta,p}$ we define $v : (0, \infty) \to \mathbf{E}$ by $v(\tau) = [\ln 2]^{-1}f_n$ if $2^{-n-1} \le \tau < 2^{-n}$ $(n = 0, 1, 2, \ldots)$ and $v(\tau) = 0$ if $\tau \ge 1$. Then $f = \int_0^{\infty}v(\tau)\tau^{-1}d\tau$ and, by (2.3.2):

$$\|f\|_{\theta,p} \le \frac{1}{\theta}\|\tau^{\theta}v\|_{L_*^p(\mathbf{E})} + \frac{1}{1-\theta}\|\tau^{\theta-1}v\|_{L_*^p(\mathbf{F})}.$$

But clearly we have

$$\|\tau^{\theta}v\|_{L_*^p(\mathbf{E})}^p = \sum_{n=0}^{\infty}\int_{2^{-n-1}}^{2^{-n}}\|\tau^{\theta}v(\tau)\|_{\mathbf{E}}^p\frac{d\tau}{\tau} \le \|\{2^{-n\theta}f_n\}\|_{\ell^p(\mathbf{E})}^p$$

and

$$\|\tau^{\theta-1}v\|_{L_*^p(\mathbf{F})} \le 2\|\{2^{n(1-\theta)}f_n\}\|_{\ell^p(\mathbf{F})}.$$

Reciprocally, let $f \in (\mathbf{E}, \mathbf{F})_{\theta,p}$ and define v as in Proposition 2.3.2 (b). Set $f_n = \int_{2^{-n-1}}^{2^{-n}}v(\tau)\tau^{-1}d\tau$ and observe that $f_n \in \mathbf{E}$ and $f = \sum_{n=0}^{\infty}f_n$ in $\mathbf{F}$. It remains to be shown that $|||\{f_n\}|||_{\theta,p} \leq c\|f\|_{\theta,p}$ for a constant c independent of f. We prove that $\|\{2^{-n\theta}f_n\}\|_{\ell^p(\mathbf{E})} \leq c'\|\tau^\theta v\|_{L^p_*(\mathbf{E})}$ and then we use (2.3.4); the second term in (2.3.10) is treated similarly. By using the Hölder inequality we get

$$\|2^{-n\theta}f_n\|_{\mathbf{E}} \leq 2^\theta \int_{2^{-n-1}}^{2^{-n}} \tau^\theta \|v(\tau)\|_{\mathbf{E}} \frac{d\tau}{\tau} \leq 2^\theta \left[\int_{2^{-n-1}}^{2^{-n}} \|\tau^\theta v(\tau)\|_{\mathbf{E}}^p \frac{d\tau}{\tau}\right]^{1/p} \cdot (\ln 2)^{1/p'}$$

which clearly implies the estimate we are looking for. $\quad\square$

2.4. Comparison and Duality of Interpolation Spaces

We now present some relations between the B-spaces $(\mathbf{E},\mathbf{F})_{\theta,p}$ for different values of θ and p, and then we shall prove the duality theorem.

PROPOSITION 2.4.1. *Let $\mathbf{E}$ and $\mathbf{F}$ be B-spaces with $\mathbf{E} \subset \mathbf{F}$. Denote by $\widetilde{\mathbf{E}}$ the Gagliardo completion of $\mathbf{E}$ with respect to $\mathbf{F}$ and let $\overline{\mathbf{E}}$ be the closure of $\mathbf{E}$ in $\mathbf{F}$. Then*

(a) $\mathbf{E} \subset \widetilde{\mathbf{E}} \subset (\mathbf{E},\mathbf{F})_{\theta,p} \subset \overline{\mathbf{E}} \subset \mathbf{F} \quad \forall\theta \in (0,1),\ \forall p \in [1,+\infty],\ \text{and}\ (\mathbf{E},\mathbf{F})_{\theta,p} = (\widetilde{\mathbf{E}},\overline{\mathbf{E}})_{\theta,p};$

(b) *if $1 \leq p_1 < p_2 \leq +\infty$, then $(\mathbf{E},\mathbf{F})_{\theta,p_1} \subset (\mathbf{E},\mathbf{F})_{\theta,p_2}$ for each $\theta \in (0,1)$;*

(c) *if $0 < \theta_1 < \theta_2 < 1$ and $p_1, p_2 \in [1,+\infty]$, one has $(\mathbf{E},\mathbf{F})_{\theta_1,p_1} \subset (\mathbf{E},\mathbf{F})_{\theta_2,p_2};$*

(d) $\mathbf{E}$ *is dense in each $(\mathbf{E},\mathbf{F})_{\theta,p}$ with $0 < \theta < 1$ and $1 \leq p < \infty$.*

PROOF. (a) has been shown before (see (2.2.9)). If $p_2 = +\infty$, the result of (b) follows from (2.2.5) and from (2.2.7) with $p = p_1$. To prove (b) if $p_2 < +\infty$, we use the definition (2.2.4) as well as (2.2.7) with $p = p_1$ to get that

$$\|f\|_{\theta,p_2}^{p_2} = \int_0^1 \left[\frac{K(\tau,f)}{\tau^{1-\theta}}\right]^{p_2-p_1} \left[\frac{K(\tau,f)}{\tau^{1-\theta}}\right]^{p_1} \frac{d\tau}{\tau}$$

$$\leq [c(\theta,p_1)\|f\|_{\theta,p_1}]^{p_2-p_1} \int_0^1 \left[\frac{K(\tau,f)}{\tau^{1-\theta}}\right]^{p_1} \frac{d\tau}{\tau}$$

$$= [c(\theta,p_1)]^{p_2-p_1} \|f\|_{\theta,p_1}^{p_2}.$$

(c) By virtue of (b), it suffices to show that $(\mathbf{E},\mathbf{F})_{\theta_1,\infty} \subset (\mathbf{E},\mathbf{F})_{\theta_2,1}$ if $0 < \theta_1 < \theta_2 < 1$. This is a straightforward consequence of (2.2.4) and (2.2.5):

$$\|f\|_{\theta_2,1} = \int_0^1 \frac{K(\tau,f)}{\tau^{1-\theta_1}} \tau^{\theta_2-\theta_1} \frac{d\tau}{\tau} \leq \frac{1}{\theta_2 - \theta_1} \|f\|_{\theta_1,\infty}.$$

(d) Let $f \in (\mathbf{E},\mathbf{F})_{\theta,p}$ with $p < \infty$, and let v be as in Proposition 2.3.2 (b). So $f = \int_0^\infty v(\tau)\tau^{-1}d\tau$, and $\int_\varepsilon^\infty v(\tau)\tau^{-1}d\tau$ belongs to $\mathbf{E}$ for each $\varepsilon > 0$. Thus it is enough to prove that $\|\int_0^\varepsilon v(\tau)\tau^{-1}d\tau\|_{\theta,p} \to 0$ as $\varepsilon \to 0$. For this we apply Lemma 2.3.1 with $w(\tau) = v(\tau)$ if $0 < \tau \leq \varepsilon$, $w(\tau) = 0$ if $\tau > \varepsilon$, to get that:

$$\left\|\int_0^\varepsilon v(\tau)\frac{d\tau}{\tau}\right\|_{\theta,p} \leq \frac{1}{\theta}\|\tau^\theta v\|_{L^p_*((0,\varepsilon);\mathbf{E})} + \frac{1}{1-\theta}\|\tau^{\theta-1}v\|_{L^p_*((0,\varepsilon);\mathbf{F})}.$$

Since $\tau^\theta v \in L^p_*(\mathbf{E})$ and $\tau^{\theta-1}v \in L^p_*(\mathbf{F})$ by (2.3.4) and because $p < \infty$, the r.h.s. of the preceding inequality converges to zero as $\varepsilon \to 0$. $\quad\square$

To motivate the following theorem, consider the situation where one has three B-spaces $\mathbf{E}$, $\mathbf{F}$ and $\mathbf{G}$ such that $\mathbf{E} \subset \mathbf{G} \subset \mathbf{F}$ with dense embeddings. Then $\mathbf{F}^* \subset \mathbf{G}^* \subset \mathbf{E}^*$ (cf. Section 2.1). Now consider the particular case where $\mathbf{G} = (\mathbf{E}, \mathbf{F})_{\theta,p}$ for some θ and p. If $p < \infty$ and if $\mathbf{E}$ is dense in $\mathbf{F}$, then we are in the above situation by virtue of Proposition 2.4.1 (d). Hence we shall have

$$\mathbf{F}^* \subset [(\mathbf{E}, \mathbf{F})_{\theta,p}]^* \subset \mathbf{E}^*.$$

The duality theorem states that the space $[(\mathbf{E}, \mathbf{F})_{\theta,p}]^*$ is identical with an interpolation space associated to the pair of B-spaces $\mathbf{F}^*$, $\mathbf{E}^*$.

THEOREM 2.4.2. *Let* $\mathbf{E}$ *and* $\mathbf{F}$ *be* B-*spaces such that* $\mathbf{E} \subset \mathbf{F}$ *densely. Let* $\theta \in (0,1)$, $p \in [1, \infty)$ *and* $p' = (1 - 1/p)^{-1}$. *Then*

(2.4.1) $$[(\mathbf{E}, \mathbf{F})_{\theta,p}]^* = (\mathbf{F}^*, \mathbf{E}^*)_{1-\theta,p'}.$$

If $p = \infty$, *let* $(\mathbf{E}, \mathbf{F})_{\theta,\infty}^{\circ}$ *be the closure of* $\mathbf{E}$ *in* $(\mathbf{E}, \mathbf{F})_{\theta,\infty}$. *Then*

(2.4.2) $$[(\mathbf{E}, \mathbf{F})_{\theta,\infty}^{\circ}]^* = (\mathbf{F}^*, \mathbf{E}^*)_{1-\theta,1}.$$

PROOF. We consider only the case $p < \infty$; the proof in the case $p = \infty$ is quite similar (at step (ii) below one has to use the fact that $[c_0(\mathbf{F})]^* = \ell^1(\mathbf{F}^*)$, where $c_0(\mathbf{F})$ is the subspace of $\ell^\infty(\mathbf{F})$ consisting of $\mathbf{F}$-valued sequences converging to zero at infinity; see [Tr]). Let $|| \cdot ||_{\mathbf{E}}$ and $|| \cdot ||_{\mathbf{F}}$ be admissible norms on $\mathbf{E}$ and $\mathbf{F}$ respectively. We denote by $|| \cdot ||_{\mathbf{E}^*}$ and $|| \cdot ||_{\mathbf{F}^*}$ the corresponding dual norms on $\mathbf{E}^*$ and $\mathbf{F}^*$ and by $|| \cdot ||_{1-\theta,p'}^*$ the associated norm on the interpolation space $(\mathbf{F}^*, \mathbf{E}^*)_{1-\theta,p'}$.

(i) Fix $\varphi \in (\mathbf{F}^*, \mathbf{E}^*)_{1-\theta,p'}$. By Proposition 2.3.2(b) there is a function $\psi : (0, \infty) \to \mathbf{E}$ such that $\varphi = \int_0^1 \psi(\tau)\tau^{-1}d\tau$ and

(2.4.3) $$||\tau^{1-\theta}\psi||_{L_*^{p'}(\mathbf{F}^*)} + ||\tau^{-\theta}\psi||_{L_*^{p'}(\mathbf{E}^*)} \leq c^* ||\varphi||_{1-\theta,p'}^*$$

with c^* independent of φ.

Now let $f \in (\mathbf{E}, \mathbf{F})_{\theta,p}$ and let u be as in Lemma 2.2.4. Then (in the notations of Section 2.1):

$$|\langle f, \varphi \rangle| \leq \int_0^1 |\langle f, \psi(\tau) \rangle| \frac{d\tau}{\tau} \leq \int_0^1 [|\langle u(\tau), \psi(\tau) \rangle| + |\langle f - u(\tau), \psi(\tau) \rangle|] \frac{d\tau}{\tau}$$

$$\leq \int_0^1 [||\tau^\theta u(\tau)||_{\mathbf{E}} + ||\tau^{\theta-1}(f - u(\tau))||_{\mathbf{F}}] \cdot [||\tau^{-\theta}\psi(\tau)||_{\mathbf{E}^*} + ||\tau^{1-\theta}\psi(\tau)||_{\mathbf{F}^*}] \frac{d\tau}{\tau}$$

$$\leq c \int_0^1 \frac{K(\tau, f)}{\tau^{1-\theta}} [||\tau^{-\theta}\psi(\tau)||_{\mathbf{E}^*} + ||\tau^{1-\theta}\psi(\tau)||_{\mathbf{F}^*}] \frac{d\tau}{\tau},$$

where c is independent of f by Lemma 2.2.4. After applying the Hölder inequality to the last integral and by taking into account (2.4.3), one arrives at the inequality:

$$|\langle f, \varphi \rangle| \leq c c^* ||f||_{\theta,p} ||\varphi||_{1-\theta,p'}^*.$$

This shows that $\varphi \in [(\mathbf{E}, \mathbf{F})_{\theta,p}]^*$ and that

$$||\varphi||_{[(\mathbf{E},\mathbf{F})_{\theta,p}]^*} \leq c c^* ||\varphi||_{1-\theta,p'}^*,$$

hence that $(\mathbf{F}^*, \mathbf{E}^*)_{1-\theta, p'} \subset [(\mathbf{E}, \mathbf{F})_{\theta, p}]^*$.

(ii) To prove the converse inclusion, we shall use the fact that $[\ell^p(\mathbf{F})]^* = \ell^{p'}(\mathbf{F}^*)$ if $p < \infty$. We define

$$G_\theta = \{(v, w) \in \ell^p(\mathbf{E}) \oplus \ell^p(\mathbf{F}) \mid 2^{n\theta} v_n + 2^{n(\theta-1)} w_n = 2^{m\theta} v_m + 2^{m(\theta-1)} w_m$$
$$\forall m, n \in \mathbb{N}\}.$$

Obviously G_θ is a closed subspace of $\ell^p(\mathbf{E}) \oplus \ell^p(\mathbf{F})$. Define $\Xi : G_\theta \to \mathbf{F}$ by

$$\Xi(v, w) = 2^{n\theta} v_n + 2^{n(\theta-1)} w_n \equiv f.$$

Then

$$(2.4.4) \qquad K(2^{-n}, f) \leq 2^{-n} \| 2^{n\theta} v_n \|_{\mathbf{E}} + \| 2^{n(\theta-1)} w_n \|_{\mathbf{F}}$$
$$= 2^{n(\theta-1)} [\| v_n \|_{\mathbf{E}} + \| w_n \|_{\mathbf{F}}].$$

Hence, by (2.3.9):

$$(2.4.5) \qquad \| f \|_{\theta, p} \leq 2[\| v \|_{\ell^p(\mathbf{E})} + \| w \|_{\ell^p(\mathbf{F})}] = 2\| (v, w) \|_{G_\theta}.$$

This shows that Ξ is a continuous mapping from G_θ into $[\mathbf{E}, \mathbf{F}]_{\theta, p}$. In fact Ξ is surjective: if $f \in (\mathbf{E}, \mathbf{F})_{\theta, p}$, let $u : (0, \infty) \to \mathbf{E}$ be as in Lemma 2.2.4 and set

$$v_n = 2^{-n\theta} u(2^{-n}), \qquad w_n = 2^{n(1-\theta)}[f - u(2^{-n})].$$

Then $\Xi(v, w) = f$ and $(v, w) \in G_\theta$, because

$$\| v_n \|_{\mathbf{E}} + \| w_n \|_{\mathbf{F}} = 2^{n(1-\theta)}[2^{-n} \| u(2^{-n}) \|_{\mathbf{E}} + \| f - u(2^{-n}) \|_{\mathbf{F}}]$$
$$\leq c 2^{n(1-\theta)} K(2^{-n}, f)$$

by (2.2.10), so that $\| v \|_{\ell^p(\mathbf{E})} + \| w \|_{\ell^p(\mathbf{F})} \leq 8c\| f \|_{\theta, p}$ by (2.3.9).

Now let $\varphi \in [(\mathbf{E}, \mathbf{F})_{\theta, p}]^*$. Then $\varphi \in \mathbf{E}^*$, and $\varphi \circ \Xi$ is a continuous antilinear functional on G_θ. By the Hahn-Banach theorem it can be extended to an antilinear functional on $\ell^p(\mathbf{E}) \oplus \ell^p(\mathbf{F})$ with the same norm. So there are $\Phi \in \ell^{p'}(\mathbf{E}^*)$ and $\Psi \in \ell^{p'}(\mathbf{F}^*)$ such that for all $(v, w) \in \ell^p(\mathbf{E}) \oplus \ell^p(\mathbf{F})$:

$$(2.4.6) \qquad \langle (v, w), \varphi \circ \Xi \rangle = \sum_{n=0}^{\infty} \{ \langle v_n, \Phi_n \rangle + \langle w_n, \Psi_n \rangle \}.$$

In particular, if $(v, w) \in G_\theta$ and $\Xi(v, w) = f$, then

$$(2.4.7) \qquad \langle f, \varphi \rangle = \sum_{n=0}^{\infty} \{ \langle v_n, \Phi_n \rangle + \langle w_n, \Psi_n \rangle \}.$$

If $f = 0$, i.e. if $w_n = -2^n v_n$ for all $n \in \mathbb{N}$, then $0 = \sum_{n=0}^{\infty} \langle v_n, \Phi_n - 2^n \Psi_n \rangle$. This clearly implies that $\Phi_n = 2^n \Psi_n$ for each n. Thus (2.4.7) becomes

$$\langle f, \varphi \rangle = \sum_{n=0}^{\infty} \{ \langle 2^{n\theta} v_n, 2^{n(1-\theta)} \Psi_n \rangle + \langle 2^{n(\theta-1)} w_n, 2^{n(1-\theta)} \Psi_n \rangle \}$$
$$= \sum_{n=0}^{\infty} \langle f, 2^{n(1-\theta)} \Psi_n \rangle.$$

Consequently we have $\varphi = \sum_{n=0}^{\infty} \varphi_n$, with $\varphi_n = 2^{n(1-\theta)}\Psi_n \in \mathbf{F}^*$.

Now, by using Proposition 2.3.5 in $(\mathbf{F}^*, \mathbf{E}^*)_{1-\theta,p'}$ and (2.4.5) and by setting $c(\theta) = \max\{(1-\theta)^{-1}, 2\theta^{-1}\}$, one finds that

$$\|\varphi\|_{1-\theta,p'}^* \le \frac{1}{1-\theta}\|\{2^{-n(1-\theta)}\varphi_n\}\|_{\ell^{p'}(\mathbf{F}^*)} + \frac{2}{\theta}\|\{2^{n\theta}\varphi_n\}\|_{\ell^{p'}(\mathbf{E}^*)}$$

$$= \frac{1}{1-\theta}\|\{\Psi_n\}\|_{\ell^{p'}(\mathbf{F}^*)} + \frac{2}{\theta}\|\{\Phi_n\}\|_{\ell^{p'}(\mathbf{E}^*)}$$

$$\le c(\theta)\|\varphi \circ \Xi\|_{G_\theta^*} = c(\theta) \sup_{(v,w)\in G_\theta} \frac{|\langle(v,w), \varphi \circ \Xi\rangle|}{\|(v,w)\|_{G_\theta}}$$

$$\le 2c(\theta) \sup_{f\in(\mathbf{E},\mathbf{F})_{\theta,p}} \frac{|\langle f,\varphi\rangle|}{\|f\|_{\theta,p}} = 2c(\theta)\|\varphi\|_{[(\mathbf{E},\mathbf{F})_{\theta,p}]^*}.$$

This proves $[(\mathbf{E},\mathbf{F})_{\theta,p}]^* \subset (\mathbf{F}^*, \mathbf{E}^*)_{1-\theta,p'}$. $\square$

2.5. The Reiteration Theorem

Let $\mathbf{E}$, $\mathbf{F}$ be two B-spaces with $\mathbf{E} \subset \mathbf{F}$. Let us consider a third B-space $\mathbf{G}$ such that $\mathbf{E} \subset \mathbf{G} \subset \mathbf{F}$. We say that $\mathbf{G}$ is of *class* θ, with $0 < \theta < 1$, if

(2.5.1) $$(\mathbf{E},\mathbf{F})_{\theta,1} \subset \mathbf{G} \subset (\mathbf{E},\mathbf{F})_{\theta,\infty}.$$

In particular, if $\mathbf{G} = (\mathbf{E},\mathbf{F})_{\theta,p}$ for some $p \in [1,+\infty]$, then $\mathbf{G}$ is of class θ by Proposition 2.4.1 (b). In the following lemma we study separately each of the two inclusions in (2.5.1).

LEMMA 2.5.1. *Let* $\mathbf{E}$, $\mathbf{F}$, $\mathbf{G}$ *be B-spaces with* $\mathbf{E} \subset \mathbf{G} \subset \mathbf{F}$, *and let* $\|\cdot\|_{\mathbf{E}}$, $\|\cdot\|_{\mathbf{F}}$, $\|\cdot\|_{\mathbf{G}}$ *be three admissible norms. Assume that* $\theta \in (0,1)$ *and denote by* $K(\tau, f)$ *the K-functional associated to the norms* $\|\cdot\|_{\mathbf{E}}$ *and* $\|\cdot\|_{\mathbf{F}}$. *Then*

(a) $\mathbf{G} \subset (\mathbf{E},\mathbf{F})_{\theta,\infty}$ *if and only if there is a constant* $c < \infty$ *such that*

(2.5.2) $$K(\tau, g) \le c\tau^{1-\theta}\|g\|_{\mathbf{G}} \qquad \forall \tau \in (0,1], \quad \forall g \in \mathbf{G}.$$

(b) $(\mathbf{E},\mathbf{F})_{\theta,1} \subset \mathbf{G}$ *if and only if there is a constant* $c < \infty$ *such that*

(2.5.3) $$\|e\|_{\mathbf{G}} \le c[\tau^\theta\|e\|_{\mathbf{E}} + \tau^{\theta-1}\|e\|_{\mathbf{F}}] \qquad \forall \tau \in (0,1], \quad \forall e \in \mathbf{E}.$$

PROOF. (a) is immediate in view of (2.2.5). For (b) we first show that (2.5.3) holds if $\|\cdot\|_{\mathbf{G}}$ is replaced by $\|\cdot\|_{\theta,1}$; this will imply (2.5.3) for any $\mathbf{G}$ such that $(\mathbf{E},\mathbf{F})_{\theta,1} \subset \mathbf{G}$. The needed estimate is easily obtained from (2.3.3):

$$\|e\|_{\theta,1} = \left\|\frac{1}{\ln 2}\int_\tau^{2\tau} e\frac{d\tau}{\tau}\right\|_{\theta,1}$$

$$\le \frac{1}{\ln 2} \cdot \max\left(\frac{1}{\theta}, \frac{1}{1-\theta}\right)\int_\tau^{2\tau}[\|\tau^\theta e\|_{\mathbf{E}} + \|\tau^{\theta-1}e\|_{\mathbf{F}}]\frac{d\tau}{\tau}$$

$$\le \max\left(\frac{1}{\theta}, \frac{1}{1-\theta}\right)[2^\theta\tau^\theta\|e\|_{\mathbf{E}} + \tau^{\theta-1}\|e\|_{\mathbf{F}}].$$

Reciprocally we must show that $\|f\|_{\mathbf{G}} \le c\|f\|_{\theta,1}$ for all $f \in (\mathbf{E},\mathbf{F})_{\theta,1}$ under the hypothesis (2.5.3). Since $\mathbf{E}$ is dense in $(\mathbf{E},\mathbf{F})_{\theta,1}$ we may assume $f \in \mathbf{E}$. We write $f = \int_0^\infty v(\tau)\tau^{-1}d\tau$ as in Proposition 2.3.2 (b) and then have by (2.5.3) and (2.3.4):

$$\int_0^\infty \|v(\tau)\|_{\mathbf{G}} \frac{d\tau}{\tau} \le c \int_0^\infty [\|\tau^\theta v(\tau)\|_{\mathbf{E}} + \|\tau^{\theta-1}v(\tau)\|_{\mathbf{F}}]\frac{d\tau}{\tau} \le c\|f\|_{\theta,1}.$$

So the integral $\int_0^\infty v(\tau)\tau^{-1}d\tau$ exists in $\mathbf{G}$ and $\|f\|_{\mathbf{G}} \le c\|f\|_{\theta,1}$ (use Remark 2.2.6). $\square$

We mention an estimate equivalent to (2.5.3) (modulo the constants). Since $\mathbf{E} \subset \mathbf{F}$, there is a constant $\kappa \ge 1$ such that $\|e\|_{\mathbf{F}} \le \kappa\|e\|_{\mathbf{E}}$ for all $e \in \mathbf{E}$. If (2.5.3) holds for all $0 < \tau \le 1$, then upon replacing c by $c\kappa$ we get an estimate which holds for all $\tau \in (0, \kappa]$. By taking $\tau = \|e\|_{\mathbf{E}}^{-1} \cdot \|e\|_{\mathbf{F}}$ we obtain

(2.5.4) $\|e\|_{\mathbf{G}} \le c_1\|e\|_{\mathbf{E}}^{1-\theta} \cdot \|e\|_{\mathbf{F}}^\theta \qquad \forall e \in \mathbf{E},$

where $c_1 = 2c\kappa$. Reciprocally, if (2.5.4) holds for some constant c_1, then (2.5.3) holds for all $\tau \in (0, \infty)$ if the constant c is changed into c_1. Indeed, we may use the following consequence of the concavity of the function log:

$$a^{1-\theta}b^\theta = (\tau^\theta a)^{1-\theta}(\tau^{\theta-1}b)^\theta \le (1-\theta)\tau^\theta a + \theta\tau^{\theta-1}b.$$

We also notice the following consequence of Lemma 2.5.1 (b):

COROLLARY 2.5.2. *Let* $\mathbf{E}$, $\mathbf{F}$, $\mathbf{H}$ *be Banach spaces such that* $\mathbf{E} \subset \mathbf{F}$ *and let* $T : \mathbf{E} \to \mathbf{H}$ *be a linear operator which is closable when considered as an operator from* $\mathbf{F}$ *to* $\mathbf{H}$. *Assume that a number* $0 < \theta < 1$ *is given. Then* T *extends to a continuous operator* $(\mathbf{E}, \mathbf{F})_{\theta,1} \to \mathbf{H}$ *if and only if there is a constant* $c < \infty$ *such that* $\|Te\|_{\mathbf{H}} \le c[\tau^\theta\|e\|_{\mathbf{E}} + \tau^{\theta-1}\|e\|_{\mathbf{F}}]$ *for all* $\tau \in (0, 1)$ *and all* $e \in \mathbf{E}$.

PROOF. One implication follows immediately from (2.5.3) written for $\mathbf{G} = (\mathbf{E}, \mathbf{F})_{\theta,1}$. Reciprocally, observe that $\|Te\|_{\mathbf{H}} + \|e\|_{\mathbf{F}}$ is a norm on $\mathbf{E}$ and that the completion $\mathbf{G}$ of $\mathbf{E}$ under this norm is a Banach space such that $\mathbf{E} \subset \mathbf{G} \subset \mathbf{F}$ and (2.5.3) is satisfied. $\square$

We prove now one of the most important results of real interpolation theory, the fact that the scale $(\mathbf{E}, \mathbf{F})_{\theta,p}$ has the "reiteration property" (no new spaces appear after an iteration of the method).

THEOREM 2.5.3. *Let* $\mathbf{E}$ *and* $\mathbf{F}$ *be B-spaces with* $\mathbf{E} \subset \mathbf{F}$. *Let* $\theta \in (0, 1)$, $p \in [1, +\infty]$ *and* $0 < \theta_1 < \theta_2 < 1$. *Assume that* $\mathbf{G}_1$ *and* $\mathbf{G}_2$ *are B-spaces of class* θ_1 *and* θ_2 *respectively. Then*

(2.5.5) $(\mathbf{G}_1, \mathbf{G}_2)_{\theta,p} = (\mathbf{E}, \mathbf{F})_{\sigma,p} \qquad with \ \sigma = (1-\theta)\theta_1 + \theta\theta_2.$

(2.5.5) *holds in particular if* $\mathbf{G}_1 = (\mathbf{E}, \mathbf{F})_{\theta_1,p_1}$ *and* $\mathbf{G}_2 = (\mathbf{E}, \mathbf{F})_{\theta_2,p_2}$ *with arbitrary* $p_1, p_2 \in [1, +\infty]$.

PROOF. For each of the four B-spaces involved in the theorem we choose an admissible norm. We denote by $K(\cdot, \cdot)$ the K-functional for the pair $\mathbf{E}$, $\mathbf{F}$ and by $k(\cdot, \cdot)$ that for the pair $\mathbf{G}_1$, $\mathbf{G}_2$, and we observe that $\mathbf{E} \subset \mathbf{G}_1 \subset \mathbf{G}_2 \subset \mathbf{F}$ by virtue of Proposition 2.4.1 (c). We shall use the following identity which is obtained by the change of variables $\tau \mapsto \varepsilon = \tau^{1/(\theta_2-\theta_1)}$:

(2.5.6)

$$\|g\|_{(\mathbf{G}_1,\mathbf{G}_2)_{\theta,p}} = \left\{ \int_0^1 \left[\frac{k(\tau, g)}{\tau^{1-\theta}} \right]^p \frac{d\tau}{\tau} \right\}^{1/p}$$

$$= (\theta_2 - \theta_1)^{1/p} \left\{ \int_0^1 \left[\frac{\varepsilon^{1-\theta_2} k(\varepsilon^{\theta_2-\theta_1}, g)}{\varepsilon^{1-\sigma}} \right]^p \frac{d\varepsilon}{\varepsilon} \right\}^{1/p}.$$

(i) Let $g \in \mathbf{G}_2$ and $h \in \mathbf{G}_1$. By Lemma 2.5.1(a) we then have for $\varepsilon \in (0, 1]$:

$$K(\varepsilon, g) \le K(\varepsilon, h) + K(\varepsilon, g - h)$$
$$\le c_1 \varepsilon^{1-\theta_1} \|h\|_{\mathbf{G}_1} + c_2 \varepsilon^{1-\theta_2} \|g - h\|_{\mathbf{G}_2}$$
$$\le c_3 \varepsilon^{1-\theta_2} [\varepsilon^{\theta_2-\theta_1} \|h\|_{\mathbf{G}_1} + \|g - h\|_{\mathbf{G}_2}].$$

By taking the infimum over all $h \in \mathbf{G}_1$, one obtains that

(2.5.7) $K(\varepsilon, g) \le c_3 \varepsilon^{1-\theta_2} k(\varepsilon^{\theta_2-\theta_1}, g) \qquad \forall \varepsilon \in (0, 1].$

It follows that:

$$\|g\|_{(\mathbf{E},\mathbf{F})_{\sigma,p}} = \left\{ \int_0^1 \left[\frac{K(\varepsilon, g)}{\varepsilon^{1-\sigma}} \right]^p \frac{d\varepsilon}{\varepsilon} \right\}^{1/p}$$

$$\le c_3 \left\{ \int_0^1 \left[\frac{\varepsilon^{1-\theta_2} k(\varepsilon^{\theta_2-\theta_1}, g)}{\varepsilon^{1-\sigma}} \right]^p \frac{d\varepsilon}{\varepsilon} \right\}^{1/p}$$

$$= c_3 \frac{1}{(\theta_2 - \theta_1)^{1/p}} \|g\|_{(\mathbf{G}_1,\mathbf{G}_2)_{\theta,p}}.$$

This shows that $(\mathbf{G}_1, \mathbf{G}_2)_{\theta,p} \subset (\mathbf{E}, \mathbf{F})_{\sigma,p}$.

(ii) To prove the converse inclusion, let $f \in (\mathbf{E}, \mathbf{F})_{\sigma,p}$ and set $f = \int_0^1 v(\tau) \tau^{-1} d\tau$ as in Proposition 2.3.2, with $v(\tau) \in \mathbf{E}$. We set $J(\tau) = \tau \|v(\tau)\|_{\mathbf{E}} + \|v(\tau)\|_{\mathbf{F}}$ and have

(2.5.8) $k(\varrho, f) \le \int_0^1 k(\rho, v(\tau)) \frac{d\tau}{\tau}.$

By applying (2.2.3) for $k(\cdot, \cdot)$ and then (2.5.3), one obtains the following two inequalities:

$$k(\varrho, v(\tau)) \le \|v(\tau)\|_{\mathbf{G}_2} \le c_2 \tau^{\theta_2-1} [\tau \|v(\tau)\|_{\mathbf{E}} + \|v(\tau)\|_{\mathbf{F}}] = c_2 \tau^{\theta_2-1} J(\tau)$$

and

$$k(\varrho, v(\tau)) \le \varrho \|v(\tau)\|_{\mathbf{G}_1} \le c_1 \varrho \tau^{\theta_1-1} J(\tau).$$

We set $\varrho = \varepsilon^{\theta_2 - \theta_1}$ in (2.5.8) and estimate the integrand by the first and second of these inequalities on the domain $[0, \varepsilon]$ and $(\varepsilon, 1]$ respectively :

$$\varepsilon^{1-\theta_2} k(\varepsilon^{\theta_2 - \theta_1}, f) \leq c_2 \int_0^\varepsilon \left(\frac{\tau}{\varepsilon}\right)^{\theta_2 - 1} J(\tau) \cdot \frac{d\tau}{\tau} + c_1 \int_\varepsilon^1 \left(\frac{\tau}{\varepsilon}\right)^{\theta_1 - 1} J(\tau) \cdot \frac{d\tau}{\tau}$$

$$\leq c_2 \int_0^1 \mu^{\theta_2 - 1} J(\varepsilon\mu) \frac{d\mu}{\mu} + c_1 \int_1^\infty \mu^{\theta_1 - 1} J(\varepsilon\mu) \cdot \frac{d\mu}{\mu}$$

$$\leq C \int_0^\infty [\min(\mu^{\theta_2 - 1}, \mu^{\theta_1 - 1})] J(\varepsilon\mu) \frac{d\mu}{\mu}.$$

By using (2.5.6), then the Minkowski inequality and finally the change of variables $\varepsilon \mapsto \lambda = \mu\varepsilon$, one then gets that

$$\|f\|_{(G_1, G_2)_{\theta,p}} \leq C \left\{ \int_0^1 \frac{d\varepsilon}{\varepsilon} \left[\frac{1}{\varepsilon^{1-\sigma}} \int_0^\infty \frac{d\mu}{\mu} [\min(\mu^{\theta_2 - 1}, \mu^{\theta_1 - 1})] \cdot J(\varepsilon\mu) \right]^p \right\}^{1/p}$$

$$\leq C \int_0^\infty \frac{d\mu}{\mu} [\min(\mu^{\theta_2 - 1}, \mu^{\theta_1 - 1})] \left\{ \int_0^1 \frac{d\varepsilon}{\varepsilon} \left[\frac{J(\varepsilon\mu)}{\varepsilon^{1-\sigma}} \right]^p \right\}^{1/p}$$

$$\leq C \int_0^\infty \frac{d\mu}{\mu} \mu^{1-\sigma} [\min(\mu^{\theta_2 - 1}, \mu^{\theta_1 - 1})] \cdot$$

$$\cdot \left\{ \int_0^\infty \frac{d\lambda}{\lambda} [\|\lambda^\theta v(\lambda)\|_E + \|\lambda^{\theta-1} v(\lambda)\|_F]^p \right\}^{1/p} .$$

The first integral on the r.h.s. is finite because $\theta_2 - \sigma > 0$ and $\theta_1 - \sigma < 0$. The second integral is bounded by $c\|f\|_{\theta,p}$ by (2.3.4), with c independent of f. $\square$

REMARK 2.5.4. The preceding proof shows that Theorem 2.5.3 remains valid even for $\theta_1 = 0$ or $\theta_2 = 1$ if we define the expression "G is of class θ" by the condition that (a) and (b) of Lemma 2.5.1 be satisfied. This clearly has a meaning for $\theta = 0$ or 1 and, for example, E is of class 0 while F is of class 1.

2.6. Interpolation of Operators

Let us consider two pairs G_0, G_1 and H_0, H_1 of B-spaces such that $G_0 \subset G_1$ and $H_0 \subset H_1$. Let T be a bounded operator from G_1 to H_1 such that $TG_0 \subset H_0$. By the closed graph theorem T belongs to $B(G_0, H_0)$. We shall show that T maps $(G_0, G_1)_{\theta,p}$ into $(H_0, H_1)_{\theta,p}$ for each θ and p (continuously, again by the closed graph theorem).

THEOREM 2.6.1. Let G_0, G_1, H_0, H_1 be as above, and set $G_{\theta,p} = (G_0, G_1)_{\theta,p}$ and $H_{\theta,p} = (H_0, H_1)_{\theta,p}$. Then one has for each $\theta \in (0, 1)$ and each $p \in [1, +\infty]$

$$(2.6.1) \qquad B(G_1, H_1) \cap B(G_0, H_0) \subset B(G_{\theta,p}, H_{\theta,p}).$$

Moreover, if one selects an admissible norm in each of the above four B-spaces, then there is a constant c which is independent of T such that :

$$(2.6.2) \qquad \|T\|_{G_{\theta,p} \to H_{\theta,p}} \leq c\|T\|_{G_0 \to H_0}^{1-\theta} \|T\|_{G_1 \to H_1}^\theta.$$

PROOF. Let us fix four admissible norms $\|\cdot\|_{\mathbf{G}_0}$, $\|\cdot\|_{\mathbf{G}_1}$, $\|\cdot\|_{\mathbf{H}_0}$ and $\|\cdot\|_{\mathbf{H}_1}$ and denote by $K_{\mathbf{G}}(\cdot,\cdot)$ and $K_{\mathbf{H}}(\cdot,\cdot)$ the K-functional for the pair $\mathbf{G}_0$, $\mathbf{G}_1$ and for the pair $\mathbf{H}_0$, $\mathbf{H}_1$ respectively. We write $\|T\|_{(0)}$ and $\|T\|_{(1)}$ for the norm of T in $B(\mathbf{G}_0, \mathbf{H}_0)$ and in $B(\mathbf{G}_1, \mathbf{H}_1)$ respectively (with respect to the chosen admissible norms). Without loss of generality we may assume that $T \neq 0$, so that $\lambda := \|T\|_{(0)}/\|T\|_{(1)}$ belongs to $(0, \infty)$.

We first observe that

$$K_{\mathbf{H}}(\varepsilon, Tf) = \inf_{h \in \mathbf{H}_0} [\varepsilon\|h\|_{\mathbf{H}_0} + \|Tf - h\|_{\mathbf{H}_1}] \leq \inf_{g \in \mathbf{G}_0} [\varepsilon\|Tg\|_{\mathbf{H}_0} + \|T(f-g)\|_{\mathbf{H}_1}]$$

$$\leq \|T\|_{(1)} \inf_{g \in \mathbf{G}_0} [\varepsilon\lambda\|g\|_{\mathbf{G}_0} + \|f - g\|_{\mathbf{G}_1}] = \|T\|_{(1)} K_{\mathbf{G}}(\varepsilon\lambda, f).$$

Hence (use $K_{\mathbf{G}}(\tau, f) \leq \|f\|_{\mathbf{G}_1} \leq c\|f\|_{\mathbf{G}_{\theta,p}}$ for $\tau \geq 1$ in the last step):

$$\|Tf\|_{\mathbf{G}_{\theta,p}} = \left\{ \int_0^1 \left[\frac{K_{\mathbf{H}}(\varepsilon, Tf)}{\varepsilon^{1-\theta}} \right]^p \frac{d\varepsilon}{\varepsilon} \right\}^{1/p}$$

$$\leq \|T\|_{(1)} \left\{ \int_0^\infty \left[\frac{K_{\mathbf{G}}(\varepsilon\lambda, f)}{\varepsilon^{1-\theta}} \right]^p \frac{d\varepsilon}{\varepsilon} \right\}^{1/p}$$

$$= \|T\|_{(1)} \lambda^{1-\theta} \left\{ \int_0^\infty \left[\frac{K_{\mathbf{G}}(\tau, f)}{\tau^{1-\theta}} \right]^p \frac{d\tau}{\tau} \right\}^{1/p}$$

$$\leq c(\theta, p) \|T\|_{(0)}^{1-\theta} \|T\|_{(1)}^{\theta} \|f\|_{\mathbf{G}_{\theta,p}}. \quad \square$$

The following consequences of the preceding theorem are often used in applications:

COROLLARY 2.6.2. *Under the assumptions of Theorem 2.6.1, one has*

$$B(\mathbf{G}_0, \mathbf{H}_0) \cap B(\mathbf{G}_0, \mathbf{H}_1) \subset B(\mathbf{G}_0, \mathbf{H}_{\theta,p})$$
$$B(\mathbf{G}_0, \mathbf{H}_0) \cap B(\mathbf{G}_1, \mathbf{H}_0) \subset B(\mathbf{G}_{\theta,p}, \mathbf{H}_0)$$

and

$$\|T\|_{\mathbf{G}_0 \to \mathbf{H}_{\theta,p}} \leq c\|T\|_{\mathbf{G}_0 \to \mathbf{H}_0}^{1-\theta} \|T\|_{\mathbf{G}_0 \to \mathbf{H}_1}^{\theta}$$
$$\|T\|_{\mathbf{G}_{\theta,p} \to \mathbf{H}_0} \leq c\|T\|_{\mathbf{G}_0 \to \mathbf{H}_0}^{1-\theta} \|T\|_{\mathbf{G}_1 \to \mathbf{H}_0}^{\theta}.$$

COROLLARY 2.6.3. *Let $\mathbf{E}$ and $\mathbf{F}$ be B-spaces with $\mathbf{E} \subset \mathbf{F}$. Let $\mathbf{G}$, $\mathbf{H}$ be other B-spaces satisfying $\mathbf{E} \subset \mathbf{G} \subset \mathbf{F}$ and $\mathbf{E} \subset \mathbf{H} \subset \mathbf{F}$. Then one has for each $\theta \in (0,1)$ and each $p \in [1, \infty]$:*

$$(2.6.3) \qquad\qquad (\mathbf{E}, \mathbf{G})_{\theta,p} \subset (\mathbf{H}, \mathbf{F})_{\theta,p}.$$

PROOF. The identity operator maps $\mathbf{G}$ into $\mathbf{F}$ and $\mathbf{E}$ into $\mathbf{H}$ continuously, hence it is bounded from $(\mathbf{E}, \mathbf{G})_{\theta,p}$ into $(\mathbf{H}, \mathbf{F})_{\theta,p}$ by (2.6.1). $\quad \square$

2.7. Quasi-Linearizable Couples of B-Spaces

We now consider a concrete realization of the interpolation spaces $(\mathbf{E}, \mathbf{F})_{\theta,p}$ for the case where $\mathbf{E}$ is the domain of a closed operator S in $\mathbf{F}$ having certain special properties. A particularly important case is that where S is of the form $S = -iA$ with A the generator of a C_0-semigroup.

It is useful to introduce the following notion (due to J. Peetre): if $\mathbf{E}$, $\mathbf{F}$ are two B-spaces with $\mathbf{E} \subset \mathbf{F}$, the couple $(\mathbf{E}, \mathbf{F})$ is said to be *quasi-linearizable* if there exists a family $\{V_\tau\}_{0<\tau\leq 1}$ of linear operators in $B(\mathbf{F}, \mathbf{E})$ such that:

(1) the family $\{V_\tau\}$ is bounded in $B(\mathbf{E})$ and in $B(\mathbf{F})$,
(2) the family $\{\tau V_\tau\}$ is bounded in $B(\mathbf{F}, \mathbf{E})$,
(3) the family $\{\tau^{-1}(V_\tau - I)\}$ is bounded in $B(\mathbf{E}, \mathbf{F})$.

The advantage of such a family is that one can give an explicit estimate for the K-functional:

LEMMA 2.7.1. *If the couple $(\mathbf{E}, \mathbf{F})$ is quasi-linearizable, then there is a constant $M \in (0, \infty)$ such that for each $\tau \in (0, 1]$ and each $f \in \mathbf{F}$:*

$$(2.7.1) \qquad K(\tau, f) \leq \tau ||V_\tau f||_{\mathbf{E}} + ||(I - V_\tau)f||_{\mathbf{F}} \leq MK(\tau, f).$$

In particular the expression

$$(2.7.2) \qquad \left[\int_0^1 [||\tau^\theta V_\tau f||_{\mathbf{E}}^p \frac{d\tau}{\tau} \right]^{1/p} + \left[\int_0^1 \left\| \frac{V_\tau - I}{\tau^{1-\theta}} f \right\|_{\mathbf{F}}^p \frac{d\tau}{\tau} \right]^{1/p}$$

defines an admissible norm on $(\mathbf{E}, \mathbf{F})_{\theta,p}$.

PROOF. The first inequality in (2.7.1) is evident since $V_\tau f \in \mathbf{E}$. For the second inequality, let $e \in \mathbf{E}$ and set $g = f - e$. Then:

$$\tau ||V_\tau f||_{\mathbf{E}} + ||(I - V_\tau)f||_{\mathbf{F}} \leq$$
$$\leq \tau ||V_\tau g||_{\mathbf{E}} + \tau ||V_\tau e||_{\mathbf{E}} + ||(I - V_\tau)g||_{\mathbf{F}} + ||(I - V_\tau)e||_{\mathbf{F}}$$
$$\leq \tau ||V_\tau||_{B(\mathbf{F},\mathbf{E})} ||g||_{\mathbf{F}} + ||V_\tau||_{B(\mathbf{E})} \cdot \tau ||e||_{\mathbf{E}} + ||I - V_\tau||_{B(\mathbf{F})} ||g||_{\mathbf{F}} +$$
$$+ \tau^{-1} ||I - V_\tau||_{B(\mathbf{E},\mathbf{F})} \tau ||e||_{\mathbf{E}} \leq M[\tau ||e||_{\mathbf{E}} + ||g||_{\mathbf{F}}].$$

By taking the infimum over all $e \in \mathbf{E}$, one obtains the second inequality in (2.7.1). $\square$

The next two propositions describe two methods of constructing families $\{V_\tau\}$ with the properties (1)–(3). First we consider the case where $\mathbf{E}$ is the domain of a "positive" operator S in the B-space $\mathbf{F}$ (this situation has been extensively studied by P. Grisvard and H. Komatsu, see [Tr] and references therein).

PROPOSITION 2.7.2. *Let $\mathbf{F}$ be a Banach space and let S be a linear operator in $\mathbf{F}$ such that for each $\lambda > 0$, $S + \lambda : D(S) \to \mathbf{F}$ is bijective, with bounded inverse and*

$$(2.7.3) \qquad\qquad ||(S + \lambda)^{-1}||_{B(\mathbf{F})} \leq c\lambda^{-1}$$

for some constant $c \in (0, \infty)$. Let $\mathbf{E} = D(S)$ with norm $\|e\|_{\mathbf{E}} = \|e\|_{\mathbf{F}} + \|Se\|_{\mathbf{F}}$. Then $\mathbf{E}$ is a B-subspace of $\mathbf{F}$, the couple $(\mathbf{E}, \mathbf{F})$ is quasi-linearizable and there is a constant $m > 0$ such that for $0 < \tau \leq 1$ and $f \in \mathbf{F}$

$$mK(\tau, f) \leq \tau \|f\|_{\mathbf{F}} + \tau \|S(I + \tau S)^{-1} f\|_{\mathbf{F}} \leq m^{-1} K(\tau, f).$$

In particular the following expression defines an admissible norm on $(\mathbf{E}, \mathbf{F})_{\theta, p}$:

$$(2.7.4) \qquad \|f\|_{\mathbf{F}} + \left[\int_0^1 \|\tau^\theta S(I + \tau S)^{-1} f\|_{\mathbf{F}}^p \frac{d\tau}{\tau} \right]^{1/p}.$$

PROOF. S is a closed operator (because $(S + \lambda)^{-1}$ is bounded, hence closed), so that $D(S)$ provided with the graph topology is a B-space. To see that the pair $(\mathbf{E}, \mathbf{F})$ is quasi-linearizable, we take $V_\tau = (I + \tau S)^{-1} = \tau^{-1}(S + \tau^{-1})^{-1}$. Then $V_\tau - I = -\tau S(I + \tau S)^{-1} = -S(S + \tau^{-1})^{-1}$, and it is straightforward to check that the conditions (1)–(3) are satisfied. $\square$

A simpler estimate for the K-functional may be obtained if S is "strictly positive", i.e. S satisfies the hypotheses of the preceding proposition but also S itself is a bijective map of $D(S)$ onto $\mathbf{F}$ with bounded inverse. In this case we may choose $\|e\|_{\mathbf{E}} = \|Se\|_{\mathbf{F}}$ as admissible norm on $\mathbf{E}$, and then we get the estimate $mK(\tau, f) \leq \tau \|S(I + \tau S)^{-1} f\|_{\mathbf{F}} \leq m^{-1} K(\tau, f)$. In particular, the term $\|f\|_{\mathbf{F}}$ in the expression (2.7.4) is no more needed in order to get an admissible norm on $(\mathbf{E}, \mathbf{F})_{\theta, p}$.

We consider now a one-parameter C_0-semigroup, i.e. a strongly continuous family $\{W(s)\}_{s \geq 0}$ of bounded linear operators in $\mathbf{F}$ with $W(0) = I$ and $W(s_1)W(s_2) = W(s_1 + s_2)$. Formally it is of the form $W(s) = \exp(iAs)$, where the generator A is a closed densely defined operator in $\mathbf{F}$. We assume that the reader is familiar with the elementary aspects of the theory of C_0-semigroups; we recommend [BB] for a detailed presentation of the subject and also for deeper aspects of the interpolation theory associated to such semigroups. Let $\mathbf{E}$ be the domain $D(A)$ of A equipped with the norm $\|e\|_{\mathbf{E}} = \|e\|_{\mathbf{F}} + \|Ae\|_{\mathbf{F}}$. Then $\mathbf{E}$ is a dense B-subspace of $\mathbf{F}$. This situation is in fact a particular case of that considered in Proposition 2.7.2. Indeed, it is not difficult to show that $\|W(s)\|_{B(\mathbf{F})} \leq M e^{\omega s}$ for some constants $M \geq 1$, $\omega \in \mathbb{R}$ and for all $s \geq 0$ (see [BB] or Section 3.2 of this text). Hence for $\lambda > 0$ the operator $\lambda + \omega - iA$ is a bijection of $D(A)$ onto $\mathbf{F}$ with inverse $(\lambda + \omega - iA)^{-1} = \int_0^\infty W(s) e^{-(\omega + \lambda)s} ds$, so $\|(\lambda + \omega - iA)^{-1}\|_{B(\mathbf{F})} \leq M\lambda^{-1}$. If we take $S = \omega - iA$ in Proposition 2.7.2 we obtain an explicit characterization of $(\mathbf{E}, \mathbf{F})_{\theta, p}$ in terms of the resolvent of A (note that S and A define the same graph topology on $D(A)$). However, new and interesting descriptions of the interpolation spaces (directly in terms of the group W for example) can be obtained by choosing the family $\{V_\tau\}_{0 < \tau \leq 1}$ differently than in the proof of Proposition 2.7.2. We shall present now a general method of constructing such V_τ; this will prove to be quite useful for the case of n-parameter groups treated in Chapter 3.

Let φ be a function of the form $\varphi(z) = \int_0^\infty e^{izs} \widehat{\varphi}(s) \underline{d}s$, where $\widehat{\varphi}$ is a measure on $[0, \infty)$ such that $\int_0^\infty e^{\omega s} |\widehat{\varphi}(s)| \underline{d}s < \infty$; here we assume $\omega \geq 0$, which is not a loss of generality for our purposes. Note that φ is a continuous function on the half-plane $\Im z \geq -\omega$, and it is analytic in the domain $\Im z > -\omega$. For any number $\tau \in [0, 1]$ we may define a bounded operator in $\mathbf{F}$ through the formula

$$(2.7.5) \qquad \varphi(\tau A) = \int_0^\infty W(\tau s)\widehat{\varphi}(s)\underline{d}s.$$

The functional calculus associated to the operator τA by this rule will be studied in detail in Chapter 3 for the case of n-parameter C_0-groups. Some obvious properties of this calculus make it natural to try to choose operators of the form $V_\tau = \varphi(\tau A)$ in the definition of quasi-linearizability. We shall explain this at a slightly formal level. Notice first that, if $\omega = 0$, the choice $\widehat{\varphi}(s) = \sqrt{2\pi}e^{-s}$ gives $V_\tau = (I - i\tau A)^{-1}$, so we obtain the operators used in the proof of Proposition 2.7.2 with $S = -iA$. In any case, for $0 \leq \tau \leq 1$ we have $V_\tau \mathbf{E} \subset \mathbf{E}$ and

$$(2.7.6) \qquad ||V_\tau||_{B(\mathbf{F})} + ||V_\tau||_{B(\mathbf{E})} \leq 2M \int_0^\infty e^{\omega s}|\widehat{\varphi}(s)|\underline{d}s.$$

Hence, by our hypothesis on $\widehat{\varphi}$, the condition (1) in the definition of quasi-linearizability is automatically satisfied. For condition (2) we must have $V_\tau \mathbf{F} \subset D(A)$. Formally $AV_\tau = \tau^{-1}\varphi_1(\tau A)$ with $\varphi_1(x) = x\varphi(x)$, hence for (2) it would be sufficient that φ_1 have the same properties as φ. For a rigorous argument, let us assume that $\widehat{\varphi}$ is a function of bounded variation on $\mathbb{R}$ such that $\widehat{\varphi}(s) = 0$ if $s < 0$ and $\int e^{\omega s}(|\widehat{\varphi}(s)| + |\widehat{\varphi}'(s)|)ds < \infty$ (the distributional derivative $\widehat{\varphi}'$ is a measure with support in $[0, \infty)$). A straightforward computation gives for $\sigma > 0$:

$$(2.7.7)$$
$$\frac{I - W(\tau\sigma)}{\sigma}\varphi(\tau A) = \frac{1}{\sigma}\int_0^\sigma W(\tau s)\widehat{\varphi}(s)\underline{d}s + \int_0^\infty W(\tau(s+\sigma))\frac{\widehat{\varphi}(s+\sigma) - \widehat{\varphi}(s)}{\sigma}\underline{d}s.$$

Hence the limit as $\sigma \to +0$ of the l.h.s. above exists strongly in $\mathbf{F}$. So, for $\tau > 0$ we have $\varphi(\tau A)\mathbf{F} \subset D(A)$ and $-i\tau A\varphi(\tau A) = \int W(\tau s)\widehat{\varphi}'(s)\underline{d}s$, where the integral extends over $[0, \infty)$. This ensures that condition (2) is fulfilled. Finally, we study the third condition in the definition of quasi-linearizability. Since $\varphi(\tau A) \to \varphi(0)I$ strongly on $\mathbf{F}$ as $\tau \to +0$, we must clearly have $\varphi(0) = 1$. Then, formally, we get $\tau^{-1}(V_\tau - I) = \tau^{-1}(\varphi(\tau A) - \varphi(0)) = \psi(\tau A)A$, where $\psi(x) = x^{-1}(\varphi(x) - \varphi(0))$. It is not difficult to give a rigorous meaning to this formula if ψ has properties similar to those of φ. This will be done in the context of Chapter 3 (see Section 3.4); in the next proposition we shall restrict ourselves to the case where $\widehat{\varphi}$ is the characteristic function of the interval $[0, 1]$ multiplied by $\sqrt{2\pi}$, hence $\varphi(x) = ix^{-1}(1 - e^{ix})$.

PROPOSITION 2.7.3. *Let $\mathbf{F}$ be a Banach space, $\{W(s)\}_{s\geq 0}$ a one-parameter C_0-semigroup in $\mathbf{F}$, and $\mathbf{E}$ the domain of its generator A equipped with the norm $||e||_{\mathbf{E}} = ||e||_{\mathbf{F}} + ||Ae||_{\mathbf{F}}$. Then $\mathbf{E}$ is a dense B-subspace of $\mathbf{F}$ and the couple $(\mathbf{E}, \mathbf{F})$ is quasi-linearizable.*

(a) *For each $f \in \mathbf{F}$ define the modulus of continuity of f with respect to W as the function $\omega(\tau, f)$ of $\tau \in [0, \infty)$ given by $\omega(\tau, f) = \sup_{0\leq\sigma\leq\tau} ||W(\sigma)f - f||_{\mathbf{F}}$.*

Then $\omega(\cdot, f)$ is a continuous, positive, increasing function such that $\omega(0, f) = 0$, and there is a constant $m > 0$ such that for all $\tau \in (0, 1]$ and $f \in \mathbf{F}$:

$$(2.7.8) \qquad m K(\tau, f) \le \tau \|f\|_{\mathbf{F}} + \omega(\tau, f) \le m^{-1} K(\tau, f).$$

In particular, the expression $\|f\|_{\mathbf{F}} + [\int_0^1 (\tau^{\theta-1} \omega(\tau, f))^p \tau^{-1} d\tau]^{1/p}$ is an admissible norm on $(\mathbf{E}, \mathbf{F})_{\theta, p}$ for all $\theta \in (0, 1)$, $p \in [1, \infty]$.
(b) If $M_\tau = \sup_{0 \le \sigma \le \tau} (1 + \|W(\sigma)\|)$, then for all $\tau \ge 0$ and $f \in \mathbf{F}$

$$(2.7.9) \quad \omega(\tau, f) \le M_\tau \|(W(\tau) - I)f\|_{\mathbf{F}} + M_\tau \left\| \int_0^1 (W(\tau t) - I) f \, dt \right\|.$$

The expression

$$(2.7.10) \qquad \|f\|_{\mathbf{F}} + \left[\int_0^1 \left\| \frac{W(\tau) - I}{\tau^{1-\theta}} f \right\|_{\mathbf{F}}^p \frac{d\tau}{\tau} \right]^{1/p}$$

defines an admissible norm on $(\mathbf{E}, \mathbf{F})_{\theta, p}$.

PROOF. We first prove the second inequality in (2.7.8). If $0 \le \sigma \le \tau \le 1$ and $e \in \mathbf{E}$, then by using $W(\sigma)e - e = \int_0^\sigma W(s) i A e \, ds$ we get

$$\|W(\sigma)f - f\|_{\mathbf{F}} \le \|W(\sigma)e - e\|_{\mathbf{F}} + \|(W(\sigma) - I)(f - e)\|_{\mathbf{F}}$$
$$\le \int_0^\sigma \|W(s)\|_{B(\mathbf{F})} ds \|Ae\|_{\mathbf{F}} + \|W(\sigma) - I\|_{B(\mathbf{F})} \|f - e\|_{\mathbf{F}}$$
$$\le c(\sigma \|Ae\|_{\mathbf{F}} + \|f - e\|_{\mathbf{F}})$$

for some constant c. Hence by taking the infimum over $e \in \mathbf{E}$ of the last member of these inequalities we obtain $\|W(\sigma)f - f\|_{\mathbf{F}} \le c K(\sigma, f) \le c K(\tau, f)$. Finally, the definition of $\omega(\tau, f)$ and the first inequality in (2.2.3) prove the second estimate in (2.7.8).
Now let $V_\tau = \tau^{-1} \int_0^\tau W(s) ds$. Then $V_\tau \mathbf{F} \subset D(A)$ and $i\tau A V_\tau = W(\tau) - I$ (see (2.7.7)), so

$$K(\tau, f) \le \tau \|V_\tau f\|_{\mathbf{E}} + \|(V_\tau - I)f\|_{\mathbf{F}}$$
$$= \tau \|V_\tau f\|_{\mathbf{F}} + \tau \|A V_\tau f\|_{\mathbf{F}} + \left\| \frac{1}{\tau} \int_0^\tau (W(s) - I) f \, ds \right\|_{\mathbf{F}}$$
$$= \tau \|V_\tau f\|_{\mathbf{F}} + \|W_\tau f - f\|_{\mathbf{F}} + \left\| \int_0^1 (W(\tau t) - I) f \, dt \right\|_{\mathbf{F}}.$$

Since $\|V_\tau\|_{B(\mathbf{F})}$ is bounded by a finite constant c if $0 < \tau \le 1$, we get $K(\tau, f) \le c \tau \|f\|_{\mathbf{F}} + 2\omega(\tau, f)$, hence (2.7.8) is completely proved.

We prove (2.7.9) by a similar argument. If $0 < \sigma < \tau$ then

$$\|(W(\sigma) - I)f\|_{\mathbf{F}} \leq \|(W(\sigma) - I)V_\tau f\|_{\mathbf{F}} + \|(W(\sigma) - I)(V_\tau - I)f\|_{\mathbf{F}}$$

$$\leq \left\| \int_0^\sigma W(s)iAV_\tau f\, ds \right\|_{\mathbf{F}} + \|(W(\sigma) - I\|_{B(\mathbf{F})} \cdot \left\| \int_0^1 (W(\tau t) - I)f\, dt \right\|_{\mathbf{F}}$$

$$\leq \sup_{0 \leq s \leq \sigma} \|W(s)\|_{B(\mathbf{F})} \frac{\sigma}{\tau} \|(W(\tau) - I)f\|_{\mathbf{F}} +$$

$$+ \|(W(\sigma) - I\|_{B(\mathbf{F})} \cdot \left\| \int_0^1 (W(\tau t) - I)f\, dt \right\|_{\mathbf{F}}.$$

This estimate is better than (2.7.9). Finally, the fact that (2.7.10) is an admissible norm on $(\mathbf{E}, \mathbf{F})_{\theta,p}$ follows from the integral Minkowski inequality:

$$\left\{ \int_0^1 \frac{d\tau}{\tau} \left[\tau^{\theta-1} \int_0^1 dt \|W(\tau t)f - f\|_{\mathbf{F}} \right]^p \right\}^{1/p} \leq$$

$$\leq \int_0^1 dt \left\{ \int_0^1 \frac{d\tau}{\tau} \left\| \frac{W(\tau t) - I}{\tau^{1-\theta}} f \right\|_{\mathbf{F}}^p \right\}^{1/p}$$

$$\leq \int_0^1 dt \cdot t^{1-\theta} \left\{ \int_0^1 \frac{d\sigma}{\sigma} \left\| \frac{W(\sigma) - I}{\sigma^{1-\theta}} f \right\|_{\mathbf{F}}^p \right\}^{1/p}. \quad \square$$

A natural question in the setting of Proposition 2.7.3 is to describe the interpolation spaces between the domain of an arbitrary power A^r of A ($r \geq 1$ integer) and $\mathbf{F}$. The K-functional in this case can be estimated in terms of the r-th order modulus of continuity $\omega_r(\tau, f) = \sup_{0 \leq \sigma \leq \tau} \|(W(\sigma) - I)^r f\|_{\mathbf{F}}$, see Section 3.4 in [BB] and also Section 5.4 of [BS] where one may find a rather precise comparison of the moduli of continuity of different orders (Marchaud's inequality). In §3.4.2 we shall treat these questions for the case of n-parameter C_0-groups (see the remarks following Theorem 3.4.6).

We end this section with one more example showing the utility of the notion of quasi-linearizability. Let $\mathbf{E}$ and $\mathbf{G}$ be B-subspaces of a B-space $\mathbf{F}$. Then $\mathbf{E} \cap \mathbf{G}$ is also a B-subspace of $\mathbf{F}$ (indeed, $\mathbf{E} \cap \mathbf{G}$ is a B-space for the intersection topology, cf. Section 2.1). We obviously have $(\mathbf{E} \cap \mathbf{G}, \mathbf{F})_{\theta,p} \subset (\mathbf{E}, \mathbf{F})_{\theta,p} \cap (\mathbf{G}, \mathbf{F})_{\theta,p}$ and it is natural to ask whether we have equality. In general the inclusion is strict, but we have the following result of Grisvard and Peetre (see [Gr3] and §1.19.9 of [Tr] for further references).

PROPOSITION 2.7.4. *Let* $\mathbf{E}$, $\mathbf{G}$ *be two B-subspaces of a B-space* $\mathbf{F}$. *Assume that the couple* $(\mathbf{E}, \mathbf{F})$ *is quasi-linearizable and that the family of operators* $\{V_\tau\}_{0 < \tau \leq 1}$ *which appears in the definition of quasi-linearizability may be chosen such that* $V_\tau \mathbf{G} \subset \mathbf{G}$ *and* $\|V_\tau\|_{\mathbf{G} \to \mathbf{G}} \leq c$ *for some constant c and all* $\tau \in (0, 1)$. *Then* $(\mathbf{E} \cap \mathbf{G}, \mathbf{F})_{\theta,p} = (\mathbf{E}, \mathbf{F})_{\theta,p} \cap (\mathbf{G}, \mathbf{F})_{\theta,p}$ *for all* $\theta \in (0, 1)$ *and* $p \in [1, \infty]$.

PROOF. Let $f \in (\mathbf{G}, \mathbf{F})_{\theta,p}$. By Lemma 2.2.4 there is a continuous function $u : (0, \infty) \to \mathbf{G}$ such that $\tau^\theta u(\cdot) \in L^p_*(\mathbf{G})$ and $\tau^{\theta-1}(f - u(\cdot)) \in L^p_*(\mathbf{F})$. We set

$v(\tau) = V_\tau u(\tau)$ for $0 < \tau \le 1$. Then $v(\tau) \in \mathbf{E} \cap \mathbf{G}$ and one may choose the constant $c < \infty$ such that for $0 < \tau \le 1$

$$||v(\tau)||_{\mathbf{E} \cap \mathbf{G}} \le ||v(\tau)||_{\mathbf{E}} + ||v(\tau)||_{\mathbf{G}} \le ||V_\tau f - V_\tau(f - u(\tau))||_{\mathbf{E}} + c||u(\tau)||_{\mathbf{G}}$$
$$\le ||V_\tau f||_{\mathbf{E}} + c\tau^{-1}||f - u(\tau)||_{\mathbf{F}} + c||u(\tau)||_{\mathbf{G}}.$$

The last two terms here belong to L_*^p (as functions of τ) when multiplied by τ^θ. If $f \in (\mathbf{E}, \mathbf{F})_{\theta,p}$, then the first term multiplied by τ^θ belongs to $L_*^p((0,1))$ by Lemma 2.7.1. Then

$$||f - v(\tau)||_{\mathbf{F}} = ||(1 - V_\tau)f + V_\tau(f - u(\tau))||_{\mathbf{F}} \le ||(1 - V_\tau)f||_{\mathbf{F}} + c||f - u(\tau)||_{\mathbf{F}}.$$

As above we see that after multiplication by $\tau^{\theta-1}$ the last expression gives a function in $L_*^p((0,1))$ if $f \in (\mathbf{E}, \mathbf{F})_{\theta,p}$. Let K be the K-functional associated to the couple $(\mathbf{E} \cap \mathbf{G}, \mathbf{F})$. Since for $0 < \tau < 1$

$$K(\tau, f) \le \tau||v(\tau)||_{\mathbf{E} \cap \mathbf{G}} + ||f - v(\tau)||_{\mathbf{F}}$$

we see that $\tau^{\theta-1} K(\tau, f) \in L_*^p((0,1))$ (i.e. $f \in (\mathbf{E} \cap \mathbf{G}, \mathbf{F})_{\theta,p}$) if $f \in (\mathbf{E}, \mathbf{F})_{\theta,p} \cap (\mathbf{G}, \mathbf{F})_{\theta,p}$. $\square$

The simplest non-trivial situation in which Proposition 2.7.4 may be applied is that considered in [Gr3]: take $\mathbf{G} = D(S)$ where S is as in Proposition 2.7.2 and has the supplementary property $(S + \lambda)^{-1}\mathbf{E} \subset \mathbf{E}$ and $||(S + \lambda)^{-1}||_{B(\mathbf{E})} \le c\lambda^{-1}$ for a constant $c < \infty$ and all $\lambda > 0$. Then we may take $V_\tau = (I + \tau S)^{-1}$. For example, let $\mathbf{F} = L^q(\mathbb{R}^n)$ for some $q \in [1, \infty]$ and $\mathbf{E} = W^{1,q}(\mathbb{R}^n)$ the Sobolev space of functions $f \in L^q(\mathbb{R}^n)$ with first order derivatives in $L^q(\mathbb{R}^n)$. For an arbitrary locally Lipschitz function $\varphi : \mathbb{R}^n \to (0, \infty)$ such that $|\nabla \varphi| \le c\varphi$ for a constant c, one may take $S = \varphi(Q)$ and get the following result ($0 < \theta < 1$, $1 \le p \le \infty$):

$$(\{f \in W^{1,q}(\mathbb{R}^n) \mid \varphi f \in L^q(\mathbb{R}^n)\}, L^q(\mathbb{R}^n))_{1-\theta,p} = \{f \in B_q^{\theta,p}(\mathbb{R}^n) \mid \varphi^\theta f \in L^q(\mathbb{R}^n)\}.$$

Here $B_q^{\theta,p}$ are Besov spaces that will be precisely defined in Chapter 4.

2.8. Friedrichs Couples

2.8.1. By a *Hilbert space* we always mean a complex Hilbert space. If $\mathbf{E}$ is a Hilbert space, we denote by $\langle \cdot, \cdot \rangle_{\mathbf{E}}$ and $|| \cdot ||_{\mathbf{E}}$ the associated scalar product (antilinear in the first variable) and norm respectively; the index will be omitted in some special cases or if there is no risk of confusion.

The adjoint $\mathbf{E}^*$ of a Hilbert space $\mathbf{E}$ is also a Hilbert space, and there is a canonical isometric isomorphism $j_{\mathbf{E}} : \mathbf{E} \to \mathbf{E}^*$, called the *Riesz isomorphism*, given by

(2.8.1) $$j_{\mathbf{E}}(f)(g) \equiv \langle g, j_{\mathbf{E}}(f) \rangle = \langle g, f \rangle_{\mathbf{E}},$$

where we have used the convention (2.1.2). In particular, if $\mathbf{E}$ is identified with $\mathbf{E}^*$ through $j_{\mathbf{E}}$, we shall have $\langle g, f \rangle = \langle g, f \rangle_{\mathbf{E}}$. However, we shall make this identification only in some very special cases. Note that $j_{\mathbf{E}}$ is always a positive (hence symmetric) operator $\mathbf{E} \to \mathbf{E}^*$. On the other hand we shall always identify $\mathbf{E}$ with $\mathbf{E}^{**}$ as explained at the beginning of §2.1 ($\mathbf{E}$ being reflexive).

A *Friedrichs couple* is a pair $(\mathbf{E}, \mathbf{F})$ of Hilbert spaces such that $\mathbf{E} \subset \mathbf{F}$ continuously and densely. One frequently meets Friedrichs couples constructed in the following manner. Let $\mathbf{F}$ and $\mathbf{G}$ be two Hilbert spaces and $T : D(T) \subset \mathbf{F} \to \mathbf{G}$ a closed densely defined linear operator. Let $\mathbf{E}$ be the domain $D(T)$ of T provided with the *graph norm*:

$$(2.8.2) \qquad \|f\|_{\mathbf{E}} = \left(\|f\|_{\mathbf{F}}^2 + \|Tf\|_{\mathbf{G}}^2 \right)^{1/2}.$$

Then $\mathbf{E}$ is a Hilbert space and $(\mathbf{E}, \mathbf{F})$ a Friedrichs couple.

In fact any Friedrichs couple $(\mathbf{E}, \mathbf{F})$ can be obtained by this method, as follows from the following theorem due to K. Friedrichs: *there is a unique positive self-adjoint operator Λ in $\mathbf{F}$ having $\mathbf{E}$ as domain and such that $\|e\|_{\mathbf{E}} = \|\Lambda e\|_{\mathbf{F}}$ for all $e \in \mathbf{E}$; the operator Λ is strictly positive,* i.e. we have $\Lambda \geq a$ for some constant $a > 0$. We shall say that Λ is the *Friedrichs operator* associated to the couple $(\mathbf{E}, \mathbf{F})$. Let us sketch a simple proof of this result (see also Section VI.2.6 in [K1]). By identifying $\mathbf{F}^* = \mathbf{F}$ by means of the Riesz isomorphism, we get continuous and dense embeddings $\mathbf{E} \subset \mathbf{F} \subset \mathbf{E}^*$. Let $j_{\mathbf{E}}$ be the Riesz isomorphism of $\mathbf{E}$ onto $\mathbf{E}^*$. Then $D := j_{\mathbf{E}}^{-1}(\mathbf{F})$ is a dense subspace of $\mathbf{E}$ (hence of $\mathbf{F}$) and $J := j_{\mathbf{E}}|_D$, considered as an operator in $\mathbf{F}$, is symmetric and strictly positive (because $\langle e, Je \rangle_{\mathbf{F}} = \langle e, j_{\mathbf{E}}e \rangle = \|e\|_{\mathbf{E}}^2 \geq a\|e\|_{\mathbf{F}}^2$ for a constant $a > 0$ and all $e \in D$). Since $JD = \mathbf{F}$, it follows that J is self-adjoint in $\mathbf{F}$ and we may take $\Lambda = \sqrt{J}$.

The operator Λ which is canonically associated to the Friedrichs couple $(\mathbf{E}, \mathbf{F})$ allows us to define new *Hilbert spaces* $[\mathbf{E}, \mathbf{F}]_\theta$ for $\theta \in (0, 1)$ in the following way: $[\mathbf{E}, \mathbf{F}]_\theta$ is the domain in $\mathbf{F}$ of the operator $\Lambda^{1-\theta}$, and

$$(2.8.3) \qquad \|f\|_{[\mathbf{E},\mathbf{F}]_\theta} := \|\Lambda^{1-\theta} f\|_{\mathbf{F}}.$$

The following two relations can easily be deduced from the preceding definition: if $0 < \theta_1 < \theta_2 < 1$, then

$$(2.8.4) \qquad \mathbf{E} \subset [\mathbf{E}, \mathbf{F}]_{\theta_1} \subset [\mathbf{E}, \mathbf{F}]_{\theta_2} \subset \mathbf{F}$$

and

$$(2.8.5) \qquad [[\mathbf{E}, \mathbf{F}]_{\theta_1}, [\mathbf{E}, \mathbf{F}]_{\theta_2}]_\theta = [\mathbf{E}, \mathbf{F}]_\sigma, \qquad \sigma = (1 - \theta)\theta_1 + \theta\theta_2.$$

The equality (2.8.5) holds at the Hilbert space level (i.e. the norms are equal too). The preceding relations are quite similar to those obtained in Proposition 2.4.1 (c) and Theorem 2.5.3 respectively. This fact is not surprising because, if $[\mathbf{E}, \mathbf{F}]_\theta$ is viewed as a H-space (see Section 2.1), then it is identical with the real interpolation space $(\mathbf{E}, \mathbf{F})_{\theta,2}$ of order $(\theta, 2)$ associated to $\mathbf{E}$ and $\mathbf{F}$:

PROPOSITION 2.8.1. *Let $(\mathbf{E}, \mathbf{F})$ be a Friedrichs couple and $0 < \theta < 1$. Then the H-space $[\mathbf{E}, \mathbf{F}]_\theta$ coincides with $(\mathbf{E}, \mathbf{F})_{\theta,2}$.*

PROOF. The proof which follows contains more information than needed here; this will be useful later on. Clearly we may take $S = \Lambda$ in Proposition 2.7.2. Taking into account the strict positivity of Λ and the remark which follows Proposition 2.7.2, we see that there is a constant $m > 0$ such that the K-functional of the couple $(\mathbf{E}, \mathbf{F})$ satisfies the estimate

$$(2.8.6) \qquad mK(\tau, f) \leq ||\tau\Lambda(I + \tau\Lambda)^{-1}f||_{\mathbf{F}} \leq m^{-1}K(\tau, f)$$

for all $0 < \tau \leq 1$ and $f \in \mathbf{F}$. The change of variables $r = \tau^{-1}$ will then give

$$(2.8.7) \quad \left[\int_1^\infty ||r^{1-\theta}\frac{\Lambda}{\Lambda + r}f||_{\mathbf{F}}^p \frac{dr}{r}\right]^{1/p} \quad \text{and} \quad \left[\sum_{n=0}^\infty [2^{n(1-\theta)}||\frac{\Lambda}{\Lambda + 2^n}f||_{\mathbf{F}}]^p\right]^{1/p}$$

as admissible norms on $(\mathbf{E}, \mathbf{F})_{\theta,p}$ (see (2.3.9)). Let E be the spectral measure of the self-adjoint operator Λ in $\mathbf{F}$. Note that the support of E is included in $[a, \infty)$ for some $a > 0$. Then for $p = 2$ we get:

$$\int_1^\infty ||r^{1-\theta}\frac{\Lambda}{\Lambda + r}f||_{\mathbf{F}}^2 \frac{dr}{r} = \int_a^\infty ||E(d\lambda)f||^2 \int_1^\infty \frac{r^{2-2\theta}\lambda^2}{(\lambda + r)^2}\frac{dr}{r}$$

and the last integral is of order $\lambda^{2-2\theta}$ if $\lambda \geq a$. Hence $||f||_{\theta,2} \sim ||\Lambda^{1-\theta}f||_{\mathbf{F}}$. $\square$

If $(\mathbf{E}, \mathbf{F})$ is a Friedrichs couple, then one can construct a new Friedrichs couple $(\mathbf{F}^*, \mathbf{E}^*)$ involving the adjoint spaces in the following way: the embedding $\mathbf{F}^* \subset \mathbf{E}^*$ is just the adjoint of the inclusion map $\mathbf{E} \subset \mathbf{F}$. In this situation, we always identify $\mathbf{F}^*$ with a subspace of $\mathbf{E}^*$ (but provide it with the Hilbert structure adjoint to that of $\mathbf{F}$). It is then clear that one has the following: if $\mathbf{G}$ is a third Hilbert space such that $\mathbf{E} \subset \mathbf{G} \subset \mathbf{F}$, and if $\mathbf{E}$ is dense in $\mathbf{G}$, then the adjoint $\mathbf{G}^*$ is identified with a subspace of $\mathbf{E}^*$ and one has $\mathbf{F}^* \subset \mathbf{G}^* \subset \mathbf{E}^*$, with dense embeddings.

The preceding considerations apply in particular if $\mathbf{G} = [\mathbf{E}, \mathbf{F}]_\theta$. In this case the duality theorem (Theorem 2.4.2) can be improved, namely

$$(2.8.8) \qquad ([\mathbf{E}, \mathbf{F}]_\theta)^* = [\mathbf{F}^*, \mathbf{E}^*]_{1-\theta},$$

with identical Hilbert structures. In fact, a direct proof of (2.8.8), starting from the definition (2.8.3), is an easy exercise.

The fact that Hilbert space structures have been fixed on the spaces of the Friedrichs couple $(\mathbf{E}, \mathbf{F})$ allows one to define the *continuous Sobolev scale of Hilbert spaces* $\{\mathbf{E}^s\}_{s\in\mathbb{R}}$ by the following procedure. If $s \geq 0$ we take $\mathbf{E}^s = D(\Lambda^s)$ with norm $||f||_{\mathbf{E}^s} := ||\Lambda^s f||_{\mathbf{F}}$. If $s < 0$, then $\mathbf{E}^s$ is the completion of $\mathbf{F}$ with respect to the norm $||f||_{\mathbf{E}^s} := ||\Lambda^s f||_{\mathbf{F}}$. Note that $\mathbf{E}^1 = \mathbf{E}$, $\mathbf{E}^0 = \mathbf{F}$, $\mathbf{E}^s = [\mathbf{E}, \mathbf{F}]_\theta$ if $\theta = 1 - s$ and $0 < s < 1$. Clearly $\mathbf{E}^s \subset \mathbf{E}^t$ continuously and densely if $s > t$. So we may introduce the vector spaces $\mathbf{E}^\infty := \cap_{s\in\mathbb{R}}\mathbf{E}^s$, $\mathbf{E}^{-\infty} := \cup_{s\in\mathbb{R}}\mathbf{E}^s$. $\mathbf{E}^\infty$ is dense in each $\mathbf{E}^s$ and is invariant under all the operators $\Lambda^t, t \in \mathbb{R}$. For each $s \in \mathbb{R}$ the operator $\Lambda : \mathbf{E}^\infty \to \mathbf{E}^\infty$ extends to a bijective isometry $\Lambda_s : \mathbf{E}^{s+1} \to \mathbf{E}^s$ and Λ_s, when considered as operator in $\mathbf{E}^s$ with domain $\mathbf{E}^{s+1}$, is self-adjoint and positive. Hence $(\mathbf{E}^{s+1}, \mathbf{E}^s)$ is a Friedrichs couple with Λ_s as Friedrichs operator. There is no possible confusion if we keep the notation Λ for all the operators Λ_s. More generally, for each $\sigma > 0$ and $s \in \mathbb{R}$, the operator Λ^σ has a unique extension to a bijective isometry of $\mathbf{E}^{s+\sigma}$ onto $\mathbf{E}^s$ and this operator (again denoted just by Λ^σ) is self-adjoint and positive when considered as an operator in $\mathbf{E}^s$ with domain $E^{s+\sigma}$, hence it is the Friedrichs operator associated to the Friedrichs couple $(\mathbf{E}^{s+\sigma}, \mathbf{E}^s)$. These assertions become particularly obvious in a spectral representation of the operator Λ (there is a positive measure space (M, μ) and a strictly positive Borel function f on M such that Λ is unitarily equivalent to the operator of multiplication by f in $L^2(M, \mu)$). As a consequence, one may prove without any difficulty that $[\mathbf{E}^{s_0}, \mathbf{E}^{s_1}]_\theta = \mathbf{E}^s$ as Hilbert spaces if $s_0 > s_1$, $0 < \theta < 1$ and $s = (1 - \theta)s_0 + \theta s_1$.

If one identifies $\mathbf{F}^*$ with $\mathbf{F}$ through the Riesz isomorphism, then for any $s \in \mathbb{R}$ there is a canonical identification $(\mathbf{E}^s)^* = \mathbf{E}^{-s}$, namely the anti-duality between $\mathbf{E}^s$ and $\mathbf{E}^{-s}$ is defined by the condition $\langle f, g \rangle = \langle f, g \rangle_{\mathbf{F}}$ if $f \in \mathbf{E}^s \cap \mathbf{F}$ and $g \in \mathbf{E}^{-s} \cap \mathbf{F}$. Then $\mathbf{E}^{-1} = \mathbf{E}^*$ and for $0 < s < 1$: $\mathbf{E}^{-s} = [\mathbf{F}, \mathbf{E}^*]_s$ (see (2.8.8)). Moreover, $(\mathbf{E}, \mathbf{E}^*)$ will be a Friedrichs couple and $\mathbf{E}^s = [\mathbf{E}, \mathbf{E}^*]_{(1-s)/2}$ if $-1 < s < 1$.

It is remarkable that the spaces $\mathbf{E}^s$ with $|s| > 1$ are quite sensitive to a change in the operator Λ, while the spaces $\mathbf{E}^s$ with $|s| \leq 1$ are rather stable. Let us keep fixed the Hilbert structure of $\mathbf{F}$ and identify $\mathbf{F}^* = \mathbf{F}$. If $|s| < 1$, then $\mathbf{E}^s = [\mathbf{E}, \mathbf{E}^*]_{(1-s)/2} = (\mathbf{E}, \mathbf{E}^*)_{(1-s)/2,2}$ as topological vector space, hence the TVS structure of $\mathbf{E}^s$ depends only on the TVS structure of $\mathbf{E}$, not on its Hilbert space structure. On the other hand, if one replaces the given scalar product on $\mathbf{E}$ by a new but equivalent one and if one denotes by $\mathbf{E}_1$ the Hilbert space obtained in this way (so $\mathbf{E} = \mathbf{E}_1$ as H-spaces), then it could happen that $\mathbf{E}^s \cap \mathbf{E}_1^s = \{0\}$ for all $s > 1$ (e.g. two positive self-adjoint operators Λ and Λ_1 may have the same form domain although $D(\Lambda) \cap D(\Lambda_1) = \{0\}$).

So it is natural to introduce the notion of *topological Friedrichs couple* as a pair $(\mathbf{E}, \mathbf{F})$ consisting of a Hilbert space $\mathbf{F}$ and a dense H-subspace $\mathbf{E}$ of $\mathbf{F}$. The preceding discussion shows that the H-spaces $\mathbf{E}^s$ are well defined for $-1 \leq s \leq 1$, but have no meaning if $|s| > 1$. We shall keep the notation $[\mathbf{E}, \mathbf{F}]_\theta \equiv (\mathbf{E}, \mathbf{F})_{\theta,2}$ even if $\mathbf{E}$ is only a H-space.

2.8.2. In this subsection we fix a topological Friedrichs couple $(\mathbf{E}, \mathbf{F})$ and summarize some of the properties of the B-spaces which can be associated to it by real interpolation. We shall identify $\mathbf{F}^* = \mathbf{F}$ through the Riesz isomorphism (hence the Hilbert structure of $\mathbf{F}$ is important) and this gives us dense continuous embeddings $\mathbf{E} \subset \mathbf{F} \subset \mathbf{E}^*$, where $\mathbf{E}$ and $\mathbf{E}^*$ are H-spaces. For $-1 < s < 1$ and $1 \leq p \leq \infty$ we define $\mathbf{E}^{s,p} = (\mathbf{E}, \mathbf{E}^*)_{(1-s)/2,p}$; for $p = \infty$ we denote by $\overset{\circ}{\mathbf{E}}{}^{s,\infty}$ the closure of $\mathbf{E}$ in $\mathbf{E}^{s,\infty}$. We shall abbreviate $\mathbf{E}^s = \mathbf{E}^{s,2}$. Since, according to the preceding subsection, we have $\mathbf{E}^0 = \mathbf{F}$, it is natural to put $\mathbf{E}^1 = \mathbf{E}$ and $\mathbf{E}^{-1} = \mathbf{E}^*$. The next properties follow immediately from the general theory of the preceding sections:

(1) $\mathbf{E}^{s,p}$ are B-spaces such that $\mathbf{E} \subset \mathbf{E}^{s,p} \subset \mathbf{E}^*$; $\mathbf{E}^s$ are H-spaces.

(2) If $p < \infty$, then $\mathbf{E}$ is dense in $\mathbf{E}^{s,p}$ and $(\mathbf{E}^{s,p})^* = \mathbf{E}^{-s,p'}$, where $p' = p/(p-1)$; in particular $\mathbf{E}^{s,p}$ are reflexive spaces if $1 < p < \infty$. If $p = \infty$, then $(\overset{\circ}{\mathbf{E}}{}^{s,\infty})^* = E^{-s,1}$.

(3) The scale $\{\mathbf{E}^{s,p}\}$ is totally ordered, namely $\mathbf{E}^{s,p} \subset \mathbf{E}^{t,q}$ if $s > t$ and $p, q \in [1, \infty]$, while $\mathbf{E}^{s,p} \subset \mathbf{E}^{s,q}$ if $1 \le p \le q \le \infty$.

(4) The scale $\{\mathbf{E}^{s,p}\}$ has the reiteration property, i.e. if $1 > s > t > -1$, $0 < \theta < 1$ and $p, q, r \in [1, \infty]$, then

$$(\mathbf{E}^{s,p}, \mathbf{E}^{t,q})_{\theta,r} = \mathbf{E}^{u,r} \qquad \text{with } u = (1 - \theta)s + \theta t.$$

This also holds if $s = 1, p = 2$ (i.e. $\mathbf{E}^{s,p} = \mathbf{E}$) or $t = -1, q = 2$ (i.e. $\mathbf{E}^{t,q} = \mathbf{E}^*$), cf. Remark 2.5.4. In particular $\mathbf{E}^{s,p} = (\mathbf{E}, \mathbf{F})_{1-s,p}$ if $0 < s < 1, 1 \le p \le \infty$.

We shall present now a Littlewood-Paley type description of the spaces $\mathbf{E}^{s,p}$; this should be considered as a preparation and motivation for the general theory of Sections 3.5 and 3.6. Assume that a strictly positive self-adjoint operator Λ is given in $\mathbf{F}$ such that $D(\Lambda) = \mathbf{E}$ (so we assume $\Lambda \ge \text{const.} > 0$; this assumption, sometimes inconvenient in applications, is not really necessary, see [BGM2] for example). If we denote by $||f||_{\mathbf{F}} \equiv ||f||$ the norm of $\mathbf{F}$, then $||f||_{\mathbf{E}} := ||\Lambda f||$ and $||f||_{\mathbf{E}^*} := ||\Lambda^{-1}f||$ are admissible and dual norms on $\mathbf{E}, \mathbf{E}^*$. So $\mathbf{E}$ and $\mathbf{E}^*$ become Hilbert spaces and we may apply the theory developed at the end of §2.8.1 to the triplet $\mathbf{E} \subset \mathbf{F} \subset \mathbf{E}^*$. We see that Λ extends to a strictly positive self-adjoint operator in $\mathbf{E}^*$ with domain $\mathbf{F}$, its square Λ^2 having domain $\mathbf{E}$ (in $\mathbf{E}^*$). In particular Λ, when considered as an operator in $\mathbf{F}$ or $\mathbf{E}^*$, is the Friedrichs operator associated to the Friedrichs couple $(\mathbf{E}, \mathbf{F})$ or $(\mathbf{F}, \mathbf{E}^*)$ respectively, while Λ^2 as operator in $\mathbf{E}^*$ is the Friedrichs operator associated to the Friedrichs couple $(\mathbf{E}, \mathbf{E}^*)$. A straightforward interpolation argument shows that for $\sigma \in (-2, 2)$ and $\max(-1, \sigma - 1) < s < \min(1, \sigma + 1)$ the operator Λ^σ is an isomorphism of $\mathbf{E}^{s,p}$ onto $\mathbf{E}^{s-\sigma,p}$.

PROPOSITION 2.8.2. *Let* $(\mathbf{E}, \mathbf{F})$ *be a topological Friedrichs couple,* Λ *a self-adjoint strictly positive operator in* $\mathbf{F}$ *with* $D(\Lambda) = \mathbf{E}$ *and* E *the spectral measure of* Λ. *Then for each* $s \in (-1, 1)$, $p \in [1, \infty]$ *or* $s \in [-1, 1]$, $p = 2$, *the following two expressions are admissible norms on* $\mathbf{E}^{s,p}$:

$$(2.8.9) \quad \left[\int_0^\infty ||r^s E((r, 2r))f||_{\mathbf{F}}^p \frac{dr}{r} \right]^{1/p}; \quad \left[\sum_{k \in \mathbb{Z}} ||2^{ks} E((2^k, 2^{k+1}))f||_{\mathbf{F}}^p \right]^{1/p}.$$

Before giving the proof we make several remarks:

(a) Only the behaviour at infinity of the integral and the sum matters. Indeed, since $\Lambda \ge a > 0$, we have $E((r, 2r)) = 0$ if $r < a/2$. If $a > 1$, it is easily seen that one may replace $\int_0^\infty$ by $\int_1^\infty$ and $\sum_{k \in \mathbb{Z}}$ by $\sum_{k=0}^\infty$.

(b) The proof which follows gives other interesting admissible norms. For example, if $s > 0$ or $s < 0$, then one may replace $E((r, 2r))$ by $E((r, \infty))$ or $E((0, r))$ respectively (see [BGM2]).

(c) If the spaces $\mathbf{E}^{s,p}$ are defined for all $s \in \mathbb{R}$ by real interpolation using the spaces $\mathbf{E}^s$ associated to the operator Λ in §2.8.1, then (2.8.9) are admissible norms for all $s \in \mathbb{R}$; this is clear from the next proof (see also Section 3.5 below, or [BGM2] for the more elementary case considered here).

PROOF. We first consider the case $0 < s < 1$ and we set $\theta = 1 - s$. Then $\mathbf{E}^{s,p} = (\mathbf{E}, \mathbf{F})_{\theta,p}$ for which (2.8.7) furnishes two admissible norms. Starting with

(2.8.6) we shall now give a new estimate for the K-functional. Let $f \in \mathbf{F}$ and $\mu(d\lambda) = ||E(d\lambda)f||^2$. Then:

$$K^2(\tau, f) \sim ||\tau\Lambda(I + \tau\Lambda)^{-1}f||^2 = \int_0^\infty \frac{\tau^2\lambda^2}{1 + \tau^2\lambda^2}\mu(d\lambda) = \int_0^\infty \frac{\lambda^2}{r^2 + \lambda^2}\mu(d\lambda)$$

where $r = \tau^{-1}$ and $0 < \tau < 1$. On the interval $(0, r]$ we have $\lambda^2/(2r^2) \leq \lambda^2/(r^2 + \lambda^2) \leq \lambda^2/r^2$, while on (r, ∞) we have $1/2 \leq \lambda^2/(r^2 + \lambda^2) \leq 1$. Hence

(2.8.10) $$K(\tau, f) \sim r^{-1}||\Lambda E((0, r])f|| + ||E((r, \infty))f||.$$

This immediately gives $K(\tau, f) \geq c||E((r, 2r])f||$ for some constant $c > 0$ and all τ, f. On the other hand, for $0 < s < 1$ we may estimate the L_*^p-norms in the r-variable as follows:

$$\left\|r^s||r^{-1}\Lambda E((0, r])f||\right\|_{L_*^p} = \left\|r^s\sum_{k=1}^\infty \frac{\Lambda}{r}E((2^{-k}r, 2^{-k+1}r])f\right\|_{L_*^p(\mathbf{F})}$$

$$\leq \sum_{k=1}^\infty 2^{-k+1}||r^sE((2^{-k}r, 2 \cdot 2^{-k}r])f||_{L_*^p(\mathbf{F})}$$

$$= \sum_{k=1}^\infty 2^{-k+1+ks}||r^sE((r, 2r])f||_{L_*^p(\mathbf{F})} = 2^s(1 - 2^{s-1})^{-1}||r^sE((r, 2r])f||_{L_*^p(\mathbf{F})};$$

$$\left\|r^s||E((r, \infty))f||\right\|_{L_*^p} = \left\|r^s\sum_{k=0}^\infty E((r2^k, 2 \cdot r2^k])f\right\|_{L_*^p(\mathbf{F})}$$

$$\leq \sum_{k=0}^\infty ||r^sE((r2^k, 2 \cdot r2^k])f||_{L_*^p(\mathbf{F})} = \sum_{k=0}^\infty 2^{-ks}||r^sE((r, 2r])f||_{L_*^p(\mathbf{F})}$$

$$= (1 - 2^{-s})^{-1}||r^sE((r, 2r])f||_{L_*^p(\mathbf{F})}.$$

By using (2.2.4) it is clear that the first expression in (2.8.9) is an admissible norm on $(\mathbf{E}, \mathbf{F})_{\theta, p}$. For the second one the argument is similar, cf. Lemma 2.3.4.

Now assume that $-1 < s \leq 0$ (we do not consider the simple cases $s = \pm 1$, $p = 2$). Then, according to a remark made before the statement of the proposition, $\Lambda^{-\sigma}$ is an isomorphism of $\mathbf{E}^{s,p}$ onto $\mathbf{E}^{s+\sigma,p}$ if $-s < \sigma < 1$. By using the result obtained before for $\mathbf{E}^{s+\sigma,p}$ and by taking into account that $||\Lambda^{-\sigma}E((r, 2r])f|| \sim r^{-\sigma}||E((r, 2r])f||$, the proof is finished. $\square$

Let us state an immediate consequence of Proposition 2.7.4 (which may be viewed as some kind of non-commutative interpolation result). Assume that S is a self-adjoint operator in $\mathbf{F}$ and that *one* of the following conditions is satisfied:

(i) $S \geq 0$, $(S + \lambda)^{-1}\mathbf{E} \subset \mathbf{E}$ and $||(S + \lambda)^{-1}||_{B(\mathbf{E})} \leq c\lambda^{-1}$ for all $\lambda > 0$ and some constant c;

(ii) $S \geq 0$, $e^{-tS}\mathbf{E} \subset \mathbf{E}$ and $||e^{-tS}||_{B(\mathbf{E})} \leq c$ for $t > 0$ and a constant c;

(iii) $e^{itS}\mathbf{E} \subset \mathbf{E}$ and $||e^{itS}||_{B(\mathbf{E})} \leq c$ for $t > 0$ and a constant c.

Then $\mathbf{E} \cap D(S)$, equipped with the intersection topology, is a dense H-subspace of $\mathbf{F}$ and $[\mathbf{E} \cap D(S), \mathbf{F}]_{1-\theta} = \mathbf{E}^\theta \cap D(|S|^\theta)$ if $0 < \theta < 1$.

We end this subsection with some considerations in connection with the B-space $B(\mathbf{E}, \mathbf{E}^*)$. We have seen in Section 2.1 that it makes sense to speak about symmetric or positive linear operators $\mathbf{E} \to \mathbf{E}^*$ and that such an operator is necessarily continuous. If $\mathbf{G}, \mathbf{K}$ are B-spaces such that $\mathbf{E} \subset \mathbf{G}$ densely and $\mathbf{K} \subset \mathbf{E}^*$, then we have defined a canonical continuous embedding $B(\mathbf{G}, \mathbf{K}) \subset B(\mathbf{E}, \mathbf{E}^*)$. In particular $B(\mathbf{E}^{s,p}, \mathbf{E}^{t,q}) \subset B(\mathbf{E}, \mathbf{E}^*)$ if $p < \infty$ and $B(\overset{\circ}{\mathbf{E}}{}^{s,\infty}, \mathbf{E}^{t,q}) \subset B(\mathbf{E}, \mathbf{E}^*)$. Also $B(\mathbf{E})$, $B(\mathbf{F})$, $B(\mathbf{E}^*)$ and $B(\mathbf{E}^*, \mathbf{E})$ are B-subspaces of $B(\mathbf{E}, \mathbf{E}^*)$; moreover, they are algebras for the natural composition product, which is not the case for $B(\mathbf{E}, \mathbf{E}^*)$. Although the product ST of two elements of $B(\mathbf{E}, \mathbf{E}^*)$ makes no sense in general, it can often be defined as a limit s-$\lim_{n\to\infty} S\varrho_n T$, where $\{\varrho_n\}_{n\in\mathbb{N}}$ is a mollifying sequence, i.e. $\varrho_n \in B(\mathbf{E}^*, \mathbf{E})$, $\|\varrho_n\|_{B(\mathbf{E})} + \|\varrho_n\|_{B(\mathbf{E}^*)} \le$ const. and $\|\varrho_n f - f\|_{\mathbf{E}} \to 0$ as $n \to \infty$ for $f \in \mathbf{E}$. By the density of $\mathbf{E}$ in $\mathbf{E}^*$, we shall have $\varrho_n f \to f$ in $\mathbf{E}^*$ for all $f \in \mathbf{E}^*$ and also $\|\varrho_n\|_{B(\mathbf{G})} \le$ const., $\varrho_n f \to f$ in $\mathbf{G}$ for all f in the closure of $\mathbf{E}$ in $\mathbf{G}$ if $\mathbf{G}$ is an interpolation space between $\mathbf{E}$ and $\mathbf{E}^*$. One may take, for example, $\varrho_n = n(n + \Lambda^2)^{-1}$, where Λ is as above.

Assume that $\mathbf{G}, \mathbf{K}$ are B-spaces, $\mathbf{K}$ is reflexive, and that $\mathbf{E} \subset \mathbf{G}$, $\mathbf{K} \subset \mathbf{G}^*$, both embeddings being dense. Then $\mathbf{G}^* \subset \mathbf{E}^*$, $\mathbf{E} \subset \mathbf{K}^*$ and the second embedding is dense. Hence $B(\mathbf{G}, \mathbf{K})$ and $B(\mathbf{K}^*, \mathbf{G}^*)$ are B-subspaces of $B(\mathbf{E}, \mathbf{E}^*)$. If T belongs to $B(\mathbf{E}, \mathbf{E}^*)$ and T^* denotes its adjoint in this space (note that $\mathbf{E}^{**} = \mathbf{E}$), then we clearly have $T \in B(\mathbf{G}, \mathbf{K}) \Rightarrow T^* \in B(\mathbf{K}^*, \mathbf{G}^*)$. This observation gives a way of proving that an operator $T \in B(\mathbf{E}, \mathbf{E}^*)$ induces a continuous operator between some B-subspaces of $\mathbf{E}^*$. For example, if $T : \mathbf{E} \to \mathbf{E}^*$ is symmetric and $T \in B(\mathbf{G}, \mathbf{K})$, then $T \in B(\mathbf{K}^*, \mathbf{G}^*)$ too. Later on, we shall often use this technique without further comments.

2.8.3. Let $\mathbf{F}$ be a Hilbert space identified with its adjoint space and let $\mathbf{E}$ be a dense B-subspace of $\mathbf{F}$, so that we have canonically $\mathbf{E} \subset \mathbf{F} \subset \mathbf{E}^*$. We shall prove here a quadratic interpolation estimate for operator norms and then we shall show that $\mathbf{F}$ is an interpolation space between $\mathbf{E}$ and $\mathbf{E}^*$.

We recall that the *spectral radius* of an operator $S \in B(\mathbf{E})$ is the number $|S|_{\mathbf{E}} = \sup\{|\lambda| \mid \lambda \in \mathbb{C} \text{ and } S - \lambda \text{ is a homeomorphism } \mathbf{E} \to \mathbf{E}\}$, i.e. $|S|_{\mathbf{E}}$ is the radius of the smallest disc in $\mathbb{C}$ with center at zero which contains the spectrum of the element S of the algebra $B(\mathbf{E})$. Note that $|S|_{\mathbf{E}}$ does not depend on the choice of an admissible norm on $\mathbf{E}$. However, if $\|\cdot\|_{\mathbf{E}}$ is such a norm, then one has $|S|_{\mathbf{E}} = \lim_{k\to\infty} \|S^k\|_{B(\mathbf{E})}^{1/k}$.

We do not assume that $\mathbf{E}$ is reflexive, hence the embedding $\mathbf{E} \subset \mathbf{E}^*$ is not dense in general and the weak*-topology on $\mathbf{E}^*$ (i.e. the weak topology on $\mathbf{E}^*$ associated to the duality of $\mathbf{E}^*$ with $\mathbf{E}$) is strictly weaker than the weak topology (which is given by the duality of $\mathbf{E}^*$ with $\mathbf{E}^{**}$). Note, however, that $\mathbf{E}$ is always weak*-dense in $\mathbf{E}^*$, hence if an operator $T \in B(\mathbf{E})$ has an extension to an operator $\widetilde{T} \in B(\mathbf{E}^*)$ which is weak*-continuous, then this extension is unique (if $\mathbf{E}$ is reflexive, any operator in $B(\mathbf{E}^*)$ is weak*-continuous). Let $T \in B(\mathbf{E})$ and let $T^* \in B(\mathbf{E}^*)$ be its adjoint. Then T has a (unique) weak*-continuous extension to $\mathbf{E}^*$ if and only if $T^*\mathbf{E} \subset \mathbf{E}$. (*Proof*: If $T^*\mathbf{E} \subset \mathbf{E}$, let $T' = T^*|_{\mathbf{E}}$ considered as an operator with values in $\mathbf{E}$. By the closed graph theorem $T' \in B(\mathbf{E})$, hence $T'^* \in B(\mathbf{E}^*)$ and is weak*-continuous; it is clear that T'^* is an extension of T. Reciprocally, if $T \subset \widetilde{T} \in B(\mathbf{E}^*)$ and $\widetilde{T}$ is weak*-continuous, then there is $S \in B(\mathbf{E})$ such that $S^* = \widetilde{T}$. For $e, e' \in \mathbf{E}$ we have $\langle Se, e' \rangle = \langle e, \widetilde{T}e' \rangle = \langle e, Te' \rangle = \langle T^*e, e' \rangle$, so $Se = T^*e$ and hence $T^*\mathbf{E} \subset \mathbf{E}$).

PROPOSITION 2.8.3. *Let $T \in B(\mathbf{E})$ such that $T^*\mathbf{E} \subset \mathbf{E}$. Denote by the same symbol T the unique weak*-continuous extension of T to $\mathbf{E}^*$. Then $T\mathbf{F} \subset \mathbf{F}$ and $||T||_{B(\mathbf{F})} \leq |T^*T|_{\mathbf{E}}^{1/2}$. If $||\cdot||_{\mathbf{E}}$ is an admissible norm on $\mathbf{E}$ and $||\cdot||_{\mathbf{E}^*}$ is the dual norm on $\mathbf{E}^*$, then we also have $||T||_{B(\mathbf{F})} \leq ||T||_{B(\mathbf{E})}^{1/2}||T||_{B(\mathbf{E}^*)}^{1/2}$. In particular, if $T \in B(\mathbf{E})$ is symmetric as an operator $\mathbf{E} \to \mathbf{E}^*$, then $||T||_{B(\mathbf{F})} \leq ||T||_{B(\mathbf{E})}$.*

PROOF. In this proof we abbreviate $||\cdot||_{\mathbf{F}} = ||\cdot||$. The operator $S = T^*T : \mathbf{E} \to \mathbf{E}$ is bounded and for $e \in \mathbf{E}$: $||Te||^2 = \langle Se, e \rangle \leq ||Se|| \cdot ||e||$. We assume $||e|| = 1$ and prove by induction $||Te||^{2k} \leq ||S^k e||$ for $k = 2^n$, $n = 0, 1, 2, \ldots$ The case $n = 0$ has been proved, and the induction hypothesis leads to:

$$||Te||^{4k} \leq ||S^k e||^2 = \langle S^k e, S^k e \rangle = \langle S^{2k} e, e \rangle \leq ||S^{2k} e||,$$

so the assertion is proved. Passing to the limit $k \to \infty$ in the inequality $||Te|| \leq ||S^k e||^{1/2k}$, we get $||Te|| \leq \liminf_{k \to \infty} ||S^k e||^{1/2k}$. Now $||S^k e|| \leq c||S^k e||_{\mathbf{E}} \leq c||S^k||_{B(\mathbf{E})}||e||_{\mathbf{E}}$ for a constant c; since $\lim_{k \to \infty} \lambda^{1/k} = 1$ if $\lambda \in (0, \infty)$ and $\mathbf{E}$ is dense in $\mathbf{F}$, we obtain the first assertion of the proposition. Then $|S|_{\mathbf{E}} \leq ||S||_{B(\mathbf{E})} \leq ||T'||_{B(\mathbf{E})} \cdot ||T||_{B(\mathbf{E})}$ with $T' = T^*|_{\mathbf{E}}$. Finally $||T'||_{B(\mathbf{E})} = ||T'^*||_{B(\mathbf{E}^*)}$ and T'^* is just the weak*-continuous extension of T to $\mathbf{E}^*$ which we decided to denote by the same letter T. If T is symmetric, then T^* is just this extension of T to $\mathbf{E}^*$. $\square$

COROLLARY 2.8.4. *For all $e \in \mathbf{E}$ one has $||e||_{\mathbf{F}} \leq ||e||_{\mathbf{E}}^{1/2} \cdot ||e||_{\mathbf{E}^*}^{1/2}$.*

PROOF. Let $Tu = e\langle e, u \rangle$, so $T \in B(\mathbf{E})$ is a symmetric operator with $||T||_{B(\mathbf{E})} = ||e||_{\mathbf{E}}||e||_{\mathbf{E}^*}$ and $||T||_{B(\mathbf{F})} = ||e||_{\mathbf{F}}^2$. $\square$

The next result is essentially due to J.-L.Lions and J.Peetre (see Theorem 4.1 in [LP]). However, their argument for the case of non-reflexive spaces $\mathbf{E}$ seems to us rather incomplete, so we give all the details of the proof.

THEOREM 2.8.5. *Let $\mathbf{F}$ be a Hilbert space identified with its adjoint space and let $\mathbf{E}$ be a dense B-subspace of $\mathbf{F}$. Consider the triplet $\mathbf{E} \subset \mathbf{F} \subset \mathbf{E}^*$ associated to these objects. Then $(\mathbf{E}, \mathbf{E}^*)_{1/2,2} = \mathbf{F}$ as B-spaces. Let $\widetilde{\mathbf{E}}$ be the Gagliardo completion of $\mathbf{E}$ in $\mathbf{F}$ (hence in $\mathbf{E}^*$) and $\overline{\mathbf{E}}$ the closure of $\mathbf{E}$ (hence of $\mathbf{F}$) in $\mathbf{E}^*$, so that $\mathbf{E} \subset \widetilde{\mathbf{E}} \subset \mathbf{F} \subset \overline{\mathbf{E}} \subset \mathbf{E}^*$. Then we also have $(\widetilde{\mathbf{E}}, \overline{\mathbf{E}})_{1/2,2} = \mathbf{F}$.*

PROOF. (i) We first prove that $(\mathbf{E}, \mathbf{E}^*)_{1/2,2} \subset \mathbf{F}$ (this is due to Lions and Peetre; note that Corollary 2.8.4, Lemma 2.5.1 (b) and (2.5.4) together imply that $(\mathbf{E}, \mathbf{E}^*)_{1/2,1} \subset \mathbf{F}$). Let $f \in \mathbf{E}$ and let u, v be the functions constructed in Lemma 2.2.4 (where $\mathbf{F}$ has to be replaced by $\mathbf{E}^*$). Then by Proposition 2.3.2(b) we have $f = \int_0^\infty v(\tau) \tau^{-1} d\tau$ and:

$$||f||_{\mathbf{F}}^2 = \int_0^\infty \langle f, v(\tau) \rangle \frac{d\tau}{\tau} = \int_0^\infty \langle u(\tau), v(\tau) \rangle \frac{d\tau}{\tau} + \int_0^\infty \langle f - u(\tau), v(\tau) \rangle \frac{d\tau}{\tau}$$
$$\leq ||\tau^{1/2} u||_{L^2_*(\mathbf{E})} ||\tau^{-1/2} v||_{L^2_*(\mathbf{E}^*)} + ||\tau^{-1/2}(f - u(\cdot))||_{L^2_*(\mathbf{E}^*)} ||\tau^{1/2} v||_{L^2_*(\mathbf{E})}.$$

The last member here is $\leq c\|f\|^2_{1/2,2}$, as a consequence of (2.2.10) and (2.3.4). Hence $\|f\|_{\mathbf{F}} \leq c'\|f\|_{1/2,2}$ for $f \in \mathbf{E}$. Since $\mathbf{E}$ is dense in $(\mathbf{E}, \mathbf{E}^*)_{1/2,2}$, the embedding $(\mathbf{E}, \mathbf{E}^*)_{1/2,2} \subset \mathbf{F}$ has been proved.

(ii) If the space $\mathbf{E}$ is reflexive, the embedding $\mathbf{F} \subset (\mathbf{E}, \mathbf{E}^*)_{1/2,2}$ follows immediately by taking adjoints in the inclusion proved above, cf. (2.4.1). For the general case we argue as follows. The space $\mathbf{F}$ being reflexive, the Gagliardo completion of $\mathbf{E}$ with respect to $\mathbf{F}$ coincides with its Gagliardo completion with respect to $\mathbf{E}^*$. Then (2.2.9) gives $(\mathbf{E}, \mathbf{E}^*)_{1/2,2} = (\mathbf{E}, \overline{\mathbf{E}})_{1/2,2} = (\widetilde{\mathbf{E}}, \overline{\mathbf{E}})_{1/2,2} = (\widetilde{\mathbf{E}}, \mathbf{E}^*)_{1/2,2}$. Since $(\mathbf{E}, \overline{\mathbf{E}})_{1/2,2} \subset \mathbf{F}$ densely and $\mathbf{F}^* = \mathbf{F}$, we get by taking the adjoints and using (2.4.1): $\mathbf{F} \subset (\mathbf{E}, \overline{\mathbf{E}})^*_{1/2,2} = (\overline{\mathbf{E}}^*, \mathbf{E}^*)_{1/2,2}$. Finally, we note that $\overline{\mathbf{E}}^* = \widetilde{\mathbf{E}}$, as has been shown in a more general setting just before Lemma 2.1.2. $\square$

The following example clarifies the preceding proof. Let $J = [0,1] \subset \mathbb{R}$ and $\mathbf{E} = C(J)$, $\mathbf{F} = L^2(J)$. Then $\mathbf{E}^* = M(J)$ is the space of integrable Borel measures on J, $\widetilde{\mathbf{E}} = L^\infty(J)$ and $\overline{\mathbf{E}} = L^1(J)$. We have $C(J) \subset L^\infty(J) \subset L^2(J) \subset L^1(J) \subset M(J)$ with all embeddings strict, and $L^1(J)^* = L^\infty(J)$.

2.8.4. If $(\mathbf{E}, \mathbf{F})$ is a Friedrichs couple, then the B-spaces $(\mathbf{E}, \mathbf{F})_{\theta,1}$ have a geometrical property which will play an important role in our proof of the existence and completeness of local wave operators in Chapter 7. We shall summarize here the necessary notions and results. Details and proofs may be found in [MP] and [Pi]; see also [DF] for an up-to-date presentation.

Let $\mathbf{G}$ be a Banach space and $q > 0$ a real number. One says that $\mathbf{G}$ is of *cotype* q if there are real numbers $\alpha > 0$ and $c > 0$ such that for an arbitrary finite family of vectors $g_1, \ldots, g_N$ in $\mathbf{G}$ the next inequality holds:

$$(2.8.11) \qquad \left[\sum_{n=1}^N \|g_n\|^q_{\mathbf{G}}\right]^{1/q} \leq c\left[2^{-N} \sum_\varepsilon \left\|\sum_{n=1}^N \varepsilon_n g_n\right\|^\alpha_{\mathbf{G}}\right]^{1/\alpha}.$$

Here $\varepsilon = (\varepsilon_1, \ldots, \varepsilon_N)$ runs over all possible choices of signs $\varepsilon_n = \pm 1$. Notice that this property depends only on the B-space structure of $\mathbf{G}$. A rather deep estimate due to Kahane and Pisier (a vector version of Khintchin's inequality) shows that if the preceding property holds for some $\alpha > 0$, then it holds for all $\alpha > 0$. It is easy to show that $q \geq 2$ if $\mathbf{G} \neq \{0\}$. The spaces of cotype 2 are, geometrically speaking, very nice: they are not far from being Hilbert spaces (it is obvious that if $\mathbf{G}$ is a Hilbert space, then (2.8.11) is valid with $q = 2, \alpha = 2, c = 1$, the inequality being in fact an equality). Note that a closed subspace of a space of cotype q is also of cotype q.

Spaces of cotype 2 are important to us because of the following *factorization theorem* established by G.Pisier in [Pi]: *Let $\mathbf{G}_1$, $\mathbf{G}_2$ be two Banach spaces, one of them having the bounded approximation property, and such that $\mathbf{G}_1^*$ and $\mathbf{G}_2$ are of cotype 2; then each continuous operator $T : \mathbf{G}_1 \to \mathbf{G}_2$ factorizes through a Hilbert space* (i.e. there are a Hilbert space $\mathbf{H}$ and bounded operators $T_1 : \mathbf{G}_1 \to \mathbf{H}$ and $T_2 : \mathbf{H} \to \mathbf{G}_2$ such that $T = T_2 T_1$). We recall that $\mathbf{G}$ has the *bounded approximation property* if for each compact set $\mathbf{K} \subset \mathbf{G}$ there is a net $\{P_\lambda\}$ of finite rank operators in $\mathbf{G}$ such that $\|P_\lambda\|_{B(\mathbf{G})} \leq$ const. and $\lim P_\lambda g = g \ \forall g \in \mathbf{K}$.

In order to apply the factorization theorem in the setting that interests us, we first prove:

LEMMA 2.8.6. *If* $(\mathbf{E}, \mathbf{F})$ *is a Friedrichs couple and* $0 < \theta < 1$, $1 \leq p \leq 2$, *then* $(\mathbf{E}, \mathbf{F})_{\theta,p}$ *is of cotype 2 and has the bounded approximation property.*

PROOF. Since $\mathbf{F}$ is a Hilbert space, (2.8.11) with $\mathbf{G} = \mathbf{F}$ and $c = 1$ becomes an equality for $q = \alpha = 2$. So, by the Kahane-Pisier estimate mentioned above, (2.8.11) holds for $\alpha = p$, i.e. there is a constant $c = c(p) < \infty$ such that for all $g_1, \ldots, g_N \in \mathbf{F}$ ($\| \cdot \|_{\mathbf{F}} \equiv \| \cdot \|$):

$$\Big[\sum_{n=1}^{N} \|g_n\|^2 \Big]^{p/2} \leq c^p 2^{-N} \sum_{\varepsilon} \Big\| \sum_{n=1}^{N} \varepsilon_n g_n \Big\|^p.$$

For each n let $\{f_n(k)\}_{k \in \mathbb{N}}$ be a sequence in $\mathbf{F}$. We take $g_n = f_n(k)$, multiply each of the obtained estimates by a number $\mu_k > 0$ and sum over all k. Since $2/p \geq 1$ we can use the Minkowski inequality in order to get:

$$(2.8.12) \quad \Big[\sum_{n=1}^{N} \Big(\sum_{k=0}^{\infty} \|f_n(k)\|^p \mu_k \Big)^{2/p} \Big]^{1/2} \leq \Big[\sum_{k=0}^{\infty} \Big(\sum_{n=1}^{N} \|f_n(k)\|^2 \Big)^{p/2} \mu_k \Big]^{1/p}$$

$$\leq c \Big[2^{-N} \sum_{\varepsilon} \sum_{k=0}^{\infty} \sum_{n=1}^{N} \Big\| \sum_{n=1}^{N} \varepsilon_n f_n(k) \Big\|^p \mu_k \Big]^{1/p}.$$

Now let $f_n \in (\mathbf{E}, \mathbf{F})_{\theta,p}$. We shall use the framework of Proposition 2.8.2 assuming, without loss of generality, that $\Lambda \geq 2$. Let $\|f\|_{\theta,p}$ be the second norm from (2.8.9). In (2.8.12) we take $\mu_k = 2^{ksp}$ with $s = 1 - \theta$ and $f_n(k) = E((2^k, 2^{k+1}]) f_n$. We get

$$\Big[\sum_{n=1}^{N} \|f_n\|_{\theta,p}^2 \Big]^{1/2} \leq c \Big[2^{-N} \sum_{\varepsilon} \Big\| \sum_{n=1}^{N} \varepsilon_n f_n \Big\|_{\theta,p}^p \Big]^{1/p}.$$

This proves that $(\mathbf{E}, \mathbf{F})_{\theta,p}$ is of cotype 2.

Now we prove that $(\mathbf{E}, \mathbf{F})_{\theta,p}$ has the bounded approximation property. Let $\mathbf{H}_k = E((2^k, 2^{k+1}]) \mathbf{F}$ equipped with the norm $\|f\|_k = 2^{ks} \|f\|$, so that $\mathbf{H}_k$ is a Hilbert space. By Proposition 2.8.2 $(\mathbf{E}, \mathbf{F})_{\theta,p}$ is isomorphic with the space $\mathbf{H} = \oplus_{k=0}^{\infty} \mathbf{H}_k$ equipped with the ℓ^p-norm given by $[\sum_{k=0}^{\infty} \|f(k)\|_k^p]^{1/p}$. If for each k we choose a finite rank orthogonal projection P_k in $\mathbf{H}_k$ such that $P_k = 0$ except for a finite number of k, then $P = \oplus_{k=0}^{\infty} P_k$ is a finite rank operator in $\mathbf{H}$ such that $\|P\|_{B(\mathbf{H})} \leq 1$. The family of such projections P in $\mathbf{H}$ is a net for the natural order relation (given by the inclusion of their ranges) and $\lim Ph = h$ for all $h \in \mathbf{H}$. $\square$

THEOREM 2.8.7. *Let* $(\mathbf{E}_1, \mathbf{F}_1)$ *and* $(\mathbf{E}_2, \mathbf{F}_2)$ *be Friedrichs couples. Define* $\mathbf{G}_j = (\mathbf{E}_j, \mathbf{F}_j)_{1/2,1}$, *so that* $\mathbf{G}_j^* = (\mathbf{F}_j^*, \mathbf{E}_j^*)_{1/2,\infty}$, *and let* $\mathbf{G}_j^{*\circ}$ *be the closure of* $\mathbf{F}_j^*$ *in* $\mathbf{G}_j^*$. *Then for each bounded operator* $V : \mathbf{G}_1^{*\circ} \to \mathbf{G}_2$ *there is a Hilbert space* $\mathbf{H}$ *(identified with its dual) and there are bounded operators* $U_j : \mathbf{G}_j^{*\circ} \to \mathbf{H}$ *such that* $V = U_2^* U_1$ *(recall that* $(\mathbf{G}_2^{*\circ})^* = \mathbf{G}_2$).

PROOF. The spaces $(\mathbf{G}_1^{*\circ})^* = \mathbf{G}_1$ and $\mathbf{G}_2$ are of cotype 2 and $\mathbf{G}_2$ has the bounded approximation property by Lemma 2.8.6. Hence by Pisier's theorem there are a Hilbert space $\mathbf{H}$ and bounded operators $U_1 : \mathbf{G}_1^{*\circ} \to \mathbf{H}$ and $T_2 : \mathbf{H} \to \mathbf{G}_2$ such that $V = T_2 U_1$. We identify $\mathbf{H}^* = \mathbf{H}$, hence $T_2^* : \mathbf{G}_2^* \to \mathbf{H}$. Let $U_2 = T_2^*|_{\mathbf{G}_2^{*\circ}}$, then $U_2^* = T_2$. $\square$

2.8.5. We end this section with some considerations concerning quadratic forms. We recall that a *sesquilinear form* on a (complex) vector space $\mathbf{E}$ is a map $S : \mathbf{E} \times \mathbf{E} \to \mathbb{C}$ anti-linear in the first argument and linear in the second argument. The *quadratic form* associated to S is the map $Q : \mathbf{E} \to \mathbb{C}$ given by $Q(e) = S(e, e)$. Q determines S uniquely because of the *polarization identity*:

$$(2.8.13) \qquad\qquad S(e, f) = \frac{1}{4} \sum_{k=0}^{3} i^k Q(i^k e + f).$$

This shows in particular that S is symmetric (i.e. $\overline{S(e, f)} = S(f, e)$) if and only if Q is real. The positive quadratic forms can be characterized intrinsically, as follows from the next result of Jordan and Von Neumann.

LEMMA 2.8.8. *Let $\mathbf{E}$ be a (complex) vector space. A function $Q : \mathbf{E} \to [0, \infty)$ is a quadratic form if and only if the following two conditions are fulfilled:*
(i) $Q(e + f) + Q(e - f) \leq 2Q(e) + 2Q(f)$, $\quad \forall e, f \in \mathbf{E}$;
(ii) $Q(\lambda e) = |\lambda|^2 Q(e)$, $\qquad\qquad\qquad\qquad \forall \lambda \in \mathbb{C}, e \in \mathbf{E}$.

PROOF. By changing e, f into $(e + f)/2$, $(e - f)/2$ in (i) we see that the inequality becomes an equality. Let S be defined by (2.8.13). By straightforward purely algebraic computations one sees that S is a symmetric sesquilinear form on $\mathbf{E}$ if we equip $\mathbf{E}$ with the structure of a vector space over the field of complex rational numbers (see §1.5 in [Y]) and $S(e, e) = Q(e)$. Hence it remains to show that $\lambda \mapsto S(e, \lambda f)$ is continuous, $\lambda \in \mathbb{C}$. For $\varepsilon > 0$ rational we get by using (i), (ii) and the positivity of Q that $|S(e, f)| = |S(\varepsilon e, \varepsilon^{-1} f)| \leq \frac{1}{4} \sum_{k=0}^{3} Q(i^k \varepsilon e + \varepsilon^{-1} f) \leq \varepsilon^2 Q(e) + \varepsilon^{-2} Q(f)$. By taking the minimum of the last member over $\varepsilon > 0$ we obtain $|S(e, f)| \leq 2[Q(e)Q(f)]^{1/2}$. Hence $|S(e, \lambda f) - S(e, \mu f)| = |S(e, (\lambda - \mu)f| \leq c|\lambda - \mu| \to 0$ if $\lambda \to \mu$. $\square$

We shall always identify a sesquilinear form with the quadratic form associated to it; so for us the words "sesquilinear" and "quadratic" are synonymous. If $\mathbf{E}$ is a B-space, there is a further identification of continuous sesquilinear (or quadratic) forms on $\mathbf{E}$ with continuous linear operators $S : \mathbf{E} \to \mathbf{E}^*$ given by $S(e, f) = \langle e, Sf \rangle$. This has already been noticed in Section 2.1 in connection with the space $B(\mathbf{E}, \mathbf{E}^*)$.

It is often convenient to allow a positive quadratic form to take the value $+\infty$. So a *positive quadratic form* on a (complex) vector space $\mathbf{F}$ is a map $Q : \mathbf{F} \to [0, \infty]$ such that (i) and (ii) of Lemma 2.8.8 are fulfilled (with the convention $0 \cdot \infty = 0$). Then the set $\mathbf{E} = \{e \in \mathbf{F} \mid Q(e) < \infty\}$ is a vector subspace of $\mathbf{F}$, called the *domain of Q*, and the restriction of Q to $\mathbf{E}$ is a quadratic form in the usual sense, in particular $Q(e) = S(e, e)$ for a (symmetric) sesquilinear form S on $\mathbf{E}$. We say that a positive quadratic form is *definite* if $Q(e) = 0 \Rightarrow e = 0$. Then $Q(f)^{1/2}$ defines a gauge on $\mathbf{F}$; gauges of this form will be called *quadratic*. Let $\mathbf{F}$ be a TVS and Q a positive definite quadratic form on $\mathbf{F}$. We say that Q is *closed*, or *coercive*, or *closable*, if the gauge associated to it has the corresponding property. For example, a coercive form on a Banach space $\mathbf{F}$ is a positive quadratic form on $\mathbf{F}$ such that $Q(f) \geq a||f||_{\mathbf{F}}^2$ for a constant $a > 0$ and all $f \in \mathbf{F}$. Note that if $\mathbf{F}$ is locally convex, Hausdorff and sequentially complete, then Q is closed if and only if it is lower semicontinuous (Proposition 2.1.1). It is an easy consequence of Lemma 2.8.8 that *the lower semicontinuous regularization of a coercive quadratic gauge on a Hausdorff sequentially complete TVS $\mathbf{F}$ is also a (coercive) quadratic gauge*. In other terms, *if Q is a coercive quadratic form on $\mathbf{F}$, then there is a largest lower semicontinuous (in particular closed) coercive quadratic form Q_* on $\mathbf{F}$ such that $Q_* \leq Q$.* When $\mathbf{F}$ is a Hilbert space this fact has been proved by B. Simon [Sim1] by rather different methods.

Let Q be a positive continuous quadratic form on a B-space $\mathbf{E}$ and $S \in B(\mathbf{E}, \mathbf{E}^*)$ the symmetric (positive) linear continuous operator associated to it: $Q(e) = \langle e, Se \rangle$. By the Cauchy-Schwarz inequality we have $|\langle e, Sf \rangle|^2 \leq Q(e)Q(f)$, in particular $Sf = 0$ if and only if $Q(f) = 0$. So Q is definite if and only if S is injective. Now *assume $\mathbf{E}$ reflexive; then Q is coercive if and only if S is bijective* (this is the Lax-Milgram lemma; notice that there exists a continuous coercive form on $\mathbf{E}$ if and only if $\mathbf{E}$ is hilbertizable). For the proof, let $|| \cdot ||$ be an admissible norm on $\mathbf{E}$ such that $|e| \equiv Q(e)^{1/2} \leq ||e||$ and denote also by $|| \cdot ||$ the adjoint norm on $\mathbf{E}^*$. We have $||e|| = \sup |\langle e, \varphi \rangle|$, where φ runs over the unit ball of $\mathbf{E}^*$. If S is bijective, then $S^{-1} : \mathbf{E}^* \to \mathbf{E}$ is continuous and each $\varphi \in \mathbf{E}^*$ with $||\varphi|| \leq 1$ is of the form $\varphi = Sf$ with $|f| \leq ||f|| = ||S^{-1}Sf|| \leq ||S^{-1}||$. Since $|\langle e, \varphi \rangle| = |\langle e, Sf \rangle| \leq |e| \cdot |f| \leq |e| \cdot ||S^{-1}||$, we get $||e|| \leq ||S^{-1}|| \cdot |e|$, so Q is coercive. Reciprocally, if $Q(e) \geq a||e||^2$, then $a||e||^2 \leq \langle e, Se \rangle \leq ||e|| \cdot ||Se||$, hence $a||e|| \leq ||Se||$ for all $e \in \mathbf{E}$. So $S : \mathbf{E} \to \mathbf{E}^*$ is continuous and injective with closed range. Then $S^* : \mathbf{E}^{**} = \mathbf{E} \to \mathbf{E}^*$ has dense range. But S is symmetric, so S is bijective. $\square$

Assume now that $(\mathbf{E}, \mathbf{F})$ is a topological Friedrichs couple and let Q be a positive continuous coercive quadratic form on $\mathbf{E}$; denote by $S \in B(\mathbf{E}, \mathbf{E}^*)$ the positive bijective operator associated to Q. After the identification $\mathbf{F}^* = \mathbf{F}$ we obtain continuous dense embeddings $\mathbf{E} \subset \mathbf{F} \subset \mathbf{E}^*$. Since S is a homeomorphism of $\mathbf{E}$ onto $\mathbf{E}^*$ and $\mathbf{F}$ is dense in $\mathbf{E}^*$, the set $D = \{e \in \mathbf{E} \mid Se \in \mathbf{F}\}$ is a dense subspace of $\mathbf{E}$, hence of $\mathbf{F}$. Let $\widehat{S}$ be the restriction of S to D, considered as an operator in $\mathbf{F}$ with domain D. Then $\widehat{S}$ is symmetric, densely defined and $\widehat{S}D = \mathbf{F}$. Hence $\widehat{S}$ *is a self-adjoint strictly positive operator in $\mathbf{F}$ ($\widehat{S} \geq$ const. > 0 because $\widehat{S}$ is positive and surjective).* We shall say that $\widehat{S}$ *is the self-adjoint operator associated to Q in $\mathbf{F}$* (the preceding construction is another version of Friedrichs' theorem). *The domain of the square root $\widehat{S}^{1/2}$ is equal to $\mathbf{E}$ and $Q(e) = ||\widehat{S}^{1/2}e||_{\mathbf{F}}^2 \; \forall e \in \mathbf{E}$.* Indeed, D is dense in both B-spaces $\mathbf{E}$ and $D(\widehat{S}^{1/2})$ and for $e \in D$ we have $Q(e) = \langle e, Se \rangle = \langle e, \widehat{S}e \rangle_{\mathbf{F}} = ||\widehat{S}^{1/2}e||_{\mathbf{F}}^2$; since Q is continuous on $\mathbf{E}$ and the last term above is continuous on $D(\widehat{S}^{1/2})$), the assertion follows.

If Q is a positive quadratic form on a Hilbert space $\mathbf{F}$ and if Q is densely defined, closed and coercive, then one may apply the preceding result with $\mathbf{E}$ equal to the domain of Q: one gets a self-adjoint strictly positive operator H in $\mathbf{F}$

such that the domain of Q is equal to the domain of $H^{1/2}$ and $Q(f) = ||H^{1/2}f||_{\mathbf{F}}^2$ on their common domain. Reciprocally, if such an operator H is given and one defines Q by the preceding formula, then one gets a positive, closed, densely defined, coercive quadratic form on $\mathbf{F}$.

If H is an arbitrary self-adjoint operator in $\mathbf{F}$, then one can still associate to it two dense B-subspaces of $\mathbf{F}$, namely its *domain* $D(H) = D(|H|)$ and its *form domain*, which is the domain $D(|H|^{1/2})$ of the square root of the modulus of H. It is easily seen that the sesquilinear form $\langle e, Hf \rangle_{\mathbf{F}}$, $(e, f \in D(H))$ has a (unique) extension to a continuous sesquilinear form on the form domain of H; this extension is denoted again by $\langle e, Hf \rangle_{\mathbf{F}}$, so now e and f are allowed to be in $D(|H|^{1/2})$. One has continuous embeddings

$$D(H) \subset D(|H|^{1/2}) \subset \mathbf{F} = \mathbf{F}^* \subset D(|H|^{1/2})^* \subset D(H)^*,$$

the continuous symmetric operator $H : D(H) \to \mathbf{F}$ extends by duality to a continuous symmetric map $H : \mathbf{F} \to D(H)^*$, and by interpolation we get the continuous symmetric map $H : D(|H|^{1/2}) \to D(|H|^{1/2})^*$ whose quadratic form is just the form $\langle e, Hf \rangle_{\mathbf{F}}$ defined above.

If the operator H is bounded from below, i.e. if there is $a \in \mathbb{R}$ such that $H \geq a$, then the preceding situation is immediately reduced to the first one (the case of strictly positive H) by considering for example $H - a + 1$ in place of H. On the other hand, if H is not semibounded, then the consideration of the form domain of H does not seem very useful, since $|H|$ could be a rather inaccessible operator. For this reason we shall prefer in later chapters to work with (topological) Friedrichs couples and with self-adjoint operators constructed by the following natural extension of the preceding techniques. In particular, quadratic forms rarely appear explicitly in our arguments.

Let $(\mathbf{E}, \mathbf{F})$ be a topological Friedrichs couple with $\mathbf{F}^*$ and $\mathbf{F}$ identified, so that $\mathbf{E} \subset \mathbf{F} \subset \mathbf{E}^*$. To each operator $S \in B(\mathbf{E}, \mathbf{E}^*)$ we associate an operator $\widehat{S}$ in $\mathbf{F}$ by the following procedure: $D(\widehat{S}) = \{e \in \mathbf{E} \mid Se \in \mathbf{F}\}$ and $\widehat{S} = S|_{D(\widehat{S})}$ considered as an operator with values in $\mathbf{F}$. Of course, one could have $D(\widehat{S}) = \{0\}$. If S is symmetric, then $\widehat{S}$ is symmetric (as an operator in $\mathbf{F}$).

LEMMA 2.8.9. *Let $S \in B(\mathbf{E}, \mathbf{E}^*)$ be symmetric. If $S - z$ is a bijective map $\mathbf{E} \to \mathbf{E}^*$ for some complex number z, then this property holds for all $z \in \mathbb{C} \setminus \mathbb{R}$. In this case $\widehat{S}$ is a self-adjoint operator in $\mathbf{F}$ and its domain is a dense subspace of $\mathbf{E}$.*

PROOF. (1) Since $(S - z)^* = S - \bar{z}$, it is sufficient to show that the bijectivity of $S - z_0 : \mathbf{E} \to \mathbf{E}^*$ for some $z_0 \in \mathbb{C}$ with $\Im z_0 \geq 0$ implies the bijectivity of $S - z :$ $\mathbf{E} \to \mathbf{E}^*$ for all z with $\Im z > 0$. We have $(S - z_0)^{-1}(S - z) = I + (z - z_0)(S - z_0)^{-1}$ as elements of $B(\mathbf{E})$. If $|z - z_0|$ is small, then the r.h.s. is invertible in $B(\mathbf{E})$, hence $S - z$ is a homeomorphism of $\mathbf{E}$ onto $\mathbf{E}^*$. The proof of the first assertion of the lemma can be completed by a standard analytic continuation argument.

(ii) Let $z \in \mathbb{C} \setminus \mathbb{R}$ and $T = S - z$. Then $D(\widehat{T}) = D(\widehat{S})$ and $\widehat{T} = \widehat{S} - z$. Since T is a homeomorphism of $\mathbf{E}$ onto $\mathbf{E}^*$ and $\mathbf{F}$ is dense in $\mathbf{E}^*$, it follows that $D(\widehat{S})$ is a dense subspace of $\mathbf{E}$, hence of $\mathbf{F}$. So $\widehat{S}$ is a symmetric densely defined operator in $\mathbf{F}$ and $(\widehat{S} - z)D(\widehat{S}) = \mathbf{F}$ for all non-real z. This is equivalent to the self-adjointness of $\widehat{S}$. $\square$

If the operator S is bounded from below, i.e. $\langle e, Se \rangle \geq a||e||_{\mathbf{F}}^2$ for some $a \in \mathbb{R}$ and all $e \in \mathbf{E}$, then $\widehat{S} \geq a$, and from the results presented before it follows that the form domain of $\widehat{S}$ is equal to $\mathbf{E}$ (we proved this for $a > 0$, and one may easily reduce the general case to this situation). But *if S is not a semibounded operator, then this is not true in general*. The next example clarifies this phenomenon.

Let $\mathbf{G}$ be a Hilbert space and N an unbounded self-adjoint operator in $\mathbf{G}$ with $N \geq 1$. We identify $\mathbf{G}^* = \mathbf{G}$ and denote $\mathbf{G}^s = D(N^s)$ with norm $||g||_s = ||N^s g||$ for $s \in \mathbb{R}$ (see §2.8.1). Let $\mathbf{F} = \mathbf{G} \oplus \mathbf{G}$ and for $0 \leq s \leq 1$ let $\mathbf{E}_s = \mathbf{G}^s \oplus \mathbf{G}^{1-s}$. Then $\mathbf{E}_s^* = \mathbf{G}^{-s} \oplus \mathbf{G}^{s-1}$, so $\mathbf{E}_s \subset \mathbf{F} \subset \mathbf{E}_s^*$. Define $S : \mathbf{E}_s \to \mathbf{E}_s^*$ by $S(g \oplus h) = Nh \oplus Ng$. Then S is an isomorphism and $D(\widehat{S}) = \mathbf{G}^1 \oplus \mathbf{G}^1$. Clearly $|\widehat{S}| = (\widehat{S}^2)^{1/2}$ is the operator $|\widehat{S}|(g \oplus h) = Ng \oplus Nh$, so its form domain is $D(|\widehat{S}|^{1/2}) = \mathbf{G}^{1/2} \oplus \mathbf{G}^{1/2} = \mathbf{E}_{1/2} \neq \mathbf{E}_s$ if $s \neq 1/2$.

One may reformulate the preceding example in more hilbertian terms. Let H be a self-adjoint operator in a Hilbert space $\mathbf{F}$ such that zero belongs to its resolvent set. If $\mathbf{E}$ is the form domain of H, then $(\mathbf{E}, \mathbf{F})$ is a topological Friedrichs couple and H extends to a symmetric isomorphism of $\mathbf{E}$ onto $\mathbf{E}^*$. One may ask whether there are other dense H-subspaces $\mathbf{E}$ of $\mathbf{F}$ having this property. If H is semibounded, the answer is no. But if H is not semibounded, then there are many such $\mathbf{E}$ (take $H = \widehat{S}$, whose spectrum is $\sigma(N) \cup \sigma(-N)$, and $\mathbf{E} = \mathbf{E}_s$, $0 \leq s \leq 1$; $\sigma(T)$ denotes the spectrum of the operator T).

CHAPTER 3

C_0-Groups and Functional Calculi

This chapter deals with strongly continuous n-parameter groups W of bounded operators in a Banach space $\mathbf{F}$. The principal topics are certain scales of Banach spaces and the functional calculus that can be associated to such groups.

The first two sections are of an introductory nature and contain mainly definitions and straightforward consequences of them. The functional calculus associated to a C_0-group will appear frequently in later chapters. Furthermore, in contrast to other texts on C_0-groups, it also plays an important technical role in the presentation of the theory of such groups given here. So we begin this chapter by introducing and studying some algebras of functions on $\mathbb{R}^n$ that are suitable for defining the functional calculus of a C_0-group having some definite growth behaviour at infinity (Section 3.1). C_0-groups, their generators and the most elementary functional calculus associated to them are discussed in Section 3.2. The analysis of the functional calculus will be taken up again in Sections 3.6 and 3.7 where more refined tools will allow us to obtain deeper results.

The n-component generator $A = (A_1, \ldots, A_n)$, formally defined by the condition that $W(x) = \exp[i(A_1 x_1 + \cdots + A_n x_n)]$, $x \in \mathbb{R}^n$, allows one to define in a natural way a scale $\{\mathbf{F}_k\}_{k \in \mathbb{Z}}$ of Banach spaces. In the case $n = 1$, for example, one has $\mathbf{F}_k = D(A^k)$ if $k \geq 0$ (equipped with the graph norm), while for $k < 0$ the space $\mathbf{F}_k$ is the completion of $\mathbf{F}$ with respect to the norm $\|(A - z)^{-k} f\|_{\mathbf{F}}$ (z is a fixed complex number with sufficiently large imaginary part). If $n > 1$, the definition of $\mathbf{F}_k$ is somewhat more involved; in any case the spaces $\mathbf{F}_k$ are the natural abstract analogue of the spaces introduced by S.L. Sobolev for the case of the translation group in $L^p(\mathbb{R}^n)$. For this reason the scale $\{\mathbf{F}_k\}_{k \in \mathbb{Z}}$ will be called the Sobolev scale. In Section 3.3 we give simple properties of this scale; for example we show that, if $\mathbf{F}$ is reflexive, then $(\mathbf{F}_k)^* = (\mathbf{F}^*)_{-k}$, where the adjoint space $\mathbf{F}^*$ is equipped with the C_0-group adjoint to W. We also specify an alternative definition of $\mathbf{F}_k$ for $k \geq 0$ in terms of moduli of continuity of the functions $x \mapsto W(x)f$, $f \in \mathbf{F}$; this is interesting in view of the definition of the Besov scale given in Section 3.4. If φ is a function on $\mathbb{R}^n$ that is admissible for the functional calculus defined by W, the operator associated to φ is naturally interpretable as a function of the generator A of the group; so we shall denote this operator by $\varphi(A)$. The action of $\varphi(A)$ in the scale $\{\mathbf{F}_k\}$ is studied in considerable detail in

W. O. Amrein et al., *C0-Groups, Commutator Methods and Spectral Theory of N-Body Hamiltonians*, Modern Birkhäuser Classics, DOI: 10.1007/978-3-0348-0733-3_3, © Springer Basel 1996

Section 3.3.

We point out an important problem on which we have only partial results. The group W induces a C_0-group in each of the spaces $\mathbf{F}_k$. So, as above, one may associate to each $\mathbf{F}_k$ a Sobolev scale $\{(\mathbf{F}_k)_m\}_{m \in \mathbb{Z}}$. It is easy to see that $(\mathbf{F}_k)_m = \mathbf{F}_{k+m}$ if k and m have the same sign. We prove in Section 3.3 that the preceding equality holds for all k, $m \in \mathbb{Z}$ if $n = 1$. In Section 3.7 we show that this stays true for any n if $\mathbf{F}$ is a Hilbert space. In the remaining cases we do not know whether the above equality holds or not. If W is the translation group in $\mathbf{F} = L^p(\mathbb{R}^n)$, with $1 < p < \infty$, then the answer is positive, as shown by G. Bourdaud in [Bd] (his proof involves the theorem of Marcinkiewicz on Fourier multipliers which holds only for $p \in (1, +\infty)$, even if $n = 1$; however, if $n = 1$, our results show that the equality under discussion is true also in $L^1(\mathbb{R}^n)$ and in $L^\infty(\mathbb{R}^n)$).

Section 3.4 is devoted to the study of a second scale of spaces $\{\mathbf{F}_{s,p} \mid s \in \mathbb{R}, p \in [1, +\infty]\}$ associated to W, called the Besov scale (indeed, if W is the translation group in $\mathbf{F} = L^r(\mathbb{R}^n)$, these spaces coincide with those defined by O.V. Besov). We define the spaces $\mathbf{F}_{s,p}$ by real interpolation, starting from the Sobolev scale. The main technical point is that, although the Sobolev scale does not behave well under interpolation [1], we are able to show that the following weaker property holds: if $m_1 < m < m_2$ are integers and $\theta = (m_2 - m)(m_2 - m_1)^{-1}$, then

$$(3.0.1) \qquad (\mathbf{F}_{m_2}, \mathbf{F}_{m_1})_{\theta,1} \subset \mathbf{F}_m \subset (\mathbf{F}_{m_2}, \mathbf{F}_{m_1})_{\theta,\infty}.$$

This shows that the Besov scale has the extremely useful reiteration property $(\mathbf{F}_{t,q}, \mathbf{F}_{s,p})_{\theta,r} = \mathbf{F}_{(1-\theta)t+\theta s,r}$ for all $s < t$, $\theta \in (0,1)$ and $p,q,r \in [1, +\infty]$. Also $\mathbf{F}_{k,1} \subset \mathbf{F}_k \subset \mathbf{F}_{k,\infty}$ for any $k \in \mathbb{Z}$. If $s > 0$, then we give a more explicit description of $\mathbf{F}_{s,p}$ in terms of moduli of continuity (which is a Besov-type description).

The results given for $s > 0$ are due to H. Triebel (if $n \geq 2$; see [Tr] and references therein). However, the case $s \leq 0$ seems to be treated here for the first time, and the extension of the theory of Triebel to these values of s is not straightforward for the following reason. At first sight one could try to define $\mathbf{F}_{s,p} = (\mathbf{F}_k)_{s-k,p}$ if $s \leq 0$, where k is some integer less than s (taking the results of Triebel for granted). This definition should be independent of k (otherwise one would have a different scale for each k, which would complicate the picture very much). For example the space $(\mathbf{F}_{-1})_{2,p}$ should be the same as $\mathbf{F}_{1,p}$. Now $(\mathbf{F}_{-1})_{2,p} = ((\mathbf{F}_{-1})_3, \mathbf{F}_{-1})_{2/3,p}$ and $\mathbf{F}_{1,p} = (\mathbf{F}_2, \mathbf{F}_0)_{1/2,p}$, so one should have $((\mathbf{F}_{-1})_3, \mathbf{F}_{-1})_{2/3,p} = (\mathbf{F}_2, \mathbf{F}_0)_{1/2,p}$. Because of (3.0.1), the preceding equality is true if $(\mathbf{F}_{-1})_3 = \mathbf{F}_2$; however, as said before, the equality of $(\mathbf{F}_{-1})_3$ and $\mathbf{F}_2$ is not known in general. Of course a weaker relation would be sufficient, for example $\mathbf{F}_{2,1} \subset (\mathbf{F}_{-1})_3 \subset \mathbf{F}_{2,\infty}$; in Section 3.6 we shall prove this for polynomially bounded groups (i.e. $\|W(x)\|_{B(\mathbf{F})} \leq c\langle x \rangle^r$) in reflexive spaces; so in such a case the definition $\mathbf{F}_{s,p} = (\mathbf{F}_k)_{s-k,p}$ leads to the same spaces as our definition given in Section 3.4. At any rate, the reduction to the theory of Triebel requires a considerable amount of technique (except in the case $n = 1$ which is easy).

---

[1]For example, if W is the translation group in $\mathbf{F} = C_\infty(\mathbb{R})$, then $\mathbf{F}_k = C_\infty^k(\mathbb{R})$ if $k \in \mathbb{N}$, and it is known that $C_\infty^1(\mathbb{R})$ is not an interpolation space between $C_\infty(\mathbb{R})$ and $C_\infty^2(\mathbb{R})$ (see the end of Section 2.1 for a reference).

When compared with the Sobolev scale, the Besov scale is very convenient because it is stable under real interpolation. This is the main reason why one may get better bounds for the operators $\varphi(A)$ when considered in $\mathbf{F}_{s,p}$ rather than in $\mathbf{F}$. In order to prove such bounds, we develop in Section 3.5 a version of the Littlewood-Paley dyadic decomposition method that holds in a framework covering the functional calculus for polynomially bounded multi-parameter C_0-groups in an arbitrary Banach space. The theory of Section 3.5 was especially suggested to us by the approach of H.S. Shapiro to some questions of approximation theory [Sh1], [Sh2], [Sh3]. An abstract version of Calderón's formula (which was mentioned after Theorem 1.3.8; see also §34 in [C2]) plays an important role in our arguments, although it does not appear explicitly (see, however, the proof of Lemma 3.5.8). As references for the classical Littlewood-Paley method we recommend [Bd], [FJW], [P1] and [St1]. In Section 3.6 we show that $\varphi(A)$ is a bounded operator in each $\mathbf{F}_{s,p}$ if φ is a symbol of class $S^0(\mathbb{R}^n)$ (see Theorem 3.6.9 for a precise formulation). This result has several important consequences, for example it implies that $\mathbf{F}_{k+m,1} \subset (\mathbf{F}_k)_m \subset \mathbf{F}^0_{k+m,\infty}$ for any $k, m \in \mathbb{Z}$ and any reflexive Banach space $\mathbf{F}$. To illustrate the advantages of the Besov spaces $\mathbf{F}_{s,p}$ over the Sobolev spaces $\mathbf{F}_k$, we mention the following simple example. Let W be the translation group in one of the spaces $L^1(\mathbb{R})$, $L^\infty(\mathbb{R})$ or $C_\infty(\mathbb{R})$; then $\varphi(A)$ is a continuous operator in the respective space $\mathbf{F}$ if and only if φ is the Fourier transform of an integrable measure, while $\varphi(A)$ is bounded in each $\mathbf{F}_{s,p}$ for an arbitrary symbol of class $S^0(\mathbb{R})$ (the Fourier transform of a symbol is not a measure in general).

In the final Section 3.7 we consider the case where $\mathbf{F}$ is a Hilbert space. Then one can improve several of the results obtained before; for example a theorem of N. Mandache states that $\varphi(A)$ is a bounded operator in $\mathbf{F}$ if $\varphi \in BC^\infty(\mathbb{R})$ and W is a polynomially bounded C_0-group.

3.1. Submultiplicative Functions and Algebras Associated to them

Let X be an euclidean space. In this section we shall consider algebras of bounded continuous functions on X that are suitable for the functional calculus for generators of representations in Banach spaces of the additive group associated to X. We set $n = \dim X$.

3.1.1. We begin with the following classical result:

PROPOSITION 3.1.1. *Let* $h : X \to [0, \infty)$ *be Borel and submultiplicative, i.e.* $h(x+y) \leq h(x)h(y)$ *for all* $x, y \in X$. *Then there are constants* $M \geq 0$ *and* $\omega \geq 0$ *such that*

$$(3.1.1) \qquad\qquad h(x) \leq M e^{\omega|x|} \qquad \forall x \in X.$$

PROOF. If Ω is a Borel subset of X, we denote by $|\Omega|$ its Fourier measure. (i) For $k \in \mathbb{N}$, let

$$\Omega_k = \{x \in X \mid |x| \leq k \text{ and } h(x) + h(-x) \leq \sqrt{k}\}.$$

Ω_k is a Borel set and $X = \bigcup_{k \in \mathbb{N}} \Omega_k$, so we may choose an integer k such that $|\Omega_k| > 0$. We denote by Ω this Borel set Ω_k and by χ_Ω its characteristic function. χ_Ω is a bounded Borel function of compact support, so that $\psi(v) := \langle \chi_\Omega, T(v)\chi_\Omega \rangle_{L^2(X)}$ is a continuous function of $v \in X$, with $\psi(0) > 0$ ($T(v)$ is defined in Section 1.2). Hence there is a neighbourhood V of $v = 0$ in X on which $\psi(v) > 0$. Now:

$$\psi(v) = \int_X \chi_\Omega(x)\chi_\Omega(x-v)\underline{d}x = \int_\Omega \chi_\Omega(x-v)\underline{d}x = |\Omega \cap (v+\Omega)|.$$

So, if $v \in V$, there is $x \in \Omega \cap (v + \Omega)$, i.e. $x \in \Omega$ and $x - v \equiv y \in \Omega$. Since $\Omega = -\Omega$, we also have $-y \in \Omega$, so that $v = x - y \in \Omega + \Omega$. This shows that $\Omega + \Omega$ contains a neighbourhood of 0.

(ii) The hypothesis of submultiplicativity of h implies that

$$h(x+y) \le h(x)h(y) \le k \text{ if } x, y \in \Omega.$$

Thus, since $\Omega + \Omega$ contains a neighbourhood of 0, there is $\varepsilon > 0$ such that $h(v) \le k$ for all $v \in X$ with $|v| \le \varepsilon$. If $x \in X$, write $x_0 = \varepsilon x |x|^{-1}$ (so that $|x_0| = \varepsilon$) and let ℓ be the largest integer less than or equal to $|x|\varepsilon^{-1}$. Then $x = \ell x_0 + y$, with $0 < |y| < \varepsilon$, hence $h(x) \le h(x_0)^\ell h(y) \le k^{\ell+1} \le kk^{|x|/\varepsilon}$. So (3.1.1) holds with $M = k$ and $\omega = \varepsilon^{-1} \ln k$. $\square$

We recall from Section 1.2 that a measure μ on X is said to be integrable if $|\mu|(X) < \infty$. The space of integrable measures is a vector subspace of $\mathscr{S}^*(X)$ and is an algebra for the convolution product (one has $|\mu * \nu| \le |\mu| * |\nu|$). We shall say that μ is a *rapidly decreasing measure* if $\int_X \langle x \rangle^r |\mu|(dx)$ is finite for each $r \in \mathbb{R}$. On various occasions we shall symbolically write $f(x)\underline{d}x$ for a measure $\mu(dx)$, and then the measure $|\mu|(dx)$ will be denoted by $|f(x)|\underline{d}x$. For example the Dirac measure at a point x_0 will be denoted δ_{x_0} or $\delta_{x_0}(x)\underline{d}x$. If μ is absolutely continuous (with respect to Fourier measure), then f is just the Radon-Nikodym derivative of μ with respect to $\underline{d}x$.

The following result will be needed below:

LEMMA 3.1.2. *Let $w : X \to [0, \infty)$ be a lower semicontinuous function.*

(a) *If ν is a measure on X, then*

$$(3.1.2) \qquad \int_X w(x)|\nu|(dx) = \sup\{|\langle f, \nu \rangle| \mid f \in \mathscr{S}(X), |f| \le w\}.$$

Moreover, if ν is an arbitrary tempered distribution on X and if the l.h.s. of (3.1.2) is defined by its r.h.s., then $\int_X w(x)|\nu|(dx) < \infty$ only if ν is a measure.

(b) *Let $\{\varphi_j\}_{j \in \mathbb{N}}$ be a sequence of functions in $L^\infty(X)$ converging pointwise to a function $\varphi : X \to \mathbb{C}$ and such that $\sup_{j \in \mathbb{N}} \|\varphi_j\|_{L^\infty(X)} < \infty$. Then*

$$(3.1.3) \qquad \int_X w(x)|\widehat{\varphi}|(dx) \le \liminf_{j \to \infty} \int_X w(x)|\widehat{\varphi_j}|(dx).$$

PROOF. (a) We first prove (3.1.2) for functions $w \in C_0(X)$. For this we use the Riesz representation theorem (apply Theorem 6.19 of [Ru] to the measure $\mu = w\nu$ to obtain the first identity below; the second identity holds by the assumed continuity of w):

$$(3.1.4) \qquad \int_X w(x)|\nu|(dx) = \sup_{f \in C_0(X), |f| < 1} |\langle f, w\nu \rangle| = \sup_{g \in C_0(X), |g| < w} |\langle g, \nu \rangle|.$$

Then (3.1.2) follows from (3.1.4) by observing that a function g in $C_0(X)$ may be expressed as a limit in $L^\infty(X)$ of a sequence $\{g_k\}$ belonging to $C_0^\infty(X)$. The case where w is lower semicontinuous can be reduced to the preceding situation because, by the lower semicontinuity of w, there is a sequence $\{w_k\}_{k \in \mathbb{N}}$ of functions in $C_0(X)$ such that $0 \leq w_k \leq w$ and $w_k(x) \to w(x)$ for each $x \in X$ as $k \to \infty$ (see Theorem 12.7.8 in [D]). The proof of the last assertion of (a) is an easy exercise.

(b) Let $N = \liminf_{j \to \infty} \int_X w(x)|\widehat{\varphi_j}|(dx)$ and choose a subsequence $\{\varphi_{j_k}\}$ of $\{\varphi_j\}$ such that $\lim_{k \to \infty} \int_X w(x)|\widehat{\varphi_{j_k}}|(dx) = N$. If $f \in \mathscr{S}(X)$, $|f| \leq w$, then one has by the Lebesgue dominated convergence theorem and (3.1.2):

$$|\langle f, \widehat{\varphi} \rangle| = |\langle \widehat{f}, \varphi \rangle| = \lim_{k \to \infty} |\langle \widehat{f}, \varphi_{j_k} \rangle| = \lim_{k \to \infty} |\langle f, \widehat{\varphi_{j_k}} \rangle|$$

$$\leq \lim_{k \to \infty} \int_X w(x)|\widehat{\varphi_{j_k}}|(dx) = N.$$

By using (3.1.2) again, one obtains the validity of (3.1.3). $\square$

A function $w : X \to \mathbb{R}$ will be called *submultiplicative* if $w(x+y) \leq w(x)w(y)$ for all $x, y \in X$. Let $w : X \to \mathbb{R}$ be submultiplicative, Borel and such that $w(x) \geq 1$ for all $x \in X$ [2]. We say that a measure μ on X is w-*integrable* if $\int_X w(x)|\mu|(dx) < \infty$. The space of w-integrable measures is an algebra for the convolution product because:

$$(3.1.5) \qquad \int_X w(x)|\mu * \nu|(dx) \leq \int_X w(x)(|\mu| * |\nu|)(dx)$$

$$= \iint_{X \times X} w(x+y)|\mu|(dx)|\nu|(dy)$$

$$\leq \iint_{X \times X} w(x)w(y)|\mu|(dx)|\nu|(dy).$$

We can now introduce the algebras of functions that will be useful in our later developments:

DEFINITION 3.1.3. Let $w : X \to \mathbb{R}$ be a lower semicontinuous submultiplicative function such that $w(x) \geq 1$ for all $x \in X$. Then $\mathscr{M}_w(X)$ is the set of all

---

[2] The examples that will be of interest for us are $w(x) = \exp(\omega|x|)$ and $w(x) = 2^{r/2}\langle x \rangle^r$, where ω, r are positive numbers; observe that in the second example the submultiplicative property is satisfied as a consequence of the inequality $\langle x + y \rangle \leq \sqrt{2}\langle x \rangle \langle y \rangle$.

$\varphi \in \mathscr{S}^*(X)$ such that the Fourier transform $\widehat{\varphi}$ is a w-integrable measure. For $\varphi \in \mathscr{M}_w(X)$ we set

$$(3.1.6) \qquad ||\varphi||_{\mathscr{M}_w} = \int_X w(x) |\widehat{\varphi}|(dx).$$

Some simple properties of the spaces $\mathscr{M}_w$ are described in the next proposition. Before stating it, let us mention that we call *Banach algebra* a complex algebra which is also a Banach space for a norm satisfying $||\varphi\psi|| \leq ||\varphi|| \cdot ||\psi||$. If the algebra has a unit element 1, then we call it a *unital Banach algebra*, although $||1|| \neq 1$ in general (however it is easy to see that there is an equivalent norm $||| \cdot |||$ on the algebra which satisfies both $|||\varphi\psi||| \leq |||\varphi||| \cdot |||\psi|||$ and $|||1||| = 1$, for example $|||\varphi||| = \sup\{||\varphi\xi|| \mid ||\xi|| \leq 1\}$, see page 173 of [KR]). A *unital homomorphism* between two unital Banach algebras is a linear, multiplicative mapping that sends 1 onto 1. A *Banach $*$-algebra* is a Banach algebra equipped with an involution $\varphi \mapsto \varphi^*$ such that $||\varphi^*|| = ||\varphi||$ for each element φ.

We recall from Section 1.1 that $BC(X)$ is a commutative unital Banach $*$-algebra for the usual operations of addition, multiplication and conjugation of functions and with norm $||\varphi||_{BC} = \sup_{x \in X} |\varphi(x)|$. If $k \in \mathbb{N}$, then $BC^k(X)$ is a subalgebra of $BC(X)$ which is invariant under conjugation and contains the constants; when equipped with the norm $|| \cdot ||_{BC^k}$ introduced in (1.1.5), it is a unital Banach $*$-algebra.

The algebra $\mathscr{M}_w$ has similar properties:

PROPOSITION 3.1.4. *Let w and $\mathscr{M}_w(X)$ be as in Definition 3.1.3. Then:*

(a) $\mathscr{M}_w(X)$ *is a subalgebra of $BC(X)$ that contains the constant functions. Equipped with the norm (3.1.6), it is a unital Banach algebra continuously embedded in $BC(X)$; in fact*

$$(3.1.7) \qquad ||\varphi||_{BC} \leq ||\varphi||_{\mathscr{M}_w} \qquad \forall \varphi \in \mathscr{M}_w.$$

(b) *The norm $|| \cdot ||_{\mathscr{M}_w}$ on $\mathscr{M}_w$ may be expressed as follows:*

$$(3.1.8) \qquad ||\varphi||_{\mathscr{M}_w} = \sup\{|\langle f, \varphi \rangle| \mid f \in \mathscr{S}(X), |\widehat{f}| \leq w\}.$$

(c) $\mathscr{M}_w(X)$ *is invariant under conjugation if and only if $w(-x) \leq cw(x)$ for some constant c and all $x \in X$. It is a Banach $*$-algebra if and only if $w(-x) = w(x)$ for all $x \in X$.*

(d) *If w_1, w_2 are two functions satisfying the assumptions of Definition 3.1.3 and if $w = \sup(w_1, w_2)$, then w is submultiplicative and*

$$\mathscr{M}_w(X) = \mathscr{M}_{w_1}(X) \cap \mathscr{M}_{w_2}(X).$$

PROOF. (a) Each w-integrable measure is integrable (because we assumed $w(x) \geq 1$), hence $\mathscr{M}_w(X) \subset BC(X)$ and $||\varphi||_{BC} \leq \int_X |\widehat{\varphi}|(dx) \leq ||\varphi||_{\mathscr{M}_w}$. The Fourier transform of a convolution is the product of the Fourier transforms of the factors; thus the inequality $||\varphi\psi||_{\mathscr{M}_w} \leq ||\varphi||_{\mathscr{M}_w} ||\psi||_{\mathscr{M}_w}$ follows from (3.1.5). We have $||1||_{\mathscr{M}_w} = w(0) < \infty$ (but different from 1 in general) because the Fourier transform of $\varphi(x) \equiv 1$ is the Dirac measure at the origin. For a proof of the completeness of $\mathscr{M}_w(X)$ for the norm (3.1.6), we refer to [HP], p. 142.

(b) This follows from Lemma 3.1.2 (a) by observing that $\langle f, \varphi \rangle = \langle \widehat{f}, \widehat{\varphi} \rangle$.

(c) Let J be the operator of conjugation of functions on X: $J\varphi = \overline{\varphi}$. If $\mathcal{M}_w(X)$ is invariant under conjugation, then J is closed as an operator in $\mathcal{M}_w(X)$ (if $\{\varphi_n\} \in \mathcal{M}_w(X)$ is such that $\{\varphi_n\}$ and $\{\overline{\varphi_n}\}$ are Cauchy in $\mathcal{M}_w(X)$, then each of these sequences has a limit in $\mathcal{M}_w(X)$, say $\varphi_n \to \varphi$ and $\overline{\varphi_n} \to \psi$; but $\varphi_n \to \varphi$ in $BC(X)$ by (3.1.7), hence $\overline{\varphi_n} \to \overline{\varphi}$ in $BC(X)$, so that $\psi = \overline{\varphi}$). Thus, by the closed graph theorem, J is a bounded operator in $\mathcal{M}_w(X)$, i.e. there is a constant $c < \infty$ such that $||\overline{\varphi}||_{\mathcal{M}_w} \leq c||\varphi||_{\mathcal{M}_w}$ for all $\varphi \in \mathcal{M}_w(X)$. Since $(\mathcal{F}\overline{\varphi})(x) = (\overline{\mathcal{F}\varphi})(-x)$, we get that

$$(3.1.9) \qquad \int_X w(-x)\mu(dx) \leq c \int_X w(x)\mu(dx)$$

for all w-integrable positive measures μ. If we take for μ the Dirac measure δ_{x_0} at x_0, we find that $w(-x_0) \leq cw(x_0)$ for all $x_0 \in X$.

Reciprocally, if $w(-x) \leq cw(x)$ for all $x \in X$, then $||\overline{\varphi}||_{\mathcal{M}_w} \leq c||\varphi||_{\mathcal{M}_w}$ for all $\varphi \in \mathcal{M}_w$, so that J maps $\mathcal{M}_w(X)$ into itself.

The last statement in (c) is simple to check, and so are the results of (d). $\qquad \square$

REMARK 3.1.5. The simplest situation which occurs in applications and in which $\mathcal{M}_w$ is not stable under conjugation is $w(x) = \sup(1, e^x), X = \mathbb{R}$.

The algebra $\mathcal{M}_w(X)$ should be thought of as the algebra of multipliers for all continuous representations W of the additive group X in Banach spaces such that $||W(x)|| \leq w(x)$. More precisely, if A is the generator of W (see Sections 3.2 and 3.3 for the precise definition), then $\varphi(A) = \int_X W(x)\widehat{\varphi}(dx)$ is a bounded operator for all $\varphi \in \mathcal{M}_w$, and in general $\mathcal{M}_w(X)$ is the optimal class of functions for which this happens (e.g. $\mathcal{M}_w(X)$ with $w(x) \equiv 1$ is exactly the set of functions φ such that $\varphi(P)$ is a continuous operator in $L^1(X)$; here W is the translation group). Of course, under special assumptions, the algebra of functions φ for which $\varphi(A)$ is bounded is much larger (if W is a unitary representation in a Hilbert space one may take any $\varphi \in BC(X)$; see Section 3.7 for more general results).

3.1.2. For some questions that appear in the preceding context, the norm topology on $\mathcal{M}_w(X)$ is too strong. For example, if we denote by e_y the function in $\mathcal{M}_w(X)$ given as $e_y(x) = e^{i\langle x,y\rangle}$, then $\widehat{e_y} = \delta_y$ (the Dirac measure with support $\{y\}$), hence for $y_1 \neq y_2$:

$$(3.1.10) \qquad ||e_{y_1} - e_{y_2}||_{\mathcal{M}_w} = \int_X w(x)|\delta_{y_1}(x) - \delta_{y_2}(x)|dx$$
$$= \int_X w(x)[\delta_{y_1}(x) + \delta_{y_2}(x)]dx$$
$$= w(y_1) + w(y_2) \geq 2.$$

So, if $y_j \to y$ in X, then e_{y_j} does not converge to e_y in $\mathcal{M}_w(X)$ (if the sequence is not eventually constant). However, $\varphi(e_{y_j}) = W(y_j) \to W(y)$ strongly on the Banach space $\mathbf{F}$ on which W acts (if the representation is strongly continuous). It will be convenient for later investigations to have a notion of "feebly convergent sequence" in $\mathcal{M}_w(X)$ such that $e_{y_j} \to e_y$ feebly if $y_j \to y$ in X and such that $\varphi_j(A) \to \varphi(A)$ strongly in $\mathbf{F}$ if $\varphi_j \to \varphi$ feebly. There is a natural topology on $\mathscr{M}_w(X)$ whose convergent sequences have these properties (it is given by the seminorms $\varphi \mapsto \int_X \theta(x)\widehat{\varphi}(dx)$ with θ continuous and such that $\lim_{|x|\to\infty} w(x)^{-1}|\theta(x)| = 0$); however this topology is not as easy to handle as the notion introduced in the following definition:

DEFINITION 3.1.6. Let $\{\varphi_k\}_{k\in\mathbb{N}}$ and φ be elements of $\mathscr{M}_w(X)$. Then the sequence $\{\varphi_k\}$ is said to be *feebly convergent* in $\mathscr{M}_w(X)$ to φ if the following three conditions are satisfied:
(a) $\lim_{k\to\infty} \varphi_k(x) = \varphi(x)$ for each $x \in X$,
(b) $\sup_{k\in\mathbb{N}} \|\varphi_k\|_{\mathscr{M}_w} < \infty$,
(c) $\lim_{R\to\infty} \sup_{k\in\mathbb{N}} \int_{|x|>R} w(x)|\widehat{\varphi_k}|(dx) = 0$.

The third condition is the natural concept of tightness in our context (see Chapter 1.6 in [Bi]) and is in general easy to verify. For example, each of the following two conditions implies (c):
(c′) there is a $\psi \in \mathscr{M}_w(X)$ such that $|\widehat{\varphi_k}| \le |\widehat{\psi}|$ (as measures) for each $k \in \mathbb{N}$,
(c″) there is a positive locally bounded Borel function $\varrho : X \to \mathbb{R}$ such that $\lim_{|x|\to\infty} \varrho(x) = \infty$ and $\int_X \varrho(x)w(x)|\widehat{\varphi_k}|(dx) \le c < \infty$ for each $k \in \mathbb{N}$.
(In fact it is easily shown that (c) and (c″) are equivalent provided that (b) is satisfied).
Obviously $\|\varphi_k - \varphi\|_{\mathscr{M}_w} \to 0$ implies that $\varphi_k \to \varphi$ feebly in $\mathscr{M}_w(X)$. Also if $\varphi_k \to \varphi$ feebly in $\mathscr{M}_w(X)$, then $\|\varphi\|_{\mathscr{M}_w} \le \liminf_{k\to\infty} \|\varphi_k\|_{\mathscr{M}_w}$, by Lemma 3.1.2 (b).

EXAMPLE 3.1.7. If $\varphi \in \mathscr{M}_w(X)$ and $\eta \in \mathscr{S}(X)$ with $\eta(0) = 1$, then for each $\varepsilon > 0$ the function φ_ε defined as $\varphi_\varepsilon(x) = \eta(\varepsilon x)\varphi(x)$ is continuous and rapidly decreasing at infinity, and φ_ε converges pointwise to φ as $\varepsilon \to 0$. If $\varphi \notin C_\infty(X)$, then clearly φ_ε cannot converge to φ in the norm of $\mathscr{M}_w(X)$ as $\varepsilon \to 0$. However, under convenient additional assumptions on η, φ_ε converges feebly in $\mathscr{M}_w(X)$. To see this, we assume in addition to $\int_X \widehat{\eta}dx \equiv \eta(0) = 1$ that $\widehat{\eta} \in C_0^\infty(X)$, $\widehat{\eta} \ge 0$ and $\widehat{\eta}(-x) = \widehat{\eta}(x)$ for all $x \in X$. For $\varepsilon > 0$, we define $\widehat{\eta_\varepsilon}$ by $\widehat{\eta_\varepsilon}(x) = \varepsilon^{-n}\widehat{\eta}(\varepsilon^{-1}x)$ and observe that $\widehat{\varphi_\varepsilon} = \widehat{\eta_\varepsilon} * \widehat{\varphi}$ is a C^∞-function with $|\widehat{\varphi_\varepsilon}| \le \widehat{\eta_\varepsilon} * |\widehat{\varphi}|$. Let $N > 0$ be such that $\text{supp}\,\widehat{\eta} \subseteq \{x \in X \mid |x| \le N\}$. Then we have

$$(\widehat{\eta_\varepsilon} * w)(x) = \int_X \widehat{\eta}(y)w(x - \varepsilon y)\underline{dy} \le w(x) \cdot \sup_{|z|\le\varepsilon N} w(z).$$

Hence, since $\widehat{\eta}(-z) = \widehat{\eta}(z)$, we get that

$$(3.1.11) \quad \|\varphi_\varepsilon\|_{\mathscr{M}_w} = \int_X w(y)|\widehat{\varphi_\varepsilon}(y)|\underline{dy} \le \int_X w(y)\underline{dy}\int_X \widehat{\eta_\varepsilon}(y - x)|\widehat{\varphi}|(dx)$$

$$= \int_X (\widehat{\eta_\varepsilon} * w)(x)|\widehat{\varphi}|(dx) \le \sup_{|z|\le\varepsilon N} w(z)\|\varphi\|_{\mathscr{M}_w}.$$

In particular we have $\varphi_\varepsilon \in \mathscr{M}_w(X)$ and $\|\varphi_\varepsilon\|_{\mathscr{M}_w} \le c < \infty$ for $0 < \varepsilon \le 1$. To see that $\varphi_\varepsilon \to \varphi$ feebly in $\mathscr{M}_w(X)$ it remains to check that the condition (c) in Definition 3.1.6 is satisfied. The validity of (c) is a consequence of the following inequality which is easily obtained by an argument similar to the preceding one: if $0 < \varepsilon \leq 1$, then

$$(3.1.12) \qquad \int_{|y| \geq R} w(y) |\widehat{\varphi_\varepsilon}(y)| dy \leq \sup_{|z| \leq N} w(z) \int_{|x| \geq R - N} w(x) |\widehat{\varphi}|(dx).$$

Some less trivial facts on feeble convergence are described in the next proposition.

PROPOSITION 3.1.8. *Let w and $\mathscr{M}_w(X)$ be as in Definition 3.1.3.*

(a) *Let $\mathbf{F}$ be a Banach space and $\Psi : X \to \mathbf{F}$ a continuous function satisfying $\|\Psi(x)\|_{\mathbf{F}} \leq cw(x)$ for all $x \in X$. Then, if $\varphi_k \to \varphi$ feebly in $\mathscr{M}_w(X)$, the sequence $\{\int_X \Psi(x) \widehat{\varphi_k}(dx)\}_{k \in \mathbb{N}}$ converges to $\int_X \Psi(x) \widehat{\varphi}(dx)$ in $\mathbf{F}$.*

(b) *The vector space generated by the exponentials (i.e. by the set of functions of the form $\varphi(x) = \sum_k c_k e^{i \langle x, y_k \rangle}$, with $y_k \in X$) is feebly sequentially dense in $\mathscr{M}_w(X)$.*

PROOF. (a) We may assume without loss of generality that $\varphi = 0$. Moreover, by a simple approximation argument, we may assume that Ψ is of class C_0^∞. Then $\|\widehat{\Psi}(\cdot)\|_{\mathbf{F}}$ belongs to $L^1(X)$. Since $|\varphi_k(x)| \leq \|\varphi_k\|_{BC} \leq \|\varphi_k\|_{\mathscr{M}_w} \leq M < \infty$ for all $k \in \mathbb{N}$ and all $x \in X$, we get the claimed result by using the hypothesis (a) of Definition 3.1.6 and the Lebesgue dominated convergence theorem :

$$\left\| \int_X \Psi(x) \widehat{\varphi_k}(dx) \right\|_{\mathbf{F}} = \left\| \int_X \widehat{\Psi}(x) \varphi_k(x) dx \right\|_{\mathbf{F}} \to 0 \text{ as } k \to \infty.$$

(b) For $k = 1, 2, 3, \ldots$, consider a partition of $X \cong \mathbb{R}^n$ into disjoint cubes of length k^{-1} and centers at the points of the lattice $k^{-1} \mathbb{Z}^n$:

$$X = \bigcup_{m \in \mathbb{Z}^n} \Delta_{k,m}, \qquad \Delta_{k,m} \cap \Delta_{k,m'} = \varnothing \quad \text{if } m \neq m',$$

where $\Delta_{k,m}$ is a (half-open) cube with edges of length k^{-1} and center $a_{k,m} = k^{-1}m$ $(m \in \mathbb{Z}^n)$. Let $\varphi \in \mathscr{M}_w(X)$ and

$$\varphi_k(x) = \sum_{|a_{k,m}| \leq k} \widehat{\varphi}(\Delta_{k,m}) \exp[i \langle a_{k,m}, x \rangle].$$

It is rather straightforward to check that the sequence $\{\varphi_k\}$ is feebly convergent in $\mathscr{M}_w(X)$ to φ. $\square$

Let Ω be an open subset of $\mathbb{R}^m$ and $F : \Omega \to \mathscr{M}_w(X)$. We say that F is *feebly continuous* at $x \in \Omega$ if for each sequence of points $x_k \in \Omega$ converging to x one has $F(x_k) \to F(x)$ feebly in $\mathscr{M}_w(X)$. Then one may define in the usual way the feeble derivatives $(\partial_j F)(x)$ $(j = 1, \ldots, m)$ as limits of quotients (if they exist), the notion of feebly C^1-functions, then higher order feeble derivatives $\partial^\alpha F$ and the notion of *feebly C^k-functions* for $k \in \mathbb{N}$. The following facts are useful:

LEMMA 3.1.9. *Let $F : \Omega \to \mathscr{M}_w(X)$.*

(a) *If F is feebly continuous, then it is locally norm bounded.*

(b) *If F is feebly of class C^k $(k \geq 1)$, then it is of class C^{k-1} in norm.*

PROOF. (a) Let $x \in \Omega$ and let $\{x_k\}_{k \in \mathbb{N}}$ be any sequence in Ω converging to x. Then $\sup_{k \in \mathbb{N}} \|F(x_k)\|_{\mathscr{M}_w} < \infty$ since F is feebly continuous (cf. (b) of Definition 3.1.6). Now assume that F is not norm bounded near x. Then one can find a sequence $\{x_k\}_{k \in \mathbb{N}}$ in Ω with $x_k \to x$ and $\|F(x_k)\|_{\mathscr{M}_w} > k$, a contradiction.

(b) If F is feebly of class C^1 and $f \in \mathscr{S}(X)$, then the function $x \mapsto \langle f, F(x) \rangle$ is of class C^1, with $\partial_j \langle f, F(x) \rangle = \langle f, \partial_j F(x) \rangle$ (this can be checked by means of the Lebesgue dominated convergence theorem, by using the norm boundedness of a feebly convergent sequence and (3.1.7) to obtain a bound on the integrand). In particular we have:

$$\langle f, F(y) - F(x) \rangle = \sum_{j=1}^{m} (y_j - x_j) \int_0^1 \langle f, \partial_j F((1 - \varrho)x + \varrho y) \rangle d\varrho.$$

If $f \in \mathscr{S}(X)$ is such that $|\widehat{f}| \leq w$, we get by using (3.1.8) that

$$|\langle f, F(y) - F(x) \rangle| \leq \sum_{j=1}^{m} |y_j - x_j| \sup_{0 \leq \varrho \leq 1} \|\partial_j F((1 - \varrho)x + \varrho y)\|_{\mathscr{M}_w}$$

$$\leq c|y - x|,$$

where (by the result of (a)) c is some finite constant. By taking the supremum over all $f \in \mathscr{S}(X)$ with $|\widehat{f}| \leq w$, we obtain that $\|F(y) - F(x)\|_{\mathscr{M}_w} \leq c|y - x|$. This shows that a feebly C^1-function is Lipschitz in norm. As in Lemma 5.A.2 one can then show by recursion that a feebly C^k-function is of class C^{k-1} in norm. $\square$

As an example, consider the function $F : X \to \mathscr{M}_w(X)$ defined by $F(y) = e_y$ ($y \in X$). It is feebly continuous but not norm continuous. Formally its derivative with respect to y_j (in some orthonormal basis of X) is the function $x \mapsto ix_j e_y(x)$, which is unbounded and hence does not belong to $\mathscr{M}_w(X)$. However the following situation will be interesting:

PROPOSITION 3.1.10. *Let $\ell \in \mathbb{N}$ and let $\varphi \in \mathscr{M}_w(X)$ be such that $\varphi_{(\alpha)}(x) \equiv x^\alpha \varphi(x)$ belongs to $\mathscr{M}_w(X)$ for each multi-index α with $|\alpha| \leq \ell$ (i.e. the distributional derivatives $\widehat{\varphi}^{(\alpha)}$ of $\widehat{\varphi}$ for $|\alpha| \leq \ell$ are w-integrable measures). Then the function $F : X \to \mathscr{M}_w(X)$ defined as $F(y) = e_y \varphi$ ($y \in X$) is feebly of class C^ℓ, and $\partial^\alpha F(y) = i^{|\alpha|} e_y \varphi_{(\alpha)}$. In particular, if $\widehat{\varphi} \in C_0^\infty(X)$, then F is of class C^∞ in norm.*

PROOF. We give the proof for $\ell = 1$. The result for $\ell > 1$ follows recursively from that for $\ell = 1$. We denote by $\{v_k\}_{k=1}^n$ an orthonormal basis of X. By taking into account the identity $e_{y+z} = e_y e_z$, it is easily shown that it suffices to prove that F is feebly differentiable at zero.

(i) We first make the additional assumption that $\widehat{\varphi} \in C^1(X)$. By using the inequality $w(y + z) \leq w(y)w(z)$, one obtains that for $0 < |\tau| \leq 1$:

$$(3.1.13) \quad \|(e_{\tau v_k} - 1)\varphi\|_{\mathscr{M}_w} = \int_X w(x)|\widehat{\varphi}(x - \tau v_k) - \widehat{\varphi}(x)|dx$$

$$= \int_X w(x)dx \left| \int_0^\tau \partial_k \widehat{\varphi}(x - \lambda v_k)d\lambda \right| \leq \left| \int_0^\tau d\lambda \int_X w(y + \lambda v_k)|\partial_k \widehat{\varphi}(y)|dy \right|$$

$$\leq \sup_{|u| \leq 1} w(u) \cdot |\tau| \cdot \|x_k \varphi\|_{\mathscr{M}_w},$$

where $x_k\varphi$ denotes the function $x \mapsto x_k\varphi(x)$ which belongs to $\mathscr{M}_w(X)$ by hypothesis. It follows that the functions $\tau^{-1}(e_{\tau v_k} - 1)\varphi$ are in $\mathscr{M}_w(X)$, with norms uniformly bounded in $0 < |\tau| \leq 1$, and obviously they converge pointwise as $\tau \to 0$ to $ix_k\varphi(x)$. To obtain their feeble convergence, it remains to show that:

$$\lim_{R \to \infty} \sup_{0 < |\tau| \leq 1} \frac{1}{|\tau|} \int_{|x| > R} w(x)|\widehat{\varphi}(x - \tau v_k) - \widehat{\varphi}(x)|dx = 0.$$

This is obtained by noticing that one has as above

$$(3.1.14) \quad \int_{|x| > R} w(x)|\widehat{\varphi}(x - \tau v_k) - \widehat{\varphi}(x)|dx$$

$$\leq \sup_{|u| \leq 1} w(u) \cdot |\tau| \cdot \int_{|y| > R-1} w(y)|\partial_k \widehat{\varphi}(y)|dy.$$

(ii) If $\widehat{\varphi}$ is not of class C^1, the proof is similar. We let η and φ_ε be as in Example 3.1.7 and observe that $(e_{\tau v_k} - 1)\varphi_\varepsilon \to (e_{\tau v_k} - 1)\varphi$ feebly in $\mathscr{M}_w(X)$ by the result of Example 3.1.7. A bound for the norm in $\mathscr{M}_w(X)$ of the function $\tau^{-1}(e_{\tau v_k} - 1)\varphi$, uniformly in $0 < |\tau| \leq 1$, can be obtained by using successively Lemma 3.1.2 (b), the inequality (3.1.13) with φ replaced by φ_ε, and the inequality (3.1.11) :

$$\|(e_{\tau v_k} - 1)\varphi\|_{\mathscr{M}_w} \leq \liminf_{\varepsilon \to 0} \|(e_{\tau v_k} - 1)\varphi_\varepsilon\|_{\mathscr{M}_w}$$

$$\leq c|\tau| \cdot \liminf_{\varepsilon \to 0} \|x_k \varphi_\varepsilon\|_{\mathscr{M}_w} \leq c_1|\tau| \cdot \|x_k \varphi\|_{\mathscr{M}_w}.$$

Finally, for $R > 1$, let $\xi_R \in C^\infty(X)$ be such that $0 \leq \xi_R \leq 1$, $\xi_R(x) = 1$ if $|x| \geq R$ and $\xi_R(x) = 0$ if $|x| \leq R - 1$. We observe that $\xi_R w$ is a lower semicontinuous function. So we get by using Lemma 3.1.2 (b) and (3.1.14) that

$$\int_{|x| > R} w(x)|\widehat{\varphi}(x - \tau v_k) - \widehat{\varphi}(x)|dx$$

$$\leq \int_X \xi_R(x)w(x)|\widehat{\varphi}(x - \tau v_k) - \widehat{\varphi}(x)|dx$$

$$\leq \liminf_{\varepsilon \to 0} \int_X \xi_R(x)w(x)|\widehat{\varphi_\varepsilon}(x - \tau v_k) - \widehat{\varphi_\varepsilon}(x)|dx$$

$$\leq \sup_{|u| \leq 1} w(u) \cdot |\tau| \cdot \liminf_{\varepsilon \to 0} \int_{|y| > R-2} w(y)|\partial_k \widehat{\varphi_\varepsilon}(y)|dy.$$

Now, by (3.1.12):

$$\int_{|y|>R-2} w(y)|\partial_k\widehat{\varphi_\varepsilon}(y)|\underline{dy} = \int_{|y|>R-2} w(y)|[\mathcal{F}(x_k\varphi)_\varepsilon](y)|\underline{dy}$$

$$\leq \sup_{|z|\leq N} w(z) \cdot \int_{|y|>R-N-2} w(y)|(\partial_k\widehat{\varphi_\varepsilon})(y)|\underline{dy}.$$

Hence

$$\sup_{0<|\tau|\leq 1} \frac{1}{|\tau|} \int_{|x|>R} w(x)|\widehat{\varphi}(x-\tau v_k) - \widehat{\varphi}(x)|\underline{dx} \to 0 \text{ as } R \to \infty. \quad \square$$

3.1.3. In the Littlewood-Paley type estimates which will be presented in Section 3.5 a certain action of the dilation group of X in $\mathcal{M}_w(X)$ will play an important role. Let

$$(3.1.15) \qquad\qquad \varphi^\sigma(x) = \varphi(\sigma x).$$

If φ is a continuous function, this makes sense for all real $\sigma \geq 0$; then $\varphi^0(x) \equiv \varphi(0)$ is a constant function and $\widehat{\varphi^0} = \varphi(0)\delta$ (δ denotes the Dirac measure with support at the point $x = 0$). If $\sigma > 0$, one may take for φ in (3.1.15) an arbitrary distribution, and then

$$(3.1.16) \qquad (\mathcal{F}\varphi^\sigma)(x) = \sigma^{-n}\widehat{\varphi}(\sigma^{-1}x) \qquad (n = \dim X).$$

If $\widehat{\varphi}$ is a measure μ, then the r.h.s. of (3.1.16) is just the measure $\Omega \mapsto \mu(\sigma^{-1}\Omega)$ (defined on the Borel subsets Ω of X).

If one replaces in the proof of Proposition 3.1.4 (c) the operator of conjugation J by the operator $J_\sigma : \varphi \mapsto \varphi^\sigma$ (with $\sigma \geq 0$ fixed), one finds that the Banach space $\mathcal{M}_w(X)$ is invariant under the operation $\varphi \mapsto \varphi^\sigma$ if and only if there is a constant $a = a(\sigma)$ such that :

$$(3.1.17) \qquad\qquad w(\sigma x) \leq a(\sigma)w(x) \qquad \forall x \in X.$$

If (3.1.17) is satisfied, we have

$$(3.1.18) \qquad\qquad ||\varphi^\sigma||_{\mathcal{M}_w} = \int_X w(\sigma x)|\widehat{\varphi}|(dx) \leq a(\sigma)||\varphi||_{\mathcal{M}_w}.$$

Let us define $a_w(\sigma) = \sup_{x\in X} w(\sigma x)w(x)^{-1}$. We obviously have $1 \leq a_w(\sigma) \leq +\infty$ and $w(\sigma x) \leq a_w(\sigma)w(x)$ for all x in X. In particular $a_w(\sigma\tau) \leq a_w(\sigma)a_w(\tau)$, so the set $\{\sigma \in [0,\infty) \mid a_w(\sigma) < \infty\}$ is stable under multiplication and contains 0 and 1.

We shall be interested only in the following two consequences of these general arguments. Firstly, if $w(x) = e^{\omega|x|}$ with $\omega > 0$, then $a(\sigma) < \infty$ if and only if $0 \leq \sigma \leq 1$, so these are the only numbers for which $\mathcal{M}_w(X)$ is stable under the operation $\varphi \mapsto \varphi^\sigma$. Secondly, assume that (3.1.17) holds for all $\sigma \geq 0$ with $a(\sigma) \leq a\langle\sigma\rangle^r$, where a and r are some finite constants. Then, by writing $x = |x|\widehat{x}$ with $|\widehat{x}| = 1$, one obtains from (3.1.17) that $w(x) \leq a\langle x\rangle^r w(\widehat{x})$ for each $x \in X$; hence $w(x) \leq b\langle x\rangle^r$ with $b = a\sup_{|y|=1} w(y) < \infty$, so that w is polynomially bounded.

The algebras that correspond to the above two extreme cases are the most important ones in our applications, so we denote them by special symbols:

DEFINITION 3.1.11. Let $\omega > 0$ and $r \geq 0$ be real numbers. Then we set
$\mathcal{M}^{(\omega)}(X) = \mathcal{M}_w(X)$ with $w(x) = e^{\omega|x|}$,
$\mathcal{M}^r(X) = \mathcal{M}_w(X)$ with $w(x) = 2^{r/2}\langle x \rangle^r$,
$\mathcal{M}(X) \equiv \mathcal{M}^0(X) = \mathcal{M}_w(X)$ with $w(x) = 1$.

To summarize, $\mathcal{M}^{(\omega)}(X)$ and $\mathcal{M}^r(X)$ are subalgebras of $BC(X)$ which contain the constant functions and are stable under conjugation. When equipped with their natural norms (3.1.6), they are abelian, unital Banach $*$-algebras continuously embedded in $BC(X)$. There are two important differences between them:

a) $\mathcal{M}^{(\omega)}(X)$ consists of real analytic functions on X, in particular it does not contain any function (not identically zero) equal to zero on some open, nonempty subset of X; on the other hand:

$$(3.1.19) \qquad \mathscr{S}(X) \subset \mathcal{M}^r(X) \subset BC^{[r]}(X)$$

continuously, where $[r]$ is the integer part of r.

(b) $\mathcal{M}^{(\omega)}(X)$ is stable under the operation of dilation $\varphi \mapsto \varphi^\sigma$ if and only if $0 \leq \sigma \leq 1$; in this case one has

$$(3.1.20) \qquad \|\varphi^\sigma\|_{\mathcal{M}^{(\omega)}} \leq \|\varphi\|_{\mathcal{M}^{(\omega)}} \text{ for } 0 \leq \sigma \leq 1, \quad \varphi \in \mathcal{M}^{(\omega)}(X).$$

On the other hand, $\mathcal{M}^r(X)$ is stable under all dilations and

$$(3.1.21) \qquad \|\varphi^\sigma\|_{\mathcal{M}^r} \leq \max(1, \sigma^r)\|\varphi\|_{\mathcal{M}^r} \text{ for } 0 \leq \sigma < \infty, \quad \varphi \in \mathcal{M}^r(X).$$

Observe that the mappings $\sigma \mapsto \varphi^\sigma \in \mathcal{M}_w(X)$ are *not* norm continuous in general. However, if $\varphi \in \mathcal{M}^{(\omega)}(X)$, then $[0,1] \ni \sigma \mapsto \varphi^\sigma \in \mathcal{M}^{(\omega)}(X)$ is feebly continuous, and so is $[0,\infty) \ni \sigma \mapsto \varphi^\sigma \in \mathcal{M}^r(X)$ if $\varphi \in \mathcal{M}^r(X)$.

Finally let us mention the obvious inclusion relations

$$(3.1.22)$$
$$\mathcal{M}^{(\omega_2)}(X) \subset \mathcal{M}^{(\omega_1)}(X) \subset \mathcal{M}^{r_2}(X) \subset \mathcal{M}^{r_1}(X) \text{ if } 0 < \omega_1 < \omega_2, \ 0 \leq r_1 < r_2,$$

where each inclusion is strict. We have $\mathcal{M}_w(X) \subset \mathcal{M}(X)$ for each w satisfying the hypotheses of Definition 3.1.3.

3.2. C_0-Groups: Continuity Properties and Elementary Functional Calculus

3.2.1. We begin with the definition of multiparameter C_0-groups and then give some of their basic properties. Throughout this section $\mathbf{F}$ is a fixed Banach space.

DEFINITION 3.2.1. Let $W = \{W(x) \mid x \in \mathbb{R}^n\}$ be a family of bounded operators in $\mathbf{F}$. We say that W is a C_0-*representation* of $\mathbb{R}^n$ in $\mathbf{F}$ if
   (i) $W(0) = I$ and $W(x + y) = W(x)W(y)$ for all $x, y \in \mathbb{R}^n$,
   (ii) the mapping $W : \mathbb{R}^n \to B(\mathbf{F})$ is strongly continuous.

A C_0-representation of $\mathbb{R}^n$ in $\mathbf{F}$ will also be called a *n-parameter C_0-group* in $\mathbf{F}$. If $\mathbf{F}$ is a Hilbert space and if $W(x)$ is a unitary operator in $\mathbf{F}$ for each $x \in \mathbb{R}^n$, then we say that W is a *unitary representation of $\mathbb{R}^n$ in $\mathbf{F}$*.

The preceding definition makes sense if $\mathbb{R}^n$ is replaced by any (finite dimensional) euclidean space X and we shall later on use such a generalization without further comments. In fact our treatment will be explicitly invariant and we shall avoid the arguments based on reduction to the case $n = 1$ (a procedure which is used in various other texts on this subject, e.g. in [Tr]). The only place where the canonical basis of $\mathbb{R}^n$ is used is in the definition of the generators $A_1, \ldots, A_n$ of W and their powers $A^\alpha = A_1^{\alpha_1} \ldots A_n^{\alpha_n}$ (see Section 3.3). The definition of the operators $\varphi(A)$ given later on in this section depends only on the euclidean structure of X.

The continuity assumption (ii) in Definition 3.2.1 is not always satisfied in our applications; as examples, consider the translation group in $\mathbf{F} = L^\infty(\mathbb{R})$ or the group W defined by $W(x)T = U(x)TU(x)^{-1}$ in $B(\mathbf{H})$, where $\mathbf{H}$ is a Hilbert space and $U(\cdot)$ a C_0-representation in $\mathbf{H}$. In Definition 3.2.6 we shall introduce a weaker notion of continuity and specify a class of representations that covers these situations. However, in most cases (ii) is easy to verify and a consequence of apparently much weaker assumptions, as the following results will show. But let us first point out the essential estimate:

PROPOSITION 3.2.2. (a) *Let $\{W(x)\}_{x \in \mathbb{R}^n}$ be a family of bounded operators in $\mathbf{F}$ such that (i) of Definition 3.2.1 is satisfied. Assume that the function $x \mapsto \|W(x)\|_{B(\mathbf{F})}$ is Borel. Then there are constants $M \geq 1$ and $\omega \geq 0$ such that*

$$(3.2.1) \qquad \|W(x)\|_{B(\mathbf{F})} \leq M e^{\omega |x|} \qquad \forall x \in \mathbb{R}^n.$$

(b) *If W is a C_0-representation of $\mathbb{R}^n$ in $\mathbf{F}$, then $x \mapsto \|W(x)\|_{B(\mathbf{F})}$ is lower semicontinuous, hence Borel, so the estimate (3.2.1) is satisfied.*

PROOF. (a) This is an immediate consequence of Proposition 3.1.1 with $h(x) = \|W(x)\|_{B(\mathbf{F})}$.

(b) The lower semicontinuity follows from the relation

$$\|W(x)\|_{B(\mathbf{F})} = \sup\{\|W(x)f\|_{\mathbf{F}} \mid \|f\|_{\mathbf{F}} \leq 1\}$$

and the fact that the least upper bound of a family of continuous functions is lower semicontinuous.

A more elementary proof of (3.2.1) under the hypotheses of (b) is as follows: By the principle of uniform boundedness we have $\sup_{|x| \leq 1} \|W(x)\|_{B(\mathbf{F})} < \infty$, and it suffices to repeat the argument at the end of part (ii) of the proof of Proposition 3.1.1. $\square$

We shall need one more notion related to the growth of the norm $\|W(x)\|_{B(\mathbf{F})}$ as $|x| \to \infty$. We recall the notation $\langle x \rangle = (1 + |x|^2)^{1/2}$.

DEFINITION 3.2.3. A C_0-representation W of $\mathbb{R}^n$ in a Banach space $\mathbf{F}$ is said to be of *polynomial growth* (or *polynomially bounded*) if there are constants $M \geq$

1 and $r \geq 0$ such that

$$(3.2.2) \qquad \|W(x)\|_{B(\mathbf{F})} \leq M\langle x \rangle^r \qquad \forall x \in \mathbb{R}^n.$$

The next result shows that in some cases measurability implies continuity.

PROPOSITION 3.2.4. *Let $\mathbf{F}$ be a Banach space. Assume that $\mathbf{F}^*$ is separable and that $W : \mathbb{R}^n \to B(\mathbf{F})$ is weakly Borel and satisfies $W(0) = I$ and $W(x+y) = W(x)W(y)$ for all $x, y \in \mathbb{R}^n$. Then W is a C_0-representation of $\mathbb{R}^n$ in $\mathbf{F}$.*

PROOF. The separability of $\mathbf{F}^*$ implies the separability of $\mathbf{F}$ (Theorem III.7 in [RS]). We denote by $\{e_j\}_{j \in \mathbb{N}}$ a countable dense sequence in the unit ball of $\mathbf{F}$ and by $\{e_k^*\}_{k \in \mathbb{N}}$ a countable dense sequence in the unit ball of $\mathbf{F}^*$.

(i) We first show that W satisfies (3.2.1). Since

$$\|W(x)\|_{B(\mathbf{F})} = \sup_{j,k \in \mathbb{N}} |\langle W(x)e_j, e_k^* \rangle|,$$

the function $x \mapsto \|W(x)\|_{B(\mathbf{F})}$ is Borel, and we may apply Proposition 3.2.2 (a).

(ii) For $\varphi \in C_0^\infty(\mathbb{R}^n)$ set (as a weak integral in $B(\mathbf{F})$):

$$(3.2.3) \qquad W[\varphi] = \int_{\mathbb{R}^n} \varphi(-y) W(y) \underline{dy}.$$

Then, for any $x \in \mathbb{R}^n$:

$$W(x)W[\varphi] = \int_{\mathbb{R}^n} \varphi(-y)W(x+y)\underline{dy} = \int_{\mathbb{R}^n} \varphi(x-y)W(y)\underline{dy} \equiv (\varphi * W)(x).$$

By using the third expression and the Lebesgue dominated convergence theorem, one sees that $x \mapsto W(x)W[\varphi] \in B(\mathbf{F})$ is norm continuous. Thus, to obtain the strong continuity of W, it suffices to show that the set $\mathscr{D} = \{W[\varphi]f \mid f \in \mathbf{F}, \varphi \in C_0^\infty(\mathbb{R}^n)\}$ is dense in $\mathbf{F}$.

(iii) To obtain the density of $\mathscr{D}$ in $\mathbf{F}$, it suffices to exhibit a sequence $\{\varphi_\ell\}_{\ell \in \mathbb{N}}$ in $C_0^\infty(\mathbb{R}^n)$ such that $W[\varphi_\ell] \to I$ weakly as $\ell \to \infty$. For this, choose $\varrho \in C_0^\infty(\mathbb{R}^n)$ such that $\varrho \geq 0$, $\varrho(-x) = \varrho(x)$ and $\int_{\mathbb{R}^n} \varrho(x)\underline{dx} = 1$, and set $\varrho_\varepsilon(x) = \varepsilon^{-n}\varrho(x/\varepsilon)$, with $\varepsilon > 0$. Then

$$(\varrho_\varepsilon * W)(x) - W(x) = \int_{\mathbb{R}^n} \varrho(y)[W(x+\varepsilon y) - W(x)]\underline{dy}.$$

We set $\psi_{jk}(y) = \langle W(y)e_j, e_k^* \rangle$. By (3.2.1) we have

$$(3.2.4) \qquad |\psi_{jk}(y)| \leq M e^{2\omega} \qquad \forall j, k \in \mathbb{N}, \forall |y| < 2.$$

Now

$$(3.2.5)$$

$$\int_{|x| \leq 1} dx \sum_{j,k=0}^{\infty} 2^{-j-k} |\langle [\varrho_\varepsilon * W(x) - W(x)]e_j, e_k^* \rangle|$$

$$\leq \sum_{j,k=0}^{\infty} 2^{-j-k} \int_{\mathbb{R}^n} \varrho(y)\underline{dy} \int_{|x| \leq 1} |\psi_{jk}(x+\varepsilon y) - \psi_{jk}(x)|dx.$$

Since ψ_{jk} is a locally integrable function, we have

$$\lim_{\varepsilon \to 0} \int_{|x| \leq 1} |\psi_{jk}(x + \varepsilon y) - \psi_{jk}(x)| dx \to 0$$

for each $j, k \in \mathbb{N}$ and $y \in \mathbb{R}^n$. By using (3.2.4) and the Lebesgue dominated convergence theorem, one obtains that the r.h.s. of (3.2.5) converges to zero as $\varepsilon \to 0$. Thus there is a sequence $\{\varepsilon_\ell\}_{\ell \in \mathbb{N}}$ such that $\varepsilon_\ell \to 0$ and

$$(3.2.6) \qquad \sum_{j,k=0}^{\infty} 2^{-j-k} |\langle [(\varrho_{\varepsilon_\ell} * W)(x) - W(x)] e_j, e_k^* \rangle| \to 0$$

as $\ell \to \infty$ for almost all $|x| \leq 1$. If we fix one of these vectors x, then we get by using again the local boundedness of W that $(\varrho_{\varepsilon_\ell} * W)(x) \to W(x)$ weakly as $\ell \to \infty$. Hence $W[\varrho_{\varepsilon_\ell}] \equiv W(-x) \cdot [\varrho_{\varepsilon_\ell} * W](x) \to I$ weakly as $\ell \to \infty$. $\square$

REMARK. The following example shows that the separability of $\mathbf{F}^*$ is essential. Let $\mathbf{F}$ be the space of all functions $f : \mathbb{R} \to \mathbb{C}$ such that $\|f\|_{\mathbf{F}} := \left(\sum_{x \in \mathbb{R}} |f(x)|^2 \right)^{1/2} < \infty$. Then $\mathbf{F}$ is a Hilbert space and the family $\{e_z\}_{z \in \mathbb{R}}$ with $e_z(x) = 1$ if $x = z$ and $e_z(x) = 0$ if $x \neq z$ is an orthonormal basis in $\mathbf{F}$. Let W be the natural representation of the translation group: $(W(x)f)(y) = f(y - x)$. Then $W(x)e_a = e_{a+x}$ if $a, x \in \mathbb{R}$, so $\langle W(x)e_a, e_b \rangle = 1$ if $x = b - a$ and $\langle W(x)e_a, e_b \rangle = 0$ otherwise. This implies that $x \mapsto \langle W(x)f, g \rangle$ is Borel for all $f, g \in \mathbf{F}$. But $\|W(x)e_a - e_a\|_{\mathbf{F}} = \sqrt{2}$ if $x \neq 0$, so W is not continuous.

We shall need only the following consequence of Proposition 3.2.4:

PROPOSITION 3.2.5. *Let $\mathbf{E}$ and $\mathbf{F}$ be reflexive Banach spaces such that $\mathbf{E} \subset \mathbf{F}$. Let W be a C_0-representation of $\mathbb{R}^n$ in $\mathbf{F}$. Then*
 (a) *the family $\{W(x)^* \mid x \in \mathbb{R}^n\}$ is a C_0-representation of $\mathbb{R}^n$ in $\mathbf{F}^*$,*
 (b) *if $W(x)\mathbf{E} \subset \mathbf{E}$ for each $x \in \mathbb{R}^n$, then the restriction $W_{\mathbf{E}}(x)$ of $W(x)$ to $\mathbf{E}$ belongs to $B(\mathbf{E})$ and $W_{\mathbf{E}}$ is a C_0-representation of $\mathbb{R}^n$ in $\mathbf{E}$.*

PROOF. (a) The strong continuity of $\{W(x)\}$ in $\mathbf{F}$ implies the weak*-continuity of $\{W(x)^*\}$ in $\mathbf{F}^*$. Thus, since $\mathbf{F}$ is reflexive, $\{W(x)^*\}$ is weakly continuous in $\mathbf{F}^*$. So, for fixed $f \in \mathbf{F}^*$, the closed subspace of $\mathbf{F}^*$ generated by the family of vectors $\{W(x)^* f\}_{x \in \mathbb{R}^n}$ is separable, so that $W(x)^* f$ is strongly continuous by Proposition 3.2.4.

(b) We first observe that $W_{\mathbf{E}}(x)$ belongs to $B(\mathbf{E})$ by the closed graph theorem. It is clear that $W_{\mathbf{E}}(x + y) = W_{\mathbf{E}}(x) W_{\mathbf{E}}(y)$. To prove the strong continuity of $W_{\mathbf{E}}$, fix a vector $e \in \mathbf{E}$ and denote by $\mathbf{E}_0$ and $\mathbf{F}_0$ the closed subspace of $\mathbf{E}$ and $\mathbf{F}$ respectively generated by the family of vectors $\{W(x)e\}_{x \in \mathbb{R}^n}$. Then $\mathbf{E}_0 \subset \mathbf{F}_0$ continuously and densely. Consequently $\mathbf{F}_0^* \subset \mathbf{E}_0^*$ continuously and densely. Also it is clear that both $\mathbf{E}_0$ and $\mathbf{F}_0$ are invariant under each $W(x)$.

Since $x \mapsto W(x)e$ is strongly continuous in $\mathbf{F}$, the Banach space $\mathbf{F}_0$ is separable. As a closed subspace of $\mathbf{F}$, it is also reflexive (see Section 2.1), so its adjoint $\mathbf{F}_0^*$ is separable (again by Theorem III.7 of [RS]). So, since $\mathbf{F}_0^* \subset \mathbf{E}_0^*$ continuously and densely, $\mathbf{E}_0^*$ is separable. Thus the strong continuity of $x \mapsto W(x)e$ in $\mathbf{E}$ (or equivalently in $\mathbf{E}_0$) follows from Proposition 3.2.4 provided that we show that

$\langle W(x)f, g \rangle$ is Borel for each $f \in \mathbf{E}_0$ and each $g \in \mathbf{E}_0^*$. Since $\mathbf{F}_0^*$ is dense in $\mathbf{E}_0^*$, it suffices to consider $g \in \mathbf{F}_0^*$. But then $\langle W(x)f, g \rangle$ is even continuous in x. $\square$

In the situation considered in Proposition 3.2.5 it often happens that W is a unitary representation of $\mathbb{R}^n$ in a Hilbert space $\mathbf{F}$ but $W_{\mathbf{E}}$ has polynomial or even exponential growth. Several examples will be given in Chapter 4.

Assume that a C_0-representation W of $\mathbb{R}^n$ in $\mathbf{F}$ is given. It is convenient to define the *dual representation* W^* *of* $\mathbb{R}^n$ *in* $\mathbf{F}^*$ by

$$(3.2.7) \qquad\qquad W^*(x) = W(-x)^*.$$

The advantage of inserting the minus sign on the r.h.s. is that, if $\mathbf{F}$ is a Hilbert space, W is unitary and $\mathbf{F}^*$ is identified with $\mathbf{F}$ through the Riesz isomorphism, then $W = W^*$. Proposition 3.2.5 (a) shows that, *if* $\mathbf{F}$ *is reflexive, then* W^* *is a* C_0-*representation*. This is not true in general in the non-reflexive case (e.g. consider the situation where $\mathbf{F} = L^1(\mathbb{R}^n)$ and W is the translation group). However, it is always true that $W^*(0) = I$, $W^*(x + y) = W^*(x)W^*(y)$ and $W^* : \mathbb{R}^n \to B(\mathbf{F}^*)$ is weak*-continuous (i.e. $x \mapsto \langle f, W^*(x)g \rangle$ is continuous for all $f \in \mathbf{F}$, $g \in \mathbf{F}^*$). Also, if (3.2.1) is fulfilled, then:

$$(3.2.8) \qquad\qquad \|W^*(x)\|_{B(\mathbf{F}^*)} = \|W(-x)\|_{B(\mathbf{F})} \le Me^{\omega|x|}.$$

The examples cited before Proposition 3.2.2 and the preceding discussion of dual representations exhibit important situations in which the continuity assumption (ii) in Definition 3.2.1 is not satisfied. In order to isolate a manageable class of representations that covers these situations, suppose that we have a family $W = \{W(x) \mid x \in \mathbb{R}^n\}$ of bounded operators in a Banach space satisfying (i) of Definition 3.2.1. Then we introduce a linear subspace Γ_W of $\mathbf{F}^*$ as follows:

$$(3.2.9) \qquad \Gamma_W = \{\varphi \in \mathbf{F}^* \mid x \mapsto \varphi(W(x)f) \text{ is continuous } \forall f \in \mathbf{F}\}.$$

If (3.2.1) is satisfied, then Γ_W is a closed subspace of $\mathbf{F}^*$. In many cases (for example in the situations cited above) it is strictly smaller than $\mathbf{F}^*$. However, many of the results valid for C_0-groups have natural analogues for the class of representations introduced in the next definition:

DEFINITION 3.2.6. A family $W = \{W(x) \mid x \in \mathbb{R}^n\}$ of bounded operators in $\mathbf{F}$ satisfying (i) of Definition 3.2.1 is said to be a C_{w}-*representation of* $\mathbb{R}^n$ *in* $\mathbf{F}$ (or a n-parameter C_{w}-*group*) if Γ_W is a determining subspace for $\mathbf{F}$ (see [HP]), in other terms if for all $f \in \mathbf{F}$:

$$(3.2.10) \qquad\qquad \|f\|_{\mathbf{F}} = \sup\{|\varphi(f)| \mid \varphi \in \Gamma_W, \|\varphi\|_{\mathbf{F}^*} \le 1\}.$$

Let us point out that (3.2.1) is satisfied for a C_{w}-representation, because Proposition 3.2.2 (a) applies; indeed one has

$$\|W(x)\|_{B(\mathbf{F})} = \sup\{|\varphi(W(x))f| \mid \varphi \in \Gamma_W, \|\varphi\|_{\mathbf{F}^*} \le 1, f \in \mathbf{F}, \|f\|_{\mathbf{F}} \le 1\},$$

so $x \mapsto \|W(x)\|_{B(\mathbf{F})}$ is lower semicontinuous and hence Borel. Moreover, the set of seminorms $f \mapsto |\varphi(f)|$ on $\mathbf{F}$ with $\varphi \in \Gamma_W$ defines a topology on $\mathbf{F}$ which is locally convex and Hausdorff and which we shall call the W-*topology* on $\mathbf{F}$.

As an example, consider a Banach space $\mathbf{G}$ and let $\mathbf{F} = \mathbf{G}^*$. Then $\mathbf{F}^* = \mathbf{G}^{**}$ and $\mathbf{G}$ is a determining subspace for $\mathbf{F}$. More generally, let $\mathbf{G}$, $\mathbf{H}$ be two Banach spaces and $\mathbf{F} = B(\mathbf{G}, \mathbf{H}^*)$. Then the linear subspace of $\mathbf{F}^*$ generated by functionals of the form $\varphi(T) = \langle h, Tg \rangle$, $h \in \mathbf{H}$, $g \in \mathbf{G}$, $T \in \mathbf{F}$, is a determining subspace for $\mathbf{F}$. Hence, if $W_\mathbf{G}$ and $W_\mathbf{H}$ are C_0-representations of $\mathbb{R}^n$ in $\mathbf{G}$ and $\mathbf{H}$ respectively, then $\mathscr{W}(x)T := W_\mathbf{H}(x)^*TW_\mathbf{G}(x)$ defines a C_w-representation of $\mathbb{R}^n$ in $B(\mathbf{G}, \mathbf{H}^*)$.

Finally, if W is a C_0-representation in an arbitrary Banach space $\mathbf{F}$, then its dual representation W^* is a C_w-representation of $\mathbb{R}^n$ in $\mathbf{F}^*$, and the W^*-topology on $\mathbf{F}^*$ is finer than the weak* topology.

3.2.2. We now turn to the second topic of this section, the construction of a functional calculus which is naturally associated to a representation of $\mathbb{R}^n$ in $\mathbf{F}$. Generally speaking, by *functional calculus* we mean a homomorphism from an algebra of complex functions on $\mathbb{R}$ (in which multiplication is defined as the usual pointwise multiplication of functions) into the algebra $B(\mathbf{F})$. The precise definition in our case is as follows:

DEFINITION 3.2.7. Let W be a C_0-representation of $\mathbb{R}^n$ in the Banach space $\mathbf{F}$. Let $w(x) = \max(1, \|W(x)\|_{B(\mathbf{F})})$ and let $\varphi \in \mathcal{M}_w(\mathbb{R}^n)$ (see Definition 3.1.3). Then $\varphi(A)$ is the continuous operator in $\mathbf{F}$ defined by

$$(3.2.11) \qquad \varphi(A) = \int_{\mathbb{R}^n} W(x)\widehat{\varphi}(x)\underline{d}x.$$

Here $\widehat{\varphi}(x)\underline{d}x$ is the measure whose Fourier transform is φ, sometimes denoted by $\widehat{\varphi}(dx)$. We prefer the first notation in order to stress our convention concerning the factor $(2\pi)^{-n/2}$ in the Fourier transformation, viz. $\varphi(x) = \int_{\mathbb{R}^n} e^{i(x,y)}\widehat{\varphi}(x)\underline{d}x$. The integral in (3.2.11) should be understood in the weak sense: for all $f \in \mathbf{F}$ and $g \in \mathbf{F}^*$ we have:

$$(3.2.12) \qquad \langle g, \varphi(A)f \rangle = \int_{\mathbb{R}^n} \langle g, W(x)f \rangle \widehat{\varphi}(x)\underline{d}x.$$

Remember that we identify $\mathbf{F} \subset \mathbf{F}^{**}$ by defining $\langle g, f \rangle \equiv \overline{\langle f, g \rangle}$ if $g \in \mathbf{F}^*$ and $f \in \mathbf{F}$ and that the anti-duality is linear in the second argument. The integral in (3.2.12) is absolutely convergent because of the continuity of the integrand and because $\widehat{\varphi}$ is a w-integrable measure. Clearly we get:

$$(3.2.13) \qquad |\langle g, \varphi(A)f \rangle| \leq \|\varphi\|_{\mathcal{M}_w} \cdot \|f\|_\mathbf{F} \cdot \|g\|_{\mathbf{F}^*},$$

from which the existence and the continuity of the operator $\varphi(A)$ follow.

Let us add a comment concerning the notation $\varphi(A)$. Let us choose φ such that $\widehat{\varphi}(x)\underline{d}x$ is the Dirac measure at the point $z \in \mathbb{R}^n$, i.e. let $\varphi(x) = e^{i(x,z)}$. Then $\varphi(A) = W(z)$. On the other hand, in the next section we shall define a collection of n commuting (in general unbounded) operators $A_1, \ldots, A_n$ in $\mathbf{F}$ such that $W(z) = \exp[i(A_1 z_1 + \cdots + A_n z_n)]$ in a natural sense. In the notation $A \cdot z \equiv A_1 z_1 + \cdots + A_n z_n$, the equation (3.2.11) can be written as

$$(3.2.14) \qquad \varphi(A) = \int_{\mathbb{R}^n} e^{iA \cdot x}\widehat{\varphi}(x)\underline{d}x,$$

so that $\varphi(A)$ has an obvious interpretation in terms of functions of A. If $n = 1$ and W is a unitary group in a Hilbert space $\mathbf{F}$, then by Stone's theorem there is a unique self-adjoint operator A in $\mathbf{F}$ such that $W(x) = e^{iA \cdot x}$ $(x \in \mathbb{R})$; then $\varphi(A)$ can be defined by the usual functional calculus for self-adjoint operators, and it is easy to see that it coincides with the operator defined by (3.2.14) if $\varphi \in \mathcal{M}(\mathbb{R}^n)$.

The next proposition summarizes the main properties of the functional calculus introduced in Definition 3.2.7.

PROPOSITION 3.2.8. *The correspondence $\varphi \mapsto \varphi(A)$ defines a unital homomorphism from $\mathcal{M}_w(\mathbb{R}^n)$ into $B(\mathbf{F})$. Its range is commutative, and one has*

$$(3.2.15) \qquad \|\varphi(A)\|_{B(\mathbf{F})} \leq \|\varphi\|_{\mathcal{M}_w}$$

and

$$(3.2.16) \qquad \|\varphi(A)f\|_{\mathbf{F}} \leq \int_{\mathbb{R}^n} \|W(x)f\|_{\mathbf{F}} \, |\widehat{\varphi}(x)| \, \underline{d}x.$$

PROOF. (i) We first check the multiplicativity of the correspondence $\varphi \mapsto \varphi(A)$. Let $\varphi, \psi \in \mathcal{M}_w(\mathbb{R}^n)$. Then $\varphi\psi$ is the inverse Fourier transform of the convolution $\widehat{\varphi} * \widehat{\psi}$ (in the sense of measures). So, for $f \in \mathbf{F}$ and $g \in \mathbf{F}^*$:

$$
\begin{aligned}
\langle g, (\varphi\psi)(A)f \rangle &= \int_{\mathbb{R}^n} \langle g, W(z)f \rangle (\mathcal{F}\varphi\psi)(z) \underline{d}z \\
&= \iint_{\mathbb{R}^n \times \mathbb{R}^n} \langle g, W(x+y)f \rangle \widehat{\varphi}(x) \underline{d}x \cdot \widehat{\psi}(y) \underline{d}y \\
&= \int_{\mathbb{R}^n} \left[\int_{\mathbb{R}^n} \langle W(x)^*g, W(y)f \rangle \widehat{\psi}(y) \underline{d}y \right] \cdot \widehat{\varphi}(x) \underline{d}x \\
&= \int_{\mathbb{R}^n} \langle g, W(x)\psi(A)f \rangle \widehat{\varphi}(x) \underline{d}x = \langle g, \varphi(A)\psi(A)f \rangle.
\end{aligned}
$$

(ii) The remaining assertions are obvious consequences of the definition (3.2.12) and of the estimate (3.2.13); for the estimate (3.2.16) also observe that the function $x \mapsto W(x)f$ is continuous and bounded by $w(x)\|f\|_{\mathbf{F}}$, so that the integral $\int_{\mathbb{R}^n} W(x)f\widehat{\varphi}(x)\underline{d}x$ exists strongly in $\mathbf{F}$. $\square$

In the next proposition we relate the functional calculus to the notion of feeble convergence introduced in Definition 3.1.6.

PROPOSITION 3.2.9. *Let w be as in Definition 3.2.7.*
(a) *If $\{\varphi_k\}_{k \in \mathbb{N}}$ is a sequence in $\mathcal{M}_w(\mathbb{R}^n)$ which is feebly convergent to $\varphi \in \mathcal{M}_w(\mathbb{R}^n)$, then $\varphi_k(A) \to \varphi(A)$ as $k \to \infty$ strongly on $\mathbf{F}$.*
(b) *The homomorphism $\varphi \mapsto \varphi(A)$ is uniquely characterized by the preceding property: there is a unique linear mapping $\Phi : \mathcal{M}_w(\mathbb{R}^n) \to B(\mathbf{F})$ such that $\Phi(\varphi_k) \to \Phi(\varphi)$ strongly on $\mathbf{F}$ if $\varphi_k \to \varphi$ feebly in $\mathcal{M}_w$ and $\Phi(e_z) = W(z)$ for all $z \in \mathbb{R}^n$ (where e_z denotes the function $e_z(x) = e^{i\langle x, z \rangle}$).*

PROOF. By virtue of the properties of the function $x \mapsto W(x)f$ pointed out in part (ii) of the preceding proof, all results are immediate consequences of Proposition 3.1.8. $\square$

The algebra $\mathscr{M}_w(\mathbb{R}^n)$ is somewhat too precise for our purposes; it will be sufficient to consider functions φ belonging to one of the algebras $\mathscr{M}^{(\omega)}(\mathbb{R}^n)$ $(\omega > 0)$ or $\mathscr{M}^r(\mathbb{R}^n)$ $(r \geq 0)$ introduced in Definition 3.1.11. More specifically, only the following two situations will occur:

(1) W is an arbitrary C_0-group: then there are constants $M \geq 1$ and $\omega > 0$ such that $||W(x)||_{B(\mathbf{F})} \leq Me^{\omega|x|}$. Clearly we shall have $\mathscr{M}^{(\omega)}(\mathbb{R}^n) \subset \mathscr{M}_w(\mathbb{R}^n)$ and $||\varphi(A)||_{B(\mathbf{F})} \leq M||\varphi||_{\mathscr{M}^{(\omega)}}$ if $\varphi \in \mathscr{M}^{(\omega)}(\mathbb{R}^n)$.

(2) W is a polynomially bounded C_0-group: then there are constants $M \geq 1$ and $r \geq 0$ such that $||W(x)||_{B(\mathbf{F})} \leq M\langle x\rangle^r$. So we shall have $\mathscr{M}^r(\mathbb{R}^n) \subset \mathscr{M}_w(\mathbb{R}^n)$ and $||\varphi(A)||_{B(\mathbf{F})} \leq 2^{-r/2}M||\varphi||_{\mathscr{M}^r}$ for $\varphi \in \mathscr{M}^r(\mathbb{R}^n)$.

In §3.1.3 we associated to a function φ defined on $\mathbb{R}^n$ a family $\{\varphi^\sigma\}_{\sigma \geq 0}$ of scaled functions by setting $\varphi^\sigma(x) = \varphi(\sigma x)$. If both φ and φ^σ belong to $\mathscr{M}_w(\mathbb{R}^n)$, then $\varphi(A)$ and $\varphi^\sigma(A)$ are continuous operators in $\mathbf{F}$, and we have (as a consequence of (3.2.11) and (3.1.16)):

$$(3.2.17) \qquad \varphi^\sigma(A) = \int_{\mathbb{R}^n} W(\sigma x)\widehat{\varphi}(x)\underline{d}x.$$

In view of (3.2.14) it is natural to use the notation $\varphi(\sigma A)$ for this operator:

$$(3.2.18) \qquad \varphi^\sigma(A) \equiv \varphi(\sigma A).$$

We mention that $\sigma \mapsto \varphi(\sigma A) \in B(\mathbf{F})$ is *not* norm-continuous in general; if $\varphi(0) = 0$, the norms $||\varphi(\sigma A)||_{B(\mathbf{F})}$ need not converge to zero as $\sigma \to 0$ (as an example, consider the situation where A is an unbounded self-adjoint operator with spectrum $\sigma(A) = \mathbb{R}$ or $\sigma(A) = [0, \infty)$; then, for $\sigma \neq 0$, one has $||\varphi(\sigma A)|| = \sup_{\lambda \in \sigma(A)} |\varphi(\sigma\lambda)| = \sup_{\mu \in \sigma(A)} |\varphi(\mu)|$).

The following facts will play a role further on:

(1') If W is as in (1) above and $\varphi \in \mathscr{M}^{(\omega)}(\mathbb{R}^n)$, then for each $0 \leq \sigma \leq 1$ we have $\varphi^\sigma \in \mathscr{M}^{(\omega)}(\mathbb{R}^n)$ and

$$(3.2.19) \qquad \begin{cases} ||\varphi(\sigma A)||_{B(\mathbf{F})} \leq M||\varphi||_{\mathscr{M}^{(\omega)}} \\ (0, 1] \ni \sigma \mapsto \varphi(\sigma A)f \in \mathbf{F} \text{ is continuous } (\forall f \in \mathbf{F}) \\ \lim_{\sigma \to 0} \varphi(\sigma A) = \varphi(0)I \text{ strongly on } \mathbf{F}. \end{cases}$$

(2') If W is as in (2) above and $\varphi \in \mathscr{M}^r(\mathbb{R}^n)$, then for each $0 \leq \sigma < \infty$ we have $\varphi^\sigma \in \mathscr{M}^r(\mathbb{R}^n)$ and

$$(3.2.20) \qquad \begin{cases} ||\varphi(\sigma A)||_{B(\mathbf{F})} \leq 2^{-r/2}M \cdot \max(1, \sigma^r) \cdot ||\varphi||_{\mathscr{M}^r} \\ (0, \infty) \ni \sigma \mapsto \varphi(\sigma A)f \in \mathbf{F} \text{ is continuous } (\forall f \in \mathbf{F}) \\ \lim_{\sigma \to 0} \varphi(\sigma A) = \varphi(0)I \text{ strongly on } \mathbf{F}. \end{cases}$$

These assertions are immediate consequences of (3.1.20), (3.1.21), (3.2.16) and (3.2.17).

One of the advantages of the functional calculus constructed above is that it makes sense for any C_w-representation of $\mathbb{R}^n$. In fact, if W is a C_w-group, then the associated function w is lower semicontinuous, so that $\mathscr{M}_w(\mathbb{R}^n)$ is well defined. Hence (3.2.12) allows us to define $\varphi(A)$ for all $\varphi \in \mathscr{M}_w(\mathbb{R}^n)$, and Proposition 3.2.8 remains valid. The statements (3.2.19) and (3.2.20) remain valid if the continuity of $\sigma \mapsto \varphi(\sigma A)f$ and the limit $\lim_{\sigma \to 0} \varphi(\sigma A)f$ are understood in the W-topology on $\mathbf{F}$. Moreover, Proposition 3.2.9 (a) remains true on condition that the convergence $\varphi_k(A) \to \varphi(A)$ is understood as convergence of $\varphi_k(A)f$ to $\varphi(A)f$ in the W-topology, for each $f \in \mathbf{F}$.

As a special case, let us consider the dual representation W^* in $\mathbf{F}^*$ associated to a C_0-group W in $\mathbf{F}$ (see (3.2.7)). We recall that the W^*-topology on $\mathbf{F}^*$ is finer than the weak* topology. If one writes $\varphi(A^*) \equiv \int_{\mathbb{R}^n} W^*(x)\widehat{\varphi}(x)\underline{dx}$ for the functional calculus associated to W^* by the method described above, one easily sees that

$$(3.2.21) \qquad\qquad \varphi(A)^* = \overline{\varphi}(A^*).$$

It should be remarked here that $\overline{\varphi} \in \mathcal{M}_{\widetilde{w}}(\mathbb{R}^n)$ if $\varphi \in \mathcal{M}_w(\mathbb{R}^n)$, where $\widetilde{w}(x) = w(-x)$. Finally, if $\mathbf{F}$ is not reflexive, then under conditions similar to those in (3.2.19) or (3.2.20) the correspondence $\sigma \mapsto \varphi(\sigma A^*) \in B(\mathbf{F}^*)$ is continuous in general only for the W^*-topology (and weaker topologies) on $\mathbf{F}^*$.

It is interesting to observe that if w is a fixed function satisfying the conditions of Definition 3.1.3, then any unital homomorphism Φ from $\mathcal{M}_w(\mathbb{R}^n)$ into $B(\mathbf{F})$ is *essentially* of the form described in this subsection. Indeed, let $e_y(x) = e^{i\langle x,y \rangle}$; then $e_y \in \mathcal{M}_w(\mathbb{R}^n)$ for each $y \in \mathbb{R}^n$, $e_0 = 1$ and $e_{y+z} = e_y e_z$. Hence, if we define $W(x) = \Phi(e_x)$, we get a representation of $\mathbb{R}^n$ in $\mathbf{F}$. It will turn out to be more fertile to adopt the point of view of the functional calculus rather than that of a representation of $\mathbb{R}^n$.

A question which will be extensively treated in Sections 3.6 and 3.7 concerns the possibility of extending the functional calculus $\varphi \mapsto \varphi(A)$ to a larger class of functions φ. Strongly related to this problem is that of obtaining estimates on $\|\varphi(A)\|_{B(\mathbf{F})}$ directly in terms of φ, rather than in terms of $\widehat{\varphi}$. In studying these problems, one is confronted with difficulties due either to the geometry of $\mathbf{F}$ or to the properties of the group W. If $\mathbf{F}$ is geometrically nice, then the behaviour of W is relatively unimportant. For example, assume $\mathbf{F}$ is a Hilbert space. Then:

(1) if W is a unitary group, then we have $\|\varphi(A)\|_{B(\mathbf{F})} \leq \sup_{x \in X} |\varphi(x)|$ because of the spectral theorem for self-adjoint operators;

(2) if W is a bounded C_0-group, we have $\|\varphi(A)\|_{B(\mathbf{F})} \leq c \sup_{x \in X} |\varphi(x)|$ because W will be similar to a unitary group (Vidav theorem, see [Da]);

(3) if W is polynomially bounded, one has a rich functional calculus due to a result of N. Mandache which will be described in Section 3.7.

If $\mathbf{F}$ is a Hilbert space and W is not polynomially bounded, then the functional calculus may be quite restricted, as is shown by the following example.

Let $\mathbf{H} = L^2(\mathbb{R})$ and $W(x) = \exp(iQx)$ ($x \in \mathbb{R}$). Let P be the self-adjoint realization of $-i\frac{d}{dx}$ in $\mathbf{H}$ and let $\mathbf{F}$ be the domain of e^P. $\mathbf{F}$ is a Hilbert space with the norm $\|f\|_{\mathbf{F}} = (\|f\|_{\mathbf{H}}^2 + \|e^P f\|_{\mathbf{H}}^2)^{1/2}$. Since $W(-x)PW(x) = P + xI$, we have $W(-x)e^P W(x) = e^x e^P$. Thus $W(x) \in B(\mathbf{F})$ and $\|W(x)\|_{B(\mathbf{F})} = \max\{1, e^x\}$, so that W is an exponentially growing group in $\mathbf{F}$. If $f \in \mathbf{F}$, then the function f has a holomorphic extension into the strip $-1 < \Im x < 0$; in fact

$$f(x) = \int_{-\infty}^{\infty} e^{ixy} \widehat{f}(y)\underline{dy} \qquad (x \in \mathbb{R}),$$

and $y \mapsto \exp(ixy)\widehat{f}(y)$ defines a function in $L^1(\mathbb{R})$ if $\Im x \in (-1, 0)$ and $f \in \mathbf{F} \equiv D(e^P)$, i.e. if $\int_{-\infty}^{\infty} |e^y \widehat{f}(y)|^2 \underline{dy} < \infty$. This implies that the functional calculus is very restricted in this case. If for example $\psi \in C_0^{\infty}(\mathbb{R})$ and $f \in \mathbf{F}$, then $\psi(Q)f$ has a holomorphic extension into the strip $-1 < \Im x < 0$ if and only if $f = 0$ or $\psi = 0$; so for $\psi \in C_0^{\infty}(\mathbb{R})$, we have $\psi(Q)\mathbf{F} \subset \mathbf{F}$ if and only if $\psi = 0$. The same phenomenon occurs for the dilation group in the Sobolev space $\mathscr{H}^1(\mathbb{R})$ (cf. Chapter 4), which can be seen explicitly by using the Mellin transformation.

In order to see the kind of difficulties which occur when $\mathbf{F}$ is not a Hilbert space, let us consider the translation group in one dimension. Let $P = -i\frac{d}{dx}$ be its generator. In Section 1.2 we have defined $\varphi(P) : \mathscr{S}(\mathbb{R}) \to \mathscr{S}^*(\mathbb{R})$ for any $\varphi \in \mathscr{S}^*(\mathbb{R})$. If $\widehat{\varphi}$ is an integrable measure on $\mathbb{R}$, then (1.2.13) shows that this operator coincides with that defined by the rules of the present section. In the case of $\mathbf{F} = L^r(\mathbb{R})$, $1 \leq r \leq \infty$, the problem of obtaining a bound for $\|\varphi(P)\|_{B(\mathbf{F})}$ directly in terms of φ is quite deep and has been much studied (see [St1], [St2]). For example, a remarkable result of Marcinkiewicz states that for each $r \in (1, \infty)$ there is a constant $c(r) < \infty$ such that $\|\varphi(P)\|_{B(L^r(\mathbb{R}))} \leq c(r)[\sup |\varphi(x)| + \sup |x\varphi'(x)|]$. On the negative side, it is known that $\varphi(P)$ is bounded in $L^1(\mathbb{R})$ or $L^{\infty}(\mathbb{R})$ if and only if φ is the Fourier transform of an integrable measure (see Theorems I.3.19 and I.3.20 in [SW]). So the estimate $\|\varphi(P)\|_{B(L^1(\mathbb{R}))} \leq \|\varphi\|_{\mathscr{M}^0(\mathbb{R})}$ is the best possible. The preceding assertions also hold when $\mathbf{F} = C_{\infty}(\mathbb{R})$, in which case they are very easy to prove: let us show that, *if $S : C_{\infty}(\mathbb{R}) \to C_{\infty}(\mathbb{R})$ is a bounded, linear operator which commutes with translations, then there is $\varphi \in \mathscr{M}^0(\mathbb{R})$ such that $S = \varphi(P)$ and $\|S\| = \|\varphi\|_{\mathscr{M}^0(\mathbb{R})}$.* Indeed, note that $f \mapsto (Sf)(0)$ is a continuous linear form on $C_{\infty}(\mathbb{R})$, i.e. is a Radon measure on $\mathbb{R}$ which we denote by μ. Then:

$$(Sf)(x) = (e^{iPx}Sf)(0) = (Se^{iPx}f)(0)$$
$$= \int_{\mathbb{R}} (e^{iPx}f)(y)\mu(dy) = \int_{\mathbb{R}} f(x+y)\mu(dy) = (\varphi(P)f)(x)$$

with $\varphi(P) = \int_{\mathbb{R}} e^{iPy}\mu(dy)$.

3.3. The Discrete Sobolev Scale Associated to a C_0-Group

3.3.1. In this section we fix a C_0-representation W of $\mathbb{R}^n$ in a Banach space $\mathbf{F}$ and begin by defining spaces of vectors in $\mathbf{F}$ which are smooth in a certain sense with respect to the action of this group. More precisely, for each $m \in \mathbb{N} \cup \{\infty\}$ we introduce the space

$$(3.3.1) \quad \mathbf{F}_m = \{f \in \mathbf{F} \mid \mathbb{R}^n \ni x \mapsto W(x)f \in \mathbf{F} \text{ is strongly of class } C^m\}.$$

Observe that this definition may be applied to an arbitrary representation W of $\mathbb{R}^n$ in $\mathbf{F}$ (without any continuity assumption). Occasionally we shall use the spaces $\mathbf{F}_m$ in such a general context. Then the continuity of W in $\mathbf{F}$ is equivalent with the statement $\mathbf{F} = \mathbf{F}_0$. If m is finite, we introduce in $\mathbf{F}_m$ the norm:

$$(3.3.2) \qquad \|f\|_{\mathbf{F}_m} = \Big(\sum_{|\alpha| \leq m} \|\partial^{\alpha} W(0)f\|_{\mathbf{F}}^2 \Big)^{1/2},$$

where the derivatives are meant with respect to the canonical orthonormal basis of $\mathbb{R}^n$. We choose this specific form of the norm because we are mostly interested in the situation where $\mathbf{F}$ is a Hilbert space, in which case $\mathbf{F}_m$ is also a Hilbert space. We first show that it suffices to require weak differentiability in the definition (3.3.1) of $\mathbf{F}_m$.

LEMMA 3.3.1. *Let $m \geq 1$ be an integer and $f \in \mathbf{F}$. Then f belongs to $\mathbf{F}_m$ if and only if the function $x \mapsto W(x)f \in \mathbf{F}$ is weakly of class C^{m-1} and for each multi-index α with $|\alpha| = m - 1$ the function $x \mapsto f_\alpha(x) \equiv \partial^\alpha W(x)f$ has weak partial derivatives $\partial_j f_\alpha(y)$ $(j = 1, \ldots, n)$ at some point $y \in \mathbb{R}^n$. In this case one has for all $|\alpha| \leq m$ and all $x, y \in \mathbb{R}^n$:*

$$(3.3.3) \qquad \partial_x^\alpha W(x)f = W(x - y)\partial_y^\alpha W(y)f.$$

PROOF. For $x, y, \varepsilon \in \mathbb{R}^n$ we have:

$$(3.3.4) \qquad W(x + \varepsilon) - W(x) = W(x - y)[W(y + \varepsilon) - W(y)].$$

Thus, if the first order derivatives of $W(\cdot)f$ exist weakly at some point $y \in \mathbb{R}^n$, they exist weakly at each $x \in \mathbb{R}^n$, and they are strongly continuous functions of x. Then $W(\cdot)f$ is strongly of class C^1 (this follows from Newton's formula; see Corollary 5.A.3 for details). If $m > 1$, it suffices to repeat this argument iteratively for the higher derivatives, by observing that (3.3.4) implies (3.3.3). $\square$

If α is a multi-index and $f \in \mathbf{F}_{|\alpha|}$, we set

$$(3.3.5) \qquad A^\alpha f = (-i)^{|\alpha|}\partial^\alpha W(x)f|_{x=0}.$$

This defines A^α as a linear operator from $\mathbf{F}_{|\alpha|}$ to $\mathbf{F}$. In terms of these operators one may rewrite the expression (3.3.2) for the norm in $\mathbf{F}_m$ and the relation (3.3.3) as follows:

$$(3.3.6) \qquad \|f\|_{\mathbf{F}_m} = \left[\sum_{|\alpha| \leq m} \|A^\alpha f\|_{\mathbf{F}}^2 \right]^{1/2},$$

$$(3.3.7) \qquad \partial^\alpha W(x)f = i^{|\alpha|}W(x)A^\alpha f = i^{|\alpha|}A^\alpha W(x)f \quad (f \in \mathbf{F}_{|\alpha|}).$$

Clearly (3.3.6) implies that, if β is a multi-index, then $A^\beta \in B(\mathbf{F}_{m+|\beta|}, \mathbf{F}_m)$ for each $m \in \mathbb{N}$, with

$$(3.3.8) \qquad \|A^\beta\|_{\mathbf{F}_{m+|\beta|} \to \mathbf{F}_m} \leq 1.$$

Similarly (3.3.6) implies that $\|f\|_{\mathbf{F}_{m-1}} \leq \|f\|_{\mathbf{F}_m}$ if $m \geq 1$ and $f \in \mathbf{F}_m$; in particular the following embeddings are continuous :

$$(3.3.9) \qquad \cdots \subset \mathbf{F}_m \subset \mathbf{F}_{m-1} \subset \cdots \subset \mathbf{F}_2 \subset \mathbf{F}_1 \subset \mathbf{F}_0 = \mathbf{F}.$$

Since the multiple derivatives in ∂^α may be calculated successively in any order, it is clear that, if $\alpha = \beta + \gamma$ and $f \in \mathbf{F}_{|\alpha|}$, then $f \in \mathbf{F}_{|\gamma|}$, $A^\gamma f \in \mathbf{F}_{|\beta|}$ and

$$(3.3.10) \qquad A^\alpha f = A^\beta(A^\gamma f).$$

If $|\alpha| = 1$, we necessarily have $\partial_x^\alpha = \partial_{x_j}$ for some $j \in \{1, \ldots, n\}$. In this case we write A_j for the operator A^α. The collection $A = (A_1, \ldots, A_n)$ is called the *generator* of the C_0-group W. The relation (3.3.10) implies that $A_j A_k f = A_k A_j f$ if $f \in \mathbf{F}_2$ and that $A^\alpha f = A_1^{\alpha_1} \cdot \ldots \cdot A_n^{\alpha_n} f$ if $f \in \mathbf{F}_{|\alpha|}$.

Next, let T be a bounded operator in $\mathbf{F}$ commuting with W. Then obviously T maps $\mathbf{F}_m$ into itself, and one has

$$(3.3.11) \qquad \qquad ||T||_{B(\mathbf{F}_m)} \leq ||T||_{B(\mathbf{F})}.$$

Two special cases ($T = W(x)$ and $T = \varphi(A)$) occur in the following two propositions.

PROPOSITION 3.3.2. *Let* $m \in \mathbb{N}$. *Then*
(a) $\mathbf{F}_m$ *is a Banach space (it is a Hilbert space if* $\mathbf{F}$ *is a Hilbert space).*
(b) $W(x)$ *leaves* $\mathbf{F}_m$ *invariant and induces in* $\mathbf{F}_m$ *a bounded operator satisfying*

$$(3.3.12) \qquad \qquad ||W(x)||_{B(\mathbf{F}_m)} \leq ||W(x)||_{B(\mathbf{F})}.$$

Moreover, $\{W(x)|_{\mathbf{F}_m}\}_{x \in \mathbb{R}^n}$ *is a* C_0-*group in* $\mathbf{F}_m$.

PROOF. The proof of (b) is easy, and we omit it. For (a), assume that $\{f_k\}$ is a Cauchy sequence in $\mathbf{F}_m$. Then for each $|\alpha| \leq m$ there is $f^\alpha \in \mathbf{F}$ such that $\partial^\alpha W(0) f_k \to f^\alpha$ as $k \to \infty$. So (3.3.3) implies that $\partial^\alpha W(x) f_k \to W(x) f^\alpha$ uniformly in x on any compact subset of $\mathbb{R}^n$. Clearly one gets that $W(x) f^0$ is of class C^m, so $f^0 \in \mathbf{F}_m$, and $\partial^\alpha W(0) f^0 = f^\alpha$. So f_k converges to f^0 in $\mathbf{F}_m$, i.e. $\mathbf{F}_m$ is a Banach space. $\square$

In the next proposition we describe the behaviour of the functional calculus of Section 3.2 in relation with the scale of Banach spaces $\{\mathbf{F}_m\}_{m \in \mathbb{N}}$. We shall also express in mathematical terms the intuition that the operators A^α should be equal to $\varphi(A)$ with $\varphi(x) = x^\alpha$ if such functions were allowed by the functional calculus (later on a more general functional calculus will be developed; in particular $\varphi(A)$ will be defined for functions φ that are Fourier transforms of distributions with compact support).

PROPOSITION 3.3.3. *Let* $k, m \in \mathbb{N}$ *and* $w(x) = \max(1, ||W(x)||_{B(\mathbf{F})})$.
(a) *If* $\varphi \in \mathscr{M}_w(\mathbb{R}^n)$, *then the operator* $\varphi(A)$ *defined by* (3.2.11) *leaves* $\mathbf{F}_m$ *invariant and defines a bounded operator in* $\mathbf{F}_m$ *with*

$$||\varphi(A)||_{B(\mathbf{F}_m)} \leq ||\varphi(A)||_{B(\mathbf{F})} \leq ||\varphi||_{\mathscr{M}_w}.$$

This operator $\varphi(A)|_{\mathbf{F}_m}$ *coincides with the operator associated by Definition 3.2.7 to the* C_0-*group* $W_{\mathbf{F}_m} \equiv W|_{\mathbf{F}_m}$ *in* $\mathbf{F}_m$.
(b) *Assume that* φ *is such that, for each multi-index* α *with* $|\alpha| \leq k$, *the function* $\varphi_{(\alpha)}(x) \equiv x^\alpha \varphi(x)$ *belongs to* $\mathscr{M}_w(\mathbb{R}^n)$. *Then the operator* $\varphi(A)$ *belongs to* $B(\mathbf{F}_m, \mathbf{F}_{m+k})$, *and one has* $A^\alpha \varphi(A) = \varphi_{(\alpha)}(A)$ *on* $\mathbf{F}$ *and* $\varphi(A) A^\alpha = \varphi_{(\alpha)}(A)$ *on* $\mathbf{F}_{|\alpha|}$ *for* $|\alpha| \leq k$.

PROOF. (a) This is simple to check by taking into account Definition 3.2.7, Proposition 3.2.8 and the inequality (3.3.12).
(b) We have $W(x) \varphi(A) = \varphi(A) W(x) = (e_x \varphi)(A)$, where $e_x(y) = e^{i(x,y)}$. Under the assumptions made on φ, the function $x \mapsto e_x \varphi$ is feebly of class C^k

with $\partial_x^\alpha(e_x\varphi) = i^{|\alpha|}e_x\varphi_{(\alpha)}$ (see Proposition 3.1.10). Thus Proposition 3.2.9 (a) implies that $W(x)\varphi(A)$ is strongly of class C^k in $\mathbf{F}$, with

$$(3.3.13) \qquad A^\alpha\varphi(A)f \equiv (-i)^{|\alpha|}\partial^\alpha W(0)\varphi(A)f = \varphi_{(\alpha)}(A)f \quad \forall f \in \mathbf{F}.$$

Furthermore

$$(3.3.14) \qquad \|\varphi(A)f\|_{\mathbf{F}_{m+k}}^2 = \sum_{|\beta|\leq m+k} \|A^\beta\varphi(A)f\|_{\mathbf{F}}^2$$

$$\leq \sum_{|\gamma|\leq m}\sum_{|\alpha|\leq k} \|A^\gamma A^\alpha\varphi(A)f\|_{\mathbf{F}}^2$$

$$= \sum_{|\alpha|\leq k} \|\varphi_{(\alpha)}(A)f\|_{\mathbf{F}_m}^2$$

$$\leq \sum_{|\alpha|\leq k} \|\varphi_{(\alpha)}\|_{\mathcal{M}_w}^2\|f\|_{\mathbf{F}_m}^2. \quad \square$$

The result of Proposition 3.3.3 (b) will be frequently used in the following way. Observe that

$$(3.3.15) \qquad \mathbf{F}_\infty = \bigcap_{m=0}^\infty \mathbf{F}_m.$$

Let $\varphi : \mathbb{R}^n \to \mathbb{C}$ be such that $\widehat{\varphi} \in C_0^\infty(\mathbb{R}^n)$ and $\varphi(0) = 1$. Then $\varphi(\sigma A) \in B(\mathbf{F}_m, \mathbf{F}_{m'})$ for all $m, m' \in \mathbb{N}$ and $\sigma > 0$, by Proposition 3.3.3 (b). Thus, if $f \in \mathbf{F}_m$ for some $m \in \mathbb{N}$, then $\varphi(\sigma A)f \in \mathbf{F}_\infty$ for each $\sigma > 0$, and $\varphi(\sigma A)f \to f$ strongly in $\mathbf{F}_m$ as $\sigma \to 0$ (by virtue of (3.2.19) applied in $\mathbf{F}_m$). In particular we have obtained the denseness of $\mathbf{F}_\infty$ in each $\mathbf{F}_m$, $m \in \mathbb{N}$.

The preceding argument can also be applied to get the following important density result which is a generalization of Nelson's self-adjointness criterion (Theorem VIII.11 in [RS]).

THEOREM 3.3.4. *Let $m \in \mathbb{N}$ and let $\mathbf{E}$ be a vector subspace of $\mathbf{F}_m$ which is invariant under W, i.e. such that $W(x)\mathbf{E} \subset \mathbf{E}$ for each $x \in \mathbb{R}^n$. If $\mathbf{E}$ is dense in $\mathbf{F}$, then it is also dense in $\mathbf{F}_m$.*
In particular $\mathbf{F}_\infty$ is a dense subspace of each of the Banach spaces $\mathbf{F}_m$.

PROOF. We denote by $\overline{\mathbf{E}}$ the closure of $\mathbf{E}$ in $\mathbf{F}_m$. Let $\psi : \mathbb{R}^n \to \mathbb{R}$ be such that $\widehat{\psi} \in C_0^\infty(\mathbb{R}^n)$. If $e \in \mathbf{E}$, then $x \mapsto W(x)e \in \mathbf{E}$ is continuous in the norm of $\mathbf{F}_m$, hence

$$\psi(A)e = \int_{\mathbb{R}^n} \widehat{\psi}(x)W(x)e\underline{d}x \in \overline{\mathbf{E}}.$$

This shows that $\psi(A)\mathbf{E} \subset \overline{\mathbf{E}}$. Since $\mathbf{E}$ is dense in $\mathbf{F}$ and $\psi(A) \in B(\mathbf{F}, \mathbf{F}_m)$ by Proposition 3.3.3 (b), we then have $\psi(A)\mathbf{F} \subset \overline{\mathbf{E}}$.

Now let $f \in \mathbf{F}_m$. Choose $\varphi : \mathbb{R}^n \to \mathbb{R}$ such that $\varphi(0) = 1$ and $\widehat{\varphi} \in C_0^\infty(\mathbb{R}^n)$. Then $\varphi(\sigma A)f \in \overline{\mathbf{E}}$ for each $\sigma > 0$ by the preceding considerations, and $\varphi(\sigma A)f$ converges to f strongly in $\mathbf{F}_m$ by virtue of (3.2.19) (applied in $\mathbf{F}_m$). This shows that $\overline{\mathbf{E}} = \mathbf{F}_m$. $\square$

It follows from Proposition 3.3.2 and Theorem 3.3.4 that, in the case $n = 1$, the operator $A = A_1$ is closed and densely defined in $\mathbf{F}$ on the domain $D(A) = \mathbf{F}_1$ (and so are its powers A^m with domain $D(A^m) = \mathbf{F}_m$, $m \geq 2$). If $n > 1$, the operators A_j are *not* closed on $\mathbf{F}_1$ in the most interesting cases. This fact is the origin of many difficulties in the case of a multi-parameter group.

We now mention a *Taylor expansion formula* which expresses the fact that, formally, we have $W(x) = \exp(iA \cdot x)$ with $A \cdot x = A_1 x_1 + \cdots + A_n x_n$. Let $m \geq 1$ be an integer and $f \in \mathbf{F}_m$. Then the function $x \mapsto W(x)f \in \mathbf{F}$ is of class C^m and

(3.3.16)

$$W(x)f = \sum_{|\alpha| < m} \frac{(ix)^\alpha}{\alpha!} A^\alpha f + \int_0^1 m(1-\tau)^{m-1} W(\tau x) d\tau \sum_{|\alpha| = m} \frac{(ix)^\alpha}{\alpha!} A^\alpha f$$

$$\equiv \sum_{|\alpha| < m} \frac{(ix)^\alpha}{\alpha!} A^\alpha f + \sum_{|\alpha| = m} \frac{(ix)^\alpha}{\alpha!} W_m(x) A^\alpha f.$$

This follows from the equations (1.1.8) and (3.3.7). $W_m(x)$ is the operator defined by $W_m(x) = \int_0^1 W(\tau x) m(1 - \tau)^{m-1} d\tau$. Then $W_m : \mathbb{R}^n \to B(\mathbf{F})$ is strongly continuous, $W_m(0) = I$ and

(3.3.17) $$\|W_m(x)\|_{B(\mathbf{F})} \leq \sup_{0 < \tau < 1} \|W(\tau x)\|_{B(\mathbf{F})}.$$

On several occasions we shall have to estimate expressions of the form $[W(x) - I]^\ell f$ as $x \to 0$. At a formal level we proceed as follows. For integers $\ell \geq m \geq 1$ and $\sigma \in \mathbb{R}$ we write

$$(e^{i\sigma} - 1)^\ell = (i\sigma)^m \frac{(e^{i\sigma} - 1)^\ell}{(i\sigma)^m} = (i\sigma)^m \int_{\mathbb{R}} e^{i\sigma\tau} \chi_{\ell,m}(\tau) d\tau.$$

One can easily show (cf. below) that this holds with $\chi_{\ell,m}$ a function of compact support and that $\chi_{\ell,m} = (-1)^{\ell-m} \chi_\ell^{(\ell-m)}$, where $\chi_\ell \equiv \chi_{\ell,\ell}$. If we (formally) replace here σ by $A \cdot x = A_1 x_1 + \cdots + A_n x_n$ and use the formula $(a_1 + \cdots + a_n)^m = \sum_{|\alpha| = m} \frac{m!}{\alpha!} a^\alpha$, we get

$$[W(x) - I]^\ell = \sum_{|\alpha| = m} \frac{m!}{\alpha!} (ix)^\alpha A^\alpha \int_{\mathbb{R}} W(\tau x) \chi_{\ell,m}(\tau) d\tau.$$

The functions $\chi_\ell : \mathbb{R} \to \mathbb{R}$ are explicitly given by the expression

$$\chi_\ell(\tau) = \sum_{j=0}^{\ell} (-1)^{\ell-j} \binom{\ell}{j} \chi_{[0,j]}(\tau) \frac{(j - \tau)^{\ell-1}}{(\ell - 1)!}.$$

Here $\chi_{[0,j]}$ is the characteristic function of the interval $[0, j]$. Below we set $\chi_1 \equiv \chi_{[0,1]}$, so $\chi_{[0,j]}(\tau) = \chi_1(\tau/j)$. It is natural to denote by χ_0 the Dirac measure at 0.

In the next proposition we give a rigorous derivation of the preceding representation of $[W(x) - I]^\ell$ and then we describe some properties of the functions χ_ℓ ($\ell \geq 1$ integer). We shall, exceptionally, define the convolution of two functions on $\mathbb{R}$ in terms of Lebesgue (not Fourier) measure: $(\varphi * \psi)(\tau) = \int_{\mathbb{R}} \varphi(\tau - \sigma)\psi(\sigma)d\sigma$. Dirac measure on $\mathbb{R}$ with support $\{j\}$ is denoted by δ_j and we set $\delta_0 \equiv \delta \equiv \chi_0$. The one-dimensional translation group will be denoted by $\{e^{i\sigma P}\}$, more precisely $(e^{i\sigma P}\varphi)(\tau) = \varphi(\tau + \sigma)$. In particular, $e^{-ijP}\delta = \delta_j$.

PROPOSITION 3.3.5. *If $\ell \geq m \geq 1$ are integers, then one has the following identity in $B(\mathbf{F}_m, \mathbf{F})$:*

(3.3.18)

$$[W(x) - I]^\ell = \sum_{|\alpha|=m} \frac{(ix)^\alpha A^\alpha}{\alpha!} \sum_{j=0}^{\ell} (-1)^{\ell-j} \binom{\ell}{j} j^m W_m(jx)$$

$$= (-1)^{\ell-m} \sum_{|\alpha|=m} \frac{m!}{\alpha!} (ix)^\alpha A^\alpha \int_{\mathbb{R}} W(\tau x) \chi_\ell^{(\ell-m)}(\tau)d\tau.$$

The functions χ_ℓ ($\ell \geq 1$) have the following properties:
(a) $\chi_1 = \chi_{[0,1]}$ *and if $\ell \geq 2$ then $\chi_\ell = \chi_1 * \cdots * \chi_1$ (ℓ factors).*
(b) $0 \leq \chi_\ell \leq 1$, supp $\chi_\ell = [0, \ell]$ *and $\int_{\mathbb{R}} \chi_\ell(\tau)d\tau = 1$.*
(c) *On each interval $[j, j+1]$ ($j = 0, 1, \ldots, \ell - 1$) χ_ℓ is a polynomial of degree $\ell - 1$; we have $\chi_\ell(\tau) = \tau^{\ell-1}/(\ell-1)!$ if $0 \leq \tau \leq 1$ and $\chi_\ell(\tau) = (\ell-\tau)^{\ell-1}/(\ell-1)!$ if $\ell - 1 \leq \tau \leq \ell$.*
(d) *If $0 \leq k \leq \ell$ is an integer, then $\chi_\ell^{(k)} = (I - e^{-iP})^k \chi_{\ell-k}$ with $\chi_0 = \delta$. In particular $\chi_\ell^{(\ell)} = \sum_{j=0}^{\ell} (-1)^j \binom{\ell}{j} \delta_j$ is a measure (binomial distribution) with support $\{0, 1, \ldots, \ell\}$ and $\chi_\ell^{(\ell-1)}$ is a function of bounded variation. If $\ell \geq 2$ then $\chi_\ell \in C_0^{\ell-2}$ and $\chi_\ell^{(\ell-2)}$ is a Lipschitz function.*

PROOF. (i) We have by (3.3.16):

$$[W(x) - 1]^\ell = \sum_{j=0}^{\ell} (-1)^{\ell-j} \binom{\ell}{j} W(jx)$$

$$= \sum_{j=0}^{\ell} \sum_{|\alpha|<m} (-1)^{\ell-j} \binom{\ell}{j} j^{|\alpha|} \frac{(ix)^\alpha}{\alpha!} A^\alpha +$$

$$+ \sum_{j=0}^{\ell} \sum_{|\alpha|=m} (-1)^{\ell-j} \binom{\ell}{j} \frac{(ix)^\alpha}{\alpha!} j^m A^\alpha W_m(jx).$$

To obtain the first equality in (3.3.18), it suffices to show that for each integer $m \in [0, \ell]$ one has

$$\sum_{j=0}^{\ell} (-1)^{\ell-j} \binom{\ell}{j} j^m = \delta_{m\ell} \cdot \ell!$$

For this observe that

$$\left(x \frac{d}{dx}\right)^m (x - 1)^\ell|_{x=1} = x^m \left(\frac{d}{dx}\right)^m (x - 1)^\ell|_{x=1} = \delta_{m\ell} \cdot \ell!$$

But

$$\left(x\frac{d}{dx}\right)^m (x-1)^\ell|_{x=1} = \sum_{j=0}^{\ell}(-1)^{\ell-j}\binom{\ell}{j}\left(x\frac{d}{dx}\right)^m x^j|_{x=1}$$

$$= \sum_{j=0}^{\ell}(-1)^{\ell-j}\binom{\ell}{j}j^m x^j|_{x=1}$$

$$= \sum_{j=0}^{\ell}(-1)^{\ell-j}\binom{\ell}{j}j^m.$$

(ii) In the expression that we have obtained so far for $[W(x)-I]^\ell$ we insert now the definition of $W_m(jx)$. After a change of variables we get

$$[W(x)-I]^\ell = \sum_{|\alpha|=m}\frac{m!}{\alpha!}(ix)^\alpha A^\alpha \int_{\mathbb{R}}W(\tau x)\chi_{\ell,m}(\tau)d\tau$$

with

$$\chi_{\ell,m}(\tau) = \sum_{j=0}^{\ell}(-1)^{\ell-j}\binom{\ell}{j}\chi_{[0,j]}(\tau)\frac{(j-\tau)^{m-1}}{(m-1)!}.$$

Now let us consider the preceding representation in the particular case $\mathbf{F}=\mathbb{C}$, $n=1$, $A=1$. We get

$$\int_{\mathbb{R}}e^{ix\tau}\chi_{\ell,m}(\tau)d\tau = \frac{(e^{ix}-1)^\ell}{(ix)^m} = (ix)^{\ell-m}\left[\frac{e^{ix}-1}{ix}\right]^\ell.$$

The last equality implies $\chi_{\ell,m}=(-1)^{\ell-m}\chi_\ell^{(\ell-m)}$ with $\chi_{\ell,\ell}\equiv\chi_\ell$. Hence (3.3.18) is completely proved.

(iii) It remains to show that the functions χ_ℓ have the properties (a)-(d). Property (c) is obvious by the definition of χ_ℓ. Property (a) follows from $\int e^{ix\tau}\chi_\ell(\tau)d\tau = [(e^{ix}-1)/(ix)]^\ell$, and (a) clearly implies (b). For $\ell=1,2$ the assertion (d) is proved by a simple calculation; in particular $\chi_1'=\delta_0-\delta_1$. If $\ell\geq 3$ then $\chi_{\ell-1}$ is a continuous function and $\chi_\ell=\chi_{\ell-1}*\chi_1$; hence

$$\chi_\ell'=\chi_{\ell-1}*\chi_1'=\chi_{\ell-1}*\delta_0-\chi_{\ell-1}*\delta_1=(I-e^{-iP})\chi_{\ell-1}.$$

This implies (d) for any k by induction. $\square$

In connection with the second equality in (3.3.18) note that $y\mapsto W(y)\in B(\mathbf{F}_m,\mathbf{F})$ is strongly of class C^m and

$$\frac{d^m}{d\tau^m}W(\tau x) = \sum_{|\alpha|=m}\frac{m!}{\alpha!}(ix)^\alpha A^\alpha W(\tau x) \equiv (ix\cdot A)^m W(\tau x)$$

where $x\cdot A=x_1A_1+\cdots+x_nA_n$. Hence the second equality in (3.3.18) is obvious if $m=0$ and one obtains it for all $m\leq\ell$ by integration by parts.

We can generalize (3.3.18) as follows. Let θ be a function on $\mathbb{R}^n$ such that its Fourier transform $\widehat{\theta}$ is a measure of compact support. By integrating the first and the last member of (3.3.18) with respect to this measure we get

(3.3.19)
$$\int_{\mathbb{R}^n} [W(x) - I]^\ell \widehat{\theta}(x) \underline{d}x = (-1)^{\ell-m} \sum_{|\alpha|=m} \frac{m!}{\alpha!} A^\alpha \int_{\mathbb{R}} \theta^{(\alpha)}(\tau A) \chi_\ell^{(\ell-m)}(\tau) d\tau.$$

If $\widehat{\theta}$ is the Dirac measure at x then (3.3.19) reduces to (3.3.18). As above we have $\sum_{|\alpha|=m} \frac{m!}{\alpha!} A^\alpha \theta^{(\alpha)}(\tau A) = (\frac{d}{d\tau})^m \theta(\tau A)$, hence for $m = 0$ the preceding identity is obvious; one then obtains (3.3.19) for any m by integrating by parts.

3.3.2. We shall now extend the scale $\{\mathbf{F}_m\}_{m \in \mathbb{N}}$ to negative integer values of the index m. The procedure closely follows one of the standard ways of defining Sobolev spaces with negative index in distribution theory. More specifically, we shall introduce for each integer $m \geq 1$ a norm $||\cdot||_{\mathbf{F}_{-m}}$ on $\mathbf{F}$ that is weaker than $||\cdot||_{\mathbf{F}}$, and then we shall define $\mathbf{F}_{-m}$ as the completion of $\mathbf{F}$ with respect to this new norm. If $m \geq 1$ is an integer and $f \in \mathbf{F}$, we set

(3.3.20)
$$||f||_{\mathbf{F}_{-m}} = \inf\left\{ \left[\sum_{|\alpha| \leq m} ||f_\alpha||_{\mathbf{F}}^2 \right]^{1/2} \;\middle|\; f_\alpha \in \mathbf{F}_{|\alpha|}, \, f = \sum_{|\alpha| \leq m} A^\alpha f_\alpha \right\}$$
$$= \inf\left\{ \left[\sum_{0 \neq |\alpha| \leq m} ||f_\alpha||_{\mathbf{F}}^2 + ||f - \sum_{0 \neq |\alpha| \leq m} A^\alpha f_\alpha||_{\mathbf{F}}^2 \right]^{1/2} \;\middle|\; f_\alpha \in \mathbf{F}_{|\alpha|} \right\}.$$

By taking $f_0 = f$ and $f_\alpha = 0$ for $|\alpha| > 0$, we get the inequality $||f||_{\mathbf{F}_{-m}} \leq ||f||_{\mathbf{F}}$, so $||f||_{\mathbf{F}_{-m}}$ is finite for each $f \in \mathbf{F}$. It is also obvious that $||\cdot||_{\mathbf{F}_{-m}}$ is a seminorm on $\mathbf{F}$, and we shall see in the next proposition that it is a norm. The special form $(\sum_{|\alpha| \leq m} ||f_\alpha||_{\mathbf{F}}^2)^{1/2}$ used to define $||f||_{\mathbf{F}_{-m}}$ is convenient because it ensures that $||\cdot||_{\mathbf{F}_{-m}}$ is a Hilbertian norm if $\mathbf{F}$ is a Hilbert space.

PROPOSITION 3.3.6. *For each integer $m \geq 1$, $||\cdot||_{\mathbf{F}_{-m}}$ is a norm on $\mathbf{F}$; if $\mathbf{F}$ is a Hilbert space, then $||\cdot||_{\mathbf{F}_{-m}}$ is a Hilbertian norm. If $1 \leq k \leq m$ are integers, then*

(3.3.21)
$$||f||_{\mathbf{F}_{-m}} \leq ||f||_{\mathbf{F}_{-k}} \leq ||f||_{\mathbf{F}} \qquad \forall f \in \mathbf{F}.$$

PROOF. (i) (3.3.21) is evident. In order to check that the seminorm $||\cdot||_{\mathbf{F}_{-m}}$ is a norm, we must show that $||f||_{\mathbf{F}_{-m}} = 0$ implies that $f = 0$. If f is such that $||f||_{\mathbf{F}_{-m}} = 0$, then there are sequences $\{f_{\alpha,j}\}_{j \in \mathbb{N}}$, $|\alpha| \leq m$ satisfying

$$f_{\alpha,j} \in \mathbf{F}_{|\alpha|}, \, f = \sum_{|\alpha| \leq m} A^\alpha f_{\alpha,j} \text{ and } \lim_{j \to \infty} \sum_{|\alpha| \leq m} ||f_{\alpha,j}||_{\mathbf{F}}^2 = 0.$$

If $\varphi \in \mathscr{S}(\mathbb{R}^n)$ is such that $\widehat{\varphi} \in C_0^\infty(\mathbb{R}^n)$, then Proposition 3.3.3 (b) implies that $\varphi(A)f = \sum_{|\alpha| \leq m} \varphi_{(\alpha)}(A)f_{\alpha,j}$ for each $j \in \mathbb{N}$. Since $\varphi_{(\alpha)}(A) \in B(\mathbf{F})$ by Proposition 3.3.3, we get that $\varphi(A)f = 0$. Now choose $\psi \in \mathscr{S}(\mathbb{R}^n)$ such that

$\psi(0) = 1$ and $\widehat{\psi} \in C_0^\infty(\mathbb{R}^n)$. Then $\psi(\varepsilon A)f = 0$ for each $\varepsilon > 0$, and $\psi(\varepsilon A)f \to \psi(0)f = f$ as $\varepsilon \to 0$ by (3.2.19). This shows that $f = 0$.

(ii) Now assume that $\mathbf{F}$ is a Hilbert space. To see that $||\cdot||_{\mathbf{F}_{-m}}$ is the norm associated to a scalar product, we use the Jordan-von Neumann theorem: it is enough to show that the condition (i) in Lemma 2.8.8 is satisfied. For this, we let f and g be represented as $f = \sum_{|\alpha| \leq m} A^\alpha f_\alpha$ and $g = \sum_{|\alpha| \leq m} A^\alpha g_\alpha$ with $f_\alpha, g_\alpha \in \mathbf{F}_{|\alpha|}$ and apply the parallelogram identity for $||\cdot||_{\mathbf{F}}$ to obtain that

$$||f + g||_{\mathbf{F}_{-m}}^2 + ||f - g||_{\mathbf{F}_{-m}}^2 \leq \sum_{|\alpha| \leq m} (||f_\alpha + g_\alpha||_{\mathbf{F}}^2 + ||f_\alpha - g_\alpha||_{\mathbf{F}}^2)$$

$$= 2 \sum_{|\alpha| \leq m} (||f_\alpha||_{\mathbf{F}}^2 + ||g_\alpha||_{\mathbf{F}}^2).$$

By taking the infimum over all such representations of f and g, we see that

$$||f + g||_{\mathbf{F}_{-m}}^2 + ||f - g||_{\mathbf{F}_{-m}}^2 \leq 2(||f||_{\mathbf{F}_{-m}}^2 + ||g||_{\mathbf{F}_{-m}}^2). \quad \square$$

REMARK 3.3.7. It suffices to consider in (3.3.20) decompositions of f such that each f_α with $|\alpha| \neq 0$ belongs to $\mathbf{F}_\infty$. More generally, let $k \in \mathbb{N} \cup \infty$. Then one has for $f \in \mathbf{F}_k$:

(3.3.22)
$$||f||_{\mathbf{F}_{-m}} = \inf\Big\{ \Big[\sum_{|\alpha| \leq m} ||f_\alpha||_{\mathbf{F}}^2 \Big]^{1/2} \mid f = \sum_{|\alpha| \leq m} A^\alpha f_\alpha, f_0 \in \mathbf{F}_k, f_\alpha \in \mathbf{F}_\infty \text{ if } |\alpha| \neq 0 \Big\}.$$

To see this, it suffices to show that, if $f \in \mathbf{F}_k$, $f = \sum_{|\alpha| \leq m} A^\alpha f_\alpha$ with $f_\alpha \in \mathbf{F}_{|\alpha|}$ and $\varepsilon > 0$, one may write $f = \sum_{|\alpha| \leq m} A^\alpha g_\alpha$ with $g_0 \in \mathbf{F}_k$, $g_\alpha \in \mathbf{F}_\infty$ if $|\alpha| \neq 0$ and $||g_\alpha||_{\mathbf{F}} \leq ||f_\alpha||_{\mathbf{F}} + \varepsilon$. Now by Theorem 3.3.4 we may choose for each α with $1 \leq |\alpha| \leq m$ a vector $g_\alpha \in \mathbf{F}_\infty$ such that $||f_\alpha - g_\alpha||_{\mathbf{F}_{|\alpha|}} \leq \varepsilon M^{-1}$, where M denotes the total number of multi-indices β satisfying $|\beta| \leq m$. If we set

$$g_0 = f_0 + \sum_{1 \leq |\alpha| \leq m} A^\alpha (f_\alpha - g_\alpha) \equiv f - \sum_{1 \leq |\alpha| \leq m} A^\alpha g_\alpha,$$

we have $g_0 \in \mathbf{F}_k$, $f = \sum_{|\alpha| \leq m} A^\alpha g_\alpha$ and, by (3.3.8):

$$||g_0||_{\mathbf{F}} \leq ||f_0||_{\mathbf{F}} + \sum_{1 \leq |\alpha| \leq m} ||f_\alpha - g_\alpha||_{\mathbf{F}_{|\alpha|}} \leq ||f_0||_{\mathbf{F}} + \varepsilon. \quad \square$$

We may now define the spaces with negative index. Let $m \geq 1$ be an integer; then

(3.3.23) $\mathbf{F}_{-m} = $ completion of $\mathbf{F}$ under the norm $||\cdot||_{\mathbf{F}_{-m}}$.

Thus we have constructed a family $\{\mathbf{F}_m\}_{m \in \mathbb{Z}}$ of Banach spaces with $\mathbf{F}_0 = \mathbf{F}$ and with the following continuous and dense embeddings ($m \geq 1$):

(3.3.24)
$$\cdots \subset \mathbf{F}_{m+1} \subset \mathbf{F}_m \subset \cdots \subset \mathbf{F}_1 \subset \mathbf{F} \subset \mathbf{F}_{-1} \subset \cdots \subset \mathbf{F}_{-m} \subset \mathbf{F}_{-m-1} \subset \cdots$$

The family $\{\mathbf{F}_m\}_{m\in\mathbb{Z}}$ will be called the *(discrete) Sobolev scale associated to the C_0-group W in $\mathbf{F}$.* We set

$$(3.3.25) \qquad \mathbf{F}_{-\infty} = \bigcup_{m\in\mathbb{Z}} \mathbf{F}_m.$$

$\mathbf{F}_{-\infty}$ is a vector space, and it has a natural topology (the inductive limit topology, i.e. the finest locally convex topology such that each of the embeddings $j_m :$ $\mathbf{F}_m \to \mathbf{F}_{-\infty}$ is continuous). Similarly we may consider on $\mathbf{F}_\infty$ the natural Fréchet space structure defined by the family of norms $\{||\cdot||_{\mathbf{F}_m}\}_{m\in\mathbb{Z}}$. Then $\mathbf{F}_\infty$ and $\mathbf{F}_{-\infty}$ are locally convex Hausdorff topological vector spaces with $\mathbf{F}_\infty \subset \mathbf{F}_{-\infty}$ continuously and densely. We shall, however, avoid the use of these topologies. We observe that $\mathbf{F}_\infty$ is dense in each $\mathbf{F}_m$, $m \in \mathbb{Z}$, and it is invariant under $W(x)$ and A_j. In the following propositions we specify the behaviour of the scale $\{\mathbf{F}_m\}_{m\in\mathbb{Z}}$ under these operators. We mention that the norms $||\cdot||_{\mathbf{F}_m}$ are extended to gauges on $\mathbf{F}_{-\infty}$ by the rule $||f||_{\mathbf{F}_m} = \infty$ if $f \notin \mathbf{F}_m$.

PROPOSITION 3.3.8. (a) *For each $x \in \mathbb{R}^n$ there is a unique linear operator $W(x)$: $\mathbf{F}_{-\infty} \to \mathbf{F}_{-\infty}$ with the following properties:*

(i) *the restriction of $W(x)$ to $\mathbf{F}_\infty$ coincides with the restriction to $\mathbf{F}_\infty$ of the original operator $W(x)$ in $B(\mathbf{F})$,*

(ii) *$W(x)$ leaves each $\mathbf{F}_m$ invariant ($m \in \mathbb{Z}$) and induces in each $\mathbf{F}_m$ a continuous operator.*

(b) *The family of operators $\{W(x)\}_{x\in\mathbb{R}^n}$ forms a representation in $\mathbf{F}_{-\infty}$ of the additive group $\mathbb{R}^n$, and for each $m \in \mathbb{Z}$ and $f \in \mathbf{F}_m$, the function $x \mapsto W(x)f \in \mathbf{F}_m$ is strongly continuous. In particular, the restriction $\{W(x)|_{\mathbf{F}_m}\}_{x\in\mathbb{R}^n}$ of W to $\mathbf{F}_m$ is a C_0-group in $\mathbf{F}_m$. Moreover one has:*

$$(3.3.26) \qquad ||W(x)||_{B(\mathbf{F}_m)} \leq ||W(x)||_{B(\mathbf{F})} \quad \forall x \in \mathbb{R}^n, \forall m \in \mathbb{Z}.$$

PROOF. In view of Proposition 3.3.2 (b), it suffices to consider the case where $m < 0$. So let $m = -k$ with $k \geq 1$. If $f \in \mathbf{F}$ and $\varepsilon > 0$, we can find for each $|\alpha| \leq k$ a vector $f_\alpha \in \mathbf{F}_{|\alpha|}$ such that $f = \sum_{|\alpha|\leq k} A^\alpha f_\alpha$ and $[\sum_{|\alpha|\leq k} ||f_\alpha||_{\mathbf{F}}^2]^{1/2} \leq ||f||_{\mathbf{F}_{-k}} + \varepsilon$. Now $W(x)f = \sum_{|\alpha|\leq k} W(x)A^\alpha f_\alpha = \sum_{|\alpha|\leq k} A^\alpha \overline{W}(x)f_\alpha$, hence $||W(x)f||_{\mathbf{F}_{-k}} \leq ||W(x)||_{B(\mathbf{F})}(||f||_{\mathbf{F}_{-k}} + \varepsilon)$. This proves (3.3.26) for $m = -k$.

The denseness of $\mathbf{F}_\infty$ in $\mathbf{F}_{-k}$ implies that $W(x)$ has a unique continuous extension from $\mathbf{F}_\infty$ to $\mathbf{F}_{-k}$. Finally we have for $f \in \mathbf{F}$:

$$||W(x)f - W(y)f||_{\mathbf{F}_{-k}} \leq ||W(x)f - W(y)f||_{\mathbf{F}}.$$

Together with (3.3.26) and the denseness of $\mathbf{F}$ in $\mathbf{F}_{-k}$, this gives the strong continuity of W in $\mathbf{F}_{-k}$. $\square$

PROPOSITION 3.3.9. (a) *For each $j = 1,\ldots,n$ there is a unique linear operator $A_j : \mathbf{F}_{-\infty} \to \mathbf{F}_{-\infty}$ with the following properties:*

(i) *the restriction of A_j to $\mathbf{F}_\infty$ coincides with the restriction to $\mathbf{F}_\infty$ of the original operator $A_j \in B(\mathbf{F}_1, \mathbf{F})$,*

(ii) *for each $m \in \mathbb{Z}$, A_j maps $\mathbf{F}_m$ into $\mathbf{F}_{m-1}$ continuously.*

One has $A_j A_k = A_k A_j$ on $\mathbf{F}_{-\infty}$. If α is a multi-index and $A^\alpha : \mathbf{F}_{-\infty} \to \mathbf{F}_{-\infty}$ is defined as $A^\alpha = A_1^{\alpha_1} \cdot \ldots \cdot A_n^{\alpha_n}$, then $A^\alpha \in B(\mathbf{F}_m, \mathbf{F}_{m-|\alpha|})$ for each $m \in \mathbb{Z}$, and $\|A^\alpha\|_{\mathbf{F}_m \to \mathbf{F}_{m-|\alpha|}} \le 1$.

(b) Let $m \in \mathbb{Z}$, $\ell \ge 1$ an integer and $f \in \mathbf{F}_{\ell+m}$. Then the function $x \mapsto W(x)f$ is of class C^ℓ when considered with values in $\mathbf{F}_m$, and $\partial^\alpha W(x)f = i^{|\alpha|}W(x)A^\alpha f = i^{|\alpha|}A^\alpha W(x)f$ for $|\alpha| \le \ell$. In particular, the Taylor expansion (3.3.16) and the identity (3.3.18) hold in $\mathbf{F}_m$. [3]

PROOF. (a) Since $\mathbf{F}_\infty$ is dense in each $\mathbf{F}_m$, it suffices to show that

$$\|A_j f\|_{\mathbf{F}_{m-1}} \le \|f\|_{\mathbf{F}_m}$$

for all $f \in \mathbf{F}_\infty$ and each $m \in \mathbb{Z}$. For $m \ge 1$ this follows from (3.3.8). For $m \le 0$, let $f \in \mathbf{F}_\infty$ and consider a representation $f = \sum_{|\alpha| \le |m|} A^\alpha f_\alpha$ with $f_\alpha \in \mathbf{F}_\infty$. Then $A_j f = \sum_{|\alpha| \le |m|} A^{\alpha+\gamma_j} f_\alpha$, where $\gamma_j = (0,0,\ldots,0,1,0,\ldots,0)$ is the multi-index such that $|\gamma_j| = 1$ and its only non-zero entry is at the j-th place. Then one has

$$\|A_j f\|_{\mathbf{F}_{-|m|-1}} \le \Big[\sum_{|\alpha| \le |m|} \|f_\alpha\|_{\mathbf{F}}^2 \Big]^{1/2}.$$

By taking the infimum over all families $\{f_\alpha\}$ in $\mathbf{F}_\infty$ with $f = \sum_{|\alpha| \le |m|} A^\alpha f_\alpha$, one obtains that $\|A_j f\|_{\mathbf{F}_{-|m|-1}} \le \|f\|_{\mathbf{F}_{-|m|}}$ (see Remark 3.3.7).

(b) In view of Proposition 3.3.5 and the considerations that led to (3.3.16) it is enough to show that the function $x \mapsto W(x)f \in \mathbf{F}_m$ is of class C^ℓ with $\partial^\alpha W(x)f = i^{|\alpha|}W(x)A^\alpha f$ if $m \le 0$. For this, choose a sequence $\{f_k\}$ in $\mathbf{F}_\infty$ with $f_k \to f$ in $\mathbf{F}_{m+\ell}$. Then $\partial^\alpha W(x)f_k = i^{|\alpha|}W(x)A^\alpha f_k$; consequently for each $|\alpha| \le \ell$ the derivative of order α of the function $x \mapsto W(x)f_k \in \mathbf{F}_m$ converges uniformly on each compact subset of $\mathbb{R}^n$ to the continuous function $x \mapsto i^{|\alpha|}W(x)A^\alpha f \in \mathbf{F}_m$. This implies the required result. $\square$

PROPOSITION 3.3.10. Let $w(x) = \max(1, \|W(x)\|_{B(\mathbf{F})})$.

(a) If $\varphi \in \mathcal{M}_w(\mathbb{R}^n)$, there is a unique linear operator $\varphi(A) : \mathbf{F}_{-\infty} \to \mathbf{F}_{-\infty}$ whose restriction to $\mathbf{F}_\infty$ coincides with the restriction to $\mathbf{F}_\infty$ of the original operator $\varphi(A)$ in $B(\mathbf{F})$. $\varphi(A)$ leaves each $\mathbf{F}_m$ invariant and satisfies $\|\varphi(A)\|_{B(\mathbf{F}_m)} \le \|\varphi\|_{\mathcal{M}_w}$ for each $m \in \mathbb{Z}$, and the restriction of $\varphi(A)$ to $\mathbf{F}_m$ coincides with the operator in $\mathbf{F}_m$ associated by (3.2.11) to the C_0-group $\{W(x)|_{\mathbf{F}_m}\}_{x \in \mathbb{R}^n}$. The correspondence $\mathcal{M}_w(\mathbb{R}^n) \ni \varphi \mapsto \varphi(A)$ with values in the algebra of linear mappings in $\mathbf{F}_{-\infty}$ is a unital homomorphism.

(b) Let $k \in \mathbb{N}$ and assume that, for each multi-index α with $|\alpha| \le k$, the function $\varphi_{(\alpha)}(x) = x^\alpha \varphi(x)$ belongs to $\mathcal{M}_w(\mathbb{R}^n)$. Then $\varphi(A) \in B(\mathbf{F}_m, \mathbf{F}_{m+k})$ for each $m \in \mathbb{Z}$, and $A^\alpha \varphi(A) = \varphi(A)A^\alpha$ on $\mathbf{F}_{-\infty}$ if $|\alpha| \le k$. Moreover, there are constants $C_m \in (0, \infty)$ $(m \in \mathbb{Z})$ such that for each φ satisfying the preceding assumptions

(3.3.27)

$$\|\varphi(A)\|_{\mathbf{F}_m \to \mathbf{F}_{m+k}} \le C_m \sum_{|\alpha| \le k} \|\varphi_{(\alpha)}(A)\|_{B(\mathbf{F})} \le C_m \sum_{|\alpha| \le k} \|\varphi_{(\alpha)}\|_{\mathcal{M}_w}.$$

---

[3] Observe that the meaning of m here is different from that in (3.3.16) and (3.3.18).

PROOF. (a) For $m \in \mathbb{Z}$, define $\varphi(A) : \mathbf{F}_m \to \mathbf{F}_m$ by (3.2.11). Then all assertions are easy to check by taking into account the results of Proposition 3.3.8 and of Proposition 3.2.8 in $B(\mathbf{F}_m)$.

(b) If $m \geq 0$, then by Proposition 3.3.3 (b) we have $\varphi(A) \in B(\mathbf{F}_m, \mathbf{F}_{m+k})$ and $A^\alpha \varphi(A) = \varphi(A)A^\alpha = \varphi_{(\alpha)}(A)$ as operator identities on $\mathbf{F}_\infty$, hence also on $\mathbf{F}_{-\infty}$ by a continuity argument. (3.3.14) implies the validity of (3.3.27) with $C_m = 1$.

It remains to show that $\|\varphi(A)f\|_{\mathbf{F}_{-m+k}} \leq \kappa \|f\|_{\mathbf{F}_{-m}}$ for $f \in \mathbf{F}_\infty$, $m \geq 1$ and κ given by the middle expression in (3.3.27). Equivalently we must show that, if $f \in \mathbf{F}_\infty$ and $f = \sum_{|\alpha| \leq m} A^\alpha f_\alpha$ with $f_\alpha \in \mathbf{F}_\infty$, then (see Remark 3.3.7)

$$(3.3.28) \qquad \| \sum_{|\alpha| \leq m} A^\alpha \varphi(A) f_\alpha \|_{\mathbf{F}_{-m+k}} \leq \kappa (\sum_{|\alpha| \leq m} \|f_\alpha\|_{\mathbf{F}}^2)^{1/2}.$$

We distinguish two cases:

(1) If $m \leq k$, then (for some constant c_m depending on m):

$$\| \sum_{|\alpha| \leq m} A^\alpha \varphi(A) f_\alpha \|_{\mathbf{F}_{-m+k}}^2 = \sum_{|\beta| \leq k-m} \| \sum_{|\alpha| \leq m} A^{\alpha+\beta} \varphi(A) f_\alpha \|_{\mathbf{F}}^2$$

$$\leq c_m \sum_{|\alpha| \leq m} \sum_{|\beta| \leq k-m} \|\varphi_{(\alpha+\beta)}(A) f_\alpha\|_{\mathbf{F}}^2$$

$$\leq c_m \sum_{|\alpha| \leq m} (\sum_{|\gamma| \leq k} \|\varphi_{(\gamma)}(A)\|_{B(\mathbf{F})}^2) \|f_\alpha\|_{\mathbf{F}}^2,$$

and (3.3.27) (with $C_{-m} = c_m$) follows by taking into account Proposition 3.3.3 (a).

(2) If $m > k$, consider for each multi-index α with $k < |\alpha| \leq m$ a decomposition $\alpha = \beta + \gamma$ with $|\gamma| = k$ and $|\beta| \leq m - k$, so that $A^\alpha \varphi(A) = A^\beta \varphi_{(\gamma)}(A)$. Then we have

$$\varphi(A)f \equiv \sum_{|\alpha| \leq m} A^\alpha \varphi(A) f_\alpha = \sum_{|\beta| \leq m-k} A^\beta g_\beta,$$

with $g_\beta = \sum_{|\alpha| \leq m} \sum_{|\gamma| \leq k} b_{\alpha\gamma} \varphi_{(\gamma)}(A) f_\alpha$ for some constants $b_{\alpha\gamma}$ ($b_{\alpha\gamma} = 0$ or 1). So, for some constant c_{km}:

$$\|\varphi(A)f\|_{\mathbf{F}_{-m+k}}^2 \leq \sum_{|\beta| \leq m-k} \|g_\beta\|_{\mathbf{F}}^2$$

$$\leq c_{km} \sum_{|\alpha| \leq m} (\sum_{|\gamma| \leq k} \|\varphi_{(\gamma)}(A)\|_{B(\mathbf{F})}^2) \|f_\alpha\|_{\mathbf{F}}^2.$$

So we have (3.3.27) with $C_{-m} = \sup_{0 \leq k \leq m} c_{km}$. $\square$

REMARK 3.3.11. The following formalism will be useful in proving properties of the spaces $\mathbf{F}_k$ for negative k. If $m > 0$ is an integer, consider the Banach space $\mathbb{F}_m = \oplus_{|\alpha| \leq m} \mathbf{F}$ (a finite direct sum of copies of $\mathbf{F}$) with norm

$$\|f\|_{\mathbb{F}_m} = \Big[\sum_{|\alpha| = m} \|f_\alpha\|_{\mathbf{F}}^2 \Big]^{1/2} \qquad \text{for } f = \{f_\alpha\}_{|\alpha| \leq m} \in \mathbb{F}_m.$$

Let $\mathbb{A}_m : \mathbb{F}_m \to \mathbf{F}_{-m}$ be the linear operator defined by $\mathbb{A}_m f = \sum_{|\alpha| \le m} A^\alpha f_\alpha$ if $f = \{f_\alpha\}_{|\alpha| \le m}$. The range $R(\mathbb{A}_m)$ of $\mathbb{A}_m$ is the set of $f \in \mathbf{F}_{-m}$ that may be written as $f = \sum_{|\alpha| \le m} A^\alpha f_\alpha$ with $f_\alpha \in \mathbf{F}$. Clearly $\mathbf{F} \subset R(\mathbb{A}_m)$. By Proposition 3.3.9 (a), $\mathbb{A}_m$ is continuous. Hence its kernel $N(\mathbb{A}_m)$ is a closed subspace of $\mathbb{F}_m$, and the quotient Banach space $\mathbb{F}_m/N(\mathbb{A}_m)$ is well defined. $\mathbb{A}_m$ induces a natural mapping from $\mathbb{F}_m/N(\mathbb{A}_m)$ onto $R(\mathbb{A}_m)$, and this mapping is an isometric isomorphism if $R(\mathbb{A}_m)$ is equipped with the norm

$$\|f\|_{R(\mathbb{A}_m)} = \inf\left\{ \left[\sum_{|\alpha| \le m} \|f_\alpha\|_{\mathbf{F}}^2\right]^{1/2} \,\middle|\, f = \sum_{|\alpha| \le m} A^\alpha f_\alpha, f_\alpha \in \mathbf{F} \right\}.$$

In the next proposition we show that the range of $\mathbb{A}_m$ is in fact equal to $\mathbf{F}_{-m}$ and that $\|f\|_{R(\mathbb{A}_m)} = \|f\|_{\mathbf{F}_{-m}}$. This proposition is the analogue in our context of the fact that each tempered distribution is a finite sum of derivatives of continuous functions.

PROPOSITION 3.3.12. *If $m \ge 1$ is an integer, then:*

$$(3.3.29) \quad \mathbf{F}_{-m} = \left\{ f \in \mathbf{F}_{-\infty} \,\middle|\, f = \sum_{|\alpha| \le m} A^\alpha f_\alpha \text{ for some } f_\alpha \in \mathbf{F} \right\},$$

$$(3.3.30) \quad \|f\|_{\mathbf{F}_{-m}} = \inf\left\{ \left[\sum_{|\alpha| \le m} \|f_\alpha\|_{\mathbf{F}}^2\right)\right]^{1/2} \,\middle|\, f = \sum_{|\alpha| \le m} A^\alpha f_\alpha, f_\alpha \in \mathbf{F} \right\}.$$

PROOF. Since $\mathbf{F}_{-m}$ is the completion of $\mathbf{F}$ in the norm $\|\cdot\|_{\mathbf{F}_{-m}}$ given by (3.3.20) and since $\mathbf{F} \subset R(\mathbb{A}_m)$, it suffices to show that $\|f\|_{R(\mathbb{A}_m)} = \|f\|_{\mathbf{F}_{-m}}$ for each f in $\mathbf{F}$. So let $f \in \mathbf{F}$. It is clear that $\|f\|_{R(\mathbb{A}_m)} \le \|f\|_{\mathbf{F}_{-m}}$. To obtain the opposite inequality, consider a representation $f = \sum_{|\alpha| \le m} A^\alpha f_\alpha$ with $f_\alpha \in \mathbf{F}$. We must show that

$$(3.3.31) \qquad\qquad \|f\|_{\mathbf{F}_{-m}} \le \left[\sum_{|\alpha| \le m} \|f_\alpha\|_{\mathbf{F}}^2 \right]^{1/2}.$$

For this, let us choose a function $\varphi : \mathbb{R}^n \to \mathbb{C}$ such that $\varphi(0) = 1$ and $\widehat{\varphi} \in C_0^\infty(\mathbb{R}^n)$, and define $f_\alpha(\varepsilon) = \varphi(\varepsilon A)f_\alpha$ if $\alpha \ne 0$ and $f_0(\varepsilon) = \varphi(\varepsilon A)f_0 + [I - \varphi(\varepsilon A)]f$, with $\varepsilon > 0$. Then $f_0(\varepsilon) \in \mathbf{F}$, $f_\alpha(\varepsilon) \in \mathbf{F}_\infty$ if $\alpha \ne 0$ (by Proposition 3.3.3 (b)) and $f = \sum_{|\alpha| \le m} A^\alpha f_\alpha(\varepsilon)$. So $\|f\|_{\mathbf{F}_{-m}} \le [\sum_{|\alpha| \le m} \|f_\alpha(\varepsilon)\|_{\mathbf{F}}^2]^{1/2}$, which implies (3.3.31) because $\|f_\alpha(\varepsilon) - f_\alpha\|_{\mathbf{F}} \to 0$ as $\varepsilon \to 0$ for each α. $\square$

PROPOSITION 3.3.13. *If $\mathbf{F}$ is a reflexive Banach space, then so is $\mathbf{F}_m$ for each $m \in \mathbb{Z}$.*

PROOF. We shall use some properties of reflexive spaces given in Section 2.1. For $m > 0$, let $\mathbb{F}_m$ and $\mathbb{A}_m$ be as in Remark 3.3.11. $\mathbb{F}_m$ is a finite direct sum of reflexive Banach spaces, hence $\mathbb{F}_m$ itself is reflexive. So the quotient space $\mathbb{F}_m/N(\mathbb{A}_m)$ is reflexive. Since $\mathbb{A}_m$ is an isometric isomorphism of this quotient space onto $\mathbf{F}_{-m}$, the Banach space $\mathbf{F}_{-m}$ is reflexive if $m > 0$.

To obtain the reflexivity of $\mathbf{F}_m$ for $m > 0$, we define an operator $\underline{A} : \mathbf{F}_m \to \mathbb{F}_m$ by $\underline{A}f = \{A^\alpha f\}_{|\alpha| \le m}$. $\underline{A}$ is an isometric operator from $\mathbf{F}_m$ onto a closed (hence reflexive) subspace of $\mathbb{F}_m$. So $\mathbf{F}_m$ is reflexive. $\square$

In applications it often happens that a topological vector space $\mathscr{F}$ is already constructed such that $\mathbf{F} \subset \mathscr{F}$ and the operators A_j have continuous extensions to $\mathscr{F}$ (in many situations $\mathscr{F}$ could be the space of tempered distributions). One can then give the following characterization of the spaces $\{\mathbf{F}_m\}_{m \in \mathbb{Z}}$:

PROPOSITION 3.3.14. *Let $\mathscr{F}$ be a Hausdorff quasi-complete locally convex vector space. Assume that a family $\{W(x)\}_{x \in \mathbb{R}^n}$ of continuous linear operators in $\mathscr{F}$ is given such that*
  (1) *$W(0) = I$ and $W(x + y) = W(x)W(y)$ for all $x, y \in \mathbb{R}^n$,*
  (2) *for each $f \in \mathscr{F}$ the function $x \mapsto W(x)f \in \mathscr{F}$ is weakly continuous,*
  (3) *for each $f \in \mathscr{F}$ and $j = 1, \dots, n$, the limit $\lim_{\varepsilon \to 0}(i\varepsilon)^{-1}[W(\varepsilon e_j) - I]f \equiv A_j f$ exists weakly and defines a continuous operator A_j in $\mathscr{F}$, where e_j ($j = 1, \dots, n$) denotes the j-th vector of the canonical basis of $\mathbb{R}^n$.*

If α is a multi-index, define $A^\alpha : \mathscr{F} \to \mathscr{F}$ by $A^\alpha = A_1^{\alpha_1} \cdot \ldots \cdot A_n^{\alpha_n}$. Let $\mathbf{F}$ be a Banach space continuously embedded in $\mathscr{F}$, invariant under each $W(x)$ and such that the mapping $x \mapsto W(x)f \in \mathbf{F}$ is continuous for each $f \in \mathbf{F}$. Then:
  (a) *The restrictions $W(x)|_{\mathbf{F}}$ define a C_0-group in $\mathbf{F}$.*
  (b) *For $m \geq 1$, the Banach space $\mathbf{F}_m$ associated to the C_0-group $W|_{\mathbf{F}}$ in $\mathbf{F}$ coincides (isometrically) with the space defined as follows:*

$$(3.3.32) \qquad \begin{cases} \mathbf{F}_m = \{f \in \mathscr{F} \mid A^\alpha f \in \mathbf{F} \text{ for each } |\alpha| \leq m\} \\ \|f\|_{\mathbf{F}_m} = \left[\sum_{|\alpha| \leq m} \|A^\alpha f\|_{\mathbf{F}}^2\right]^{1/2}. \end{cases}$$

Moreover, the restriction of the operator $A^\alpha : \mathscr{F} \to \mathscr{F}$ to $\mathbf{F}_{|\alpha|}$ coincides with the operator defined by (3.3.5).
  (c) *For $m \leq -1$, let*

$$(3.3.33)$$
$$\begin{cases} \mathbf{F}_m = \{f \in \mathscr{F} \mid \exists \{f_\alpha\}_{|\alpha| \leq -m}, \; f_\alpha \in \mathbf{F}, f = \sum_{|\alpha| \leq -m} A^\alpha f_\alpha\} \\ \|f\|_{\mathbf{F}_m} = \inf\{[\sum_{|\alpha| \leq -m} \|f_\alpha\|_{\mathbf{F}}^2]^{1/2} \mid f_\alpha \in \mathbf{F}, \; f = \sum_{|\alpha| \leq -m} A^\alpha f_\alpha\}. \end{cases}$$

Then $\mathbf{F}_m$ is a Banach space continuously embedded in $\mathscr{F}$. $\mathbf{F}$ is a dense subspace of $\mathbf{F}_m$, and the restriction of the norm (3.3.33) to $\mathbf{F}$ coincides with the norm defined by (3.3.20). In particular the space $\mathbf{F}_m$ defined by (3.3.33) is canonically identified with that defined by (3.3.23).

PROOF. For the purpose of the proof we denote by $\mathbf{F}_m$ the spaces defined by (3.3.1) or (3.3.23) and by $\mathbf{F}_m^+$ those defined by (3.3.32) or (3.3.33). We let $A_{\mathbf{F}}^\alpha : \mathbf{F}_{|\alpha|} \to \mathbf{F}$ be the operators given by (3.3.5) and denote by $A_{\mathscr{F}}^\alpha : \mathscr{F} \to \mathscr{F}$ those introduced in the statement of the present proposition.

(a) We just have to show that, for fixed $x \in \mathbb{R}^n$, $W(x)|_{\mathbf{F}}$ is a continuous operator in $\mathbf{F}$, or (by the closed graph theorem) that $W(x)|_{\mathbf{F}}$ is a closed operator in $\mathbf{F}$. Now if $\{f_k\} \in \mathbf{F}$ is such that $f_k \to f$ and $W(x)f_k \to g$ in $\mathbf{F}$, then $f_k \to f$ and $W(x)f_k \to g$ in $\mathscr{F}$, hence $g = W(x)f$ because $W(x)$ is continuous in $\mathscr{F}$ and the topology of $\mathscr{F}$ is Hausdorff.

(b) We first observe that $x \mapsto W(x)f \in \mathscr{F}$ is weakly C^∞ for each $f \in \mathscr{F}$ and $(-i\partial_x)^\alpha W(x)f = W(x)A_{\mathscr{F}}^\alpha f$. Also Taylor's formula (3.3.16) holds weakly in $\mathscr{F}$.

Now assume that $f \in \mathbf{F}$ is such that $x \mapsto W(x)f \in \mathbf{F}$ is of class C^m for some $m \geq 1$. Then (by arguments as in (a)), we get from (3.3.5) that $A_{\mathbf{F}}^\alpha f = A_{\mathscr{F}}^\alpha f$ if $|\alpha| \leq m$. Hence we have $\mathbf{F}_m \subset \mathbf{F}_m^+$ isometrically.

For the opposite inclusion, let $f \in \mathbf{F}_m^+$. Then $\mathrm{f}_\alpha \equiv A_{\mathscr{F}}^\alpha f \in \mathbf{F}$ for each $|\alpha| \leq m$, and Taylor's formula (3.3.16) gives

$$W(x+y)f = \sum_{|\alpha| \leq m} \frac{(iy)^\alpha}{\alpha!} W(x)f_\alpha + \sum_{|\alpha|=m} \frac{(iy)^\alpha}{\alpha!} [W_m(y) - I]W(x)f_\alpha.$$

Then the converse to Taylor's theorem (Proposition 1.1.1) implies that $x \mapsto W(x)f$ is of class C^m in $\mathbf{F}$.

(c) We here need a functional calculus for the group W in $\mathscr{F}$. If $\varphi : \mathbb{R}^n \to \mathbb{C}$ is such that $\widehat{\varphi} \in C_0^\infty(\mathbb{R}^n)$, we define [4] $\varphi(A_{\mathscr{F}})f = \int_{\mathbb{R}^n} W(x)f\widehat{\varphi}(x)dx$ for each $f \in \mathscr{F}$, and we use the following simple properties of the linear operator $\varphi(A_{\mathscr{F}}) : \mathscr{F} \to \mathscr{F}$:

(1) $\varphi(A_{\mathscr{F}})A_{\mathscr{F}}^\alpha = A_{\mathscr{F}}^\alpha \varphi(A_{\mathscr{F}}) = \varphi_{(\alpha)}(A_{\mathscr{F}})$, where $\varphi_{(\alpha)}(x) = x^\alpha \varphi(x)$ (this follows from the continuity of $A_{\mathscr{F}}$, the formula $A_{\mathscr{F}}^\alpha W(x)f = (-i\partial_x^\alpha)^\alpha W(x)f$ and an integration by parts),

(2) if $f \in \mathbf{F}$, then $\varphi_{(\alpha)}(A_{\mathscr{F}})f = \varphi_{(\alpha)}(A_{\mathbf{F}})f \in \mathbf{F}_\infty$, and $\varphi(\varepsilon A_{\mathscr{F}})f \to f$ in $\mathbf{F}$ as $\varepsilon \to 0$ provided that $\varphi(0) = 1$.

Now let $m \leq -1$ and $f \in \mathbf{F}_m^+$. For $|\alpha| \leq -m$ choose $f_\alpha \in \mathbf{F}$ such that $f = \sum_{|\alpha| \leq -m} A_{\mathscr{F}}^\alpha f_\alpha$ and consider $\varphi(\varepsilon A_{\mathscr{F}})f = \sum_{|\alpha| \leq -m} A_{\mathscr{F}}^\alpha \varphi(\varepsilon A_{\mathscr{F}})f_\alpha \in \mathbf{F}_\infty$, with $\varphi(0) = 1$. Then $\|\varphi(\varepsilon A_{\mathscr{F}})f - f\|_{\mathbf{F}_m^+}^2 \leq \sum_{|\alpha| \leq -m} \|\varphi(\varepsilon A_{\mathbf{F}})f_\alpha - f_\alpha\|_{\mathbf{F}}^2 \to 0$ as $\varepsilon \to 0$. So $\mathbf{F}_\infty$ is dense in $\mathbf{F}_m^+$. Finally assume that $f \in \mathbf{F} \subset \mathbf{F}_m^+$; obviously $\|f\|_{\mathbf{F}_m^+} \leq \|f\|_{\mathbf{F}_m}$ (the l.h.s. is defined in (3.3.33) and the r.h.s. in (3.3.20)). The opposite inequality is obtained by observing that, if $f \in \mathbf{F}$ is written as $f = \sum_{|\alpha| \leq -m} A_{\mathscr{F}}^\alpha f_\alpha$ with $f_\alpha \in \mathbf{F}$, then $f = \sum_{|\alpha| \leq -m} A_{\mathbf{F}}^\alpha f_\alpha'(\varepsilon)$ with $f_0'(\varepsilon) = [I - \varphi(\varepsilon A_{\mathbf{F}})]f + \varphi(\varepsilon A_{\mathbf{F}})f_0 \in \mathbf{F}$, $f_\alpha'(\varepsilon) = \varphi(\varepsilon A_{\mathbf{F}})f_\alpha \in \mathbf{F}_\infty \subset \mathbf{F}_{|\alpha|}$ if $\alpha \neq 0$, and $f_\alpha'(\varepsilon) \to f_\alpha$ in $\mathbf{F}$ for each α as $\varepsilon \to 0$. $\square$

REMARK 3.3.15. The definitions (3.3.32) and (3.3.33) involve only a Hausdorff topological linear vector space $\mathscr{F}$ and a family $\{A_1, \ldots, A_n\}$ of commuting continuous linear operators in $\mathscr{F}$ (i.e. the group W is not used in these definitions). It is quite easy to show that, in this situation, each $\mathbf{F}_m$ is a Banach space continuously embedded in $\mathscr{F}$, with $\cdots \subset \mathbf{F}_2 \subset \mathbf{F}_1 \subset \mathbf{F} \subset \mathbf{F}_{-1} \subset \mathbf{F}_{-2} \subset \ldots$; the embeddings in this scale are contractions (and not dense in general), and one has $A^\alpha \mathbf{F}_m \subset \mathbf{F}_{m-|\alpha|}, \|A^\alpha f\|_{\mathbf{F}_{m-|\alpha|}} \leq \|f\|_{\mathbf{F}_m}$ for all $m \in \mathbb{Z}$, $\alpha \in \mathbb{N}^n$ (we set

---

[4] Let us explain the meaning of the integral. Consider a weakly continuous function $\phi : \mathbb{R}^n \to \mathscr{F}$ of compact support ("weak" refers to the topology $\sigma(\mathscr{F}, \mathscr{F}^*)$). Then $\int \langle \phi(x), g \rangle dx$ is well defined for each $g \in \mathscr{F}^*$. Let us show that there is a unique $f \in \mathscr{F}$ such that $\langle f, g \rangle = \int \langle \phi(x), g \rangle dx \ \forall g \in \mathscr{F}^*$ (then we shall identify $\int \phi(x)dx$ with f). Fix a cube Δ in $\mathbb{R}^n$ such that $\mathrm{supp}\,\phi \subset \Delta$; for each $k \in \mathbb{N}$ consider a partition $\{\Delta_j\}$ of Δ into sub-cubes with edges of order k^{-1}, and let $x_j \in \Delta_j$. If $f_k = \sum_j \phi(x_j)|\Delta_j|$ then $\langle f_k, g \rangle$ is a Riemann sum which converges to $\int \langle \phi(x), g \rangle dx$ as $k \to \infty$. In particular, $\{f_k\}$ is a weak Cauchy sequence. Observe that $|\Delta|^{-1} f_k$ belongs to the convex hull of the range M of ϕ. Since M is weakly compact, the weak closure of its convex hull is a weakly compact subset of $\mathscr{F}$ (by Krein's theorem, cf. Theorem 3 in §IV.5 of [Bo2]). So f_k will converge weakly to some f in $\mathscr{F}$. The uniqueness is obvious.

$\mathbf{F}_0 = \mathbf{F}$). The fact that $\mathbf{F}_m$ is complete can be checked by the arguments used in the proof of Proposition 3.3.12.

REMARK 3.3.16. Any C_0-group W in a Banach space $\mathbf{F}$ can be obtained as the restriction to $\mathbf{F}$ of a family $\{W(x)\}_{x \in \mathbb{R}^n}$ of operators in a topological vector space $\mathscr{F}$ as described in Proposition 3.3.14. In fact there is a choice for $\mathscr{F}$ which is in some sense minimal, namely $\mathscr{F} = \mathbf{F}_{-\infty}$ (given by (3.3.25); cf. Propositions 3.3.8, 3.3.9, 3.3.12 and 3.3.18).

3.3.3. In this subsection we first establish two new descriptions of the spaces $\mathbf{F}_m$ for $m \geq 1$ and then we derive convex type inequalities involving the norms of the spaces $\mathbf{F}_m$ ($m \in \mathbb{Z}$) or some new seminorms naturally associated to the family of operators $\{A^\alpha\}$.

We shall need the following simple results, which will be also helpful in other circumstances.

LEMMA 3.3.17. Let $\varphi \in \mathscr{S}(\mathbb{R}^n)$ such that $\widehat{\varphi}$ is of compact support and denote by $\Gamma(\varphi)$ the set of $x \in \mathbb{R}^n$ of the form $x = \tau y$ with $0 \leq \tau \leq 1$ and $y \in \operatorname{supp}\widehat{\varphi}$. Then

(a) One has $\varphi(A)\mathbf{F}_{-\infty} \subset \mathbf{F}_{+\infty}$ and $\lim_{\varepsilon \to 0} \varphi(\varepsilon A) = \varphi(0)I$ strongly in $B(\mathbf{F}_m)$ for each $m \in \mathbb{Z}$. Moreover, the map $\tau \mapsto \varphi(\tau A) \in B(\mathbf{F}_m)$ is strongly continuous $(0 \leq \tau < \infty)$.

(b) For each integer $k \geq 1$ the following identity holds on $\mathbf{F}_{-\infty}$:

$$\varphi(A) = \sum_{|\alpha|<k} \frac{\varphi^{(\alpha)}(0)}{\alpha!}A^\alpha + \sum_{|\alpha|=k} \frac{1}{\alpha!} \int_0^1 \varphi^{(\alpha)}(\tau A)k(1-\tau)^{k-1}d\tau \cdot A^\alpha.$$

(c) For each multi-index α define $J_\alpha^\varphi = -\frac{1}{\alpha!}\int_0^1 \varphi^{(\alpha)}(\tau A)k(1-\tau)^{k-1}d\tau$; the integral exists strongly in each $B(\mathbf{F}_m)$ ($m \in \mathbb{Z}$) and the following estimates holds:

$$\|J_\alpha^\varphi\|_{B(\mathbf{F}_m)} \leq \frac{1}{\alpha!} \sup_{x \in \Gamma(\varphi)} \|W(x)\|_{B(\mathbf{F})}\|Q^\alpha\widehat{\varphi}\|_{L^1(\mathbb{R}^n)}.$$

(d) Let Γ be a non-empty open cone with vertex at the origin in $\mathbb{R}^n$ and R a strictly positive number; set $\Gamma_R = \{x \in \Gamma \mid |x| < R\}$. Then for each integer $k \geq 1$ and each multi-index α with $|\alpha| < k$, one may find a function $\varphi \in \mathscr{S}(\mathbb{R}^n)$ with $\varphi^{(\alpha)}(0) = \alpha!$, $\varphi^{(\beta)}(0) = 0$ if $|\beta| < k$, $\beta \neq \alpha$, and $\operatorname{supp}\widehat{\varphi} \subset \Gamma_R$. For such a function we have on $\mathbf{F}_{-\infty}$:

$$(3.3.34) \qquad A^\alpha = \varphi(A) + \sum_{|\beta|=k} J_\beta^\varphi A^\beta = \varphi(A) + \sum_{|\beta|=k} A^\beta J_\beta^\varphi.$$

$\varphi(A)$ is a regularizing operator (i.e. $\varphi(A)\mathbf{F}_{-\infty} \subset \mathbf{F}_{+\infty}$) and each J_β^φ is regularity preserving (i.e. $J_\beta^\varphi\mathbf{F}_m \subset \mathbf{F}_m \ \forall m \in \mathbb{Z}$).

PROOF. (a) is an immediate consequence of Proposition 3.3.10. For (b), we start with the Taylor expansion (3.3.16) (see Proposition 3.3.9 (b)), multiply it by $\widehat{\varphi}$ and integrate over x. To prove (c) we first use (3.2.19) and then estimate J_α^φ as follows (by using (3.3.26)):

$$\alpha! \, ||J_\alpha^\varphi||_{B(\mathbf{F}_m)} \leq \int_0^1 ||\varphi^{(\alpha)}(\tau A)||_{B(\mathbf{F}_m)} k(1-\tau)^{k-1} d\tau$$

$$\leq \sup_{0\leq\tau\leq 1} ||\varphi^{(\alpha)}(\tau A)||_{B(\mathbf{F}_m)} \leq \sup_{0\leq\tau\leq 1} \int_{\mathbb{R}^n} ||W(\tau y)||_{B(\mathbf{F})} |y^\alpha \widehat{\varphi}(y)| \underline{dy}.$$

It remains to check the existence of a function φ with the properties stated in (d). It is sufficient to show that for any non-empty ball $B \subset \mathbb{R}^n$ there is $\psi_\alpha \in C_0^\infty(B)$ such that $\int x^\beta \psi_\alpha(x) dx = 0$ if $|\beta| < k$, $\beta \neq \alpha$ and $\int x^\alpha \psi_\alpha(x) dx = 1$ (then we take $\widehat{\varphi}$ equal to a multiple of ψ_α, the ball B being chosen such that the convex hull of B and the origin is contained in Γ_R). Let V be the finite-dimensional vector space of polynomials of degree less than k, V' its dual and W the subspace of V' consisting of the linear forms of the form $L_\psi(f) = \int f(x)\psi(x) dx$ with $\psi \in C_0^\infty(B)$. If $W \neq V'$, then there is $f \in V$, $f \neq 0$, such that $L_\psi(f) = 0$ for all $\psi \in C_0^\infty(B)$ (because $V'' = V$). But this implies $f|_B = 0$, hence $f = 0$ (f being a polynomial); so $W = V'$. Since the monomials $\{x^\beta\}_{|\beta|<k}$ form a basis of V, we may choose a family $\{\psi_\alpha\}_{|\alpha|<k}$ in $C_0^\infty(B)$ such that $\{L_{\psi_\alpha}\}$ is the dual basis of V', i.e. $\int x^\beta \psi_\alpha(x) dx = \delta_{\alpha\beta}$. $\square$

As a first application of Lemma 3.3.17 we obtain a regularity result in the scale $\{\mathbf{F}_m\}_{m\in\mathbb{Z}}$ which implies a new characterization of $\mathbf{F}_m$ for $m \geq 1$.

PROPOSITION 3.3.18. (a) *Let* $m \in \mathbb{Z}$ *and* $k \in \mathbb{N}$. *If* $f \in \mathbf{F}_{-\infty}$ *and* $A^\alpha f \in \mathbf{F}_m$ *for each multi-index* α *with* $|\alpha| = k$, *then we have* $A^\alpha f \in \mathbf{F}_m$ *for all* α *such that* $|\alpha| \leq k$. *Moreover, for each integer* $\ell \leq m$ *there is a constant* $C > 0$ *such that for all* $f \in \mathbf{F}_{-\infty}$

(3.3.35)
$$C \sum_{|\alpha|\leq k} ||A^\alpha f||_{\mathbf{F}_m} \leq ||f||_{\mathbf{F}_\ell} + \sum_{|\alpha|=k} ||A^\alpha f||_{\mathbf{F}_m} \leq C^{-1} \sum_{|\alpha|\leq k} ||A^\alpha f||_{\mathbf{F}_m}.$$

(b) *If* $m \geq 1$ *is an integer, then* $\mathbf{F}_m$ *coincides with the set of* $f \in \mathbf{F}_{-\infty}$ *such that* $A^\alpha f \in \mathbf{F}$ *for* $|\alpha| = m$. *The gauge* $|| \cdot ||_{\mathbf{F}_m}$ *given by* (3.3.6) *on* $\mathbf{F}_{-\infty}$ *is equivalent to the gauge*

(3.3.36) $$||f||'_{\mathbf{F}_m} = ||f||_{\mathbf{F}} + \sum_{|\alpha|=m} ||A^\alpha f||_{\mathbf{F}}.$$

PROOF. (a) Let $f \in \mathbf{F}_{-\infty}$ such that $A^\beta f \in \mathbf{F}_m$ if $|\beta| = k$ and let α be a multi-index with $|\alpha| < k$. Choose φ as in Lemma 3.3.17 (d), with $\Gamma = \mathbb{R}^n$. Then by using (3.3.34) and the properties $\varphi(A)\mathbf{F}_{-\infty} \subset \mathbf{F}_{+\infty}$ and $J_\beta^\varphi \in B(\mathbf{F}_m)$, we obtain $A^\alpha f \in \mathbf{F}_m$ and

$$||A^\alpha f||_{\mathbf{F}_m} \leq ||\varphi(A)f||_{\mathbf{F}_m} + C_1 \sum_{|\beta|=k} ||A^\beta f||_{\mathbf{F}_m}.$$

Then (3.3.35) is a consequence of the fact that $\varphi(A) \in B(\mathbf{F}_\ell, \mathbf{F}_m)$ for arbitrary ℓ, m.

(b) The equivalence of the gauges $|| \cdot ||_{\mathbf{F}_m}$ and $|| \cdot ||'_{\mathbf{F}_m}$ is a particular case of (3.3.35) (one may replace $||f||_{\mathbf{F}}$ in (3.3.36) by $||f||_{\mathbf{F}_\ell}$ with $\ell \leq 0$ arbitrary). Now let $f \in \mathbf{F}_{-\infty}$ such that $A^\alpha f \in \mathbf{F}$ for $|\alpha| = m$. By part (a) of the proposition, this will hold for all $|\alpha| \leq m$. We may finish the proof by applying Proposition 3.3.14 (b) in the context of the Remark 3.3.16, but we prefer to give a more direct argument. We choose a function $\varphi \in \mathscr{S}(\mathbb{R}^n)$ with $\varphi(0) = 1$ and $\widehat{\varphi}$ of compact support, and we set $f_j = \varphi(j^{-1}A)f$. By Lemma 3.3.17 (a) we have $f_j \in \mathbf{F}_\infty$ and, if $|\alpha| \leq m$, $A^\alpha f_j = \varphi(j^{-1}A)A^\alpha f \to A^\alpha f$ in $\mathbf{F}$ as $j \to \infty$, because $A^\alpha f \in \mathbf{F}$ for such α. Hence $||f_j - f||_{\mathbf{F}} \to 0$ and $\{f_j\}$ is Cauchy in $\mathbf{F}_m$. Since $\mathbf{F}_m$ is a B-subspace of $\mathbf{F}$, we obtain $f \in \mathbf{F}_m$. $\square$

Notice that the proof of part (b) of the proposition gives slightly more: if $m \geq 1$, $k \geq 0$ and we have $A^\alpha f \in \mathbf{F}_k$ for some $f \in \mathbf{F}_{-\infty}$ and all α with $|\alpha| = m$, then $f \in \mathbf{F}_{k+m}$. We do not know whether this holds for $k < 0$ too (we shall prove it later on in some particular situations, cf. §3.3.4).

We shall now deduce from Proposition 3.3.18 a characterization of $\mathbf{F}_m$ for $m \geq 1$ in terms of the group W itself. It is interesting to point out first a consequence of the identity (3.3.18). Let $\theta : \mathbb{R}^n \to \mathbb{C}$ be such that $\widehat{\theta}$ is a sufficiently rapidly decreasing measure (e.g. $\widehat{\theta}$ is a measure of compact support). Using (3.3.18) and Proposition 3.3.9 (b), we get as an identity on $\mathbf{F}_{-\infty}$ for each integer $m \geq 1$ and $\varepsilon > 0$:

$$(3.3.37) \quad \int_{\mathbb{R}^n} \left[\frac{W(\varepsilon y) - I}{\varepsilon} \right]^m \widehat{\theta}(y)dy = \sum_{|\alpha|=m} \frac{m!}{\alpha!} A^\alpha \int_{\mathbb{R}} \theta^{(\alpha)}(\tau \varepsilon A)\chi_m(\tau)d\tau.$$

Now let $f \in \mathbf{F}_{-\infty}$ be such that $A^\alpha f \in \mathbf{F}_k$ for all $|\alpha| = m$ and some fixed $k \in \mathbb{Z}$. Then, by using (b) of Proposition 3.3.5, we obtain (strongly in $\mathbf{F}_k$):

$$(3.3.38) \quad \lim_{\varepsilon \to +0} \int_{\mathbb{R}^n} \left[\frac{W(\varepsilon y) - I}{\varepsilon} \right]^m f\widehat{\theta}(y)dy = \sum_{|\alpha|=m} \frac{m!}{\alpha!} \theta^{(\alpha)}(0)A^\alpha f.$$

The special case of (3.3.38) when $\widehat{\theta}$ is the Dirac measure with support $\{x\}$ will allow us to prove the following result (for the case $n = 1$ one may find several versions of the results of our Propositions 3.3.19 and 3.3.25 (b) in Sections 2.2 and 3.4 of [BB]):

PROPOSITION 3.3.19. *Let $f \in \mathbf{F}$ and let $m \geq 1$ be an integer. Then $f \in \mathbf{F}_m$ if and only if $\lim_{\varepsilon \to +0} \varepsilon^{-m}[W(\varepsilon x) - I]^m f$ exists in $\mathbf{F}$ for each $x \in \mathbb{R}^n$. Moreover, the norm (3.3.2) of $\mathbf{F}_m$ is equivalent to that defined by*

$$(3.3.39) \quad ||f||''_{\mathbf{F}_m} = ||f||_{\mathbf{F}} + \sup_{|x| \leq 1} \left\| \left[\frac{W(x) - I}{|x|} \right]^m f \right\|_{\mathbf{F}}.$$

PROOF. (i) By using (3.3.18) and Proposition 3.3.9 (b) we get for an arbitrary $f \in \mathbf{F}_{-\infty}$:

$$(3.3.40) \quad \left[\frac{W(\varepsilon x) - I}{i\varepsilon} \right]^m f = \sum_{|\alpha|=m} \frac{m!}{\alpha!} x^\alpha \int_{\mathbb{R}} W(\tau \varepsilon x)\chi_m(\tau)d\tau \cdot A^\alpha f.$$

If $f \in \mathbf{F}_m$, then $A^\alpha f \in \mathbf{F}$ for each $|\alpha| = m$, hence the r.h.s. of (3.3.40) is strongly convergent in $\mathbf{F}$ as $\varepsilon \to 0$.

(ii) If $f \in \mathbf{F}$ and the expression (3.3.40) is strongly convergent in $\mathbf{F}$ as $\varepsilon \to 0$, then the limit (given by (3.3.38) with $\widehat{\theta}$ equal to the Dirac measure at x) belongs to $\mathbf{F}$, i.e. $\sum_{|\alpha|=m} (\alpha!)^{-1} x^\alpha A^\alpha f \in \mathbf{F}$ for each $x \in \mathbb{R}^n$. We now apply Proposition 1.1.2 (with $\ell > m$) to the polynomials $P(x) \equiv \sum_{|\alpha|=m} (\alpha!)^{-1} x^\alpha A^\alpha f$. This shows that $A^\alpha f = \sum_{y \in G} C_y^\alpha P(y)$, so $A^\alpha f \in \mathbf{F}$ for $|\alpha| = m$. Hence $f \in \mathbf{F}_m$ by Proposition 3.3.18 (b).

(iii) Now we prove the equivalence of the norm $|| \cdot ||_{\mathbf{F}_m}''$ with $|| \cdot ||_{\mathbf{F}_m}'$ (given by (3.3.36)). The inequality $|| \cdot ||_{\mathbf{F}_m}'' \le c_1 || \cdot ||_{\mathbf{F}_m}'$ for some finite constant c_1 is a straightforward consequence of (3.3.40). For the converse inequality it suffices to observe that, for $|\alpha| = m$:

$$||A^\alpha f||_{\mathbf{F}} \le \sum_{y \in G} |C_y^\alpha| \cdot ||P(y)||_{\mathbf{F}} \le c_2 \sum_{y \in G} \sup_{0 < \varepsilon < 1} \left\| \left[\frac{W(\varepsilon y) - I}{\varepsilon} \right]^m f \right\|_{\mathbf{F}}. \qquad \square$$

If $\sigma > 0$ is a real number, one may apply Lemma 3.3.17 to the group $\{W(\sigma x)\}_{x \in \mathbb{R}^n}$ with generator σA. Then we get:

(3.3.41)

$$\varphi(\sigma A) = \sum_{|\alpha| < k} \frac{\varphi^{(\alpha)}(0)}{\alpha!} \sigma^{|\alpha|} A^\alpha + \sum_{|\alpha|=k} \frac{\sigma^k}{\alpha!} \int_0^1 \varphi^{(\alpha)}(\tau \sigma A) k (1-\tau)^{k-1} d\tau \cdot A^\alpha.$$

Let $J_{\alpha,\sigma} = -\frac{1}{\alpha!} \int_0^1 \varphi^{(\alpha)}(\tau \sigma A) k (1-\tau)^{k-1} d\tau$. If we assume that $\operatorname{supp} \widehat{\varphi} \subset \Gamma_R$, where Γ_R is as in Lemma 3.3.17 (d), then part (c) of the lemma gives the following estimate for the norm of $J_{\alpha,\sigma}$ in $\mathbf{F}_m$ ($m \in \mathbb{Z}$):

(3.3.42)

$$||J_{\alpha,\sigma}||_{B(\mathbf{F}_m)} \le \frac{1}{\alpha!} \sup_{x \in \Gamma(\varphi)} ||W(\sigma x)||_{B(\mathbf{F})} ||Q^\alpha \widehat{\varphi}||_{L^1}$$

$$\le \frac{1}{\alpha!} \sup_{x \in \Gamma_{\sigma R}} ||W(x)||_{B(\mathbf{F})} ||Q^\alpha \widehat{\varphi}||_{L^1}.$$

If φ has all the properties required in part (d) of Lemma 3.3.17, we obtain a representation of the operator A^α (on $\mathbf{F}_{-\infty}$) of the form:

(3.3.43)
$$A^\alpha = \sigma^{-|\alpha|} \varphi(\sigma A) + \sigma^{k-|\alpha|} \sum_{|\beta|=k} J_{\beta,\sigma} A^\beta.$$

Before giving some consequences of these formulas we prove two technical estimates that will be useful later on again.

LEMMA 3.3.20. *Let* $\varphi \in \mathscr{S}(\mathbb{R}^n)$ *such that* $\widehat{\varphi}$ *has compact support and let* k, m *be integers with* $k \ge 0$. *Then there is a constant* $C < \infty$ *such that*

(3.3.44)
$$||\varphi(\sigma A)||_{\mathbf{F}_m \to \mathbf{F}_{m+k}} \le C \sigma^{-k}, \qquad 0 < \sigma \le 1.$$

If $\varphi(0) = 1$ *and* $\varphi^{(\beta)}(0) = 0$ *for* $1 \le |\beta| < k$, *then we also have*

(3.3.45)
$$||I - \varphi(\sigma A)||_{\mathbf{F}_m \to \mathbf{F}_{m-k}} \le C \sigma^k, \qquad 0 < \sigma \le 1.$$

PROOF. In the following estimates we shall be more careful than needed for the proof of the lemma. Observe the $[\varphi^\sigma]_{(\alpha)} = \sigma^{-|\alpha|}[\varphi_{(\alpha)}]^\sigma$. Then, by using (3.3.27), we get for all $\sigma > 0$:

$$||\varphi(\sigma A)||_{\mathbf{F}_m \to \mathbf{F}_{m+k}} \leq C_m \sum_{|\alpha| \leq k} ||[\varphi^\sigma]_{(\alpha)}(A)||_{B(\mathbf{F})} = C_m \sum_{|\alpha| \leq k} \sigma^{-|\alpha|}||\varphi_{(\alpha)}(\sigma A)||_{B(\mathbf{F})}$$

$$\leq C_m \sum_{|\alpha| \leq k} \sigma^{-|\alpha|}||\widehat{\varphi}^{(\alpha)}||_{L^1(\mathbb{R}^n)} \sup_{x \in \sigma\Gamma(\varphi)} ||W(x)||_{B(\mathbf{F})}$$

where $\Gamma(\varphi)$ has the same meaning as in Lemma 3.3.17. This clearly proves (3.3.44). For the second part of the lemma we use (3.3.43) which holds with $\alpha = 0$:

$$I - \varphi(\sigma A) = \sigma^k \sum_{|\beta|=k} A^\beta J_{\beta,\sigma}.$$

By using Proposition 3.3.9 (a) and (3.3.42) we get

$$||I - \varphi(\sigma A)||_{\mathbf{F}_m \to \mathbf{F}_{m-k}} \leq \sigma^k \sum_{|\beta|=k} ||J_{\beta,\sigma}||_{B(\mathbf{F}_m)}$$

$$\leq \sigma^k \sum_{|\beta|=k} \frac{1}{\beta!} ||Q^\beta \widehat{\varphi}||_{L^1(\mathbb{R}^n)} \sup_{x \in \sigma\Gamma(\varphi)} ||W(x)||_{B(\mathbf{F})}. \quad \square$$

Now we point out a (logarithmic) convexity property of the family of gauges $\{||\cdot||_{\mathbf{F}_m}\}_{m \in \mathbb{Z}}$.

PROPOSITION 3.3.21. *Let* $m_1 < m < m_2$ *be integers. Then there is a constant* $C < \infty$ *such that for all real* $\sigma > 0$ *and all* $f \in \mathbf{F}_{-\infty}$

(3.3.46) $$||f||_{\mathbf{F}_m} \leq C\sigma^{-(m-m_1)}||f||_{\mathbf{F}_{m_1}} + C\sigma^{m_2-m}||f||_{\mathbf{F}_{m_2}}$$

and, with $\theta \equiv (m - m_1)(m_2 - m_1)^{-1}$:

(3.3.47) $$||f||_{\mathbf{F}_m} \leq C||f||_{\mathbf{F}_{m_1}}^{1-\theta}||f||_{\mathbf{F}_{m_2}}^\theta.$$

PROOF. Let φ be as in the second part of Lemma 3.3.20 with $k = m_2 - m_1$. Then there is a constant C_1 such that for all $f \in \mathbf{F}_{-\infty}$ and $0 < \sigma \leq 1$:

$$||f||_{\mathbf{F}_m} \leq ||\varphi(\sigma A)f||_{\mathbf{F}_m} + ||[I - \varphi(\sigma A)]f||_{\mathbf{F}_m}$$

$$\leq ||\varphi(\sigma A)||_{\mathbf{F}_{m_1} \to \mathbf{F}_m}||f||_{\mathbf{F}_{m_1}} + ||I - \varphi(\sigma A)||_{\mathbf{F}_{m_2} \to \mathbf{F}_m}||f||_{\mathbf{F}_{m_2}}$$

$$\leq C_1\sigma^{-(m-m_1)}||f||_{\mathbf{F}_{m_1}} + C_1\sigma^{m_2-m}||f||_{\mathbf{F}_{m_2}}.$$

This proves (3.3.46). By taking $\sigma = [||f||_{\mathbf{F}_{m_1}}||f||_{\mathbf{F}_{m_2}}^{-1}]^{1/(m_2-m_1)}$ (we assume $0 < ||f||_{\mathbf{F}_{m_2}} < \infty$, otherwise the estimates (3.3.46) and (3.3.47) are obvious; then $0 < \sigma \leq 1$) we get (3.3.47) with $C = 2C_1$. $\quad \square$

We mention that (3.3.47) implies (3.3.46). To see this, it suffices to set $\tau = \sigma^{\theta(m-m_2)}$ in the following inequality which holds for all $a, b \geq 0$, $\tau > 0$ and $\theta \in (0,1)$ by the concavity of the function log:

$$a^{1-\theta}b^\theta = [\tau^{1/(1-\theta)}a]^{1-\theta}[\tau^{-1/\theta}b]^\theta \leq (1-\theta)\tau^{1/(1-\theta)}a + \theta\tau^{-1/\theta}b.$$

Finally, we are going to prove more precise estimates of the form (3.3.46), (3.3.47) involving the family of semi-gauges (i.e. seminorms that are allowed to take the value $+\infty$) on $\mathbf{F}_{-\infty}$ defined by

$$(3.3.48) \qquad |f|_m^{(k)} = \sum_{|\alpha|=k} \|A^\alpha f\|_{\mathbf{F}_m},$$

with the convention $|f|_m^{(k)} = \infty$ if $A^\alpha f \notin \mathbf{F}_m$ for some α with $|\alpha| = k$. Note that $|\cdot|_m^{(0)}$ is just the gauge $\|\cdot\|_{\mathbf{F}_m}$ on $\mathbf{F}_{-\infty}$.

PROPOSITION 3.3.22. *Let Γ be a non-empty open cone in $\mathbb{R}^n$ with vertex at the origin and set for $\sigma > 0$:*

$$(3.3.49) \qquad M_\sigma = \sup_{\substack{x \in \Gamma \\ |x| < \sigma}} \|W(x)\|_{B(\mathbf{F})}.$$

For each integer $\ell \geq 1$ there is a constant C_ℓ, depending only on n and ℓ, such that for all $f \in \mathbf{F}_{-\infty}$, $m \in \mathbb{Z}$, $k \in \{0, 1, \dots, \ell\}$ and $\sigma > 0$:

$$(3.3.50) \qquad |f|_m^{(k)} \leq C_\ell M_\sigma \{\sigma^{-k}|f|_m^{(0)} + \sigma^{\ell-k}|f|_m^{(\ell)}\}.$$

If $\|W(x)\|_{B(\mathbf{F})} \leq M < \infty$ for some constant M and all $x \in \Gamma$, then:

$$(3.3.51) \qquad |f|_m^{(k)} \leq C_\ell M(|f|_m^{(0)})^{1-(k/\ell)}(|f|_m^{(\ell)})^{k/\ell}.$$

PROOF. Note first that $M_\sigma \geq 1$, hence (3.3.50) is satisfied for $k = 0, \ell$ if $C_\ell \geq 1$. Now fix a multi-index α with $|\alpha| < \ell$ and choose $\varphi \in \mathscr{S}(\mathbb{R}^n)$ with supp $\widehat{\varphi} \subset \Gamma_1$ and $\varphi^{(\alpha)}(0) = \alpha!$, $\varphi^{(\beta)}(0) = 0$ if $|\beta| < \ell$, $\beta \neq \alpha$ (see Lemma 3.3.17 (d)). Then, if we take into account (3.3.43) (with k replaced by ℓ) and (3.3.42), we get for all $f \in \mathbf{F}_{-\infty}$ and $\sigma > 0$:

(3.3.52)

$$\|A^\alpha f\|_{\mathbf{F}_m} \leq \sigma^{-|\alpha|}\|\varphi(\sigma A)f\|_{\mathbf{F}_m} + \sigma^{\ell-|\alpha|}M_\sigma \sum_{|\beta|=\ell} \frac{1}{\beta!}\|Q^\beta\widehat{\varphi}\|_{L^1}\|A^\beta f\|_{\mathbf{F}_m}.$$

Clearly $\|\varphi(\sigma A)\|_{B(\mathbf{F}_m)} \leq M_\sigma\|\widehat{\varphi}\|_{L^1}$. Since supp $\widehat{\varphi}$ is included in the unit ball, we also have $\|Q^\beta\widehat{\varphi}\|_{L^1} \leq \|\widehat{\varphi}\|_{L^1}$. Hence

$$(3.3.53)\|A^\alpha f\|_{\mathbf{F}_m} \leq \|\widehat{\varphi}\|_{L^1}M_\sigma\{\sigma^{-|\alpha|}\|f\|_{\mathbf{F}_m} + \sigma^{\ell-|\alpha|}\sum_{|\beta|=\ell} \frac{1}{\beta!}\|A^\beta f\|_{\mathbf{F}_m}\}.$$

So we have (3.3.50). Then (3.3.51) follows by minimizing the r.h.s. of (3.3.50) over $\sigma \in (0, \infty)$ (under the assumption $M_\sigma \leq M$). $\square$

3.3.4.   We shall now discuss a question that appears naturally in the present context. Let $\{\mathbf{F}_m\}_{m\in\mathbb{Z}}$ be the scale of Banach spaces associated to a C_0-group W in a Banach space $\mathbf{F} \equiv \mathbf{F}_0$. Since W induces in each $\mathbf{F}_m$ a C_0-group, one may apply the definitions of §3.3.1 and §3.3.2 to the C_0-group W induced in one of these spaces $\mathbf{F}_k$ to obtain a scale $\{(\mathbf{F}_k)_m\}_{m\in\mathbb{Z}}$. Does this lead to new spaces, or can each $(\mathbf{F}_k)_m$ be identified with one of the spaces of the original scale $\{\mathbf{F}_\ell\}_{\ell\in\mathbb{Z}}$? It turns out that in non-pathological cases (in fact in all cases that are known to us) no new spaces appear. For example we shall prove in Section 3.7 that $(\mathbf{F}_k)_m = \mathbf{F}_{k+m}$ for all $k, m \in \mathbb{Z}$ if $\mathbf{F}$ is a Hilbert space. The same result will be established below for the case where $\mathbf{F}$ is a Banach space and W is a *one-parameter* C_0-group.

Let us consider more closely the spaces $(\mathbf{F}_k)_m$. If $m \geq 1$, then $(\mathbf{F}_k)_m$ is a Banach space continuously and densely embedded in $\mathbf{F}_k$, namely the subspace of vectors $f \in \mathbf{F}_k$ such that $x \mapsto W(x)f$ is of class C^m in $\mathbf{F}_k$. In particular, by taking into account Lemma 3.3.1, one gets that

$$(3.3.54) \qquad (\mathbf{F}_k)_m = \mathbf{F}_{k+m} \quad \text{if } k, m \geq 0,$$

with equivalent norms. For a general $k \in \mathbb{Z}$ we have as a consequence of Proposition 3.3.9:

$$(3.3.55) \qquad \mathbf{F}_{k+m} \subset (\mathbf{F}_k)_m \subset \mathbf{F}_k \quad \text{if } k \in \mathbb{Z}, m \geq 0,$$

where both embeddings are continuous and dense (the denseness of the first embedding follows from Theorem 3.3.4 applied to the C_0-group W in $\mathbf{F}_k$ by observing that $\mathbf{F}_{k+m}$ is dense in $\mathbf{F}_k$ and invariant under W). One can also use Proposition 3.3.14 (b) (with $\mathscr{I} = \mathbf{F}_{-\infty}$ and with $\mathbf{F}$ replaced by $\mathbf{F}_k$) to find that for any $k \in \mathbb{Z}$ and $m \geq 0$:

$$(3.3.56) \qquad (\mathbf{F}_k)_m = \{f \in \mathbf{F}_k \mid A^\alpha f \in \mathbf{F}_k \text{ for all } |\alpha| \leq m\},$$

with

$$\|f\|_{(\mathbf{F}_k)_m} = \Big[\sum_{|\alpha|\leq m} \|A^\alpha f\|_{\mathbf{F}_k}^2\Big]^{1/2}.$$

This again leads to (3.3.55) and to the continuity of the first embedding.

The preceding considerations show that the question of whether one has equality in the relation $(\mathbf{F}_k)_m \subset (\mathbf{F}_k)_m$ is a *regularity problem*: the equality will be true if the conditions $f \in \mathbf{F}_k$ and $A^\alpha f \in \mathbf{F}_k$ for all $|\alpha| \leq m$ imply that $f \in \mathbf{F}_{k+m}$. We shall prove the validity of this implication in several cases; in general, the best we can do is to show that $(\mathbf{F}_k)_m$ is contained in some slightly larger space than $\mathbf{F}_{k+m}$ (see Section 3.4).

If m is negative, the space $(\mathbf{F}_k)_m$ is defined in terms of an operation of completion, hence it is a rather abstract object which a priori might not be comparable with the spaces of the scale $\{\mathbf{F}_\ell\}_{\ell\in\mathbb{Z}}$. However, Proposition 3.3.14 (c) (with

$\mathcal{F} = \mathbf{F}_{-\infty}$ and with $\mathbf{F}$ replaced by $\mathbf{F}_k$) leads to a relatively concrete realization of $(\mathbf{F}_k)_m$ as a subspace of $\mathbf{F}_{-\infty}$: if $k \in \mathbb{Z}$ and $m < 0$, then

(3.3.57)
$$\begin{cases} (\mathbf{F}_k)_m = \Big\{ f \in \mathbf{F}_{-\infty} \; \Big| \; f = \sum_{|\alpha| \leq -m} A^\alpha f_\alpha \text{ for some } f_\alpha \in \mathbf{F}_k \Big\} \\ \|f\|_{(\mathbf{F}_k)_m} = \inf\Big\{ \Big[\sum_{|\alpha| \leq -m} \|f_\alpha\|^2_{\mathbf{F}_k} \Big]^{1/2} \; \Big| \; f = \sum_{|\alpha| \leq -m} A^\alpha f_\alpha, \, f_\alpha \in \mathbf{F}_k \Big\}. \end{cases}$$

This implies together with Proposition 3.3.9 (a) that one has the following continuous and dense embeddings:

(3.3.58) $\mathbf{F}_k \subset (\mathbf{F}_k)_m \subset \mathbf{F}_{k+m}$ if $k \in \mathbb{Z}$, $m \leq 0$

(the continuity of the second embedding can be checked by arguments similar to those used in the proof of Proposition 3.3.10 (b)).

As before, we have equality in the relation $(\mathbf{F}_k)_m \subset \mathbf{F}_{k+m}$ in all non-pathological cases, and we shall see in Section 3.4 that in all cases $(\mathbf{F}_k)_m$ contains spaces that are slightly smaller than $\mathbf{F}_{k+m}$. In the present situation (m negative) the question of equality in the relation $(\mathbf{F}_k)_m \subset \mathbf{F}_{k+m}$ is equivalent to an *existence problem* (essentially dual to the regularity problem that we met for positive m): one has to show that each $f \in \mathbf{F}_{k+m}$ can be written as $f = \sum_{|\alpha| \leq -m} A^\alpha f_\alpha$ for some $f_\alpha \in \mathbf{F}_k$. In particular, it is clear that (3.3.57) and Proposition 3.3.12 imply that

(3.3.59) $(\mathbf{F}_k)_m = \mathbf{F}_{k+m}$ if $k, m \leq 0$,

with equivalent norms.

Since $(\mathbf{F}_k)_m = \mathbf{F}_{k+m}$ if k and m have equal signs, one may construct the scale $\{\mathbf{F}_\ell\}_{\ell \in \mathbb{Z}}$ by a recursion procedure: $\mathbf{F}_2 = (\mathbf{F}_1)_1$, $\mathbf{F}_3 = (\mathbf{F}_2)_1$, etc. and $\mathbf{F}_{-2} = (\mathbf{F}_{-1})_{-1}$, $\mathbf{F}_{-3} = (\mathbf{F}_{-2})_{-1}$, etc.

We shall now prove that $(\mathbf{F}_k)_m = \mathbf{F}_{k+m}$ for all $k, m \in \mathbb{Z}$ if W is a one-parameter C_0-group. We remark that (for all aspects of the theory presented until now or developed further on) the case $n = 1$ is considerably more elementary than the case $n > 1$ because, if $n = 1$, then the generator A of the C_0-group has a non-empty resolvent set, so we have at our disposal the resolvent $R = (A + z)^{-1}$ (for some $z \in \mathbb{C}$) which satisfies $AR + zR = I$. Thus, if $n = 1$, we can work with the function $\varphi(\lambda) = \lambda(\lambda + z)^{-1}$ which is a symbol of degree 0 with very simple properties (namely its Fourier transform is a measure). The advantages of this will be apparent in the proof of the next result.

THEOREM 3.3.23. *Let $\mathbf{F}$ be a Banach space and $\{W(x)\}_{x \in \mathbb{R}}$ a one-parameter C_0-group in $\mathbf{F}$. Let $\omega \geq 0$ be such that (3.2.1) is satisfied. Then*
   (a) *The generator A of W (defined by the convention $W(x) = e^{iAx}$) is a closed, densely defined operator in $\mathbf{F}$ with domain $\mathbf{F}_1$.*
   (b) *For each $z \in \mathbb{C}$ with $|\Im z| > \omega$, the mapping $A + z : \mathbf{F}_1 \to \mathbf{F}$ is a homeomorphism.*

(c) *Let $m \geq 1$ and $z \in \mathbb{C}$ with $|\Im z| > \omega$. Then $D(A^m) = \mathbf{F}_m$, the norm $\|\cdot\|_{\mathbf{F}_m}$ on $\mathbf{F}_m$ is equivalent to that defined by $\|(A+z)^m f\|_{\mathbf{F}}$, and the norm $\|\cdot\|_{\mathbf{F}_{-m}}$ on $\mathbf{F}$ is equivalent to that defined by $\|(A+z)^{-m} f\|_{\mathbf{F}}$.*

(d) *For any $k, m \in \mathbb{Z}$, one has $(\mathbf{F}_k)_m = \mathbf{F}_{k+m}$, with equivalent norms. In particular, the domain of the generator of the C_0-group W in $\mathbf{F}_k$ is equal to $\mathbf{F}_{k+1}$ for each $k \in \mathbb{Z}$. Moreover, if the canonical extension of A to $\mathbf{F}_{-\infty}$ is denoted by the same symbol A, then for $|\Im z| > \omega$ the map $A + z : \mathbf{F}_{-\infty} \to \mathbf{F}_{-\infty}$ is bijective and $(A+z)^{-m}$ induces an isomorphism of $\mathbf{F}_k$ onto $\mathbf{F}_{k+m}$.*

PROOF. The result of (a) has already been pointed out after the proof of Theorem 3.3.4. For (b), assume that $z \in \mathbb{C}$ with $\Im z > \omega$ or $\Im z < -\omega$ and set

$$(3.3.60) \quad R = -i \int_0^{\pm\infty} e^{izx} W(x) dx \quad \text{(for } \pm \Im z > \omega \text{ respectively)}.$$

Then $R \in B(\mathbf{F})$, and a short calculation shows that $R\mathbf{F} \subset \mathbf{F}_1$ and $(A+z)R = I$. In particular $A + z$ maps $\mathbf{F}_1 \equiv D(A)$ onto $\mathbf{F}$. Since R commutes with the C_0-group W, the identity $(A+z)R = I$ leads to $R(A+z)f = f$ if $f \in \mathbf{F}_1$. This implies that $R\mathbf{F} = \mathbf{F}_1$, that $A + z : \mathbf{F}_1 \to \mathbf{F}$ is invertible and that $R = (A+z)^{-1}$.

(c) The identity $D(A^m) = \mathbf{F}_m$ follows by recursion from (a) and the relation $\mathbf{F}_k = (\mathbf{F}_{k-1})_1$ (see (3.3.54)). To compare the indicated norms, we observe that for each $k \in \{1, \ldots, m\}$ there are constants d_j (depending on k and z) such that one has on $\mathbf{F}_m$:

$$(3.3.61) \qquad A^k = \sum_{j=0}^{k} d_j R^{m-j}(A+z)^m \qquad (0 \leq k \leq m).$$

This identity implies that $\|f\|_{\mathbf{F}_m} \leq c\|(A+z)^m f\|_{\mathbf{F}}$, whereas the inequality $\|(A+z)^m f\|_{\mathbf{F}}^2 \leq c\sum_{k=0}^{m} \|A^k f\|_{\mathbf{F}}^2 = c\|f\|_{\mathbf{F}_m}^2$ is evident.

Next, if $f \in \mathbf{F}$ has the form $f = \sum_{k=0}^{m} A^k f$ with $f_k \in \mathbf{F}_k$, then by (3.3.61) there are constants γ_{jk} such that

$$R^m f = \sum_{j,k=0}^{m} \gamma_{jk} R^{m-j} f_k.$$

Thus $\|R^m f\|_{\mathbf{F}}^2 \leq c\sum_{k=0}^{m} \|f_k\|_{\mathbf{F}}^2$, so that $\|R^m f\|_{\mathbf{F}} \leq c^{1/2}\|f\|_{\mathbf{F}_{-m}}$. For the opposite inequality we write $f = (A+z)^m R^m f = \sum_{k=0}^{m} \delta_k A^k R^m f$, so that $\|f\|_{\mathbf{F}_{-m}} \leq c\|R^m f\|_{\mathbf{F}}$ because $R^m f \in \mathbf{F}_k$ for each $k = 0, \ldots, m$.

(d) We set $S = A + z$. By Proposition 3.3.9 (a) we have $S \in B(\mathbf{F}_{k+1}, \mathbf{F}_k)$ for each $k \in \mathbb{Z}$. By using the equivalent norms given in (c), one sees that $R \in B(\mathbf{F}_k, \mathbf{F}_{k+1})$ for each $k \in \mathbb{Z}$. Clearly $RSf = SRf = f$ for $f \in \mathbf{F}_\infty$, hence $RS = SR = I$ as maps from $\mathbf{F}_{-\infty}$ into $\mathbf{F}_{-\infty}$, because $\mathbf{F}_\infty$ is dense in $\mathbf{F}_k$.

We now prove that $(\mathbf{F}_k)_1 = \mathbf{F}_{k+1}$ for each $k \in \mathbb{Z}$. By virtue of (3.3.55) it suffices to show that $(\mathbf{F}_k)_1 \subset \mathbf{F}_{k+1}$ continuously. If $f \in (\mathbf{F}_k)_1$, then $f \in \mathbf{F}_k$ and $Af \in \mathbf{F}_k$ by (3.3.56). So $Sf \in \mathbf{F}_k$; hence $f = RSf \in \mathbf{F}_{k+1}$ and $\|f\|_{\mathbf{F}_{k+1}} \leq \|R\|_{\mathbf{F}_k \to \mathbf{F}_{k+1}}\|(A+z)f\|_{\mathbf{F}_k} \leq c\|f\|_{(\mathbf{F}_k)_1}$.

We now show by induction on m that $(\mathbf{F}_k)_m = \mathbf{F}_{k+m}$ if $m \geq 2$. Suppose we know that $(\mathbf{F}_k)_{m-1} = \mathbf{F}_{k+m-1}$. Then, by using also the result of the preceding paragraph, we have $\mathbf{F}_{k+m} = (\mathbf{F}_{k+m-1})_1 = [(\mathbf{F}_k)_{m-1}]_1$. On the other hand, by applying the result of the preceding paragraph (i.e. $\mathbf{F}_{\ell+1} = (\mathbf{F}_\ell)_1$ for all $\ell \in \mathbb{Z}$) to the C_0-group induced by W in $\mathbf{F}_k$, we get that $(\mathbf{F}_k)_{\ell+1} = [(\mathbf{F}_k)_\ell]_1$; in particular (take $\ell = m - 1$): $(\mathbf{F}_k)_m = [(\mathbf{F}_k)_{m-1}]_1$. This completes the induction argument.

Next we show that $(\mathbf{F}_k)_{-1} = \mathbf{F}_{k-1}$ for each $k \in \mathbb{Z}$. In view of (3.3.58), we must show that $\mathbf{F}_{k-1} \subset (\mathbf{F}_k)_{-1}$ continuously. If $f \in \mathbf{F}_{k-1}$, then $Rf \in \mathbf{F}_k$. So $f = SRf = A(Rf) + zRf \in (\mathbf{F}_k)_{-1}$ by (3.3.57), and $\|f\|_{(\mathbf{F}_k)_{-1}} \le \|Rf\|_{\mathbf{F}_k} + |z| \cdot \|Rf\|_{\mathbf{F}_k} \le c\|R\|_{\mathbf{F}_{k-1} \to \mathbf{F}_k}\|f\|_{\mathbf{F}_{k-1}}$.

Now one obtains the relation $(\mathbf{F}_k)_{-m} = \mathbf{F}_{k-m}$ for $m \ge 2$ by an induction argument similar to that used above. $\square$

REMARK 3.3.24. If the space $\mathbf{F}$ in the preceding theorem is a Hilbert space (identified with its dual $\mathbf{F}^*$), then one has the following expression for the norms $\|\cdot\|_{\mathbf{F}_1}$ and $\|\cdot\|_{\mathbf{F}_{-1}}$:

$$(3.3.62) \quad \|f\|_{\mathbf{F}_1} = \|(I + A^*A)^{1/2}f\|_{\mathbf{F}}, \quad \|f\|_{\mathbf{F}_{-1}} = \|(I + AA^*)^{-1/2}f\|_{\mathbf{F}}.$$

The first equality is obvious on $D(A^*A)$; it holds on $\mathbf{F}_1$ because $D(A^*A)$ is a core for A and for $(I + A^*A)^{1/2}$. To verify the second equality, we start from

$$(3.3.63) \quad \|f\|_{\mathbf{F}_{-1}} = \inf_{g \in D(A)} [\|f - Ag\|_{\mathbf{F}}^2 + \|g\|_{\mathbf{F}}^2]^{1/2}.$$

We fix $f \in \mathbf{F}$ and show that there is a vector $g_0 \in D(A)$ for which the minimum in (3.3.63) is attained. If such a vector g_0 exists, one must have

$$\|f - Ag_0 - Ah\|_{\mathbf{F}}^2 + \|g_0 + h\|_{\mathbf{F}}^2 \ge \|f - Ag_0\|_{\mathbf{F}}^2 + \|g_0\|_{\mathbf{F}}^2 \quad \forall h \in D(A)$$

or equivalently

$$2\Re[\langle Ag_0 - f, Ah \rangle + \langle g_0, h \rangle] + \|Ah\|^2 + \|h\|^2 \ge 0 \quad \forall h \in D(A).$$

This last condition is satisfied for $g_0 = A^*(I + AA^*)^{-1}f \in D(A)$, because with this choice the square bracket is zero for each $h \in D(A)$. By taking $g = g_0$ in (3.3.63), one obtains the second expression in (3.3.62).

3.3.5. If the Banach space $\mathbf{F}$ is reflexive, then several aspects of the theory that we have developed so far can be improved. An example is the following improvement of Proposition 3.3.19:

PROPOSITION 3.3.25. (a) *Assume that there is Hausdorff topology $\mathscr{T}$ on $\mathbf{F}$ which is weaker than the initial topology and such that*

(1) $\varphi(A) : \mathbf{F} \to \mathbf{F}$ *is continuous for the $\mathscr{T}$-topology if $\widehat{\varphi} \in C_0^\infty(\mathbb{R}^n)$,*

(2) *any norm-bounded sequence in $\mathbf{F}$ contains a subsequence which is convergent in the $\mathscr{T}$-topology.*

Let $m \ge 1$ and $f \in \mathbf{F}$. Then $f \in \mathbf{F}_m$ if and only if

$$(3.3.64) \quad \liminf_{\varepsilon \to +0} \varepsilon^{-m}\|[W(\varepsilon x) - I]^m f\|_{\mathbf{F}} < \infty \quad \forall x \in \mathbb{R}^n.$$

(b) *Let $\mathbf{F}$ be a reflexive Banach space, $m \ge 1$ and $f \in \mathbf{F}$. Then $f \in \mathbf{F}_m$ if and only if (3.3.64) is satisfied.*

PROOF. (b) This follows from (a) by taking for $\mathscr{T}$ the weak topology of $\mathbf{F}$.

(a) The "only if" part is an immediate consequence of Proposition 3.3.19. To prove the "if" part, let $f \in \mathbf{F}$ and assume that (3.3.64) holds. It follows easily from (2) that, for fixed $x \in \mathbb{R}^n$, one may find a sequence $\{\varepsilon_k\}$ of positive numbers such that $\varepsilon_k \to 0$ and a vector $g \in \mathbf{F}$ such that $g_k \equiv \varepsilon_k^{-m}[W(\varepsilon_k x) - I]^m f$ defines a bounded sequence of vectors in $\mathbf{F}$ and $g_k \to g$ in the $\mathscr{T}$-topology.

Now choose $\varphi : \mathbb{R}^n \to \mathbb{R}$ such that $\varphi(0) = 1$ and $\widehat{\varphi} \in C_0^\infty(\mathbb{R}^n)$, and let $\sigma > 0$. Then, by (1), we have $\varphi(\sigma A)g_k \to \varphi(\sigma A)g$ in the $\mathscr{T}$-topology as $k \to \infty$. Also, by (3.3.40) and Proposition 3.3.5, g_k converges in $\mathbf{F}_{-m}$ to $h \equiv \sum_{|\alpha|=m} i^m (m!/\alpha!)x^\alpha A^\alpha f$. Hence $\varphi(\sigma A)g_k$ converges in $\mathbf{F}$ to $\varphi(\sigma A)h$ (because $\varphi(\sigma A) : \mathbf{F}_{-m} \to \mathbf{F}$ is continuous). As the $\mathscr{T}$-topology is weaker than the norm topology on $\mathbf{F}$, we get that

$$(3.3.65) \qquad \varphi(\sigma A)g = \varphi(\sigma A)h \equiv \varphi(\sigma A)i^m \sum_{|\alpha|=m} \frac{m!}{\alpha!} x^\alpha A^\alpha f \quad \forall \sigma > 0.$$

By letting $\sigma \to 0$ in (3.3.65), one obtains that $g = h$ in $\mathbf{F}_{-m}$. Hence $h \in \mathbf{F}$, i.e. $\sum_{|\alpha|=m}(\alpha!)^{-1}x^\alpha A^\alpha f \in \mathbf{F}$ for each $x \in \mathbb{R}^n$. Proposition 1.1.2 allows one to conclude that $f \in \mathbf{F}_m$. $\square$

The preceding proposition leads to a strengthening of Lemma 3.3.1:

COROLLARY 3.3.26. *Assume that* $\mathbf{F}$ *is a reflexive Banach space. Let* $m \geq 1$ *and let* $f \in \mathbf{F}$ *be such that the function* $x \mapsto W(x)f \in \mathbf{F}$ *is of class* C^{m-1} *and its derivatives of order* $m - 1$ *are locally Lipschitz. Then* $f \in \mathbf{F}_m$.

PROOF. We use (3.3.18) to obtain that

$$\left\| \left[\frac{W(x) - I}{|x|} \right]^m f \right\|_{\mathbf{F}} \leq C(x) \sum_{|\alpha|=m-1} \frac{(m-1)!|x^\alpha|}{\alpha!|x|^{m-1}} \left\| \frac{W(x) - I}{|x|} A^\alpha f \right\|_{\mathbf{F}}$$

where $C(x) = \sup_{|z| \leq (m-1)|x|} \|W(z)\|_{B(\mathbf{F})}$. The hypotheses made on f imply that the r.h.s. is a locally bounded function of x. $\square$

Let us illustrate Proposition 3.3.25 by some examples. We shall consider the translation group T (given by (1.2.12)) successively in the spaces $L^p(\mathbb{R})$ with $1 < p < \infty$, $L^1(\mathbb{R})$, $C_\infty(\mathbb{R})$ and $L^\infty(\mathbb{R})$.

(1) Let $\mathbf{F} = L^p(\mathbb{R})$, $1 < p < \infty$. Then part (b) of Proposition 3.3.25 applies; so if $f \in L^p(\mathbb{R})$, then its distributional derivative f' belongs to $L^p(\mathbb{R})$ if and only if $\|f(\cdot + y) - f\|_{L^p(\mathbb{R})} \leq c|y|$ for some constant c.

(2) Let $\mathbf{F} = L^1(\mathbb{R})$. A priori it is not clear whether part (a) of Proposition 3.3.25 can be used or not. In fact it can not, because its conclusion is not true: by rather standard distributional arguments one can show that $\|f(\cdot + y) - f\|_{L^1(\mathbb{R})} \leq c|y|$ (with $f \in L^1(\mathbb{R})$) if and only if the distributional derivative f' of f is an integrable measure on $\mathbb{R}$.

(3) Let $\mathbf{F} = C_\infty(\mathbb{R})$. Here the situation is similar to that in (2): it is simple to check that the estimate $\|f(\cdot + y) - f\|_{C_\infty(\mathbb{R})} \leq c|y|$ holds if and only if $f' \in L^\infty(\mathbb{R})$, i.e. if and only if f is Lipschitz.

(4) Let $\mathbf{F} = L^\infty(\mathbb{R})$. This case is of a different nature than the preceding ones because the translation group is not of class C_0 in $L^\infty(\mathbb{R})$. It is clear that $x \mapsto T(x)f \in L^\infty(\mathbb{R})$ is continuous if and only if f is a uniformly continuous function. However, $L^\infty(\mathbb{R})$ is the adjoint space of $L^1(\mathbb{R})$ and the translation group in $L^\infty(\mathbb{R})$ is the dual of the translation group in $L^1(\mathbb{R})$ (cf. (3.2.7)). Thus, if we take for $\mathscr{T}$ in Proposition 3.3.25 (a) the weak* topology on $L^\infty(\mathbb{R})$, then the assumptions (1) and (2) of that proposition are clearly satisfied. As in case (3) above, the condition $\|f(\cdot + y) - f\|_{L^\infty(\mathbb{R})} \leq c|y|$ is equivalent to the condition that $f' \in L^\infty(\mathbb{R})$. Moreover, if $f' \in L^\infty(\mathbb{R})$, then $y^{-1}[f(\cdot + y) - f]$ converges to f' in the weak* topology of $L^\infty(\mathbb{R})$ as $y \to 0$, but the convergence takes place in the norm of $L^\infty(\mathbb{R})$ if and only if f' is a uniformly continuous function. This example shows in particular that, for a dual group in a non-reflexive Banach space $\mathbf{F}$, the weak* and the strong domain of the generator may be quite different (none of them is norm dense in the present example).

If W is a C_0-group in a reflexive Banach space $\mathbf{F}$, then by Proposition 3.2.5 the adjoint $\mathbf{F}^*$ is equipped with the dual C_0-group W^* defined by $W^*(x) = W(-x)^*$. We denote the generator of W^* by $A^* = (A_1^*, \ldots, A_n^*)$. This notation is justified because $\varphi(A)^* = \overline{\varphi}(A^*)$ if $\varphi \in \mathcal{M}_w(\mathbb{R}^n)$, by (3.2.21). Furthermore we have $A^{*\alpha} \subset (A^\alpha)^*$ in the sense of unbounded operators, for each multi-index $\alpha \in \mathbb{N}^n$. More precisely, let $\{(\mathbf{F}^*)_m\}$, $m \in \mathbb{Z} \cup \{-\infty, +\infty\}$, be the scale of Banach spaces associated to the C_0-group W^* in $\mathbf{F}^*$. Then one has for each $m \in \mathbb{N}$, $|\alpha| \leq m$, $f \in \mathbf{F}_m$ and $g \in (\mathbf{F}^*)_m$:

$$(3.3.66) \qquad\qquad \langle A^\alpha f, g \rangle = \langle f, A^{*\alpha} g \rangle.$$

To see this, it suffices to calculate the derivative $(i\partial_x)^\alpha$ at $x = 0$ of the equation $\langle W(x)f, g \rangle = \langle f, W(x)^* g \rangle = \langle f, W^*(-x)g \rangle$.

As a last result in this section we shall show that there is a canonical identification of the adjoint space $(\mathbf{F}_m)^*$ of $\mathbf{F}_m$ with $(\mathbf{F}^*)_{-m}$. We mention that, even if $\mathbf{F}$ is a Hilbert space and $\mathbf{F}^*$ is identified with $\mathbf{F}$ through the Riesz isomorphism, then $(\mathbf{F}_m)^*$ is not the same as $\mathbf{F}_{-m}$ if W is not unitary (for $n = 1$ this is easy to see by using Remark 3.3.24: $\mathbf{F}_{-1}$ is the closure of $\mathbf{F}$ in the norm $\|(I + AA^*)^{-1/2}f\|_{\mathbf{F}}$, whereas $(\mathbf{F}_1)^*$ is the closure of $\mathbf{F}$ in the norm $\|(I + A^*A)^{-1/2}f\|_{\mathbf{F}}$).

We first consider the case $m > 0$. Then $\mathbf{F}_m \subset \mathbf{F}$ continuously and densely, hence $\mathbf{F}^* \subset (\mathbf{F}_m)^*$ continuously. If $\mathbf{F}$ is reflexive, then so is $\mathbf{F}_m$ by Proposition 3.3.13, hence the embedding $\mathbf{F}^* \subset (\mathbf{F}_m)^*$ is also dense (see Section 2.1). So the norm on $(\mathbf{F}_m)^*$ is uniquely determined by its restriction to $\mathbf{F}^*$, and we must prove that $\|g\|_{(\mathbf{F}_m)^*} = \|g\|_{(\mathbf{F}^*)_{-m}}$ if $g \in \mathbf{F}^*$.

LEMMA 3.3.27. *Assume that* $\mathbf{F}$ *is reflexive. Let* $m \geq 0$ *and* $g \in \mathbf{F}^*$. *Then* $\|g\|_{(\mathbf{F}^*)_{-m}} = \|g\|_{(\mathbf{F}_m)^*}$. *Moreover there is a family* $\{h_\alpha\}_{|\alpha| \leq m}$ *of vectors in* $\mathbf{F}^*$ *such that* $g = \sum_{|\alpha| \leq m} A^{*\alpha} h_\alpha$ *and* $\|g\|_{(\mathbf{F}^*)_{-m}} = (\sum_{|\alpha| \leq m} \|h_\alpha\|_{\mathbf{F}^*}^2)^{1/2}$.

PROOF. (i) We use the Banach space $\mathbb{F}_m$ associated to $\mathbf{F}$ by Remark 3.3.11 and the Banach space $\mathbb{F}_m^*$ associated to $\mathbf{F}^*$ by the same construction. Both $\mathbb{F}_m$ and $\mathbb{F}_m^*$ are reflexive, and $\mathbb{F}_m$ is the dual of $\mathbb{F}_m^*$. We denote by $\langle\langle \cdot, \cdot \rangle\rangle$ the duality between these two spaces; if $\mathfrak{f} = \{f_\alpha\}_{|\alpha| \leq m} \in \mathbb{F}_m$ and $\mathfrak{g} = \{g_\alpha\}_{|\alpha| \leq m} \in \mathbb{F}_m^*$, then

$$\langle\langle \mathfrak{f}, \mathfrak{g} \rangle\rangle = \sum_{|\alpha| \leq m} \langle f_\alpha, g_\alpha \rangle,$$

$$\|\mathfrak{f}\|_{\mathbb{F}_m} = \Big(\sum_{|\alpha| \leq m} \|f_\alpha\|_{\mathbb{F}}^2 \Big)^{1/2}, \quad \|\mathfrak{g}\|_{\mathbb{F}_m^*} = \Big(\sum_{|\alpha| \leq m} \|g_\alpha\|_{\mathbb{F}^*}^2 \Big)^{1/2}.$$

To each vector $f \in \mathbf{F}_m$ one can associate a vector $\underline{A}f \equiv \{A^\alpha f\}_{|\alpha| \leq m}$ in $\mathbb{F}_m$. We observe that $\|\underline{A}f\|_{\mathbb{F}_m} = \|f\|_{\mathbf{F}_m}$. Also, if $\mathfrak{g} \equiv \{g_\alpha\}_{|\alpha| \leq m} \in \mathbb{F}_m^*$ is such that $g_\alpha \in (\mathbf{F}^*)_{|\alpha|}$, then by (3.3.66):

$$\langle\langle \underline{A}f, \mathfrak{g} \rangle\rangle = \sum_{|\alpha| \leq m} \langle A^\alpha f, g_\alpha \rangle = \langle f, \sum_{|\alpha| \leq m} A^{*\alpha} g_\alpha \rangle.$$

In particular we have

$$\Big\| \sum_{|\alpha| \leq m} A^{*\alpha} g_\alpha \Big\|_{(\mathbf{F}_m)^*} = \sup_{\|f\|_{\mathbf{F}_m} \leq 1} |\langle\langle \underline{A}f, \mathfrak{g} \rangle\rangle|.$$

(ii) Now let $g \in \mathbf{F}^*$ and consider a decomposition $g = \sum_{|\alpha| \leq m} A^{*\alpha} g_\alpha$ with $g_0 \in \mathbf{F}^*$ and $g_\alpha \in (\mathbf{F}^*)_\infty$ if $\alpha \neq 0$. Denote by $\mathfrak{g} \equiv \{g_\alpha\}_{|\alpha| \leq m}$ the associated element of $\mathbb{F}_m^*$. Then, by the results of (i):

$$(3.3.67) \qquad \Big(\sum_{|\alpha| \leq m} \|g_\alpha\|_{\mathbb{F}^*}^2 \Big)^{1/2} = \sup_{\|\mathfrak{e}\|_{\mathbb{F}_m} \leq 1} |\langle\langle \mathfrak{e}, \mathfrak{g} \rangle\rangle|$$

$$\geq \sup_{\|f\|_{\mathbb{F}_m} \leq 1} |\langle\langle \underline{A}f, \mathfrak{g} \rangle\rangle| = \|g\|_{(\mathbf{F}_m)^*}.$$

In particular, by using (3.3.22) in $\mathbf{F}^*$, we see that $\|g\|_{(\mathbf{F}^*)_{-m}} \geq \|g\|_{(\mathbf{F}_m)^*}$.

(iii) To obtain the opposite inequality, let g and $\mathfrak{g}$ be as in (ii). By the Hahn-Banach theorem, the restriction $\mathfrak{g}|_{\mathbb{G}}$ of the functional $\mathfrak{g}$ to the subspace $\mathbb{G} \equiv \{\underline{A}f \mid f \in \mathbf{F}_m\}$ has an extension from $\mathbb{G}$ to $\mathbb{F}_m$ with the same norm. In other terms there is an element $\mathfrak{h} = \{h_\alpha\}_{|\alpha| \leq m}$ of $\mathbb{F}_m^*$ (i.e. $h_\alpha \in \mathbf{F}^*$) such that

$$(3.3.68)$$

$$\|\mathfrak{h}\|_{\mathbb{F}_m^*} \equiv \Big(\sum_{|\alpha| \leq m} \|h_\alpha\|_{\mathbf{F}^*}^2 \Big)^{1/2}$$

$$= \|\mathfrak{g}|_{\mathbb{G}}\| = \sup_{\|f\|_{\mathbf{F}_m} \leq 1} |\langle\langle \underline{A}f, \mathfrak{g} \rangle\rangle| = \|g\|_{(\mathbf{F}_m)^*}$$

and

$$(3.3.69) \qquad \langle\langle \underline{A}f, \mathfrak{g} \rangle\rangle = \langle\langle \underline{A}f, \mathfrak{h} \rangle\rangle \qquad \forall f \in \mathbf{F}_m.$$

We shall see below that (3.3.69) implies that $g = \sum_{|\alpha| \leq m} A^{*\alpha} h_\alpha$. By virtue of Proposition 3.3.12, the relation (3.3.68) then shows that $\|g\|_{(\mathbf{F}^*)_{-m}} \leq \|g\|_{(\mathbf{F}_m)^*}$.

It remains to check that $\sum_{|\alpha| \leq m} A^{*\alpha} h_\alpha = g$ (for instance as elements of $(\mathbf{F}^*)_{-m}$). For this we choose a function $\varphi : \mathbb{R}^n \to \mathbb{C}$ such that $\varphi(0) = 1$ and

$\widehat{\varphi} \in C_0^\infty(\mathbb{R}^n)$. If $\varepsilon > 0$, $\varphi(\varepsilon A)$ maps $\mathbf{F}_m$ into $\mathbf{F}_\infty$ and $\overline{\varphi}(\varepsilon A^*)$ maps $\mathbf{F}^*$ into $(\mathbf{F}^*)_\infty$. Hence we have for all $f \in \mathbf{F}_m$:

$$
\begin{aligned}
0 = \langle\langle \underline{A}\varphi(\varepsilon A)f, \mathfrak{g} - \mathfrak{h} \rangle\rangle &= \sum_{|\alpha| \leq m} \langle \varphi(\varepsilon A) A^\alpha f, g_\alpha - h_\alpha \rangle \\
&= \sum_{|\alpha| \leq m} \langle A^\alpha f, \overline{\varphi}(\varepsilon A^*)(g_\alpha - h_\alpha) \rangle \\
&= \langle f, \sum_{|\alpha| \leq m} A^{*\alpha}\overline{\varphi}(\varepsilon A^*)(g_\alpha - h_\alpha) \rangle \\
&= \langle f, \overline{\varphi}(\varepsilon A^*)g - \sum_{|\alpha| \leq m} A^{*\alpha}\overline{\varphi}(\varepsilon A^*)h_\alpha \rangle.
\end{aligned}
$$

Since $\overline{\varphi}(\varepsilon A^*)g - \sum_{|\alpha| \leq m} A^{*\alpha}\overline{\varphi}(\varepsilon A^*)h_\alpha \in \mathbf{F}^*$ and $\mathbf{F}_m$ is dense in $\mathbf{F}$, we get

$$
\overline{\varphi}(\varepsilon A^*)g = \sum_{|\alpha| \leq m} A^{*\alpha}\overline{\varphi}(\varepsilon A)h_\alpha.
$$

As $\varepsilon \to 0$, $\varphi(\varepsilon A^*)g$ converges to g in $\mathbf{F}^*$ (hence in $(\mathbf{F}^*)_{-m}$) and

$$
\sum_{|\alpha| \leq m} A^{*\alpha}\overline{\varphi}(\varepsilon A^*)h_\alpha \text{ converges to } \sum_{|\alpha| \leq m} A^{*\alpha}h_\alpha \text{ in } (\mathbf{F}^*)_{-m}. \quad \square
$$

Lemma 3.3.27 shows (together with the remarks preceding it) that $(\mathbf{F}_m)^* = (\mathbf{F}^*)_{-m}$ for each $m > 0$ if $\mathbf{F}$ is reflexive. If $m < 0$, we have $\mathbf{F} \subset \mathbf{F}_m$ continuously and densely, hence $(\mathbf{F}_m)^* \subset \mathbf{F}^*$ continuously and densely (if $\mathbf{F}$ is reflexive). On the other hand, we also have $(\mathbf{F}^*)_{-m} \subset \mathbf{F}^*$ continuously and densely. The next theorem shows that the two subspaces $(\mathbf{F}_m)^*$ and $(\mathbf{F}^*)_{-m}$ of $\mathbf{F}^*$ are identical, with equal norms.

THEOREM 3.3.28. *If $\mathbf{F}$ is a reflexive Banach space equipped with a C_0-representation of $\mathbb{R}^n$, then $(\mathbf{F}_m)^* = (\mathbf{F}^*)_{-m}$ (as Banach spaces) for each $m \in \mathbb{Z}$.*

PROOF. For $m > 0$ the result has been established above. In particular we have $[(\mathbf{F}^*)_m]^* = (\mathbf{F}^{**})_{-m} \equiv \mathbf{F}_{-m}$ if $m > 0$. Thus $(\mathbf{F}_{-m})^* = [(\mathbf{F}^*)_m]^{**} = (\mathbf{F}^*)_m$ if $m > 0$. $\square$

3.4. Besov Spaces Associated to a C_0-Group

We continue to work with a fixed C_0-representation W of $\mathbb{R}^n$ in a Banach space $\mathbf{F}$. In the preceding section we associated to the couple $(W, \mathbf{F})$ a scale

(3.4.1) $\cdots \subset \mathbf{F}_2 \subset \mathbf{F}_1 \subset \mathbf{F} = \mathbf{F}_0 \subset \mathbf{F}_{-1} \subset \mathbf{F}_{-2} \subset \ldots$

consisting of Banach spaces; each embedding in (3.4.1) is continuous and dense. For various practical purposes it is natural to try to extend this discrete scale to a continuous one by using the complex or real interpolation method: for each couple of successive integers m, $m + 1$ one may define the complex or real interpolating spaces $[\mathbf{F}_{m+1}, \mathbf{F}_m]_\theta$ or $(\mathbf{F}_{m+1}, \mathbf{F}_m)_{\theta,p}$ respectively, where $0 < \theta < 1$ and $1 \leq p \leq \infty$. Since each non-integer real number s is situated between two uniquely defined successive integers m, $m + 1$, one can define spaces $\mathbf{F}_s^{(c)}$, $\mathbf{F}_{s,p}^{(r)}$ by complex or real interpolation for all $s \in \mathbb{R} \backslash \mathbb{Z}$ and $1 \leq p \leq \infty$ (use the parameter θ defined by $s = \theta m + (1-\theta)(m+1)$). It turns out that the scales defined in this way have the very useful reiteration property, for example $[\mathbf{F}_s^{(c)}, \mathbf{F}_t^{(c)}]_\mu = \mathbf{F}_{(1-\mu)s+\mu t}^{(c)}$ if $0 < \mu < 1$, $t < s$ and the numbers s, t, $(1-\mu)s + \mu t$ are not integers. Then the question arises whether the spaces $\mathbf{F}_m$, $m \in \mathbb{Z}$, fit naturally into one of the preceding scales or not. For example, one would be tempted to define $\mathbf{F}_s^{(c)} = \mathbf{F}_m$ if $s = m \in \mathbb{Z}$ and hope to get the reiteration property for all $0 < \mu < 1$, $t < s$; it turns out that the full reiteration property for the scale defined in this way is satisfied only in very few cases, for example if $\mathbf{F}$ is a Hilbert space (cf. Section 3.7). In most other cases the scale $\{\mathbf{F}_m\}_{m \in \mathbb{Z}}$ does not have good interpolation properties (for example $[\mathbf{F}_{m+1}, \mathbf{F}_{m-1}]_{1/2} \neq \mathbf{F}_m$ and $(\mathbf{F}_{m+1}, \mathbf{F}_{m-1})_{1/2,p} \neq \mathbf{F}_m$ for any $p \in [1, \infty]$), and consequently one gets a rather heavy formalism which, moreover, does not allow one to obtain natural results in various estimates (see part II of [ABG1]).

It is possible to complete the scales $\{\mathbf{F}_s^{(c)}\}$ and $\{\mathbf{F}_{s,p}^{(r)}\}$ (defined for non-integer values of s) such as to be defined for all real numbers s and such that the reiteration property is satisfied. The spaces introduced in this way for integer values of s are different from the original spaces $\mathbf{F}_m$ but in fact considerably more useful in many applications; this explains the abundance of spaces that appear in the harmonic analysis of the translation group (Sobolev spaces, potential spaces, Besov spaces, Triebel-Lizorkin spaces, etc.; see [P1], [St1], [FJW]).

For the kind of applications we have in mind it appears that the scale constructed by real interpolation is appropriate. From now on we shall denote these spaces by $\mathbf{F}_{s,p}$ and view them as B-spaces (see Section 2.1). We shall first define the scale $\{\mathbf{F}_{s,p}\}$ and establish its basic properties. Then we shall study admissible norms on $\mathbf{F}_{s,p}$ that can be introduced directly in terms of the C_0-group.

3.4.1. In order to apply interpolation theory in the scale $\{\mathbf{F}_m\}_{m \in \mathbb{Z}}$, we need to know the K-functional $K_{m_2 m_1}$ for each couple $(\mathbf{F}_{m_2}, \mathbf{F}_{m_1})$ with $m_1, m_2 \in \mathbb{Z}$, $m_1 < m_2$. We claim that

(3.4.2)
$$K_{m_2 m_1}(\tau, f) \sim \tau \|\varphi(\tau^{1/(m_2-m_1)}A)f\|_{\mathbf{F}_{m_2}} + \|[\varphi(\tau^{1/(m_2-m_1)}A) - I]f\|_{\mathbf{F}_{m_1}},$$

where φ is any function in $\mathscr{S}(\mathbb{R}^n)$ satisfying $\varphi(0) = 1$, $\varphi^{(\beta)}(0) = 0$ if $1 \leq |\beta| < m_2 - m_1$ and (if the C_0-group W is not polynomially bounded) $\widehat{\varphi} \in C_0^\infty(\mathbb{R}^n)$. The validity of (3.4.2) follows from Lemma 2.7.1 by observing that the couple $(\mathbf{F}_{m_2}, \mathbf{F}_{m_1})$ is quasi-linearizable: if we set $V_\tau = \varphi(\tau^{1/(m_2-m_1)}A)$, then the conditions (1)-(3) in the definition of a quasi-linearizable couple (Section 2.7) are satisfied as a consequence of Lemma 3.3.20. We can now prove the basic result:

THEOREM 3.4.1. *Let $m_1 < m < m_2$ be integers and denote by $\theta = (m_2 - m) \cdot (m_2 - m_1)^{-1}$ the unique number in $(0,1)$ such that $m = \theta m_1 + (1 - \theta)m_2$. Then*

(3.4.3)
$$(\mathbf{F}_{m_2}, \mathbf{F}_{m_1})_{\theta,1} \subset \mathbf{F}_m \subset (\mathbf{F}_{m_2}, \mathbf{F}_{m_1})_{\theta,\infty}.$$

PROOF. We apply the criteria given in Lemma 2.5.1. For the first inclusion in (3.4.3) we must show that:

$$(3.4.4) \quad \|f\|_{\mathbf{F}_m} \le c[\tau^\theta \|f\|_{\mathbf{F}_{m_2}} + \tau^{\theta-1}\|f\|_{\mathbf{F}_{m_1}}] \quad \forall \tau \in (0,1], \ \forall f \in \mathbf{F}_\infty.$$

This follows from (3.3.46) with $\sigma = \tau^{1/(m_2-m_1)}$.

For the second inclusion in (3.4.3), we must verify that

$$(3.4.5) \quad K_{m_2 m_1}(\tau, f) \le c\tau^{1-\theta}\|f\|_{\mathbf{F}_m} \quad \forall \tau \in (0,1], \ \forall f \in \mathbf{F}_\infty.$$

This inequality is easily obtained from (3.4.2) after noticing that, by Lemma 3.3.20 (for $0 < \tau \le 1$):

$$\|\varphi(\tau^{1/(m_2-m_1)}A)\|_{\mathbf{F}_m \to \mathbf{F}_{m_2}} \le c[\tau^{-1/(m_2-m_1)}]^{m_2-m} = c\tau^{-\theta}$$

and

$$\|\varphi(\tau^{1/(m_2-m_1)}A) - I\|_{\mathbf{F}_m \to \mathbf{F}_{m_1}} \le c[\tau^{1/(m_2-m_1)}]^{m-m_1} = c\tau^{1-\theta}. \quad \square$$

In the terminology of Section 2.5, the preceding theorem states that, if $m_1 < m < m_2$ are integers, then $\mathbf{F}_m$ is of class $\theta \equiv (m_2-m)(m_2-m_1)^{-1}$ for the couple $(\mathbf{F}_{m_2}, \mathbf{F}_{m_1})$. So, if ℓ_k, m_k $(k = 1, 2)$ are integers with $\ell_1 < m_1 < m_2 < \ell_2$, then $\mathbf{F}_{m_1}$ and $\mathbf{F}_{m_2}$ are of class θ_1 and θ_2 respectively for the couple $(\mathbf{F}_{\ell_2}, \mathbf{F}_{\ell_1})$ with

$$\theta_1 = \frac{\ell_2 - m_1}{\ell_2 - \ell_1}, \qquad \theta_2 = \frac{\ell_2 - m_2}{\ell_2 - \ell_1}.$$

Note that $\theta_2 < \theta_1$. Then Theorem 2.5.3 implies that, for any $\theta \in (0,1)$ and any $p \in [1, \infty]$:

$$(3.4.6) \quad (\mathbf{F}_{m_2}, \mathbf{F}_{m_1})_{\theta,p} = (\mathbf{F}_{\ell_2}, \mathbf{F}_{\ell_1})_{\sigma,p} \text{ with } \sigma = (1-\theta)\theta_2 + \theta\theta_1.$$

Now observe that

$$\sigma\ell_1 + (1-\sigma)\ell_2 = \theta m_1 + (1-\theta)m_2.$$

So, if $s \in \mathbb{R}$, one may define for each $p \in [1, \infty]$ a unique B-space $\mathbf{F}_{s,p}$ by choosing any integers m_1, m_2 with $m_1 < s < m_2$ and setting $\mathbf{F}_{s,p} = (\mathbf{F}_{m_2}, \mathbf{F}_{m_1})_{\theta,p}$ with θ such that $s = \theta m_1 + (1-\theta)m_2$, i.e. $\theta = (m_2 - s)(m_2 - m_1)^{-1}$. This justifies the following definition:

DEFINITION 3.4.2. (a) Let $\mathbf{F}$ be a Banach space equipped with a C_0-representation W of $\mathbb{R}^n$. Then the *Besov scale* associated to $(W, \mathbf{F})$ is the collection of B-spaces $\{\mathbf{F}_{s,p}\}_{s\in\mathbb{R}, p\in[1,\infty]}$ defined by

$$(3.4.7) \quad \mathbf{F}_{s,p} = (\mathbf{F}_{m_2}, \mathbf{F}_{m_1})_{\theta,p} \quad \text{with } m_1 < s < m_2, \quad \theta = \frac{m_2 - s}{m_2 - m_1}.$$

The closure of $\mathbf{F}_\infty$ in $\mathbf{F}_{s,\infty}$ is denoted by $\mathbf{F}_{s,\infty}^\circ$.

(b) If W is an arbitrary representation of $\mathbb{R}^n$ in $\mathbf{F}$, we define $\mathbf{F}_{s,p}$ for $s > 0$, $p \in [1, \infty]$ by the same rule but with $m_1 \ge 0$ (cf. the remark following (3.3.1)).

The properties of the Besov scale follow from general facts concerning real interpolation. In particular we have:

THEOREM 3.4.3. *Let* $\{\mathbf{F}_{s,p}\}_{s\in\mathbb{R},p\in[1,\infty]}$ *be the Besov scale associated to a couple* $(W,\mathbf{F})$.

(a) *If* $s < t$, *then* $\mathbf{F}_{t,q} \subset \mathbf{F}_{s,p}$ *continuously for all* p, q. *If* $p < \infty$, *the embedding is dense; if* $p = \infty$, *then* $\mathbf{F}_{t,q} \subset \mathbf{F}_{s,\infty}^{\circ}$ *continuously and densely. If* $s = t$ *but* $1 \le p \le q < \infty$, *then* $\mathbf{F}_{t,p} \subset \mathbf{F}_{t,q} \subset \mathbf{F}_{t,\infty}^{\circ}$ *continuously and densely.*

(b) *Let* $s < t$ *and* $1 \le p, q, r \le \infty$. *Then for* $0 < \theta < 1$:

$$(\mathbf{F}_{t,q}, \mathbf{F}_{s,p})_{\theta,r} = \mathbf{F}_{(1-\theta)t+\theta s, r}.$$

(c) *One has the following relation between the Sobolev and Besov scales:*

$$(3.4.8) \qquad \mathbf{F}_{k,1} \subset \mathbf{F}_k \subset \mathbf{F}_{k,\infty}^{\circ}$$

continuously and densely for all $k \in \mathbb{Z}$. *In particular, if* $\ell, m \in \mathbb{Z}$ *and* $\ell < m$, *then for* $0 < \theta < 1$, $1 \le p \le \infty$:

$$(3.4.9) \qquad (\mathbf{F}_m, \mathbf{F}_\ell)_{\theta,p} = \mathbf{F}_{(1-\theta)m+\theta\ell, p}.$$

(d) *Assume that* $\mathbf{F}$ *is reflexive and denote by* $\mathbf{F}_{s,p}^* \equiv (\mathbf{F}^*)_{s,p}$ *the Besov spaces associated to the couple* $(W^*, \mathbf{F}^*)$, *where* W^* *is the dual representation of* W *defined by* (3.2.7). *Then* $(\mathbf{F}_{s,p})^* = \mathbf{F}_{-s,p'}^*$ *if* $1 \le p < \infty$ *and* $(p')^{-1} = 1 - p^{-1}$, *and* $(\mathbf{F}_{s,\infty}^\circ)^* = \mathbf{F}_{-s,1}^*$.

PROOF. (a) and (b) follow from Proposition 2.4.1 and Theorem 2.5.3 respectively. (c) holds by (3.4.3) and because $\mathbf{F}_\infty$ is contained (in fact densely) in each of the spaces appearing in (3.4.8). (d) is obtained from Theorem 2.4.2 by remembering that $(\mathbf{F}_m)^* = (\mathbf{F}^*)_{-m}$ if $m \in \mathbb{Z}$ (cf. Theorem 3.3.28). $\square$

Except for some special cases (see Theorem 3.7.8), the spaces $\mathbf{F}_m$ with $m \in \mathbb{Z}$ do not belong to the family $\{\mathbf{F}_{s,p}\}$ of Besov spaces, although they are very close to the spaces $\mathbf{F}_{m,p}$ (see Theorem 3.4.1). Also they generate the same interpolation spaces as the spaces $\mathbf{F}_{m,p}$; more precisely one has

$$(3.4.10) \qquad (\mathbf{F}_m, \mathbf{F}_{s,p})_{\theta,q} = \mathbf{F}_{\theta s+(1-\theta)m, q} = (\mathbf{F}_{m,p_1}, \mathbf{F}_{s,p})_{\theta,q} \quad \text{if } s < m$$

$$(3.4.11) \qquad (\mathbf{F}_{s,p}, \mathbf{F}_m)_{\theta,q} = \mathbf{F}_{\theta m+(1-\theta)s, q} = (\mathbf{F}_{s,p}, \mathbf{F}_{m,p_1})_{\theta,q} \quad \text{if } s > m,$$

where $0 < \theta < 1$ and $p, q, p_1 \in [1, \infty]$. These identities follow from Theorem 2.5.3 (for example if $s < m$, choose $m_1, m_2 \in \mathbb{Z}$ with $m_1 < s < m < m_2$; then Theorem 3.4.1 states that $\mathbf{F}_m$ is of class $\theta_m \equiv (m_2 - m)(m_2 - m_1)^{-1}$ for the couple $(\mathbf{F}_{m_2}, \mathbf{F}_{m_1})$, and the first identity in (3.4.10) is identical with (2.5.5) for $\mathbf{G}_1 = \mathbf{F}_m$, $\mathbf{G}_2 = \mathbf{F}_{s,p}$, $\mathbf{E} = \mathbf{F}_{m_2}$ and $\mathbf{F} = \mathbf{F}_{m_1}$).

The following examples may help to arrive at a better understanding of the relation between the spaces $\mathbf{F}_m$ and $\mathbf{F}_{m,p}$ for integer m. Let $\mathbf{F} = C_\infty(\mathbb{R})$ be the space of functions $f : \mathbb{R} \to \mathbb{C}$ that are continuous and converge to zero at infinity, equipped with the translation group T. Clearly $\mathbf{F}_1 = C_\infty^1(\mathbb{R})$ is the space of $f \in C_\infty(\mathbb{R})$ which are of class C^1 and whose derivative tends to zero at infinity. Weiss and Zygmund [WZ] have constructed a function f such that $f \in \mathbf{F}_{1,p}$ for *all* $p > 2$ but the set of points where f is differentiable is of Lebesgue measure zero. In particular $\mathbf{F}_{1,p} \not\subset \mathbf{F}_1$ for any $p > 2$. On the other hand, one may show that if $g \in \mathbf{F}_{1,2}$, then g is absolutely continuous (hence differentiable almost everywhere) and $g' \in L^2_{loc}(\mathbb{R})$ [5]. In fact, the result of [WZ] is more precise, since their function satisfies $||(T(\varepsilon) - I)^2 f||_{\mathbf{F}} = O(\varepsilon |\log \varepsilon|^{-1/2})$ as $\varepsilon \to +0$ (compare with Proposition 3.4.6 below). They also show that from $||(T(\varepsilon) - I)^2 g|| = O(\varepsilon |\log \varepsilon|^{-\alpha})$ with $\alpha > 1/2$, it follows that g is absolutely continuous and $g' \in L^q_{loc}(\mathbb{R})$ for all $q < \infty$. We mention that Shapiro [Sh4] has an example of a function $f \in \mathbf{F}$ such that $||(T(\varepsilon) - I)^2 f|| = O(\varepsilon |\log \varepsilon|^{-1/2})$ and f is increasing and singular on an interval. Finally, we remark that the relation between the spaces $\mathbf{F}_m$ and $\mathbf{F}_{m,p}$ for the case of $\mathbf{F} = L^r(\mathbb{R}^n)$, $1 < r < \infty$, equipped with the translation group, is rather well understood (see [P1] for a résumé of the main results). For example, let $m \geq 1$ be an arbitrary integer; then $\mathbf{F}_{m,r} \subset \mathbf{F}_m \subset \mathbf{F}_{m,2}$ if $1 < r \leq 2$ and $\mathbf{F}_{m,2} \subset \mathbf{F}_m \subset \mathbf{F}_{m,r}$ if $2 \leq r \leq \infty$, all embeddings being optimal on the Besov scale.

To end this subsection, we consider the properties of certain operators in the scale $\{\mathbf{F}_{s,p}\}$.

PROPOSITION 3.4.4. *Let $s \in \mathbb{R}$ and $p \in [1, \infty]$. Then $W(x)\mathbf{F}_{s,p} \subset \mathbf{F}_{s,p}$ for each $x \in \mathbb{R}^n$, and the restriction of $W(x) : \mathbf{F}_{-\infty} \to \mathbf{F}_{-\infty}$ to $\mathbf{F}_{s,p}$ forms a group of bounded operators in $\mathbf{F}_{s,p}$. If $p < \infty$ this group is a C_0-group; if $p = \infty$ its restriction to $\mathbf{F}^\circ_{s,\infty}$ is a C_0-group. Moreover, $\mathbf{F}^\circ_{s,\infty}$ coincides with the set of vectors $f \in \mathbf{F}_{s,\infty}$ such that the function $x \mapsto W(x)f \in \mathbf{F}_{s,\infty}$ is continuous.*

PROOF. By interpolation (Theorem 2.6.1) and by taking into account Proposition 3.3.8 (b), one sees that $W(x)\mathbf{F}_{s,p} \subset \mathbf{F}_{s,p}$ and that

$$(3.4.12) \qquad \sup_{|x| \leq 1} ||W(x)||_{B(\mathbf{F}_{s,p})} < \infty.$$

Next we observe that, if $f \in \mathbf{F}_\infty$, then $W(x)f \to f$ as $x \to 0$ in each $\mathbf{F}_m$ $(m \in \mathbb{Z})$ by Proposition 3.3.8 (b), hence also in $\mathbf{F}_{s,p}$. By the denseness of $\mathbf{F}_\infty$ in $\mathbf{F}_{s,p}$ (if $p < \infty$) as well as in $\mathbf{F}^\circ_{s,\infty}$ and by (3.4.12), one obtains the strong continuity of W in $\mathbf{F}_{s,p}$ $(p < \infty)$ and in $\mathbf{F}^\circ_{s,\infty}$. Now assume that $f \in \mathbf{F}_{s,\infty}$ is such that the map $W(\cdot)f : \mathbb{R} \to \mathbf{F}_{s,\infty}$ is continuous. Let $\varphi \in \mathscr{S}(\mathbb{R}^n)$ be such that $\widehat{\varphi}$ is of compact support and $\varphi(0) = 1$. Then for each $\varepsilon > 0$ we have $\varphi(\varepsilon A)f \in \mathbf{F}_\infty$ (cf. Lemma 3.3.17 (a)) and $\varphi(\varepsilon A)f = \int_{\mathbb{R}^n} W(\varepsilon x)f \cdot \widehat{\varphi}(x)\underline{d}x$ converges in the norm of $\mathbf{F}_{s,\infty}$ to $\int_{\mathbb{R}^n} W(0)f \cdot \widehat{\varphi}(x)\underline{d}x = f$ as $\varepsilon \to 0$ (by the dominated convergence theorem). So $f \in \mathbf{F}^\circ_{s,\infty}$. $\square$

Let us point out that the group induced by W in $\mathbf{F}_{s,\infty}$ is not strongly continuous in general. As an example, let $\mathbf{F} = C_\infty(\mathbb{R})$ and let A be the operator of multiplication by the variable in $C_\infty(\mathbb{R})$. Then for $s > 0$, $\mathbf{F}_{s,\infty}$ is the set of all $f \in \mathbf{F}$ such that $||f||_{s,\infty} \equiv \sup \langle x \rangle^s |f(x)| < \infty$ (this follows from the results of Section 3.6). Then f belongs to the domain $D(A)$ of A in $\mathbf{F}_{s,\infty}$ if and only if $||f||_{s,\infty}$ and $||xf(\cdot)||_{s,\infty}$ are finite, and the set of these functions f is not dense in $\mathbf{F}_{s,\infty}$ (for example the function $g(x) = \langle x \rangle^{-s}$ cannot be approximated in $\mathbf{F}_{s,\infty}$ by a sequence belonging to $D(A)$).

---

[5]This is a consequence of the inclusion $C(\mathbb{R}) \subset L^2_{loc}(\mathbb{R})$ and of the fact that $[L^2(\mathbb{R})]_{1,2}$ is the Sobolev space $\mathscr{H}^1(\mathbb{R})$ (the last assertion is a trivial special case of Proposition 3.7.7).

PROPOSITION 3.4.5. (a) *Let $s \in \mathbb{R}$ and $p \in [1, \infty]$. Then, for any multi-index α, the operator A^α belongs to $B(\mathbf{F}_{s,p}, \mathbf{F}_{s-|\alpha|,p})$.*

(b) *Let $s, t \in \mathbb{R}$ with $s < t$ and $p, q \in [1, \infty]$. Assume that $\varphi \in \mathscr{S}(\mathbb{R}^n)$ with $\hat{\varphi} \in C_0^\infty(\mathbb{R}^n)$. Then there is a constant c such that*

$$(3.4.13) \qquad \|\varphi(\sigma A)\|_{\mathbf{F}_{s,p} \to \mathbf{F}_{t,q}} \le c\sigma^{-(t-s)} \qquad \forall \sigma \in (0, 1].$$

If φ also satisfies $\varphi(0) = 1$ and $\varphi^{(\beta)}(0) = 0$ for each multi-index β with $1 \le |\beta| < t - s + 1$, then one may choose c such that

$$(3.4.14) \qquad \|\varphi(\sigma A) - I\|_{\mathbf{F}_{t,q} \to \mathbf{F}_{s,p}} \le c\sigma^{t-s} \qquad \forall \sigma \in (0, 1].$$

PROOF. (a) Let m_1, m_2 be integers with $m_1 < s < m_2$. By Proposition 3.3.9 we have $A^\alpha \in B(\mathbf{F}_{m_1}, \mathbf{F}_{m_1-|\alpha|}) \cap B(\mathbf{F}_{m_2}, \mathbf{F}_{m_2-|\alpha|})$. So the assertion (a) follows by interpolation (Theorem 2.6.1).

(b) These results follow from those of Lemma 3.3.20 by interpolation. We indicate the proof of (3.4.14); that of (3.4.13) is similar but easier. Let k_0 be the least integer $\ge t - s$ and set $k = k_0 + 1$. Then, for any multi-index β, we have $|\beta| < t - s + 1$ if and only if $|\beta| < k$. So our hypotheses allow us to use Lemma 3.3.20. It follows that, if we set $\Phi_\sigma = I - \varphi(\sigma A)$, then for each integer m there is a constant C such that $\|\Phi_\sigma\|_{\mathbf{F}_m \to \mathbf{F}_m} \le C$ and $\|\Phi_\sigma\|_{\mathbf{F}_m \to \mathbf{F}_{m-k}} \le C\sigma^k$ for $0 < \sigma \le 1$. By interpolation (cf. (3.4.7) and Theorem 2.6.1) we see that for each real t and each $q \in [1, \infty]$ there is a constant C such that the norm of Φ_σ as an operator $\mathbf{F}_{t,q} \to \mathbf{F}_{t,q}$ is $\le C$, while its norm as an operator $\mathbf{F}_{t,q} \to \mathbf{F}_{t-k,q}$ is $\le C\sigma^k$, for $0 < \sigma \le 1$. Now, since $t - k < s < t$ we may interpolate again (see Corollary 2.6.2 and Theorem 3.4.3 (b)) and obtain that the norm of Φ_σ as an operator $\mathbf{F}_{t,q} \to \mathbf{F}_{s,p}$ is $\le c\sigma^{t-s}$ for some constant c and all $\sigma \in (0, 1]$. $\square$

3.4.2. It is convenient to have a description of the spaces $\mathbf{F}_{s,p}$, $s > 0$, in terms of smoothness properties of the functions $x \mapsto W(x)f \in \mathbf{F}$, like those which appeared in the definition of the spaces $\mathbf{F}_m$ for $m \in \mathbb{N}$. For this we introduce some new gauges on $\mathbf{F}$ and show that they give admissible norms on the spaces $\mathbf{F}_{s,p}$. The following proposition should be compared with Proposition 3.3.19 which it generalizes in a certain sense.

THEOREM 3.4.6. *Let $s > 0$ be a real number, $m > s$ an integer and $p \in [1, \infty]$. Then the expression*

$$(3.4.15) \qquad \|f\|_{s,p}^{(m)} = \|f\|_{\mathbf{F}} + \left[\int_{|x| \le 1} \left\| \frac{[W(x) - I]^m}{|x|^s} f \right\|_{\mathbf{F}}^p \frac{dx}{|x|^n} \right]^{1/p}$$

defines a gauge on $\mathbf{F}$, and the associated normed space coincides as a B-space with $\mathbf{F}_{s,p}$. In particular $\| \cdot \|_{s,p}^{(m)}$ is an admissible norm on $\mathbf{F}_{s,p}$. The domain of integration $|x| \le 1$ in (3.4.15) may be replaced by $|x| \le b$ with an arbitrary $b > 0$.

PROOF. Let us denote by $K(\cdot,\cdot)$ the K-functional for the couple $(\mathbf{F}_m, \mathbf{F})$. We set $\theta = m^{-1}(m-s)$, so that $\mathbf{F}_{s,p} = (\mathbf{F}_m, \mathbf{F})_{\theta,p}$, with

$$(3.4.16) \qquad \|f\|_{\mathbf{F}_{s,p}} = \left\{ \int_0^1 \left[\frac{K(\tau,f)}{\tau^{1-\theta}} \right]^p \frac{d\tau}{\tau} \right\}^{1/p}$$

$$\equiv m^{1/p} \left\{ \int_0^1 \left[\frac{K(\varrho^m, f)}{\varrho^s} \right]^p \frac{d\varrho}{\varrho} \right\}^{1/p}.$$

(i) By using (3.3.18) and (3.2.1), one sees that there is a constant c such that for all $f \in \mathbf{F}$, all $e \in \mathbf{F}_m$ and all $x \in \mathbb{R}^n$ with $|x| \leq 1$:

$$\|[W(x) - I]^m f\|_{\mathbf{F}} \leq \|[W(x) - I]^m e\|_{\mathbf{F}} + \|[W(x) - I]^m (f-e)\|_{\mathbf{F}}$$
$$\leq c|x|^m \|e\|_{\mathbf{F}_m} + c\|f - e\|_{\mathbf{F}}.$$

By taking into account the definition (2.2.1) of $K(\tau, f)$, we obtain that

$$(3.4.17) \qquad \|[W(x) - I]^m f\|_{\mathbf{F}} \leq cK(|x|^m, f) \quad \text{if } |x| \leq 1.$$

We insert (3.4.17) into (3.4.15). By setting $\varrho = |x|$ and by using also (2.2.6), one gets that for some constant c_1 and all $f \in \mathbf{F}$:

$$\|f\|_{s,p}^{(m)} \leq c_1 \|f\|_{\mathbf{F}_{s,p}} + c_1 \left\{ \int_0^1 \left[\frac{K(\varrho^m, f)}{\varrho^s} \right]^p \frac{d\varrho}{\varrho} \right\}^{1/p} \leq 2c_1 \|f\|_{\mathbf{F}_{s,p}}.$$

(ii) To prove the opposite inequality, we shall construct in (iii) below a function $\varphi \in \mathscr{S}(\mathbb{R}^n)$ such that $\varphi(0) = 1$, $\varphi^{(\beta)}(0) = 0$ if $1 \leq |\beta| < m$, $\widehat{\varphi} \in C_0^\infty(\mathbb{R}^n)$ and such that for some constant c_2, all $f \in \mathbf{F}$ and all $\varrho \in (0,1]$:

$$(3.4.18)$$
$$\|[\varphi(\varrho A) - I]f\|_{\mathbf{F}} + \sum_{|\alpha|=m} \varrho^m \|A^\alpha \varphi(\varrho A) f\|_{\mathbf{F}} \leq c_2 \int_{|x| \leq 1} \|[W(\varrho x) - I]^m f\|_{\mathbf{F}} \underline{d}x.$$

By applying the Hölder inequality, one deduces from (3.4.18) that, for some constant c_3, all $f \in \mathbf{F}$ and all $\varrho \in (0,1]$:

$$(3.4.19) \quad \|[\varphi(\varrho A) - I]f\|_{\mathbf{F}} + \sum_{|\alpha|=m} \varrho^m \|A^\alpha \varphi(\varrho A) f\|_{\mathbf{F}} \leq$$

$$\leq c_3 \left\{ \int_{|x| \leq 1} \|[W(\varrho x) - I]^m f\|_{\mathbf{F}}^p \underline{d}x \right\}^{1/p}.$$

Now, by using first (3.4.2), then Proposition 3.3.18 (b) and finally (3.2.19) and (3.4.19), one gets that

$$K(\varrho^m, f) \leq c_4 \varrho^m \|\varphi(\varrho A) f\|_{\mathbf{F}_m} + c_4 \|[\varphi(\varrho A) - I]f\|_{\mathbf{F}}$$

$$\leq c_5 \varrho^m \|\varphi(\varrho A) f\|_{\mathbf{F}} + c_5 \varrho^m \sum_{|\alpha|=m} \|A^\alpha \varphi(\varrho A) f\|_{\mathbf{F}} + c_4 \|[\varphi(\varrho A) - I]f\|_{\mathbf{F}}$$

$$\leq c_6 \varrho^m \|f\|_{\mathbf{F}} + c_6 \left\{ \int_{|x| \leq 1} \|[W(\varrho x) - I]^m f\|_{\mathbf{F}}^p \underline{d}x \right\}^{1/p}.$$

Together with (3.4.16) this leads to

$$\|f\|_{\mathbf{F}_{s,p}} \leq c_7 \left\{ \int_0^1 \varrho^{(m-s)p} \frac{d\varrho}{\varrho} \right\}^{1/p} \|f\|_{\mathbf{F}}$$

$$+ c_7 \left\{ \int_0^1 \frac{d\varrho}{\varrho} \int_{|x| \leq 1} dx \frac{\|[W(\varrho x) - I]^m f\|_{\mathbf{F}}^p}{\varrho^{sp}} \right\}^{1/p}.$$

Since $m > s$, the first integral on the r.h.s. is finite. After a change of variables $x \mapsto y = \varrho x$ in the last integral, one finds that $\|f\|_{\mathbf{F}_{s,p}} \leq c_8 \|f\|_{s,p}^{(m)}$ for all $f \in \mathbf{F}$.

(iii) We now exhibit a function φ with the properties stated at the beginning of (ii). We start with a function $\xi \in \mathscr{S}(\mathbb{R}^n)$ and associate to it and to each multi-index α with $|\alpha| = m$ the functions $\psi, \psi_\alpha \in \mathscr{S}(\mathbb{R}^n)$ defined as follows:

$$(3.4.20) \qquad \psi(x) = \sum_{j=1}^m (-1)^{j-1} \binom{m}{j} j^m \xi(jx),$$

$$(3.4.21) \qquad \psi_\alpha(x) = \sum_{j=1}^m (-1)^{j-1} \binom{m}{j} x^\alpha \xi(jx).$$

Now we set

$$(3.4.22)\ \varphi(x) = \psi(0) - \int_{\mathbb{R}^n} (1 - e^{ix \cdot y})^m \widehat{\psi}(y) dy = \sum_{j=1}^m (-1)^{j-1} \binom{m}{j} \psi(jx).$$

It is clear from the first expression for $\varphi(x)$ that $\varphi(0) = \psi(0) = (-1)^{m-1} m! \xi(0)$ (cf. the proof of Proposition 3.3.5) and $\varphi^{(\beta)}(0) = 0$ if $1 \leq |\beta| < m$. We shall prove below that for $|\alpha| = m$ we have

$$(3.4.23) \qquad \varphi_{(\alpha)}(x) \equiv x^\alpha \varphi(x) = - \int_{\mathbb{R}^n} (1 - e^{ix \cdot y})^m \widehat{\psi_\alpha}(y) dy.$$

By choosing ξ such that $\xi(0) = (-1)^{m-1}/m!$ and $\operatorname{supp} \widehat{\xi} \subset \{x \mid |x| \leq 1/m\}$ we see that $\widehat{\varphi} \in C_0^\infty(\mathbb{R}^n)$, $\varphi(0) = 1$, $\varphi^{(\beta)}(0) = 0$ if $1 \leq |\beta| < m$ and

$$I - \varphi(\varrho A) = \int_{\mathbb{R}^n} [I - W(\varrho y)]^m \widehat{\psi}(y) dy,$$

$$\varrho^m A^\alpha \varphi(\varrho A) = - \int_{\mathbb{R}^n} [I - W(\varrho y)]^m \widehat{\psi_\alpha}(y) dy \quad \text{if } |\alpha| = m,$$

where $\widehat{\psi}, \widehat{\psi_\alpha}$ are of class C_0^∞ with support in $|y| \leq 1$. This clearly implies (3.4.18).

One may prove (3.4.23) by a straightforward computation but the following formalism makes things more natural (in particular it explains the choice of ψ and ψ_α). Let δ_j be the Dirac measure on $\mathbb{R}$ with support $\{j\}$ and set $\mu = \sum_{j=1}^m (-1)^{j-1} \binom{m}{j} \delta_j$. Then μ is a measure on $\mathbb{R}$ with compact (finite) support included in $(0, \infty)$ and we have $\widetilde{\mu}(\tau) := \int_0^\infty e^{i\tau\sigma} \mu(d\sigma) = 1 - (1 - e^{i\tau})^m$ for all $\tau \in \mathbb{R}$. For $\sigma > 0$ let $\mathfrak{I}(\sigma)$ be the linear operator in $C(\mathbb{R}^n)$ given by $(\mathfrak{I}(\sigma)f)(x) = f(\sigma x)$. For an arbitrary real r we set $\mathfrak{I}_r = \int_0^\infty \mathfrak{I}(\sigma)\sigma^r \mu(d\sigma)$, i.e. $\mathfrak{I}_r : C(\mathbb{R}^n) \to C(\mathbb{R}^n)$ is the linear operator given by $(\mathfrak{I}_r f)(x) = \int_0^\infty f(\sigma x)\sigma^r \mu(d\sigma)$; we abbreviate $\mathfrak{I}_0 = \mathfrak{I}$.

If f, g are continuous functions on $\mathbb{R}^n$ and g is the inverse Fourier transform of a measure $\widehat{g}$, then $f = \mathfrak{J}g$ is clearly equivalent to

(3.4.24)
$$f(x) = \int_0^\infty g(\sigma x)\mu(d\sigma) = \int_{\mathbb{R}^n} \widetilde{\mu}(x \cdot y)\widehat{g}(y)\underline{dy} = g(0) - \int_{\mathbb{R}^n} (1 - e^{ix \cdot y})^m \widehat{g}(y)\underline{dy}.$$

Now observe that for an arbitrary multi-index α and all $\sigma > 0$ we have $Q^\alpha \mathfrak{J}(\sigma) = \sigma^{-|\alpha|}\mathfrak{J}(\sigma)Q^\alpha$ (as operators on $C(\mathbb{R}^n)$), hence $Q^\alpha \mathfrak{J}_r = \mathfrak{J}_{r-|\alpha|}Q^\alpha$ for all $r \in \mathbb{R}$. Since the operators $\mathfrak{J}_r$ for different values of r commute, we get:

$$Q^\alpha \mathfrak{J}\mathfrak{J}_{|\alpha|} = \mathfrak{J}_{-|\alpha|}Q^\alpha \mathfrak{J}_{|\alpha|} = \mathfrak{J}_{-|\alpha|}\mathfrak{J}Q^\alpha = \mathfrak{J}\mathfrak{J}_{-|\alpha|}Q^\alpha.$$

In particular, if $m \geq 1$ is an integer and we define $\varphi = \mathfrak{J}\mathfrak{J}_m\xi$, then for $|\alpha| = m$ we shall have $Q^\alpha \varphi = \mathfrak{J}\mathfrak{J}_{-m}Q^\alpha \xi$. Clearly (3.4.20), (3.4.21), (3.4.22) are equivalent to $\psi = \mathfrak{J}_m\xi$, $\psi_\alpha = \mathfrak{J}_{-m}Q^\alpha \xi$ and $\varphi = \mathfrak{J}\psi$ respectively. Then the relation $Q^\alpha \varphi = \mathfrak{J}\psi_\alpha$ and the fact that $\psi_\alpha(0) = 0$ imply (3.4.23) (cf. (3.4.24)). $\square$

We mention that the preceding proof gives more than stated in Theorem 3.4.6: we have in fact a natural extension of Proposition 2.7.3 to the context of n-parameter C_0-groups. To state these results, let us define the *m-th order modulus of continuity of* $f \in \mathbf{F}$ *with respect to* W by $\omega_m(\varrho, f) = \sup_{|x| \leq \varrho}\|(W(x) - I)^m f\|_{\mathbf{F}}$, where $0 < \varrho \leq 1$. From (3.4.17) and (2.2.2) we then get $\omega_m(\varrho, f) \leq cK(\varrho^m, f)$. In part (ii) of the proof we obtained

$$K(\varrho^m, f) \leq c\varrho^m\|\varphi(\varrho A)f\|_{\mathbf{F}} + c\int_{|x| \leq 1} \|(W(\varrho x) - I)^m f\|_{\mathbf{F}}dx,$$

and $\|\varphi(\varrho A)f\|_{\mathbf{F}} \leq c\|f\|_{\mathbf{F}}$. In particular, we have the following extension of (2.7.8): there is a constant $c > 0$ such that for $0 < \varrho \leq 1$ and $f \in \mathbf{F}$:

$$c^{-1}K(\varrho^m, f) \leq \varrho^m\|f\|_{\mathbf{F}} + \omega_m(\varrho, f) \leq cK(\varrho^m, f).$$

The next result is the so-called *reduction theorem*.

THEOREM 3.4.7. *Let* $s \in (0, \infty)$, $p \in [1, +\infty]$ *and* ℓ *an integer such that* $0 \leq \ell < s$. *Then one has, in the sense of equivalence of gauges on* $\mathbf{F}$:

(3.4.25)
$$\|f\|_{\mathbf{F}_{s,p}} \sim \|f\|_{\mathbf{F}} + \sum_{|\alpha|=\ell} \|A^\alpha f\|_{\mathbf{F}_{s-\ell,p}}.$$

PROOF. The inequality

$$\|f\|_{\mathbf{F}} + \sum_{|\alpha|=\ell} \|A^\alpha f\|_{\mathbf{F}_{s-\ell,p}} \leq c_1\|f\|_{\mathbf{F}_{s,p}} \qquad \forall f \in \mathbf{F}$$

follows from Proposition 3.4.5 (a). It remains to prove for example that for any integer $m > s$:

$$\|f\|_{s,p}^{(m)} \leq c_2[\|f\|_{\mathbf{F}} + \sum_{|\alpha|=\ell} \|A^\alpha f\|_{\mathbf{F}_{s-\ell,p}}] \qquad \forall f \in \mathbf{F}.$$

To estimate the second term on the r.h.s. of (3.4.15), we write $[W(x) - I]^m = [W(x) - I]^\ell[W(x) - I]^{m-\ell}$ and use the expansion (3.3.18) for $[W(x) - I]^\ell$. We obtain that for all $|x| \leq 1$ and all $f \in \mathbf{F}$

$$\frac{||[W(x) - I]^m f||_{\mathbf{F}}}{|x|^s} \leq c_3 \sum_{|\alpha|=\ell} \frac{||[W(x) - I]^{m-\ell} A^\alpha f||_{\mathbf{F}}}{|x|^{s-\ell}}$$

so that

$$\left\{ \int_{|x| \leq 1} \left[\frac{||[W(x) - I]^m f||_{\mathbf{F}}}{|x|^s} \right]^p \cdot \frac{dx}{|x|^n} \right\}^{1/p} \leq c_3 \sum_{|\alpha|=\ell} ||A^\alpha f||_{s-\ell,p}^{(m-\ell)}. \quad \square$$

COROLLARY 3.4.8. (a) *Let* $s > t > 0$ *be such that* $\ell \equiv s-t$ *is an integer. Then* $f \in \mathbf{F}_{s,p}$ *if and only if* $f \in \mathbf{F}_{t,p}$ *and* $A^\alpha f \in \mathbf{F}_{t,p}$ $\forall \alpha$ *with* $|\alpha| = \ell$. *Moreover:*

$$(3.4.26) \qquad ||f||_{\mathbf{F}_{s,p}} \sim ||f||_{\mathbf{F}_{t,p}} + \sum_{|\alpha|=\ell} ||A^\alpha f||_{\mathbf{F}_{t,p}}.$$

(b) *Let* $s = k + \tau$ *with* $k \in \mathbb{N}$ *and* $0 < \tau \leq 1$. *Then* $f \in \mathbf{F}_{s,p}$ *if and only if* $f \in \mathbf{F}_k$ *and* $A^\alpha f \in \mathbf{F}_{\tau,p}$ *for each multi-index* α *with* $|\alpha| = k$.

One may reformulate part (a) of Corollary 3.4.8 in the following terms. Let $t > 0$ a real number, $\ell \geq 1$ an integer and $s = \ell + t$. The B-space $\mathbf{F}_{t,p}$ is equipped with the representation of $\mathbb{R}^n$ induced by W. Then we have $(\mathbf{F}_{t,p})_\ell = \mathbf{F}_{s,p}$.

We next point out an anisotropic variant of the gauge (3.4.15); this rather deep result and the proof we give are due to Triebel (see §1.13.4 in [Tr]).

THEOREM 3.4.9. *Let* $\{v_1, \ldots, v_n\}$ *be a basis of* $\mathbb{R}^n$, $0 < s < \infty$ *and* $1 \leq p \leq \infty$. *If* $m > s$ *is an integer, then the following expression defines an admissible norm on* $\mathbf{F}_{s,p}$:

$$(3.4.27) \qquad |||f|||_{s,p}^{(m)} = ||f||_{\mathbf{F}} + \sum_{j=1}^n \left[\int_0^1 \left\| \frac{[W(\tau v_j) - I]^m f}{\tau^s} \right\|_{\mathbf{F}}^p \frac{d\tau}{\tau} \right]^{1/p}.$$

PROOF. (i) We denote by $x_1, \ldots, x_n$ the coordinates of the vector $x \in \mathbb{R}^n$ in the basis $\{v_1, \ldots, v_n\}$, i.e. $x = \sum_{j=1}^n x_j v_j$, and let $W_j(\tau) = W(\tau v_j)$. Then (with an obvious definition of $T_j(x)$):

$(3.4.28)$
$$W(x) - I = W_1(x_1) \ldots W_n(x_n) - I$$

$$= W_1(x_1) - I + \sum_{j=2}^n W_1(x_1) \ldots W_{j-1}(x_{j-1})[W_j(x_j) - I]$$

$$= \sum_{j=1}^n T_j(x)[W_j(x_j) - I].$$

We set $k = mn$ and then have

$$[W(x) - I]^k = \sum_{k=k_1+\cdots+k_n} \frac{k!}{k_1! \ldots k_n!} \prod_{j=1}^{n} [T_j(x)]^{k_j} [W_j(x_j) - I]^{k_j}$$

$$\equiv \sum_{j=1}^{n} U_j(x)[W_j(x_j) - I]^m$$

for some operators $U_j(x)$, because in each term of the first sum there is at least one k_j with $k_j \geq m$. Hence there is a locally bounded function $q : \mathbb{R}^n \to \mathbb{R}$ such that:

(3.4.29) $$\|[W(x) - I]^k f\|_{\mathbf{F}} \leq q(x) \sum_{j=1}^{n} \|[W_j(x_j) - I]^m f\|_{\mathbf{F}}.$$

(ii) The preceding inequality implies that

(3.4.30)

$$\int_{|x|\leq 1} \left\| \frac{[W(x) - I]^k}{|x|^s} f \right\|_{\mathbf{F}}^p \frac{dx}{|x|^n} \leq c_1 \sum_{j=1}^{n} \int_{|x|\leq 1} \left\| \frac{[W_j(x_j) - I]^m}{|x|^s} f \right\|_{\mathbf{F}}^p \frac{dx}{|x|^n}.$$

The n terms on the r.h.s. are all of the same form. We estimate for example the one with $j = 1$. For this we set $\varrho = x_1$ and $x' = (x_2, \ldots, x_n) \in \mathbb{R}^{n-1}$. Then there are constants $a > 0$ and $c_2 < \infty$ so that the term with $j = 1$ in (3.4.30) is less than

(3.4.31) $$c_2 \int_{-a}^{a} d\varrho \left\| \frac{[W_1(\varrho) - I]^m f}{|\varrho|^s} \right\|_{\mathbf{F}}^p \int_{\mathbb{R}^{n-1}} \frac{dx'}{(\varrho^2 + |x'|^2)^{n/2}}.$$

The last integral is equal to $c_3 \varrho^{-1}$ for some $c_3 > 0$. Since $W_1(-\varrho) - I = -W_1(-\varrho)[W_1(\varrho) - I]$, it follows that the expression (3.4.31) is majorized by

$$c_4 \int_{0}^{a} \frac{d\varrho}{\varrho} \left\| \frac{[W_1(\varrho) - I]^m f}{\varrho^s} \right\|_{\mathbf{F}}^p.$$

Thus we have $\|f\|_{s,p}^{(m)} \leq c_5 \|f\|_{s,p}^{(k)} \leq c_6 \||f\||_{s,p}^{(m)}$ for all $f \in \mathbf{F}$ (use Theorem 3.4.6).

(iii) We shall now show that for each $v \neq 0$ in $\mathbb{R}^n$ there are constants c and b such that

(3.4.32)

$$\left[\int_{0}^{1} \left\| \frac{[W(\tau v) - I]^k f}{\tau^s} \right\|_{\mathbf{F}}^p \frac{d\tau}{\tau} \right]^{1/p} \leq c \left[\int_{|x|\leq b} \left\| \frac{[W(x) - I]^m f}{|x|^s} \right\|_{\mathbf{F}}^p \frac{dx}{|x|^n} \right]^{1/p}.$$

Then, by applying Theorem 3.4.6 to each of the one-parameter groups $\{W(\tau v_j)\}$ $(j = 1, \ldots, n)$ and by using (3.4.32), one gets that $\||f\||_{s,p}^{(m)} \leq c_7 \|f\|_{s,p}^{(k)} \leq c_8 \|f\|_{s,p}^{(m)}$.

To prove (3.4.32), choose n disjoint open subsets $\Omega_1, \ldots, \Omega_n$ of the unit sphere S^{n-1} in $\mathbb{R}^n$ such that, for each choice of vectors $e_1 \in \Omega_1, \ldots, e_n \in \Omega_n$, the family

$\{e_1, \ldots, e_n\}$ is a basis of $\mathbb{R}^n$ and $v = \sum_{j=1}^n z_j e_j$ with $0 < z_j \le b$. By using (3.4.29) with $x = \tau v$ and $\{v_j\}$ replaced by the basis $\{e_j\}$, one obtains that

$$\int_0^1 \left\| \frac{[W(\tau v) - I]^k f}{\tau^s} \right\|_{\mathbf{F}}^p \frac{d\tau}{\tau} \le c_9 \sum_{j=1}^n \int_0^b \left\| \frac{[W(\varrho e_j) - I]^m f}{\varrho^s} \right\|_{\mathbf{F}}^p \frac{d\varrho}{\varrho}.$$

We divide this inequality by $\prod_{j=1}^n |\Omega_k|$ ($|\Omega_k|$ denotes the spherical Lebesgue measure of Ω_k) and then integrate over $\Omega_1 \times \cdots \times \Omega_n$ with respect to the spherical product measure $de_1 \ldots de_n$. This leads to

$$\int_0^1 \left\| \frac{[W(\tau v) - I]^k f}{\tau^s} \right\|_{\mathbf{F}}^p \frac{d\tau}{\tau} \le c_{10} \sum_{j=1}^n \frac{1}{|\Omega_j|} \int_{\Omega_j} d\omega \int_0^b \frac{d\varrho}{\varrho} \left\| \frac{[W(\varrho \omega) - I]^m f}{\varrho^s} \right\|_{\mathbf{F}}^p,$$

where $d\omega$ means spherical Lebesgue measure. Since $\varrho^{-1} d\varrho d\omega = |x|^{-n} dx$, the preceding inequality implies (3.4.32). $\square$

Let us add some comments on the significance of the preceding theorem. Let $\{e_1, \ldots, e_n\}$ be the canonical orthonormal basis of $\mathbb{R}^n$ and $W_j(\tau) = W(\tau e_j) = \exp(i\tau A_j)$. For each $j = 1, \ldots, n$, denote by $\{\mathbf{F}_m^{(j)}\}_{m \in \mathbb{Z}}$ the Sobolev scale and by $\{\mathbf{F}_{s,p}^{(j)}\}_{s \in \mathbb{R}, 1 \le p \le \infty}$ the Besov scale associated to the C_0-group W_j in $\mathbf{F}$. Then Theorem 3.4.9 states that

$$(3.4.33) \qquad \mathbf{F}_{s,p} = \bigcap_{j=1}^n \mathbf{F}_{s,p}^{(j)} \qquad \text{if } s > 0, \ 1 \le p \le \infty,$$

as an identity between B-spaces (we recall that, if $\mathbf{E}_1, \ldots, \mathbf{E}_n$ are B-spaces continuously embedded in a Hausdorff topological vector space $\mathscr{E}$, then the vector subspace $\mathbf{E}_1 \cap \ldots \cap \mathbf{E}_n$ of $\mathscr{E}$ has a canonical B-space structure defined by the norm $\| \cdot \|_{\mathbf{E}_1} + \cdots + \| \cdot \|_{\mathbf{E}_n}$, where $\| \cdot \|_{\mathbf{E}_j}$ is an admissible norm on $\mathbf{E}_j$).

The identity (3.4.33) is a quite remarkable fact, and its analogue for the discrete Sobolev scale is *not valid* in general Banach space. For example, if $\mathbf{F} = C_\infty(\mathbb{R}^2)$ or $\mathbf{F} = L^1(\mathbb{R}^2)$ and W is the translation group, then $\mathbf{F}_2 \subset \mathbf{F}_2^{(1)} \cap \mathbf{F}_2^{(2)}$ strictly (see Remark 2 in §1.13.4 of [Tr]). On the other hand, if W is the translation group in $\mathbf{F} = L^p(\mathbb{R}^n)$ with $1 < p < \infty$, then $\mathbf{F}_m = \cap_{j=1}^n \mathbf{F}_m^{(j)}$ for each $m \in \mathbb{N}$; this is a quite deep result of Marcinkiewicz and Mihlin (see Remark 2 in §2.3.3 and Remark 4 in §2.2.4 of [Tr]). In Section 3.7 we shall prove such a result for the discrete Sobolev scale of a Hilbert space.

Our purpose now is to understand the difference between the spaces $\mathbf{F}_m$ and $\mathbf{F}_{m,1}$ for $m \ge 1$; they should be quite near to each other in a certain sense. The next proposition shows that this is indeed true, the distinction between $f \in \mathbf{F}_m$ and $f \in \mathbf{F}_{m,1}$ being rather subtle. For the case $n = 1$ the following result is due to H. Komatsu (see Theorem 4.4 in II of [Km]).

PROPOSITION 3.4.10. *Let $\ell, m \in \mathbb{N}$ with $\ell > m \ge 1$, and let $f \in \mathbf{F}$. Then $f \in \mathbf{F}_{m,1}$ if and only if $\int_0^1 \|\tau^{-m}[W(\tau x) - I]^\ell f\|_{\mathbf{F}} \tau^{-1} d\tau < \infty$ for each $x \in \mathbb{R}^n$, and $f \in \mathbf{F}_m$ if and only if $\lim_{\varepsilon \to +0} \int_\varepsilon^1 \tau^{-m}[W(\tau x) - I]^\ell f \tau^{-1} d\tau$ exists in $\mathbf{F}$ for each $x \in \mathbb{R}^n$.*

Thus the difference between $\mathbf{F}_m$ and $\mathbf{F}_{m,1}$ lies in the fact that, for $f \in \mathbf{F}_{m,1}$, the function $\tau^{-m}[W(\tau x) - I]^\ell f$ is absolutely integrable in $\mathbf{F}$ at $\tau = 0$ if $\ell > m$, whereas if $f \in \mathbf{F}_m$ but $f \notin \mathbf{F}_{m,1}$, then the integral of this function at $\tau = 0$ exists only as an improper integral in $\mathbf{F}$.

PROOF. The characterization of $\mathbf{F}_{m,1}$ follows from Theorem 3.4.9. For that of $\mathbf{F}_m$, we use the identity (3.3.18) with m replaced by $m + 1$. If we write $x \cdot A = x_1 A_1 + \cdots + x_n A_n$ and observe that $\sum_{|\alpha|=k} \frac{k!}{\alpha!}(ix)^\alpha A^\alpha = (ix \cdot A)^k$ for any $k \in \mathbb{N}$, we obtain (on $\mathbf{F}_{-\infty}$):

$$[W(x) - I]^\ell = (-1)^{\ell-m-1}(ix \cdot A)^{m+1} \int_{\mathbb{R}} W(\sigma x)\chi_\ell^{(\ell-m-1)}(\sigma)d\sigma.$$

We replace x by τx, divide both sides by τ^{m+1} and then integrate over $\tau \in [\varepsilon, 1]$. Since $\int_\varepsilon^1 (ix \cdot A)W(\sigma\tau x)d\tau = \sigma^{-1}[W(\sigma x) - W(\varepsilon\sigma x)]$ we get

$$\int_\varepsilon^1 [I - W(\tau x)]^\ell \frac{d\tau}{\tau^{m+1}} = (-ix \cdot A)^m \int_{\mathbb{R}} [W(\varepsilon\sigma x) - W(\sigma x)]\chi_\ell^{(\ell-m-1)}(\sigma)\frac{d\sigma}{\sigma}.$$

From (c) of Proposition 3.3.5 we get $\chi_\ell^{(\ell-m-1)}(\sigma) = \sigma^m/m!$ if $0 \le \sigma \le 1$, hence there are no problems of convergence for the last integral. Moreover, the derivative of order m of the function $\chi_\ell^{(\ell-m-1)}(\sigma)/\sigma$ is a measure of compact support. Since $(-ix \cdot A)^m W(\sigma x) = (-1)^m \frac{d^m}{d\sigma^m}W(\sigma x)$, we obtain after integrating by parts m times:

(3.4.34)
$$\int_\varepsilon^1 [I - W(\tau x)]^\ell \frac{d\tau}{\tau^{m+1}} = (-ix \cdot A)^m \int_{\mathbb{R}} W(\varepsilon\sigma x)\chi_\ell^{(\ell-m-1)}(\sigma)\frac{d\sigma}{\sigma} -$$
$$- \int_{\mathbb{R}} W(\sigma x)\frac{d^m}{d\sigma^m}\left[\frac{\chi_\ell^{(\ell-m-1)}(\sigma)}{\sigma}\right]d\sigma.$$

The last term here is an operator $B(x) \in B(\mathbf{F}_j)$ for all $j \in \mathbb{Z}$. If we set $a = \int_{\mathbb{R}} \chi_\ell^{(\ell-m-1)}(\sigma)\sigma^{-1}d\sigma$, then a is a finite number; let us prove that a is not zero. By using (d) of Proposition 3.3.5 and the fact that $\chi_\ell(\sigma)$ is zero for $\sigma \le 0$, we obtain

$$a = \int_0^\infty \chi_{m+1}(\sigma)[(I - e^{iP})^{\ell-m-1}u](\sigma)d\sigma,$$

where u is the function defined on $(0, \infty)$ by $u(\sigma) = \sigma^{-1}$. We have $(e^{iP} - I)^k u = \int_{\mathbb{R}} e^{i\tau P}(iP)^k u\chi_k(\tau)d\tau$; this is formally a particular case of (3.3.18) (take $A = P$) and is an easy consequence of the Taylor formula. Since $[(iP)^k u](\sigma) = u^{(k)}(\sigma) = (-1)^k k!\sigma^{-k-1}$, by taking $k = \ell - m - 1$ we get

$$a = (\ell - m - 1)! \int_0^\infty \int_0^\infty \chi_{m+1}(\sigma)\chi_{\ell-m-1}(\tau)\frac{d\sigma d\tau}{(\sigma + \tau)^{\ell-m}} > 0.$$

We clearly have $\lim_{\varepsilon \to 0} \int_{\mathbb{R}} W(\varepsilon \sigma x) \chi_\ell^{(\ell-m-1)}(\sigma) \sigma^{-1} d\sigma = a$ strongly in $B(\mathbf{F}_j)$ for each $j \in \mathbb{Z}$. So, if $f \in \mathbf{F}$ and $(x \cdot A)^m f$ belongs to $\mathbf{F}$ too, then the next limit exists in $\mathbf{F}$ and the following equality holds:

$$(3.4.35) \qquad \lim_{\varepsilon \to 0} \int_\varepsilon^1 [I - W(\tau x)]^\ell f \frac{d\tau}{\tau^{m+1}} = a(-ix \cdot A)^m f + B(x)f.$$

Reciprocally, let $f \in \mathbf{F}$ such that the l.h.s. limit above exists in $\mathbf{F}$ (note that the integral $\int_0^1 [I - W(\tau x)]^\ell f \tau^{-1-m} d\tau$ is always absolutely convergent in $\mathbf{F}_{-m-1}$; indeed $\|[W(y) - I]^\ell f\|_{\mathbf{F}_{-m-1}} \leq c|y|^{m+1}$ by (3.3.18)). Then $(x \cdot A)^m f \in \mathbf{F}$ for all $x \in \mathbb{R}^n$, so $A^\alpha f \in \mathbf{F}$ for $|\alpha| = m$ (apply Proposition 1.1.2 as in (ii) of the proof of Proposition 3.3.19), hence $f \in \mathbf{F}_m$ (cf. Proposition 3.3.18 (b)). $\square$

REMARK 3.4.11. It is clear from the preceding proof that $f \in \mathbf{F}_m$ if and only if $f \in \mathbf{F}$ and, for each x in $\mathbb{R}^n$, $\int_{\varepsilon_k}^1 \tau^{-m}[W(\tau x) - I]^\ell f \tau^{-1} d\tau$ has a weak limit in $\mathbf{F}$ for some sequence $\{\varepsilon_k\}$ converging to zero (in fact it is sufficient that this property be satisfied for a finite number of points x in $\mathbb{R}^n$). In particular, if $\mathbf{F}$ is a reflexive Banach space, $f \in \mathbf{F}$ and if there are integers $\ell > m \geq 1$ such that $\left\| \int_\varepsilon^1 [W(\tau x) - I]^\ell f \tau^{-m-1} d\tau \right\|_{\mathbf{F}} \leq c(x) < \infty$ for each $\varepsilon \in (0,1]$ and each $x \in \mathbb{R}^n$, then $f \in \mathbf{F}_m$.

We mention the following consequence of (3.4.34) (this is also a direct consequence of (3.3.19)). Let $\xi(dx) \equiv \xi(x)dx$ be a measure on $\mathbb{R}^n \setminus \{0\}$ which is homogeneous of degree zero. If ξ is a function this means $\xi(x) = \xi(x/|x|)$ for $x \neq 0$, while in general we have $\xi(x)dx = r^{n-1}dr \cdot \xi(\omega)d\omega$ (in polar coordinates $x = r\omega$, $r = |x|$) where $\xi(\omega)d\omega$ is a measure on the unit sphere S^{n-1} of $\mathbb{R}^n$. We take $x = \omega \in S^{n-1}$ in (3.4.34), then we integrate with respect to $\xi(\omega)d\omega$ over S^{n-1} and finally we let $\varepsilon \to 0$. We get:

$$\lim_{\varepsilon \to 0} \int_{\varepsilon < |x| < 1} \frac{[I - W(x)]^\ell}{|x|^m} f \xi(x) \frac{dx}{|x|^n} = a(-i)^m \sum_{|\alpha|=m} \frac{m!}{\alpha!} \int_{S^{n-1}} \omega^\alpha \xi(\omega)d\omega \cdot A^\alpha f$$
$$+ \int_{S^{n-1}} B(\omega) f \xi(\omega)d\omega.$$

Here f is an arbitrary element of $\mathbf{F}_{-\infty}$ and the limit exists in $\mathbf{F}_j$ ($j \in \mathbb{Z}$) if f and $A^\alpha f$ belong to $\mathbf{F}_j$ for all $|\alpha| = m$. The formula (3.4.35) with $x = \omega \in S^{n-1}$ is a particular case of this one (ξ is the Dirac measure with support at ω). Now let $\{\xi_\alpha\}_{|\alpha|=m}$ be a family of measures on S^{n-1} such that the integral $\int_{S^{n-1}} \omega^\alpha \xi_\beta(\omega)d\omega$ equals 1 if $\alpha = \beta$ and equals 0 otherwise ($|\alpha| = |\beta| = m$). If $f \in \mathbf{F}$ and if for each $\xi = \xi_\beta$ the limit in the l.h.s. of the preceding formula exists in $\mathbf{F}$ (weakly and along a subsequence $\varepsilon_k \to 0$ is sufficient in fact), then $A^\beta f \in \mathbf{F}$ for all $|\beta| = m$, hence $f \in \mathbf{F}_m$ (Proposition 3.3.18 (b)). This is an improved form of the second part of Proposition 3.4.10.

If W is a one-parameter group, the results that we have proven in this subsection in $\mathbf{F}_{s,p}$ with $s > 0$ have analogues for negative values of s, essentially as a consequence of Theorem 3.3.23 (d). We collect a few typical facts in the next proposition:

PROPOSITION 3.4.12. *Let W be a C_0-representation of $\mathbb{R}$ in a Banach space $\mathbf{F}$, and let $1 \le p \le \infty$. Then*
  (a) *For any $k, m \in \mathbb{Z}$ one has $\mathbf{F}_{k+m,1} \subset (\mathbf{F}_k)_m = \mathbf{F}_{k+m} \subset \mathbf{F}_{k+m,\infty}^0$.*
  (b) *Let $s \in \mathbb{R}$ and let $k \in \mathbb{Z}$. Then*

$$(3.4.36) \qquad \mathbf{F}_{s,p} = (\mathbf{F}_k)_{s-k,p}.$$

If $k < s$, then for each integer $\ell > s - k$ the following expression defines an admissible norm on $\mathbf{F}_{s,p}$:

$$(3.4.37) \qquad \|f\|_{\mathbf{F}_k} + \left[\int_0^1 \left\| \frac{[W(x)-I]^\ell f}{x^{s-k}} \right\|_{\mathbf{F}_k}^p \frac{dx}{x} \right]^{1/p}.$$

  (c) *Let $s, t \in \mathbb{R}$ be such that $\ell \equiv s - t$ is an integer. Then $\mathbf{F}_{s,p} = (\mathbf{F}_{t,p})_\ell$ and, if $\ell > 0$:*

$$(3.4.38) \qquad \mathbf{F}_{s,p} = \{ f \in \mathbf{F}_{t,p} \mid A^\alpha f \in \mathbf{F}_{t,p} \ \forall |\alpha| = \ell \}.$$

PROOF. (a) is immediate by (3.4.8) and Theorem 3.3.23 (d). The second part of (b) and (c) follow from Theorem 3.4.6 and Corollary 3.4.8 respectively by using the relation (3.4.36) and the fact that $(A + z)^\ell$ is an isomorphism of $\mathbf{F}_{t,p}$ onto $\mathbf{F}_{s,p}$ if $\Im z$ is large enough (Theorem 3.3.23).

To prove (3.4.36), let m_1, m_2 be integers satisfying $m_1 < s - k < m_2$. Then, by Definition 3.4.2 (applied in the scale $\{(\mathbf{F}_k)_m\}_{m \in \mathbb{Z}}$), we have $(\mathbf{F}_k)_{s-k,p} = ((\mathbf{F}_k)_{m_2}, (\mathbf{F}_k)_{m_1})_{\theta,p}$ with $\theta = [m_2 - (s - k)](m_2 - m_1)^{-1}$. Since $(\mathbf{F}_k)_m = \mathbf{F}_{k+m}$, this means that $(\mathbf{F}_k)_{s-k,p} = (\mathbf{F}_{k+m_2}, \mathbf{F}_{k+m_1})_{\theta,p}$. But $(\mathbf{F}_{k+m_2}, \mathbf{F}_{k+m_1})_{\theta,p} = \mathbf{F}_{s,p}$ if θ is as given above, again by (3.4.7). $\square$

3.5. Littlewood-Paley Estimates

In the preceding section we have considered a Banach space $\mathbf{F}$ equipped with a C_0-representation of $\mathbb{R}^n$ and defined the associated Besov spaces $\mathbf{F}_{s,p}$ ($s \in \mathbb{R}$, $p \in [1, +\infty]$) by interpolation. Then for $s > 0$ we obtained a characterization of these spaces in terms of smoothness properties of the functions $x \mapsto W(x)f$ (Theorem 3.4.6). For polynomially bounded C_0-groups one can define other admissible norms on $\mathbf{F}_{s,p}$ in terms of the functional calculus associated to these C_0-groups. These norms appear rather naturally because our original expression (3.4.2) for the K-functionals leading to the interpolation spaces $\mathbf{F}_{s,p}$ involved directly certain operators from the functional calculus for the given C_0-group.

It turns out that for many developments the use of the functional calculus rather than the C_0-group itself allows more flexibility, in addition to a certain theoretical simplicity. For instance there are situations where a representation of $\mathbb{R}^n$ has restricted continuity properties (we think in particular of a C_w-representation as introduced in Definition 3.2.6) but the associated functional calculus is just as easy to handle as that associated to a C_0-group. The fact that a functional calculus is a morphism of algebras contributes to the facility of its use.

In the present section we shall introduce an abstract notion of functional calculus. It covers the functional calculus associated to polynomially bounded

C_{w}-groups and is very well adapted to a certain type of estimates that are generally called *Littlewood-Paley estimates*. The principal result is Theorem 3.5.11 which contains estimates of this type and which will be the basis, in the next section, for new admissible norms on the Besov spaces $\mathbf{F}_{s,p}$ of Section 3.4.

Let X be an euclidean space and set $X^0 = X \setminus \{0\}$. We shall consider algebras $\mathcal{N}$ of bounded complex functions on X that are continuous except possibly at the origin of X. The advantage of not requiring continuity at the origin is that these algebras may contain functions that are homogeneous of degree zero and non-constant on X^0 (e.g. the functions $x^\alpha |x|^{-|\alpha|}$ for any multi-index α).

It is convenient to consider the functions in such algebras to be defined on X^0 rather than on X. We define $BC(X^0)$ to be the set of all bounded continuous functions from X^0 to $\mathbb{C}$, equipped with the norm

$$(3.5.1) \qquad \|\varphi\|_{BC(X^0)} = \sup_{x \in X^0} |\varphi(x)|.$$

Each bounded function φ that is defined and continuous on all of X ($\varphi \in BC(X)$) naturally defines an element of $BC(X^0)$ by taking its restriction to X^0. Keeping this convention in mind, we shall occasionally view $BC(X)$ as a subset of $BC(X^0)$.

We recall that a sequence $\{\varphi_k\}_{k \in \mathbb{N}}$ of functions in $C_0^\infty(X)$ is $C_0^\infty(X)$-convergent to a function $\varphi \in C_0^\infty(X)$ (written $\varphi_k \to \varphi$ in $C_0^\infty(X)$) if there is a compact subset Ω of X such that $\operatorname{supp} \varphi_k \subset \Omega$ for each $k \in \mathbb{N}$ and such that for each multi-index α, $\varphi_k^{(\alpha)}(x) \to \varphi^{(\alpha)}(x)$ as $k \to \infty$, uniformly in $x \in X$.

DEFINITION 3.5.1. Let $r \geq 0$ be a real number. A LP-*algebra of order* r on X is a Banach algebra $\mathcal{N}$, with norm $\| \cdot \|_{\mathcal{N}}$, consisting of bounded continuous functions from X^0 to $\mathbb{C}$, and satisfying the following conditions:

(α) There is a constant $c < \infty$ such that

$$(3.5.2) \qquad \sup_{x \in X^0} |\varphi(x)| \leq c \|\varphi\|_{\mathcal{N}} \qquad \forall \varphi \in \mathcal{N},$$

(β) $\mathcal{N}$ contains $C_0^\infty(X)$ and all constant functions. Moreover, if $\varphi_k \to \varphi$ in $C_0^\infty(X)$, then $\|\varphi_k - \varphi\|_{\mathcal{N}} \to 0$ as $k \to \infty$.

(γ) If $\varphi \in \mathcal{N}$ and $\sigma > 0$, then $\varphi^\sigma \in \mathcal{N}$, and there is a constant $N < \infty$ such that

$$(3.5.3) \qquad \|\varphi^\sigma\|_{\mathcal{N}} \leq N \langle \sigma \rangle^r \|\varphi\|_{\mathcal{N}} \qquad \forall \varphi \in \mathcal{N}, \forall \sigma > 0.$$

Examples of LP-algebras of order r are the algebra $\mathcal{M}^r(X)$ introduced in Definition 3.1.11 and, if r is an integer, the algebra $BC^r(X^0)$. For another example, consider for some $k \in \mathbb{N}$ the set $\mathcal{N}_k$ of functions $\varphi : X^0 \to \mathbb{C}$ that are of class C^k on X^0 and satisfy $\|\varphi\|_{\mathcal{N}_k} < \infty$, where

$$\|\varphi\|_{\mathcal{N}_k} = \sup_{x \in X^0} \sum_{|\alpha| \leq k} \frac{1}{\alpha!} |x|^\alpha |\varphi^{(\alpha)}(x)|.$$

Then $\mathcal{N}_k$ is a LP-algebra of order $r = 0$ on X.

REMARK 3.5.2. There is an important class of algebras that are not LP-algebras, namely the algebras $\mathscr{M}^{(\omega)}(X)$ with $\omega > 0$ introduced in Definition 3.1.11. In fact, we know that $\mathscr{M}^{(\omega)}(X)$ is not invariant under the operation of dilation $\varphi \mapsto \varphi^\sigma$ if $\sigma > 1$. Moreover, since $\mathscr{M}^{(\omega)}(X)$ contains only analytic functions, the only function in it that vanishes on a set of non-zero Lebesgue measure is the function identically equal to zero. Analogously, if $X = \mathbb{R}$, the algebra $\mathscr{M}_w(\mathbb{R})$ with $w(x) = \max(1, e^{\omega x})$ has similar properties. In applications, this excludes for example the consideration of the dilation group $\exp(ixD)$ in $\mathscr{H}^1(\mathbb{R})$; in fact it can be shown that, if $\varphi(D)$ — which exists as a bounded operator in $\mathscr{H}(X)$ for each $\varphi \in L^\infty(\mathbb{R})$ by the functional calculus for the self-adjoint operator D — leaves $\mathscr{H}^1(\mathbb{R})$ invariant, then φ cannot vanish on a set of non-zero Lebesgue measure unless $\varphi \equiv 0$ (see the end of Section 3.2).

REMARK 3.5.3. Let $\mathscr{N}$ be a LP-algebra on X. It is easy to see from simple examples that for most $\varphi \in \mathscr{N}$ the mapping $\sigma \mapsto \varphi^\sigma \in \mathscr{N}$ is *not* norm-continuous. However, the norm-continuity of the preceding mapping is satisfied if $\varphi \in C_0^\infty(X)$, because $\sigma \mapsto \varphi^\sigma \in C_0^\infty(X)$ is continuous on $(0, \infty)$ and the embedding $C_0^\infty(X) \subset \mathscr{N}$ is continuous by the condition (β) in Definition 3.5.1. Since $\mathscr{N}$ is a unital algebra, it follows that, *if $\varphi \in C^\infty(X)$ and $\varphi(x)$ is constant in a neighbourhood of infinity, then $\varphi \in \mathscr{N}$ and $\sigma \mapsto \varphi^\sigma \in \mathscr{N}$ is norm continuous on $0 < \sigma < \infty$*; however, norm continuity at $\sigma = 0$ holds only if φ is a constant: $\varphi^0(x) = \varphi(0)$ is a constant, hence

$$\lim_{\sigma \to 0} \|\varphi^\sigma - \varphi^0\|_{BC(X^0)} = \lim_{\sigma \to 0} \sup_{x \in X^0} |\varphi(\sigma x) - \varphi(0)| = \sup_{y \neq 0} |\varphi(y) - \varphi(0)|,$$

which is different from zero unless φ is a constant function.

In the following definition we introduce the notion of a zero of order ℓ ($\ell > 0$) at the origin for functions in a LP-algebra.

DEFINITION 3.5.4. (a) A function $\theta : X^0 \to \mathbb{C}$ is said to be a *tauberian function* if $\theta \in C_0^\infty(X^0)$ and if for each $x \in X^0$ there is a number $\lambda > 0$ such that $\theta(\lambda x) \neq 0$.

(b) Let $\mathscr{N}$ be a LP-algebra and $\ell > 0$ a real number. A function $\varphi \in \mathscr{N}$ is said to have a *zero of order ℓ at the origin in the $\mathscr{N}$-sense* (written $\varphi(x) = O_{\mathscr{N}}(|x|^\ell)$ as $x \to 0$) if there exists a tauberian function θ such that $\|\varphi^\sigma \theta\|_{\mathscr{N}} \leq c\sigma^\ell$ for some constant c and all $\sigma \in (0, 1)$.

As a typical example of a tauberian function one may take a function $\theta \in C_0^\infty(X)$ such that $\theta(x) > 0$ if $|x| \in (a, b)$ for some $0 < a < b < \infty$ and $\theta(x) = 0$ if $|x| \notin (a, b)$. If θ is any tauberian function, we define for $x \in X^0$:

$$(3.5.4) \qquad\qquad \varrho(x) = \int_0^\infty |\theta(\sigma x)|^2 \frac{d\sigma}{\sigma}.$$

Then ϱ is homogeneous of degree zero and of class C^∞ on X^0, and it satisfies $\varrho(x) > 0$ for each $x \in X^0$. Furthermore θ/ϱ and $\bar{\theta}/\varrho$ belong to $C_0^\infty(X)$, hence to $\mathscr{N}$.

LEMMA 3.5.5. *Let $\mathcal{N}$ be a LP-algebra, $\varphi \in \mathcal{N}$ with $\varphi(x) = O_{\mathcal{N}}(|x|^{\ell})$ as $x \to 0$ for some $\ell > 0$. Then*

(a) *if θ_0 is any element of $\mathcal{N}$ with compact support in X^0, there is a constant $c_0 < \infty$ such that $\|\varphi^{\tau}\theta_0\|_{\mathcal{N}} \leq c_0 \tau^{\ell}$ for all $\tau \in (0,1)$,*

(b) *given any real numbers a, b with $0 < a < b < \infty$, there is a constant $c_1 < \infty$ such that $\sup\{|\varphi(x)| \mid a\tau < |x| < b\tau\} \leq c_1 \tau^{\ell}$ for all $\tau \in (0,1)$.*

PROOF. (i) By hypothesis there is a tauberian function θ such that $\|\varphi^{\sigma}\theta\|_{\mathcal{N}} \leq c_0 \sigma^{\ell}$ if $0 < \sigma < 1$. We define ϱ by (3.5.4) and let $\eta = |\theta|^2/\varrho$. Then $\eta \in C_0^{\infty}(X^0)$, and

$$(3.5.5) \qquad \int_0^{\infty} \eta^{\sigma}(x)\frac{d\sigma}{\sigma} = \frac{1}{\varrho(x)}\int_0^{\infty}|\theta(\sigma x)|^2\frac{d\sigma}{\sigma} = 1 \quad \forall x \in X^0.$$

Also, for some constant c_2:

$$\|\varphi^{\sigma}\eta\|_{\mathcal{N}} = \|\varphi^{\sigma}|\theta|^2/\varrho\|_{\mathcal{N}} \leq \|\varphi^{\sigma}\theta\|_{\mathcal{N}}\|\overline{\theta}/\varrho\|_{\mathcal{N}} \leq c_2\sigma^{\ell} \qquad \forall\sigma \in (0,1).$$

(ii) Let $\theta_0 \in \mathcal{N}$ with compact support in X^0. Choose $0 < r < s < \infty$ such that $\operatorname{supp}\theta_0 \cup \operatorname{supp}\eta \subset \{x \in X \mid r < |x| < s\}$. If $x \in X$ is such that $\theta_0(x)\eta(\sigma x) \neq 0$, then $|x| \in (r,s)$ and $\sigma|x| \in (r,s)$, hence also $\sigma \in (R^{-1}, R)$ with $R = sr^{-1}$. By using this fact, we get from (3.5.5) that

$$\theta_0 = \int_{R^{-1}}^{R} \theta_0 \eta^{\sigma}\frac{d\sigma}{\sigma},$$

where the integral exists in the norm of $\mathcal{N}$ as a Riemann integral of a continuous $\mathcal{N}$-valued function (see Remark 3.5.3). Hence, for $0 < \tau < 1$:

$$(3.5.6) \qquad \begin{aligned} \|\varphi^{\tau}\theta_0\|_{\mathcal{N}} &\leq \int_{R^{-1}}^{R} \|\varphi^{\tau}\eta^{\sigma}\theta_0\|_{\mathcal{N}}\frac{d\sigma}{\sigma} \\ &\leq 2\ln R\|\theta_0\|_{\mathcal{N}} \sup_{R^{-1}<\sigma<R} \|\varphi^{\tau}\eta^{\sigma}\|_{\mathcal{N}} \\ &= c_3 \sup_{R^{-1}<\sigma<R} \|(\varphi^{\tau/\sigma}\eta)^{\sigma}\|_{\mathcal{N}} \\ &\leq c_3 N\langle R\rangle^r \sup_{R^{-1}<\sigma<R} \|\varphi^{\tau/\sigma}\eta\|_{\mathcal{N}} \\ &\leq c_4 \sup_{R^{-1}<\sigma<R} \left(\frac{\tau}{\sigma}\right)^{\ell} \equiv c_0\tau^{\ell}. \end{aligned}$$

This proves (a). To obtain (b) we choose r and s such that they also satisfy $r < a$ and $s > b$. Then one may take θ_0 such that $\theta_0(x) = 1$ for $|x| \in (a,b)$. So (3.5.2) and (3.5.6) imply that

$$\sup_{\tau a<|x|<\tau b}|\varphi(x)| = \sup_{a<|y|<b}|\varphi(\tau y)| \leq c\|\varphi^{\tau}\theta_0\|_{\mathcal{N}} \leq c_1\tau^{\ell} \qquad \text{if } \tau \in (0,1). \qquad \square$$

We shall now give two examples of functions that have a zero of order ℓ at the origin in the $\mathcal{N}$-sense.

EXAMPLE 3.5.6. Let $\mathscr{N}$ be a LP-algebra on X. Assume that $\varphi \in \mathscr{N}$ is of the form $\varphi(x) = \varrho(x)\psi(x)$ with $\psi \in \mathscr{N}$ and $\varrho \in C^\infty(X^0)$ homogeneous of degree ℓ (i.e. $\varrho(\lambda x) = \lambda^\ell \varrho(x)$ for $x \neq 0$ and $\lambda > 0$). Then $\varphi(x) = O_\mathscr{N}(|x|^\ell)$ as $x \to 0$.

PROOF. Let θ be any function in $C_0^\infty(X^0)$. Then $\varrho\theta \in C_0^\infty(X^0) \subset \mathscr{N}$, hence for $\sigma \in (0, 1)$:

$$\|\varphi^\sigma\theta\|_\mathscr{N} = \|\varrho^\sigma\psi^\sigma\theta\|_\mathscr{N} = \sigma^\ell\|\psi^\sigma\varrho\theta\|_\mathscr{N} \leq \sigma^\ell\|\psi^\sigma\|_\mathscr{N}\|\varrho\theta\|_\mathscr{N}$$

$$\leq 2^{r/2}N\|\psi\|_\mathscr{N}\|\varrho\theta\|_\mathscr{N}\sigma^\ell. \quad \square$$

EXAMPLE 3.5.7. Let $r \geq 0$ be a real number and let $\ell \geq 1$ be an integer. Let $\varphi \in BC^\ell(X)$ be such that $\varphi^{(\alpha)} \in \mathscr{M}^r(X)$ for each multi-index α with $|\alpha| \leq \ell$ and such that $\varphi(x) = O(|x|^\ell)$ as $x \to 0$ in the usual sense (hence $\varphi^{(\alpha)}(0) = 0$ if $|\alpha| \leq \ell - 1$). Then $\varphi(x) = O_{\mathscr{M}^r(X)}(|x|^\ell)$ as $x \to 0$.

PROOF. By using Taylor's formula (1.1.8) we get for all $x \in X$ and $\sigma > 0$:

$$\varphi(\sigma x) = \sum_{|\alpha|=\ell} \frac{\sigma^\ell x^\alpha}{\alpha!} \int_0^1 \varphi^{(\alpha)}(\tau\sigma x)\ell(1-\tau)^{\ell-1}d\tau.$$

Let $\theta \in \mathscr{M}^r(X)$ be such that $\theta_{(\alpha)} \in \mathscr{M}^r(X)$ for $|\alpha| = \ell$, where $\theta_{(\alpha)}(x) = x^\alpha\theta(x)$. Upon multiplying the preceding identity by $\theta(x)$ we get:

$$\|\varphi^\sigma\theta\|_{\mathscr{M}^r} \leq \sum_{|\alpha|=\ell} \frac{\sigma^\ell}{\alpha!} \int_0^1 \|\varphi^{(\alpha)\sigma\tau}\|_{\mathscr{M}^r}\|\theta_{(\alpha)}\|_{\mathscr{M}^r}\ell(1-\tau)^{\ell-1}d\tau.$$

By taking into account (3.1.21) we obtain for $0 < \sigma \leq 1$:

$$(3.5.7) \qquad \|\varphi^\sigma\theta\|_{\mathscr{M}^r} \leq \sigma^\ell \sum_{|\alpha|=\ell} \frac{1}{\alpha!}\|\varphi^{(\alpha)}\|_{\mathscr{M}^r} \cdot \|Q^\alpha\theta\|_{\mathscr{M}^r}. \quad \square$$

In the context of the preceding example it is seen that the factor θ in $\|\varphi^\sigma\theta\|_{\mathscr{M}^r}$ is essential, because

$$\|\varphi^\sigma\|_{\mathscr{M}^r} = 2^{r/2}\int_X \langle\sigma x\rangle^r|\widehat{\varphi}|(dx) \geq 2^{r/2}\int_X |\widehat{\varphi}|(dx) > 0 \text{ unless } \varphi \equiv 0.$$

For the remainder of this section we consider a fixed LP-algebra $\mathscr{N}$ of order r on X, a Banach space $\mathbf{E}$ and a norm-continuous unital homomorphism of $\mathscr{N}$ into $B(\mathbf{E})$ (i.e. a representation of the algebra $\mathscr{N}$ in $\mathbf{E}$). If $\varphi \in \mathscr{N}$, we denote by the symbol $\varphi(A)$ the operator in $B(\mathbf{E})$ associated to φ by the above homomorphism (the reason for this notation is, as already explained in relation with (3.2.11), that in applications $\varphi(A)$ can be naturally interpreted in terms of the functional calculus associated to a finite set A of unbounded operators in $\mathbf{E}$). If $\sigma > 0$, we shall write $\varphi(\sigma A)$ for the operator in $B(\mathbf{E})$ associated to $\varphi^\sigma \in \mathscr{N}$ (so $\varphi(\sigma A) \equiv \varphi^\sigma(A)$). Notice that the mapping $\sigma \mapsto \varphi(\sigma A) \in B(\mathbf{E})$ is not norm-continuous in general; however, in many of our applications it will be strongly continuous (even at $\sigma = 0$ when this makes sense, i.e. when φ is continuous at $x = 0$). Also, *if $\varphi \in C^\infty(X)$ and $\varphi(x)$ is constant in a neighbourhood of infinity, then $\sigma \mapsto \varphi(\sigma A) \in B(\mathbf{E})$ is norm-continuous on $0 < \sigma < \infty$* by Remark

3.5.3. Finally let us mention that in all non-trivial cases $||\varphi(\sigma A)||_{B(\mathbf{E})}$ is either a constant independent of $\sigma \in (0,1)$ or bounded below by a strictly positive constant (consider the case where $\mathcal{N} = BC(\mathbb{R})$ and A is a self-adjoint operator in a Hilbert space $\mathbf{E}$).

The next lemma contains the principal technical estimates.

LEMMA 3.5.8. *Let ξ and $\widetilde{\xi}$ be functions of class C^∞ on X such that $\xi(x) = 0$ in a neighbourhood of the origin, $\xi(x) = 1$ in a neighbourhood of infinity and $\widetilde{\xi} \in C_0^\infty(X)$ with $\widetilde{\xi}(x) = 1$ inside a ball containing the support of $\nabla\xi$. Define $\eta \in C_0^\infty(X^0)$ by $\eta(x) = x \cdot \nabla\xi(x) = \sum_{j=1}^n x_j \partial_j \xi(x)$.*

Let $\varphi \in \mathcal{N}$ be such that $\varphi(x) = O_\mathcal{N}(|x|^\ell)$ as $x \to 0$ for some real number $\ell > r$. Then there is a constant $c > 0$ such that for all $f \in \mathbf{E}$ and all $\varepsilon \in (0,1)$:

(3.5.8)
$$c||\varphi(\varepsilon A)f||_\mathbf{E} \le ||\xi(\varepsilon A)f||_\mathbf{E} + \int_\varepsilon^1 \left(\frac{\varepsilon}{\tau}\right)^\ell ||\eta(\tau A)f||_\mathbf{E} \frac{d\tau}{\tau} + \varepsilon^\ell ||\widetilde{\xi}(A)f||_\mathbf{E}.$$

If $||\xi(\tau A)f||_\mathbf{E} \to 0$ as $\tau \to 0$, then one also has

$$(3.5.9)\ c||\varphi(\varepsilon A)f||_\mathbf{E} \le \int_0^1 \min\left(1, \left(\frac{\varepsilon}{\tau}\right)^\ell\right) \cdot ||\eta(\tau A)f||_\mathbf{E} \frac{d\tau}{\tau} + \varepsilon^\ell ||\widetilde{\xi}(A)f||_\mathbf{E}.$$

REMARK. The main point of these results lies in the fact that ξ and η have zeroes of infinite order at the origin, whereas φ has a zero of finite order ℓ.

PROOF. (i) If $0 < a < b < \infty$, we have for any $x \in X^0$:

(3.5.10)
$$\xi(bx) - \xi(ax) = \int_a^b \frac{d}{d\sigma}\xi(\sigma x)d\sigma = \int_a^b \eta(\sigma x)\frac{d\sigma}{\sigma}.$$

We take $a = 1$, $b \to \infty$ and $x = \varepsilon y$ ($0 < \varepsilon < 1$, $y \in X^0$) to obtain that

$$1 = \xi(\varepsilon y) + \int_1^\infty \eta(\varepsilon\sigma y)\frac{d\sigma}{\sigma} = \xi(\varepsilon y) + \int_\varepsilon^\infty \eta(\tau y)\frac{d\tau}{\tau}.$$

Upon multiplying this identity by $\varphi(\varepsilon y)$, we get that

$$(3.5.11)\qquad \varphi^\varepsilon(y) = \varphi^\varepsilon(y)\xi^\varepsilon(y) + \int_\varepsilon^\infty \varphi^\varepsilon(y)\eta^\tau(y)\frac{d\tau}{\tau} \qquad \forall y \in X^0.$$

Since $\eta \in C_0^\infty(X^0)$, the function $\tau \mapsto \varphi^\varepsilon\eta^\tau \in \mathcal{N}$ is norm-continuous. Also, by (3.5.3) and Lemma 3.5.5 (a):

$$(3.5.12)\quad ||\varphi^\varepsilon\eta^\tau||_\mathcal{N} = ||(\varphi^{\varepsilon/\tau}\eta)^\tau||_\mathcal{N} \le N\langle\tau\rangle^r ||\varphi^{\varepsilon/\tau}\eta||_\mathcal{N} \le cN\langle\tau\rangle^r \left(\frac{\varepsilon}{\tau}\right)^\ell.$$

Hence, because $\ell > r$, $\int_\varepsilon^\infty \varphi^\varepsilon\eta^\tau\tau^{-1}d\tau$ is a norm-convergent integral in $\mathcal{N}$. Denote by δ_x the linear form $\psi \mapsto \psi(x)$ on $\mathcal{N}$. Then $\{\delta_x \mid x \in X^0\}$ is a family of linear continuous forms on $\mathcal{N}$ which separates the points of $\mathcal{N}$. Hence we get as an identity in $\mathcal{N}$:

$$\varphi^\varepsilon = \varphi^\varepsilon\xi^\varepsilon + \int_\varepsilon^\infty \varphi^\varepsilon\eta^\tau \frac{d\tau}{\tau}.$$

Since $\mathcal{N} \ni \psi \mapsto \psi(A) \in B(\mathbf{E})$ is a norm-continuous homomorphism, this implies that

$$\varphi(\varepsilon A) = \varphi(\varepsilon A)\xi(\varepsilon A) + \int_\varepsilon^\infty \varphi(\varepsilon A)\eta(\tau A)\frac{d\tau}{\tau},$$

where the integral exists (as a Riemann integral) in the norm of $B(\mathbf{E})$. By taking into account that $||\varphi(\varepsilon A)||_{B(\mathbf{E})} \le c_1||\varphi^\varepsilon||_{\mathcal{N}} \le c_1 N\langle\varepsilon\rangle^r ||\varphi||_{\mathcal{N}}$, one sees that there is a constant c_2 such that for all $f \in \mathbf{E}$ and all $\varepsilon \in (0,1)$:

$$(3.5.13) \qquad ||\varphi(\varepsilon A)f||_{\mathbf{E}} \le c_2||\xi(\varepsilon A)f||_{\mathbf{E}} + \int_\varepsilon^\infty ||\varphi(\varepsilon A)\eta(\tau A)f||_{\mathbf{E}}\frac{d\tau}{\tau}.$$

We consider separately the contributions to the integral in (3.5.13) for $\tau \in (\varepsilon, 1)$ and for $\tau \ge 1$. Let $\theta_0 \in C_0^\infty(X^0)$ be such that $\eta = \eta\theta_0$. Then

$$||\varphi(\varepsilon A)\eta(\tau A)f||_{\mathbf{E}} \le ||\varphi(\varepsilon A)\theta_0(\tau A)||_{B(\mathbf{E})}||\eta(\tau A)f||_{\mathbf{E}} \le c_1||\varphi^\varepsilon \theta_0^\tau||_{\mathcal{N}}||\eta(\tau A)f||_{\mathbf{E}}.$$

Now for $\varepsilon < \tau < 1$ we have as in (3.5.12): $||\varphi^\varepsilon \theta_0^\tau||_{\mathcal{N}} \le c_3(\varepsilon/\tau)^\ell$. So the contribution for $\tau \in (\varepsilon, 1)$ to the integral in (3.5.13) is majorized by the second term on the r.h.s. of (3.5.8).

The argument for $\tau \ge 1$ is similar. If $\tau \ge 1$ we have $\eta(\tau x) = \eta(\tau x)\tilde{\xi}(x)$ for all $x \in X^0$, hence:

$$||\varphi(\varepsilon A)\eta(\tau A)f||_{\mathbf{E}} \le c_1||\varphi^\varepsilon \eta^\tau||_{\mathcal{N}}||\tilde{\xi}(A)f||_{\mathbf{E}}.$$

(3.5.12) implies that $||\varphi^\varepsilon \eta^\tau||_{\mathcal{N}} \le c_4\varepsilon^\ell\tau^{r-\ell}$ for $\tau \ge 1$. So, since $\ell > r$, the contribution for $\tau \ge 1$ to the integral in (3.5.13) can be majorized by the last term on the r.h.s. of (3.5.8).

(ii) To prove (3.5.9) we observe that (3.5.10) implies that

$$(3.5.14) \qquad \xi^\varepsilon = \xi^\tau + \int_\tau^\varepsilon \eta^\sigma\frac{d\sigma}{\sigma}.$$

Consequently

$$||\xi(\varepsilon A)f||_{\mathbf{E}} \le ||\xi(\tau A)f||_{\mathbf{E}} + \int_\tau^\varepsilon ||\eta(\sigma A)f||_{\mathbf{E}}\frac{d\sigma}{\sigma}.$$

By inserting this inequality into (3.5.8) and letting $\tau \to 0$, one arrives at (3.5.9). $\square$

In relation with (3.5.9) the following observations are interesting. Let ξ_1, ξ_2 be two functions having the same properties as the function ξ in the preceding lemma. Then there is a number $\lambda > 0$ such that $\xi_2(\lambda x) = 1$ for all $x \in \mathrm{supp}\,\xi_1$. Hence $\xi_1 = \xi_1\xi_2^\lambda$, so that for some constant c_1, all $\varepsilon \in (0,1)$ and all $f \in \mathbf{E}$:

$$||\xi_1(\varepsilon A)f||_{\mathbf{E}} = ||\xi_1(\varepsilon A)\xi_2(\varepsilon\lambda A)f||_{\mathbf{E}} \le c_1||\xi_2(\varepsilon\lambda A)f||_{\mathbf{E}}.$$

Similarly there are a number $\mu > 0$ and a constant c_2 such that $||\xi_2(\varepsilon A)f||_{\mathbf{E}} \le c_2||\xi_1(\varepsilon\mu A)f||_{\mathbf{E}}$ for all $\varepsilon \in (0,1)$ and all $f \in \mathbf{E}$. In particular the two quantities $||\xi_1(\varepsilon A)f||_{\mathbf{E}}$ and $||\xi_2(\varepsilon A)f||_{\mathbf{E}}$ have the same behaviour as $\varepsilon \to +0$; for example $||\xi_1(\varepsilon A)f||_{\mathbf{E}} \to 0$ as $\varepsilon \to 0$ if and only if $||\xi_2(\varepsilon A)f|| \to 0$ as $\varepsilon \to 0$.

These considerations justify the following definition:

DEFINITION 3.5.9. Let $\mathbf{E}$ be a Banach space. Given a norm-continuous unital homomorphism of a LP-algebra $\mathcal{N}$ into $B(\mathbf{E})$, let $\mathbf{E}_{\mathcal{N}}$ be the set of vectors $f \in \mathbf{E}$ such that $\|\xi(\varepsilon A)f\|_{\mathbf{E}} \to 0$ as $\varepsilon \to 0$ for some (and hence each) function $\xi \in C^{\infty}(X)$ vanishing in a neighbourhood of $x = 0$ and satisfying $\xi(x) = 1$ in a neighbourhood of infinity.

It is clear that $\mathbf{E}_{\mathcal{N}}$ is a closed subspace of $\mathbf{E}$. *If the functional calculus is associated to a C_0-group W in $\mathbf{E}$, then $\mathbf{E}_{\mathcal{N}} = \mathbf{E}$ by (3.2.19).* In general $\mathbf{E}_{\mathcal{N}}$ is strictly smaller than $\mathbf{E}$. We illustrate this with the following two examples:

(A) Let $\mathbf{E} = BC(\mathbb{R})$ and $A = Q$ the operator of multiplication by the variable $x \in \mathbb{R}$. Then $\varphi(Q)$ is multiplication by φ, where $\varphi \in BC(\mathbb{R}) \equiv \mathcal{N}$. If $\xi \in C^{\infty}(\mathbb{R})$ is such that $\xi(x) = 1$ for $|x| \geq R$, then $\|\xi(\varepsilon Q)f\|_{BC} \geq \sup_{|x| > R\varepsilon^{-1}} |f(x)|$, which tends to zero as $\varepsilon \to 0$ if and only if $f \in C_{\infty}(\mathbb{R})$. So $\mathbf{E}_{\mathcal{N}} = C_{\infty}(\mathbb{R})$.

(B) We take again $\mathbf{E} = BC(\mathbb{R})$ but $A = P$, and $\mathcal{N} = \mathcal{M}_1(\mathbb{R}) \equiv \mathcal{M}_w(\mathbb{R})$ for $w(x) \equiv 1$. Then $\mathbf{E}_{\mathcal{N}}$ is the subspace of uniformly continuous functions. In fact it is easily shown that $\varphi(P)f$ is in $BC^{\infty}(\mathbb{R})$ if $\varphi \in C_0^{\infty}(\mathbb{R})$ and $f \in BC(\mathbb{R})$ (because $P^k\varphi(P)f = \varphi_k(P)f$ where $\varphi_k(x) = x^k\varphi(x)$). If $\varphi \in C_0^{\infty}(\mathbb{R})$ is such that $\varphi(x) = 1$ near $x = 0$, we may take $\xi = 1 - \varphi$ and write $f = \varphi(\varepsilon P)f + \xi(\varepsilon P)f$; then $\|\xi(\varepsilon P)f\|_{BC} \to 0$ as $\varepsilon \to 0$ implies that f is a (uniform) limit of uniformly continuous functions, so it is uniformly continuous .

Before giving the abstract Littlewood-Paley estimates, we prove a version of an inequality due to Hardy (see Theorem 319 in [HLP]):

LEMMA 3.5.10. *Let χ, h be two positive Borel functions on $(0, \infty)$ and $s \in \mathbb{R}$. For $\sigma > 0$, set*

$$h_0(\sigma) = \int_0^{\infty} \chi\left(\frac{\sigma}{\tau}\right)h(\tau)\frac{d\tau}{\tau}, \quad g(\sigma) = \sigma^{-s}h(\sigma) \text{ and } g_0(\sigma) = \sigma^{-s}h_0(\sigma).$$

Then one has for each $p \in [1, \infty]$:

$$(3.5.15) \qquad \left[\int_0^{\infty} |g_0(\sigma)|^p \frac{d\sigma}{\sigma}\right]^{1/p} \leq \int_0^{\infty} \chi(\lambda)\frac{d\lambda}{\lambda^{1+s}}\left[\int_0^{\infty} |g(\sigma)|^p \frac{d\sigma}{\sigma}\right]^{1/p}.$$

PROOF. We write L_*^p for $L^p((0, \infty); \sigma^{-1}d\sigma)$. Then

$$g_0(\sigma) = \int_0^{\infty} \left(\frac{\tau}{\sigma}\right)^s \chi\left(\frac{\sigma}{\tau}\right)\tau^{-s}h(\tau)\frac{d\tau}{\tau} = \int_0^{\infty} \lambda^{-s}\chi(\lambda)g\left(\frac{\sigma}{\lambda}\right)\frac{d\lambda}{\lambda}.$$

So, by the integral Minkowski inequality:

$$\|g_0\|_{L_*^p} \leq \int_0^{\infty} \lambda^{-s}\chi(\lambda)\left\|g\left(\frac{\cdot}{\lambda}\right)\right\|_{L_*^p}\frac{d\lambda}{\lambda} = \int_0^{\infty} \lambda^{-s}\chi(\lambda)\|g\|_{L_*^p}\frac{d\lambda}{\lambda}. \quad \square$$

THEOREM 3.5.11. *Let ξ and $\widetilde{\xi}$ be functions of class C^{∞} on X such that $\xi(x) = 0$ in a neighbourhood of the origin, $\xi(x) = 1$ in a neighbourhood of infinity and $\widetilde{\xi} \in C_0^{\infty}(X)$ with $\widetilde{\xi}(x) = 1$ inside a ball containing the support of $\nabla\xi$. Let θ be a tauberian function and let $\widetilde{\theta} \in C_0^{\infty}(X)$ be such that $\widetilde{\theta}(x) = 1$ inside a ball containing the support of θ.*

Let $\mathscr{N}$ be a LP-algebra of order r on X and let $\varphi \in \mathscr{N}$ be such that $\varphi(x) = O_{\mathscr{N}}(|x|^{\ell})$ as $x \to 0$ for some real number $\ell > r$. Then

(a) *There is a constant $c > 0$ such that for all $f \in \mathbf{E}$ and all $\varepsilon \in (0,1)$:*

(3.5.16)
$$c\|\varphi(\varepsilon A)f\|_{\mathbf{E}} \le \|\xi(\varepsilon A)f\|_{\mathbf{E}} + \int_{\varepsilon}^{1} \left(\frac{\varepsilon}{\tau}\right)^{\ell} \|\xi(\tau A)f\|_{\mathbf{E}} \frac{d\tau}{\tau} + \varepsilon^{\ell}\|\widetilde{\xi}(A)f\|_{\mathbf{E}}.$$

Furthermore, for all $f \in \mathbf{E}_{\mathscr{N}}$ and all $\varepsilon \in (0,1)$:

(3.5.17) $c\|\varphi(\varepsilon A)f\|_{\mathbf{E}} \le \int_{0}^{1} \min\left(1, \left(\frac{\varepsilon}{\tau}\right)^{\ell}\right) \|\theta(\tau A)f\|_{\mathbf{E}} \frac{d\tau}{\tau} + \varepsilon^{\ell}\|\widetilde{\theta}(A)f\|_{\mathbf{E}}.$

(b) *For each real $s < \ell$ there is a constant $c_s < \infty$ such that for all $f \in \mathbf{E}$ and all $p \in [1, +\infty]$:*

(3.5.18)
$$\left[\int_{0}^{1} \|\varepsilon^{-s}\varphi(\varepsilon A)f\|_{\mathbf{E}}^{p} \frac{d\varepsilon}{\varepsilon}\right]^{1/p} \le c_s\|\widetilde{\xi}(A)f\|_{\mathbf{E}} + c_s \left[\int_{0}^{1} \|\varepsilon^{-s}\xi(\varepsilon A)f\|_{\mathbf{E}}^{p} \frac{d\varepsilon}{\varepsilon}\right]^{1/p}.$$

If $0 < s < \ell$, there is a constant d_s such that for all $f \in \mathbf{E}_{\mathscr{N}}$ and all $p \in [1, +\infty]$:

(3.5.19)
$$\left[\int_{0}^{1} \|\varepsilon^{-s}\varphi(\varepsilon A)f\|_{\mathbf{E}}^{p} \frac{d\varepsilon}{\varepsilon}\right]^{1/p} \le d_s\|\widetilde{\theta}(A)f\|_{\mathbf{E}} + d_s \left[\int_{0}^{1} \|\varepsilon^{-s}\theta(\varepsilon A)f\|_{\mathbf{E}}^{p} \frac{d\varepsilon}{\varepsilon}\right]^{1/p}.$$

PROOF. (a) Let $\xi_1 \in C^{\infty}(X)$ be such that $\xi_1(x) = 0$ in a neighbourhood of $x = 0$, $\xi_1(x) = 1$ in a neighbourhood of infinity, $\operatorname{supp}\xi_1 \subset \{x \in X \mid \xi(x) = 1\}$ and $\operatorname{supp}\nabla\xi_1 \in \{x \in X \mid \widetilde{\xi}(x) = 1\}$. We set $\eta_1(x) = x \cdot \nabla\xi_1(x)$. Then (3.5.8) implies that

(3.5.20)
$$c\|\varphi(\varepsilon A)f\|_{\mathbf{E}} \le \|\xi_1(\varepsilon A)f\|_{\mathbf{E}} + \int_{\varepsilon}^{1} \left(\frac{\varepsilon}{\tau}\right)^{\ell} \|\eta_1(\tau A)f\|_{\mathbf{E}} \frac{d\tau}{\tau} + \varepsilon^{\ell}\|\widetilde{\xi}(A)f\|_{\mathbf{E}}.$$

Now $\xi_1 = \xi_1\xi$ and $\eta_1 = \eta_1\xi$. Hence we have for all $\varepsilon \in (0,1]$:

$$\|\xi_1(\varepsilon A)f\|_{\mathbf{E}} \le \|\xi_1(\varepsilon A)\|_{B(\mathbf{E})}\|\xi(\varepsilon A)f\|_{\mathbf{E}} \le c'\|\xi(\varepsilon A)f\|_{\mathbf{E}}.$$

By estimating similarly $\|\eta_1(\varepsilon A)f\|_{\mathbf{E}}$, one obtains (3.5.16) from (3.5.20).
(3.5.17) follows similarly from (3.5.9). We define $\xi_2 : X \to \mathbb{R}$ by

$$\xi_2(x) = \frac{1}{\varrho(x)} \int_{0}^{1} |\theta(\sigma x)|^2 \frac{d\sigma}{\sigma} = \frac{1}{\varrho(x)} \int_{0}^{|x|} \left|\theta\left(\tau \frac{x}{|x|}\right)\right|^2 \frac{d\tau}{\tau},$$

with $\varrho(x)$ given by (3.5.4), and we set $\eta_2(x) = x \cdot \nabla\xi_2(x)$. Then $\eta_2 = |\theta|^2/\varrho \equiv \theta(\overline{\theta}/\varrho)$. So (3.5.9) holds with η and $\widetilde{\xi}$ replaced by η_2 and $\overline{\theta}$ respectively, and it suffices to observe that $\overline{\theta}/\varrho \in C_0^{\infty}(X) \subset \mathscr{N}$, so that

$$\|\eta_2(\tau A)f\|_{\mathbf{E}} \le c''\|\theta(\tau A)f\|_{\mathbf{E}} \quad \text{for } \tau \in (0,1].$$

(b) The estimate (3.5.18) is a consequence of (3.5.16). This is easily seen by applying Lemma 3.5.10 with $h(\tau) = \|\xi(\tau A)f\|_{\mathbf{E}}$ if $\tau \le 1$, $h(\tau) = 0$ if $\tau > 1$ and $\chi(\tau) = \tau^\ell \chi_{[0,1]}(\tau)$ (where $\chi_{[0,1]}$ denotes the characteristic function of the interval $[0,1]$) and by observing that

$$\int_0^\infty \lambda^\ell \chi_{[0,1]}(\lambda) \frac{d\lambda}{\lambda^{1+s}} = \int_0^1 \lambda^{\ell-s-1} d\lambda = \frac{1}{\ell-s} \quad \text{if } s < \ell.$$

Similarly (3.5.19) can be obtained from (3.5.17) by using Lemma 3.5.10 with $h(\tau) = \|\theta(\tau A)f\|_{\mathbf{E}}$ if $\tau \le 1$, $h(\tau) = 0$ if $\tau > 1$ and $\chi(\tau) = \min(1, \tau^\ell)$, since

$$\int_0^\infty \min(1, \lambda^\ell) \frac{d\lambda}{\lambda^{1+s}} = \frac{1}{\ell-s} + \frac{1}{s} \quad \text{if } 0 < s < \ell. \quad \square$$

The techniques developed in this section will be used in the next one in the framework of C_0-representations. We now present an application of a slightly different nature which will be important in Chapter 7.

Let A be a closed (not necessarily densely defined) operator in the Banach space $\mathbf{E}$ and let $r \ge 0$ be a real number. We say that A has a (LP)-functional calculus of order r if there are a LP-algebra $\mathscr{N}$ of order r on $\mathbb{R}$, a norm-continuous unital homomorphism $\Phi : \mathscr{N} \to B(\mathbf{E})$ and a real number $\lambda > 0$ such that: (i) $A - i\lambda : D(A) \to \mathbf{E}$ is bijective; (ii) if $\varrho_\lambda(x) = (x - \lambda i)^{-1}$, then $\varrho_\lambda \in \mathscr{N}$ and $\Phi(\varrho_\lambda) = (A - \lambda i)^{-1}$. Observe that $(A - \lambda i)^{-1} \in B(\mathbf{E})$ by the closed graph theorem. We shall use the notation $\Phi(\varphi) = \varphi(A)$ for any $\varphi \in \mathscr{N}$, if there is no danger of confusion. If A is the generator of a C_0-group such that $\|e^{iAx}\|_{B(\mathbf{E})} \le M\langle x \rangle^r$, then it clearly has these properties (take $\mathscr{N} = \mathscr{M}^r(\mathbb{R})$); but often it is easier to associate to A a (LP)-functional calculus than to show that it generates a group.

We keep the assumptions of the definition and derive some consequences. First we show that $\varrho_\lambda \in \mathscr{N}$ for all $\lambda > 0$ and $\|\varrho_\lambda\|_{\mathscr{N}} \le c\lambda^{-1-r}$ if $0 < \lambda \le 1$, $\|\varrho_\lambda\|_{\mathscr{N}} \le c\lambda^{-1}$ if $\lambda \ge 1$, where c is a finite constant. For this we use the fact that $\varrho_\lambda^\sigma \in \mathscr{N}$ if $0 < \sigma < \infty$ and the obvious identity $\varrho_\lambda^\sigma = \sigma^{-1} \varrho_{\lambda\sigma^{-1}}$, in particular $\varrho_\lambda = \lambda^{-1} \varrho_1^{\lambda^{-1}}$. Now we prove that $A - \lambda i : D(A) \to \mathbf{E}$ is bijective for all $\lambda > 0$ and that $\|(A - \lambda i)^{-1}\|_{B(\mathbf{E})} \le c \max(\lambda^{-1-r}, \lambda^{-1})$ for (another) constant c. We know that this holds for some $\lambda > 0$. For any $\mu > 0$, let $R_\mu = \Phi(\varrho_\mu)$, in particular $R_\lambda = (A - \lambda i)^{-1}$. From $\varrho_\lambda - \varrho_\mu = (\lambda - \mu)i\varrho_\lambda \varrho_\mu$ we get $R_\mu = R_\lambda[I + i(\mu - \lambda)R_\mu]$, hence the range of R_μ is included in $R_\lambda \mathbf{E} = D(A)$ and $(A - \mu i)R_\mu = (A - \lambda i)R_\lambda[I + i(\mu - \lambda)R_\mu] + (\lambda i - \mu i)R_\mu = I$. Similarly $R_\mu = [I + i(\mu - \lambda)R_\mu]R_\lambda$ implies $R_\mu(A - \mu i)f = f$ if $f \in D(A)$. Hence $R_\mu = (A - \mu i)^{-1}$ and the norm-continuity of Φ gives the desired estimate.

Observe that $x(x - \lambda i)^{-1} = 1 + i\lambda(x - \lambda i)^{-1}$ is a function in $\mathscr{N}$; since $A(A - \lambda i)^{-1} = I + i\lambda(A - \lambda i)^{-1}$, the image through Φ of the preceding function is just $A(A - \lambda i)^{-1}$. But $\mathscr{N}$ is an algebra, hence $x \mapsto x^\ell(x - \lambda i)^{-\ell}$ is a function in $\mathscr{N}$ for any $\ell \in \mathbb{N}$, and its image through Φ is $[A(A - \lambda i)^{-1}]^\ell$. According to Example 3.5.6, the function $x^\ell(x - \lambda i)^{-\ell}$ has a zero of order ℓ at the origin in the $\mathscr{N}$-sense. Hence, if $\ell > r$, we can apply Theorem 3.5.11 (a), in which we set $\varepsilon = \lambda^{-1}$ (notice then $[\varepsilon A(\varepsilon A - i)^{-1}]^\ell = [A(A - \lambda i)^{-1}]^\ell$). So, if $\xi \in C^\infty(\mathbb{R})$, $\xi(x) = 0$ (resp. 1) near zero (resp. ∞) and if $\tilde{\xi} \in C_0^\infty(\mathbb{R})$ with $\tilde{\xi}(x) = 1$ on an interval containing supp ξ', then there is a constant $0 < c < \infty$ such that for all $f \in \mathbf{E}$ and $\lambda \geq 1$:

(3.5.21)

$$c||[A(A - \lambda i)^{-1}]^\ell f||_{\mathbf{E}} \leq \left\|\xi\left(\frac{A}{\lambda}\right)f\right\|_{\mathbf{E}} + \int_1^\lambda \left(\frac{\tau}{\lambda}\right)^\ell \left\|\xi\left(\frac{A}{\tau}\right)f\right\|_{\mathbf{E}} \frac{d\tau}{\tau} + \frac{1}{\lambda^\ell}||\widetilde{\xi}(A)f||_{\mathbf{E}}.$$

If $||\xi(\frac{A}{\lambda})f||_{\mathbf{E}} \to 0$ as $\lambda \to \infty$, then we have the better estimate:

(3.5.22)

$$c||[A(A - \lambda i)^{-1}]^\ell f||_{\mathbf{E}} \leq \int_1^\infty \min\left(1, \left(\frac{\tau}{\lambda}\right)^\ell\right)\left\|\theta\left(\frac{A}{\tau}\right)f\right\|_{\mathbf{E}} \frac{d\tau}{\tau} + \frac{1}{\lambda^\ell}||\widetilde{\theta}(A)f||_{\mathbf{E}}.$$

Here $\theta \in C_0^\infty(\mathbb{R})$ is an arbitrary function equal to zero in a neighbourhood of zero, but such that $\theta(x_\pm) \neq 0$ for some $x_+ > 0$ and some $x_- < 0$; and $\widetilde{\theta} \in C_0^\infty(\mathbb{R})$ has to be equal to 1 on an interval containing supp θ.

Assume that the function $x \mapsto \xi(x)x^{-1}$ belongs to $\mathcal{N}$ (this is a rather trivial assumption); then $||\xi(\frac{A}{\lambda})f||_{\mathbf{E}} \to 0$ as $\lambda \to \infty$, for all $f \in \overline{D(A)}$ = closure of the domain of A. In fact, we may write $\xi(x) = \eta(x)x(x - i)^{-1}$ with $\eta \in \mathcal{N}$, hence $||\xi(\varepsilon A)f||_{\mathbf{E}} = ||\eta(\varepsilon A)\varepsilon A(\varepsilon A - i)^{-1}f||_{\mathbf{E}} \leq c||\varepsilon A(\varepsilon A - i)^{-1}f||_{\mathbf{E}}$ if $0 < \varepsilon \leq 1$. Then, if $f \in D(A)$, we have $A(A - \lambda i)^{-1}f = (A - \lambda i)^{-1}Af \to 0$ as $\lambda \to \infty$.

We shall now deduce from (3.5.21) and (3.5.22) two estimates which will be needed in Chapter 7. Let $\mathbf{F}$ be a second Banach space and $T \in B(\mathbf{F}, \mathbf{E})$. We write $|| \cdot ||$ for the norm in $B(\mathbf{F}, \mathbf{E})$ and clearly have:

(3.5.23)

$$c||[A(A - \lambda i)^{-1}]^\ell T|| \leq \left\|\xi\left(\frac{A}{\lambda}\right)T\right\| + \int_1^\lambda \left(\frac{\tau}{\lambda}\right)^\ell \left\|\xi\left(\frac{A}{\tau}\right)T\right\| \frac{d\tau}{\tau} + \frac{1}{\lambda^\ell}||\widetilde{\xi}(A)T||$$

(3.5.24)

$$c||[A(A - \lambda i)^{-1}]^\ell T|| \leq \int_1^\infty \min\left(1, \left(\frac{\tau}{\lambda}\right)^\ell\right)\left\|\theta\left(\frac{A}{\tau}\right)T\right\| \frac{d\tau}{\tau} + \frac{1}{\lambda^\ell}||\widetilde{\theta}(A)T||.$$

All the terms involved in these formulas are continuous functions of λ and τ. For each $s \in [0, \ell)$ we clearly get:

(3.5.25) $\quad c\displaystyle\int_1^\infty \lambda^s ||[A(A - \lambda i)^{-1}]^\ell T|| \frac{d\lambda}{\lambda} \leq$

$$\leq \frac{\ell - s + 1}{\ell - s}\int_1^\infty \lambda^s \left\|\xi\left(\frac{A}{\lambda}\right)T\right\| \frac{d\lambda}{\lambda} + \frac{1}{\ell - s}||\widetilde{\xi}(A)T||$$

(3.5.26) $\quad c\displaystyle\int_1^\infty \lambda^s ||[A(A - \lambda i)^{-1}]^\ell T|| \frac{d\lambda}{\lambda} \leq$

$$\leq \int_1^\infty \left[\frac{\lambda^s - 1}{s} + \frac{\lambda^s}{\ell - s}\right]\left\|\theta\left(\frac{A}{\lambda}\right)T\right\| \frac{d\lambda}{\lambda} + \frac{1}{\ell - s}||\widetilde{\theta}(A)T||.$$

If $s = 0$, the term $(\lambda^s - 1)s^{-1}$ in (3.5.26) has to be replaced by $\ln s$. Moreover, the estimates (3.5.24) and (3.5.26) hold only if the range of T is included in the set of $f \in \mathbf{E}$ such that $\xi(\frac{A}{\lambda})f \to 0$ as $\lambda \to \infty$.

The choice of the integer ℓ in the last two estimates is dictated not only by the condition $s < \ell$ but also by our initial assumption $\ell > r$. Since r could be quite large, this is often inconvenient in concrete calculations. Fortunately, the second condition can be eliminated by using the next result, due to Komatsu (cf. Proposition 1.2 in II of [Km]):

LEMMA 3.5.12. *Let A be a closed densely defined operator in the Banach space* **E** *such that for each $\lambda > 0$ the number $i\lambda$ belongs to the resolvent set of A and such that $||(A - i\lambda)^{-1}||_{B(\mathbf{E})} \leq c\lambda^{-1}$ for all $\lambda > 0$ and some finite constant c. Let $T \in B(\mathbf{F}, \mathbf{E})$ and $s \geq 0$ real. If $\int_1^\infty \lambda^s ||[A(A - i\lambda)^{-1}]^\ell T|| \frac{d\lambda}{\lambda} < \infty$ for some integer $\ell > s$, then this holds for each integer $\ell > s$.*

Before giving the proof, we make some comments. Since

$$||A(A - i\lambda)^{-1}||_{B(\mathbf{E})} = ||I + i\lambda(A - i\lambda)^{-1}||_{B(\mathbf{E})} \leq \text{const.},$$

the finiteness of the integral for $\ell = m$ implies its finiteness for all $\ell \geq m$. Hence the non-trivial fact is that from a large ℓ we can go to a smaller ℓ, namely to the smallest integer $> s$. The non-triviality of the assertion can be clarified if one uses $\varepsilon = \lambda^{-1}$ as variable: then $A(A - i\lambda)^{-1} = \varepsilon A(\varepsilon A - i)^{-1}$, and the point is that the function $[x(x - i)^{-1}]^\ell$ vanishes at zero more rapidly than $[x(x - i)^{-1}]^{\ell-1}$ (here $\ell \geq 2$), so it is not a priori obvious that the finiteness of the integral for some ℓ implies its finiteness for $\ell - 1$. So, like Theorem 3.5.11, this lemma has a tauberian character.

Observe that we assumed $||(A - i\lambda)^{-1}||_{B(\mathbf{E})} \leq c\lambda^{-1}$ even if $0 < \lambda < 1$. Now, if A has a (LP)-functional calculus of order r, then as $\lambda \to 0$ only an estimate $O(\lambda^{-1-r})$ is a priori available. However, this assumption is easily satisfied and we give now a typical example. Let **E** be the Sobolev space $\mathscr{H}^r(\mathbb{R})$ and A the operator of multiplication by $\langle x \rangle = \sqrt{1 + x^2}$. It is trivial to check that $||(A - i\lambda)^{-1}||_{B(\mathbf{E})} \leq c(1 + \lambda)^{-1}$ for all $\lambda > 0$. However $||e^{iAt}||_{B(\mathbf{E})} \leq c\langle t \rangle^r$ and this is optimal. One may also consider directly the functional calculus associated to A: for any $\varphi \in BC^r(\mathbb{R})$ (here we assume $r \in \mathbb{N}$ for simplicity), $\varphi(A)$ can be defined as multiplication by $\varphi(\langle x \rangle)$. The estimate $||\varphi(\sigma A)||_{B(\mathbf{E})} \leq c\langle \sigma \rangle^r$ cannot be improved (as $\sigma \to \infty$) in the class $BC^r(\mathbb{R})$, but we have $||\varphi(\sigma A)||_{B(\mathbf{E})} \leq c$ ($\forall \sigma > 0$) if $\varphi \in S^0(\mathbb{R})$. This shows the advantage of choosing a smaller (LP)-algebra for the functional calculus. Now let A be multiplication by x (in the same space). Then $||(A - i\lambda)^{-1}||_{B(\mathbf{E})} \leq c \max(\lambda^{-1-r}, \lambda^{-1})$ and this can not be improved; moreover, $||\varphi(\sigma A)||_{B(\mathbf{E})}$ can be made of order σ^r as $\sigma \to \infty$ even if $\varphi \in S^0(\mathbb{R})$.

PROOF OF LEMMA 3.5.12. By using the identity $\lambda A(A - i\lambda)^{-1} = \lambda + i\lambda^2(A - i\lambda)^{-1}$, it is easy to prove that for any $\lambda > 0$ and $\ell \in \mathbb{N}$:

$$\frac{d}{d\lambda}[\lambda A(A - i\lambda)^{-1}]^\ell = \frac{\ell}{\lambda^2}[\lambda A(A - i\lambda)^{-1}]^{\ell+1}.$$

The hypothesis $\lambda ||(A - i\lambda)^{-1}|| \leq c$ if $\lambda > 0$ implies that $||\lambda A(A - i\lambda)^{-1}|| \to 0$ as $\lambda \to 0$. Hence for $\ell \geq 1$:

$$[A(A - i\lambda)^{-1}]^\ell = \frac{\ell}{\lambda^\ell} \int_0^\lambda t^\ell [A(A - it)^{-1}]^{\ell+1} \frac{dt}{t}$$

with a norm-convergent integral. This implies for $0 \leq s < \ell$:

$$\int_1^\infty \lambda^s \||[A(A-i\lambda)^{-1}]^\ell T\|| \frac{d\lambda}{\lambda} \leq \frac{\ell}{\ell-s} \int_0^1 t^\ell \||[A(A-it)^{-1}]^{\ell+1} T\|| \frac{dt}{t}$$

$$+ \frac{\ell}{\ell-s} \int_1^\infty t^s \||[A(A-it)^{-1}]^{\ell+1} T\|| \frac{dt}{t},$$

which easily gives the assertion of the lemma. □

3.6. Polynomially Bounded C_0-Groups

In this section we consider a fixed C_0-representation W of $\mathbb{R}^n$ in a Banach space $\mathbf{F}$ and we assume that it is of polynomial growth, more precisely that there are constants $M \geq 1$ and $r \geq 0$ such that

(3.6.1) $\|W(x)\|_{B(\mathbf{F})} \leq M\langle x\rangle^r \qquad \forall x \in \mathbb{R}^n.$

We saw in Section 3.2 that a rather rich functional calculus can be associated to such a representation: $\varphi(A)$ is well defined for any $\varphi \in \mathcal{M}^r(\mathbb{R}^n)$, and the LP-algebra $\mathcal{N} \equiv \mathcal{M}^r(\mathbb{R}^n)$ is quite large, e.g. it contains $\mathcal{S}(\mathbb{R}^n)$, in particular all functions of class $C_0^\infty(\mathbb{R}^n)$. The operators $\varphi(A)$, for $\varphi \in \mathcal{M}^r(\mathbb{R}^n)$, are bounded in $\mathbf{F}$. One of the purposes here is to show that one may naturally define $\varphi(A)$ as a densely defined unbounded operator in $\mathbf{F}$ for any $\varphi \in C_{\text{pol}}^\infty(\mathbb{R}^n)$. The properties of these operators will be determined in §3.6.2.

In the first part of this section we use the general Littlewood-Paley estimates of Section 3.5 to obtain new admissible norms on the Besov spaces $\mathbf{F}_{s,p}$, $s > 0$, associated to polynomially bounded C_0-groups. In the last part of this section (§3.6.3) we consider the interpolation spaces $\mathbf{F}_{s,p}$ for negative values of s and show that various results obtained for $s > 0$ remain valid for $s < 0$ if $\mathbf{F}$ is a reflexive Banach space or if W is a one-parameter C_0-group.

3.6.1. We first indicate how the Littlewood-Paley estimates of Theorem 3.5.11 lead to new expressions for the K-functional for the couple $(\mathbf{F}_m, \mathbf{F})$ with $m > r$. We use the LP-algebra $\mathcal{N} = \mathcal{M}^r(\mathbb{R}^n)$.

LEMMA 3.6.1. *Let $m > r$ be an integer and denote by $K(\cdot, \cdot)$ the K-functional for the couple $(\mathbf{F}_m, \mathbf{F})$. Let $\theta \in C_0^\infty(\mathbb{R}^n)$ be a tauberian function, in particular $0 \notin \text{supp}\,\theta$. Then there is a constant $c > 1$ such that for all $\varepsilon \in (0,1)$ and all $f \in \mathbf{F}$:*

(3.6.2) $c^{-1}[\varepsilon^m \|f\|_{\mathbf{F}} + \|\theta(\varepsilon A)f\|_{\mathbf{F}}] \leq K(\varepsilon^m, f)$

$$\leq c\varepsilon^m \|f\|_{\mathbf{F}} + c\int_0^1 \min\left(1, \left(\frac{\varepsilon}{\tau}\right)^m\right) \|\theta(\tau A)f\|_{\mathbf{F}} \frac{d\tau}{\tau}.$$

PROOF. Let $\psi : \mathbb{R}^n \to \mathbb{R}$ be the function $\psi(x) = \exp(-x^{2m})$. We have $\psi \in \mathcal{S}(\mathbb{R}^n)$, $\psi(0) = 1$ and $\psi^{(\beta)}(0) = 0$ if $1 \leq |\beta| < 2m$. Hence, by (3.4.2):

(3.6.3) $K(\varepsilon^m, f) \sim \varepsilon^m \|\psi(\varepsilon A)f\|_{\mathbf{F}_m} + \|[\psi(\varepsilon A) - I]f\|_{\mathbf{F}}.$

(i) From (2.2.3) we see that $\varepsilon^m \|f\|_{\mathbf{F}} \leq c_1 K(\varepsilon^m, f)$. Since $0 \notin \text{supp}\,\theta$, we have $1 - \psi(x) > 0$ on $\text{supp}\,\theta$. Thus, if we set $\varrho = \theta/(1-\psi)$, then $\varrho \in C_0^\infty(\mathbb{R}^n)$.

Hence $||\varrho(\varepsilon A)||_{B(\mathbf{F})} \le c_2 < \infty$ for all $\varepsilon \in (0,1]$ by (3.2.19). So we obtain the first inequality in (3.6.2) after observing that for $\varepsilon \in (0,1]$:

$$||\theta(\varepsilon A)f||_{\mathbf{F}} \le ||\varrho(\varepsilon A)||_{B(\mathbf{F})}||[I - \psi(\varepsilon A)]f||_{\mathbf{F}} \le c_2||[I - \psi(\varepsilon A)]f||_{\mathbf{F}} \le c_3 K(\varepsilon^m, f).$$

(ii) We have $\psi(x) - 1 = O_{\mathscr{M}^r(\mathbb{R}^n)}(|x|^m)$ as $x \to 0$ by Example 3.5.7. So (3.5.17) applied to the function $\varphi = \psi - 1$ shows that $||[\psi(\varepsilon A) - I]f||_{\mathbf{F}}$ is majorized by the r.h.s. of (3.6.2).

By using Proposition 3.3.18 (b) and (3.2.19) we have for $\varepsilon \in (0,1]$:

$$\varepsilon^m ||\psi(\varepsilon A)f||_{\mathbf{F}_m} \le c_4 \varepsilon^m ||\psi(\varepsilon A)f||_{\mathbf{F}} + c_4 \varepsilon^m \sum_{|\alpha|=m} ||A^\alpha \psi(\varepsilon A)f||_{\mathbf{F}}$$

$$\le c_5 \varepsilon^m ||f||_{\mathbf{F}} + c_4 \sum_{|\alpha|=m} ||\psi_{(\alpha)}(\varepsilon A)f||_{\mathbf{F}}.$$

Again Example 3.5.7 shows that $\psi_{(\alpha)}(x) = O_{\mathscr{M}^r(\mathbb{R}^n)}(|x|^m)$ as $x \to 0$; by using (3.5.17) for the functions $\varphi = \psi_{(\alpha)}(x)$, one finds that $\varepsilon^m ||\psi(\varepsilon A)f||_{\mathbf{F}_m}$ is bounded by the r.h.s. of (3.6.2). $\square$

THEOREM 3.6.2. *Let $\mathbf{F}$ be a Banach space equipped with a polynomially bounded C_0-representation of $\mathbb{R}^n$. Let $\theta \in C_0^\infty(\mathbb{R}^n)$ be a tauberian function (hence $0 \notin \operatorname{supp}\theta$) and let $\widetilde{\theta} \in C_0^\infty(\mathbb{R}^n)$ be such that $\widetilde{\theta}(x) = 1$ inside a ball containing the support of θ. Let $s > 0$ and $p \in [1, +\infty]$. Then*

(a) *The following expression defines an admissible norm on $\mathbf{F}_{s,p}$:*

$$(3.6.4) \qquad ||\widetilde{\theta}(A)f||_{\mathbf{F}} + \left[\int_0^1 ||\tau^{-s}\theta(\tau A)f||_{\mathbf{F}}^p \frac{d\tau}{\tau}\right]^{1/p}.$$

(b) *Let $a > 1$ and assume that $\theta(x) > 0$ if $a^{-1} < |x| < a$ and $\theta(x) = 0$ otherwise. Then the following expression is an admissible norm on $\mathbf{F}_{s,p}$:*

$$(3.6.5) \qquad ||\widetilde{\theta}(A)f||_{\mathbf{F}} + \left[\sum_{j=0}^\infty ||a^{js}\theta(a^{-j}A)f||_{\mathbf{F}}^p\right]^{1/p}.$$

PROOF. (a) We fix an integer $m > \max(r,s)$ and denote by $K(\cdot,\cdot)$ the K-functional for the couple $(\mathbf{F}_m, \mathbf{F})$. By using the first inequality in (3.6.2) we get that

$$||\widetilde{\theta}(A)f||_{\mathbf{F}} + \left[\int_0^1 ||\tau^{-s}\theta(\tau A)f||_{\mathbf{F}}^p \frac{d\tau}{\tau}\right]^{1/p} \le$$

$$\le c_1||f||_{\mathbf{F}} + c\left[\int_0^1 \left|\frac{K(\tau^m, f)}{\tau^s}\right|^p \frac{d\tau}{\tau}\right]^{1/p} \le c_2||f||_{\mathbf{F}_{s,p}} \qquad \forall f \in \mathbf{F}.$$

To obtain the opposite inequality, we use the second inequality in (3.6.2) to get that

$$||f||_{\mathbf{F}_{s,p}} = m^{1/p}\left[\int_0^1 \left|\frac{K(\varepsilon^m, f)}{\varepsilon^s}\right|^p \frac{d\varepsilon}{\varepsilon}\right]^{1/p} \leq cm^{1/p}\left[\int_0^1 \varepsilon^{(m-s)p}\frac{d\varepsilon}{\varepsilon}\right]^{1/p}||f||_{\mathbf{F}} +$$

$$+ cm^{1/p}\left[\int_0^1 \frac{d\varepsilon}{\varepsilon}\left|\varepsilon^{-s}\int_0^1 \min\left(1, \left(\frac{\varepsilon}{\tau}\right)^m\right)||\theta(\tau A)f||_{\mathbf{F}}\frac{d\tau}{\tau}\right|^p\right]^{1/p}.$$

The second term on the r.h.s. can be majorized by applying Lemma 3.5.10 exactly as we did in the passage from (3.5.17) to (3.5.19) (see the last paragraph in the proof of Theorem 3.5.11). One gets that

$$||f||_{\mathbf{F}_{s,p}} \leq c_3||\widetilde{\theta}(A)f||_{\mathbf{F}} + c_3||[I - \widetilde{\theta}(A)]f||_{\mathbf{F}} + c_3\left[\int_0^1 ||\varrho^{-s}\theta(\varrho A)f||_{\mathbf{F}}^p \frac{d\varrho}{\varrho}\right]^{1/p}.$$

The r.h.s. of this last inequality is of the form (3.6.4) except for the additional term $c_3||[I - \widetilde{\theta}(A)]f||_{\mathbf{F}}$; to see that this term can be majorized by a multiple of (3.6.4) we first apply (3.5.17) with $\varphi = 1 - \widetilde{\theta}$ and $\varepsilon = 1$, and then we use the Hölder inequality ($p' \equiv (1 - p^{-1})^{-1}$):

$$||[I - \widetilde{\theta}(A)]f||_{\mathbf{F}} \leq c_4||\widetilde{\theta}(A)f||_{\mathbf{F}} + c_4\int_0^1 ||\theta(\tau A)f||_{\mathbf{F}}\frac{d\tau}{\tau}$$

$$\leq c_4||\widetilde{\theta}(A)f||_{\mathbf{F}} + c_4\left[\int_0^1 \tau^{p's}\frac{d\tau}{\tau}\right]^{1/p'}\left[\int_0^1 ||\tau^{-s}\theta(\tau A)f||_{\mathbf{F}}^p\frac{d\tau}{\tau}\right]^{1/p}.$$

(b) For $j \in \mathbb{Z}$, define θ_j by $\theta_j(x) = \theta(a^j x)$. We have $\operatorname{supp}\theta_j = \{x \in \mathbb{R}^n \mid a^{-j-1} \leq |x| \leq a^{-j+1}\}$, $\theta_j(x) > 0$ if x is in the interior of $\operatorname{supp}\theta_j$, $\operatorname{supp}\theta_j \cap \operatorname{supp}\theta_k = \varnothing$ if $|j - k| \geq 2$ and $\operatorname{supp}\theta_j \cap \operatorname{supp}\theta_{j+1} = \{x \in \mathbb{R}^n \mid a^{-j-1} \leq |x| \leq a^{-j}\}$. So, if η is defined as $\eta(x) = \theta(x)\left[\sum_{j\in\mathbb{Z}}\theta_j(x)\right]^{-1}$ for $x \neq 0$ and $\eta(0) = 0$, then $\eta \in C_0^\infty(\mathbb{R}^n)$, $\operatorname{supp}\eta = \operatorname{supp}\theta$ and $\sum_{k\in\mathbb{Z}}\eta(a^k x) = 1$ for all $x \neq 0$. Moreover there are functions $\xi_1, \xi_2 \in C_0^\infty(\mathbb{R}^n)$ such that $\eta = \xi_1\theta$ and $\theta = \xi_2\eta$. By (3.2.19) there are constants c_1, c_2 such that for all $j \geq 0$:

$$||\eta(a^{-j}A)f||_{\mathbf{F}} \leq ||\xi_1(a^{-j}A)||_{B(\mathbf{F})}||\theta(a^{-j}A)f||_{\mathbf{F}} \leq c_1||\theta(a^{-j}A)f||_{\mathbf{F}}$$

and $||\theta(a^{-j}A)f||_{\mathbf{F}} \leq c_2||\eta(a^{-j}A)f||_{\mathbf{F}}$. Thus it suffices to show that that the expression (3.6.5) with θ replaced by η or, equivalently, under the assumption that $\sum_{k\in\mathbb{Z}}\theta(a^k x) = 1$ for all $x \neq 0$, defines an admissible norm on $\mathbf{F}_{s,p}$.

Now let $k \in \mathbb{N}$ and $\tau \in (a^{-k-1}, a^{-k})$. Then $\operatorname{supp}\theta^\tau \subset \operatorname{supp}\theta_{-k} \cap \operatorname{supp}\theta_{-k-1}$, hence

$$\theta^\tau(x) \equiv \theta(\tau x) = \sum_{j\in\mathbb{Z}}\theta(\tau x)\theta_j(x) = \theta(\tau x)\theta_{-k}(x) + \theta(\tau x)\theta_{-k-1}(x).$$

So we have (we consider the case $p < \infty$)

$$\int_0^1 ||\tau^{-s}\theta(\tau A)f||_{\mathbf{F}}^p \frac{d\tau}{\tau} = \sum_{k=0}^{\infty} \int_{a^{-k-1}}^{a^{-k}} ||\tau^{-s}\theta(\tau A)f||_{\mathbf{F}}^p \frac{d\tau}{\tau}$$

$$\leq c_3 \sum_{k=0}^{\infty} a^{(k+1)sp}[||\theta(a^{-k}A)f||_{\mathbf{F}}^p + ||\theta(a^{-k-1}A)f||_{\mathbf{F}}^p]$$

$$\leq c_3(1 + a^{sp}) \sum_{k=0}^{\infty} ||a^{ks}\theta(a^{-k}A)f||_{\mathbf{F}}^p.$$

This implies that the norm (3.6.4) is weaker than that given by (3.6.5). To arrive at the opposite conclusion, we observe that supp $\theta^\sigma \subset [a^{-2}, a]$ for each $\sigma \in [1, a]$ and that the function $\varphi \equiv \theta^{a^2} + \theta^a + \theta + \theta^{1/a}$ is strictly positive on $[a^{-2}, a]$. Let $\psi \in \mathscr{S}(\mathbb{R})$ be such that $\psi(x) = 1/\varphi(x)$ for all $x \in [a^{-2}, a]$. Then by (3.2.20) there is a constant c_4 such that for all $\sigma \in [1, a]$, all $\tau \in (0, 1]$ and all $f \in \mathbf{F}$:

(3.6.6) $\quad ||\theta^\sigma(\tau A)f||_{\mathbf{F}} \leq ||\theta^\sigma(\tau A)||_{B(\mathbf{F})} \cdot ||\psi(\tau A)||_{B(\mathbf{F})} \cdot ||\varphi(\tau A)f||_{\mathbf{F}}$

$\qquad\qquad\qquad\qquad \leq c_4||\varphi(\tau A)f||_{\mathbf{F}}.$

Now let again $k \in \mathbb{N}$ and $\tau \in (a^{-k-1}, a^{-k})$. Set $\sigma = \tau^{-1}a^{-k}$; then $\sigma \in [1, a]$ and $\theta^\sigma(\tau x) = \theta(a^{-k}x)$. Hence, by (3.6.6):

(3.6.7) $\quad ||\theta(a^{-k}A)f||_{\mathbf{F}} = ||\theta^\sigma(\tau A)f||_{\mathbf{F}} \leq c_4 \sum_{j=-1}^{2} ||\theta(a^j\tau A)f||_{\mathbf{F}}.$

So (we again consider the case $p < \infty$):

$$\sum_{k=0}^{\infty} a^{ksp}||\theta(a^{-k}A)f||_{\mathbf{F}}^p = \frac{1}{\ln a} \sum_{k=0}^{\infty} a^{ksp} \int_{a^{-k-1}}^{a^{-k}} ||\theta(a^{-k}A)f||_{\mathbf{F}}^p \frac{d\tau}{\tau}$$

$$\leq c_5 \sum_{k=0}^{\infty} \sum_{j=-1}^{2} \int_{a^{-k-1}}^{a^{-k}} \tau^{-sp}||\theta(a^j\tau A)f||_{\mathbf{F}}^p \frac{d\tau}{\tau}$$

$$\leq c_5 a^{2sp} \sum_{k=0}^{\infty} \sum_{j=-1}^{2} \int_{a^{j-k-1}}^{a^{j-k}} \varrho^{-sp}||\theta(\varrho A)f||_{\mathbf{F}}^p \frac{d\varrho}{\varrho}$$

$$\leq c_6 \int_0^1 ||\varrho^{-s}\theta(\varrho A)f||_{\mathbf{F}}^p \frac{d\varrho}{\varrho} + c_6 \int_1^{a^2} ||\theta(\varrho A)f||_{\mathbf{F}}^p d\varrho.$$

The fact that the norm (3.6.5) is weaker than that given by (3.6.4) now follows after observing that there is a constant c_7 such that

$$||\theta(\varrho A)f||_{\mathbf{F}} \leq ||\theta(\varrho A)||_{B(\mathbf{F})}||\tilde{\theta}(A)f||_{\mathbf{F}} \leq c_7||\tilde{\theta}(A)f||_{\mathbf{F}} \qquad \forall \varrho \in [1, a^2]. \quad \square$$

3.6.2. We now develop a functional calculus for the class $C_{\mathrm{pol}}^\infty(\mathbb{R}^n)$. We assume that W satisfies (3.6.1). If $s \in \mathbb{R}$ and $k \in \mathbb{N}$, we denote by $C_{(s)}^k(\mathbb{R}^n)$ the Banach space of all functions φ in $C^k(\mathbb{R}^n)$ for which the following expression is finite:

$$(3.6.8) \qquad \|\varphi\|_{C_{(s)}^k} := \sup_{x \in \mathbb{R}^n} \langle x \rangle^s \sum_{|\alpha| \leq k} |\varphi^{(\alpha)}(x)|.$$

We denote by $\overset{\circ}{C}_{(s)}^k(\mathbb{R}^n)$ the completion of $\mathscr{S}(\mathbb{R}^n)$ in $C_{(s)}^k(\mathbb{R}^n)$. One may check without difficulty that a function φ of class C^k on $\mathbb{R}^n$ belongs to $\overset{\circ}{C}_{(s)}^k(\mathbb{R}^n)$ if and only if $\langle x \rangle^s \varphi^{(\alpha)}(x) \to 0$ as $|x| \to \infty$ for each multi-index α with $|\alpha| \leq k$. In particular

$$C_{(s)}^k(\mathbb{R}^n) \subset \overset{\circ}{C}_{(s')}^k(\mathbb{R}^n) \text{ if } s' < s \text{ and } C_{\mathrm{pol}}^k(\mathbb{R}^n) = \bigcup_{s \in \mathbb{R}} \overset{\circ}{C}_{(s)}^k(\mathbb{R}^n).$$

The definition of $\varphi(A)$ for $\varphi \in C_{\mathrm{pol}}^\infty(\mathbb{R}^n)$ will be based on the following fact (we use the notation $\underline{n} = [n/2] + 1$):

LEMMA 3.6.3. *Let* k, $\ell \geq 0$ *be even integers. Then there is a constant* c *such that for all* ξ, $\eta \in \mathscr{S}(\mathbb{R}^n)$:

$$(3.6.9) \qquad |\langle \xi, \eta \rangle| \leq c \|\xi\|_{C_{(\underline{n}-k)}^\ell} \|\breve{\eta}\|_{C_{(\underline{n}-\ell)}^k}.$$

PROOF. We have

$$|\langle \xi, \eta \rangle| \leq \|\langle Q \rangle^{-\ell} \langle P \rangle^k \xi\|_{L^2(\mathbb{R}^n)} \|\langle Q \rangle^\ell \langle P \rangle^{-k} \eta\|_{L^2(\mathbb{R}^n)}$$
$$= \|\langle Q \rangle^{-\ell} \langle P \rangle^k \xi\|_{L^2(\mathbb{R}^n)} \|\langle P \rangle^\ell \langle Q \rangle^{-k} \breve{\eta}\|_{L^2(\mathbb{R}^n)}$$
$$\leq c \|\langle Q \rangle^{-\ell} \langle P \rangle^k \xi\|_{L^2(\mathbb{R}^n)} \|\langle Q \rangle^{-k} \langle P \rangle^\ell \breve{\eta}\|_{L^2(\mathbb{R}^n)},$$

where the last inequality is obvious because $\langle P \rangle^\ell$ is a differential operator if ℓ is even. Now (3.6.9) is obtained by observing that, for $s \in \mathbb{R}$ and even $m \in \mathbb{N}$:

$$\|\langle Q \rangle^{-s} \langle P \rangle^m \psi\|_{L^2(\mathbb{R}^n)} \leq \|\langle \cdot \rangle^{-\underline{n}}\|_{L^2(\mathbb{R}^n)} \|\langle Q \rangle^{\underline{n}-s} \langle P \rangle^m \psi\|_{L^\infty(\mathbb{R}^n)}$$
$$\leq c(m) \|\langle \cdot \rangle^{-\underline{n}}\|_{L^2(\mathbb{R}^n)} \|\psi\|_{C_{(\underline{n}-s)}^m}. \qquad \square$$

COROLLARY 3.6.4. *Let* k *be the smallest even integer in* $(\underline{n} + r, \infty)$. *Then for each even* $\ell \in \mathbb{N}$ *there is a constant* c_ℓ *such that for all* $m \in \mathbb{Z}$, $f \in \mathbf{F}_\infty$ *and* $\varphi \in \mathscr{S}(\mathbb{R}^n)$:

$$(3.6.10) \qquad \|\varphi(A)f\|_{\mathbf{F}_m} \leq c_\ell \|f\|_{\mathbf{F}_{m+\ell}} \|\varphi\|_{C_{(\underline{n}-\ell)}^k}.$$

PROOF. Let $m \in \mathbb{Z}$, $f \in \mathbf{F}_\infty$, $f^* \in (\mathbf{F}_m)^*$ and $\varphi \in \mathscr{S}(\mathbb{R}^n)$. Then

$$(3.6.11) \qquad \langle f^*, \varphi(A)f \rangle = \langle \langle W(\cdot)f, f^* \rangle, \widehat{\varphi} \rangle,$$

where the outer bracket on the r.h.s. means antiduality in the sense of distributions. We take $\eta = \widehat{\varphi}$ in (3.6.9) and remark that the inequality (3.6.9) will remain true for all $\xi \in \overset{\circ}{C}^\ell_{(\underline{n}-k)}(\mathbb{R}^n)$. Since $\langle W(\cdot)f, f^* \rangle \in C^\ell_{(-r)}(\mathbb{R}^n) \subset \overset{\circ}{C}^\ell_{(\underline{n}-k)}(\mathbb{R}^n)$ for each $\ell \geq 0$, we may take $\xi = \langle W(\cdot)f, f^* \rangle$. Then

$$\|\xi\|_{C^\ell_{(\underline{n}-k)}} \leq c\|f\|_{\mathbf{F}_{m+\ell}}\|f^*\|_{(\mathbf{F}_m)^*}.$$

So

$$|\langle f^*, \varphi(A)f \rangle| \leq c\|f\|_{\mathbf{F}_{m+\ell}}\|f^*\|_{(\mathbf{F}_m)^*}\|\varphi\|_{C^k_{(\underline{n}-\ell)}}$$

which implies (3.6.10). $\square$

Let $f \in \mathbf{F}_\infty$ and $\varphi \in C^\infty_{\mathrm{pol}}(\mathbb{R}^n)$; choose k be as in Corollary 3.6.4. Then $\varphi \in \overset{\circ}{C}^k_{(s)}(\mathbb{R}^n)$ for some $s \in \mathbb{R}$. By the definition of $\overset{\circ}{C}^k_{(s)}(\mathbb{R}^n)$, there is a sequence $\{\varphi_j\}_{j\in\mathbb{N}}$ in $\mathscr{S}(\mathbb{R}^n)$ such that $\|\varphi - \varphi_j\|_{C^k_{(s)}} \to 0$ as $j \to \infty$. By Corollary 3.6.4, the sequence $\{\varphi_j(A)f\}$ is Cauchy in each $\mathbf{F}_m$ $(m \in \mathbb{Z})$, and if its limit is denoted by $\varphi(A)f$, then for even $\ell \geq \underline{n} - s$:

$$(3.6.12) \qquad \|\varphi(A)f\|_{\mathbf{F}_m} \leq c\|f\|_{\mathbf{F}_{m+\ell}}\|\varphi\|_{C^k_{(s)}}.$$

So $\varphi(A)$ defines an operator in $B(\mathbf{F}_{m+\ell}, \mathbf{F}_m)$ for each $m \in \mathbb{Z}$ and some integer ℓ depending on φ.

One may rephrase this definition in a more intrinsic way as follows: for each fixed $f \in \mathbf{F}_\infty$ and $m \in \mathbb{Z}$, the linear mapping $\mathscr{S}(\mathbb{R}^n) \ni \varphi \mapsto \varphi(A)f \in \mathbf{F}_m$ is continuous when $\mathscr{S}(\mathbb{R}^n)$ is equipped with the norm $\|\cdot\|_{C^k_{(s)}}$; hence it has a unique continuous extension to $\overset{\circ}{C}^k_{(s)}$ which we again denote by $\varphi(A)f$. In particular this proves the independence of $\varphi(A)f$ of the approximating sequence $\{\varphi_j\}$ in $\mathscr{S}(\mathbb{R}^n)$ chosen above (clearly $\varphi(A)f$ does not depend on s either). Finally, if $f \in \mathbf{F}_\infty$ and $f^* \in \mathbf{F}^*$, then

$$(3.6.13) \qquad \langle f^*, \varphi(A)f \rangle = \langle\langle W(\cdot)f, f^* \rangle, \widehat{\varphi}\rangle,$$

where the r.h.s. means antiduality of distributions (for $\varphi \in \mathscr{S}(\mathbb{R}^n)$, this is just (3.6.11); since both sides in (3.6.13) are continuous functions of φ for the norm of $C^k_{(s)}$, (3.6.13) will hold for any $\varphi \in \overset{\circ}{C}^k_{(s)}$).

The preceding results may be rewritten by introducing the following space:

$$(3.6.14) \qquad B(\mathbf{F}_{-\infty}) = \bigcap_{m\in\mathbb{Z}} \bigcup_{k\in\mathbb{Z}} B(\mathbf{F}_m, \mathbf{F}_k).$$

It is the set of linear operators $T : \mathbf{F}_{-\infty} \to \mathbf{F}_{-\infty}$ such that for each $m \in \mathbb{Z}$ there is $k \in \mathbb{Z}$ with $T \in B(\mathbf{F}_m, \mathbf{F}_k)$, and it has the structure of an algebra. We have shown above that $\varphi(A) \in B(\mathbf{F}_{-\infty})$ if $\varphi \in C^\infty_{\mathrm{pol}}(\mathbb{R}^n)$, and the correspondence $\varphi \mapsto \varphi(A)$ defines a unital homomorphism from $C^\infty_{\mathrm{pol}}(\mathbb{R}^n)$ into $B(\mathbf{F}_{-\infty})$.

By interpolation (see Theorem 2.6.1) one obtains from the inequality (3.6.12) that $\varphi(A) \in B(\mathbf{F}_{t+\ell,p}, \mathbf{F}_{t,p})$ for each $t \in \mathbb{R}$, each $p \in [1, \infty]$ and some $\ell \in \mathbb{N}$ depending on φ. One may also derive a somewhat more explicit expression for $\varphi(A)$ by observing that, if $\varphi \in C^\infty_{\mathrm{pol}}(\mathbb{R}^n)$, then the distribution $\widehat{\varphi}$ is rapidly w-vanishing at infinity (see Theorem 1.4.5); hence, if $N > 0$ is fixed, there are an integer $k \in \mathbb{N}$ and a family of continuous functions $\{\varphi_\alpha\}_{|\alpha| \leq k}$ such that $|\widehat{\varphi_\alpha}(x)| \leq c\langle x \rangle^{-N}$ and $\varphi(x) = \sum_{|\alpha| \leq k} x^\alpha \varphi_\alpha(x)$ (see Corollary 1.4.4 (c)). If one assumes that $N > n + r$, then $\varphi_\alpha \in \mathscr{M}^r(\mathbb{R}^n)$ for each $|\alpha| \leq k$, so that $\varphi_\alpha(A) \in B(\mathbf{F}_m)$ for each $m \in \mathbb{Z}$; then one has

$$(3.6.15) \qquad \varphi(A) = \sum_{|\alpha| \leq k} A^\alpha \varphi_\alpha(A) \in B(\mathbf{F}_{m+k}, \mathbf{F}_m) \qquad \forall m \in \mathbb{Z}.$$

We next prove a Taylor expansion formula for $\varphi(A)$. The result is particularly useful when the derivatives of higher order of φ produce bounded operators.

PROPOSITION 3.6.5. *Let* $\varphi \in C^\infty_{\mathrm{pol}}(\mathbb{R}^n)$ *and* $j \in \mathbb{N} \setminus \{0\}$.
(a) *The following identity holds on* $\mathbf{F}_{-\infty}$:

$$(3.6.16)$$
$$\varphi(A) = \sum_{|\beta| < j} \frac{\varphi^{(\beta)}(0)}{\beta!} A^\beta + \sum_{|\beta| = j} \frac{j}{\beta!} \int_0^1 \varphi^{(\beta)}(\tau A)(1 - \tau)^{j-1} d\tau \cdot A^\beta,$$

where the integral exists strongly in $B(\mathbf{F}_{m+\ell}, \mathbf{F}_m)$ *for some* $\ell \in \mathbb{N}$ *and each* $m \in \mathbb{Z}$.
(b) *If* $\varphi^{(\beta)} \in \mathscr{M}^r(\mathbb{R}^n)$ *for each multi-index* β *with* $|\beta| = j$, *e.g. if* $\varphi \in S^{j-\varepsilon}(\mathbb{R}^n)$ *for some* $\varepsilon > 0$, *then the integral in* (3.6.16) *exists strongly in* $B(\mathbf{F}_m)$ *for each* $m \in \mathbb{Z}$. *In particular* $\varphi(A) \in B(\mathbf{F}_m, \mathbf{F}_{m-j})$ *for each* $m \in \mathbb{Z}$ *and* $\varphi(A) \in B(\mathbf{F}_{t,p}, \mathbf{F}_{t-j,p})$ *for each* $t \in \mathbb{R}$ *and each* $p \in [1, \infty]$.

PROOF. (a) For $\varphi \in \mathscr{S}(\mathbb{R}^n)$, the identity (3.6.16) follows from the Taylor expansion (3.3.16). By a limiting argument one can then extend (3.6.16) to all $\varphi \in C^\infty_{\mathrm{pol}}(\mathbb{R}^n)$. More precisely, let $\varphi \in C^\infty_{\mathrm{pol}}(\mathbb{R}^n)$. Choose an integer $k \geq \underline{n} + r + j$, an even $\ell \in \mathbb{N}$ and a sequence $\{\varphi_s\}_{s \in \mathbb{N}}$ in $\mathscr{S}(\mathbb{R}^n)$ such that $\varphi \in \overset{\circ}{C}^k_{(\underline{n}-\ell)}(\mathbb{R}^n)$ and $\varphi_s \to \varphi$ in $C^k_{(\underline{n}-\ell)}(\mathbb{R}^n)$. Let $f \in \mathbf{F}_\infty$ (clearly it is sufficient to show that (3.6.16) holds on $\mathbf{F}_\infty$). Then $\varphi_s(A)f \to \varphi(A)f$ strongly in each $\mathbf{F}_m$ by (3.6.10), and $\varphi_s^{(\alpha)}(0) \to \varphi^{(\alpha)}(0)$ for each $|\alpha| \leq k$. Furthermore, for $|\beta| = j$, we have $\varphi^{(\beta)} \in \overset{\circ}{C}^{k-j}_{(\underline{n}-\ell)}(\mathbb{R}^n)$ and $\varphi_s^{(\beta)} \to \varphi^{(\beta)}$ in $C^{k-j}_{(\underline{n}-\ell)}(\mathbb{R}^n)$. Since $||\psi^\tau||_{C^a_{(b)}} \leq ||\psi||_{C^a_{(b)}}$ if $0 < \tau \leq 1$, one easily obtains by using (3.6.10) and the Lebesgue dominated convergence theorem that

$$\int_0^1 \varphi_s^{(\beta)}(\tau A)f \cdot (1 - \tau)^{j-1} d\tau \to \int_0^1 \varphi^{(\beta)}(\tau A)f \cdot (1 - \tau)^{j-1} d\tau \quad \text{in each } \mathbf{F}_m.$$

(b) It suffices to observe that one has $||\varphi^{(\beta)}(\sigma A)||_{B(\mathbf{F}_m)} \leq c||\varphi^{(\beta)}||_{\mathscr{M}^r}$ for each $m \in \mathbb{Z}$ and each $\sigma \in (0, 1]$. $\square$

We shall now consider the case where $\varphi \in S^k(\mathbb{R}^n)$ for some $k \in \mathbb{N}$. It is known that $\varphi(A)$ need not be bounded in $\mathbf{F}$ if $\varphi \in S^0(\mathbb{R}^n)$. For example, if $\mathbf{F} = L^1(\mathbb{R}^n)$ and W is the translation group, then $\varphi(P)$ is continuous as an operator from the real Hardy space $\mathbb{H}^1(\mathbb{R}^n)$ into $L^1(\mathbb{R}^n)$ but *not* continuous as an operator in $L^1(\mathbb{R}^n)$ for arbitrary $\varphi \in S^0(\mathbb{R}^n)$ (see the end of our Section 3.2 and Chapter 7 in [Me]). So in general one cannot expect $\varphi(A)$ to be bounded from $\mathbf{F}_m$ to $\mathbf{F}_{m-k}$

if $\varphi \in S^k(\mathbb{R}^n)$. We shall see below that the Besov scale has better properties in this respect. We shall need the preliminary facts given in the following three lemmas.

LEMMA 3.6.6. *Let $\varphi \in S^k(\mathbb{R}^n)$ for some real $k \geq 0$. Then, for each $\nu > 0$, $\varphi(A) \in B(\mathbf{F}_{k+\nu,\infty}, \mathbf{F})$.*

PROOF. We may assume without loss of generality that $0 < \nu < 1$. Let $\eta \in C_0^\infty(\mathbb{R}^n)$ be such that $\eta(x) = 1$ in some neighbourhood of the origin. Let $f \in \mathbf{F}_\infty$ and $f^* \in \mathbf{F}^*$. By using first (3.6.13) and then Proposition 1.3.7 and the fact that $(1 - \eta)\widehat{\varphi} \in \mathscr{S}(\mathbb{R}^n)$ (see Lemma 1.3.5), one gets that

$$|\langle f^*, \varphi(A)f\rangle| \leq |\langle\langle\eta(\cdot)W(\cdot)f, f^*\rangle, \widehat{\varphi}\rangle| + |\langle\langle W(\cdot)f, f^*\rangle, (1 - \eta)\widehat{\varphi}\rangle|$$
$$\leq c_1 ||\langle\eta(\cdot)W(\cdot)f, f^*\rangle||_{BC^{k+\nu}} +$$
$$+ ||f||_{\mathbf{F}}||f^*||_{\mathbf{F}^*} \int_{\mathbb{R}^n} ||W(x)||_{B(\mathbf{F})}|1 - \eta(x)| \cdot |\widehat{\varphi}(x)|dx.$$

We observe that the last integral is finite and that

$$||\langle\eta(\cdot)W(\cdot)f, f^*\rangle||_{BC^{k+\nu}} \leq c_2 ||f||_{\mathbf{F}_{k+\nu,\infty}}||f^*||_{\mathbf{F}^*}$$

(this can be checked by using the definition (1.3.7) of $|| \cdot ||_{BC^\mu}$, Corollary 3.4.8 (b) and Theorem 3.4.6). So $|\langle f^*, \varphi(A)f\rangle| \leq c_3||f^*||_{\mathbf{F}^*}||f||_{k+\nu,\infty}$, which implies the result of the lemma. $\square$

LEMMA 3.6.7. *Let $\theta, \varphi \in C^j(\mathbb{R}^n \setminus \{0\})$ for some $j \in \mathbb{N}$. Assume that there is a number $k \in \mathbb{R}$ such that*

(3.6.17) $$\mathscr{I}_k \equiv \int_{\mathbb{R}^n} |x|^k \left[|\theta(x)| + \sum_{|\alpha|\leq j} |x|^{|\alpha|-j}|\theta^{(\alpha)}(x)|\right] dx < \infty$$

and such that $|\varphi^{(\alpha)}(x)| \leq c|x|^{k-|\alpha|}$ for some constant c, all $x \in \mathbb{R}^n \setminus \{0\}$ and each multi-index α with $|\alpha| \leq j$. Set $\theta^\varepsilon(x) = \theta(\varepsilon x)$. Then there is a constant C such that

(3.6.18) $$|(\mathscr{F}\theta^\varepsilon\varphi)(x)| \leq C\varepsilon^{-k-n}\langle\varepsilon^{-1}x\rangle^{-j} \qquad \forall x \in \mathbb{R}^n, \forall \varepsilon > 0.$$

PROOF. Let $\ell \in \mathbb{N}$. Then $|x|^\ell \leq n^\ell \sum_{i=1}^n |x_i|^\ell \leq n^\ell \sum_{|\alpha|=\ell} |x^\alpha|$. Hence, for

$\ell \leq j$:

$$|x|^\ell |(\mathcal{F}\theta^\varepsilon \varphi)(x)| \leq n^\ell \sum_{|\alpha|=\ell} |x^\alpha (\mathcal{F}\theta^\varepsilon \varphi)(x)|$$

$$= n^\ell \sum_{|\alpha|=\ell} |(-i)^{|\alpha|} \mathcal{F}(\partial^\alpha \theta^\varepsilon \varphi)(x)| \leq n^\ell \sum_{|\alpha|=\ell} ||\partial^\alpha \theta^\varepsilon \varphi||_{L^1(X)}$$

$$\leq n^\ell \sum_{|\alpha|=\ell} \sum_{\beta \leq \alpha} \varepsilon^{|\alpha-\beta|} \int_{\mathbb{R}^n} |\theta^{(\alpha-\beta)}(\varepsilon x)| \, |\varphi^{(\beta)}(x)| \underline{dx}$$

$$\leq n^\ell \sum_{|\alpha|=\ell} \sum_{\beta \leq \alpha} \varepsilon^{|\alpha|-|\beta|} \int_{\mathbb{R}^n} |\theta^{(\alpha-\beta)}(\varepsilon x)| \, c \, |x|^{k-|\beta|} \underline{dx}$$

$$= c n^\ell \sum_{|\alpha|=\ell} \sum_{\beta \leq \alpha} \varepsilon^{\ell-k-n} \int_{\mathbb{R}^n} |\theta^{(\alpha-\beta)}(y)| \, |y|^{k-|\beta|} \underline{dy}$$

$$\leq c_\ell \varepsilon^{\ell-k-n} \sum_{|\gamma| \leq \ell} \int_{\mathbb{R}^n} |\theta^{(\gamma)}(y)| \, |y|^{k-\ell+|\gamma|} \underline{dy}.$$

We apply this inequality for $\ell = 0$ and $\ell = j$ and obtain that

$$|(\mathcal{F}\theta^\varepsilon \varphi)(x)| \leq c_0 \mathcal{J}_k \varepsilon^{-k-n}$$

and

$$|(\mathcal{F}\theta^\varepsilon \varphi)(x)| \leq c_j \mathcal{J}_k \varepsilon^{-k-n} |\varepsilon^{-1} x|^{-j}.$$

These two inequalities imply (3.6.18). $\square$

LEMMA 3.6.8. *Let $j > n + r$ be an integer and φ, $\theta \in C_{\mathrm{pol}}^j(\mathbb{R}^n)$ satisfying the conditions of Lemma 3.6.7 for some $k \in \mathbb{R}$. Then $\theta^\varepsilon \varphi \in \mathcal{M}^r(\mathbb{R}^n)$ if $\varepsilon > 0$, and there is a constant c such that for $\varepsilon \in (0,1)$ and $m \in \mathbb{Z}$:*

(3.6.19) $||\theta(\varepsilon A)\varphi(A)||_{B(\mathbf{F}_m)} \leq c \varepsilon^{-k}.$

PROOF. We have $\theta^\varepsilon \varphi \in \mathcal{M}^r(\mathbb{R}^n)$ by (3.6.18) because $j - r > n$. Then

$$||\theta(\varepsilon A)\varphi(A)||_{B(\mathbf{F}_m)} = ||(\theta^\varepsilon \varphi)(A)||_{B(\mathbf{F}_m)}$$

$$\leq c' ||\theta^\varepsilon \varphi||_{\mathcal{M}^r} = c' 2^{r/2} \int_{\mathbb{R}^n} \langle x \rangle^r |(\mathcal{F}\theta^\varepsilon \varphi)(x)| \underline{dx}$$

$$\leq c'' 2^{r/2} \int_{\mathbb{R}^n} \langle x \rangle^r \varepsilon^{-k-n} \langle \varepsilon^{-1} x \rangle^{-j} \underline{dx}$$

$$= c'' 2^{r/2} \varepsilon^{-k} \int_{\mathbb{R}^n} \frac{\langle \varepsilon y \rangle^r}{\langle y \rangle^j} \underline{dy} \leq c \varepsilon^{-k} \quad \text{if } 0 < \varepsilon \leq 1. \quad \square$$

THEOREM 3.6.9. *Let $\mathbf{F}$ be a Banach space equipped with a polynomially bounded C_0-representation of $\mathbb{R}^n$. Let $\varphi \in S^k(\mathbb{R}^n)$ for some $k \in \mathbb{R}$.*

Then $\varphi(A) \in B(\mathbf{F}_{s,p}, \mathbf{F}_{s-k,p})$ whenever $s > 0$, $s - k > 0$ and $1 \leq p \leq \infty$. If $n = 1$ or if $\mathbf{F}$ is reflexive, then $\varphi(A) \in B(\mathbf{F}_{s,p}, \mathbf{F}_{s-k,p})$ for each $s \in \mathbb{R}$.

PROOF. (i) Let $\eta \in C_0^\infty(\mathbb{R}^n)$ be such that $\eta(x) = 1$ in some neighbourhood of the origin. Then $\eta\varphi \in C_0^\infty(\mathbb{R}^n)$, hence $\eta(A)\varphi(A) \in B(\mathbf{F}_{s,p}, \mathbf{F}_{t,p})$ for all $s, t \in \mathbb{R}$. Hence it suffices to prove that $[I - \eta(A)]\varphi(A) \in B(\mathbf{F}_{s,p}, \mathbf{F}_{s-k,p})$, in other words it suffices to prove the theorem under the additional assumption that $\varphi(x) = 0$ in some neighbourhood of the origin. For this we fix a function $\theta \in C_0^\infty(\mathbb{R}^n)$ such that $\theta(x) > 0$ if $1/2 < |x| < 2$ and $\theta(x) = 0$ otherwise and observe that, if $f \in \mathbf{F}$ and $0 < \tau < 1$, then $\|\theta(\tau A)^2\varphi(A)f\|_{\mathbf{F}} \leq c\tau^{-k}\|\theta(\tau A)f\|_{\mathbf{F}}$ by Lemma 3.6.8.

We first assume that $k \geq 0$ and fix a number ν satisfying $0 < \nu < s - k$. By using first Theorem 3.6.2 (a) and then Lemma 3.6.6, one sees that

(3.6.20)
$$\|\varphi(A)f\|_{\mathbf{F}_{s-k,p}} \leq c_1\|\varphi(A)f\|_{\mathbf{F}} + c_1\left[\int_0^1 \|\tau^{-s+k}\theta^2(\tau A)\varphi(A)f\|_{\mathbf{F}}^p \frac{d\tau}{\tau}\right]^{1/p}$$
$$\leq c_2\|f\|_{\mathbf{F}_{k+\nu,\infty}} + c_1 C\left[\int_0^1 \|\tau^{-s}\theta(\tau A)f\|_{\mathbf{F}}^p \frac{d\tau}{\tau}\right]^{1/p}.$$

Each term on the r.h.s. is bounded by $c_3\|f\|_{\mathbf{F}_{s,p}}$ (use $s > k + n$ and (3.6.4)). So we have $\varphi(A) \in B(\mathbf{F}_{s,p}, \mathbf{F}_{s-k,p})$ if $k \geq 0$ and $s - k > 0$.

The proof in the case $k < 0$, $s > 0$ is almost identical; it suffices to replace $c_2\|f\|_{\mathbf{F}_{k+\nu,\infty}}$ on the r.h.s. of (3.6.20) by $c_2\|f\|_{\mathbf{F}}$, which is justified since φ belongs to $\mathscr{M}^r(\mathbb{R}^n)$ if $\varphi \in S^{-\delta}(\mathbb{R}^n)$ for some $\delta > 0$.

Note that, since $\varphi(A)$ maps $\mathbf{F}_\infty$ into $\mathbf{F}_\infty$ and $\mathbf{F}_{t,\infty}^\circ$ is the closure of $\mathbf{F}_\infty$ in $\mathbf{F}_{t,\infty}$, we also have $\varphi(A) \in B(\mathbf{F}_{s,\infty}^\circ, \mathbf{F}_{s-k,\infty}^\circ)$.

(ii) If $n = 1$, choose an integer ℓ such that $s - \ell > 0$ and $s - \ell - k > 0$. Then, by the preceding result, we have $\varphi(A) \in B((\mathbf{F}_\ell)_{s-\ell,p}, (\mathbf{F}_\ell)_{s-\ell-k,p})$. It now suffices to observe that $(\mathbf{F}_\ell)_{s-\ell,p} = \mathbf{F}_{s,p}$ and $(\mathbf{F}_\ell)_{s-\ell-k,p} = \mathbf{F}_{s-k,p}$ by (3.4.36).

(iii) Now assume $\mathbf{F}$ to be reflexive. By applying the results of (i) to the C_0-group W^* in $\mathbf{F}^*$, we see that

$$\overline{\varphi}(A^*) \in B(\mathbf{F}_{k-s,q}^*, \mathbf{F}_{-s,q}^*) \cap B(\mathbf{F}_{k-s,\infty}^{*\circ}, \mathbf{F}_{-s,\infty}^{*\circ})$$

if $1 \leq q \leq \infty$ and $k - s > 0$, $-s > 0$. This implies together with Theorem 3.4.3 (d) that $\varphi(A) \in B(\mathbf{F}_{s,p}, \mathbf{F}_{s-k,p})$ for each $p \in [1, \infty]$ if $s < 0$ and $s - k < 0$. So this inclusion holds if s and $s - k$ have the same sign. If s and $s - k$ are of opposite signs (or equal to zero), one obtains the same inclusion by choosing a number $t > |k|$ and by interpolating between the relations $\varphi(A) \in B(\mathbf{F}_{t,p}, \mathbf{F}_{t-k,p})$ and $\varphi(A) \in B(\mathbf{F}_{-t,p}, \mathbf{F}_{-t-k,p})$ (use Theorems 2.6.1 and 3.4.3 (b)). $\square$

COROLLARY 3.6.10. *For $k \in \mathbb{R}$ denote by $\langle A \rangle^k$ the operator $\varphi(A)$ associated to the function $\varphi(x) = (1 + |x|^2)^{k/2}$. Then $\langle A \rangle^k$ is an isomorphism of $\mathbf{F}_\infty$ onto itself, and $\langle A \rangle^k\langle A \rangle^\ell = \langle A \rangle^{k+\ell}$ if $k, \ell \in \mathbb{R}$. If $s > 0$, $s - k > 0$ and $1 \leq p \leq \infty$, then $\langle A \rangle^k$ is an isomorphism of $\mathbf{F}_{s,p}$ onto $\mathbf{F}_{s-k,p}$ with inverse $\langle A \rangle^{-k}$. If $n = 1$ or if $\mathbf{F}$ is reflexive, then $\langle A \rangle^k$ is an isomorphism of $\mathbf{F}_{-\infty}$ onto itself, one has $\langle A \rangle^k\langle A \rangle^\ell = \langle A \rangle^{k+\ell}$ on $\mathbf{F}_{-\infty}$, and $\langle A \rangle^k : \mathbf{F}_{s,p} \to \mathbf{F}_{s-k,p}$ is an isomorphism with inverse $\langle A \rangle^{-k}$ for all $k, s \in \mathbb{R}$ and $1 \leq p \leq \infty$.*

3.6.3. One remarks from Theorem 3.6.9 and Corollary 3.6.10 that stronger results can be obtained if the Banach space $\mathbf{F}$ is reflexive or if W is a one-parameter C_0-group (the statements hold in $\mathbf{F}_{s,p}$ for all $s \in \mathbb{R}$ and not only for $s > 0$). In the last part of this section we shall show similarly that the expressions (3.6.4) and (3.6.5) determine admissible norms on $\mathbf{F}_{s,p}$ for *all* $s \in \mathbb{R}$ if one of the above additional conditions is satisfied. We begin with some preliminary results which are the analogue for a reflexive Banach space of those obtained in Proposition 3.4.12 for a one-parameter group.

LEMMA 3.6.11. *Assume that* $\mathbf{F}$ *is reflexive, and let* $k, m \in \mathbb{Z}$. *Then*

$$(3.6.21) \qquad \mathbf{F}_{k+m} \subset (\mathbf{F}_k)_m \subset \mathbf{F}^\circ_{k+m,\infty} \qquad \textit{if } m \geq 0$$

$$(3.6.22) \qquad \mathbf{F}_{k+m,1} \subset (\mathbf{F}_k)_m \subset \mathbf{F}_{k+m} \qquad \textit{if } m \leq 0.$$

PROOF. (i) We first prove (3.6.22). If $k \leq 0$ or $m = 0$, this is obvious because $(\mathbf{F}_k)_m = \mathbf{F}_{k+m}$. So assume that $k \geq 1$ and $m < 0$. We set $\ell = -m > 0$ and have on $\mathbf{F}_{-\infty}$ (see Corollary 3.6.10) :

$$(3.6.23)$$
$$f = \langle A \rangle^{2\ell} \langle A \rangle^{-2\ell} f = \sum_{|\alpha| \leq \ell} A^\alpha \cdot \frac{\ell!}{\alpha!(\ell - |\alpha|)!} A^\alpha \langle A \rangle^{-2\ell} f \equiv \sum_{|\alpha| \leq \ell} A^\alpha f_\alpha,$$

with an obvious definition of f_α. If $f \in \mathbf{F}_{k-\ell,1}$, then $f_\alpha \in \mathbf{F}_{k+\ell-|\alpha|,1}$ by Theorem 3.6.9 (observe that the function $\varphi(x) \equiv x^\alpha \langle x \rangle^{-2\ell}$ belongs to $S^{|\alpha|-2\ell}(\mathbb{R}^n)$). Hence $f_\alpha \in \mathbf{F}_k$ for each $|\alpha| \leq \ell$. By using (3.3.29) in the Banach space $\mathbf{F}_k$, one sees that $f \in (\mathbf{F}_k)_{-\ell} \equiv (\mathbf{F}_k)_m$. This proves the first inclusion in (3.6.22). The second one was already pointed out in (3.3.58).

(ii) Now let $k \in \mathbb{Z}$ and $m \geq 0$. By writing $\mathbf{F}^*_\ell$ for $(\mathbf{F}^*)_\ell$ and by using (3.6.22) in the Banach space $\mathbf{F}^*$, we have dense embeddings

$$\mathbf{F}^*_{-k-m,1} \subset (\mathbf{F}^*_{-k})_{-m} \subset \mathbf{F}^*_{-k-m}.$$

Hence

$$(3.6.24) \qquad (\mathbf{F}^*_{-k-m})^* \subset [(\mathbf{F}^*_{-k})_{-m}]^* \subset (\mathbf{F}^*_{-k-m,1})^*.$$

Theorems 3.3.28 and 3.4.3 (d) show that (3.6.24) is identical with

$$\mathbf{F}_{k+m} \subset (\mathbf{F}_k)_m \subset \mathbf{F}_{k+m,\infty}.$$

Since $\mathbf{F}_\infty$ is a dense subspace of $(\mathbf{F}_k)_m$, we obtain (3.6.21). $\square$

REMARK 3.6.12. An alternative proof of (3.6.22) may be obtained by using instead of (3.6.23) the relation

$$(3.6.25) \qquad f = \varphi_0(A)f + \sum_{|\alpha|=\ell} A^\alpha \varphi_\alpha(A)f$$

where $\varphi_0 \in \mathscr{S}(\mathbb{R}^n)$ and $\varphi_\alpha \in S^{-\ell}(\mathbb{R}^n)$. (3.6.25) follows from Theorem 1.3.8 which implies (together with Corollary 1.3.4) the existence of functions φ_0, φ_α ($|\alpha| = \ell$) such that $\widehat{\varphi_0} \in C_0^\infty(\mathbb{R}^n)$, $\widehat{\varphi_\alpha} \in L^1(\mathbb{R}^n)$ and of compact support, and

$$(3.6.26) \qquad 1 = \varphi_0(x) + \sum_{|\alpha|=\ell} x^\alpha \varphi_\alpha(x) \qquad \forall x \in \mathbb{R}^n.$$

THEOREM 3.6.13. *Let W be a polynomially bounded C_0-representation of $\mathbb{R}^n$ in a reflexive Banach space $\mathbf{F}$.*
 (a) *For any $k, m \in \mathbb{Z}$ one has $\mathbf{F}_{k+m,1} \subset (\mathbf{F}_k)_m \subset \mathbf{F}_{k+m,\infty}^\circ$.*
 (b) *Let $s \in \mathbb{R}$, $k \in \mathbb{Z}$ and $p \in [1, \infty]$. Then*

$$(3.6.27) \qquad \mathbf{F}_{s,p} = (\mathbf{F}_k)_{s-k,p}.$$

If $k < s$, then for each integer $\ell > s - k$ the following expression defines an admissible norm on $\mathbf{F}_{s,p}$:

$$(3.6.28) \qquad \|f\|_{\mathbf{F}_k} + \left[\int_{|x| \le 1} \left\| \frac{[W(x) - I]^\ell f}{|x|^{s-k}} \right\|_{\mathbf{F}_k}^p \frac{dx}{|x|^n} \right]^{1/p}.$$

PROOF. If $m \ge 0$, the result of (a) is obtained from the first inclusion in (3.4.8) and (3.6.21). If $m < 0$ we use (3.6.22) and the second inclusion in (3.4.8).

The second part of (b) follows from (3.6.27) and Theorem 3.4.6. To prove (3.6.27), we choose integers m_1, m_2, ℓ_1 and ℓ_2 such that $m_1 < s - k < m_2$ and $\ell_1 < m_1 + k < m_2 + k < \ell_2$. By (3.4.7) we then have $(\mathbf{F}_k)_{s-k,p} = ((\mathbf{F}_k)_{m_2}, (\mathbf{F}_k)_{m_1})_{\theta,p}$ with $\theta = [m_2 - (s - k)](m_2 - m_1)^{-1}$. By the result of (a), $(\mathbf{F}_k)_{m_i}$ is of class θ_i for the pair of B-spaces $(\mathbf{F}_{\ell_2}, \mathbf{F}_{\ell_1})$, with $\theta_i = (\ell_2 - k - m_i)(\ell_2 - \ell_1)^{-1}$ (see (2.5.1)). By applying the reiteration theorem (Eq. (2.5.5)) one finds that $((\mathbf{F}_k)_{m_2}, (\mathbf{F}_k)_{m_1})_{\theta,p} = (\mathbf{F}_{\ell_2}, \mathbf{F}_{\ell_1})_{\sigma,p}$ with $\sigma = (1 - \theta)\theta_2 + \theta\theta_1 \equiv (\ell_2 - s)(\ell_2 - \ell_1)^{-1}$, so that the last interpolation space is identical with $\mathbf{F}_{s,p}$. $\square$

THEOREM 3.6.14. *Let W be a polynomially bounded C_0-representation of $\mathbb{R}^n$ in a Banach space $\mathbf{F}$. If $n = 1$ or if $\mathbf{F}$ is reflexive, then the conclusions of Theorem 3.6.2 are valid for each $s \in \mathbb{R}$ and $1 \le p \le \infty$.*

REMARK. We observe that $\widetilde{\theta}(A)f$ and $\theta(\tau^{-1}A)f$ are well defined for any $f \in \mathbf{F}_{-\infty}$ and belong to $\mathbf{F}_\infty$. Hence the expressions (3.6.4) and (3.6.5) make sense for all $f \in \mathbf{F}_{-\infty}$, although they could be equal to $+\infty$ because the integral in (3.6.4) or the series in (3.6.5) could be divergent at infinity.

PROOF. We prove the validity of the statement of part (a) of Theorem 3.6.2 for $s \le 0$. The proof of part (b) is very similar.

Let $s \le 0$ and choose an even integer $k > 0$ such that $s + k > 0$. By (3.4.36) or (3.6.27) we have $\mathbf{F}_{s,p} = (\mathbf{F}_{-k})_{s+k,p}$. So Theorem 3.6.2 (applied for the Banach space $\mathbf{F}_{-k}$) implies that

$$(3.6.29) \quad \|f\|_{\mathbf{F}_{s,p}} \sim \|\widetilde{\theta}(A)f\|_{\mathbf{F}_{-k}} + \left[\int_0^1 \|\tau^{-s-k}\theta(\tau A)f\|_{\mathbf{F}_{-k}}^p \frac{d\tau}{\tau} \right]^{1/p}.$$

To obtain the equivalence with the norm (3.6.4), it is enough to prove the existence of a constant $c > 0$ such that for all $f \in \mathbf{F}_{-\infty}$:

$$(3.6.30) \qquad c^{-1}||\widetilde{\theta}(A)f||_{\mathbf{F}_{-k}} \leq ||\widetilde{\theta}(A)f||_{\mathbf{F}} \leq c||\widetilde{\theta}(A)f||_{\mathbf{F}_{-k}}$$

and

$$(3.6.31)$$
$$c^{-1}||\tau^{-k}\theta(\tau A)f||_{\mathbf{F}_{-k}} \leq ||\theta(\tau A)f||_{\mathbf{F}} \leq c||\tau^{-k}\theta(\tau A)f||_{\mathbf{F}_{-k}} \quad \forall \tau \in (0,1].$$

The first inequality in (3.6.30) holds because $k > 0$. Next let $\eta \in C_0^\infty(\mathbb{R}^n)$ be such that $\eta\widetilde{\theta} = \widetilde{\theta}$ (hence also $\eta\theta = \theta$). Then Lemma 3.3.20 implies that $||\eta(\tau A)||_{\mathbf{F}_{-k}\to\mathbf{F}} \leq c_1\tau^{-k}$ for all $\tau \in (0,1]$, so that

$$||\theta(\tau A)f||_{\mathbf{F}} = ||\eta(\tau A)\theta(\tau A)f||_{\mathbf{F}} \leq c_1\tau^{-k}||\theta(\tau A)f||_{\mathbf{F}_{-k}}.$$

This proves the second inequality in (3.6.31). The second inequality in (3.6.30) is obtained by the same argument. Finally, to obtain the first inequality in (3.6.31), let χ be a function in $C_0^\infty(\mathbb{R}^n \setminus \{0\})$ such that $\chi\theta = \theta$ and set $\xi(x) = |x|^{-k}\chi(x)$. Observe that $\xi \in C_0^\infty(\mathbb{R}^n)$ and $\chi = \sum_{|\alpha|=k} c_\alpha x^\alpha \xi(x)$ for some constants $c_\alpha \geq 0$ (remember that k is even). Then

$$||\chi(\tau A)||_{\mathbf{F}\to\mathbf{F}_{-k}} \leq \sum_{|\alpha|=k} c_\alpha\tau^k ||A^\alpha||_{\mathbf{F}\to\mathbf{F}_{-k}}||\xi(\tau A)||_{B(\mathbf{F})} \leq c_2\tau^k$$

for all $\tau \in (0,1]$, by Proposition 3.3.9 (a) and (3.2.19). Hence

$$||\theta(\tau A)f||_{\mathbf{F}_{-k}} = ||\chi(\tau A)\theta(\tau A)f||_{\mathbf{F}_{-k}} \leq c_2\tau^k||\theta(\tau A)f||_{\mathbf{F}}. \qquad \square$$

3.7. C_0-Groups in Hilbert Spaces

If $\mathbf{F}$ is a Hilbert space, then several of the results obtained in the preceding sections can be improved. More precisely, one can associate a bounded operator $\varphi(A)$ in $\mathbf{F}$ to a much larger class of functions φ and one can show that $\mathbf{F}_k = \mathbf{F}_{k,2}$ for all $k \in \mathbb{Z}$ (i.e. the Sobolev scale is part of the Besov scale). The estimate (3.7.23) and Proposition 3.7.7 are due to N. Mandache [Ma]; in the proof of the next proposition we use the representation (3.7.5) which is also due to him.

We recall (see Definition 3.1.11) that $\varphi \in \mathcal{M}^{(\omega_1)}(\mathbb{R}^n)$ means that φ is the inverse Fourier transform of a measure $\widehat{\varphi}$ on $\mathbb{R}^n$ which satisfies $\int_{\mathbb{R}^n} e^{\omega_1|x|}|\widehat{\varphi}|(dx) < \infty$. If $\varphi \in \mathcal{M}^{(\omega_1)}(\mathbb{R}^n)$, it has an analytic extension to the strip $\Sigma_{\omega_1} = \{x + iy \mid x, y \in \mathbb{R}^n, |y| < \omega_1\}$ of $\mathbb{C}^n$ and a bounded continuous extension to the closure of this strip, namely:

$$\varphi(x + iy) = \int_{\mathbb{R}^n} e^{i(x+iy)k}\widehat{\varphi}(k)\underline{d}k.$$

PROPOSITION 3.7.1. *Let $\mathbf{F}$ be a Hilbert space and W a C_0-representation of $\mathbb{R}^n$ in $\mathbf{F}$ such that $||W(x)||_{B(\mathbf{F})} \leq Me^{\omega|x|}$ for some $M \geq 1$ and $\omega \geq 0$. Let*

$\omega_1 > \omega$. Then there is a constant $C < \infty$ such that for each $\varphi \in \mathscr{M}^{(\omega_1)}(\mathbb{R}^n)$ and each $m \in \mathbb{Z}$:

$$(3.7.1) \qquad \|\varphi(A)\|_{B(\mathbf{F}_m)} \leq C \sup_{\substack{x,y \in \mathbb{R}^n \\ |y| < \omega_1}} |\varphi(x+iy)|.$$

PROOF. (i) If $\psi : \mathbb{R}^n \to \mathbb{C}$ is the Fourier transform of an integrable measure $\widehat{\psi}$, then the operator $\psi(P) : \mathscr{S}(\mathbb{R}^n) \to \mathscr{S}^*(\mathbb{R}^n)$, introduced in Section 1.2, is explicitly given by the integral formula:

$$(3.7.2) \qquad [\psi(P)g](x) = \int_{\mathbb{R}^n} g(x+y)\widehat{\psi}(y)\underline{d}y, \qquad g \in \mathscr{S}(\mathbb{R}^n).$$

By using also Plancherel's theorem, one sees that

$$(3.7.3) \qquad \|\psi(P)g\|_{L^2(\mathbb{R}^n)} = \|\psi(Q)\widehat{g}\|_{L^2(\mathbb{R}^n)} \leq \sup_{x \in \mathbb{R}^n} |\psi(x)| \cdot \|g\|_{L^2(\mathbb{R}^n)}.$$

Now assume that g is a bounded continuous $\mathbf{F}_m$-valued function. Then (3.7.2) makes sense and defines a new $\mathbf{F}_m$-valued function with the same properties. Moreover, (3.7.3) still holds if the norms are taken in $L^2(\mathbb{R}^n; \mathbf{F}_m)$ (with the convention that the norm of a non-square integrable function is infinite). In fact, by a standard approximation argument, we may assume that $g(x) = \sum_{j=1}^{N} g_j(x)f_j$, where $g_j : \mathbb{R}^n \to \mathbb{C}$ are in $\mathscr{S}(\mathbb{R}^n)$ and $\{f_1, \ldots, f_N\}$ is an orthonormal set in $\mathbf{F}_m$. Then the vector-valued version of (3.7.3) follows immediately from the scalar version.

(ii) Now let $\varphi : \mathbb{R}^n \to \mathbb{C}$ be as in the statement of the theorem. Then (3.7.2) allows us to define $\varphi(P)g$ for a much larger class of functions $g : \mathbb{R}^n \to \mathbf{F}_m$, e.g. it is enough to assume g continuous and $\|g(x)\|_{\mathbf{F}_m} \leq C \exp(\omega_1|x|)$, for some constant C depending on g. Notice the following fact: if $v \in \mathbb{R}^n$ and $|v| \leq \omega_1$, then $\varphi(P)e^{(v,Q)}g = e^{(v,Q)}\varphi(P-iv)g$ for $g : \mathbb{R}^n \to \mathbf{F}_m$ bounded and continuous (to prove this, just observe that the Fourier transform of the function $x \mapsto \varphi(x-iv)$ is the measure $e^{(v,k)}\widehat{\varphi}(k)$).

To each vector $f \in \mathbf{F}_m$ one can associate a function $\Theta_f : \mathbb{R}^n \to \mathbf{F}_m$ by setting $\Theta_f(x) = W(x)f$. We clearly have:

$$(3.7.4) \qquad \varphi(P)\Theta_f = \Theta_{\varphi(A)f}.$$

(iii) We fix a C^∞-function η with support in the unit ball of $\mathbb{R}^n$ and satisfying $\int_{\mathbb{R}^n} \eta(x)\underline{d}x = 1$. Then, by (3.7.4) we obtain the obvious but fundamental identity due to N. Mandache:

$$(3.7.5) \qquad \varphi(A)f = \int_{\mathbb{R}^n} \eta(x)W(-x)[\varphi(P)\Theta_f](x)\underline{d}x.$$

By using (3.3.12) one then gets that

$$(3.7.6) \quad \|\varphi(A)f\|_{\mathbf{F}_m} \leq \sup_{|y| \leq 1} \|W(y)\|_{B(\mathbf{F})} \cdot \int_{\mathbb{R}^n} |\eta(x)| \cdot \|[\varphi(P)\Theta_f](x)\|_{\mathbf{F}_m}\underline{d}x$$

$$\leq M e^{\omega} \left[\int_{|x| \leq 1} \underline{d}x\right]^{1/2} \|\eta(Q)\varphi(P)\Theta_f\|_{L^2(\mathbb{R}^n, \mathbf{F}_m)}.$$

Let $\chi_0, \chi_1, \ldots, \chi_N$ be a partition of unity on $\mathbb{R}^n$ consisting of smooth functions such that $0 \leq \chi_k(x) \leq 1$ for each k and such that χ_0 has a compact support, whereas the support of χ_k $(k \neq 0)$ is contained in a cone $\Gamma_k = \{x \in \mathbb{R}^n \mid (x, e_k) \geq \omega_1^{-1}(\omega + \varepsilon)|x|\}$, where e_k is a unit vector in $\mathbb{R}^n$ and ε a fixed number with $0 < \varepsilon < \omega_1 - \omega$. Then

$$(3.7.7) \qquad \eta(Q)\varphi(P)\Theta_f = \eta(Q)\varphi(P)\chi_0(Q)\Theta_f +$$

$$+ \sum_{k=1}^{N} \eta(Q)\varphi(P)e^{\omega_1 Q \cdot e_k}\chi_k(Q)e^{-\omega_1 Q \cdot e_k}\Theta_f$$

$$= \sum_{k=0}^{N} \eta_k(Q)\varphi(P - iv_k)\widetilde{\chi}_k(Q)\Theta_f,$$

where $\eta_0 = \eta$, $\eta_k(x) = \eta(x)\exp[\omega_1(x, e_k)]$, $v_0 = 0$, $v_k = \omega_1 e_k$ for $1 \leq k \leq N$, $\widetilde{\chi}_0 = \chi_0$ and $\widetilde{\chi}_k(x) = \chi_k(x)\exp[-\omega_1(x, e_k)]$ for $1 \leq k \leq N$. Since η has compact support, each $\eta_k(Q)$ is a bounded operator in $L^2(\mathbb{R}^n; \mathbf{F}_m)$. So, by using the vector-valued version of (3.7.3), with $\psi(x) = \varphi(x - iv_k)$, and by taking into account (3.3.12) and the fact that $|\widetilde{\chi}_k(x)| \leq e^{-(\omega+\varepsilon)|x|}$ for $k \geq 1$, one finds that

$$(3.7.8)$$
$$\|\eta_k(Q)\varphi(P - iv_k)\widetilde{\chi}_k(Q)\Theta_f\|_{L^2(\mathbb{R}^n; \mathbf{F}_m)} \leq C \sup_{x \in \mathbb{R}^n} |\varphi(x - iv_k)| \cdot \|f\|_{\mathbf{F}_m}$$

for a finite constant C. (3.7.1) now follows from (3.7.6)-(3.7.8). $\square$

The preceding theorem will be used to associate a bounded operator $\varphi(A)$ in $\mathbf{F}$ to each function $\varphi : \mathbb{R}^n \to \mathbb{C}$ having a bounded, holomorphic extension to a strip $\Sigma_{\omega(\varphi)}$ for some $\omega(\varphi) > \omega$. We denote by $BH^\omega(\mathbb{R}^n)$ the space of all these functions (we stress the fact that $\omega(\varphi)$ depends on φ) and begin with some preparatory remarks concerning this space. Obviously $BH^\omega(\mathbb{R}^n)$ is an algebra with unit for the usual operations. We shall say that a subset $\mathcal{N} \subset BH^\omega(\mathbb{R}^n)$ is *bounded* if there are constants $\omega_0 > \omega$ and $m < \infty$ such that each function $\varphi \in \mathcal{N}$ has a holomorphic extension to Σ_{ω_0} and $|\varphi(z)| \leq m$ for all $z \in \Sigma_{\omega_0}$. A sequence $\{\varphi_k\}_{k \in \mathbb{N}}$ of elements of $BH^\omega(\mathbb{R}^n)$ is said to be *strongly convergent* to $\varphi \in BH^\omega(\mathbb{R}^n)$ if it is bounded and $\lim_{k \to \infty} \varphi_k(x) = \varphi(x)$ for each $x \in \mathbb{R}^n$. We shall prove in a moment that $BH^\omega(\mathbb{R}^n) \cap \mathcal{M}^{(\omega)}(\mathbb{R}^n)$ is dense in $BH^\omega(\mathbb{R}^n)$ with respect to this notion of convergence.

It is clear that the Fourier transform $\widehat{\varphi}$ of a function $\varphi \in BH^\omega(\mathbb{R}^n)$ should be exponentially decaying in some sense. Since $\widehat{\varphi}$ is not a measure in general (e.g. if $n = 1$ and $\varphi(x) = x\langle x \rangle^{-1}$ with $\omega(\varphi) = 1/2$, then $\widehat{\varphi}$ is not a measure in any neighbourhood of zero), we shall have to interpret this decay in a generalized sense. The following construction will be convenient. Let $\mathcal{N}$ be a bounded subset of $BH^\omega(\mathbb{R}^n)$ and let ω_0, m be as in the definition of boundedness. We fix a number $\sigma \in (0, \omega_0^{-1})$ and associate to each $\varphi \in \mathcal{N}$ a function ϕ by setting $\phi(x) = \varphi(x)\langle \sigma x \rangle^{-2\underline{n}}$, where $\underline{n} = [n/2] + 1$. As φ varies over $\mathcal{N}$, the associated functions ϕ vary over another bounded subset of $BH^\omega(\mathbb{R}^n)$ (with the same ω_0 but with m replaced by $m[1 - \sigma^2\omega_0^2]^{-2\underline{n}}$). The function ϕ has better decay properties than φ, namely $|\phi(x + iy)| \leq b_0|\varphi(x + iy)|\langle x \rangle^{-2\underline{n}}$ if $|y| < \omega_0$, where

b_0 is a constant depending only on σ, ω_0 and n. Since $2\underline{n} > n$, we see that $\widehat{\phi} : \mathbb{R}^n \to \mathbb{C}$ is a continuous function, and we can use Cauchy's theorem to get that for $|y| < \omega_0$ and all $w \in \mathbb{R}^n$:

$$(3.7.9) \qquad \widehat{\phi}(w) = \int_{\mathbb{R}^n} e^{-i(w,x)}\phi(x)dx = \int_{\mathbb{R}^n} e^{-i(w,x+iy)}\phi(x+iy)dx.$$

In particular, there is a constant b, depending only on σ, ω_0 and n, such that $|\widehat{\phi}(w)| \le mbe^{(w,y)}$. By taking $y = -\omega_1 w|w|^{-1}$ with $\omega_1 < \omega_0$ and then letting $\omega_1 \to \omega_0$, one obtains that

$$(3.7.10) \qquad\qquad |\widehat{\phi}(w)| \le mbe^{-\omega_0|w|} \qquad \forall w \in \mathbb{R}^n.$$

In particular, $\widehat{\phi}$ is a continuous exponentially decaying function on $\mathbb{R}^n$, and

$$(3.7.11) \qquad \widehat{\varphi}(w) = \sum_{|\alpha| \le \underline{n}} \frac{n!}{\alpha!(n-|\alpha|)!}(i\sigma)^{2|\alpha|}\partial^{2\alpha}\widehat{\phi}(w).$$

We notice a useful consequence of the preceding remarks: if $\{\varphi_k\}_{k \in \mathbb{N}}$ is a sequence of elements of $\mathcal{N}$ which converges strongly to some $\varphi \in \mathcal{N}$, then $|\widehat{\phi}_k(w)| \le \text{const.}\, e^{-\omega_0|w|}$ and $\lim_{k\to\infty} \widehat{\phi}_k(w) = \widehat{\phi}(w)$ uniformly in $w \in \mathbb{R}^n$.

Let us now prove the density of $BH^\omega(\mathbb{R}^n) \cap \mathcal{M}^{(\omega)}(\mathbb{R}^n)$ in $BH^\omega(\mathbb{R}^n)$. If $\varphi \in BH^\omega(\mathbb{R}^n)$ and $0 < \varepsilon < 1$, we define $\varphi_\varepsilon(x) = \varphi(x)e^{-\varepsilon|x|^2}$. Clearly $\mathcal{N} = \{\psi \mid \psi = \varphi$ or $\psi = \varphi_\varepsilon$ for some $\varepsilon \in (0,1)\}$ is a bounded subset of $BH^\omega(\mathbb{R}^n)$ and $\lim_{\varepsilon \to 0} \varphi_\varepsilon = \varphi$ strongly in $BH^\omega(\mathbb{R}^n)$. So it suffices to check that $\varphi_\varepsilon \in \mathcal{M}^{(\omega)}(\mathbb{R}^n)$. This a consequence of the following inequality which can be obtained by using Cauchy's theorem as in (3.7.9) and an argument as in the derivation of (3.7.10):

$$(3.7.12) \qquad\qquad |\widehat{\varphi_\varepsilon}(w)| \le (2\varepsilon)^{-n/2}e^{\omega_0^2 - \omega_0|w|} \sup_{z \in \Sigma_{\omega_0}} |\varphi(z)|.$$

This inequality implies that $\varphi_\varepsilon \in \mathcal{M}^{(\omega_1)}(\mathbb{R}^n)$ for each $\omega_1 < \omega_0$.

PROPOSITION 3.7.2. *Let* $\mathbf{F}$ *be a Hilbert space equipped with a C_0-representation* W *of* $\mathbb{R}^n$ *such that* $\|W(x)\|_{B(\mathbf{F})} \le Me^{\omega|x|}$ *for some* $\omega \ge 0$. *Then*
(a) *There is a unique linear mapping* $BH^\omega(\mathbb{R}^n) \ni \varphi \mapsto \varphi(A) \in B(\mathbf{F})$ *having the following properties:*
(1) $\varphi(A) = \int_{\mathbb{R}^n} W(x)\widehat{\varphi}(x)dx$ *if* $\varphi \in \mathcal{M}^{(\omega)}(\mathbb{R}^n)$,
(2) *if* $\varphi_k \to \varphi$ *strongly in* $BH^\omega(\mathbb{R}^n)$, *then* $\varphi_k(A) \to \varphi(A)$ *strongly in* $B(\mathbf{F})$. *The correspondence* $\varphi \mapsto \varphi(A)$ *is a homomorphism of algebras, and* $\varphi(A)^* = \overline{\varphi}(A^*)$, *where* A^* *is the generator of the adjoint representation* $W^*(x) = W(-x)^*$ *in* $\mathbf{F}^*$.

(b) *The inequality (3.7.1) is satisfied for each* $\varphi \in BH^\omega(\mathbb{R}^n)$, *and if* $|\varphi(z)| \le c(1+|z|)^{-r}$ *for some* $r \in \mathbb{N}$ *and all* $z \in \Sigma_{\omega(\varphi)}$ *(with* $\omega(\varphi) > \omega$*), then* $\varphi(A) \in B(\mathbf{F}_m, \mathbf{F}_{m+r})$ *for all* $m \in \mathbb{Z}$.

PROOF. (i) Let φ be a fixed function in $BH^\omega(\mathbb{R}^n)$. Define φ_ε by $\varphi_\varepsilon(x) = \varphi(x)e^{-\varepsilon|x|^2}$ $(0 < \varepsilon < 1)$ and let ϕ, ϕ_ε be the functions associated to φ and φ_ε by the rule given before (3.7.9). From (3.7.9) one gets as in the derivation of (3.7.10) that $|\partial^\alpha \widehat{\phi}_\varepsilon(x)| \le c(\alpha, \varepsilon)e^{-\omega_0|x|}$ for any multi-index α. For f, $g \in \mathbf{F}_\infty$

set $\theta(x) = \langle g, W(x)f \rangle$. Then one obtains by using (3.7.11) with φ replaced by φ_ε and an integration by parts that

$$\langle g, \varphi_\varepsilon(A)f \rangle = \int_{\mathbb{R}^n} \widehat{\varphi_\varepsilon}(x)\theta(x)\underline{dx} = \sum_{|\alpha| \leq \underline{n}} \frac{n!}{\alpha!(\underline{n}-|\alpha|)!}(i\sigma)^{2|\alpha|} \int_{\mathbb{R}^n} \widehat{\phi_\varepsilon}(x)\partial^{2\alpha}\theta(x)\underline{dx}.$$

We let $\varepsilon \to 0$ in this identity; by the remarks made after (3.7.11), one may apply the Lebesgue dominated convergence theorem to get that

$$\lim_{\varepsilon \to 0}\langle g, \varphi_\varepsilon(A)f \rangle \equiv \sum_{|\alpha| \leq \underline{n}} \frac{n!}{\alpha!(\underline{n}-|\alpha|)!}\sigma^{2|\alpha|} \cdot \int_{\mathbb{R}^n} \widehat{\phi}(x)\langle g, W(x)A^{2\alpha}f \rangle \underline{dx}.$$

On the other hand we have $||\varphi_\varepsilon(A)||_{B(\mathbf{F}_m)} \leq c$ for some constant c independent of $\varepsilon \in (0,1)$ and $m \in \mathbb{Z}$ (see (3.7.1)). Since $\mathbf{F}_\infty$ is dense in each $\mathbf{F}_m$ and $\mathbf{F}_m^* = \mathbf{F}_{-m}$, it follows that $\{\varphi_\varepsilon(A)\}$ is weakly convergent in $B(\mathbf{F}_m)$ as $\varepsilon \to 0$ for each $m \in \mathbb{Z}$. We denote the limit by $\varphi(A)$; it satisfies (3.7.1) and has the following representation for $f \in \mathbf{F}_\infty$:

$$(3.7.13) \qquad \varphi(A)f = \sum_{|\alpha| \leq \underline{n}} \frac{n!}{\alpha!(\underline{n}-|\alpha|)!}\sigma^{2|\alpha|} \int_{\mathbb{R}^n} \widehat{\phi}(x)W(x)A^{2\alpha}f\underline{dx}$$

(observe that $|\widehat{\phi}(x)| \leq ce^{-\omega_0|x|}$ by (3.7.10)).

(ii) We have obtained a linear mapping $BH^\omega(\mathbb{R}^n) \ni \varphi \mapsto \varphi(A) \in B(\mathbf{F})$ such that (3.7.1) and the property (1) in the statement of the proposition are satisfied. We now prove (2). If $\varphi_k \to \varphi$ strongly in $BH^\omega(\mathbb{R}^n)$, then $||\varphi_k(A)f - \varphi(A)f||_\mathbf{F} \to 0$ as $k \to \infty$ for each $f \in \mathbf{F}_\infty$ by (3.7.13) and the remark made after (3.7.11). By (3.7.1) and the definition of strong convergence in $BH^\omega(\mathbb{R}^n)$ we have $||\varphi_k(A)||_{B(\mathbf{F})} \leq c < \infty$ for each k. Hence $\varphi_k(A) \to \varphi(A)$ strongly in $B(\mathbf{F})$.

(iii) The uniqueness and the algebraic properties of the mapping $\varphi \mapsto \varphi(A)$ follow from the density of $BH^\omega(\mathbb{R}^n) \cap \mathcal{M}^{(\omega)}(\mathbb{R}^n)$ in $BH^\omega(\mathbb{R}^n)$. Finally, if $|\varphi(z)| \leq c(1+|z|)^{-r}$ for $z \in \Sigma_{\omega(\varphi)}$, then $\varphi_{(\alpha)}(A) = A^\alpha\varphi(A)$ belongs to $B(\mathbf{F})$ for each multi-index α with $|\alpha| \leq r$. So $\varphi(A) \in B(\mathbf{F}_m, \mathbf{F}_{m+r})$ by (3.3.27) (the proof of (3.3.27) given in Section 3.3 applies also in the situation considered here). $\quad\square$

Some particular elements of $BH^\omega(\mathbb{R}^n)$ are the functions $\varphi(x) = \langle \sigma x \rangle^{-\lambda} = (1 + \sigma^2 x^2)^{-\lambda/2}$, with $\lambda \geq 0$ and $0 \leq \sigma < \omega^{-1}$. Indeed, since $1 + \sigma^2(x+iy)^2 = \sigma^2(\sigma^{-2} - y^2) + \sigma^2(x^2 + 2ixy)$, φ has a holomorphic extension to the strip Σ_{ω_1} with $\omega_1 > \omega$ if $\sigma \leq \omega_1^{-1}$ (in fact the Fourier transform of φ can be explicitly calculated, see Chapter V.3 of [St1]). If we denote the corresponding operator $\varphi(A)$ by $\langle \sigma A \rangle^{-\lambda}$, then we have:

COROLLARY 3.7.3. *Let $\mathbf{F}$ and W be as in Proposition 3.7.2, and let $\lambda \geq 0$. Then for each multi-index α with $0 \leq |\alpha| \leq \lambda$ and for $0 < \sigma < \omega^{-1}$, the operator $A^\alpha\langle \sigma A \rangle^{-\lambda}$ belongs to $B(\mathbf{F}_m, \mathbf{F}_{m+r})$ for each $m \in \mathbb{Z}$ and each integer $r \leq \lambda - |\alpha|$.*

REMARK 3.7.4. If $0 \leq |\alpha| < \lambda$, then the function $\varphi(x) = x^\alpha\langle \sigma x \rangle^{-\lambda}$ is the Fourier transform of a rather explicit function that is exponentially decreasing at infinity and belongs to $L^1(\mathbb{R}^n)$. So $A^\alpha\langle \sigma A \rangle^{-\lambda} \in B(\mathbf{F})$ if $0 \leq |\alpha| < \lambda$ and $\mathbf{F}$

is an arbitrary Banach space. This is not true if $|\alpha| = \lambda$; for example if W is the translation group in $\mathbf{F} = L^p(\mathbb{R}^n)$, then $A^\alpha \langle \sigma A \rangle^{-\lambda} \in B(\mathbf{F})$ for $|\alpha| = \lambda$ if and only if $1 < p < \infty$.

THEOREM 3.7.5. *Let W be a C_0-representation of $\mathbb{R}^n$ in a Hilbert space $\mathbf{F}$. Then $(\mathbf{F}_k)_m = \mathbf{F}_{k+m}$ for all $k, m \in \mathbb{Z}$.*

PROOF. By virtue of the results of §3.3.4, it is enough to show that $(\mathbf{F}_k)_m \subset \mathbf{F}_{k+m}$ and $\mathbf{F}_{k-m} \subset (\mathbf{F}_k)_{-m}$ if $k \in \mathbb{Z}$ and $m > 0$. For this we use the following identity which is obtained from the binomial expansion for $(1 + \sigma^2 x^2)^m$:

$$(3.7.14) \qquad f = \sum_{|\alpha| \leq m} \frac{\sigma^{2|\alpha|} m!}{\alpha!(m - |\alpha|)!} A^\alpha \langle \sigma A \rangle^{-2m} A^\alpha f \qquad (f \in \mathbf{F}_{-\infty}).$$

First let $f \in (\mathbf{F}_k)_m$. Then $A^\alpha f \in \mathbf{F}_k$ for each $|\alpha| \leq m$ by (3.3.56). Since $A^\alpha \langle \sigma A \rangle^{-2m} \in B(\mathbf{F}_k, \mathbf{F}_{k+m})$ for $|\alpha| \leq m$ and $\sigma < \omega^{-1}$ (see Corollary 3.7.3), we have $f \in \mathbf{F}_{k+m}$. Next let $f \in \mathbf{F}_{k-m}$. Then, again by Corollary 3.7.3, $\langle \sigma A \rangle^{-2m} A^\alpha f \in \mathbf{F}_k$ if $|\alpha| \leq m$, and consequently $f \in (\mathbf{F}_k)_{-m}$ by (3.3.57). $\square$

As a preparation for the proof of the next theorem we establish some properties of the interpolation spaces associated to one-parameter C_0-groups.

LEMMA 3.7.6. *Let $\mathbf{F}$ be a Hilbert space equipped with a one-parameter C_0-group W. Assume that for each $f \in \mathbf{F}$ the function $x \mapsto \|W(x)f\|_{\mathbf{F}}^2$ is of class C^2. Then $(\mathbf{F}_2, \mathbf{F})_{1/2,2} = \mathbf{F}_1$.*

PROOF. (i) We shall exhibit in (ii) a pair of self-adjoint operators (S, T) such that $D(S) = D(A)$, the operators T, $[A, T]$ and $[S, T]$ belong to $B(\mathbf{F})$, T maps $D(A)$ into $D(A)$ and $A = S - iT$. It follows that $D(S^2) = D(A^2)$ and

$$A^2 = S^2 + 2iTS + i[S, T] - T^2, \qquad S^2 = A^2 - 2iTA - i[A, T] - T^2.$$

Hence $\mathbf{F}_1 \equiv D(A) = D(S)$, $\mathbf{F}_2 = D(A^2) = D(S^2)$ with

$$\|f\|_{\mathbf{F}_1} \sim \|f\|_{\mathbf{F}} + \|Sf\|_{\mathbf{F}}, \quad \|f\|_{\mathbf{F}_2} \sim \|f\|_{\mathbf{F}} + \|Sf\|_{\mathbf{F}} + \|S^2 f\|_{\mathbf{F}} \sim \|(I + S^2)f\|_{\mathbf{F}}.$$

By setting $\Lambda = I + S^2$, one obtains from Proposition 2.8.1 that $(\mathbf{F}_2, \mathbf{F})_{1/2,2} = D(\Lambda^{1/2}) = D(S)$ with

$$\|f\|_{(\mathbf{F}_2, \mathbf{F})_{1/2,2}} \sim \|(I + S^2)^{1/2} f\|_{\mathbf{F}} \sim \|f\|_{\mathbf{F}} + \|Sf\|_{\mathbf{F}} \sim \|f\|_{\mathbf{F}_1},$$

which proves the assertion of the lemma.

(ii) The function $x \mapsto W(x)^* W(x) \in B(\mathbf{F})$ is weakly of class C^2 by assumption. We set

$$T = -\frac{1}{2} \frac{d}{dx} [W(x)^* W(x)]|_{x=0}, \qquad T_2 = \frac{1}{2i} \frac{d^2}{dx^2} [W(x)^* W(x)]|_{x=0}.$$

Both T and T_2 belong to $B(\mathbf{F})$, and T is self-adjoint. If $f \in D(A)$, we have

(3.7.15)
$$\langle f, Tf \rangle_{\mathbf{F}} = -\frac{1}{2}\frac{d}{dx}\langle W(x)f, W(x)f \rangle_{\mathbf{F}}|_{x=0}$$
$$= -\frac{1}{2}[\langle iAf, f \rangle_{\mathbf{F}} + \langle f, iAf \rangle_{\mathbf{F}}] = \Im\langle f, Af \rangle_{\mathbf{F}}.$$

Let $S = A - iT$. (3.7.15) implies that $\langle f, Sf \rangle = \langle Sf, f \rangle$ for $f \in D(A)$. Hence, by the polarization principle, S is a symmetric operator on the domain $D(S) \equiv D(A)$. To see that it is self-adjoint, we show that $S \pm i\lambda$ are surjective if $\lambda \in \mathbb{R}$ is large enough. So let $\lambda > M\|T\|_{B(\mathbf{F})} + \omega$. Then $A \pm i\lambda$ map $D(A)$ onto $\mathbf{F}$ and $\|(A \pm i\lambda)^{-1}\|_{B(\mathbf{F})} \leq M(\lambda - \omega)^{-1} < 1/\|T\|_{B(\mathbf{F})}$ (see the beginning of the proof of Theorem 3.3.23). The identity $S \pm i\lambda = [I - iT(A \pm i\lambda)^{-1}](A \pm i\lambda)$ implies that $S \pm i\lambda$ map $D(A)$ onto $\mathbf{F}$ (observe that $I - iT(A \pm i\lambda)^{-1}$ have inverses in $B(\mathbf{F})$, so they map $\mathbf{F}$ onto $\mathbf{F}$).

Finally we observe that, for $f, g \in D(A)$:

$$\langle f, T_2 g \rangle_{\mathbf{F}} = -\frac{1}{i}\frac{d}{dx}\langle W(x)f, TW(x)g \rangle_{\mathbf{F}}|_{x=0} = \langle Af, Tg \rangle_{\mathbf{F}} - \langle f, TAg \rangle_{\mathbf{F}}.$$

So $[A, T]$, defined as a sesquilinear form on $D(A)$, has a bounded extension in $B(\mathbf{F})$. Consequently the sesquilinear form $[S, T]$, defined on $D(A) \equiv D(S)$, has an extension belonging to $B(\mathbf{F})$, and

$$\langle Sf, Tg \rangle_{\mathbf{F}} = \langle f, (TA + T_2 + iT^2)g \rangle_{\mathbf{F}} \quad \text{if } f, g \in D(A) \equiv D(S).$$

Thus $Tg \in D(S^*) = D(S)$ if $g \in D(S)$, which proves that T maps $D(A) \equiv D(S)$ into itself. $\square$

PROPOSITION 3.7.7. *If* $\mathbf{F}$ *is a Hilbert space equipped with a one-parameter* C_0-*group* W, *then* $(\mathbf{F}_2, \mathbf{F})_{1/2,2} = \mathbf{F}_1$.

PROOF. Let $\varphi \in C_0^\infty(\mathbb{R})$ be such that $\varphi \geq 0$ and $\varphi \not\equiv 0$. Let $\mathbf{G}$ be the Hilbert space obtained by introducing on the set $\mathbf{F}$ the following scalar product:

(3.7.16)
$$\langle f, g \rangle_{\mathbf{G}} = \int_{\mathbb{R}} \varphi(y)\langle W(y)f, W(y)g \rangle_{\mathbf{F}} dy.$$

Then $\|f\|_{\mathbf{G}} \sim \|f\|_{\mathbf{F}}$, so that $\mathbf{G} = \mathbf{F}$ as B-spaces. W defines a C_0-group in $\mathbf{G}$ the generator of which is identical with A; consequently $\mathbf{G}_1 = \mathbf{F}_1$ and $\mathbf{G}_2 = \mathbf{F}_2$ as B-spaces. Now observe that

$$\|W(x)f\|_{\mathbf{G}}^2 = \int_{\mathbb{R}} \varphi(y)\|W(x+y)f\|_{\mathbf{F}}^2 dy = \int_{\mathbb{R}} \varphi(y-x)\|W(y)f\|_{\mathbf{F}}^2 dy.$$

Thus the function $x \mapsto \|W(x)f\|_{\mathbf{G}}^2$ is of class C^∞. By using Lemma 3.7.6, we get as an identity between B-spaces:

$$(\mathbf{F}_2, \mathbf{F})_{1/2,2} = (\mathbf{G}_2, \mathbf{G})_{1/2,2} = \mathbf{G}_1 = \mathbf{F}_1. \quad \square$$

To understand the significance of the next theorem, the reader should return to the discussion following the proof of Theorem 3.4.9. We also use some of the notations introduced there.

THEOREM 3.7.8. *Let* $\mathbf{F}$ *be a Hilbert space equipped with a C_0-representation* W *of* $\mathbb{R}^n$. *For* $j = 1, \ldots, n$, *let* $\{\mathbf{F}_m^{(j)}\}$ *be the Sobolev scale associated to the one-parameter C_0-group* $\{\exp(i\tau A_j)\}_{\tau \in \mathbb{R}}$. *Then* $\mathbf{F}_{k,2} = \mathbf{F}_k$ *for each* $k \in \mathbb{Z}$, *and*

$$(3.7.17) \qquad \mathbf{F}_k = \bigcap_{j=1}^n \mathbf{F}_k^{(j)} \quad \text{if } k \in \mathbb{N}$$

(as identities between B-spaces).

REMARK. If $\mathbf{F}_k^{(j)}$ is identified with $D(A_j^k)$ as in Theorem 3.3.23 (c), then the last identity in the preceding theorem may be written as

$$(3.7.18) \qquad \mathbf{F}_k = \bigcap_{j=1}^n D(A_j^k) \quad \text{(as } B\text{-spaces)}.$$

PROOF. (i) We first treat the case $k = 1$. In this case the identity (3.7.17) is evident from the definitions given at the beginning of §3.3.1. Then, by using successively (3.7.17), Proposition 3.7.7, (3.4.9) and (3.4.33), one gets that

$$(3.7.19) \qquad \mathbf{F}_1 = \bigcap_{j=1}^n \mathbf{F}_1^{(j)} = \bigcap_{j=1}^n (\mathbf{F}_2^{(j)}, \mathbf{F})_{1/2,2} = \bigcap_{j=1}^n \mathbf{F}_{1,2}^{(j)} = \mathbf{F}_{1,2}.$$

(ii) Now let $k \in \mathbb{Z}$. We use the identity $\mathbf{F}_{1,2} = (\mathbf{F}_2, \mathbf{F})_{1/2,2}$ and apply (3.7.19) to the C_0-group W in $\mathbf{F}_{k-1}$ (instead of $\mathbf{F}$) to obtain that $(\mathbf{F}_{k-1})_1 = ((\mathbf{F}_{k-1})_2, \mathbf{F}_{k-1})_{1/2,2}$. By Theorem 3.7.5, this is equivalent to

$$(3.7.20) \qquad \mathbf{F}_k = (\mathbf{F}_{k+1}, \mathbf{F}_{k-1})_{1/2,2} = \mathbf{F}_{k,2}.$$

(3.7.20) implies that $\mathbf{F}_k^{(j)} = \mathbf{F}_{k,2}^{(j)}$ for $j = 1, \ldots, n$, hence (by using also (3.4.33)):

$$\mathbf{F}_k = \mathbf{F}_{k,2} = \bigcap_{j=1}^n \mathbf{F}_{k,2}^{(j)} = \bigcap_{j=1}^n \mathbf{F}_k^{(j)} \quad \text{if } k \geq 1. \quad \square$$

The preceding theorem justifies the notation $\mathbf{F}_s = \mathbf{F}_{s,2}$ for all $s \in \mathbb{R}$ if $\mathbf{F}$ is a Hilbert space equipped with a C_0-representation of $\mathbb{R}^n$. Then $\{\mathbf{F}_s\}_{s \in \mathbb{R}}$ is the *continuous Sobolev scale* associated to $(\mathbf{F}, W)$.

We now consider polynomially bounded C_0-groups in a Hilbert space. According to Section 3.6 we then have a rich functional calculus in $\mathbf{F}_{-\infty}$: to each $\varphi \in C_{\mathrm{pol}}^\infty(\mathbb{R}^n)$ one can associate a linear operator $\varphi(A) : \mathbf{F}_{-\infty} \to \mathbf{F}_{-\infty}$ which leaves $\mathbf{F}_\infty$ invariant. Since a Hilbert space is reflexive, we also have the following boundedness criterion in the Besov scale (see Theorem 3.6.9): if $\varphi \in S^0(\mathbb{R}^n)$, then $\varphi(A) \in B(\mathbf{F}_{s,p})$ for each $s \in \mathbb{R}$ and each $p \in [1, \infty]$. Together with the preceding theorem, this shows that $\varphi(A) \in B(\mathbf{F}_k)$ for all $k \in \mathbb{Z}$, in particular $\varphi(A) \in B(\mathbf{F})$. We shall now improve this criterion. We use the norms $\| \cdot \|_{BC^\sigma}$ defined in (1.3.7) and the following fact:

LEMMA 3.7.9. *Let $\sigma \geq 0$ and $\delta > 0$ be real numbers. There is a constant $c = c(\sigma, \delta)$ such that*

$$(3.7.21) \quad ||\langle Q \rangle^{-\sigma} \varphi(P) \langle Q \rangle^{\sigma}||_{B(L^2(\mathbb{R}^n))} \leq c ||\varphi||_{BC^{\sigma+\delta}} \quad \forall \varphi \in \mathscr{S}(\mathbb{R}^n).$$

PROOF. We denote the norm in $L^2(\mathbb{R}^n)$ by $|| \cdot ||$ and that in $\mathscr{H}^\sigma(\mathbb{R}^n)$ by $|| \cdot ||_{\mathscr{H}^\sigma}$. By taking the adjoint and applying a Fourier transformation, we get

$$||\langle Q \rangle^{-\sigma} \varphi(P) \langle Q \rangle^{\sigma}||_{B(L^2(\mathbb{R}^n))} = ||\langle P \rangle^{\sigma} \overline{\varphi}(Q) \langle P \rangle^{-\sigma}||_{B(L^2(\mathbb{R}^n))} = ||\overline{\varphi}(Q)||_{B(\mathscr{H}^\sigma(\mathbb{R}^n))}.$$

So it suffices to show that

$$(3.7.22) \quad ||\psi(Q)||_{B(\mathscr{H}^\sigma(\mathbb{R}^n))} \leq c ||\psi||_{BC^{\sigma+\delta}} \quad \forall \psi \in \mathscr{S}(\mathbb{R}^n).$$

If σ is an integer, then

$$||g||_{\mathscr{H}^\sigma} \sim \sum_{|\alpha| \leq \sigma} ||P^\alpha g|| = \sum_{|\alpha| \leq \sigma} ||g^{(\alpha)}||,$$

and (3.7.22) is immediate (it holds even for $\delta = 0$). If $\sigma = k + \varepsilon$ with $k \in \mathbb{N}$ and $0 < \varepsilon < 1$, the norm in $\mathscr{H}^\sigma(\mathbb{R}^n)$ is equivalent to the following expression (use (3.4.25) and (3.4.15) with $A = P$ and $p = 2$) :

$$||g||_{\mathscr{H}^\sigma} \sim \sum_{|\alpha| \leq k} ||g^{(\alpha)}|| + \sum_{|\alpha| = k} \left[\int_{|x| \leq 1} \left\| \frac{e^{iP \cdot x} - I}{|x|^\varepsilon} g^{(\alpha)} \right\|^2 \frac{dx}{|x|^n} \right]^{1/2}.$$

We replace g by $\psi(Q)f$ and assume that $\delta < 1 - \varepsilon$. The first sum is then bounded by $c ||\psi||_{BC^k} ||f||_{\mathscr{H}^\sigma}$. The second sum is bounded by

$$c \sum_{|\beta| + |\gamma| = k} \left[\int_{\mathbb{R}^n} dy \int_{|x| \leq 1} \frac{dx}{|x|^n} \frac{|\psi^{(\beta)}(x+y) f^{(\gamma)}(x+y) - \psi^{(\beta)}(y) f^{(\gamma)}(y)|^2}{|x|^{2\varepsilon}} \right]^{1/2} \leq$$

$$\leq c_1 ||\psi||_{BC^k} \cdot ||f||_{\mathscr{H}^\sigma} +$$

$$+ c_1 \sum_{|\beta| \leq k} \sup_{\substack{x, y \in \mathbb{R}^n \\ x \neq y}} \frac{|\psi^{(\beta)}(x+y) - \psi^{(\beta)}(y)|}{|x|^{\varepsilon + \delta}} \left[\int_{|x| \leq 1} \frac{dx}{|x|^{n-2\delta}} \right]^{1/2} ||f||_{\mathscr{H}^\sigma}.$$

This implies (3.7.22). $\square$

THEOREM 3.7.10. *Let $\mathbf{F}$ be a Hilbert space equipped with a C_0-representation W of $\mathbb{R}^n$ such that $||W(x)||_{B(\mathbf{F})} \leq M \langle x \rangle^r$ for some real $r \geq 0$.*

(a) *If $\varphi \in BC^\infty(\mathbb{R}^n)$, then $\varphi(A) \in B(\mathbf{F}_{s,p})$ for all $s \in \mathbb{R}$ and $p \in [1, \infty]$. In particular $\varphi(A) \in B(\mathbf{F})$.*

(b) *Let σ be a real number such that $\sigma > r + n/2$. Then for each $s \in \mathbb{R}$ and $p \in [1, \infty]$ there is a constant C such that for all $\varphi \in BC^\infty(\mathbb{R}^n)$:*

$$(3.7.23) \quad ||\varphi(A)||_{B(\mathbf{F}_{s,p})} \leq C ||\varphi||_{BC^\sigma}.$$

PROOF. (i) The result of (a) is a direct consequence of (3.7.23). To prove (b), let $\sigma > r + n/2$ be fixed. We first show that there is a constant c_0 such that for all $m \in \mathbb{Z}$ and all $\varphi \in \mathscr{S}(\mathbb{R}^n)$:

$$(3.7.24) \qquad \|\varphi(A)\|_{B(\mathbf{F}_m)} \leq c_0 \|\varphi\|_{BC^\sigma}.$$

For this we use the representation (3.7.5) for $\varphi(A)$; by the Cauchy-Schwarz inequality, we obtain that for $f \in \mathbf{F}_\infty$ and any $\sigma' \in \mathbb{R}$:

$$(3.7.25)$$

$$\|\varphi(A)f\|_{\mathbf{F}_m} \leq 2^{r/2} M \|\langle \cdot \rangle^{\sigma'} \eta\|_{L^2(\mathbb{R}^n)} \cdot \|\langle Q \rangle^{-\sigma'} \varphi(P)\Theta_f\|_{L^2(\mathbb{R}^n;\mathbf{F}_m)}$$

$$= c_1 \|\langle Q \rangle^{-\sigma'} \varphi(P)\langle Q \rangle^{\sigma'} \cdot \langle Q \rangle^{-\sigma'} \Theta_f\|_{L^2(\mathbb{R}^n;\mathbf{F}_m)}$$

$$\leq c_1 \|\langle Q \rangle^{-\sigma'} \varphi(P)\langle Q \rangle^{\sigma'}\|_{B(L^2(\mathbb{R}^n))} \cdot \|\langle Q \rangle^{-\sigma'} \Theta_f\|_{L^2(\mathbb{R}^n;\mathbf{F}_m)}.$$

We assume $\sigma' \in (r + n/2, \sigma)$. Then there is a constant c_2 such that

$$\|\langle Q \rangle^{-\sigma'} \Theta_f\|_{L^2(\mathbb{R}^n;\mathbf{F}_m)} \leq c_2 \|f\|_{\mathbf{F}_m}$$

for each $m \in \mathbb{Z}$ and each $f \in \mathbf{F}_\infty$. Now (3.7.24) follows from (3.7.25) upon using the result of Lemma 3.7.9.

(ii) Next let $\varphi \in BC^\infty(\mathbb{R}^n)$. For $0 < \varepsilon \leq 1$, define φ_ε by $\varphi_\varepsilon(x) = e^{-\varepsilon|x|^2}\varphi(x)$. We have $\varphi \in C_{(\nu)}^k(\mathbb{R}^n)$ for each $k \in \mathbb{N}$ and $\nu < 0$, whereas $\varphi_\varepsilon \in \mathscr{S}(\mathbb{R}^n) \subset C_{(\nu)}^k(\mathbb{R}^n)$. By using the fact that $|\partial^\alpha e^{-\varepsilon|x|^2}| \leq c(\alpha)$ for all $\varepsilon \in (0,1)$ and all $x \in \mathbb{R}^n$, it is easy to check that $\|\varphi - \varphi_\varepsilon\|_{C_{(\nu)}^k} \to 0$ as $\varepsilon \to 0$. Hence, by the definition of $\varphi(A)$ given in relation with (3.6.12), and by (3.7.24), we have for $f \in \mathbf{F}_\infty$:

$$(3.7.26) \quad \|\varphi(A)f\|_{\mathbf{F}_m} = \lim_{\varepsilon \to 0} \|\varphi_\varepsilon(A)f\|_{\mathbf{F}_m} \leq c_0 \sup_{0 < \varepsilon \leq 1} \|\varphi_\varepsilon\|_{BC^\sigma} \|f\|_{\mathbf{F}_m}.$$

By using the property $|\partial^\alpha e^{-\varepsilon x^2}| \leq c(\alpha)$ one finds that $\|\varphi_\varepsilon\|_{BC^\sigma} \leq c_3 \|\varphi\|_{BC^\sigma}$ for some constant c_3 depending on σ, all $\varepsilon \in (0,1]$ and all $\varphi \in BC^\infty(\mathbb{R}^n)$. So $\|\varphi(A)\|_{B(\mathbf{F}_m)} \leq c_4 \|\varphi\|_{BC^\sigma}$ for some constant c_4 depending on σ and for each $\varphi \in BC^\infty(\mathbb{R}^n)$. This inequality implies (3.7.23) by interpolation. $\square$

By virtue of (3.7.23) and Theorem 3.6.9, we have

COROLLARY 3.7.11. *Let* $s \in \mathbb{R}$, $p \in [1,\infty]$, $\sigma > r + n/2$ *and* $k \in \mathbb{R}$. *Then there is a constant* c *such that for all* $\varphi \in BC^\infty(\mathbb{R}^n)$:

$$(3.7.27) \qquad \|\varphi(A)\|_{B(\mathbf{F}_{s,p}, \mathbf{F}_{s+k,p})} \leq c\|\langle \cdot \rangle^{-k}\varphi\|_{BC^\sigma}.$$

REMARK 3.7.12. (a) We mention the following result which can be checked by using the preceding theorem and the methods of the proof of Proposition 3.7.2. Let $\{\varphi_k\}_{k \in \mathbb{N}}$ be a sequence in $BC^\infty(\mathbb{R}^n)$ such that $\lim_{k \to \infty} \varphi_k(x) = \varphi(x)$ for some $\varphi \in BC^\infty(\mathbb{R}^n)$ and each $x \in \mathbb{R}^n$, and assume that $\sup_{k \in \mathbb{N}, x \in \mathbb{R}^n} |\partial^\alpha \varphi_k(x)| < \infty$ for each multi-index α. Then $\lim_{k \to \infty} \varphi_k(A) = \varphi(A)$ strongly in each $B(\mathbf{F}_{s,p})$, $s \in \mathbb{R}$, $1 \leq p < \infty$.

(b) It is now easy to associate a bounded operator $\psi(A)$ in each $\mathbf{F}_{s,p}$ to any $\psi \in BC^\sigma(\mathbb{R}^n)$ with $\sigma > r + n/2$. By virtue of (3.7.23) it is enough to construct a sequence $\{\psi_k\}_{k\in\mathbb{N}}$ in $BC^\infty(\mathbb{R}^n)$ such that $\|\psi_k - \psi\|_{BC^{\sigma'}} \to 0$ as $k \to \infty$ for each $\sigma' < \sigma$. For this let $\theta \in C_0^\infty(\mathbb{R}^n)$ be such that $\int_{\mathbb{R}^n} \theta(y)\underline{dy} = 1$ and set $\psi_k(x) = \int_{\mathbb{R}^n} \theta(y)\psi(x - k^{-1}y)\underline{dy}$. So $\psi_k = \widehat{\theta}(k^{-1}P)\psi$, and one obtains the preceding convergence property by applying the general theory of Section 3.4 to the C_0-group $\{e^{iP\cdot x}\}$ in the Banach space of bounded uniformly continuous functions on $\mathbb{R}^n$ (for example one may use (3.4.14) with $p = q = \infty$, $t = \sigma$, $s = \sigma'$ and $\sigma = k^{-1}$).

REMARK 3.7.13. Let $\mathbf{F}$ be a Hilbert space equipped with a polynomially bounded C_0-representation W of $\mathbb{R}^n$ and let $k \in \mathbb{R}$. Then $\langle A \rangle^k$ is an isomorphism of $\mathbf{F}_s$ onto $\mathbf{F}_{s-k}$ for each $s \in \mathbb{R}$ (cf. Corollary 3.6.10 and Theorem 3.7.8). In particular, $\langle A \rangle^k$ is an isomorphism of $\mathbf{F}_k$ onto $\mathbf{F}$ for each $k \in \mathbb{Z}$. If $k = 2$ we get that $f \in \mathbf{F}_2$ if and only if $f \in \mathbf{F}$ and $A^2 f = (A_1^2 + \cdots + A_n^2)f \in \mathbf{F}$ (compare with Proposition 3.3.18 (b)). Such an assertion is not true if $\mathbf{F}$ is a Banach space; for example, it is false if $\mathbf{F} = C_\infty(\mathbb{R}^n)$ and W is the translation group ($n \geq 2$). Indeed, there is a function $f \in C_0^1(\mathbb{R}^n)$ such that $\Delta f \in C_0^0(\mathbb{R}^n)$ but $\partial^\alpha f \notin C^0(\mathbb{R}^n)$ for each multi-index α with $|\alpha| = 2$ (see page 248 in Volume I of [H]).

CHAPTER 4

Some Examples of C_0-Groups

In this chapter we present several examples of C_0-groups which are frequently used. We shall consider Banach spaces $\mathbf{F}$ embedded in $\mathscr{S}^*(\mathbb{R}^n)$. We know (see Section 1.2) two n-parameter groups acting in $\mathscr{S}^*$, namely $\{e^{i(x,P)}\}$ and $\{e^{i(x,Q)}\}$. If the Banach space $\mathbf{F}$ is invariant under one of these groups, one may define the associated Sobolev and Besov scales according to the theory of Chapter 3. This situation is described in Section 4.1. In Section 1.2 we also introduced the dilation group $e^{iD\tau}$, where $D = (P \cdot Q + Q \cdot P)/4$. It is interesting to consider the more general one-parameter group induced by a generator of the form $A = [P \cdot F(Q) + F(Q) \cdot P]/2$, where F is a vector field on $\mathbb{R}^n$. This is the topic of Section 4.2.

4.1. Weighted Sobolev and Besov Spaces

Let $X = \mathbb{R}^n$ and let $\mathbf{F}$ be a Banach space embedded in $\mathscr{S}^*(X)$ such that $e^{iQ \cdot x}\mathbf{F} \subset \mathbf{F}$ for each $x \in X$, where $Q \cdot x = (x, Q)$. If the group $\{e^{iQ \cdot x}\}$ in $\mathbf{F}$ is of class C_0, the discrete Sobolev scale $\{\mathbf{F}_k \mid k \in \mathbb{Z}\}$ and the Besov scale $\{\mathbf{F}_{s,p} \mid s \in \mathbb{R}, 1 \leq p \leq \infty\}$ are defined according to the rules of Sections 3.3 and 3.4. An explicit description of $\mathbf{F}_k$ is contained in Proposition 3.3.14, while $\mathbf{F}_{s,p}$ is obtained from the scale $\{\mathbf{F}_k\}$ by real interpolation according to Definition 3.4.2 (a). We remark that, in order for $\{e^{iQ \cdot x}\}$ to be a C_0-group in $\mathbf{F}$, it is enough that

(i) $\mathscr{S}(X) \subset \mathbf{F}$ densely and
(ii) there is a constant $c < \infty$ such that $||e^{iQ \cdot x}||_{B(\mathbf{F})} \leq c$ for $|x| \leq 1$.

For example one may take $\mathbf{F} = L^r(X)$ with $1 \leq r < \infty$. On the other hand, the group induced by $\{e^{iQ \cdot x}\}$ in $\mathbf{F}$ could be only of class C_w (this is the case if $\mathbf{F} = L^\infty(X)$); however, according to our conventions at the beginning of Section 3.1 and in Definition 3.4.2 (b), the spaces $\mathbf{F}_k$ and $\mathbf{F}_{s,p}$ are well defined for $k \geq 0$ and $s > 0$ (of course $\mathbf{F}_0$ will be different from $\mathbf{F}$ in such a case).

In the situations described above, it is usual to say that $\mathbf{F}_k$ and $\mathbf{F}_{s,p}$ are *weighted* $\mathbf{F}$-*spaces*. For example, if $\mathbf{F} - L^r(X), 1 \leq r < \infty$, the space $L^r_{s,p}(X) = [L^r(X)]_{s,p}$ is called a *weighted* L^r-*space*. We note that $L^r_k(X)$ is identical (as a B-space) with $L^r(X; \langle x \rangle^k dx)$.

W. O. Amrein et al., *C₀-Groups, Commutator Methods and Spectral Theory of N-Body Hamiltonians*, Modern Birkhäuser Classics,
DOI: 10.1007/978-3-0348-0733-3_4, © Springer Basel 1996

One may consider in the same manner the case where **F** is invariant under the translation group $\{e^{iP\cdot x}\}$. We shall denote the associated Sobolev and Besov scales with upper indices, i.e. by $\{\mathbf{F}^k\}$ and $\{\mathbf{F}^{s,p}\}$. If $\mathbf{F} = L^r(X)$, the spaces $\mathbf{F}^k$ are often denoted in the literature by $W_r^k(X)$ (e.g. in [P1]); these are the spaces originally introduced by Sobolev. The spaces $[L^r(X)]^{s,p}$, often denoted $B_r^{s,p}(X)$, are the standard Besov spaces ([P1]).

Now let us assume that the Banach space **F** is invariant under both groups $\{e^{iQ\cdot x}\}$ and $\{e^{iP\cdot x}\}$. In this situation we shall systematically denote by $\mathbf{F}_k$ and $\mathbf{F}_{s,p}$ the Sobolev and Besov spaces defined in terms of $\{e^{iQ\cdot x}\}$ and (as already said above) by $\mathbf{F}^k$ and $\mathbf{F}^{s,p}$ those associated to the group $\{e^{iP\cdot x}\}$. We claim that $\mathbf{F}^k$ and $\mathbf{F}^{s,p}$ are invariant under $e^{iQ\cdot x}$ and that $\mathbf{F}_k$ and $\mathbf{F}_{s,p}$ are invariant under $e^{iP\cdot x}$. This follows from the commutation relation:

$$(4.1.1) \qquad e^{iP\cdot y}e^{iQ\cdot x} = e^{iQ\cdot x}e^{i(P+x)\cdot y} = e^{i(Q+y)\cdot x}e^{iP\cdot y}$$

and the general fact that the group with generator A defines the same scales as the group with generator $A + a$, for any $a \in \mathbb{R}^n$. So, if μ and ν stand either for an integer k or for a couple (s, p), as above, then one obtains four types of spaces $(\mathbf{F}^\mu)_\nu$ and four types of spaces $(\mathbf{F}_\nu)^\mu$. The study of the relations between these spaces is outside the scope of this text; however a special case will be treated below.

The most important case is that where $\mathbf{F} = L^2(X) = \mathcal{H}(X) \equiv \mathcal{H}$. This case is particularly nice because $\mathcal{H}$ is a Hilbert space, the representations $\{e^{iQ\cdot x}\}$ and $\{e^{iP\cdot x}\}$ in $\mathcal{H}$ are unitary and the Fourier transformation is a unitary operator in $\mathcal{H}$ which intertwines these two representations. Let $\{\mathcal{H}_{s,p}(X)\}$ be the Besov scale associated to $\{e^{iQ\cdot x}\}$ and set $\mathcal{H}_s(X) = \mathcal{H}_{s,2}(X)$. Here s is an arbitrary real number, and there is no ambiguity of notation if $s = k$ is an integer, because of Theorem 3.7.8 (see the remark after the proof of that theorem). Similarly let $\mathcal{H}^{s,p}(X)$ and $\mathcal{H}^s(X) \equiv \mathcal{H}^{s,2}(X)$ be the spaces associated to the translation group in $\mathcal{H}$. According to the discussion above, one may consider *weighted Sobolev or Besov spaces* of the type $(\mathcal{H}^{s,p})_{t,q}$ or $(\mathcal{H}_{t,q})^{s,p}$. Further on (e.g. in Sections 7.4, 7.6 and 9.4) the spaces

$$(4.1.2) \qquad \mathcal{H}_{t,q}^s \equiv \mathcal{H}_{t,q}^s(X) = [\mathcal{H}^s(X)]_{t,q}$$

will play an important role. For this reason we shall describe them in more detail.

We denote by $\|\cdot\|$ the L^2-norm on $\mathcal{H}$, given by (1.2.2), and we identify $\mathcal{H}$ with its adjoint $\mathcal{H}^*$. So far the space $\mathcal{H}^s(X)$ has been treated as an H-space, but from now on we specify on it the following admissible norm:

$$(4.1.3) \qquad \|f\|_{\mathcal{H}^s(X)} = \|\langle P \rangle^s f\|.$$

Then $[\mathcal{H}^s(X)]^* = \mathcal{H}^{-s}(X)$ as Hilbert spaces. The spaces $\mathcal{H}_{t,q}^s$ are B-spaces; they are reflexive if $1 < q < \infty$ but non-reflexive if $q = 1$ or $q = \infty$. $\mathscr{S}(X)$ is dense in $\mathcal{H}_{t,q}^s$ if $q < \infty$ but not dense in $\mathcal{H}_{t,\infty}^s$; we denote by $\overset{\circ}{\mathcal{H}}_{t,\infty}^s$ the closure of

$\mathscr{S}(X)$ in $\mathscr{H}^s_{t,\infty}$. Then we have (see Theorem 3.4.3):

(4.1.4) $(\mathscr{H}^s_{t,q})^* = \mathscr{H}^{-s}_{-t,q'}$ if $1 \le q < \infty$, $q'^{-1} = 1 - q^{-1}$

(4.1.5) $(\overset{\circ}{\mathscr{H}^s_{t,\infty}})^* = \mathscr{H}^{-s}_{-t,1}.$

For fixed s, the family $\{\mathscr{H}^s_{t,q}\}$ is well behaved with respect to interpolation (see Theorem 3.4.3). Finally one has the following Paley-Littlewood description of these spaces: let $\theta \in C^\infty_0(X)$ be such that $\theta(x) > 0$ if $0 < a < |x| < b < \infty$ and $\theta(x) = 0$ otherwise, and let $\widetilde{\theta} \in C^\infty_0(X)$ be such that $\widetilde{\theta}(x) = 1$ if $|x| < b$. Then

(4.1.6) $||\widetilde{\theta}(Q)f||_{\mathscr{H}^s} + \left[\int_1^\infty ||r^t\theta(r^{-1}Q)f||^q_{\mathscr{H}^s} \frac{dr}{r} \right]^{1/q}$

is an admissible norm on $\mathscr{H}^s_{t,q}$ (see Theorems 3.6.2 and 3.6.14). If $a = 1/2$ and $b = 2$ then another admissible norm is given by the dyadic analogue of (4.1.6):

(4.1.7) $||\widetilde{\theta}(Q)f||_{\mathscr{H}^s} + \left[\sum_{j=0}^\infty ||2^{jt}\theta(2^{-j}Q)f||^q_{\mathscr{H}^s} \right]^{1/q}.$

The only H-spaces in the scale $\{\mathscr{H}^s_{t,q}\}$ are the spaces $\mathscr{H}^s_{t,2}$. We denote by $\mathscr{H}^s_t$ the Hilbert space obtained by fixing the following admissible norm on $\mathscr{H}^s_{t,2}$:

(4.1.8) $||f||_{\mathscr{H}^s_t(X)} = ||\langle P\rangle^s\langle Q\rangle^t f||.$

We first show that

(4.1.9) $[\mathscr{H}^s_t(X)]^* = \mathscr{H}^{-s}_{-t}(X)$ isometrically.

For this we use the convention that, if $\mathbf{E}$ is a Banach space such that $\mathscr{S}(X) \subset \mathbf{E} \subset \mathscr{S}^*(X)$ continuously and densely, then we identify its adjoint $\mathbf{E}^*$ with the space of all $g \in \mathscr{S}^*(X)$ that have a finite dual norm

$$||g||_{\mathbf{E}^*} = \sup\{|\langle f,g\rangle| \mid f \in \mathscr{S}(X), ||f||_{\mathbf{E}} \le 1\}.$$

Now, for $\mathbf{E} = \mathscr{H}^s_t(X)$, we get that

$\sup\{|\langle f,g\rangle| \mid f \in \mathscr{S}(X), ||f||_{\mathscr{H}^s_t(X)} \le 1\} =$
$= \sup\{|\langle\langle F\rangle^s\langle Q\rangle^t f, \langle F\rangle^{-s}\langle Q\rangle^{-t}y\rangle| \mid f \in \mathscr{S}(X), ||\langle F\rangle^s\langle Q\rangle^t f|| \le 1\} =$
$= \sup\{|\langle h, \langle P\rangle^{-s}\langle Q\rangle^{-t}g\rangle| \mid h \in \mathscr{S}(X), ||h|| \le 1\} = ||\langle P\rangle^{-s}\langle Q\rangle^{-t}g||,$

which proves (4.1.9).

It is an easy consequence of the inequality (1.3.16) that the family of norms $\{||\cdot||\}_{\mathscr{H}^s}$ defines the topology of $\mathscr{S}(X)$. In fact, we have

(4.1.10) $\mathscr{S}(X) = \bigcap_{s,t\in\mathbb{R}} \mathscr{H}^s_t(X), \qquad \mathscr{S}^*(X) = \bigcup_{s,t\in\mathbb{R}} \mathscr{H}^s_t(X),$

where the second equality follows by taking into account (4.1.9). By using Proposition 1.3.9 and the inequality (1.3.16) again one can easily verify that

$$(4.1.11) \qquad\qquad C_{\text{pol}}^{\infty}(X) = \bigcap_{s\in\mathbb{R}} \bigcup_{t\in\mathbb{R}} \mathscr{H}_t^s(X)$$

and that $\bigcap_{t\in\mathbb{R}} \bigcup_{s\in\mathbb{R}} \mathscr{H}_t^s(X)$ is the set of rapidly decreasing distributions (use Corollary 1.4.4 (c)).

Our next purpose is to show that $(\mathscr{H}^s)_t = (\mathscr{H}_t)^s$.

PROPOSITION 4.1.1. *For any* $s, t \in \mathbb{R}$, *the norms* $f \mapsto \|\langle P\rangle^s\langle Q\rangle^t f\|$ *and* $f \mapsto \|\langle Q\rangle^t\langle P\rangle^s f\|$ *are equivalent on* $\mathscr{S}(X)$. *In particular* $(\mathscr{H}^s)_t = (\mathscr{H}_t)^s$ *as B-spaces. In other terms the Fourier transformation is a topological isomorphism of* $\mathscr{H}_t^s(X)$ *onto* $\mathscr{H}_s^t(X)$.

PROOF. The first assertion will follow once we have shown that the operator $\langle P\rangle^s\langle Q\rangle^t\langle P\rangle^{-s}\langle Q\rangle^{-t}$ is bounded on $\mathscr{H}(X)$. One can prove this by elementary means, but in order to save space we deduce it as a special case of Corollary 5.5.5. Let $\mathbf{F}' = \mathscr{H}_t$, $\mathbf{F}'' = \mathscr{H}$, $A' = P$ acting in $\mathscr{H}_t$ and $A'' = P$ acting in $\mathscr{H}$. Then the operator $S = \langle Q\rangle^t$ belongs to $B(\mathbf{F}', \mathbf{F}'') \equiv B(\mathscr{H}_t, \mathscr{H})$ and is of class C^∞ with respect to the groups generated by A' and A''. Let $\varphi(x) = \langle x\rangle^s$ and $\psi(x) = \langle x\rangle^{-s}$. It clearly follows from Corollary 5.5.5 that $\varphi(A'')S\psi(A')$ belongs to $B(\mathbf{F}', \mathbf{F}'')$, which is equivalent with the boundedness in $\mathscr{H}(X)$ of $\langle P\rangle^s\langle Q\rangle^t\langle P\rangle^{-s}\langle Q\rangle^{-t}$.

The remaining two assertions of the proposition are interpretations of the first one. $\square$

In the remainder of this section we consider some properties of operators of the form $\varphi(Q)\psi(P)$, namely a commutator expansion and several compactness criteria.

PROPOSITION 4.1.2. *Let* φ *and* ψ *be symbols of class* S^a *and* S^b *respectively, with* $a, b \in \mathbb{R}$. *Let* $m \geq 1$ *be an integer such that* $m = a + \mu = b + \nu$ *with* μ, $\nu > 0$. *Then*

$$(4.1.12) \qquad \varphi(Q)\psi(P) = \sum_{|\alpha|<m} \frac{i^{|\alpha|}}{\alpha!}\psi^{(\alpha)}(P)\varphi^{(\alpha)}(Q) + R_m,$$

where R_m *is a bounded operator from* $\mathscr{H}_t^s(X)$ *to* $\mathscr{H}_{t+\mu}^{s+\nu}(X)$ *for all* $s, t \in \mathbb{R}$.

PROOF. (i) We again use results and some notations from Section 5.5. We first apply Theorem 5.5.3 with $\mathbf{F}' = \mathscr{H}^s$, $\mathbf{F}'' = \mathscr{H}^{s-b}$, $A' = Q$ acting in $\mathscr{H}^s$ and $A'' = Q$ acting in $\mathscr{H}^{s-b}$. Then $S \equiv \psi(P) \in B(\mathbf{F}', \mathbf{F}'')$. So (5.5.20) gives (4.1.12) with

$$R_m = \sum_{|\alpha|=m} \frac{i^{|\alpha|}}{\alpha!}\mathscr{I}_\alpha^R[\psi^{(\alpha)}(P)].$$

(ii) To obtain the indicated boudedness of R_m, we apply Theorem 5.5.2 with $\mathbf{F}' = \mathscr{H}^s$, $\mathbf{F}'' = \mathscr{H}^{s-b+m} \equiv \mathscr{H}^{s+\nu}$, $A' = Q$ acting in $\mathbf{F}'$ and $A'' = Q$ acting in $\mathbf{F}''$. We take $S = \psi^{(\alpha)}(P)$ ($|\alpha| = m$), so S belongs to $B(\mathbf{F}', \mathbf{F}'')$ and is of class C^∞ with respect to the groups generated by A' and A''. Then (cf. (5.5.24) and (5.5.14)) $\mathscr{I}_\alpha^R[\psi^{(\alpha)}(P)] = \mathscr{I}_{\varphi^{(\alpha)}}[S]$. Now $\varphi^{(\alpha)} \in S^{-\mu}(X) \subset \Sigma^{-\mu}(X)$ (because $\mu > 0$). So Theorem 5.5.2 (a) shows that $\mathscr{I}_{\varphi^{(\alpha)}}[S] \in B(\mathbf{F}'_t, \mathbf{F}''_{t+\mu})$. $\square$

We now describe some compactness criteria for operators of the form $\varphi(Q)\psi(P)$ or $[\varphi(Q), \psi(P)]$ in $\mathscr{H}(X)$. The first result is elementary and very useful.

PROPOSITION 4.1.3. *If* φ, $\psi \in L^2(X)$, *then the operator* $\varphi(Q)\psi(P)$ *is of Hilbert-Schmidt class in* $\mathscr{H}(X)$ *and its Hilbert-Schmidt norm is equal to*

$$||\varphi||_{L^2(X)} \cdot ||\psi||_{L^2(X)}.$$

If φ, $\psi \in C_\infty(X)$, *then* $\varphi(Q)\psi(P)$ *is a compact operator in* $\mathscr{H}(X)$.

PROOF. (i) We have $\varphi(Q)\psi(P)f = \varphi(Q)(\check{\psi} * f)$, so $\varphi(Q)\psi(P)$ is an integral operator with kernel $k(x, y) = \varphi(x)\check{\psi}(x - y)$. Its Hilbert-Schmidt norm as an operator in $\mathscr{H}(X)$ is given by the expression

$$\left[\iint_{X \times X} |k(x, y)|^2 dx\, dy\right]^{1/2} = ||\varphi||_{L^2(X)} \cdot ||\psi||_{L^2(X)}.$$

(ii) For $k \in \mathbb{N}$, define $\eta_k : X \to \mathbb{R}$ by $\eta_k(x) = 1$ if $|x| \le k$ and $\eta_k(x) = 0$ if $|x| > k$. If φ, $\psi \in C_\infty(X)$, set $\varphi_k = \eta_k\varphi$ and $\psi_k = \eta_k\psi$. We have φ_k, $\psi_k \in L^2(X)$ and $\varphi_k \to \varphi$, $\psi_k \to \psi$ in $L^\infty(X)$ as $k \to \infty$. Then $||\varphi_k(Q)\psi_k(P) - \varphi(Q)\psi(P)||_{B(\mathscr{H})} \to 0$ as $k \to \infty$, so that $\varphi(Q)\psi(P)$ is compact as the norm limit of the sequence of Hilbert-Schmidt operators $\{\varphi_k(Q)\psi_k(P)\}$. $\square$

COROLLARY 4.1.4. *If* φ, $\psi \in \mathscr{H}_t(X)$ *for some* $t > \frac{1}{2}\dim X$, *then* $\varphi(Q)\psi(P)$ *is a trace class operator in* $\mathscr{H}(X)$ *with trace norm bounded by* $C||\varphi||_{\mathscr{H}_t} \cdot ||\psi||_{\mathscr{H}_t}$, *where* C *is a constant depending only on* t *and* $\dim X$.

PROOF. We write

$$\varphi(Q)\psi(P) = [\varphi(Q)\langle Q\rangle^t \cdot \langle P\rangle^{-t}] \cdot [\langle P\rangle^t \langle Q\rangle^{-t} \langle P\rangle^{-t} \langle Q\rangle^t] \cdot [\langle Q\rangle^{-t} \cdot \langle P\rangle^t \psi(P)].$$

The first and the third factor on the r.h.s. are Hilbert-Schmidt operators by the preceding proposition and the second factor is a bounded operator (see the proof of Proposition 4.1.1). $\square$

PROPOSITION 4.1.5. *If* $s_1 \ge s_2$ *and* $t_1 \ge t_2$, *then* $\mathscr{H}_{t_1}^{s_1}(X) \subset \mathscr{H}_{t_2}^{s_2}(X)$ *continuously. If* $s_1 > s_2$ *and* $t_1 > t_2$, *then the preceding embedding is compact.*

PROOF. Let us write

(4.1.13) $\langle P\rangle^{s_2}\langle Q\rangle^{t_2} = [\langle P\rangle^{s_2}\langle Q\rangle^{t_2-t_1}\langle P\rangle^{-s_2}\langle Q\rangle^{t_1-t_2}] \cdot$

$$\cdot [\langle Q\rangle^{t_2-t_1}\langle P\rangle^{s_2-s_1}] \cdot [\langle P\rangle^{s_1}\langle Q\rangle^{t_1}].$$

The first factor on the r.h.s. is a bounded operator in $\mathscr{H}(X)$. The second factor is bounded if $s_1 \ge s_2, t_1 \ge t_2$, and it is compact if $s_1 > s_2$ and $t_1 > t_2$ by Proposition 4.1.3. In particular, (4.1.13) implies that $||f||_{\mathscr{H}_{t_2}^{s_2}(X)} \le c||f||_{\mathscr{H}_{t_1}^{s_1}(X)}$. $\square$

Let us now consider the commutator $[\varphi(Q), \psi(P)]$. It can be a compact operator even if φ or ψ do not converge to zero at infinity (then neither $\varphi(Q)\psi(P)$ nor $\psi(P)\varphi(Q)$ are compact in general). However, this happens only if φ and ψ behave in a rather regular way at infinity. For example, if $\varphi, \psi \in S^1(X)$ and $\varphi^{(\alpha)}, \psi^{(\alpha)}$ converge to zero at infinity for $|\alpha| = 1$, then $[\varphi(Q), \psi(P)]$ is compact in $\mathcal{H}(X)$ (this is a straightforward consequence of Propositions 4.1.1 and 4.1.3). But a much deeper fact is true:

THEOREM 4.1.6 (CORDES). *If $\varphi, \psi \in C^\infty(X)$ and all their derivatives of order ≥ 1 tend to zero at infinity, then $[\varphi(Q), \psi(P)]$ is a compact operator in $\mathcal{H}(X)$.*

In this theorem the functions φ and ψ could be unbounded. The result of the theorem is suggested by the expansion (4.1.12), but the rigorous proof, due to H. Cordes [Cr], is far from trivial. For the moment we shall deduce an important consequence (also due to Cordes) and give an elementary proof of a special case that often appears in applications.

DEFINITION 4.1.7. *A continuous function $f : X \to \mathbb{C}$ is of vanishing oscillation at infinity if*

$$(4.1.14) \qquad \lim_{|x| \to \infty} \sup_{|y| \leq 1} |f(x + y) - f(x)| = 0.$$

EXAMPLES 4.1.8. (a) If $f \in C^1(X)$ and all its partial derivatives of the first order vanish at infinity, then f is of vanishing oscillation at infinity.

(b) If f is continuous on X and homogeneous of degree zero outside the unit ball of X (i.e. $f(x) = f\left(\frac{x}{|x|}\right)$ if $|x| \geq 1$), then f is of vanishing oscillation at infinity. In fact a simple geometric consideration shows that

$$|f(y + x) - f(x)| \leq \mu\left(\frac{|y|}{(|x|^2 - |y|^2)^{1/2}}\right) \quad \text{if } |x| \geq 2(1 + |y|),$$

where μ is the modulus of continuity of f on the unit sphere:

$$\mu(\varepsilon) = \sup\{|f(\omega_1) - f(\omega_2)| \mid |\omega_1| = |\omega_2| = 1, |\omega_1 - \omega_2| \leq \varepsilon\}.$$

In connection with the next theorem, it is interesting to observe that the space of functions which are bounded, continuous and of vanishing oscillation at infinity is a C^*-subalgebra of $BC(X)$ which contains $C_\infty(X)$ and the constants; moreover, it is just the closure in $BC(X)$ of the set of $f \in BC^\infty(X)$ such that $f^{(\alpha)}(x) \to 0$ if $|x| \to \infty$ for each multi-index $\alpha \neq 0$:

LEMMA 4.1.9. *If $f \in BC(X)$ is of vanishing oscillation at infinity, then there is a sequence $\{f_k\}$ in $BC^\infty(X)$ such that*

$$\partial^\alpha f_k \in C_\infty(X) \text{ if } \alpha \neq 0 \text{ and } \|f_k - f\|_{L^\infty(X)} \to 0 \text{ as } k \to \infty.$$

PROOF. Since f is continuous and of vanishing oscillation at infinity, we have

$$(4.1.15) \qquad \lim_{\varepsilon \to 0} \sup_{\substack{x \in X \\ |z| \leq 1}} |f(x + \varepsilon z) - f(x)| = 0.$$

Choose $\eta \in C_0^\infty(X)$ such that $\eta(x) = 0$ if $|x| \geq 1$ and $\int_X \eta(x)dx = 1$, and for $\varepsilon \in (0,1)$ define $f_\varepsilon = \hat{\eta}(\varepsilon P)f$. Then (see e.g. (1.2.15)):

$$f_\varepsilon(x) = \int_X f(x - \varepsilon z)\eta(z)dz = \varepsilon^{-n} \int_X f(y)\eta\Big(\frac{x - y}{\varepsilon}\Big)dy.$$

This implies that f_ε is of class C^∞ and bounded (because $f \in L^\infty(X)$). Furthermore we have

$$f_\varepsilon(x) - f(x) = \int_X [f(x - \varepsilon z) - f(x)]\eta(z)dz,$$

so that $\|f_\varepsilon - f\|_{L^\infty(X)} \to 0$ as $\varepsilon \to 0$ by virtue of (4.1.15).

Next let α be a multi-index with $|\alpha| \geq 1$. Then $\int_X (\partial^\alpha \eta)(z)dz = 0$, so that

$$\partial^\alpha f_\varepsilon(x) = \varepsilon^{-|\alpha|} \int_X f(x - \varepsilon z)(\partial^\alpha \eta)(z)dz$$

$$= \varepsilon^{-|\alpha|} \int_X [f(x - \varepsilon z) - f(x)](\partial^\alpha \eta)(z)dz.$$

By taking into account (4.1.14), one sees that $\partial^\alpha f_\varepsilon \in C_\infty(X)$. $\square$

THEOREM 4.1.10 (CORDES). *If* $\varphi, \psi : X \to \mathbb{C}$ *are bounded, continuous and of vanishing oscillation at infinity, then* $[\varphi(Q), \psi(P)]$ *is a compact operator in* $\mathscr{H}(X)$.

PROOF. Choose $\varphi_k, \psi_k \in BC^\infty(X)$ such that $\partial^\alpha \varphi_k, \partial^\alpha \psi_k \in C_\infty(X)$ if $|\alpha| \geq 1$ and $\varphi_k \to \varphi$, $\psi_k \to \psi$ in $L^\infty(X)$ as $k \to \infty$ (see Lemma 4.1.9). Then $[\varphi_k(Q), \psi_k(P)]$ is compact for each $k \in \mathbb{N}$ by Theorem 4.1.6. The result of the present theorem follows because $[\varphi_k(Q), \psi_k(P)]$ converges to $[\varphi(Q), \psi(P)]$ in the norm of $B(\mathscr{H})$ as $k \to \infty$. $\square$

Theorem 4.1.10 is especially important in the case where one of the two functions φ and ψ belongs to $C_\infty(X)$. We now give a direct elementary proof of this result:

PROOF OF THEOREM 4.1.10 IN THE CASE $\psi \in C_\infty(X)$. Since $\mathscr{S}(X)$ is dense in $C_\infty(X)$, we may assume that $\psi \in \mathscr{S}(X)$.

(i) We have (see (1.2.13)):

$$[\varphi(Q), \psi(P)] = \int_X [\varphi(Q), e^{i(y,P)}]\hat{\psi}(y)dy$$

$$= \int_X \{\varphi(Q) - e^{i(y,P)}\varphi(Q)e^{-i(y,P)}\}e^{i(y,P)}\hat{\psi}(y)dy$$

$$= \int_X \{\varphi(Q) - \varphi(Q + y)\}\hat{\psi}(y)e^{i(y,P)}dy \equiv \int_X \Psi(y)e^{i(y,P)}dy.$$

Here $\Psi : X \to B(\mathcal{H})$ is defined by $\Psi(y) = \widehat{\psi}(y)(\varphi(Q) - \varphi(Q+y))$. We notice the following properties of Ψ:

(α) For fixed $y \in X$, $\Psi(y)$ is multiplication by a function $\theta_y \in C_\infty(X)$, viz. $\theta_y(x) = \widehat{\psi}(y)(\varphi(x) - \varphi(x+y))$;

(β) $\|\Psi(y)\|_{B(\mathcal{H})} \leq 2|\widehat{\psi}(y)| \cdot \|\varphi\|_{L^\infty(X)}$, which is rapidly decreasing at infinity;

(γ) Ψ is norm-continuous (because φ is bounded and uniformly continuous on X, cf. (4.1.15)).

(ii) We show now that for any Ψ with properties (α)-(γ) the operator $S \equiv \int_X \Psi(y)e^{i(y,P)}dy$ is compact in $\mathcal{H}$. We fix a function $\eta \in \mathscr{S}(X)$ with $\int_X \eta(x)dx = 1$ and set, for $\varepsilon > 0$: $\eta_\varepsilon(x) = \varepsilon^{-n}\eta(\varepsilon^{-1}x)$. Then

$$(4.1.16)\, S_\varepsilon \equiv \int_X (\eta_\varepsilon * \Psi)(x)e^{i(x,P)}dx = \int_X dx \int_X dy\, \eta_\varepsilon(x-y)\Psi(y)e^{i(x,P)}$$
$$= \int_X \Psi(y)\widetilde{\eta}_\varepsilon(P)e^{i(y,P)}dy = \int_X \theta_y(Q)\widetilde{\eta}_\varepsilon(P)e^{i(y,P)}dy.$$

By Proposition 4.1.3, $\theta_y(Q)\widetilde{\eta}_\varepsilon(P)$ is a compact operator for each $y \in X$ and each $\varepsilon > 0$, and it is norm-continuous as a function of y, by (γ). Since $y \mapsto e^{i(y,P)}$ is strongly continuous, an easy argument shows that the integrand of the last integral in (4.1.16) is a rapidly decreasing norm-continuous function of y with values in the Banach space of compact operators on $\mathcal{H}$. Hence S_ε is a compact operator. Finally we have:

$$\|S_\varepsilon - S\|_{B(\mathcal{H})} = \left\|\int_X dx \int_X dy[\Psi(y-x) - \Psi(y)]\eta_\varepsilon(x)e^{i(y,P)}\right\|_{B(\mathcal{H})}$$
$$\leq \int_X dz\, |\eta(z)| \int_X dy\|\Psi(y-\varepsilon z) - \Psi(y)\|_{B(\mathcal{H})},$$

which converges to zero as $\varepsilon \to 0$ by the Lebesgue dominated convergence theorem (for each fixed $z \in X$ the norm of $\|\Psi(\cdot - \varepsilon z) - \Psi(\cdot)\|_{B(\mathcal{H})}$ in $L^1(X)$ is bounded by $2\int_X \|\Psi(y)\|_{B(\mathcal{H})}dx$ and converges to zero as $\varepsilon \to 0$, by (β) and (γ)). $\square$

4.2. C_0-Groups Associated to Vector Fields

In Section 1.2 we introduced the one-parameter group $W(\tau) = \exp(2iD\tau)$ in $\mathscr{S}^*(X)$ induced by the dilations of the euclidean space X, see (1.2.17). Since $W(-\tau)PW(\tau) = e^\tau P$, it is easily seen (by using a Fourier transformation) that W induces a C_0-group in each Sobolev space $\mathcal{H}^s(X)$ and that $\|W(\tau)\|_{B(\mathcal{H}^s)} = \max\{1, e^{\tau s}\}$. For $s \neq 0$, the group is of exponential growth in $\mathcal{H}^s(X)$, which is rather inconvenient because it allows only a very restricted functional calculus for D. We now discuss a considerably more general situation.

DEFINITION 4.2.1. Let $\varphi : \mathbb{R} \times X \to X$ and, for $\tau \in \mathbb{R}$, define $\varphi_\tau : X \to X$ by $\varphi_\tau(x) = \varphi(\tau, x)$. We say that φ is a C^∞ flow on X if φ is a function of class C^∞ and

(1) $\varphi(0, x) = x$ for all $x \in X$,

(2) $\varphi_\tau \circ \varphi_\nu = \varphi_{\tau+\nu}$ for all $\tau, \nu \in \mathbb{R}$.

If φ is a C^∞ flow on X, then it is clear that $\varphi_\tau : X \to X$ is a C^∞ diffeomorphism with $\varphi_{-\tau}$ as its inverse. For $\tau \in \mathbb{R}$ we set

$$(4.2.1) \qquad J_\tau(x) = \det \nabla \varphi_\tau(x).$$

Clearly $J_\tau : X \to \mathbb{R}$ is of class C^∞, $J_0(x) = 1$ and $J_\tau(x) > 0$ for each $x \in X$.

Now we may define $W(\tau)$ by

$$(4.2.2) \qquad [W(\tau)f](x) = \{J_\tau(x)\}^{1/2} f(\varphi_\tau(x))$$

for any distribution f on X. We shall say that the family $\{W(\tau)\}_{\tau \in \mathbb{R}}$, thus defined, is *the C_0-group associated to the flow φ*. One can easily prove the following statements [1]:

    (a) $W(0) = I$, $W(\tau)W(\nu) = W(\tau + \nu)$ on all $\mathcal{D}^*(X)$,

    (b) $W(\tau)$ leaves $\mathcal{D}(X)$ and $\mathcal{H}(X)$ invariant,

    (c) $\{W(\tau)\}_{\tau \in \mathbb{R}}$ is a (strongly continuous) unitary group in $\mathcal{H}(X)$.

In general $\mathcal{S}(X)$ and $\mathcal{H}^k(X)$ are not invariant under $W(\tau)$ (because φ_τ could grow too rapidly at infinity).

In order to calculate the generator of the C_0-group W introduced in (4.2.2) we recall some facts concerning systems of ordinary differential equations. Let us define the *generator of the flow φ* as the C^∞ vector field $F : X \to X$ given by

$$(4.2.3) \qquad F(x) = \frac{d}{d\tau}\varphi(\tau, x)|_{\tau=0}.$$

Since

$$\frac{d}{d\tau}\varphi(\tau, x) = \frac{d}{d\nu}\varphi_{\tau+\nu}(x)|_{\nu=0} = \frac{d}{d\nu}\varphi_\nu(\varphi_\tau(x))|_{\nu=0} = F(\varphi_\tau(x)),$$

we see that, for each $x \in X$, the function $\mathbb{R} \ni \tau \mapsto \varphi_\tau(x) \in X$ is a solution of the following system

$$(4.2.4) \qquad \begin{cases} \frac{d}{d\tau}\varphi_\tau(x) = F(\varphi_\tau(x)) \\ \varphi_\tau(x)|_{\tau=0} = x. \end{cases}$$

The uniqueness theorem for ordinary differential equations shows that F determines φ uniquely. Moreover, there is a bijective correspondence between C^∞ flows φ on X and completely integrable C^∞ vector fields $F : X \to X$, i.e. C^∞ mappings $F : X \to X$ such that, for each $x \in X$, the unique solution of the system $dx(\tau)/d\tau = F(x(\tau))$, $x(0) = x$ exists for all $\tau \in \mathbb{R}$. We refer to Theorem 1.8.8 and §2.10 of [N] for more details.

Now observe that

$$\frac{d}{d\tau}\nabla\varphi_\tau(x) = \nabla F \circ \varphi_\tau(x) = \nabla F(\varphi_\tau(x)) \cdot \nabla\varphi_\tau(x),$$

---

[1] $\mathcal{D}^*(X)$ denotes the set of continuous anti-linear functionals (distributions) on the space of test functions $\mathcal{D}(X)$ (see e.g. Section V.4 of [RS]).

where the dot on the r.h.s. means multiplication in $L(X)$ (algebra of linear operators on X). Thus the function $\mathbb{R} \ni \tau \mapsto \nabla\varphi_\tau(x) \in L(X)$ is the solution of the linear system

(4.2.5)
$$\begin{cases} \frac{d}{d\tau}\nabla\varphi_\tau(x) = \nabla F(\varphi_\tau(x)) \cdot \nabla\varphi_\tau(x) \\ \nabla\varphi_0(x) = I_X \end{cases}$$

where I_X denotes the identity mapping on X. With respect to some orthonormal basis $(v_1, \ldots, v_n)$, the above system reads

(4.2.6)
$$\begin{cases} \frac{d}{d\tau}\partial_k\varphi_{\tau,j}(x) = \sum_{\ell=1}^n [\partial_\ell F_j(\varphi_\tau(x))] \cdot [\partial_k\varphi_{\tau,\ell}(x)] \\ \partial_k\varphi_{0,j}(x) = \delta_{jk}. \end{cases}$$

One can express $J_\tau \equiv \det \nabla\varphi_\tau$ in terms of the generator F of the flow φ. If we define

(4.2.7)
$$\operatorname{div} F(x) = \operatorname{Tr} \nabla F(x) = \sum_{j=1}^n \partial_j F_j(x),$$

then

(4.2.8)
$$J_\tau(x) = \exp\left\{ \int_0^\tau (\operatorname{div} F)(\varphi_\nu(x))d\nu \right\}.$$

The following technical lemma will be useful.

LEMMA 4.2.2. *Let φ be a C^∞ flow on X, F its generator and assume that all derivatives $\partial^\alpha F$ of order $|\alpha| \geq 1$ are bounded. Then:*

(a) *If α is a multi-index with $|\alpha| \geq 1$, then there is a constant $c = c(\alpha) \in (0, \infty)$ such that:*

(4.2.9)
$$|\partial^\alpha \varphi_\tau(x)| \leq ce^{c|\tau|} \quad \forall x \in X, \forall \tau \in \mathbb{R}.$$

(b) *Let $c = \sup\{\|\nabla F(x)\|_{L(X)} \mid x \in X\}$; then*

(4.2.10)
$$e^{-c|\tau|}|x - y| \leq |\varphi_\tau(x) - \varphi_\tau(y)| \leq e^{c|\tau|}|x - y| \quad \forall x, y \in X, \forall \tau \in \mathbb{R}.$$

PROOF. (i) For (a) we proceed by induction on the order $|\alpha|$ of α. We prove (4.2.9) for $\tau \geq 0$; the proof for $\tau < 0$ is essentially the same after a change of variables $\tau \mapsto -\tau$. First assume that $|\alpha| = 1$ and remark that, by (4.2.5):

(4.2.11)
$$\nabla\varphi_\tau(x) = I_X + \int_0^\tau \nabla F(\varphi_\nu(x)) \cdot \nabla\varphi_\nu(x)d\nu.$$

Hence, for $c = \sup_{x\in X} \|\nabla F(x)\|_{L(X)}$:

$$\|\nabla\varphi_\tau(x)\|_{L(X)} \leq 1 + c \int_0^\tau \|\nabla\varphi_\nu(x)\|_{L(X)}d\nu.$$

By Gronwall's lemma (see Appendix 7.A) we then have

(4.2.12)
$$\|\nabla\varphi_\tau(x)\|_{L(X)} \leq e^{c\tau} \quad \forall \tau \geq 0,$$

which implies (4.2.9) for $|\alpha| = 1$ and $\tau \geq 0$.

Now let $|\alpha| > 1$. By (4.2.6) we have

$$\partial^\alpha \varphi_{0,j}(x) = 0$$

and

$$\frac{d}{d\nu} \partial^\alpha \varphi_{\nu,j}(x) = \sum_{\ell=1}^n [\partial_\ell F_j(\varphi_\nu(x))][\partial^\alpha \varphi_{\nu,\ell}(x)] + R_j,$$

where R_j is a finite sum of terms each of which is the product of a factor $\partial^\gamma F_j(\varphi_\nu(x))$ and of a finite number of factors of the form $\partial^\delta \varphi_{\nu,k}(x)$, where the multi-indices γ and δ satisfy $|\gamma| \geq 2$ and $|\delta| \leq |\alpha| - 1$. By using the hypothesis made on F (boundedness of all its derivatives) and the induction hypothesis, one obtains that

$$\sum_{j=1}^n |\partial^\alpha \varphi_{\tau,j}(x)| = \sum_{j=1}^n \left| \int_0^\tau \left[\frac{d}{d\nu} \partial^\alpha \varphi_{\nu,j}(x) \right] d\nu \right|$$

$$\leq c_1 \int_0^\tau \left[\sum_{j=1}^n |\partial^\alpha \varphi_{\nu,j}(x)| \right] d\nu + \tau \sum_{j=1}^n |R_j|$$

and

$$|R_j| \leq c_2 e^{c_2 \tau} < \infty \qquad \forall x \in X, \forall \tau \geq 0.$$

By Gronwall' s lemma, this leads to

$$\sum_{j=1}^n |\partial^\alpha \varphi_{\tau,j}(x)| \leq c_3 e^{c_3 \tau} \qquad \forall x \in X, \forall \tau \geq 0,$$

which completes the inductive proof of (4.2.9).

(ii) The second inequality in (4.2.10) is obtained by using the mean value theorem and (4.2.12). By setting $x = \varphi_{-\tau}(x')$, $y = \varphi_{-\tau}(y')$ in the second inequality in (4.2.10) one gets that

$$|x' - y'| \leq e^{c|\tau|} |\varphi_{-\tau}(x') - \varphi_{-\tau}(y')| \qquad \forall x', y' \in X,$$

which implies the first inequality in (4.2.10). $\square$

We now determine the generator of the C_0-group W.

PROPOSITION 4.2.3. *Let φ be a C^∞ flow on the euclidean space X, $F : X \to X$ its generator and $\{W(\tau)\}_{\tau \in \mathbb{R}}$ the C_0-group associated to φ by (4.2.2). Then the operator*

$$(4.2.13) \qquad \frac{1}{2}[P \cdot F(Q) + F(Q) \cdot P] = \frac{1}{2} \sum_{j=1}^n [P_j F_j(Q) + F_j(Q) P_j]$$

is essentially self-adjoint on $C_0^\infty(X)$ in $\mathcal{H}(X)$, and its closure A is the generator of W, i.e. $W(\tau) = \exp(iA\tau)$.

PROOF. Let A denote the generator of W. Then A is a self-adjoint operator in $\mathscr{H}(X)$ and, by Nelson's lemma (see e.g. Theorem 3.3.4) and the fact that $W(\tau)$ leaves $C_0^\infty(X)$ invariant, it is essentially self-adjoint on $C_0^\infty(X)$. But one can explicitly calculate the action of A on functions f in $C_0^\infty(X)$:

$$
\begin{aligned}
(iAf)(x) &= \frac{d}{d\tau}[W(\tau)f](x)|_{\tau=0} \\
&= \frac{d}{d\tau}[J_\tau(x)]^{1/2}|_{\tau=0} f(x) + \frac{d}{d\tau}f(\varphi_\tau(x))|_{\tau=0} \\
&= \frac{1}{2}[J_\tau(x)]^{-1/2}\frac{d}{d\tau}J_\tau(x)|_{\tau=0} f(x) + \nabla f(x) \cdot \frac{d}{d\tau}\varphi_\tau(x)|_{\tau=0} \\
&= \frac{1}{2}[\operatorname{div} F(x)]f(x) + [F(x) \cdot \nabla]f(x),
\end{aligned}
$$

which is identical with the action of the operator (4.2.13) on f. □

In most cases the flow φ cannot be calculated explicitly (an exception is the dilation group where $\varphi(\tau, x) = e^\tau x$ and $F(x) = x$). Therefore it is useful to have conditions directly on its generator F in order to ensure its complete integrability. This is the purpose of the following proposition.

PROPOSITION 4.2.4. *If $F : X \to X$ is of class C^∞ and all its derivatives $\partial^\alpha F$ of order $|\alpha| \geq 1$ are bounded (so that F may grow at most linearly), then F is a completely integrable vector field on X. Furthermore the C_0-group W associated to its flow φ leaves $\mathscr{S}(X)$ and $\mathscr{H}_t^s(X)$ invariant, for all $s, t \in \mathbb{R}$.*

PROOF. (i) For the complete integrability of F, it would suffice to require that $|F(x) - F(y)| \leq c|x - y|$ for some constant $c \in (0, \infty)$ and all $x, y \in X$. For details we refer to Theorem 1.8.4 and Remark 1.8.7 of [N].

(ii) To prove the last statement of the proposition, we shall use the properties (4.2.9) and (4.2.10) of the flow φ. We first show that there is a constant $C \in (0, \infty)$ such that for all $x \in X$ and all $\tau \in \mathbb{R}$:

(4.2.14) $e^{-C|\tau|} \leq J_\tau(x) \leq e^{C|\tau|}.$

Indeed, since $\operatorname{div} F$ is bounded on X by hypothesis, there is a constant $C \in (0, \infty)$ such that $|\int_0^\tau \operatorname{div} F(\varphi_\nu(x))d\nu| \leq C|\tau|$ for all $x \in X$, so that (4.2.14) follows from (4.2.8).

By the definition of J_τ and by (4.2.14) and (4.2.9), we see that, for each fixed $\tau \in \mathbb{R}$, the function $[J_\tau]^{1/2} : X \to X$ is bounded, of class C^∞ and with bounded derivatives. In particular, in order to prove that $W(\tau)$ leaves $\mathscr{H}_t^s(X)$ invariant, it suffices to show that

(4.2.15) $f \in \mathscr{H}_t^s(X) \implies f \circ \varphi_\tau \in \mathscr{H}_t^s(X),$

which will be done in (iii) below. (4.2.15) also implies that $W(\tau)$ leaves $\mathscr{S}(X)$ invariant, because $\mathscr{S}(X) = \cap_{s,t \geq 0}\mathscr{H}_t^s(X)$.

(iii) To prove (4.2.15) we first observe that, if α is a multi-index with $|\alpha| \neq 0$, then $\partial^\alpha(f \circ \varphi_\tau)$ is a finite sum of terms each of which is the product of a factor $f^{(\beta)} \circ \varphi$ (with $|\beta| \leq |\alpha|$) and of a finite number of factors of the form $\partial^\gamma \varphi_{\nu,k}(x)$

with $0 < |\gamma| \leq |\alpha|$. By taking into account (4.2.9), one sees that for each multi-index α there is a constant $c_\alpha = c_\alpha(\tau)$ such that for all $f \in \mathscr{S}(X)$ and all $x \in X$:

$$(4.2.16) \qquad |\partial^\alpha (f \circ \varphi_\tau)(x)| \leq c_\alpha \sum_{|\beta| \leq |\alpha|} |(f^{(\beta)} \circ \varphi_\tau)(x)|.$$

Now let $s \geq 0$ be an integer and $f \in \mathscr{S}(X)$. Then, by (4.2.16) and Proposition 4.1.1:

$$\|f \circ \varphi_\tau\|^2_{\mathscr{H}^s_t(X)} \leq c \sum_{|\alpha| \leq s} \|\langle Q \rangle^t \partial^\alpha (f \circ \varphi_\tau)\|^2_{\mathscr{H}(X)}$$

$$\leq c_1 \sum_{|\beta| \leq s} \|\langle Q \rangle^t f^{(\beta)} \circ \varphi_\tau\|^2_{\mathscr{H}(X)}$$

$$= c_1 \sum_{|\beta| \leq s} \int_X \langle x \rangle^{2t} |f^{(\beta)}(\varphi_\tau(x))|^2 \, dx$$

$$= c_1 \sum_{|\beta| \leq s} \int_X \langle \varphi_{-\tau}(y) \rangle^{2t} |f^{(\beta)}(y)|^2 J_{-\tau}(y) \, dy.$$

Since $J_{-\tau}$ is bounded on X and $\langle \varphi_{-\tau}(y) \rangle \leq c_2 \langle y \rangle$, we see that

$$(4.2.17) \qquad \|f \circ \varphi_\tau\|^2_{\mathscr{H}^s_t(X)} \leq c_3 \|f\|^2_{\mathscr{H}^s_t(X)} \qquad \forall f \in \mathscr{S}(X),$$

where $c_3 = c_3(\tau)$. This proves (4.2.15) if s is a non-negative integer and $t \in \mathbb{R}$.

Let us denote by T_τ the linear operator defined as $T_\tau f = f \circ \varphi_\tau$. We have shown that $T_\tau \in B(\mathscr{H}^s_t(X))$ or equivalently that $\langle Q \rangle^t T_\tau \langle Q \rangle^{-t} \in B(\mathscr{H}^s(X))$ if s is a non-negative integer. By interpolating one then obtains that $\langle Q \rangle^t T_\tau \langle Q \rangle^{-t} \in B(\mathscr{H}^s(X))$ for each real $s \geq 0$, which establishes (4.2.15) for $s \geq 0$ and $t \in \mathbb{R}$.

Finally we consider the case $s < 0$. Since $\mathscr{H}^s_t(X)^* = \mathscr{H}^{-s}_{-t}(X)$ and $T_{-\tau} \in B(\mathscr{H}^{-s}_{-t}(X))$ if $s < 0$, we know that $T^*_{-\tau} \in B(\mathscr{H}^s_t(X))$ if $s < 0$. But, for f, $g \in \mathscr{S}(X)$:

$$\int_X f(x) g(\varphi_{-\tau}(x)) \, dx = \int_X f(\varphi_\tau(y)) g(y) J_\tau(y) \, dy,$$

so that $T^*_{-\tau} = J_\tau(Q) T_\tau$. Hence, as an operator on $\mathscr{H}^s_t(X)$ with $s < 0$, T_τ is equal to $[J_\tau(Q)]^{-1} T^*_{-\tau}$, which belongs to $B(\mathscr{H}^s_t(X))$ because $[J_\tau]^{-1} \in BC^\infty(X)$. $\square$

We have seen in Section 3.6 that it is possible to develop a sufficiently general functional calculus for the generator of a C_0-group of polynomial growth. We shall now determine a class of vector fields F on X for which the C_0-group with generator (4.2.13) is of polynomial growth in $\mathscr{H}^2(X)$ (in fact in all Sobolev spaces, but $\mathscr{H}^2$ is especially interesting in the applications to Schrödinger operators). The dilation group corresponds to $F(x) = x$, and we have seen that the associated C_0-group is *not* of polynomial growth. At the end of this section we shall cite a result showing that the operator γ which corresponds to $F(x) = x \langle x \rangle^{-1}$ has the same drawback: the group $\{\exp(i\gamma\tau)\}_{\tau \in \mathbb{R}}$ is not of polynomial growth in $\mathscr{H}^1(X)$ (hence neither in $\mathscr{H}^2(X)$; we mention that the operator

γ played an important role in the proof of asymptotic completeness for N-body hamiltonians given by Sigal and Soffer).

First let us consider an arbitrary C^∞ diffeomorphism $\psi : X \to X$ such that $\det \nabla \psi(x) > 0$ for all $x \in X$, and define

$$(4.2.18) \qquad (Tf)(x) = [J(x)]^{1/2} f(\psi(x)), \qquad J(x) = \det \nabla \psi(x).$$

Then T is unitary in $\mathscr{H}(X)$, and a straightforward calculation implies the existence of a constant C (depending only on n) such that

$$(4.2.19) \quad \|T\|_{B(\mathscr{H}^1(X))} \leq 1 + C[\sup_{x \in X} |\operatorname{grad} \ln J(x)| + \sup_{x \in X} \|\nabla \psi(x)\|_{L(X)}],$$

$$(4.2.20) \quad \|T\|_{B(\mathscr{H}^2(X))} \leq C\Big[1 + \sup_{x \in X} |\operatorname{grad} \ln J(x)| + \sup_{x \in X} \|\nabla \psi(x)\|_{L(X)}\Big]^2$$

$$+ C \sum_{j,k=1}^{n} [\sup_{x \in X} |\partial_j \partial_k \ln J(x)| + \sup_{x \in X} |\partial_j \partial_k \psi(x)|].$$

In what follows, we shall study in more detail the case where ψ is a *radial diffeomorphism*, i.e. where there is a function $\varrho : [0, \infty) \to [0, \infty)$ such that $\psi(r\omega) = \varrho(r)\omega$ for all $r \geq 0$ and all $\omega \in S_X$ (the unit sphere of X). We first express the estimate (4.2.20) for the norm of T in $B(\mathscr{H}^2(X))$ in terms of the function ϱ:

LEMMA 4.2.5. *Let ψ be a radial C^∞ diffeomorphism of X : $\psi(x) = \varrho(r)\omega$, and let T be the operator associated to ψ by (4.2.18). Define θ by $\theta(r) = r^{-1}\varrho(r)$ for $r > 0$. Then θ is a function of class C^∞ on $[0, \infty)$, and one has*

$$(4.2.21) \qquad \varrho'(0) = \theta(0) \neq 0$$

$$(4.2.22) \qquad \varrho(0) = \theta'(0) = \varrho''(0) = 0$$

and, for some constant c depending only on n:

$$(4.2.23) \quad \|T\|_{B(\mathscr{H}^2(X))} \leq c\left[1 + \sup_{r>0} |\varrho'(r)| + \sup_{r>0}\left|\frac{\varrho''(r)}{\varrho'(r)}\right| + \sup_{r>0}\left|\frac{\theta'(r)}{\theta(r)}\right|\right]^2$$

$$+ c\left[\sup_{r>0} |\varrho''(r)| + \sup_{r>0}\left|\frac{d}{dr}\frac{\varrho''(r)}{\varrho'(r)}\right| + \sup_{r>0}\left|\frac{\theta''(r)}{\theta(r)}\right|\right].$$

PROOF. (i) We shall use several times the following fact. Let $h : [0, \infty) \to \mathbb{R}$ be continuously differentiable with $h(0) = 0$. If $r > 0$, then by the mean value theorem there is a number $\xi(r) \in [0, r]$ such that

$$\frac{h(r)}{r} \equiv \frac{h(r) - h(0)}{r} = h'(\xi(r)).$$

Hence

$$(4.2.24) \qquad \sup_{r>0} \frac{|h(r)|}{r} \leq \sup_{r \geq 0} |h'(r)|.$$

(ii) Clearly the conditions of the lemma imply that $\varrho \in C^\infty([0, \infty))$ and that $\varrho(0) = 0$ (because $\psi(0) = \psi(0\omega) = \varrho(0)\omega$ has to be independent of ω).

Consequently θ will also be of class C^∞ on $[0,\infty)$. If we consider r and ω_k as functions on X given by $r(x) = |x|$, $\omega_k = x_k|x|^{-1}$, then

$$(4.2.25) \qquad \partial_j r = \omega_j, \qquad \partial_j \omega_k = r^{-1}(\delta_{jk} - \omega_j \omega_k).$$

If h is a function defined on $[0,\infty)$, then $\partial_j h(r) = h'(r)\omega_j$. By using also the identity

$$(4.2.26) \qquad \varrho'(r) - \theta(r) = r\theta'(r),$$

one then finds that

$$(4.2.27) \qquad (\partial_j \psi_k)(r\omega) = \theta(r)\delta_{jk} + [\varrho'(r) - \theta(r)]\omega_j \omega_k$$

$$(4.2.28) \qquad (\partial_j \partial_k \psi_m)(r\omega) = \theta'(r)[\delta_{jk}\omega_m + \delta_{km}\omega_j + \delta_{jm}\omega_k]$$
$$+[\varrho''(r) - 3\theta'(r)]\omega_j \omega_k \omega_m.$$

(iii) The relation (4.2.26) implies that $\varrho'(0) = \theta(0)$. Thus, by (4.2.27), we have $(\partial_j \psi_k)(0) = \theta(0)\delta_{jk}$. Since ψ is a diffeomorphism, $(\partial_j \psi_k)(x)$ has an inverse for each $x \in X$, in particular we must have $\theta(0) \neq 0$. This proves (4.2.21).

(iv) (4.2.26) implies that $\varrho''(0) = 2\theta'(0)$. On the other hand, $(\partial_j \partial_k \psi_m)(0)$ must be independent of ω; if we take for example $j = 1$, $k = 2$, $m = 3$ in (4.2.28), we see that one must have $\varrho''(0) = 3\theta'(0)$. Consequently $\theta'(0) = 0$, and we have verified (4.2.22). Note that the preceding argument works if $n \geq 3$; the cases $n = 1$ and $n = 2$ are easier and are left as an exercise.

(v) To prove (4.2.23) we use (4.2.20). We first estimate the two terms involving the derivatives of ψ in (4.2.20). By using first (4.2.27) and then (4.2.24) with $h = \varrho$, one obtains that

$$\sup_{x \in X} |(\partial_j \psi_k)(x)| \leq \sup_{r \geq 0}[2|\theta(r)| + |\varrho'(r)|] \leq 3\sup_{r \geq 0}|\varrho'(r)|.$$

Next we observe that (since $\varrho(0) = 0$):

$$r\varrho'(r) - \varrho(r) = \int_0^r s\varrho''(s)ds.$$

Hence

$$(4.2.29) \quad |\theta'(r)| = \frac{|r\varrho'(r) - \varrho(r)|}{r^2} \leq \frac{1}{r^2}\int_0^r |s\varrho''(s)|ds \leq \frac{1}{2}\sup_{0 \leq s \leq r}|\varrho''(s)|.$$

Then (4.2.28) implies that

$$\sup_{x \in X}|\partial_j \partial_k \psi_m(x)| \leq 3\sup_{r \geq 0}|\theta'(r)| + \sup_{r \geq 0}|\varrho''(r)| < 3\sup_{r \geq 0}|\varrho''(r)|.$$

(vi) It remains to estimate the terms in (4.2.20) involving $\ln J(x)$. For this we first show that ($n = \dim X$):

$$(4.2.30) \qquad J(x) \equiv \det\{(\partial_j \psi_k)(x)\} = \varrho'(r)[\theta(r)]^{n-1}.$$

Indeed (4.2.27) shows that, for fixed x, the matrix $\{(\partial_j \psi_k)(x)\}$ (viewed as a mapping from $\mathbb{R}^n$ to $\mathbb{R}^n$) may be written in the form

$$(\nabla\psi)(x) = \theta(r)I + [\varrho'(r) - \theta(r)]P_\omega,$$

where I is the identity on $\mathbb{R}^n$ and P_ω the orthogonal projection of $\mathbb{R}^n$ onto the one-dimensional subspace spanned by the vector $\omega = \frac{x}{|x|}$. Thus

$$(\nabla \psi)(x) = \theta(r)[I - P_\omega] + \varrho'(r)P_\omega,$$

which implies (4.2.30) since the range of $I - P_\omega$ is of dimension $n - 1$.

From (4.2.30) we deduce that

$$(4.2.31) \qquad \partial_j \ln J(x) = \left[\frac{\varrho''(r)}{\varrho'(r)} + (n-1)\frac{\theta'(r)}{\theta(r)} \right]\omega_j,$$

$$(4.2.32)$$

$$\partial_j \partial_k \ln J(x) = \left[\frac{\varrho''(r)}{r\varrho'(r)} + (n-1)\frac{\theta'(r)}{r\theta(r)} \right]\delta_{jk} +$$

$$+ \left[\frac{d}{dr}\frac{\varrho''(r)}{\varrho'(r)} - \frac{\varrho''(r)}{r\varrho'(r)} + (n-1)\frac{d}{dr}\frac{\theta'(r)}{\theta(r)} - (n-1)\frac{\theta'(r)}{r\theta(r)} \right]\omega_j\omega_k.$$

Hence

$$|\partial_j \ln J(x)| \le \left| \frac{\varrho''(r)}{\varrho'(r)} \right| + (n-1)\left| \frac{\theta'(r)}{\theta(r)} \right|$$

and

$$|\partial_j \partial_k \ln J(x)| \le 2\left| \frac{\varrho''(r)}{r\varrho'(r)} \right| + 2(n-1)\left| \frac{\theta'(r)}{r\theta(r)} \right| + \left| \frac{d}{dr}\frac{\varrho''(r)}{\varrho'(r)} \right| + (n-1)\left| \frac{d}{dr}\frac{\theta'(r)}{\theta(r)} \right|$$

$$\le 3\sup_{r\ge 0}\left| \frac{d}{dr}\frac{\varrho''(r)}{\varrho'(r)} \right| + 3(n-1)\sup_{r\ge 0}\left| \frac{d}{dr}\frac{\theta'(r)}{\theta(r)} \right|,$$

where we have used (4.2.24) with $h = \varrho''/\varrho'$ and with $h = \theta'/\theta$. By inserting these estimates into (4.2.20) one obtains (4.2.23) after noticing that $(\theta'/\theta)' = \theta''/\theta - (\theta'/\theta)^2$. $\square$

PROPOSITION 4.2.6. *Let $F : X \to X$ be of class C^∞ and assume that all its derivatives of order ≥ 1 are bounded. Assume that $F(r\omega) = f(r)\omega$ for $r \ge 0$, $\omega \in S_X$, where $f : [0, \infty) \to \mathbb{R}$ has the following properties :*
   (a) $f(0) = f'(0) = 0, \qquad f(r) > 0 \quad \forall r > 0,$
   (b) *there are constants $c, \delta, R > 0$ such that for all $r > R$:*

$$(4.2.33) \qquad |f'(r)| \le cr^{-\delta}, \qquad f(r) \ge c^{-1}r^{1-\delta}.$$

Then, for each $s \in [-2, 2]$, the C_0-group $\{W(\tau)\}_{\tau \in \mathbb{R}}$ associated to the flow of F leaves the Sobolev space $\mathscr{H}^s(X)$ invariant and is of polynomial growth in this space.

PROOF. For each $r \ge 0$ we denote by $\varrho_\tau(r)$ the solution of the equation

$$(4.2.34) \qquad \frac{d}{d\tau}\varrho_\tau(r) = f(\varrho_\tau(r)), \qquad \varrho_0(r) = r \quad (\tau \in \mathbb{R}).$$

Since $f(0) = 0$, we have $\varrho_\tau(0) = 0$ for all τ. (4.2.4) shows that the flow φ_τ associated to the radial vector field F is also radial and given by $\varphi_\tau(r\omega) = \varrho_\tau(r)\omega$.

If we define $\theta_\tau(r) = r^{-1}\varrho_\tau(r)$ and set $\varrho'_\tau(r) = d\varrho_\tau(r)/dr$, then (4.2.2) and (4.2.30) lead to the following expression for $W(\tau)$:

$$(4.2.35) \qquad [W(\tau)g](rw) = \left(\varrho'_\tau(r)[\theta_\tau(r)]^{n-1}\right)^{1/2} g(\varrho_\tau(r)w).$$

The proof of the proposition will consist of two parts. In part (i) we show that the conclusions of the proposition are true provided that

$$(4.2.36) \qquad \sup_{r>0} \frac{d}{dr}\varrho_\tau(r) \leq c\langle\tau\rangle^k$$

for some constants c, k and all $\tau \in \mathbb{R}$, where $\langle\tau\rangle = (1+\tau^2)^{1/2}$. In part (ii) we deduce the validity of (4.2.36) from the hypotheses on the function f.

(i) By interpolation it suffices to show that $\{W(\tau)\}$ is of polynomial growth in $\mathcal{H}^2(X)$. Now the norm of $W(\tau)$ in $B(\mathcal{H}^2(X))$ is bounded by the r.h.s. of (4.2.23) with ϱ and θ replaced by ϱ_τ and θ_τ respectively. We show that the assumption (4.2.36) leads to a polynomial bound in $\langle\tau\rangle$ for each of these terms.

The relation (4.2.34) implies that

$$(4.2.37) \qquad \frac{d}{d\tau}\varrho'_\tau(r) = f'(\varrho_\tau(r))\varrho'_\tau(r), \qquad \varrho'_0(r) = 1.$$

This leads to

$$(4.2.38) \qquad \varrho'_\tau(r) = \exp\left[\int_0^\tau f'(\varrho_\nu(r))d\nu\right].$$

In particular we have $\varrho'_\tau > 0$. Furthermore

$$(4.2.39) \qquad \varrho''_\tau(r) = \varrho'_\tau(r)\int_0^\tau f''(\varrho_\nu(r))\varrho'_\nu(r)d\nu$$

and

$$(4.2.40) \qquad \frac{d}{dr}\frac{\varrho''_\tau(r)}{\varrho'_\tau(r)} = \int_0^\tau \left[f'''(\varrho_\nu(r))[\varrho'_\nu(r)]^2 + f''(\varrho_\nu(r))\varrho''_\nu(r)\right]d\nu.$$

By using the positivity of ϱ'_τ, the boundedness of f'' and f''', the assumption (4.2.36) and the identities (4.2.39) and (4.2.40), one sees that there are constants c', k' such that for all $\tau \in \mathbb{R}$:

$$(4.2.41) \quad \sup_{r>0}|\varrho'_\tau(r)| + \sup_{r>0}|\varrho''_\tau(r)| + \sup_{r>0}\left|\frac{\varrho''_\tau(r)}{\varrho'_\tau(r)}\right| + \sup_{r>0}\left|\frac{d}{dr}\frac{\varrho''_\tau(r)}{\varrho'_\tau(r)}\right| \leq c'\langle\tau\rangle^{k'}.$$

This gives the polynomial bound of all terms involving the function ϱ_τ on the r.h.s. of (4.2.23). It remains to consider the two terms involving the function θ_τ; clearly it suffices to show that there are constants c_i and k_i ($i = 1, 2, 3$) such that the following three inequalities hold for each $r > 0$ and $\tau \in \mathbb{R}$:

$$(4.2.42) \quad \theta_\tau(r) \geq c_1\langle\tau\rangle^{-k_1}, \qquad |\theta'_\tau(r)| \leq c_2\langle\tau\rangle^{k_2}, \qquad |\theta''_\tau(r)| \leq c_3\langle\tau\rangle^{k_3}.$$

To prove these inequalities, we first observe that, by differentiating the identity $\varrho_{-\tau}(\varrho_\tau(r)) = r$, one has $\varrho'_{-\tau}(\varrho_\tau(r)) \cdot \varrho'_\tau(r) = 1$, so that

$$\varrho'_\tau(r) = \frac{1}{\varrho'_{-\tau}(\varrho_\tau(r))} \geq \frac{1}{\sup_{s>0} \varrho'_{-\tau}(s)} \geq c^{-1} \langle \tau \rangle^{-k},$$

by (4.2.36). Since $\varrho_\tau(0) = 0$, this implies the first inequality in (4.2.42):

$$\theta_\tau(r) = r^{-1} \varrho_\tau(r) = r^{-1} \int_0^r \varrho'_\tau(s) ds \geq c^{-1} \langle \tau \rangle^{-k}.$$

The second inequality in (4.2.42) follows from (4.2.29) and (4.2.41):

$$|\theta'_\tau(r)| \leq \frac{1}{2} \sup_{0 \leq s \leq r} |\varrho''_\tau(s)| \leq \frac{1}{2} c' \langle \tau \rangle^{k'}.$$

For the last inequality we use the following identity which is easily checked by integrating by parts and using the fact that $\varrho_\tau(0) = 0$:

$$\varrho_\tau(r) - r\varrho'_\tau(r) + \frac{1}{2} r^2 \varrho''_\tau(r) = \frac{1}{2} \int_0^r s^2 \varrho'''_\tau(s) ds.$$

We then have

$$|\theta''_\tau(r)| \equiv \frac{2}{r^3} \left| \varrho_\tau(r) - r\varrho'_\tau(r) + \frac{1}{2} r^2 \varrho''_\tau(r) \right|$$
$$\leq \frac{1}{r^3} \int_0^r s^2 |\varrho'''_\tau(s)| ds \leq \frac{1}{3} \sup_{0 \leq s \leq r} |\varrho'''_\tau(s)|.$$

To see that $|\varrho'''_\tau(r)|$ is polynomially bounded in $\langle \tau \rangle$, uniformly in $r > 0$, it suffices to differentiate (4.2.39) with respect to r and then to use the boundedness of f'' and f''' and the polynomial bounds on $\varrho'_\tau(r)$ and $\varrho''_\tau(r)$ contained in (4.2.41).

(ii) Let us define $\Phi : (0, \infty) \to \mathbb{R}$ by

(4.2.43) $$\Phi(r) = \frac{1}{f(1)} + \int_1^r \frac{1}{f(s)} ds.$$

Then $\Phi'(r) = \frac{1}{f(r)}$. The hypothesis that f' be bounded implies together with the assumption (a) that $0 < f(r) \leq cr$ for some constant c and all $r > 0$. Hence we have

$$\lim_{r \to +\infty} \Phi(r) = +\infty, \qquad \lim_{r \to 0} \Phi(r) = -\infty.$$

Thus Φ is a strictly increasing C^∞ diffeomorphism of $(0, \infty)$ onto $\mathbb{R}$. We denote by Φ^{-1} its inverse and observe (since $\Phi(\Phi^{-1}(r)) = r$) that

(4.2.44) $$(\Phi^{-1})'(t) = \frac{1}{\Phi'(\Phi^{-1}(t))} = f(\Phi^{-1}(t)).$$

It follows that

$$\frac{d}{d\tau} \Phi^{-1}(\tau + \Phi(r)) = f(\Phi^{-1}(\tau + \Phi(r))),$$

and comparison with (4.2.34) shows that

$$(4.2.45) \qquad \varrho_\tau(r) = \Phi^{-1}(\tau + \Phi(r)).$$

Therefore

$$(4.2.46) \qquad \frac{d}{dr}\varrho_\tau(r) \equiv \varrho_\tau'(r) = (\Phi^{-1})'(\tau + \Phi(r)) \cdot \Phi'(r).$$

Upon setting $t = \Phi(r)$, one sees that (4.2.36) is equivalent to

$$(4.2.47) \quad (\Phi^{-1})'(\tau + t) \leq c\langle\tau\rangle^k f(\Phi^{-1}(t)) \equiv c\langle\tau\rangle^k (\Phi^{-1})'(t) \quad \forall \tau, t \in \mathbb{R}.$$

We have to show that the validity of (4.2.47) follows from the hypotheses of the proposition. For this we first prove that

$$(4.2.48) \qquad |\Phi(r)f'(r)| \leq \gamma$$

for some constant γ and all $r \geq 0$. Clearly $\Phi f'$ is locally bounded on $(0, \infty)$, so it suffices to analyze its behaviour at $r = 0$ and at $r = \infty$. By the condition (a) of the proposition we have $f(r) = ar^m(1 + O(r))$ for small r, for some constants $a > 0$ and $m \geq 2$. Hence $\Phi(r) \leq c_1 + c_2 r^{-m+1}$ for small r (c_1 and c_2 are constants). Also $f'(r) = a_1 r^{m-1}(1 + O(r)) \leq c_3 r^{m-1}$ for sufficiently small r, so that (4.2.48) also holds in a neighbourhood of $r = 0$. Similarly the validity of (4.2.48) for large r follows from the condition (b).

By setting $r = \Phi^{-1}(t)$ in (4.2.48) and using also the boundedness of f' (i.e. $|f'(r)| \leq \gamma_1$ for all $r \geq 0$) one finds that

$$(4.2.49) \qquad |f'(\Phi^{-1}(t))| \leq \frac{\gamma_2}{1 + |t|} \qquad \forall t \in \mathbb{R},$$

with $\gamma_2 = \gamma + \gamma_1$. Now (4.2.44) implies that

$$(\Phi^{-1})''(t) = f'(\Phi^{-1}(t)) \cdot (\Phi^{-1})'(t),$$

or equivalently (observe that $(\Phi^{-1})'(t) > 0$ for all t by (4.2.44) and the strict positivity of f on $(0, \infty)$):

$$\frac{d}{dt} \ln(\Phi^{-1})'(t) = f'(\Phi^{-1}(t)).$$

Thus, by (4.2.49):

$$(4.2.50) \qquad \left| \ln \frac{(\Phi^{-1})'(\tau + t)}{(\Phi^{-1})'(t)} \right| = \left| \int_t^{\tau+t} \left[\frac{d}{ds} \ln(\Phi^{-1})'(s) \right] ds \right|$$
$$\leq \gamma_2 \left| \int_t^{\tau+t} \frac{1}{1 + |s|} ds \right|$$
$$\leq 2\gamma_2 \ln(1 + |\tau|)$$

(the factor 2 in the last step is necessary to handle the case where t and $\tau + t$ have opposite signs). The inequality (4.2.50) implies (4.2.47) (if $(\Phi^{-1})'(\tau + t) \leq (\Phi^{-1})'(t)$, then (4.2.47) evidently holds with $k = 0$, in all other cases (4.2.47) holds with $k = 2\sqrt{2}\gamma_2$ by (4.2.50) and the fact that $1 + |\tau| \leq \sqrt{2}\langle\tau\rangle$). $\square$

We wish to point out here that the condition $f'(0) = 0$ in Proposition 4.2.6 is essential in order for the corresponding C_0-group $\{W(\tau)\}$ to be of polynomial growth in $\mathscr{H}^1(X)$ (and hence in each $\mathscr{H}^s(X)$ with $|s| \geq 1$). More precisely, we have the following result:

PROPOSITION 4.2.7. *Let* $F : X \to X$ *be of class* C^∞ *and assume that all its derivatives of order* ≥ 1 *are bounded. Assume that* $F(r\omega) = f(r)\omega$, *where the* C^∞ *function* $f : [0, \infty) \to \mathbb{R}$ *satisfies* $f'(0) \neq 0$ *and* $f(r) > 0$ *if* $r > 0$. *Then the* C_0-*group* $\{W(\tau)\}_{\tau \in \mathbb{R}}$ *associated to the flow of* F *leaves* $\mathscr{H}^1(X)$ *invariant but is not of polynomial growth in* $\mathscr{H}^1(X)$.

We omit the proof of this result (it was given in Lemma 3.10.8 of [ABG1]). As examples we may consider the operator D (but then the result is obvious) or the γ operator of Sigal and Soffer [SS] which corresponds to $f(r) = r(1 + r^2)^{-1/2}$. So $\exp(i\gamma\tau)$ is a C_0-group in each $\mathscr{H}^s(X)$, but it is not polynomially bounded at least if $|s| \geq 1$.

CHAPTER 5

Groups of Automorphisms Associated to C_0-Representations of $\mathbb{R}^n$

A representation $\mathscr{W}$ of $\mathbb{R}^n$ with values in the group of automorphisms (i.e. linear multiplicative bijections) of a Banach algebra $\mathscr{B}$ has many interesting features due to the richness of the algebraic structure which comes into play. We have no intention to present the general theory of such representations (elements of this theory may be found in [Br], [BR], [Cm], [Pd]) but rather to develop a very special aspect in view of later applications in spectral and scattering theory. More precisely, the algebras which will appear in this and in the next chapter are of the form $\mathscr{B} = B(\mathbf{F})$, where $\mathbf{F}$ is a Banach space equipped with a C_0-representation W of $\mathbb{R}^n$, while the automorphism $\mathscr{W}(x)$ of $\mathscr{B}$ is given by $\mathscr{W}(x)[S] = W(-x)SW(x)$. Unless the generator of W is bounded in $\mathbf{F}$, the family $\{\mathscr{W}(x)\}_{x\in\mathbb{R}^n}$ does not form a C_0-representation of $\mathbb{R}^n$ in the Banach space $\mathscr{B}$ but only a C_w-representation (cf. Definition 3.2.6). Nevertheless, by using its continuity in the strong operator topology, one can develop for $\mathscr{W}$ the theory obtained in Chapter 3 for C_0-groups; the only difference lies in the fact that the domain of the generator of $\mathscr{W}$ is not norm dense in $\mathscr{B}$ but only strongly dense. Of course, $\mathscr{W}$ induces a C_0-group in the subspace $\mathscr{B}_u$ of $\mathscr{B}$ consisting of all $S \in \mathscr{B}$ such that $x \mapsto \mathscr{W}(x)[S]$ is norm-continuous, and this observation will be frequently used.

For technical reasons that will become clear in the next chapter, we shall consider a slightly more general framework than indicated above. Let $\mathbf{F}'$, $\mathbf{F}''$ be two Banach spaces and W', W'' two C_0-representations of $\mathbb{R}^n$ in $\mathbf{F}'$, $\mathbf{F}''$ respectively. We denote their generators by $A' = (A'_1, \dots, A'_n)$ and $A'' = (A''_1, \dots, A''_n)$. For each of these representations one may develop the formalism of Chapter 3; in particular, the spaces $\mathbf{F}'_m$, $\mathbf{F}''_m$, $\mathbf{F}'_{s,p}$ and $\mathbf{F}''_{s,p}$ are well defined for each integer m, each real s and each $p \in [1, \infty]$.

In this context one gets a natural representation $\mathscr{W}$ of $\mathbb{R}^n$ in the Banach space $\mathscr{B} = B(\mathbf{F}', \mathbf{F}'')$ by setting for $S \in B(\mathbf{F}', \mathbf{F}'')$:

$$(5.0.1) \qquad \mathscr{W}(x)S = \mathscr{W}(x)[S] = S(x) := W''(-x)SW'(x), \quad x \in \mathbb{R}^n.$$

Then $\mathscr{W}$ is a C_w-representation of $\mathbb{R}^n$ in $\mathscr{B}$. In fact, for each $x \in \mathbb{R}^n$, $\mathscr{W}(x)$

W. O. Amrein et al., C_0-Groups, Commutator Methods and Spectral Theory of N-Body Hamiltonians, Modern Birkhäuser Classics, DOI: 10.1007/978-3-0348-0733-3_5, © Springer Basel 1996

is bounded as an operator acting in $B(\mathbf{F}', \mathbf{F}'')$: if M and ω are such that $\max\{||W'(x)||_{B(\mathbf{F}')}, ||W''(x)||_{B(\mathbf{F}'')}\} \leq M \exp(\omega|x|)$, then

$$(5.0.2) \qquad ||\mathscr{W}(x)S||_{\mathbf{F}' \to \mathbf{F}''} \leq M^2 e^{2\omega|x|} ||S||_{\mathbf{F}' \to \mathbf{F}''}.$$

Moreover, $\mathscr{W}(0)$ is the identity operator on $\mathscr{B}$, $\mathscr{W}(x + y) = \mathscr{W}(x)\mathscr{W}(y)$, and for each $S \in \mathscr{B}$ the mapping $\mathbb{R}^n \ni x \mapsto \mathscr{W}(x)S \in B(\mathbf{F}', \mathbf{F}'')$ is continuous for the strong operator topology. In particular, $\Gamma_{\mathscr{W}}$ will contain all the functionals of the form $\varphi(S) = \langle Sf, g \rangle$ with $f \in \mathbf{F}'$, $g \in (\mathbf{F}'')^*$, hence the conditions of Definition 3.2.6 are fulfilled.

The purpose of this chapter is to introduce and study some notions and objects naturally associated to the group $\mathscr{W}$. In Section 5.1 we introduce the analogue for the group $\mathscr{W}$ of the spaces $\mathbf{F}_k$ of Chapter 3 (for $k \geq 0$). Since $\mathscr{W}$ is only a C_w-group, several versions of such spaces can be defined by using different topologies on $B(\mathbf{F}', \mathbf{F}'')$. We shall denote by $C^k(A', A''; \mathbf{F}', \mathbf{F}'')$ and $C_u^k(A', A''; \mathbf{F}', \mathbf{F}'')$ the spaces obtained from the strong and the uniform (norm) topology on $B(\mathbf{F}', \mathbf{F}'')$ respectively; these notations stress the fact that we are thinking in terms of regularity (smoothness) properties of operators $S : \mathbf{F}' \to \mathbf{F}''$ with respect the action of the groups W', W''. Of course one could introduce also spaces $C_w^k(A', A''; \mathbf{F}', \mathbf{F}'')$ by using the weak operator topology on $B(\mathbf{F}', \mathbf{F}'')$, but it turns out that $C_w^k(A', A''; \mathbf{F}', \mathbf{F}'') = C^k(A', A''; \mathbf{F}', \mathbf{F}'')$. Notice that the exact analogues of the spaces $\mathbf{F}_k$ are the spaces $C_u^k(A', A''; \mathbf{F}', \mathbf{F}'')$ (more precisely: if in (3.3.1) one takes $\mathbf{F} = B(\mathbf{F}', \mathbf{F}'')$, $m = k$, $W = \mathscr{W}$, then one gets $\mathbf{F}_k = C_u^k(A', A''; \mathbf{F}', \mathbf{F}'')$).

In Section 5.2 we study the analogue of the interpolation spaces $\mathbf{F}_{s,p}$ for $\mathscr{W}$ (with $s > 0$). This time we consider only the norm-topology on $B(\mathbf{F}', \mathbf{F}'')$ and denote by $\mathscr{C}^{s,p}(A', A''; \mathbf{F}', \mathbf{F}'')$ the space which, according to the rules of Chapter 4, should be called $B(\mathbf{F}', \mathbf{F}'')_{s,p}$ (we remark that similar spaces could be defined by using the strong or the weak operator topology; however, we did not develop their theory, although they could be useful in some applications). We also introduce spaces $\mathscr{C}^{0,p}(A', A''; \mathbf{F}', \mathbf{F}'')$ and $\mathscr{C}^{k+0}(A', A''; \mathbf{F}', \mathbf{F}'')$ which have no analogues in the scale $\mathbf{F}_{s,p}$, but which are quite natural in the framework of the present chapter.

The main point of Sections 5.1 and 5.2 is to study the behaviour of the regularity classes C^k, C_u^k, $\mathscr{C}^{s,p}$ with respect to various algebraic operations (product, inverse, adjoint). In the special case where $\mathbf{F}' = \mathbf{F}''$ and $W' = W''$, these properties generalize easily to the context of Banach algebras equipped with representations of $\mathbb{R}^n$ by automorphisms. Sections 5.3 to 5.5 are devoted a different question, namely the behaviour of an operator $S : \mathbf{F}' \to \mathbf{F}''$, having a certain type of regularity with respect to $\mathscr{W}$, in relation with the Besov scales $\{\mathbf{F}'_{t,q}\}, \{\mathbf{F}''_{t,q}\}$. For example, we show that, if $S \in \mathscr{C}^{s,p}$ and $0 < t < s$, then $S\mathbf{F}'_{t,q} \subset \mathbf{F}''_{t,q}$. This means that, if S is regular with respect to $\mathscr{W}$, then it *preserves regularity of vectors* in the Besov scales associated to W' and W'' (a similar but more elementary assertion obviously holds in the discrete Sobolev scale). We shall also give conditions for an operator S to *improve regularity* in the Besov scales (i.e. such that $S\mathbf{F}'_{t,q} \subset \mathbf{F}''_{t+\tau,r}$ for some $\tau > 0$). This will allow us in Section 5.5 to present commutator expansion formulas with precise estimates on the remainder. We shall not explain here the terminology ("commutator expansions", "remainder"), because it requires some new formalism which will be introduced only in Sections 5.3 and 5.5.

5.1. Regularity and Commutators

5.1.1. We use the framework explained above and begin with some comments on the generator of the C_w-group $\mathscr{W}$ and the associated functional calculus. The generator $\mathscr{A}$ of $\mathscr{W}$ is an n-tuple $\mathscr{A} = (\mathscr{A}_1, \ldots, \mathscr{A}_n)$ of operators in the Banach space $\mathscr{B} = B(\mathbf{F}', \mathbf{F}'')$ formally given by the rule $\mathscr{W}(x) = \exp(i\mathscr{A} \cdot x)$; in general, the operators $\mathscr{A}_k$ are unbounded and not densely defined. Stated differently, we shall have:

$$(5.1.1) \qquad \frac{\partial}{\partial x_j} \mathscr{W}(x)[S] = i\mathscr{A}_j \mathscr{W}(x)[S] = i\mathscr{W}(x)\mathscr{A}_j[S]$$

for those operators S for which the derivative exists strongly (some basic facts about derivatives of operator-valued functions are collected in an appendix to this chapter and will be freely used in our arguments). We shall describe a more general and more precise version of (5.1.1) below, while the domains of powers of $\mathscr{A}$ will be studied in §5.1.2.

We explained at the end of Section 3.2 how to define a bounded functional calculus for arbitrary C_w-groups. So, for certain easily specified classes of functions $\varphi : \mathbb{R}^n \to \mathbb{C}$, a bounded linear operator $\varphi(\mathscr{A}) : \mathscr{B} \to \mathscr{B}$ is well defined, more precisely:

$$(5.1.2) \qquad \varphi(\mathscr{A})[S] = \int_{\mathbb{R}^n} W''(-x)SW'(x)\widehat{\varphi}(x)\underline{d}x,$$

where the integral exists in the strong operator topology of $\mathscr{B} = B(\mathbf{F}', \mathbf{F}'')$ if $\widehat{\varphi}$ is a measure such that $\int_{\mathbb{R}^n} \|\mathscr{W}(x)\|_{B(\mathscr{B})} |\widehat{\varphi}(x)|\underline{d}x < \infty$ (see (5.0.2)).

In order to treat unbounded functions φ (e.g. polynomials in $\mathscr{A}$), we shall now introduce a vector space $\mathscr{B}_{-\infty}$ that has the same role, in the present context, as the space $\mathbf{F}_{-\infty}$ in the context of Chapter 3. Observe first that, if $k, m \in \mathbb{N}$ and $k < m$, then we have a natural embedding $B(\mathbf{F}'_k, \mathbf{F}''_{-k}) \subset B(\mathbf{F}'_m, \mathbf{F}''_{-m})$; hence we may define

$$(5.1.3) \qquad \mathscr{B}_{-\infty} \equiv B(\mathbf{F}'_\infty, \mathbf{F}''_{-\infty}) = \bigcup_{m=0}^{\infty} B(\mathbf{F}'_m, \mathbf{F}''_{-m}).$$

We shall consider this set as a vector subspace of the space of all linear operators $\mathbf{F}'_\infty \to \mathbf{F}''_{-\infty}$. By using Proposition 3.3.8 we get a natural extension of the representation $\mathscr{W}$ to a n-parameter group of linear operators acting in $\mathscr{B}_{-\infty}$ (we keep the notation $\mathscr{W}(x) : \mathscr{B}_{-\infty} \to \mathscr{B}_{-\infty}$ for this extension). Observe that each space $B(\mathbf{F}'_k, \mathbf{F}''_m)$ and $B(\mathbf{F}'_{s,p}, \mathbf{F}''_{t,q})$ is contained in $\mathscr{B}_{-\infty}$ and is left invariant by $\mathscr{W}$ (here $k, m \in \mathbb{Z}$; $s, t \in \mathbb{R}$; $p \in [1, \infty)$ and $q \in [1, \infty]$).

We are now in a position to give explicit expressions for the generators $\mathscr{A}_j$. They have canonical extensions to linear operators $\mathscr{A}_j : \mathscr{B}_{-\infty} \to \mathscr{B}_{-\infty}$ which act as follows:

$$(5.1.4) \qquad \mathscr{A}_j[S] = SA'_j - A''_j S.$$

A simple purely algebraic calculation shows that $\mathscr{A}_1, \ldots, \mathscr{A}_n$ are pairwise commuting operators in $\mathscr{B}_{-\infty}$, so that one may define without ambiguity $\mathscr{A}^\alpha = \mathscr{A}_1^{\alpha_1} \ldots \mathscr{A}_n^{\alpha_n} : \mathscr{B}_{-\infty} \to \mathscr{B}_{-\infty}$ for each multi-index α (with the usual convention $\mathscr{A}^0 =$ identity map). Hence for each polynomial function $\varphi : \mathbb{R}^n \to \mathbb{C}$, the map $\varphi(\mathscr{A}) : \mathscr{B}_{-\infty} \to \mathscr{B}_{-\infty}$ is also well defined and, if the degree of φ is ℓ, then for any $k, m \in \mathbb{Z}$; $s, t \in \mathbb{R}$; $1 \le p < \infty$, $1 \le q \le \infty$:

$$(5.1.5) \qquad S \in B(\mathbf{F}'_k, \mathbf{F}''_m) \Longrightarrow \varphi(\mathscr{A})[S] \in B(\mathbf{F}'_{k+\ell}, \mathbf{F}''_{m-\ell}),$$

$$(5.1.6) \qquad S \in B(\mathbf{F}'_{s,p}, \mathbf{F}''_{t,q}) \Longrightarrow \varphi(\mathscr{A})[S] \in B(\mathbf{F}'_{s+\ell,p}, \mathbf{F}''_{t-\ell,q}).$$

By taking into account (3.3.7) it is now easy to give a general version of (5.1.1): if $k, m \in \mathbb{Z}$ and $S \in B(\mathbf{F}'_k, \mathbf{F}''_m)$, then

$$(5.1.7) \qquad \partial^\alpha \mathscr{W}(x)[S] = i^{|\alpha|} \mathscr{A}^\alpha \mathscr{W}(x)[S] = i^{|\alpha|} \mathscr{W}(x) \mathscr{A}^\alpha[S].$$

The derivatives on the left hand side exist in the strong operator topology of $B(\mathbf{F}'_{k+|\alpha|}, \mathbf{F}''_{m-|\alpha|})$. Moreover, for each integer $\ell \ge 1$, the function $x \mapsto \mathscr{W}(x)[S] \in B(\mathbf{F}'_{k+\ell}, \mathbf{F}''_{m-\ell})$ is of class C^ℓ in the strong operator topology and one has the Taylor expansion (use (1.1.8) and (5.1.7)):

$$(5.1.8)$$

$$\mathscr{W}(x)[S] = \sum_{|\alpha|<\ell} \frac{(ix)^\alpha}{\alpha!} \mathscr{A}^\alpha[S] + \sum_{|\alpha|=\ell} \frac{(ix)^\alpha}{\alpha!} \int_0^1 \mathscr{W}(\tau x) \mathscr{A}^\alpha[S] \ell (1-\tau)^{\ell-1} d\tau.$$

A useful consequence of this is the following identity (see the proof of Proposition 3.3.5):

$$(5.1.9)$$

$$[\mathscr{W}(x) - I]^\ell S = \sum_{|\alpha|=\ell} \frac{x^\alpha}{\alpha!} \sum_{k=1}^\ell (-1)^{\ell-k} \ell \binom{\ell}{k} (ik)^\ell \int_0^1 \mathscr{W}(\tau k x) \mathscr{A}^\alpha[S] (1-\tau)^{\ell-1} d\tau.$$

In numerous situations we shall use the following alternative notation for the operators $\mathscr{A}_j$:

$$(5.1.10) \qquad \mathrm{ad}_{A_j}(S) = A''_j S - SA'_j \equiv -\mathscr{A}_j[S],$$

hence $\mathscr{A}^\alpha = (-1)^{|\alpha|} \mathrm{ad}_A^\alpha = (-1)^{|\alpha|} \mathrm{ad}_{A_1}^{\alpha_1} \ldots \mathrm{ad}_{A_n}^{\alpha_n}$. The reason for this is that in many applications $\mathbf{F}'$ and $\mathbf{F}''$ will both be embedded in the same Banach space $\mathbf{F}$ and W', W'' will be restrictions of a C_0-representation $W(x) = e^{iAx}$ of $\mathbb{R}^n$ in $\mathbf{F}$. Then A'_j, A''_j will be restrictions of some operator A_j acting in $\mathbf{F}$ and so: $\mathrm{ad}_{A_j}(S) = A_j S - SA_j = [A_j, S]$.

5.1.2. We may now introduce and study the discrete Sobolev scales associated to the C_w-group $\mathscr{W}$ in the Banach space $B(\mathbf{F}', \mathbf{F}'')$ (see the discussion in the introduction to this chapter). We shall restrict ourselves to positive indices k, since the case $k < 0$ will not be needed further on.

DEFINITION 5.1.1. Let $k \in \mathbb{N}$.

(a) We denote by $C^k(A', A''; \mathbf{F}', \mathbf{F}'')$ the Banach space formed by all operators $S \in B(\mathbf{F}', \mathbf{F}'')$ such that the function $S(\cdot)$ given by (5.0.1) is strongly of class C^k, equipped with the norm

$$(5.1.11) \qquad \|S\|_{C^k} = \left[\sum_{|\alpha| \le k} \|\partial^\alpha S(0)\|^2_{\mathbf{F}' \to \mathbf{F}''} \right]^{1/2}.$$

(b) We denote by $C^k_u(A', A''; \mathbf{F}', \mathbf{F}'')$ the Banach space of all operators $S \in B(\mathbf{F}', \mathbf{F}'')$ such that the function $S(\cdot)$ is of class C^k in the norm of $B(\mathbf{F}', \mathbf{F}'')$, equipped with the norm (5.1.11).

If $S \in C^k(A', A''; \mathbf{F}', \mathbf{F}'')$, then $S \in C^k_u(A', A''; \mathbf{F}', \mathbf{F}'')$ means that the derivatives $\partial^\alpha S(\cdot)$ for $|\alpha| \le k$ are norm-continuous functions (cf. (5.1.8)). The argument of the proof of Proposition 3.3.2 can be used to see that $C^k(A', A''; \mathbf{F}', \mathbf{F}'')$ and $C^k_u(A', A''; \mathbf{F}', \mathbf{F}'')$ are complete for the norm $\| \cdot \|_{C^k}$. The Banach space $C^k_u(A', A''; \mathbf{F}', \mathbf{F}'')$ coincides with the space $\mathbf{F}_k$ associated by Eqs. (3.3.1) and (3.3.2) to the C_0-group $\mathscr{W}$ in $\mathbf{F} \equiv C^0_u(A', A''; \mathbf{F}', \mathbf{F}'')$.

In situations where the C_0-groups are fixed (and especially in the course of proofs) we shall write $C^k(\mathbf{F}', \mathbf{F}'')$ instead of $C^k(A', A''; \mathbf{F}', \mathbf{F}'')$. In many cases $\mathbf{F}'$ and $\mathbf{F}''$ will be embedded in a Banach space $\mathbf{F}$ and W', W'' will be restrictions of a fixed C_0-representation $W(x) = e^{iA \cdot x}$ of $\mathbb{R}^n$ in $\mathbf{F}$; in such a case we write $C^k(A; \mathbf{F}', \mathbf{F}'')$ rather than $C^k(A', A''; \mathbf{F}', \mathbf{F}'')$. If we also have $\mathbf{F}' = \mathbf{F}'' = \mathbf{F}$, we denote this space by $C^k(A; \mathbf{F})$ or just by $C^k(\mathbf{F})$.

Some simple features of the preceding definition are described in the next proposition.

PROPOSITION 5.1.2. *Let $S \in B(\mathbf{F}', \mathbf{F}'')$ and $k \in \mathbb{N} \setminus \{0\}$.*

(a) *If there is a point $y \in \mathbb{R}^n$ such that the derivatives $\partial^\alpha S(y)$ exist weakly in $B(\mathbf{F}', \mathbf{F}'')$ for each multi-index α with $|\alpha| \le k$, then $S \in C^k(\mathbf{F}', \mathbf{F}'') \cap C^{k-1}_u(\mathbf{F}', \mathbf{F}'')$, and one has*

$$(5.1.12) \qquad \partial^\alpha S(x) = W''(y - x)[\partial^\alpha S(y)]W'(x - y), \qquad x \in \mathbb{R}^n.$$

(b) *S belongs to $C^k(\mathbf{F}', \mathbf{F}'')$ if and only if $\mathrm{ad}_A^\alpha(S)$ belongs to $B(\mathbf{F}', \mathbf{F}'')$ for all multi-indices α with $|\alpha| \le k$.*

(c) *If $S \in C^k(\mathbf{F}', \mathbf{F}'')$, then for each $|\alpha| \le k$, the function $x \mapsto \mathrm{ad}_A^\alpha(S(x)) \in B(\mathbf{F}', \mathbf{F}'')$ is strongly of class $C^{k-|\alpha|}$ and one has the identity*

$$(5.1.13) \quad \partial^\alpha S(x) = (-i)^{|\alpha|} W''(-x) \, \mathrm{ad}_A^\alpha(S) W'(x) \equiv (-i)^{|\alpha|} \, \mathrm{ad}_A^\alpha(S(x)).$$

PROOF. (a) By using the fact that W' and W'' are groups and the hypothesis that $\partial^\alpha S(y)$ exists, one sees that $\partial^\alpha S(x)$ exists at each $x \in \mathbb{R}^n$ (weakly in $B(\mathbf{F}', \mathbf{F}'')$), for all α with $|\alpha| \le k$, and that $\partial^\alpha S(x)$ is given by (5.1.12). Part (b) of Lemma 5.A.2 now implies that $S(\cdot)$ is strongly C^k, whereas part (a) of that lemma shows that $S(\cdot)$ is C^{k-1} in norm.

(b) If $S(\cdot)$ is strongly C^k as a function from $\mathbb{R}^n$ to $B(\mathbf{F}', \mathbf{F}'')$, then (5.1.7) implies that $\mathrm{ad}_A^\alpha(S) \in B(\mathbf{F}', \mathbf{F}'')$ for each multi-index α with $|\alpha| \leq k$. For the converse we observe that, by (5.1.8):

$$S(x+y) - \sum_{|\alpha| \leq k} \frac{(-iy)^\alpha}{\alpha!} \mathrm{ad}_A^\alpha(S(x)) =$$

$$= k \sum_{|\alpha|=k} \frac{(-iy)^\alpha}{\alpha!} \int_0^1 \varrho^{k-1} d\varrho \{\mathrm{ad}_A^\alpha[S(x+(1-\varrho)y)] - \mathrm{ad}_A^\alpha(S(x))\}.$$

Now, by assumption, $\mathrm{ad}_A^\alpha(S(\cdot))$ is strongly continuous in $B(\mathbf{F}', \mathbf{F}'')$ for each $|\alpha| \leq k$, hence uniformly continuous on each compact subset of $\mathbb{R}^n$. By applying Proposition 1.1.1, one obtains that $S(\cdot)$ is strongly of class C^k in $B(\mathbf{F}', \mathbf{F}'')$.

(c) The identity (5.1.13) is an immediate consequence of (5.1.12), with $y = 0$, and of (5.1.7). The first assertion of (c) follows from (5.1.13). $\square$

The next result is much deeper.

THEOREM 5.1.3. *Let $S \in B(\mathbf{F}', \mathbf{F}'')$ and let $k \geq 1$ be an integer.*

(a) *S belongs to $C^k(A', A''; \mathbf{F}', \mathbf{F}'')$ or to $C_u^k(A', A''; \mathbf{F}', \mathbf{F}'')$ if and only if, for each $x \in \mathbb{R}^n$, $\lim_{\varepsilon \to +0} \varepsilon^{-k}[\mathscr{W}(\varepsilon x) - I]^k S$ exists in the strong operator topology or in the norm topology of $B(\mathbf{F}', \mathbf{F}'')$ respectively.*

(b) *If $\mathbf{F}''$ is reflexive and $\liminf_{\varepsilon \to +0} \varepsilon^{-k} \|[\mathscr{W}(\varepsilon x) - I]^k S\|_{\mathbf{F}' \to \mathbf{F}''} < \infty$ for each $x \in \mathbb{R}^n$, then $S \in C^k(A', A''; \mathbf{F}', \mathbf{F}'')$.*

PROOF. (a) The assertion concerning C_u^k follows from that concerning C^k and from Proposition 3.3.19 applied to $\mathbf{F} = C_u^0(A', A''; \mathbf{F}', \mathbf{F}'')$ (note that $C^k \subset C_u^0$ if $k \geq 1$). The case of C^k will be treated by an adaptation of the proof of Proposition 3.3.19. Instead of (3.3.40) we use (see (5.1.9)):

(5.1.14)

$$\left[\frac{\mathscr{W}(\varepsilon x) - I}{\varepsilon}\right]^k S =$$

$$= i^k \sum_{|\alpha|=k} \frac{x^\alpha}{\alpha!} \sum_{\ell=0}^k (-1)^{k-\ell} \binom{k}{\ell} \ell^k \cdot \int_0^1 k(1-\tau)^{k-1} \mathscr{W}(\tau \varepsilon \ell x) d\tau \mathscr{A}^\alpha[S].$$

If $S \in C^k(\mathbf{F}', \mathbf{F}'')$, then $\mathscr{A}^\alpha[S] \in B(\mathbf{F}', \mathbf{F}'')$ for $|\alpha| \leq k$ by Proposition 5.1.2 (b), so the r.h.s. of (5.1.14) is strongly convergent in $B(\mathbf{F}', \mathbf{F}'')$ as $\varepsilon \to 0$.

If for each $x \in \mathbb{R}^n$ the l.h.s. of (5.1.14) is strongly convergent in $B(\mathbf{F}', \mathbf{F}'')$, then the formula (5.1.14) implies that $\sum_{|\alpha|=k} (\alpha!)^{-1} x^\alpha \mathscr{A}^\alpha[S] \in B(\mathbf{F}', \mathbf{F}'')$ for each $x \in \mathbb{R}^n$. Then (5.1.8) leads to $\sum_{|\alpha| \leq k} (\alpha!)^{-1} (ix)^\alpha \mathscr{A}^\alpha[S] \in B(\mathbf{F}', \mathbf{F}'')$ for each $x \in \mathbb{R}^n$. By using Proposition 1.1.2 as in the proof of Proposition 3.3.19, one finds that $\mathscr{A}^\alpha[S] = (-1)^{|\alpha|} \mathrm{ad}_A^\alpha(S) \in B(\mathbf{F}', \mathbf{F}'')$ for each $|\alpha| \leq k$. Hence $S \in C^k(\mathbf{F}', \mathbf{F}'')$ by Proposition 5.1.2 (b).

(b) For the proof of the second part of the theorem we could use Proposition 3.3.25, but we prefer to give a direct, independent proof. Let $x \in \mathbb{R}^n$ be fixed and set $\mathcal{W}(\sigma) = \mathcal{W}(\sigma x)$ for $\sigma \in \mathbb{R}$. The hypothesis of (b) implies the existence of a sequence $\{\varepsilon_j\}$ of positive numbers such that $\varepsilon_j \to 0$ and $\varepsilon_j^{-k}[\mathcal{W}(\varepsilon_j) - I]^k S$ is weakly convergent in $B(\mathbf{F}', \mathbf{F}'')$ as $j \to \infty$ (because, by the reflexivity of $\mathbf{F}''$, the unit ball in $B(\mathbf{F}', \mathbf{F}'')$ is weakly compact). We denote the preceding weak limit by S_0.

By using the group property of $\mathcal{W}$, one sees that for any $\tau > 0$:

(5.1.15)

$$\int_0^\tau d\sigma_1 \cdots \int_0^\tau d\sigma_k \, \mathcal{W}(\sigma_1) \ldots \mathcal{W}(\sigma_k) \left[\frac{\mathcal{W}(\varepsilon_j) - I}{\varepsilon_j} \right]^k S =$$
$$= \frac{1}{\varepsilon_j} \int_0^{\varepsilon_j} d\sigma_1 \cdot \mathcal{W}(\sigma_1)[\mathcal{W}(\tau) - I] \ldots \frac{1}{\varepsilon_j} \int_0^{\varepsilon_j} d\sigma_k \cdot \mathcal{W}(\sigma_k)[\mathcal{W}(\tau) - I]S.$$

As $j \to \infty$, the l.h.s. converges weakly to $\int_0^\tau d\sigma_1 \cdots \int_0^\tau d\sigma_k \, \mathcal{W}(\sigma_1) \ldots \mathcal{W}(\sigma_k)S_0$ (apply the Lebesgue dominated convergence theorem). By the strong continuity of $\mathcal{W}$, the r.h.s. of (5.1.15) converges strongly to $[\mathcal{W}(\tau) - I]^k S$ as $j \to \infty$. Hence

$$\frac{1}{\tau^k}[\mathcal{W}(\tau) - I]^k S = \frac{1}{\tau} \int_0^\tau d\sigma_1 \, \mathcal{W}(\sigma_1) \ldots \frac{1}{\tau} \int_0^\tau d\sigma_k \, \mathcal{W}(\sigma_k)S_0.$$

So $\tau^{-k}[\mathcal{W}(\tau) - I]^k S$ is strongly convergent in $B(\mathbf{F}', \mathbf{F}'')$ as $\tau \to 0$, and the conclusion $S \in C^k(\mathbf{F}', \mathbf{F}'')$ follows from (a). $\square$

COROLLARY 5.1.4. *Assume that $\mathbf{F}''$ is reflexive. Let $k \in \{1, 2, 3, \ldots\}$ and $T \in B(\mathbf{F}', \mathbf{F}'')$. Then T belongs to $C^k(A', A''; \mathbf{F}', \mathbf{F}'')$ if and only if $x \mapsto T(x) \in B(\mathbf{F}', \mathbf{F}'')$ is of class C^{k-1} in norm and $x \mapsto \partial^\alpha T(x) \in B(\mathbf{F}', \mathbf{F}'')$ is locally Lipschitz for each multi-index α with $|\alpha| = k - 1$.*

PROOF. The "only if" part follows from Lemma 5.A.2. To obtain the "if" part, one may apply Proposition 5.1.3 (b) with $k = 1$ to the operator $S = \mathrm{ad}_A^\alpha(T)$, $|\alpha| = k - 1$, to conclude that $\mathrm{ad}_A^\alpha(S) \in B(\mathbf{F}', \mathbf{F}'')$ for $|\alpha| = k$. $\square$

We shall now study the regularity classes of products, inverses and adjoints of operators. The first result involves a third Banach space $\mathbf{F}'''$ equipped with a C_0-representation W''' of $\mathbb{R}^n$.

PROPOSITION 5.1.5. *Let $k \in \mathbb{N}$. Then:*

$$S \in C^k(\mathbf{F}', \mathbf{F}'') \text{ and } T \in C^k(\mathbf{F}'', \mathbf{F}''') \Rightarrow TS \in C^k(\mathbf{F}', \mathbf{F}''').$$

The corresponding assertion holds with C^k replaced by C_u^k. Moreover one has for each multi-index α with $|\alpha| \le k$:

(5.1.16) $$\mathrm{ad}_A^\alpha(TS) = \sum_{\beta + \gamma = \alpha} \frac{\alpha!}{\beta! \gamma!} \, \mathrm{ad}_A^\beta(T) \, \mathrm{ad}_A^\gamma(S).$$

PROOF. We observe that

$$(TS)(x) \equiv W'''(-x)TSW'(x) = W'''(-x)TW''(x)W''(-x)SW'(x) \equiv T(x)S(x).$$

Thus Leibniz' rule implies that $(TS)(\cdot)$ is strongly C^k on $\mathbb{R}^n$, with

$$(5.1.17) \qquad \partial^\alpha (TS)(x) = \sum_{\beta+\gamma=\alpha} \binom{\alpha}{\beta} \partial^\beta T(x) \cdot \partial^\gamma S(x).$$

In view of (5.1.7), this implies the validity of (5.1.16). The case of C_u^k is similar. $\square$

PROPOSITION 5.1.6. *Let $S \in B(\mathbf{F}', \mathbf{F}'')$ be a bijection of $\mathbf{F}'$ onto $\mathbf{F}''$, and let $k \in \mathbb{N}$. Define $S^{-1}(\cdot) : \mathbb{R}^n \to B(\mathbf{F}'', \mathbf{F}')$ by $S^{-1}(x) = W'(-x)S^{-1}W''(x)$. Then*
  (a) $S^{-1} \in C^k(A'', A'; \mathbf{F}'', \mathbf{F}') \Longleftrightarrow S \in C^k(A', A''; \mathbf{F}', \mathbf{F}'')$.
  (b) $S^{-1} \in C_u^k(A'', A'; \mathbf{F}'', \mathbf{F}') \Longleftrightarrow S \in C_u^k(A', A''; \mathbf{F}', \mathbf{F}'')$.
  (c) *For each non-zero multi-index β and each decomposition $\beta = \beta(1) + \cdots + \beta(r)$ of β into a sum of multi-indices $\beta(j)$, with $|\beta(j)| \geq 1$ for each j, there is a real number $c(\beta(1), \ldots, \beta(r))$ such that the following is true:*
  If $S(\cdot)$ is strongly of class C^k in $B(\mathbf{F}', \mathbf{F}'')$ and $1 \leq |\alpha| \leq k$, then

$$(5.1.18) \quad \partial^\alpha S^{-1}(x) = \sum c(\alpha(1), \ldots, \alpha(r)) \cdot S^{-1}(x) \partial^{\alpha(1)} S(x) \cdot$$
$$\cdot S^{-1}(x) \partial^{\alpha(2)} S(x) \cdot \ldots \cdot S^{-1}(x) \partial^{\alpha(r)} S(x) \cdot S^{-1}(x)$$

and

$$(5.1.19) \quad \mathrm{ad}_A^\alpha(S^{-1}) = \sum c(\alpha(1), \ldots, \alpha(r)) \cdot S^{-1} \mathrm{ad}_A^{\alpha(1)}(S) \cdot$$
$$\cdot S^{-1} \mathrm{ad}_A^{\alpha(2)}(S) \cdot \ldots \cdot S^{-1} \mathrm{ad}_A^{\alpha(r)}(S) \cdot S^{-1},$$

where the sums are over all possible decompositions $\alpha = \alpha(1) + \cdots + \alpha(r)$ with $|\alpha(j)| \geq 1$.

PROOF. We first observe that S^{-1} belongs to $B(\mathbf{F}'', \mathbf{F}')$ by the closed graph theorem.

(a) Since the statement is completely symmetric in S and S^{-1}, it suffices to show that, if S is strongly of class C^k, then so is S^{-1}. We proceed by induction on the order of the derivatives of $S^{-1}(\cdot)$. Assume that (for some $k \geq 1$) $S(\cdot)$ is strongly C^k as a function from $\mathbb{R}^n$ to $B(\mathbf{F}', \mathbf{F}'')$ and that we know that $S^{-1}(\cdot)$ is strongly C^m as a function from $\mathbb{R}^n$ to $B(\mathbf{F}'', \mathbf{F}')$, for some $m < k$ (which is certainly the case for $m = 0$). We show that this implies that $S^{-1}(\cdot)$ is strongly C^{m+1} in $B(\mathbf{F}'', \mathbf{F}')$. We shall use the following fact: if $T : \mathbb{R}^n \to B(\mathbf{F}'', \mathbf{F}')$ is strongly continuous, then

$$(5.1.20) \qquad \partial_j \{T(x)[S(x) - S]S^{-1}\}|_{x=0} = T(0) \cdot \partial_j S(0) \cdot S^{-1}$$

(strong derivative in $B(\mathbf{F}'', \mathbf{F}')$ on the l.h.s.).
  We observe that $S^{-1}(x) = [S(x)]^{-1}$, hence:

$$(5.1.21) \qquad S^{-1}(x) = S^{-1} - S^{-1}(x)[S(x) - S]S^{-1}.$$

Now let α be a multi-index with $|\alpha| = m$. By applying Leibniz' rule in (5.1.21), we get

(5.1.22)
$$\partial^\alpha S^{-1}(x) = \delta_{m0} S^{-1} - \sum_{\beta \leq \alpha} \binom{\alpha}{\beta} \partial^\beta S^{-1}(x) \cdot \partial^{\alpha-\beta}[S(x) - S] \cdot S^{-1}.$$

Since $S^{-1}(\cdot)$ is assumed to be of class C^m and $S(\cdot)$ of class C^k $(k > m)$, each term with $\beta \neq \alpha$ on the r.h.s. is strongly C^1 (because $\beta \leq \alpha$, $\beta \neq \alpha$ imply that $|\beta| < |\alpha| = m$). The term with $\beta = \alpha$ is differentiable at $x = 0$, because its partial derivatives are of the form (5.1.20), hence it is strongly C^1 by Proposition 5.1.2 (a). Thus $S^{-1}(\cdot)$ is strongly C^{m+1}.

(b) The proof is identical with that of (a) except for the replacement of strong continuity and strong differentiability by norm continuity and norm differentiability respectively.

(c) By virtue of (5.1.12) (applied to $S(\cdot)$ and to $S^{-1}(\cdot)$), it suffices to prove (5.1.18) for $x = 0$. We proceed by induction on the order $|\alpha|$ of the derivative. The argument is based on the identity (5.1.22). The term with $\beta = \alpha$ in the sum (5.1.22) vanishes if $x = 0$ $(S(x) - S = 0$ at $x = 0)$, so if $x = 0$ this sum only involves those β such that $\beta < \alpha$ (instead of $\beta \leq \alpha$). Hence, for $|\alpha| \geq 1$:

(5.1.23)
$$\partial^\alpha S^{-1}(0) = - \sum_{\beta < \alpha} \binom{\alpha}{\beta} \partial^\beta S^{-1}(0) \cdot \partial^{\alpha-\beta} S(0) \cdot S^{-1}.$$

This represents a recursion relation for the operators $\partial^\beta S^{-1}(0)$. For $|\alpha| = 1$, there is only one multi-index β with $\beta < \alpha$, namely $|\beta| = 0$, and (5.1.23) then reads

$$\partial_j S^{-1}(0) = -S^{-1} \cdot \partial_j S(0) \cdot S^{-1},$$

which is (5.1.18) for $|\alpha| = 1$ (at $x = 0$). It is now not difficult to see that the solution of the recursion relation (5.1.23) is of the form (5.1.18). $\square$

PROPOSITION 5.1.7. *Assume that $\mathbf{F}'$ and $\mathbf{F}''$ are reflexive. Then*

$$S \in C^k(A', A''; \mathbf{F}', \mathbf{F}'') \text{ if and only if } S^* \in C^k(A''^*, A'^*; \mathbf{F}''^*, \mathbf{F}'^*).$$

PROOF. Let $S^*(x) = S(x)^* = W'(x)^* S^* W''(-x)^*$. Then $S^*(\cdot)$ is weakly C^k in $B(\mathbf{F}''^*, \mathbf{F}'^*)$ if and only if $S(\cdot)$ is weakly C^k in $B(\mathbf{F}', \mathbf{F}'')$. By Proposition 5.1.2 (a) one may replace "weakly" by "strongly", which completes the proof. $\square$

5.2. Regularity of Fractional Order

5.2.1. In Section 3.4 we saw that a C_0-representation of $\mathbb{R}^n$ in a Banach space $\mathbf{F}$ leads naturally to a scale of Banach spaces $\{\mathbf{F}_{s,p}\}$, with $s \in \mathbb{R}$ and $p \in [1, \infty]$. In this section we consider the C_0-group $\mathscr{W}$ in the Banach space $\mathbf{F} = C_u^0(A', A''; \mathbf{F}', \mathbf{F}'')$ and study properties of the associated scale.

We are interested here only in the situation where the elements of these spaces have a certain degree of regularity (e.g. for which $S(\cdot)$ is Hölder continuous or differentiable). Since we insist on regularity of the operators with respect to the group, the norms (3.4.15) are especially convenient here. More precisely, let $s > 0$, $p \in [1, \infty]$ and let ℓ be an integer with $\ell > s$. Then we set for $S \in B(\mathbf{F}', \mathbf{F}'')$:

$$(5.2.1)\ ||S||_{s,p}^{(\ell)} = ||S||_{\mathbf{F}' \to \mathbf{F}''} + \left[\int_{|x| \leq 1} \left| ||x|^{-s} [\mathscr{W}(x) - I]^\ell S ||_{\mathbf{F}' \to \mathbf{F}''}^p \frac{dx}{|x|^n} \right]^{1/p} .$$

An equivalent but somewhat more explicit form of this expression is obtained by using the definition (5.0.1) of $S(x)$, viz. $S(x) = W''(-x)SW'(x)$:

(5.2.2)

$$||S||_{s,p}^{(\ell)} = ||S||_{\mathbf{F}' \to \mathbf{F}''} + \left[\int_{|x| \leq 1} \left| ||x|^{-s} \sum_{m=0}^{\ell} (-1)^m \binom{\ell}{m} S(mx) \right|\right|_{\mathbf{F}' \to \mathbf{F}''}^p \frac{dx}{|x|^n} \right]^{1/p} .$$

Observe that the integrand is a lower semicontinuous function of x, hence measurable.

If one applies Theorem 3.4.6 in the Banach space $\mathbf{F} = C_u^0(A', A''; \mathbf{F}', \mathbf{F}'')$ equipped with the C_0-group $\mathscr{W}$, one sees that $|| \cdot ||_{s,p}^{(\ell)}$ and $|| \cdot ||_{s,p}^{(\ell')}$ are equivalent gauges on $\mathbf{F}$ for any integers ℓ, $\ell' > s$, and the associated B-space is just the space $\mathbf{F}_{s,p}$ defined by interpolation between $\mathbf{F}$ and $C_u^\ell(A', A''; \mathbf{F}', \mathbf{F}'')$. However it is convenient to consider the gauge (5.2.1) also for $s = 0$; *the space that one gets in this case* (which will be shown to be independent of ℓ too) *is different from the interpolation space* $\mathbf{F}_{0,p}$. For this reason we use a special notation for the spaces $\mathbf{F}_{s,p}$ in the present context: if $s \geq 0$ and $p \in [1, \infty]$, we denote by $\mathscr{C}^{s,p}(A', A''; \mathbf{F}', \mathbf{F}'')$ the set of all operators S in $B(\mathbf{F}', \mathbf{F}'')$ for which $||S||_{s,p}^{(\ell)}$ is finite for some (and hence for each) integer $\ell > s$. Clearly $\mathscr{C}^{0,\infty}(A', A''; \mathbf{F}', \mathbf{F}'')$ is identical with $B(\mathbf{F}', \mathbf{F}'')$; $\mathscr{W}$ is not a C_0-group in this space in general, since it is only strongly continuous. On the other hand, if $s > 0$ or if $s = 0$ and $p \neq \infty$, the following lemma shows that the set $\mathscr{C}^{s,p}(A', A''; \mathbf{F}', \mathbf{F}'')$ is identical with the set of operators $S \in C_u^0(A', A''; \mathbf{F}', \mathbf{F}'')$ for which $||S||_{s,p}^{(\ell)}$ is finite for some $\ell > s$; since the restriction of $\mathscr{W}$ to $C_u^0(A', A''; \mathbf{F}', \mathbf{F}'')$ defines a C_0-representation of $\mathbb{R}^n$, the set $\mathscr{C}^{s,p}(A', A''; \mathbf{F}', \mathbf{F}'')$ is a B-space for the family of admissible norms $\{|| \cdot ||_{s,p}^{(\ell)}\}_{\ell > s}$, and for $s > 0$ one may use the theory developed in Section 3.4.

Let us mention some abbreviations that will often be used later on. We set $\mathscr{C}^{s,p}(A', A''; \mathbf{F}', \mathbf{F}'') = \mathscr{C}^{s,p}(\mathbf{F}', \mathbf{F}'')$ when we are working with a fixed pair of representations W', W''. If $\mathbf{F}'$, $\mathbf{F}''$ are B-subspaces of a Banach space $\mathbf{F}$ equipped with a C_0-representation $W(x) = e^{iA \cdot x}$ which leaves $\mathbf{F}'$ and $\mathbf{F}''$ invariant, then we may equip $\mathbf{F}'$, $\mathbf{F}''$ with the C_0-representations $W'(x) = W(x)|_{\mathbf{F}'}$ and $W''(x) = W(x)|_{\mathbf{F}''}$. In such a situation we set $\mathscr{C}^{s,p}(A', A''; \mathbf{F}', \mathbf{F}'') = \mathscr{C}^{s,p}(A; \mathbf{F}', \mathbf{F}'')$ (if the specification of A is important). Finally, we write $\mathscr{C}^{s,p}(A; \mathbf{F})$ or $\mathscr{C}^{s,p}(\mathbf{F})$ for $\mathscr{C}^{s,p}(A, A; \mathbf{F}, \mathbf{F})$.

In the next lemma we strengthen and prove some of the assertions made above. Note that the assertion (a) in the most important case $s > 0$ follows immediately from the assertion (b) together with Theorem 3.4.6. We preferred, however, to give a new proof which is of independent interest and which covers the case $s = 0$ too (the proof of part (a) of Lemma 5.2.1 is based on the identity (5.2.3) which is due, as far as we know, to Triebel; see §1.13.4 in [Tr] for references and historical comments and also Lemma 3.4.8 in [BB]).

LEMMA 5.2.1. *Let $s \geq 0$ and $p \in [1, \infty]$.*

(a) *If ℓ_1, ℓ_2 are integers such that $s < \ell_1 < \ell_2$, then $|| \cdot ||_{s,p}^{(\ell_1)}$ and $|| \cdot ||_{s,p}^{(\ell_2)}$ are equivalent gauges on $B(\mathbf{F}', \mathbf{F}'')$.*

(b) *Let $S \in B(\mathbf{F}', \mathbf{F}'')$ be such that $\lim_{\varepsilon \to 0} \varepsilon^{-n} \int_{|x| \leq \varepsilon} ||[\mathscr{W}(x) - I]^{\ell} S||_{\mathbf{F}' \to \mathbf{F}''} dx = 0$ for some integer $\ell \geq 1$. Then S belongs to $C_u^0(\mathbf{F}', \mathbf{F}'')$, i.e. the function $x \mapsto \mathscr{W}(x) S \in B(\mathbf{F}', \mathbf{F}'')$ is norm continuous. In particular, S belongs to $C_u^0(\mathbf{F}', \mathbf{F}'')$ if $||S||_{s,p}^{(\ell)} < \infty$ for some $0 \leq s < \ell$ and $1 \leq p \leq \infty$, with $p < \infty$ if $s = 0$, or if $\lim_{x \to 0} ||[\mathscr{W}(x) - I]^{\ell} S||_{\mathbf{F}' \to \mathbf{F}''} = 0$.*

PROOF. (a) It suffices to consider the case where $\ell_2 - \ell_1 = 1$, so we set $\ell_1 = \ell$ and take $\ell_2 = \ell + 1$.

(i) By (5.0.2) there is a constant c_1 such that

$$||[\mathscr{W}(x) - I]^{\ell+1} S||_{\mathbf{F}' \to \mathbf{F}''} \leq c_1 ||[\mathscr{W}(x) - I]^{\ell} S||_{\mathbf{F}' \to \mathbf{F}''} \quad \forall |x| \leq 1.$$

Thus we have $||S||_{s,p}^{(\ell+1)} \leq c_1 ||S||_{s,p}^{(\ell)}$.

(ii) We have the following algebraic identity:

$$(t^2 - 1)^{\ell} = (t - 1)^{\ell} [2 + (t - 1)]^{\ell}$$

$$= 2^{\ell} (t - 1)^{\ell} + (t - 1)^{\ell+1} \sum_{m=1}^{\ell} 2^{\ell-m} \binom{\ell}{m} (t - 1)^{m-1}.$$

It implies that (take $t = \mathscr{W}(x)$):

(5.2.3)

$$[\mathscr{W}(x) - I]^{\ell} = 2^{-\ell} [\mathscr{W}(2x) - I]^{\ell} - [\mathscr{W}(x) - I]^{\ell+1} \sum_{m=1}^{\ell} 2^{-m} \binom{\ell}{m} [\mathscr{W}(x) - I]^{m-1}.$$

So by (5.0.2) there is a constant c_2 such that for $0 < |x| \leq 1$:

$$\frac{||[\mathscr{W}(x) - I]^{\ell} S||_{\mathbf{F}' \to \mathbf{F}''}}{|x|^s} \leq 2^{s-\ell} \frac{||[\mathscr{W}(2x) - I]^{\ell} S||_{\mathbf{F}' \to \mathbf{F}''}}{|2x|^s}$$

$$+ c_2 \frac{||[\mathscr{W}(x) - I]^{\ell+1} S||_{\mathbf{F}' \to \mathbf{F}''}}{|x|^s}.$$

We set $\varphi_k(x) = |x|^{-s} ||[\mathscr{W}(x) - I]^k S||_{\mathbf{F}' \to \mathbf{F}''}$ and obtain from the preceding inequality that, for any $\varepsilon \in (0, 1]$:

$$\left[\int_{\varepsilon < |x| < 1} \varphi_\ell(x)^p \frac{dx}{|x|^n} \right]^{1/p} \leq 2^{s-\ell} \left[\int_{\varepsilon < |x| < 1} \varphi_\ell(2x)^p \frac{dx}{|x|^n} \right]^{1/p} +$$

$$+ c_2 \left[\int_{\varepsilon < |x| < 1} \varphi_{\ell+1}(x)^p \frac{dx}{|x|^n} \right]^{1/p}$$

$$\leq 2^{s-\ell} \left[\int_{\varepsilon < |y| < 1} \varphi_\ell(y)^p \frac{dy}{|y|^n} \right]^{1/p} + 2^{s-\ell} \left[\int_{1 < |y| < 2} \varphi_\ell(y)^p \frac{dy}{|y|^n} \right]^{1/p} +$$

$$+ c_2 \left[\int_{\varepsilon < |x| < 1} \varphi_{\ell+1}(x)^p \frac{dx}{|x|^n} \right]^{1/p}.$$

We observe that $2^{s-\ell} < 1$ and that the second term on the r.h.s. is majorized by $c_3 \|S\|_{\mathbf{F}' \to \mathbf{F}''}$ (again by (5.0.2)). Hence the preceding inequality implies that $\|S\|_{s,p}^{(\ell)} \leq c_4 \|S\|_{s,p}^{(\ell+1)}$ for some constant c_4 and all $S \in B(\mathbf{F}', \mathbf{F}'')$.

(b) Let $\theta \in C_0^\infty(\mathbb{R}^n)$ be such that $\theta(x) = 0$ if $|x| \geq 1$ and $\int_{\mathbb{R}^n} \theta(x) dx = 1$. For $\varepsilon > 0$, set

$$S_\varepsilon = S - \int_{\mathbb{R}^n} [I - \mathscr{W}(\varepsilon x)]^\ell S \, \theta(x) dx = \sum_{m=1}^\ell (-1)^{m+1} \binom{\ell}{m} \int_{\mathbb{R}^n} \mathscr{W}(\varepsilon m x) S \, \theta(x) dx.$$

Then:

$$\mathscr{W}(y) S_\varepsilon = \sum_{m=1}^\ell (-1)^{m+1} \binom{\ell}{m} \frac{1}{\varepsilon^n m^n} \int_{\mathbb{R}^n} \mathscr{W}(z) S \cdot \theta\left(\frac{z-y}{\varepsilon m}\right) dz.$$

Each integral in the last expression is weakly of class C^∞ as a function of y. Hence, by Lemma 5.A.2 (a), $\mathscr{W}(\cdot) S_\varepsilon$ is C^∞ in the norm of $B(\mathbf{F}', \mathbf{F}'')$. Since C_u^0 is a closed subspace of $B(\mathbf{F}', \mathbf{F}'')$, it now suffices to show that $S_\varepsilon \to S$ in norm as $\varepsilon \to 0$. But, by the assumptions made on S, this follows immediately from the estimate:

$$\|S_\varepsilon - S\|_{\mathbf{F}' \to \mathbf{F}''} \leq \frac{1}{\varepsilon^n} \int_{|y| \leq \varepsilon} \|[\mathscr{W}(y) - I]^\ell S\|_{\mathbf{F}' \to \mathbf{F}''} \cdot \left|\theta\left(\frac{y}{\varepsilon}\right)\right| dy. \quad \square$$

The identity (5.2.3) provides us with a more elementary proof of the implication $\lim_{x \to 0} \|[\mathscr{W}(x) - I]^\ell S\|_{\mathbf{F}' \to \mathbf{F}''} = 0 \Rightarrow S \in C_u^0$. Indeed, for each $\ell \geq 2$ there is a constant c such that for $|x| \leq 1$:

$$2^{\ell-1} \|[\mathscr{W}(x) - I]^{\ell-1} S\|_{\mathbf{F}' \to \mathbf{F}''} \leq \|[\mathscr{W}(2x) - I]^{\ell-1} S\|_{\mathbf{F}' \to \mathbf{F}''}$$

$$+ c \|[\mathscr{W}(x) - I]^\ell S\|_{\mathbf{F}' \to \mathbf{F}''}.$$

Together with the hypothesis made on S, this implies

$$2^{\ell-1} \limsup_{x \to 0} \|[\mathscr{W}(x) - I]^{\ell-1} S\|_{\mathbf{F}' \to \mathbf{F}''} \leq \limsup_{x \to 0} \|[\mathscr{W}(2x) - I]^{\ell-1} S\|_{\mathbf{F}' \to \mathbf{F}''}.$$

The two superior limits in this inequality are finite and equal, so they must be zero, i.e. $\lim_{x \to 0} \|[\mathscr{W}(x) - I]^{\ell-1} S\|_{\mathbf{F}' \to \mathbf{F}''} = 0$. By iterating this argument we get $S \in C_u^0$.

Let us mention some particular cases of the spaces $\mathscr{C}^{s,p}(\mathbf{F}',\mathbf{F}'')$ that will be important. If $0 \le s < 1$, one may take $\ell = 1$ in (5.2.1), in which case the condition that $S \in \mathscr{C}^{s,p}(\mathbf{F}',\mathbf{F}'')$ may be written as

$$(5.2.4) \qquad \|S\|_{\mathbf{F}'\to\mathbf{F}''} + \left[\int_{|x|\le 1}\left\|\frac{S(x)-S}{|x|^s}\right\|_{\mathbf{F}'\to\mathbf{F}''}^p \frac{dx}{|x|^n}\right]^{1/p} < \infty.$$

If $0 \le s < 2$, one may take $\ell = 2$, so that we have $S \in \mathscr{C}^{s,p}(\mathbf{F}',\mathbf{F}'')$ if and only if

$$(5.2.5) \quad \|S\|_{\mathbf{F}'\to\mathbf{F}''} + \left[\int_{|x|\le 1}\left\|\frac{S(x)+S(-x)-2S}{|x|^s}\right\|_{\mathbf{F}'\to\mathbf{F}''}^p \frac{dx}{|x|^n}\right]^{1/p} < \infty.$$

If $s = 1$, we are forced to use the form (5.2.5), i.e. the condition that $S \in \mathscr{C}^{1,p}(\mathbf{F}',\mathbf{F}'')$ cannot be expressed in terms of the first order difference $S(x) - S$.

5.2.2. We now establish some properties of the B-spaces $\mathscr{C}^{s,p}(\mathbf{F}',\mathbf{F}'')$. By applying Theorem 3.4.6 to the present situation, we get that for $s > 0$ and any integer $\ell > s$

$$(5.2.6) \qquad \mathscr{C}^{s,p}(\mathbf{F}',\mathbf{F}'') = (C_u^\ell(\mathbf{F}',\mathbf{F}''), C_u^0(\mathbf{F}',\mathbf{F}''))_{\theta,p}, \qquad \theta = 1 - \frac{s}{\ell}.$$

As a consequence we obtain some other useful relations between the B-spaces $\mathscr{C}^{s,p}(\mathbf{F}',\mathbf{F}'')$, namely:

$$(5.2.7) \qquad \mathscr{C}^{s,p}(\mathbf{F}',\mathbf{F}'') \subset \mathscr{C}^{t,q}(\mathbf{F}',\mathbf{F}'') \qquad \text{if } s > t \ge 0, \quad p,q \in [1,\infty]$$

and

$$(5.2.8) \qquad\qquad \mathscr{C}^{t,p}(\mathbf{F}',\mathbf{F}'') \subset \mathscr{C}^{t,q}(\mathbf{F}',\mathbf{F}'') \qquad \text{if } q \ge p.$$

For $t > 0$ these relations follow from Theorem 3.4.3 (a). The extension to $t \ge 0$ is easy. The inclusion (5.2.8) implies in particular that for $s \ge 0$

$$(5.2.9) \qquad \mathscr{C}^{s,1}(\mathbf{F}',\mathbf{F}'') \subset \mathscr{C}^{s,p}(\mathbf{F}',\mathbf{F}'') \subset \mathscr{C}^{s,\infty}(\mathbf{F}',\mathbf{F}'') \text{ if } p \in [1,\infty].$$

If $s \equiv k$ is an integer, we have in addition to the spaces $\mathscr{C}^{k,p}(\mathbf{F}',\mathbf{F}'')$ the two spaces $C^k(\mathbf{F}',\mathbf{F}'')$ and $C_u^k(\mathbf{F}',\mathbf{F}'')$ introduced in Definition 5.1.1. One has the following relations between these spaces:

$$(5.2.10) \qquad \mathscr{C}^{k,1}(\mathbf{F}',\mathbf{F}'') \subset C_u^k(\mathbf{F}',\mathbf{F}'') \subset C^k(\mathbf{F}',\mathbf{F}'') \subset \mathscr{C}^{k,\infty}(\mathbf{F}',\mathbf{F}'')$$

if $k \in \mathbb{N}$. In general all embeddings in (5.2.10) are strict and the spaces C_u^k, C^k and $\mathscr{C}^{k,p}$ are distinct for all $p \in [1,\infty]$.

PROOF OF (5.2.10). The case $k = 0$ is evident. If $k \ge 1$, then Theorem 3.4.3 (c) applied to the C_0-group $\mathscr{W}$ in $C_u^0(\mathbf{F}',\mathbf{F}'')$ gives the inclusions $\mathscr{C}^{k,1} \subset C_u^k \subset \mathscr{C}^{k,\infty}$. The relation $C_u^k \subset C^k$ is evident. The inclusion $C^k \subset \mathscr{C}^{k,\infty}$ is obtained by observing that, if $S \in C^k(\mathbf{F}',\mathbf{F}'')$, then

$$\|[\mathscr{W}(x) - I]^k S\|_{\mathbf{F}'\to\mathbf{F}''} \le c|x|^k \sum_{|\alpha|=k} \|\operatorname{ad}_A^\alpha(S)\|_{\mathbf{F}'\to\mathbf{F}''} \le c_1|x|^k$$

for all $|x| \le 1$ by (5.1.9). $\quad\square$

Now let $s > 0$ and let θ be a real number satisfying $0 < \theta < 1$ and $\theta \leq s$. Then, if $S \in \mathscr{C}^{s,p}(\mathbf{F}', \mathbf{F}'')$, the function $S(\cdot)$ is locally Hölder continuous of order θ. Indeed, by (5.2.7) and (5.2.8) we then have $S \in \mathscr{C}^{\theta,\infty}(\mathbf{F}', \mathbf{F}'')$; hence for each compact subset Ω of $\mathbb{R}^n$ there is a constant c_Ω such that for all $x, y \in \Omega$:

$$\|\mathscr{W}(x)S - \mathscr{W}(y)S\|_{\mathbf{F}' \to \mathbf{F}''} \leq c_\Omega \|S\|_{\theta,\infty}^{(1)} |x - y|^\theta.$$

PROPOSITION 5.2.2. *Let $s \in (0, \infty)$ and $p \in [1, \infty]$. Write $s = k + \sigma$ with $k \in \mathbb{N}$ and $0 < \sigma \leq 1$.*
(a) *If $S \in \mathscr{C}^{s,p}(\mathbf{F}', \mathbf{F}'')$ and $|\alpha| \leq k$, then $\mathrm{ad}_A^\alpha(S) \in \mathscr{C}^{s-|\alpha|,p}(\mathbf{F}', \mathbf{F}'')$.*
(b) *If $\| \cdot \|_{\mathscr{C}^{\sigma,p}}$ is an admissible norm on $\mathscr{C}^{\sigma,p}(\mathbf{F}', \mathbf{F}'')$, then*

$$\|S\|_{C^k} + \sum_{|\alpha|=k} \| \mathrm{ad}_A^\alpha(S)\|_{\mathscr{C}^{\sigma,p}}$$

defines an admissible norm on $\mathscr{C}^{s,p}(\mathbf{F}', \mathbf{F}'')$. In particular: $S \in \mathscr{C}^{s,p}(\mathbf{F}', \mathbf{F}'')$ if and only if $S \in C^k(\mathbf{F}', \mathbf{F}'')$ and $\mathrm{ad}_A^\alpha(S) \in \mathscr{C}^{\sigma,p}(\mathbf{F}', \mathbf{F}'')$ for each multi-index α with $|\alpha| = k$.

PROOF. We consider the C_0-group $\mathscr{W}$ in $\mathbf{F} = C_u^0(\mathbf{F}', \mathbf{F}'')$. If $\mathscr{A}$ denotes the generator of $\mathscr{W}$, we have $\mathscr{A}^\alpha S = (-1)^{|\alpha|} \mathrm{ad}_A^\alpha(S)$. Thus (a) follows immediately from Proposition 3.4.5 (a). Theorem 3.4.7 implies that $\|S\|_{C^k} + \sum_{|\alpha|=k} \| \mathrm{ad}_A^\alpha(S)\|_{\mathscr{C}^{\sigma,p}}$ is an admissible norm on $\mathscr{C}^{s,p}(\mathbf{F}', \mathbf{F}'')$. Thus it remains to show that

(5.2.11) $S \in C^k, \ \mathrm{ad}_A^\alpha(S) \in \mathscr{C}^{\sigma,p}$ for $|\alpha| = k \implies S \in C_u^k.$

For this we observe that $\partial^\alpha S(\cdot) \equiv (-i)^{|\alpha|} \mathrm{ad}_A^\alpha(S(\cdot))$ is locally Hölder continuous of order $\sigma/2$ by the observation made before the proposition, so that the validity of (5.2.11) follows from Lemma 5.A.2 (c) in the Appendix. $\square$

For some applications it is useful to consider, for $p = 1$, the limiting case in Proposition 5.2.2 where formally $\sigma = 0$. We set $C^{+0}(\mathbf{F}', \mathbf{F}'') \equiv \mathscr{C}^{0,1}(\mathbf{F}', \mathbf{F}'')$ and observe that the condition that $S \in C^{+0}(\mathbf{F}', \mathbf{F}'')$ amounts to a Dini type continuity condition on the function $S(\cdot)$, namely

$$(5.2.12) \qquad \int_{|x|\leq 1} \|W''(-x)SW'(x) - S\|_{\mathbf{F}' \to \mathbf{F}''} \frac{dx}{|x|^n} < \infty.$$

If $k \geq 1$ is an integer, we define $C^{k+0}(\mathbf{F}', \mathbf{F}'')$ as the space of all S in $C^k(\mathbf{F}', \mathbf{F}'')$ such that $\mathrm{ad}_A^\alpha(S) \in C^{+0}(\mathbf{F}', \mathbf{F}'')$ for each multi-index α with $|\alpha| = k$. We observe that $S \in C^{k+0}(\mathbf{F}', \mathbf{F}'')$ if and only if $S(\cdot) : \mathbb{R}^n \to B(\mathbf{F}', \mathbf{F}'')$ is strongly of class C^k and its derivatives of order k are Dini continuous in norm (i.e. (5.2.12) is satisfied with S replaced by $\mathrm{ad}_A^\alpha(S)$, $|\alpha| = k$). $C^{k+0}(\mathbf{F}', \mathbf{F}'')$ is a Banach space for the norm

(5.2.13)

$$\|S\|_{C^{k+0}} = \|S\|_{C^k} + \sum_{|\alpha|=k} \int_{|x|\leq 1} \|[\mathscr{W}(x) - I]\, \mathrm{ad}_A^\alpha(S)\|_{\mathbf{F}' \to \mathbf{F}''} \frac{dx}{|x|^n}.$$

One has the following inclusions:

$$(5.2.14) \qquad C^{k+0}(\mathbf{F}', \mathbf{F}'') \subset \mathscr{C}^{k,1}(\mathbf{F}', \mathbf{F}'') \subset C_u^k(\mathbf{F}', \mathbf{F}'') \quad \text{if } k \in \mathbb{N}.$$

PROOF OF (5.2.14). The second inclusion has already been established in (5.2.10). To obtain the first inclusion if $k \geq 1$, we use (5.1.9) which allows us to write, for some constants $c_{\alpha m}$:

$$[\mathscr{W}(x) - I]^{k+1} S = \sum_{|\alpha|=k} \sum_{m=1}^{k} c_{\alpha m} x^{\alpha} \int_0^1 (1 - \tau)^{k-1} d\tau \, \mathscr{W}(\tau m x)[\mathscr{W}(x) - I] \, \mathrm{ad}_A^{\alpha}(S).$$

Hence we have

$$||S||_{k,1}^{(k+1)} = ||S||_{\mathbf{F}' \to \mathbf{F}''} + \int_{|x| \leq 1} |||x|^{-k} [\mathscr{W}(x) - I]^{k+1} S||_{\mathbf{F}' \to \mathbf{F}''} \frac{dx}{|x|^n}$$

$$\leq ||S||_{\mathbf{F}' \to \mathbf{F}''} + cM^2 e^{2k\omega} \sum_{|\alpha|=k} \int_{|x| \leq 1} |||[\mathscr{W}(x) - I] \, \mathrm{ad}_A^{\alpha}(S)||_{\mathbf{F}' \to \mathbf{F}''} \frac{dx}{|x|^n}$$

$$\leq c_1 ||S||_{C^{k+0}}. \quad \square$$

5.2.3. We next give some results on the behaviour of the classes $\mathscr{C}^{s,p}(\mathbf{F}', \mathbf{F}'')$ under algebraic operations.

PROPOSITION 5.2.3. *Let $\mathbf{F}'$, $\mathbf{F}''$ and $\mathbf{F}'''$ be Banach spaces, W', W'' and W''' C_0-representations of $\mathbb{R}^n$ in $\mathbf{F}'$, $\mathbf{F}''$ and $\mathbf{F}'''$ respectively.*

(a) Let $s \geq 0$, $p \in [1, \infty]$ and assume that $S \in \mathscr{C}^{s,p}(\mathbf{F}', \mathbf{F}'')$ and $T \in \mathscr{C}^{s,p}(\mathbf{F}'', \mathbf{F}''')$. Then $TS \in \mathscr{C}^{s,p}(\mathbf{F}', \mathbf{F}''')$.

(b) If $k \in \mathbb{N}$, $S \in C^{k+0}(\mathbf{F}', \mathbf{F}'')$ and $T \in C^{k+0}(\mathbf{F}'', \mathbf{F}''')$, then TS belongs to $C^{k+0}(\mathbf{F}', \mathbf{F}''')$.

PROOF. We prove (a) and leave the simple proof of (b) as an exercise. With some abuse of notation, we write $\mathscr{W}(x)$ for each of the three operators $\mathscr{W}(x)$ defined in $B(\mathbf{F}', \mathbf{F}'')$, $B(\mathbf{F}', \mathbf{F}''')$ and $B(\mathbf{F}'', \mathbf{F}''')$ respectively. We set $\mathscr{W}_-(x) = \mathscr{W}(x) - I$ and $\mathscr{W}_+(x) = [\mathscr{W}(x) + I]/2$ and claim that for each $\ell \in \mathbb{N}$:

$$(5.2.15)$$

$$[\mathscr{W}(x) - I]^{\ell}(TS) = \sum_{m=0}^{\ell} \binom{\ell}{m} [\mathscr{W}_+(x)^{\ell-m} \mathscr{W}_-(x)^m T] \cdot [\mathscr{W}_+(x)^m \mathscr{W}_-(x)^{\ell-m} S].$$

(5.2.15) is evident for $\ell = 0$ and easy to check for $\ell = 1$. For general ℓ, (5.2.15) is obtained by induction; one may write $[\mathscr{W}(x) - I]^{\ell+1} = [\mathscr{W}(x) - I][\mathscr{W}(x) - I]^{\ell}$, use the induction hypothesis (5.2.15) as well as (5.2.15) with $\ell = 1$, and take into account the identity

$$\binom{\ell}{m} + \binom{\ell}{m-1} = \binom{\ell+1}{m} \qquad \text{if } m = 1, \ldots, \ell.$$

Now choose an integer $\ell > 2s$. By (5.0.2) there is a constant $c \in [1, \infty)$ such that for all $|x| \leq 1$:

$$||\mathscr{W}_+(x)||_{\mathbf{F}' \to \mathbf{F}''} + ||\mathscr{W}_+(x)||_{\mathbf{F}'' \to \mathbf{F}'''} + ||\mathscr{W}_-(x)||_{\mathbf{F}' \to \mathbf{F}''} + ||\mathscr{W}_-(x)||_{\mathbf{F}'' \to \mathbf{F}'''} \leq c.$$

Thus one gets (see (5.2.1) and denote by B the unit ball in $\mathbb{R}^n$):

$$\left\| |x|^{-s}[\mathscr{W}(x) - I]^{\ell}(TS) \right\|_{L_*^p(B;B(\mathbf{F}',\mathbf{F}'''))} \leq$$

$$\leq \ell! c^{2\ell} \Big\{ \|T\|_{\mathbf{F}'' \to \mathbf{F}'''} \sum_{0 \leq m \leq \ell/2} \left\| |x|^{-s}[\mathscr{W}(x) - I]^{\ell-m} S \right\|_{L_*^p(B;B(\mathbf{F}',\mathbf{F}''))}$$

$$+ \|S\|_{\mathbf{F}' \to \mathbf{F}''} \sum_{\ell/2 < m \leq \ell} \left\| |x|^{-s}[\mathscr{W}(x) - I]^m T \right\|_{L_*^p(B;B(\mathbf{F}'',\mathbf{F}'''))} \Big\},$$

which is finite since $\ell - m > s$ for each term in the first sum and $m > s$ for each term in the second sum. $\square$

PROPOSITION 5.2.4. *Let $s \geq 0$ and $p \in [1,\infty]$. Let $S \in B(\mathbf{F}',\mathbf{F}'')$ be a bijection of $\mathbf{F}'$ onto $\mathbf{F}''$. Then $S \in \mathscr{C}^{s,p}(\mathbf{F}',\mathbf{F}'')$ if and only if $S^{-1} \in \mathscr{C}^{s,p}(\mathbf{F}'',\mathbf{F}')$.*

PROOF. As in the proof of Proposition 5.1.6, it suffices to show that $S \in \mathscr{C}^{s,p}(\mathbf{F}',\mathbf{F}'') \Longrightarrow S^{-1} \in \mathscr{C}^{s,p}(\mathbf{F}'',\mathbf{F}')$.

(i) Let $0 \leq s < 1$. By (5.1.21) we have

$$[\mathscr{W}(x) - I]S^{-1} = -S^{-1}(x) \cdot \{[\mathscr{W}(x) - I]S\} \cdot S^{-1}.$$

Since $\|S^{-1}(x)\|_{\mathbf{F}'' \to \mathbf{F}'} \leq c\|S^{-1}\|_{\mathbf{F}'' \to \mathbf{F}'}$ for all $|x| \leq 1$, we get that

$$\|S^{-1}\|_{s,p}^{(1)} \leq c\|S^{-1}\|_{\mathbf{F}'' \to \mathbf{F}'}^2 \|S\|_{s,p}^{(1)}.$$

(ii) Let $1 \leq s < 2$, so that $s/2 < 1$. In this case we use the identity

(5.2.16)
$$[\mathscr{W}(x) - I]^2 S^{-1} = -S^{-1}(2x) \cdot \{[\mathscr{W}(x) - I]^2 S$$
$$- 2\mathscr{W}(x)[\mathscr{W}(x) - I]S \cdot S^{-1}(x) \cdot [\mathscr{W}(x) - I]S\} \cdot S^{-1}.$$

One then finds as above that

$$\|S^{-1}\|_{s,p}^{(2)} \leq c_1 \|S^{-1}\|_{\mathbf{F}'' \to \mathbf{F}'}^2 \big\{ \|S\|_{s,p}^{(2)} + \|S^{-1}\|_{\mathbf{F}'' \to \mathbf{F}'} [\|S\|_{s/2,2p}^{(2)}]^2 \big\},$$

which is finite since $S \in \mathscr{C}^{s/2,2p}(\mathbf{F}',\mathbf{F}'')$ by (5.2.7).

(iii) Now let $s \geq 2$. We write $s = k + \sigma$ with $k \in \mathbb{N}$ and $0 < \sigma \leq 1$. By Proposition 5.2.2 (b) we must show that $S^{-1} \in C^k(\mathbf{F}'',\mathbf{F}')$ and that $\mathrm{ad}_A^\alpha(S^{-1}) \in \mathscr{C}^{\sigma,p}(\mathbf{F}'',\mathbf{F}')$ for $|\alpha| = k$. The first condition is satisfied by Proposition 5.1.6 (a). For the second condition we observe that, by (5.1.19), the operator $\mathrm{ad}_A^\alpha(S^{-1})$ is a finite linear combination of terms of the form

$$\left[\prod_{j=1}^m S^{-1} \mathrm{ad}_A^{\alpha(j)}(S) \right] \cdot S^{-1},$$

with $1 \leq m \leq k$ and $1 \leq |\alpha(j)| \leq k$ if $|\alpha| = k$. By Proposition 5.2.2 and by (5.2.7), each $\mathrm{ad}_A^{\alpha(j)}(S)$ belongs to $\mathscr{C}^{\sigma,p}(\mathbf{F}',\mathbf{F}'')$ by (i) or (ii) above (since $0 < \sigma \leq 1$). By Proposition 5.2.3, we conclude that $\mathrm{ad}_A^\alpha(S^{-1}) \in \mathscr{C}^{\sigma,p}(\mathbf{F}'',\mathbf{F}')$. $\square$

The following obvious property should also be mentioned: if $\mathbf{F}'$ and $\mathbf{F}''$ are reflexive, then

$$(5.2.17) \qquad S \in \mathscr{C}^{s,p}(\mathbf{F}', \mathbf{F}'') \Longleftrightarrow S^* \in \mathscr{C}^{s,p}(\mathbf{F}''^*, \mathbf{F}'^*).$$

REMARK 5.2.5. The notion of $\mathscr{C}^{s,p}$-regularity is stable under other operations, in particular under integration. More precisely, let Ω be a measurable space, μ a measure on Ω and $\Psi : \Omega \to B(\mathbf{F}', \mathbf{F}'')$ a (strongly) integrable function such that for some $s \geq 0$, some integer $\ell > s$ and some $p \in [1, \infty]$:

$$(5.2.18) \qquad \int_\Omega \| \Psi(\omega) \|_{s,p}^{(\ell)} \mu(d\omega) < \infty.$$

Then $S = \int_\Omega \Psi(\omega)\mu(d\omega)$ belongs to $\mathscr{C}^{s,p}(\mathbf{F}', \mathbf{F}'')$.

5.2.4. For each integer $k \geq 0$ we have introduced in this chapter four distinct classes of regularity for bounded operators $S : \mathbf{F}' \to \mathbf{F}''$, namely: C^k, C_u^k, C^{k+0} and $\mathscr{C}^{k,p}$ (where $1 \leq p \leq \infty$). We shall end this section with some further comments concerning these classes and the relations between them. As the Banach spaces $\mathbf{F}', \mathbf{F}''$ and the representations W', W'' are fixed, we shall not specify them in the notations.

We recall the following inclusions, which have been established before:

$$(5.2.19) \qquad C^{k+0} \subset \mathscr{C}^{k,1} \subset C_u^k \subset C^k \subset \mathscr{C}^{k,\infty} \subset B(\mathbf{F}', \mathbf{F}'').$$

All the spaces here are B-spaces and the embeddings are continuous; the first two are also dense, but the remaining ones are not dense in general. The B-spaces $\mathscr{C}^{k,p}$, for $1 < p < \infty$, lie somewhere between $\mathscr{C}^{k,1}$ and $\mathscr{C}^{k,\infty}$, but they are (in general) distinct from C_u^k and C^k.

For $k = 0$ we have $C^0 = \mathscr{C}^{0,\infty} = B(\mathbf{F}', \mathbf{F}'')$, but C_u^0 is a closed (non trivial in general) subspace of $B(\mathbf{F}', \mathbf{F}'')$. Also $C^{+0} = \mathscr{C}^{0,1}$ (this is just the definition of C^{+0}), but for $k > 0$ the space C^{k+0} is significantly different from $\mathscr{C}^{k,1}$. For example, it follows from (5.2.19) that by real interpolation of order (θ, q), with $0 < \theta < 1$, $1 \leq q \leq \infty$, between two spaces C^k and C^m one obtains one of the spaces $\mathscr{C}^{s,p}$; but by interpolating between C^{k+0} and C^{m+0} one gets new spaces.

In the next theorem we sum up several characterizations of the preceding spaces for $k \geq 1$; most of them have been established before, but collecting them in one statement allows a better understanding of the relations between the four classes of regularity $\mathscr{C}^{k,1}, C_u^k, C^k$ and $\mathscr{C}^{k,\infty}$.

THEOREM 5.2.6. Let $S : \mathbf{F}' \to \mathbf{F}''$ be a continuous operator and k, ℓ two integers such that $1 \leq k < \ell$.

(a) One has $S \in \mathscr{C}^{k,\infty}$ if and only if one of the following conditions is satisfied:

(a_1) for each $x \in \mathbb{R}^n$ there is a constant $c(x) < \infty$ such that $\|[\mathscr{W}(\varepsilon x) - I]^\ell S\|_{\mathbf{F}' \to \mathbf{F}''} \leq c(x)\varepsilon^k \; \forall \varepsilon \in (0,1)$;

(a_2) there is $c < \infty$ such that $\|[\mathscr{W}(x) - I]^\ell S\|_{\mathbf{F}' \to \mathbf{F}''} \leq c|x|^k$ for $|x| \leq 1$.

(b) The condition $S \in C^k$ is equivalent to each of the following two conditions:

(b_1) for each $x \in \mathbb{R}^n$ the limit $\lim_{\varepsilon \to +0} \varepsilon^{-k}[\mathscr{W}(\varepsilon x) - I]^k S$ exists in the strong operator topology of $B(\mathbf{F}', \mathbf{F}'')$;

(b_2) *for each $x \in \mathbb{R}^n$ the limit $\lim_{\varepsilon \to +0} \int_\varepsilon^1 \tau^{-k}[\mathscr{W}(\tau x) - I]^\ell S \tau^{-1} d\tau$ exists in the strong operator topology of $B(\mathbf{F}', \mathbf{F}'')$.*

If $\mathbf{F}''$ is reflexive, then these conditions are also equivalent to each of the next ones:

(b_3) *for each $x \in \mathbb{R}^n$, $\liminf_{\varepsilon \to +0} \varepsilon^{-k} ||[\mathscr{W}(\varepsilon x) - I]^k S||_{\mathbf{F}' \to \mathbf{F}''} < \infty$;*

(b_4) *for each $x \in \mathbb{R}^n$ there is a finite constant $c(x)$ such that $||[\mathscr{W}(\varepsilon x) - I]^k S||_{\mathbf{F}' \to \mathbf{F}''} \le c(x)\varepsilon^k \ \forall \varepsilon \in (0, 1)$;*

(b_5) *one has $||[\mathscr{W}(x) - I]^k S||_{\mathbf{F}' \to \mathbf{F}''} \le c|x|^k$ for a finite constant c and all $x \in \mathbb{R}^n$ with $|x| \le 1$;*

(b_6) *for each $x \in \mathbb{R}^n$, $\liminf_{\varepsilon \to +0} \left\| \int_\varepsilon^1 \tau^{-k}[\mathscr{W}(\tau x) - I]^\ell S \tau^{-1} d\tau \right\|_{\mathbf{F}' \to \mathbf{F}''} < \infty$.*

(c) *The condition $S \in C_u^k$ is equivalent to each of the following two conditions:*

(c_1) *for each $x \in \mathbb{R}^n$, the limit $\lim_{\varepsilon \to +0} \varepsilon^{-k}[\mathscr{W}(\varepsilon x) - I]^k S$ exists in the norm topology of $B(\mathbf{F}', \mathbf{F}'')$;*

(c_2) *for each $x \in \mathbb{R}^n$ the limit $\lim_{\varepsilon \to +0} \int_\varepsilon^1 \tau^{-k}[\mathscr{W}(\tau x) - I]^\ell S \tau^{-1} d\tau$ exists in the norm topology of $B(\mathbf{F}', \mathbf{F}'')$.*

(d) *One has $S \in \mathscr{C}^{k,1}$ if and only if one of the following two conditions is satisfied:*

(d_1) *$\int_0^1 ||\tau^{-k}[\mathscr{W}(\tau x) - I]^\ell S||_{\mathbf{F}' \to \mathbf{F}''} \tau^{-1} d\tau < \infty$ for each $x \in \mathbb{R}^n$;*

(d_2) *$\int_{|x| \le 1} || \, |x|^{-k}[\mathscr{W}(x) - I]^\ell S||_{\mathbf{F}' \to \mathbf{F}''} |x|^{-n} dx < \infty$.*

PROOF. (a) The condition (a_2) is just the definition of the class $\mathscr{C}^{k,\infty}$, and it clearly implies (a_1). If (a_1) is satisfied, then so is the hypothesis of Lemma 5.2.1 (b) (use the dominated convergence theorem). So $S \in C_u^0$, and we may apply Theorem 3.4.9 in the Banach space $\mathbf{F} = C_u^0$, with $s = k$ and $p = \infty$, to conclude that $S \in \mathscr{C}^{k,\infty}$.

(b) The equivalence of the condition $S \in C^k$ with (b_1) (and with (b_3) if $\mathbf{F}''$ is reflexive) has been established in Theorem 5.1.3. Then, assuming $\mathbf{F}''$ reflexive, we have (b_5) $\Rightarrow$ (b_3) $\Rightarrow S \in C^k \Rightarrow$ (b_5) (for the last step, see the proof of (5.2.10)).

It remains to prove the equivalence with the integral conditions. By following the proof of Proposition 3.4.10 we see that for arbitrary $S \in B(\mathbf{F}', \mathbf{F}'')$ and $f \in \mathbf{F}'_k$:

(5.2.20)
$$\int_\varepsilon^1 \frac{[\mathscr{W}(\tau x) - I]^\ell S}{\tau^k} f \frac{d\tau}{\tau} = b \sum_{|\alpha| = k} \frac{x^\alpha}{\alpha!} \mathscr{A}^\alpha[S]f + \mathscr{S}f$$

$$+ b \sum_{|\alpha| = k} \frac{x^\alpha}{\alpha!} \int_\mathbb{R} \{ (\mathscr{W}(\varepsilon \sigma x)\chi(\sigma) - I)\mathscr{A}^\alpha[S] \} f \, d\sigma.$$

Here b is a non-zero constant, $\mathscr{S}$ is a continuous operator $\mathbf{F}' \to \mathbf{F}''$ depending on S and χ is a bounded function with compact support such that $\int_\mathbb{R} \chi(\sigma) d\sigma = 1$.

Now assume $S \in C^k$. Then all the operators that appear in (5.2.20) belong to $B(\mathbf{F}', \mathbf{F}'')$, hence the identity will hold for all $f \in \mathbf{F}'$, by the density of $\mathbf{F}'_k$ in $\mathbf{F}'$. If we fix such an f, then each of the integrals on the r.h.s. of (5.2.20) tends to zero in $\mathbf{F}''$ as $\varepsilon \to +0$; hence (b_2) is satisfied. Reciprocally, assume that (b_2)

holds. For $f \in \mathbf{F}'_k$, (5.2.20) implies that the next limit exists strongly in $\mathbf{F}''_{-k}$ and that

$$(5.2.21) \qquad \lim_{\varepsilon \to +0} \int_\varepsilon^1 \frac{[\mathscr{W}(\tau x) - I]^\ell S}{\tau^k} f \frac{d\tau}{\tau} = b \sum_{|\alpha|=k} \frac{x^\alpha}{\alpha!} \mathscr{A}^\alpha[S] f + \mathscr{S} f.$$

But, by hypothesis, the limit in the l.h.s. exists in $\mathbf{F}''$ and gives an operator in $B(\mathbf{F}', \mathbf{F}'')$. Hence $\sum_{|\alpha|=k} \frac{x^\alpha}{\alpha!} \mathscr{A}^\alpha[S]$ belongs to $B(\mathbf{F}', \mathbf{F}'')$ for each $x \in \mathbb{R}^n$. By applying Proposition 1.1.2 (see the proof of Proposition 3.3.19) we get $\mathscr{A}^\alpha[S] \in B(\mathbf{F}', \mathbf{F}'')$ for $|\alpha| \leq k$. Finally, Proposition 5.1.2 (b) implies that $S \in C^k$.

We now prove the implication $(b_6) \Rightarrow S \in C^k$ assuming $\mathbf{F}''$ reflexive ($S \in C^k \Rightarrow (b_2) \Rightarrow (b_6)$ is obvious).

Since the unit ball in $B(\mathbf{F}', \mathbf{F}'')$ is weakly compact, we may find a sequence $\varepsilon_j \to 0$ such that the sequence of operators $\int_{\varepsilon_j}^1 \tau^{-k} [\mathscr{W}(\tau x) - I]^\ell S \tau^{-1} d\tau$ is weakly convergent in $B(\mathbf{F}', \mathbf{F}'')$; let $U(x) \in B(\mathbf{F}', \mathbf{F}'')$ be its limit. Then (5.2.21) implies for $f \in \mathbf{F}_k : U(x) f = b \sum_{|\alpha|=k} \frac{x^\alpha}{\alpha!} \mathscr{A}^\alpha[S] f + \mathscr{S} f$, and we may argue as before.

(c) We have seen in Theorem 5.1.3 that $S \in C_u^k \Leftrightarrow (c_1)$. The implication $S \in C_u^k \Rightarrow (c_2)$ follows from Proposition 3.4.10 applied to the C_0-group $\mathscr{W}$ in the Banach space C_u^0. Reciprocally, if $S \in B(\mathbf{F}', \mathbf{F}'')$ and (c_2) holds, then $S \in C^k$ by part (b) of the theorem. Since $C^k \subset C_u^0$ (recall that $k \geq 1$), we get $S \in C_u^0$, hence we may use again Proposition 3.4.10.

(d) Condition (d_2) is the definition of the relation $S \in \mathscr{C}^{k,1}$. To prove that $S \in \mathscr{C}^{k,1} \Leftrightarrow (d_1)$ one proceeds as in the proof of (c) (observe that $(d_1) \Rightarrow (b_2)$). $\square$

The scale $\{\mathscr{C}^{s,p}\}_{s>0, 1 \leq p \leq \infty}$ is obtained by real interpolation starting from the scale $\{C_u^k\}_{k \in \mathbb{N}}$, cf. (5.2.6). However, from (5.2.19) and the fact that the closure of C^ℓ ($\ell \geq 1$) in $C^0 \equiv B(\mathbf{F}', \mathbf{F}'')$ is C_u^0, one obtains after an application of Theorem 2.5.3 and Proposition 2.4.1 (a):

$$(5.2.22) \qquad \mathscr{C}^{s,p} = (C^\ell, C^0)_{\theta,p}, \qquad \theta = 1 - s\ell^{-1}.$$

Finally, we mention that, *if $\mathbf{F}''$ is a reflexive space, then the Gagliardo completion of the B-space C_u^k ($k \geq 1$ integer) with respect to $B(\mathbf{F}', \mathbf{F}'')$ is the B-space C^k.* The proof of this assertion is left as an exercice (compute, by using (5.1.8), the closure in $B(\mathbf{F}', \mathbf{F}'')$ of the closed unit ball of C_u^k).

5.3. Regularity Preserving and Regularity Improving Operators

The purpose of this section is to show the following: if $S : \mathbf{F}' \to \mathbf{F}''$ has a certain degree of regularity with respect to the action of a pair of C_0-groups W' and W'', then S is *regularity preserving* in the Besov scales associated to these C_0-groups in the sense that $SF'_{t,q} \subset F''_{t,q}$ for certain values of t and q. We shall also see that S, when considered as an operator $\mathbf{F}'_{t,q} \to \mathbf{F}''_{t,q}$, has a well defined degree of regularity with respect to W' and W''; for example, if $S \in \mathscr{C}^{s,p}(\mathbf{F}', \mathbf{F}'')$, then $S \in \mathscr{C}^{s-t,p}(\mathbf{F}'_{t,q}, \mathbf{F}''_{t,q})$ for each $0 < t < s$. Finally, we shall give conditions under which S is *regularity improving* in the Besov scales, i.e. such that $SF'_{t,q} \subset F''_{t+\tau,r}$ for some $\tau > 0$.

The considerations of this and later sections are clarified by the observation that, besides $\mathscr{W}$, there are two other $C_{\mathbf{w}}$-representations $\mathscr{W}'$ and $\mathscr{W}''$ of $\mathbb{R}^n$ in the Banach space $\mathscr{B} = B(\mathbf{F}', \mathbf{F}'')$ which appear naturally in our context, namely

(5.3.1) $$\mathscr{W}'(x)[S] = SW'(x), \qquad \mathscr{W}''(x)[S] = W''(x)S.$$

Notice that, even if $\mathbf{F}' = \mathbf{F}''$ and $W' = W''$, the representation $\mathscr{W}'$, $\mathscr{W}''$ are distinct. Clearly $\mathscr{W}'$ commutes with $\mathscr{W}''$:

(5.3.2) $$\mathscr{W}'(x)\mathscr{W}''(y) = \mathscr{W}''(y)\mathscr{W}'(x) \qquad \forall x, y \in \mathbb{R}^n,$$

and one can express $\mathscr{W}$ in terms of $\mathscr{W}'$ and $\mathscr{W}''$:

(5.3.3) $$\mathscr{W}(x) = \mathscr{W}'(x)\mathscr{W}''(-x), \qquad x \in \mathbb{R}^n.$$

The relations (5.3.3) and (5.3.2) imply that for $\ell \in \mathbb{N}$:

(5.3.4)
$$(\mathscr{W}''(x) - I)^\ell = [(\mathscr{W}(-x) - I)\mathscr{W}'(x) + (\mathscr{W}'(x) - I)]^\ell$$
$$= \sum_{k=0}^\ell \binom{\ell}{k}(\mathscr{W}(-x) - I)^k[\mathscr{W}'(x)]^k(\mathscr{W}'(x) - I)^{\ell-k}.$$

Since $[\mathscr{W}'(x)]^k = \mathscr{W}'(kx)$, this identity may be written in a more explicit form as follows:

(5.3.5)
$$(W''(x)-I)^\ell S - S(W'(x) - I)^\ell =$$
$$= \sum_{k=1}^\ell \binom{\ell}{k}(\mathscr{W}(-x) - I)^k[S]W'(kx)(W'(x) - I)^{\ell-k}$$

for all $S \in B(\mathbf{F}', \mathbf{F}'')$.

As in §5.1.1 one may consider $\mathscr{W}'$ and $\mathscr{W}''$ as n-parameter groups of linear operators in the vector space $\mathscr{B}_{-\infty}$, and their generators $\mathscr{A}' = (\mathscr{A}'_1, \ldots, \mathscr{A}'_n)$ and $\mathscr{A}'' = (\mathscr{A}''_1, \ldots, \mathscr{A}''_n)$ extend to n-component linear operators $\mathscr{A}$ in $\mathscr{B}_{-\infty}$ satisfying

(5.3.6) $$\mathscr{A}'_j[S] = SA'_j, \qquad \mathscr{A}''_j[S] = A''_j S$$

and

(5.3.7) $$\mathscr{A}'_j\mathscr{A}''_k = \mathscr{A}''_k\mathscr{A}'_j, \qquad \mathscr{A}'_j\mathscr{A}'_k = \mathscr{A}'_k\mathscr{A}'_j, \qquad \mathscr{A}''_j\mathscr{A}''_k = \mathscr{A}''_k\mathscr{A}''_j.$$

For large classes of functions φ (we shall specify them more precisely when needed) $\varphi(\mathscr{A}')$ and $\varphi(\mathscr{A}'')$ are well defined linear operators in $\mathscr{B}_{-\infty}$, in fact $\varphi(\mathscr{A}')[S] = S\varphi(A')$, $\varphi(\mathscr{A}'')[S] = \varphi(A'')S$ for $S \in \mathscr{B}_{-\infty}$. For example, if $\alpha \in \mathbb{N}^n$ and $\varphi(x) = x^\alpha$, we get the operators $\mathscr{A}'^\alpha$ and $\mathscr{A}''^\alpha$, etc .

The equation (5.1.4) may now rewritten as

(5.3.8) $$\mathscr{A}_j = \mathscr{A}'_j - \mathscr{A}''_j, \qquad (j = 1, \ldots, n).$$

By the binomial theorem and (5.3.7) we then have

$$(5.3.9) \qquad \mathscr{A}^\alpha = \sum_{\beta+\gamma=\alpha} \frac{\alpha!}{\beta!\gamma!}(-1)^{|\gamma|}\mathscr{A}'^\beta\mathscr{A}''^\gamma.$$

Similar formulas can be obtained by starting from $\mathscr{A}' = \mathscr{A} + \mathscr{A}''$ or $\mathscr{A}'' = \mathscr{A}' - \mathscr{A}$. We write these identities explicitly by using the more usual notation $\mathrm{ad}_A = -\mathscr{A}$: for $S \in \mathscr{B}_{-\infty} \equiv B(\mathbf{F}'_\infty, \mathbf{F}''_{-\infty})$ and each $\alpha \in \mathbb{N}^n$ one has

$$(5.3.10) \qquad \mathrm{ad}_A^\alpha(S) = \sum_{\beta+\gamma=\alpha} \frac{\alpha!}{\beta!\gamma!}(-1)^{|\gamma|}A''^\beta SA'^\gamma,$$

$$(5.3.11) \qquad SA'^\alpha = \sum_{\beta+\gamma=\alpha} \frac{\alpha!}{\beta!\gamma!}(-1)^{|\gamma|}A''^\beta \,\mathrm{ad}_A^\gamma(S),$$

$$(5.3.12) \qquad A''^\alpha S = \sum_{\beta+\gamma=\alpha} \frac{\alpha!}{\beta!\gamma!} \mathrm{ad}_A^\beta(S)A'^\gamma.$$

The last identity readily implies the simplest form of the assertion "S regular $\implies S$ preserves regularity". In fact, by using (5.3.12) and Proposition 3.3.18, we obtain: *if* $\mathrm{ad}_A^\alpha(S) \in B(\mathbf{F}'_{|\alpha|}, \mathbf{F}'')$ *for all* $|\alpha| \leq k$, *then* $S\mathbf{F}'_m \subset \mathbf{F}''_m$ *for all integers* $0 \leq m \leq k$. In the next proposition we prove a stronger version of this result. It involves the classes $C^k(\mathbf{F}'_m, \mathbf{F}''_m)$ for $k \in \mathbb{N}$ and $m \in \mathbb{Z}$ which are well defined because W' and W'' induce C_0-groups in each of the spaces $\mathbf{F}'_m$ and $\mathbf{F}''_m$ respectively ($m \in \mathbb{Z}$).

PROPOSITION 5.3.1. *Let* $k \in \mathbb{N}$ *and* $m \in \mathbb{Z}$ *with* $|m| \leq k$. *Then*

$$(5.3.13) \qquad C^k(A', A''; \mathbf{F}', \mathbf{F}'') \subset C^{k-|m|}(A', A''; \mathbf{F}'_m, \mathbf{F}''_m).$$

Furthermore

$$S \in C^k(A', A''; \mathbf{F}', \mathbf{F}'') \implies S \in B(\mathbf{F}'_{t,q}, \mathbf{F}''_{t,q})$$

for each real t *with* $|t| < k$ *and each* $q \in [1, \infty]$.

To be more specific, the relation (5.3.13) states the following fact: if $S \in C^k(A', A''; \mathbf{F}', \mathbf{F}'')$, then $S : \mathbf{F}' \to \mathbf{F}''$ is continuous when $\mathbf{F}'$ and $\mathbf{F}''$ are equipped with the topology induced by $\mathbf{F}'_{-k}$ and $\mathbf{F}''_{-k}$ respectively. The extension by continuity of S to an operator from $\mathbf{F}'_{-k}$ to $\mathbf{F}''_{-k}$ maps $\mathbf{F}'_m$ boundedly into $\mathbf{F}''_m$ and is of class $C^{k-|m|}(A', A''; \mathbf{F}'_m, \mathbf{F}''_m)$.

PROOF. (i) We first show that, if $m \in \mathbb{Z}$ and $S \in C^{|m|}(\mathbf{F}', \mathbf{F}'')$, then $S \in B(\mathbf{F}'_m, \mathbf{F}''_m)$. In fact, the assumption $S \in C^{|m|}(\mathbf{F}', \mathbf{F}'')$ implies that $\mathrm{ad}_A^\beta(S) \in B(\mathbf{F}', \mathbf{F}'')$ if $|\beta| \leq |m|$ (see Proposition 5.1.2 (b)). So, if $m > 0$, one obtains from (5.3.12) and (3.3.6) that for some finite constants c and c':

$$\|Sf\|_{\mathbf{F}''_m} \leq \sum_{|\alpha|\leq m} \|A''^\alpha Sf\|_{\mathbf{F}''} \leq c \sum_{|\gamma|\leq m} \|A'^\gamma f\|_{\mathbf{F}'} \leq c'\|f\|_{\mathbf{F}'_m}.$$

If $m < 0$, let $f \in \mathbf{F}'$ be of the form $f = \sum_{|\alpha| \leq |m|} A'^\alpha f_\alpha$ $(f_\alpha \in \mathbf{F}'_{|\alpha|})$; then by (5.3.11) :

$$Sf = \sum_{|\beta| \leq |m|} A''^\beta \sum_{\substack{|\alpha| \leq |m| \\ \alpha \geq \beta}} \frac{\alpha!}{\beta!(\alpha - \beta)!} (-1)^{|\alpha - \beta|} \operatorname{ad}_A^{\alpha - \beta}(S) f_\alpha.$$

Since $\operatorname{ad}_A^{\alpha - \beta}(S) \in B(\mathbf{F}', \mathbf{F}'')$, this implies (see (3.3.20)) that

$$\|Sf\|_{\mathbf{F}''_{-|m|}} \leq C \Big[\sum_{|\alpha| \leq |m|} \|f_\alpha\|_{\mathbf{F}'}^2 \Big]^{1/2}$$

for some constant C. Hence $\|Sf\|_{\mathbf{F}''_{-|m|}} \leq C\|f\|_{\mathbf{F}'_{-|m|}}$.

(ii) Now assume that $|m| \leq k$ and $T \in C^k(\mathbf{F}', \mathbf{F}'')$. Then, if α is a multi-index with $|\alpha| \leq k - |m|$, we have $\operatorname{ad}_A^\alpha(T) \in C^{|m|}(\mathbf{F}', \mathbf{F}'')$ by Proposition 5.1.2 (c). Hence $\operatorname{ad}_A^\alpha(T) \in B(\mathbf{F}'_m, \mathbf{F}''_m)$ by the result of (i). So $T \in C^{k-|m|}(\mathbf{F}'_m, \mathbf{F}''_m)$ by Proposition 5.1.2 (b) (applied for the pair of Banach spaces $(\mathbf{F}'_m, \mathbf{F}''_m)$).

(iii) The last assertion of the proposition can be obtained by interpolation (Theorem 2.6.1). $\square$

We shall now prove a relation similar to (5.3.13) in the context of regularity of fractional order. For a deeper understanding of these results we refer to the comments contained in part (iii) of the proof of Theorem 5.3.3. We need the following elementary fact.

LEMMA 5.3.2. *Let* $s > 0$, $p \in [1, \infty]$ *and let* $\ell > s$ *be an integer. Then there is a finite constant* c *such that for each* $S \in B(\mathbf{F}', \mathbf{F}'')$:

$$(5.3.14) \qquad \|S\|_{\mathbf{F}'_{s,p} \to \mathbf{F}''_{s,p}} \leq c\|S\|_{s,p}^{(\ell)}.$$

PROOF. We use the norm given by (3.4.15) on $\mathbf{F}'_{s,p}$ and on $\mathbf{F}''_{s,p}$ (e.g. $\|\cdot\|_{\mathbf{F}'_{s,p}} = \|\cdot\|_{s,p}^{(\ell)}$). So

$$\|Sf\|_{\mathbf{F}''_{s,p}} = \|Sf\|_{\mathbf{F}''} + \Big[\int_{|x| \leq 1} \Big\| \frac{[W''(x) - I]^\ell}{|x|^s} Sf \Big\|_{\mathbf{F}''}^p \frac{dx}{|x|^n} \Big]^{1/p}$$

$$\leq \|S\|_{\mathbf{F}' \to \mathbf{F}''} \|f\|_{\mathbf{F}'_{s,p}} + \Big[\int_{|x| \leq 1} \Big\| \frac{[W''(x) - I]^\ell S - S[W'(x) - I]^\ell}{|x|^s} f \Big\|_{\mathbf{F}''}^p \frac{dx}{|x|^n} \Big]^{1/p}.$$

By (5.3.5) the last term here is majorized by

$$c \sum_{k=1}^\ell \binom{\ell}{k} \int_{|x| \leq 1} \Big[\big[|x|^{-s} \big\| [\mathscr{W}(x) - I]^k S \big\|_{\mathbf{F}' \to \mathbf{F}''} \big\| [W'(x) - I]^{\ell - k} f \big\|_{\mathbf{F}'} \big]^p \frac{dx}{|x|^n} \Big]^{1/p}.$$

By Lemma 5.2.1 we may assume without loss of generality that $\ell > 2s$. Then for each $k = 1, \dots, \ell$, one has either $k > s$ or $\ell - k > s$. So the preceding expression is majorized by

$$c_1 \sum_{k>s} \binom{\ell}{k} ||S||_{s,p}^{(k)} ||f||_{\mathbf{F}'} + c_1 \sum_{k<\ell-s} \binom{\ell}{k} ||S||_{\mathbf{F}'\to\mathbf{F}''} ||f||_{s,p}^{(\ell-k)} \leq$$

$$\leq c_2 ||S||_{s,p}^{(\ell)} ||f||_{\mathbf{F}'} + c_2 ||S||_{\mathbf{F}'\to\mathbf{F}''} ||f||_{\mathbf{F}'_{s,p}}. \quad \square$$

Before stating the next theorem, we recall that the group W' induces a C_0-group in each space $\mathbf{F}'_{s,p}$ $(p < \infty)$ and in $\mathbf{F}'^{\circ}_{s,\infty}$ (see Proposition 3.4.4); the group induced in $\mathbf{F}'_{s,\infty}$ is not of class C_0 in general. The group W'' has analogous properties in the scale $\{\mathbf{F}''_{s,p}\}$. So, if S is an operator mapping $\mathbf{F}'_{s,p}$ into $\mathbf{F}''_{t,q}$ with $p, q < \infty$, the statement $S \in \mathscr{C}^{u,r}(A', A''; \mathbf{F}'_{s,p}, \mathbf{F}''_{t,q})$ has a well defined meaning. If p or q is infinite, the meaning of this statement is a natural extension of the preceding one.

THEOREM 5.3.3. *Let* $s > 0$, $1 \leq p \leq \infty$ *and* $S \in \mathscr{C}^{s,p}(A', A''; \mathbf{F}', \mathbf{F}'')$. *If* $t \in (0, s)$, *then* $S\mathbf{F}'_{t,q} \subset \mathbf{F}''_{t,q}$ *for each* $q \in [1, \infty]$ *and* $S\mathbf{F}'^{\circ}_{t,\infty} \subset \mathbf{F}''^{\circ}_{t,\infty}$. *Furthermore one has* $S \in \mathscr{C}^{s-t,p}(A', A''; \mathbf{F}'_{t,q}, \mathbf{F}''_{t,q})$. *In other terms, if* $0 < t < s$ *and* $p, q \in [1, \infty]$:

$$(5.3.15) \qquad \mathscr{C}^{s,p}(A', A''; \mathbf{F}', \mathbf{F}'') \subset \mathscr{C}^{s-t,p}(A', A''; \mathbf{F}'_{t,q}, \mathbf{F}''_{t,q}).$$

If $q = \infty$, *then one may replace* $\mathbf{F}'_{t,\infty}, \mathbf{F}''_{t,\infty}$ *by* $\mathbf{F}'^{\circ}_{t,\infty}, \mathbf{F}''^{\circ}_{t,\infty}$.

PROOF. (i) By (5.2.7) we have $S \in \mathscr{C}^{t,q}(\mathbf{F}', \mathbf{F}'')$ if $0 < t < s$ and $q \in [1, \infty]$. So S maps $\mathbf{F}'_{t,q}$ boundedly into $\mathbf{F}''_{t,q}$ by (5.3.14). If $f \in \mathbf{F}'_{\infty}$, then $Sf \in \mathbf{F}''_{t,1} \subset \mathbf{F}''^{\circ}_{t,\infty}$, hence S maps $\mathbf{F}'^{\circ}_{t,\infty}$ into $\mathbf{F}''^{\circ}_{t,\infty}$.

(ii) We find it instructive to prove first a statement weaker than (5.3.15) by elementary methods; the more precise assertion (5.3.15) will be proved in step (iii) by quite different arguments. Let $q \in [p, \infty)$, $0 < t < s$ and $S \in \mathscr{C}^{s,p}(\mathbf{F}', \mathbf{F}'')$. We are going to show that $S \in \mathscr{C}^{s-t,q}(\mathbf{F}'_{t,q}, \mathbf{F}''_{t,q})$. It is enough to prove that for some integer $\ell > s$:

$$\mathscr{J} \equiv \left[\int_{|x|\leq 1} \left\| \frac{[\mathscr{W}(x) - I]^{\ell} S}{|x|^{s-t}} \right\|^q_{\mathbf{F}'_{t,q}\to\mathbf{F}''_{t,q}} \frac{dx}{|x|^n} \right]^{1/q} < \infty.$$

By (5.3.14) we have $\mathscr{J} \leq c(\mathscr{J}_1 + \mathscr{J}_2)$ with

$$\mathscr{J}_1 = \left[\int_{|x|\leq 1} \left\| \frac{[\mathscr{W}(x) - I]^{\ell} S}{|x|^{s-t}} \right\|^q_{\mathbf{F}'\to\mathbf{F}''} \frac{dx}{|x|^n} \right]^{1/q}$$

$$\mathscr{J}_2 \equiv \left[\int_{|x|\leq 1} \frac{dx}{|x|^n} \left\{ \int_{|y|\leq 1} \frac{dy}{|y|^n} \left\| \frac{[\mathscr{W}(x) - I]^{\ell}}{|x|^{s-t}} \frac{[\mathscr{W}(y) - I]^{\ell} S}{|y|^t} \right\|^q_{\mathbf{F}'\to\mathbf{F}''} \right\} \right]^{1/q}.$$

Clearly $\mathscr{J}_1 < \infty$ because $S \in \mathscr{C}^{s-t,q}(\mathbf{F}', \mathbf{F}'')$. To estimate $\mathscr{J}_2$ we define $\mathscr{J}_{2+}$ and $\mathscr{J}_{2-}$ as above but with the integral over dy restricted to the domain $\{|x| \leq |y|\}$ and $\{|y| \leq |x|\}$ respectively. Then $\mathscr{J}_2 \leq \mathscr{J}_{2+} + \mathscr{J}_{2-}$, and it suffices to show that $\mathscr{J}_{2+} < \infty$ and $\mathscr{J}_{2-} < \infty$.

For $\mathscr{J}_{2+}$ we use the fact that $\|\lfloor \mathscr{W}(y) - I \rfloor^\ell\|_{B(B(\mathbf{F}',\mathbf{F}''))} \leq c < \infty$ for all $|y| \leq 1$, so that

$$(\mathscr{J}_{2+})^q \leq c^q \int_{|x| \leq 1} \frac{dx}{|x|^n} \left\| \frac{\lfloor \mathscr{W}(x) - I \rfloor^\ell S}{|x|^s} \right\|^q_{\mathbf{F}' \to \mathbf{F}''} \cdot \int_{|y| \geq |x|} \frac{dy}{|y|^n} \left(\frac{|x|}{|y|} \right)^{tq}.$$

The integral over dy is finite and independent of x, and the integral over dx is finite because $S \in \mathscr{C}^{s,p}(\mathbf{F}',\mathbf{F}'') \subset \mathscr{C}^{s,q}(\mathbf{F}',\mathbf{F}'')$. The finiteness of $\mathscr{J}_{2-}$ is obtained by a similar estimate (interchange the roles of x and y and of t and $s - t$).

(iii) Proposition 5.3.1 and Theorem 5.3.3 are formulated in a way which is convenient for applications but which hides the true nature of their assertions (for example, it is difficult to decide whether the conclusion of Theorem 5.3.3 is optimal or not). We shall present here a more conceptual approach to these matters. We prefer now to use notations which are more suggestive, and which are consistent with those of Chapter 3. Let $\mathscr{B}$ be the B-space $B(\mathbf{F}',\mathbf{F}'')$, equipped with the C_{w}-representation $\mathscr{W}$. Then $\{\mathscr{B}_k\}_{k \in \mathbb{N}}$ is (half of) the discrete (strong) Sobolev scale associated to $\mathscr{B}$, namely $\mathscr{B}_k = C^k(\mathbf{F}',\mathbf{F}'')$, and $\{\mathscr{B}_{s,p} \mid 0 < s < \infty, 1 \leq p \leq \infty\}$ is (half of) the Besov scale associated to $\mathscr{B}$, more precisely $\mathscr{B}_{s,p} = \mathscr{C}^{s,p}(\mathbf{F}',\mathbf{F}'')$. Each of the spaces $\mathscr{B}_k$, $\mathscr{B}_{s,p}$ is equipped with the C_{w}-representation induced by $\mathscr{W}$, so we may consider the Sobolev and Besov scales associated to them. Hence the spaces $(\mathscr{B}_k)_m$ and $(\mathscr{B}_{s,p})_{t,q}$ have a well defined meaning (k, $m \in \mathbb{N}$; s, $t \in (0,\infty)$; p, $q \in [1,\infty]$). One can easily convince oneself that $(\mathscr{B}_k)_m = \mathscr{B}_{k+m}$ (in Section 3.3 we have seen that this is trivial if k, m have the same sign, which is the case here). Less trivial is the equality $(\mathscr{B}_{s,p})_{t,q} = \mathscr{B}_{s+t,q}$. To prove it, let us write the reduction theorem in the form $(\mathscr{B}_{s,p})_\ell = \mathscr{B}_{s+\ell,p}$ for $\ell \in \mathbb{N}$ (see Corollary 3.4.8). Then choosing $\ell > t$, we get:

$$(\mathscr{B}_{s,p})_{t,q} = ((\mathscr{B}_{s,p})_\ell, \mathscr{B}_{s,p})_{1-t/\ell,q}$$
$$= (\mathscr{B}_{s+\ell,p}, \mathscr{B}_{s,p})_{1-t/\ell,q} = \mathscr{B}_{(s+\ell)t/\ell+s(1-t/\ell),q} = \mathscr{B}_{s+t,q},$$

so the assertion is proved.

Now, in order to deduce Proposition 5.3.1 and the optimal form of Theorem 5.3.3 from the above two identities, let us introduce the following spaces (for $k \in \mathbb{N}$, $0 < s < \infty$ and $1 \leq p \leq \infty$):

$$\mathscr{B}^k = \{S \in B(\mathbf{F}',\mathbf{F}'') \mid S\mathbf{F}'_j \subset \mathbf{F}''_j \text{ if } 0 \leq j \leq k\},$$
$$\mathscr{B}^{s,p} = \{S \in B(\mathbf{F}',\mathbf{F}'') \mid S\mathbf{F}'_{s,p} \subset \mathbf{F}''_{s,p}\}.$$

These are naturally B-spaces continuously embedded in $\mathscr{B}$, and they are invariant under $\mathscr{W}$ which induces in each of them a C_{w}-representation of $\mathbb{R}^n$ (in the definition of $\mathscr{B}^k$ we have taken into account the fact that $\mathbf{F}_j$, for $1 \leq j \leq k - 1$, is not in general an interpolation space between $\mathbf{F}$ and $\mathbf{F}_k$). $\mathscr{B}^k$ and $\mathscr{B}^{s,p}$ coincide with the set of all bounded operators $S : \mathbf{F}' \to \mathbf{F}''$ that preserve regularity up to order k and (s,p) respectively. Hence the statement "a regular operator preserves regularity" may be written as $\mathscr{B}_k \subset \mathscr{B}^k$ and $\mathscr{B}_{s,p} \subset \mathscr{B}^{s,p}$. Now the validity of the first one of these inclusions is obvious from (5.3.12), while that of the second case is equivalent to Lemma 5.3.2. Since $\mathscr{B}^k$ and $\mathscr{B}^{s,p}$ are equipped with the groups induced by $\mathscr{W}$, we can consider the associated Sobolev and Besov scales, e.g. $(\mathscr{B}^k)_m$ and $(\mathscr{B}^{s,p})_{t,q}$ make sense for $m \in \mathbb{N}$,

$0 < t < \infty$, $1 \leq q \leq \infty$. From the relations $\mathscr{B}_k \subset \mathscr{B}^k$ and $\mathscr{B}_{s,p} \subset \mathscr{B}^{s,p}$ one obviously gets that $(\mathscr{B}_k)_m \subset (\mathscr{B}^k)_m$ and $(\mathscr{B}_{s,p})_{t,q} \subset (\mathscr{B}^{s,p})_{t,q}$. By taking into account the two identities proved further up, we shall have $\mathscr{B}_{k+m} \subset (\mathscr{B}^k)_m$ and $\mathscr{B}_{s+t,q} \subset (\mathscr{B}^{s,p})_{t,q}$. Let us change the notations, in order to make the connection with the results proved before in this section: we have $\mathscr{B}_k \subset (\mathscr{B}^m)_{k-m}$ if $0 \leq m \leq k$ are integers, and $\mathscr{B}_{s,p} \subset (\mathscr{B}^{t,q})_{s-t,p}$ if $0 < t < s$ are real and $p, q \in [1, \infty]$. The first of these embeddings is equivalent to Proposition 5.3.1. The second one completes the proof of Theorem 5.3.3, since it is equivalent to (5.3.15) (the last assertion of the theorem is obvious). $\square$

COROLLARY 5.3.4. *Assume that $\mathbf{F}', \mathbf{F}''$ are reflexive spaces. Then one has for each real t with $0 < |t| < s$ and all $p, q \in [1, \infty]$:*

$$(5.3.16) \qquad \mathscr{C}^{s,p}(\mathbf{F}', \mathbf{F}'') \subset \mathscr{C}^{s-|t|,p}(\mathbf{F}'_{t,q}, \mathbf{F}''_{t,q}).$$

PROOF. Let $S \in \mathscr{C}^{s,p}(\mathbf{F}', \mathbf{F}'')$ and $t \in (-s, 0)$. Then by (5.2.17) and Theorem 5.3.3 we have

$$S^* \in \mathscr{C}^{s,p}(\mathbf{F}''^*, \mathbf{F}'^*) \subset \mathscr{C}^{s-|t|,p}((\mathbf{F}''^*)_{-t,q'}, (\mathbf{F}'^*)_{-t,q'})$$

for each number $q' \in [1, \infty]$. Furthermore, if $q' = \infty$, we also have $S^* \in \mathscr{C}^{s-|t|,p}((\mathbf{F}''^*)^\circ_{-t,\infty}, (\mathbf{F}'^*)^\circ_{-t,\infty})$. Define q by the relation $1/q + 1/q' = 1$. By using again (5.2.17) and Theorem 3.4.3 (d), one now obtains $S \in \mathscr{C}^{s-|t|,p}(\mathbf{F}'_{t,q}, \mathbf{F}''_{t,q})$. $\square$

As a final result of this section we prove a criterion for an operator S to be regularity improving in the Besov scales:

PROPOSITION 5.3.5. *Let $s = t + \mu$ with $t, \mu > 0$ and let $1 \leq p \leq q \leq \infty$. Define r by $1/r = 1/p - 1/q$. Assume that $S \in B(\mathbf{F}', \mathbf{F}'')$ satisfies*

$$(5.3.17) \qquad \left[\int_{|x| \leq 1} \left\| \frac{[W''(x) - I]^k}{|x|^\mu} S \right\|^r_{\mathbf{F}' \to \mathbf{F}''} \frac{dx}{|x|^n} \right]^{1/r} < \infty$$

and

$$(5.3.18) \qquad \left[\int_{|x| \leq 1} \left\| \frac{[W''(x) - I]^k}{|x|^s} [\mathscr{W}(x) - I]^\ell S \right\|^p_{\mathbf{F}' \to \mathbf{F}''} \frac{dx}{|x|^n} \right]^{1/p} < \infty$$

for some integers k, ℓ with $k > \mu$ and $\ell > t$. Then $S \in B(\mathbf{F}'_{t,q}, \mathbf{F}''_{s,p})$.

PROOF. In view of Theorem 3.4.6 it suffices to show that for some constant c and all $f \in \mathbf{F}'_{t,q}$:

$$(5.3.19) \qquad \left[\int_{|x| \leq 1} \left\| \frac{[W''(x) - I]^{k+2\ell} Sf}{|x|^s} \right\|^p_{\mathbf{F}''} \frac{dx}{|x|^n} \right]^{1/p} \leq$$
$$\leq c\|f\|_{\mathbf{F}'} + c \left[\int_{|x| \leq 1} \left\| \frac{[W'(x) - I]^\ell}{|x|^t} f \right\|^q_{\mathbf{F}'} \frac{dx}{|x|^n} \right]^{1/q}.$$

For this we use (5.3.5) to write

$$\frac{[W''(x)-I]^{k+2\ell}}{|x|^s}Sf = \frac{[W''(x)-I]^k}{|x|^\mu}S\frac{[W'(x)-I]^{2\ell}}{|x|^t}f +$$

$$+ \sum_{m=1}^{2\ell}\binom{2\ell}{m}\frac{[W''(x)-I]^k}{|x|^\mu}\frac{[\mathscr{W}(-x)-I]^m S}{|x|^{a_m}}\cdot W'(mx)\frac{[W'(x)-I]^{2\ell-m}}{|x|^{b_m}}f,$$

with $a_m + b_m = t$. We choose $a_m = 0$, $b_m = t$ if $m < \ell$ and $a_m = t$, $b_m = 0$ if $m \geq \ell$. By the local boundedness of W', W'' and $\mathscr{W}$ one obtains the following inequality which holds for all $|x| \leq 1$ and some constant c_1:

(5.3.20)

$$\left\|\frac{[W''(x)-I]^{k+2\ell}}{|x|^s}Sf\right\|_{\mathbf{F}''} \leq c_1\left\|\frac{[W''(x)-I]^k}{|x|^\mu}S\right\|_{\mathbf{F}'\to\mathbf{F}''}\left\|\frac{[W'(x)-I]^\ell}{|x|^t}f\right\|_{\mathbf{F}'}$$

$$+ c_1\left\|\frac{[W''(x)-I]^k}{|x|^s}[\mathscr{W}(-x)-I]^\ell S\right\|_{\mathbf{F}'\to\mathbf{F}''}\|f\|_{\mathbf{F}'}.$$

To obtain (5.3.19), one inserts the preceding inequality into the l.h.s. of (5.3.19); the contribution from the first term on the r.h.s. of (5.3.20) is estimated by using the Hölder inequality and the hypothesis (5.3.17), whereas the contribution from the second term on the r.h.s. of (5.3.20)) is treated by using (5.3.18). □

5.4. The spaces $\mathscr{M}^r_{s,p}(\mathbb{R}^n)$

In this section we define and study some new spaces of functions on $\mathbb{R}^n$ that will be useful in Section 5.5. We use some notations and conventions concerning measures that were introduced in Sections 1.2 and 3.1. For $r \geq 0$ we have defined $\mathscr{M}^r(\mathbb{R}^n)$ to be the Banach algebra of all bounded continuous functions $\varphi : \mathbb{R}^n \to \mathbb{C}$ such that $\widehat{\varphi}$ is a measure on $\mathbb{R}^n$ satisfying

$$(5.4.1) \qquad \|\varphi\|_{\mathscr{M}^r} \equiv 2^{r/2}\int_{\mathbb{R}^n}\langle x\rangle^r|\widehat{\varphi}(x)|\,d\!\!\!\!\!\!\,x < \infty.$$

We set

$$(5.4.2) \qquad \mathscr{M}^\infty(\mathbb{R}^n) = \bigcap_{r\geq 0}\mathscr{M}^r(\mathbb{R}^n).$$

Clearly $\mathscr{M}^\infty(\mathbb{R}^n) \subset BC^\infty(\mathbb{R}^n)$, and $\varphi \in \mathscr{M}^\infty(\mathbb{R}^n)$ means that $\widehat{\varphi}$ is a rapidly decreasing measure.

It is useful to know that $\mathscr{M}^r(\mathbb{R}^n)$ is the adjoint of a Banach space $(\mathscr{M}^r)_*$. To see this, let $\overset{\circ}{C}_{(-r)}(\mathbb{R}^n)$ be the space of continuous functions $\varphi : \mathbb{R}^n \to \mathbb{C}$ such that $\langle x\rangle^{-r}\varphi(x) \to 0$ as $|x| \to \infty$ (see the beginning of §3.6.2). We define $(\mathscr{M}^r)_* = \mathcal{F}^*[\overset{\circ}{C}_{(-r)}(\mathbb{R}^n)]$ equipped with the norm

$$(5.4.3) \qquad \|\psi\|_{(\mathscr{M}^r)_*} = \sup_{x\in\mathbb{R}^n}2^{-r/2}\langle x\rangle^{-r}|\widehat{\psi}(x)|.$$

$\mathscr{S}(\mathbb{R}^n)$ is a dense subspace of $(\mathscr{M}^r)_*$, so $(\mathscr{M}^r)_*$ is a separable Banach space and its adjoint $(\mathscr{M}^r)^*_*$ is identified with a subspace of $\mathscr{S}^*(\mathbb{R}^n)$. For $g \in \mathscr{S}$ and

$f \in \mathscr{S}^*$ we have $\langle g, f \rangle = \langle \hat{g}, \hat{f} \rangle$, so by taking into account Proposition 3.1.4 (b) it is easily shown that $(\mathcal{M}^r)_*^* = \mathcal{M}^r$ (with identical norms), i.e. $\mathcal{M}^r$ is just the adjoint Banach space of $(\mathcal{M}^r)_*$.

The notion of feebly convergent sequences in $\mathcal{M}^r(\mathbb{R}^n)$ (see Definition 3.1.6) is stronger than that of weakly convergent sequences: a feebly convergent sequence in $\mathcal{M}^r(\mathbb{R}^n)$ is also weakly convergent, but the converse need not be true (for example we have $\lim_{x \to \infty} e_x = 0$ weakly in $\mathcal{M}^0(\mathbb{R}^n)$, but $\{e_{x_j}\}_{j \in \mathbb{N}}$ does not converge feebly in $\mathcal{M}^0(\mathbb{R}^n)$ if $x_j \to \infty$ as $j \to \infty$).

For each $y \in \mathbb{R}^n$, the operator $e^{iQ \cdot y}$, defined by $[e^{iQ \cdot y}\varphi](x) = e^{i(x,y)}\varphi(x) \equiv e_y(x)\varphi(x)$, is bounded as an operator in $\mathcal{M}^r(\mathbb{R}^n)$ (we recall that e_y, defined as $e_y(x) = \exp[i(x,y)]$, is an element of $\mathcal{M}^r(\mathbb{R}^n)$ and that $\mathcal{M}^r(\mathbb{R}^n)$ is an algebra). Thus the collection $\{e^{iQ \cdot y}\}_{y \in \mathbb{R}^n}$ defines a representation of $\mathbb{R}^n$ in $\mathcal{M}^r(\mathbb{R}^n)$.

Let us now consider the operators $e^{iQ \cdot y}$ in $(\mathcal{M}^r)_*$. Since $e^{iQ \cdot y}\mathcal{F}^* = \mathcal{F}^* e^{-iP \cdot y}$ and because the translation group $\{e^{-iP \cdot y}\}_{y \in \mathbb{R}^n}$ is obviously a C_0-group in $\overset{\circ}{C}_{(-r)}(\mathbb{R}^n)$, the collection $\{e^{iQ \cdot y}\}_{y \in \mathbb{R}^n}$ is a C_0-group in $(\mathcal{M}^r)_*$. It follows that the representation $\{e^{iQ \cdot y}\}_{y \in \mathbb{R}^n}$ of $\mathbb{R}^n$ in $\mathcal{M}^r(\mathbb{R}^n)$ is a C_w-representation (see the end of §3.2.1).

One may apply the theory developed in Sections 5.1 and 5.2 to the following objects: $\mathbf{F}' = (\mathcal{M}^r)_*$, $W'(x) = e^{-iQ \cdot x}$, $\mathbf{F}'' = \mathbb{C}$, $W''(x) = 1$. Then $B(\mathbf{F}', \mathbf{F}'') = \mathcal{M}^r(\mathbb{R}^n)$ (here we make the harmless abuse of considering anti-linear maps as elements of $B(\mathbf{F}', \mathbf{F}'')$ — recall that we have made the convention that the adjoint space consists of antilinear forms) and $\mathcal{W}(x) = e^{iQ \cdot x}$. In the remainder of this section we shall describe the Sobolev and Besov scales associated to the couple $(\mathcal{M}^r, \{e^{iQ \cdot x}\})$ according to the general theory of Sections 5.1, 5.2 and Chapter 3.

We first introduce the following closed subalgebra $\mathscr{L}^r(\mathbb{R}^n)$ of $\mathcal{M}^r(\mathbb{R}^n)$ in which the family $\{e^{iQ \cdot y}\}_{y \in \mathbb{R}^n}$ induces a C_0 representation of $\mathbb{R}^n$:

$$(5.4.4) \qquad \mathscr{L}^r(\mathbb{R}^n) = \{\varphi \in \mathcal{M}^r(\mathbb{R}^n) \mid \lim_{y \to 0} \|[e^{iQ \cdot y} - I]\varphi\|_{\mathcal{M}^r} = 0\}.$$

PROPOSITION 5.4.1. (a) Let $\varphi \in \mathcal{M}^r(\mathbb{R}^n)$ be such that

$$\lim_{y \to 0} \|[e^{iQ \cdot y} - I]^\ell \varphi\|_{\mathcal{M}^r} = 0$$

for some integer $\ell \geq 1$. Then $\varphi \in \mathscr{L}^r(\mathbb{R}^n)$.

(b) An element φ of $\mathcal{M}^r(\mathbb{R}^n)$ belongs to $\mathscr{L}^r(\mathbb{R}^n)$ if and only if the measure $\hat{\varphi}$ is absolutely continuous (with respect to Lebesgue measure). In particular, $\mathscr{L}^0(\mathbb{R}^n) = \mathcal{F}L^1(\mathbb{R}^n)$, and for integer $r \geq 1$: $\varphi \in \mathscr{L}^r$ if and only if φ is of class C^r and $\varphi^{(\alpha)} \in \mathcal{F}L^1(\mathbb{R}^n)$ for all $|\alpha| \leq r$.

PROOF. (a) This has been shown in a more general setting in Lemma 5.2.1 (b) (for an elementary proof, see the remark following the proof of this lemma).

(b) It is easy to show that the absolute continuity of $\hat{\varphi}$ implies that $\varphi \in \mathscr{L}^r$. Reciprocally, let W be the C_0-group in the Banach space $\mathbf{F} = \mathscr{L}^r(\mathbb{R}^n)$ defined by $W(x) = e^{iQ \cdot x}$ $(x \in \mathbb{R}^n)$. If $\varphi \in \mathbf{F}_1 \equiv (\mathscr{L}^r)_1$, then $Q_j\varphi \in \mathbf{F} = \mathscr{L}^r \subset \mathcal{M}^r$, i.e. $\partial_j\hat{\varphi}$ are measures for $j = 1, \ldots, n$. Thus, by Proposition 1.3.9, $\hat{\varphi}$ is absolutely continuous. So $(\mathscr{L}^r)_1$ consists of functions the Fourier transforms of which are absolutely continuous measures. Since $(\mathscr{L}^r)_1$ is dense in $\mathscr{L}^r$ and the limit (in the norm given by the variation) of a sequence of absolutely continuous measures is absolutely continuous, the same is true for $\mathscr{L}^r$. □

Another characterization of $\mathscr{L}^r(\mathbb{R}^n)$ is as follows (see Definition 3.5.9): $\varphi \in \mathscr{L}^r(\mathbb{R}^n)$ if and only if $\varphi \in \mathscr{M}^r(\mathbb{R}^n)$ and $\lim \|\xi(\varepsilon Q)f\|_{\mathscr{M}^r} = 0$ as $\varepsilon \to 0$ for some (and hence each) function $\xi \in C^\infty(X)$ with $\xi(x) = 0$ in a neighbourhood of $x = 0$ and $\xi(x) = 1$ in a neighbourhood of infinity. We note that the results of Proposition 5.4.1 remain true if r is allowed to assume negative values.

The space that will play the most important role further on is $\mathscr{M}_{s,p}^r$, defined for $r \geq 0$, $s > 0$ and $1 \leq p \leq \infty$ as the space of those $\varphi \in \mathscr{M}^r$ for which there is an integer $\ell > s$ such that:

$$(5.4.5) \qquad \|\varphi\|_{\mathscr{M}^r} + \left[\int_{|y| \leq 1} \||y|^{-s}[e^{iQ \cdot y} - 1]^\ell \varphi\|_{\mathscr{M}^r}^p \frac{dy}{|y|^n} \right]^{1/p} < \infty.$$

For each integer $\ell > s$, the expression (5.4.5) defines a gauge on $\mathscr{M}^r(\mathbb{R}^n)$, and all these gauges are equivalent (by Lemma 5.2.1 (a)). Notice that $\mathscr{M}_{s,p}^r(\mathbb{R}^n) \subset \mathscr{L}^r(\mathbb{R}^n)$ (by Lemma 5.2.1 (b)); in particular, the norm (5.4.5) or the Littlewood-Paley type version given below shows that $\mathscr{M}_{s,p}^0(\mathbb{R}^n) = \mathscr{F}B_1^{s,p}(\mathbb{R}^n)$, where the space $B_1^{s,p}(\mathbb{R}^n)$ is the standard Besov space introduced in Section 4.1. For definiteness, let $\|\varphi\|_{\mathscr{M}_{s,p}^r}$ be the gauge associated to the *least* integer ℓ strictly larger than s. Observe that the upper index r characterizes the local regularity of the (continuous) functions φ in $\mathscr{M}_{s,p}^r(\mathbb{R}^n)$, while the lower indices s, p characterize their decay at infinity (i.e. $\mathscr{M}_{s,p}^r(\mathbb{R}^n)$ is a sort of weighted Sobolev space or the Fourier transform of a weighted Besov space). Since $\mathscr{M}^r(\mathbb{R}^n)$ is a Banach algebra, it follows from (5.4.5) that $\mathscr{M}_{s,p}^r(\mathbb{R}^n)$ *is a subalgebra of* $\mathscr{M}^r(\mathbb{R}^n)$.

We now give a description of the spaces $\mathscr{M}_{s,p}^r(\mathbb{R}^n)$ based on the Littlewood-Paley theory. This description is sometimes more convenient than (5.4.5) (e.g. the fact that the indices s, p describe the decay at infinity of φ becomes obvious). Let θ and $\widetilde{\theta}$ be functions in $C_0^\infty(\mathbb{R}^n)$ such that $\theta(x) > 0$ if $2^{-1} < |x| < 2$ and $\theta(x) = 0$ otherwise, and $\widetilde{\theta}(x) = 1$ if $|x| < 2$. Then a continuous function $\varphi : \mathbb{R}^n \to \mathbb{C}$ belongs to $\mathscr{M}_{s,p}^r(\mathbb{R}^n)$ if and only if

$$\|\widetilde{\theta}\varphi\|_{\mathscr{M}^r} + \left[\int_0^1 \|\tau^{-s}\theta^\tau \varphi\|_{\mathscr{M}^r}^p \frac{d\tau}{\tau} \right]^{1/p} < \infty$$

where, as usual, θ^τ is the function defined by $\theta^\tau(x) = \theta(\tau x)$. The necessity of this condition is obtained by applying Theorem 3.6.2 to the C_0-group $\{e^{iQ \cdot x}\}$ in $\mathscr{L}^r(\mathbb{R}^n)$ (recall that $\mathscr{M}_{s,p}^r \subset \mathscr{L}^r$). For the sufficiency, it is enough to show that the preceding inequality implies $\varphi \in \mathscr{L}^r$ (then we may use Theorem 3.6.2 again). Let $\xi \in C^\infty(\mathbb{R}^n)$ with $\xi(x) = 0$ near the origin and $\xi(x) = 1$ near infinity. Then $\eta(x) = x \cdot \nabla\xi(x)$ defines a function $\eta \in C_0^\infty(\mathbb{R}^n \setminus \{0\})$, and it is straightforward to prove (by using the hypothesis on φ) that for any such function one has $[\int_0^1 \|\tau^{-s}\eta^\tau \varphi\|_{\mathscr{M}^r}^p \tau^{-1} d\tau]^{1/p} < \infty$. Clearly $\xi(\varepsilon x)\varphi(x) = \int_0^\varepsilon \eta(\tau x)\varphi(x)\tau^{-1} d\tau$ for $x \in \mathbb{R}^n$ (see (3.5.10)). Since the integral $\int_0^\varepsilon \eta^\tau \varphi \tau^{-1} d\tau$ exists in $\mathscr{M}^r$, we see that $\xi^\varepsilon \varphi \in \mathscr{M}^r$ and $\|\xi^\varepsilon \varphi\|_{\mathscr{M}^r} \leq \int_0^\varepsilon \|\eta^\tau \varphi\|_{\mathscr{M}^r} \tau^{-1} d\tau \to 0$ as $\varepsilon \to 0$. To conclude that

$\varphi \in \mathscr{L}^r$, it now suffices to observe that $\varphi \in \mathscr{M}^r$ (write $\varphi = \tilde{\theta}\varphi + (1-\tilde{\theta})\varphi$; we have $\tilde{\theta}\varphi \in \mathscr{M}^r$ by hypothesis and $(1-\tilde{\theta})\varphi \in \mathscr{M}^r$ by setting $\xi^\varepsilon = 1 - \tilde{\theta}$ above).

The following inclusion relations hold:

(5.4.6) $\quad \mathscr{M}^r_{s,p}(\mathbb{R}^n) \subset \mathscr{M}^r_{t,q}(\mathbb{R}^n) \qquad$ if $0 < t < s$ and $p,q \in [1,\infty]$,

(5.4.7) $\quad \mathscr{M}^r_{s,p}(\mathbb{R}^n) \subset \mathscr{M}^r_{s,q}(\mathbb{R}^n) \qquad$ if $1 \le p \le q \le \infty$

(5.4.8) $\quad \mathscr{M}^{r_1}_{s,p}(\mathbb{R}^n) \subset \mathscr{M}^{r_2}_{s,p}(\mathbb{R}^n) \qquad$ if $0 \le r_2 \le r_1 < \infty$,

(5.4.9) $\quad \mathscr{M}^r_{s,p}(\mathbb{R}^n) \subset \mathscr{L}^r(\mathbb{R}^n)$.

One may also define

(5.4.10) $$\mathscr{M}^\infty_{s,p}(\mathbb{R}^n) = \bigcap_{r \ge 0} \mathscr{M}^r_{s,p}(\mathbb{R}^n).$$

Clearly (5.4.6) and (5.4.7) hold for $r = \infty$ too.

A multiplicative property that has no analogue in the abstract formalism is given in the next proposition:

PROPOSITION 5.4.2. *Let* $r \in [1,\infty]$, $s,t > 0$ *and* $1 \le p,q,u \le \infty$ *such that* $1/p + 1/q = 1/u$. *Then*

$$\mathscr{M}^r_{s,p}(\mathbb{R}^n) \cdot \mathscr{M}^r_{t,q}(\mathbb{R}^n) \subset \mathscr{M}^r_{s+t,u}(\mathbb{R}^n),$$

where the dot means multiplication of functions defined on $\mathbb{R}^n$.

PROOF. Choose $k,m \in \mathbb{N}$ with $k > s$ and $m > t$, and set $\ell = k + m$. Since $\mathscr{M}^r(\mathbb{R}^n)$ is an algebra, we have for any $\varphi, \psi \in \mathscr{M}^r(\mathbb{R}^n)$:

$$\left\| |y|^{-s-t}[e^{iQ\cdot y} - I]^\ell \varphi\psi \right\|_{\mathscr{M}^r} \le \left\| |y|^{-s}[e^{iQ\cdot y} - I]^k \varphi \right\|_{\mathscr{M}^r} \left\| |y|^{-t}[e^{iQ\cdot y} - I]^m \psi \right\|_{\mathscr{M}^r}.$$

The assertion of the proposition follows upon applying the Hölder inequality. $\square$

A more detailed description of $\mathscr{M}^r_{s,p}(\mathbb{R}^n)$, in terms of the spaces $\mathscr{M}^r_k(\mathbb{R}^n)$ ($k \in \mathbb{N}$) belonging to the Sobolev scale, can be obtained by using the reduction theorem (Theorem 3.4.7). Since $\{e^{iQ\cdot x}\}_{x\in\mathbb{R}^n}$ is not a C_0-group but only a C_w-group in $\mathscr{M}^r(\mathbb{R}^n)$, there are two natural possibilities for $\mathscr{M}^r_k(\mathbb{R}^n)$, as explained at the beginning of this chapter. Our main emphasis will be on the spaces $C^k(-Q,0;(\mathscr{M}^r)_*,\mathbb{C})$, which we shall introduce below and denote by $\mathscr{M}^r_k(\mathbb{R}^n)$. The description of $C^k_u(-Q,0;(\mathscr{M}^r)_*,\mathbb{C})$ is quite simple by Proposition 5.4.1 and the fact that $\mathscr{L}^r(\mathbb{R}^n)$ is identical with the set of ψ in $\mathscr{M}^r(\mathbb{R}^n)$ such that $\hat{\psi}$ is a function: a function φ in $\mathscr{M}^r(\mathbb{R}^n)$ is such that $x \mapsto e^{iQ\cdot x}\varphi \in \mathscr{M}^r(\mathbb{R}^n)$ is of class C^k in norm if and only if $\partial^\alpha \hat{\varphi}$ is a function for each multi-index α with $|\alpha| \le k$ and $\sum_{|\alpha|\le k} \int_{\mathbb{R}^n} |\partial^\alpha \hat{\varphi}(x)|\langle x\rangle^r dx < \infty$.

In order to avoid the explicit use of the weak topology on $\mathscr{M}^r(\mathbb{R}^n)$ (as adjoint of $(\mathscr{M}^r)_*$), we define $\mathscr{M}^r_k(\mathbb{R}^n)$ by taking into account the criterion of Theorem 5.1.3 (b): if $k \in \mathbb{N}$, then $\mathscr{M}^r_k(\mathbb{R}^n)$ is the set of all φ in $\mathscr{M}^r(\mathbb{R}^n)$ for which the following expression is finite:

(5.4.11) $$\|\varphi\|_{\mathscr{M}^r_k} \equiv \|\varphi\|_{\mathscr{M}^r} + \sup_{|x|\le 1} |x|^{-k} \left\| [e^{iQ\cdot x} - I]^k \varphi \right\|_{\mathscr{M}^r}.$$

Clearly $\mathscr{M}_k^r(\mathbb{R}^n) \cdot \mathscr{M}_m^r(\mathbb{R}^n) \subset \mathscr{M}_{k+m}^r(\mathbb{R}^n)$ for k, $m \in \mathbb{N}$, and the general theory (Section 5.1) gives $\mathscr{M}_k^r(\mathbb{R}^n) \subset \mathscr{M}_m^r(\mathbb{R}^n)$ if $0 \leq m \leq k$. These two inclusions imply in particular that $\mathscr{M}_k^r(\mathbb{R}^n)$ is a subalgebra of $\mathscr{M}^r(\mathbb{R}^n)$. We also have $\mathscr{M}_k^{r_1}(\mathbb{R}^n) \subset \mathscr{M}_k^{r_2}(\mathbb{R}^n)$ if $0 \leq r_2 \leq r_1$; this suggests the definition

$$\mathscr{M}_k^\infty(\mathbb{R}^n) = \bigcap_{r \geq 0} \mathscr{M}_k^r(\mathbb{R}^n).$$

The general theory (see (5.2.10)) also implies the rather deep relation (both inclusions below are strict):

(5.4.12) $\mathscr{M}_{k,1}^r(\mathbb{R}^n) \subset \mathscr{M}_k^r(\mathbb{R}^n) \subset \mathscr{M}_{k,\infty}^r(\mathbb{R}^n)$ $(k \geq 1$ integer$)$.

We can now present a rather explicit description of these spaces:

THEOREM 5.4.3. (a) *Let* $k \in \mathbb{N}$. *Then* φ *belongs to* $\mathscr{M}_k^r(\mathbb{R}^n)$ *if and only if* $Q^\alpha \varphi \in \mathscr{M}^r(\mathbb{R}^n)$ *for each multi-index* α *with* $|\alpha| \leq k$.

(b) *Let* $s = k + \sigma$ *with* $k \in \mathbb{N}$ *and* $0 < \sigma \leq 1$. *Then* φ *belongs to* $\mathscr{M}_{s,p}^r(\mathbb{R}^n)$ *if and only* $Q^\alpha \varphi \in \mathscr{M}^r(\mathbb{R}^n)$ *for each* $|\alpha| < k$ *and* $Q^\alpha \varphi \in \mathscr{M}_{\sigma,p}^r(\mathbb{R}^n)$ *for each* $|\alpha| = k$.

These results are straightforward consequences of Proposition 5.1.2 (b) (see also Proposition 3.3.14) and of the reduction theorem (Proposition 5.2.2 (b)). Their non-triviality will be illustrated by a comment at the end of this section. We now consider the special case where $r = \infty$ and $p = \infty$:

COROLLARY 5.4.4. *Let* $k \in \mathbb{N}$, $0 < \sigma \leq 1$ *and* $s = k + \sigma$. *Let* $\varphi \in \mathscr{S}^*(\mathbb{R}^n)$.

(a) φ *belongs to* $\mathscr{M}_k^\infty(\mathbb{R}^n)$ *if and only if all derivatives of order* $\leq k$ *of* $\widehat{\varphi}$ *are rapidly decreasing measures.*

(b) φ *belongs to* $\mathscr{M}_{s,\infty}^\infty(\mathbb{R}^n)$ *if and only if* $\widehat{\varphi}^{(\alpha)}$ *is a rapidly decreasing function for each multi-index* α *with* $|\alpha| \leq k$ *and the following estimate holds for each* $r \in \mathbb{R}$, *some constant* c_r *and all* $|y| \leq 1$:

(i) *if* $\sigma < 1$:

(5.4.13) $$\sum_{|\alpha|=k} \int_{\mathbb{R}^n} |\widehat{\varphi}^{(\alpha)}(x+y) - \widehat{\varphi}^{(\alpha)}(x)| \langle x \rangle^r dx \leq c_r |y|^\sigma,$$

(ii) *if* $\sigma = 1$:

(5.4.14) $$\sum_{|\alpha|=k} \int_{\mathbb{R}^n} |\widehat{\varphi}^{(\alpha)}(x+y) + \widehat{\varphi}^{(\alpha)}(x-y) - 2\widehat{\varphi}^{(\alpha)}(x)| \langle x \rangle^r dx \leq c_r |y|.$$

PROOF. This is a particular case of Theorem 5.4.3 except for the fact that $\widehat{\varphi}^{(\alpha)}(x)$ is a function if $\varphi \in \mathscr{M}_{s,\infty}^\infty(\mathbb{R}^n)$ and $|\alpha| \leq k$ with $s > k$. This fact follows from Proposition 5.4.1. $\square$

We next give a result for the case where φ is a symbol in $S^{-\mu}(\mathbb{R}^n)$ for some $\mu > 0$. Some comments on the optimality of this statement will be made after the proof.

PROPOSITION 5.4.5. *Let* $\mu > 0$ *be a real number. Then*

(5.4.15) $$S^{-\mu}(\mathbb{R}^n) \subset \mathscr{M}_{\mu,\infty}^\infty(\mathbb{R}^n).$$

PROOF. Assume that $\varphi \in S^{-\mu}(\mathbb{R}^n)$. Then $\widehat{\varphi}$ is a function in $L^1(\mathbb{R}^n)$ of rapid decay at infinity (see Proposition 1.3.6). Thus $\varphi \in \mathcal{M}^r(\mathbb{R}^n)$ for each $r \geq 0$ and $\widehat{\varphi}$ is a function, in other terms $\varphi \in \mathscr{L}^r(\mathbb{R}^n)$. So, by applying Theorem 3.6.2 to the C_0-group $\{e^{iQ \cdot x}\}$ in $\mathscr{L}^r(\mathbb{R}^n)$, one sees that

$$(5.4.16) \quad \varphi \in \mathcal{M}^r_{\mu,\infty}(\mathbb{R}^n) \Longleftrightarrow \|\widetilde{\theta}(Q)\varphi\|_{\mathcal{M}^r} + \sup_{0 < \tau < 1} \|\tau^{-\mu}\theta(\tau Q)\varphi\|_{\mathcal{M}^r} < \infty,$$

where $\theta \in C_0^\infty(\mathbb{R}^n)$ is such that $\theta(x) > 0$ if $2^{-1} < |x| < 2$ and $\theta(x) = 0$ otherwise, and $\widetilde{\theta} \in C_0^\infty(\mathbb{R}^n)$ is such that $\widetilde{\theta}(x) = 1$ for $|x| \leq 2$. Since $\widetilde{\theta}\varphi \in C_0^\infty(\mathbb{R}^n)$, the finiteness of $\|\widetilde{\theta}(Q)\varphi\|_{\mathcal{M}^r}$ is obvious. To estimate $\|\theta(\tau Q)\varphi\|_{\mathcal{M}^r}$, we use Lemma 3.6.7 with $k = -\mu$ to obtain that $|\{\mathcal{F}[\theta(\tau Q)\varphi]\}(y)| \leq C\tau^{\mu-n}\langle \tau^{-1}y\rangle^{-\kappa}$ for some constant $C < \infty$ and $\kappa > n + r$. Hence (see (5.4.1))

$$\|\theta(\tau Q)\varphi\|_{\mathcal{M}^r} \leq 2^{r/2}C\tau^\mu \int_{\mathbb{R}^n} \langle y\rangle^r \langle \tau^{-1}y\rangle^{-\kappa} d(\tau^{-1}y) \leq c_r\tau^\mu \qquad \forall \tau \in (0,1),$$

where c_r is some finite constant. So $\varphi \in \mathcal{M}^r_{\mu,\infty}(\mathbb{R}^n)$ for each $r \geq 0$ by (5.4.16). $\square$

The result of Proposition 5.4.5 suggests the notation $\Sigma^{-\mu}(\mathbb{R}^n) = \mathcal{M}^\infty_{\mu,\infty}(\mathbb{R}^n)$. So $\{\Sigma^{-\mu}(\mathbb{R}^n)\}_{\mu>0}$ is a family of algebras of continuous functions on $\mathbb{R}^n$ explicitly characterized in Corollary 5.4.4. One has $\Sigma^{-\mu}(\mathbb{R}^n) \cdot \Sigma^{-\nu}(\mathbb{R}^n) \subset \Sigma^{-(\mu+\nu)}(\mathbb{R}^n)$ for any $\mu, \nu > 0$. Moreover, if $0 < \mu \leq \nu$, then $\Sigma^{-\nu}(\mathbb{R}^n) \subset \Sigma^{-\mu}(\mathbb{R}^n) \subset BC^\infty(\mathbb{R}^n)$.

The result of Proposition 5.4.5 is optimal in the sense that $S^{-\mu} \not\subset \mathcal{M}^0_{s,p}$ if $s > \mu$ or if $s = \mu$ and $p < \infty$ (we assume $\mu > 0$; if $\mu \leq 0$, then the Fourier transform of an element φ of $S^{-\mu}$ need not be a measure, hence need not belong to $\mathcal{M}^0$, cf. the discussion that follows Proposition 1.3.6; notice that only the behaviour of $\widehat{\varphi}$ at the origin matters, since $\widehat{\varphi}$ is of class $\mathscr{S}$ outside zero). Indeed, on page 50 of [P1] it is shown that a function φ of class C^∞ on $\mathbb{R}^n$ and equal to $|x|^{-\mu}$ in a neighbourhood of infinity has the property that $\widehat{\varphi} \in B_q^{\mu-n/q',\infty}(\mathbb{R}^n)$ for each $q \in [1,\infty]$ (where $q' = q/(q-1)$) and that this is optimal on the Besov scale; in particular $\widehat{\varphi} \in B_1^{\mu,\infty}$ and $\widehat{\varphi} \notin B_1^{s,p}$ if $s > \mu$ or $s = \mu$ and $p < \infty$. But we know that $\varphi \in \mathcal{M}^0_{s,p}$ $(s > 0)$ is equivalent to $\widehat{\varphi} \in B_1^{s,p}$.

We end this section with some comments which make the connection between Proposition 5.4.1, Theorem 5.4.3 and some classical results of real analysis due to Hardy, Littlewood and Plessner (see Sections 1.4.2, 2.1.2 and 2.2.3 in [BB]). The connection is obtained by rewriting our results in terms of the Fourier transform of φ. First, consider an integrable measure g on $\mathbb{R}^n$ such that $\int_{\mathbb{R}^n} |g(x+y) - g(x)|dx \to 0$ as $y \to 0$ (we denote by $g(\cdot + y)$ the measure obtained from g by a translation through y, and $|g(x+y) - g(x)|dx$ is the variation of the measure $g(\cdot + y) - g(\cdot)$). By applying Proposition 5.4.1 with $r = 0$, $\ell = 1$ and $\varphi = \breve{g}$, one finds that g is absolutely continuous, which for $n = 1$ is Plessner's theorem. Next, by taking $r = 0$ and $k = 1$ in Theorem 5.4.3 (a), one sees that an integrable measure g satisfies $\int_{\mathbb{R}^n} |g(x+y) - g(x)|dx \leq c|y|$ if and only if its distributional derivatives $\partial_1 g, \ldots, \partial_n g$ are integrable measures; if $n = 1$, this means that g is a *function* of bounded variation, and this is a theorem of Hardy and Littlewood.

We shall now state more general results that follow by the methods of this section. In fact, if $r \geq 0$, then the next theorem is a corollary of Proposition 5.4.1 and Theorem 5.4.3 (see also formula (3.3.18) and Proposition 1.3.9). On the other hand, it is clear that the positivity of r was irrelevant for the proof of these assertions (only the multiplicative properties $\mathscr{M}^r_{s,p} \cdot \mathscr{M}^r_{t,q} \subset \mathscr{M}^r_{s+t,u}$ and $\mathscr{M}^r_k \cdot \mathscr{M}^r_m \subset \mathscr{M}^r_{k+m}$ required $r \geq 0$). Hence we have proved indeed the theorem stated below (we mention that one can easily prove it by using only the identities (3.3.18) and (3.3.38) for the translation group acting on C^∞ functions and simple approximation arguments based on regularization by convolution).

THEOREM 5.4.6. *Let* $f \in \mathscr{S}^*(\mathbb{R}^n)$, $k \geq 1$ *an integer, and* $r \in \mathbb{R}$.

(a) *If* $f^{(\alpha)}$ *is a measure for each multi-index* α *with* $|\alpha| = k$, *then* $f^{(\alpha)} \in L^1_{\text{loc}}(\mathbb{R}^n)$ *for all* $|\alpha| < k$, *and for each* $y \in \mathbb{R}^n$ *one has*

$$\int_{\mathbb{R}^n} \left| \sum_{j=0}^k \binom{k}{j} (-1)^j f(x+jy) \right| \langle x \rangle^r \underline{dx} \leq$$

$$\leq 2^{|r|/2} \langle ky \rangle^{|r|} \sum_{|\alpha|=k} \frac{k!}{\alpha!} |y^\alpha| \int_{\mathbb{R}^n} |f^{(\alpha)}(x)| \langle x \rangle^r \underline{dx}.$$

(b) *Assume that* f *is a measure such that*

$$\int_{\mathbb{R}^n} \left| \sum_{j=0}^k \binom{k}{j} (-1)^j f(x+jy) \right| \langle x \rangle^r \underline{dx} \leq M|y|^k$$

for some constant $M < \infty$ *and all* $|y| \leq 1$. *Then* $f^{(\alpha)}$ *is a measure if* $|\alpha| = k$.

(c) *Assume that* f *is a measure such that* $\int_{\mathbb{R}^n} |f(x)| \langle x \rangle^r \underline{dx} < \infty$ *and*

$$\lim_{y \to 0} \int_{\mathbb{R}^n} \left| \sum_{j=0}^k \binom{k}{j} (-1)^j f(x+jy) \right| \langle x \rangle^r \underline{dx} = 0.$$

Then f *is absolutely continuous with respect to Lebesgue measure, i.e.* $f \in L^1_{\text{loc}}(\mathbb{R}^n)$.

5.5. Commutator Expansions

5.5.1. In order to explain the meaning of the term "commutator expansion", let $\varphi : \mathbb{R}^n \to \mathbb{C}$ be a complex polynomial in $\mathbb{R}^n$. Then we have a Taylor expansion

$$\varphi(x+y) = \sum_\alpha \frac{1}{\alpha!} \varphi^{(\alpha)}(x) y^\alpha$$

with a finite number of terms. This identity is, obviously, a purely algebraic fact (it follows by linearity from the binomial theorem), hence we can replace $x = (x_1, \ldots, x_n)$ and $y = (y_1, \ldots, y_n)$ by elements of an arbitrary commutative algebra with unit. By setting $x = \mathscr{A}''$ and $y = \mathscr{A}$ (in the notations introduced at the beginning of Section 5.3) and by taking into account the relation $\mathscr{A}' = \mathscr{A} + \mathscr{A}''$, we get

(5.5.1) $$\varphi(\mathscr{A}') = \sum_\alpha \frac{1}{\alpha!} \varphi^{(\alpha)}(\mathscr{A}'') \mathscr{A}^\alpha.$$

This is an identity between operators acting on the vector space $B(\mathbf{F}'_\infty, \mathbf{F}''_{-\infty})$. Hence, for any $S \in B(\mathbf{F}'_\infty, \mathbf{F}''_{-\infty})$:

(5.5.2)
$$S\varphi(A') = \sum_\alpha \frac{(-1)^{|\alpha|}}{\alpha!} \varphi^{(\alpha)}(A'') \operatorname{ad}_A^\alpha(S).$$

We call this a *left commutator expansion* because it expresses the generalized commutator $S\varphi(A') - \varphi(A'')S$ as a sum of terms of the form

$$(-1)^{|\alpha|}(\alpha!)^{-1}\varphi^{(\alpha)}(A'') \operatorname{ad}_A^\alpha(S)$$

(with $|\alpha| \geq 1$) in which all the functions of A'' stand on the left of the operators $\operatorname{ad}_A^\alpha(S)$ (in the most usual case where $A' = A''$, (5.5.2) does represent an expression of the commutator $[S, \varphi(A)]$).

By starting with $\mathscr{A}'' = \mathscr{A}' - \mathscr{A}$, one similarly obtains that

(5.5.3)
$$\varphi(\mathscr{A}'') = \sum_\alpha \frac{(-1)^{|\alpha|}}{\alpha!} \varphi^{(\alpha)}(\mathscr{A}')\mathscr{A}^\alpha,$$

or equivalently that

(5.5.4)
$$\varphi(A'')S = \sum_\alpha \frac{1}{\alpha!} \operatorname{ad}_A^\alpha(S)\varphi^{(\alpha)}(A')$$

for all $S \in B(\mathbf{F}'_\infty, \mathbf{F}''_{-\infty})$. Naturally, we call this a *right commutator expansion*. Observe that (5.3.11) and (5.3.12) are the simplest examples of left and right commutator expansions respectively.

The functions φ which are interesting for us are not polynomials but rather symbols of class $S^m(\mathbb{R}^n)$ with $m \in \mathbb{R}$. Then any finite Taylor expansion of φ contains a remainder, so the formulas cannot be as simple as above. By arguing formally, one gets from (1.1.8):

(5.5.5)
$$\varphi(\mathscr{A}') = \varphi(\mathscr{A}'' + \mathscr{A}) = \sum_{|\alpha|<m} \frac{1}{\alpha!}\varphi^{(\alpha)}(\mathscr{A}'')\mathscr{A}^\alpha + \sum_{|\alpha|=m} \frac{1}{\alpha!}\varphi_\alpha(\mathscr{A}';\mathscr{A}'')\mathscr{A}^\alpha,$$

where (see (1.1.7) and observe the slight change of notation):

(5.5.6)
$$\varphi_\alpha(\mathscr{A}';\mathscr{A}'') \equiv \varphi_\alpha(\mathscr{A}'',\mathscr{A}) = |\alpha| \int_0^1 \varphi^{(\alpha)}(\mathscr{A}'' + \tau\mathscr{A})(1-\tau)^{|\alpha|-1}d\tau$$
$$= |\alpha| \int_0^1 \varphi^{(\alpha)}((1-\tau)\mathscr{A}' + \tau\mathscr{A}'')\tau^{|\alpha|-1}d\tau.$$

Hence, for any $S \in B(\mathbf{F}'_\infty, \mathbf{F}''_{-\infty})$:

(5.5.7)
$$S\varphi(A') = \sum_{|\alpha|<m} \frac{(-1)^{|\alpha|}}{\alpha!}\varphi^{(\alpha)}(A'') \operatorname{ad}_A^\alpha(S) + \sum_{|\alpha|=m} \frac{(-1)^{|\alpha|}}{\alpha!}\varphi_\alpha(\mathscr{A}';\mathscr{A}'')[\operatorname{ad}_A^\alpha(S)].$$

This is a *left commutator expansion of order m*, and the last sum is called the *remainder of the expansion*.

Similarly, by developing $\varphi(\mathscr{A}'') = \varphi(\mathscr{A}' - \mathscr{A})$ in powers of $\mathscr{A}$, we get a *right commutator expansion of order m*:

(5.5.8) $\varphi(\mathscr{A}'') = \varphi(\mathscr{A}' - \mathscr{A})$

$$= \sum_{|\alpha|<m} \frac{(-1)^{|\alpha|}}{\alpha!} \varphi^{(\alpha)}(\mathscr{A}')\mathscr{A}^\alpha + \sum_{|\alpha|=m} \frac{(-1)^{|\alpha|}}{\alpha!} \varphi_\alpha(\mathscr{A}'';\mathscr{A}')\mathscr{A}^\alpha.$$

Here $\varphi_\alpha(\mathscr{A}'';\mathscr{A}') \equiv \varphi_\alpha(\mathscr{A}',-\mathscr{A})$ is given by the last member of (5.5.6), but with the roles of $\mathscr{A}'$ and $\mathscr{A}''$ interchanged. More explicitly, for $S \in B(\mathbf{F}'_\infty, \mathbf{F}''_{-\infty})$:

(5.5.9)

$$\varphi(A'')S = \sum_{|\alpha|<m} \frac{1}{\alpha!} \mathrm{ad}_A^\alpha(S)\varphi^{(\alpha)}(A') + \sum_{|\alpha|=m} \frac{1}{\alpha!} \varphi_\alpha(\mathscr{A}'';\mathscr{A}')[\mathrm{ad}_A^\alpha(S)].$$

The formulas (5.5.5), (5.5.8) are easy to justify for arbitrary commuting C_w-groups $\mathscr{W}'$, $\mathscr{W}''$ and $\mathscr{W}$ such that $\mathscr{W}(x) = \mathscr{W}''(-x)\mathscr{W}'(x)$. However, in our case it is easier to obtain (5.5.7) and (5.5.9) by starting from (5.1.8). For example, let us multiply (5.1.8) (with $\ell = m$) on the left by $W''(x)$ and apply the distribution $\widehat{\varphi}$ to both sides of the resulting equation (if W', W'' are arbitrary, we assume that $\widehat{\varphi}$ has compact support; if W', W'' are polynomially bounded, we assume that $\widehat{\varphi}$ is a rapidly decreasing distribution). Clearly we obtain (5.5.7).

The main purpose of this section is to show that, for certain φ and m, the remainder of a commutator expansion transforms a regular operator (with respect to W', W'') into a regularity improving operator (with respect to the Besov scale). In order to express the remainder in a convenient form, we introduce for each $\tau \in \mathbb{R}$ a C_w-group $\mathscr{W}_\tau$ acting in $B(\mathbf{F}', \mathbf{F}'')$ by setting

(5.5.10)

$$\mathscr{W}_\tau(x) = \mathscr{W}'((1-\tau)x)\mathscr{W}''(\tau x) = \mathscr{W}'(x)\mathscr{W}(-\tau x) = \mathscr{W}''(x)\mathscr{W}((1-\tau)x).$$

As usual, we extend $\mathscr{W}_\tau$ to an operator in $\mathscr{B}_{-\infty} \equiv B(\mathbf{F}'_\infty, \mathbf{F}''_{-\infty})$. The generator $\mathscr{A}_\tau$ of $\mathscr{W}_\tau$ is given by

(5.5.11) $\mathscr{A}_\tau = (1-\tau)\mathscr{A}' + \tau\mathscr{A}'' = \mathscr{A}' - \tau\mathscr{A} = \mathscr{A}'' + (1-\tau)\mathscr{A}.$

If ψ is a function from $\mathbb{R}^n$ to $\mathbb{C}$, we set $\psi(\mathscr{A}_\tau) = \int_{\mathbb{R}^n} \mathscr{W}_\tau(x)\widehat{\psi}(x)\underline{d}x$ whenever this makes sense. More explicitly, for $S \in B(\mathbf{F}'_\infty, \mathbf{F}''_{-\infty})$:

(5.5.12)

$$\psi(\mathscr{A}_\tau)[S] = \int_{\mathbb{R}^n} \mathscr{W}_\tau(x)[S]\widehat{\psi}(x)\underline{d}x = \int_{\mathbb{R}^n} W''(\tau x)SW'((1-\tau)x)\widehat{\psi}(x)\underline{d}x.$$

Then we have

$$\varphi_\alpha(\mathscr{A}';\mathscr{A}'') = |\alpha| \int_0^1 \varphi^{(\alpha)}(\mathscr{A}_\tau)\tau^{|\alpha|-1}d\tau,$$

$$\varphi_\alpha(\mathscr{A}'';\mathscr{A}') = |\alpha| \int_0^1 \varphi^{(\alpha)}(\mathscr{A}_\tau)(1-\tau)^{|\alpha|-1}d\tau.$$

In conclusion, we have to study properties of operators (acting in $\mathscr{B}_{-\infty}$) of the form $\int_0^1 \psi(\mathscr{A}_\tau)\chi(\tau)d\tau$, where $\chi(\tau)d\tau$ could be any integrable measure.

The fact that, under certain conditions, $\psi(\mathscr{A}_\tau)$ transforms regular operators into regularity improving operators may be illustrated as follows: from the relation $\mathscr{A}'' = \mathscr{A}_\tau + (\tau-1)\mathscr{A}$ (see (5.5.11)) one obtains that

$$(5.5.13) \qquad \mathscr{A}''^\alpha\psi(\mathscr{A}_\tau) = \sum_{\beta+\gamma=\alpha} \frac{\alpha!}{\beta!\gamma!}(\tau-1)^{|\beta|}\mathscr{A}^\beta\mathscr{A}_\tau^\gamma\psi(\mathscr{A}_\tau)$$

$$= \sum_{\beta+\gamma=\alpha} \frac{\alpha!}{\beta!\gamma!}(\tau-1)^{|\beta|}\psi_{(\gamma)}(\mathscr{A}_\tau)\mathscr{A}^\beta,$$

where $\psi_{(\gamma)}(x) = x^\gamma\psi(x)$. So if for example $S \in C^k(\mathbf{F}',\mathbf{F}'')$ for some $k \in \mathbb{N}$ (so that $\mathscr{A}^\beta[S] \in B(\mathbf{F}',\mathbf{F}'')$ if $|\beta| \leq k$) and if $\psi_{(\gamma)}(\mathscr{A}_\tau) \in B(B(\mathbf{F}',\mathbf{F}''))$ for all $|\gamma| \leq k$, the preceding identity implies that $\psi(\mathscr{A}_\tau)[S] \in B(\mathbf{F}',\mathbf{F}''_k)$.

In what follows we shall develop the preceding observation. We shall restrict ourselves to the case where W' and W'' are polynomially bounded C_0-groups; the general case could be dealt with in a similar way by considering a more restricted class of functions ψ.

5.5.2. For the remainder of this section *we assume the representations* W', W'' *to be polynomially bounded*. We fix a measure χ of compact support on $\mathbb{R}$ and set, for $\psi \in C_{\mathrm{pol}}^\infty(\mathbb{R}^n)$

$$(5.5.14) \qquad \mathscr{I}_\psi = \int_{\mathbb{R}} \psi(\mathscr{A}_\tau)\chi(\tau)\underline{d}\tau.$$

Clearly $\mathscr{I}_\psi : B(\mathbf{F}'_\infty,\mathbf{F}''_{-\infty}) \to B(\mathbf{F}'_\infty,\mathbf{F}''_{-\infty})$ is a well defined linear map (we refer to §3.6.2 for the meaning of (5.5.14) for $\psi \in C_{\mathrm{pol}}^\infty(\mathbb{R}^n)$; we do not insist on this point because this generality will be irrelevant).

Now let us assume that $\psi \in \mathscr{M}^\infty(\mathbb{R}^n)$, so that $\hat{\psi}$ is a rapidly decreasing measure. We shall use the notation $||S||_{\mathscr{C}^{s,p}}$ for $||S||_{s,p}^{(\ell)}$ (cf. (5.2.1)), where ℓ is the smallest integer strictly larger than s. Since $\mathscr{A}_\tau$ commutes with $\mathscr{A}$, there is a constant $C < \infty$ such that for all $S \in B(\mathbf{F}'_\infty,\mathbf{F}''_{-\infty})$, all $s > 0$ and all $1 \leq p \leq \infty$:

$$(5.5.15) \qquad ||\mathscr{I}_\psi[S]||_{\mathscr{C}^{s,p}} \leq C||S||_{\mathscr{C}^{s,p}}.$$

In other words for each $s > 0$ and $p \in [1,\infty]$, the operator $\mathscr{I}_\psi$ leaves $\mathscr{C}^{s,p}(\mathbf{F}',\mathbf{F}'')$ invariant. In particular, if $S \in \mathscr{C}^{s,p}(\mathbf{F}',\mathbf{F}'')$ then $\mathscr{I}_\psi[S] \in B(\mathbf{F}'_{t,q},\mathbf{F}''_{t,q})$ for $0 < t < s$ and $q \in [1,\infty]$ (see Theorem 5.3.3) .

We shall now give conditions under which $\mathscr{I}_\psi$ transforms an operator $S \in \mathscr{C}^{s,p}(\mathbf{F}',\mathbf{F}'')$ into a regularity improving operator.

PROPOSITION 5.5.1. *Assume that W' and W'' are polynomially bounded C_0-groups. Let $0 < t < s < \infty$, $1 \leq p \leq q \leq \infty$ and set $\mu = s - t$. Let r be defined by $1/r = 1/p - 1/q$. Then $\mathscr{I}_\psi[S] \in B(\mathbf{F}'_{t,q},\mathbf{F}''_{s,p})$ whenever $\psi \in \mathscr{M}_{\mu,r}^\infty(\mathbb{R}^n)$ and $S \in \mathscr{C}^{s,p}(A',A'';\mathbf{F}',\mathbf{F}'')$.*

PROOF. We denote the norm in $B(\mathbf{F}', \mathbf{F}'')$ by $|| \cdot ||$. Let $\psi \in \mathscr{M}_{\mu,r}^\infty(\mathbb{R}^n)$ be fixed. In (i) we shall show that there are constants c, $u < \infty$ such that for all k, $\ell \in \mathbb{N}$ and all $|x| \leq 1$:

(5.5.16)
$$\left\| [W''(x) - I]^k [\mathscr{W}(x) - I]^\ell \mathscr{I}_\psi[S] \right\| \leq$$
$$\leq c \sum_{j=0}^k \left\| [e^{iQ \cdot x} - I]^j \psi \right\|_{\mathscr{M}^u} \int_{\mathbb{R}} \left\| [\mathscr{W}((1-\tau)x) - I]^{k-j} [\mathscr{W}(x) - I]^\ell S \right\| |\chi(\tau)| \underline{d}\tau.$$

In (ii) we shall use this inequality to show that the hypotheses of the criterion for regularity improvement of Proposition 5.3.5 are satisfied for the operator $\mathscr{I}_\psi[S]$ if $S \in \mathscr{C}^{s,p}(\mathbf{F}', \mathbf{F}'')$.

(i) By taking into account (5.5.10), one obtains that
$$[\mathscr{W}''(x) - I]^k = [\mathscr{W}((\tau-1)x)\mathscr{W}_\tau(x) - I]^k$$
$$= \mathscr{W}(k(\tau-1)x)\{[\mathscr{W}_\tau(x) - I] + [I - \mathscr{W}((1-\tau)x)]\}^k$$
$$= \mathscr{W}(k(\tau-1)x) \sum_{j=0}^k \binom{k}{j} [\mathscr{W}_\tau(x) - I]^j [I - \mathscr{W}((1-\tau)x)]^{k-j}.$$

It follows that

(5.5.17) $[\mathscr{W}''(x) - I]^k [\mathscr{W}(x) - I]^\ell \psi(\mathscr{A}_\tau) =$
$$\mathscr{W}(k(\tau-1)x) \sum_{j=0}^k \binom{k}{j} [(e^{iQ \cdot x} - I)^j \psi](\mathscr{A}_\tau) \cdot [I - \mathscr{W}((1-\tau)x)]^{k-j} [\mathscr{W}(x) - I]^\ell,$$

where we have used the relation $[e^{i\mathscr{A}_\tau \cdot x} - I]^j \psi(\mathscr{A}_\tau) = [(e^{iQ \cdot x} - I)\psi](\mathscr{A}_\tau)$ which follows directly from the definitions or from the fact that $\varphi \mapsto \varphi(\mathscr{A}_\tau)$ is a homomorphism. Since the representations W' and W'' are polynomially bounded, there are a finite number u and a constant c_0 such that for $\tau \in \text{supp}\,\chi$: $||\varphi(\mathscr{A}_\tau)[S]|| \leq c_0 ||\varphi||_{\mathscr{M}^u} ||S||$, and (5.5.16) is an easy consequence of (5.5.17).

(ii) We now check the hypotheses of Proposition 5.3.5 for $\mathscr{I}_\psi[S]$. For (5.3.17) we take $k > \mu$ and $\ell = 0$ in (5.5.16). By using the integral Minkowski inequality, we get that

$$\left[\int_{|x| \leq 1} \left\| |x|^{-\mu} [W''(x) - I]^k \mathscr{I}_\psi[S] \right\|^r \frac{dx}{|x|^n} \right]^{1/r} \leq$$
$$\leq c \sum_{j=0}^k \int_{\mathbb{R}} |\chi(\tau)| d\tau \left[\int_{|x| \leq 1} \left\| |x|^{-\mu j/k} [e^{iQ \cdot x} - I]^j \psi \right\|_{\mathscr{M}^u}^r \cdot \right.$$
$$\left. \cdot \left\| |x|^{-\mu(k-j)/k} [\mathscr{W}((1-\tau)x) - I]^{k-j} S \right\|^r \frac{dx}{|x|^n} \right]^{1/r}.$$

The term with $j = k$ in the sum on the r.h.s. is finite because $\psi \in \mathscr{M}_{\mu,r}^u(\mathbb{R}^n)$ and $S \in B(\mathbf{F}', \mathbf{F}'')$. For the remaining terms we majorize the first norm under the integral by a constant (observe that $\psi \in \mathcal{M}^u_{\mu j/k, \infty}(\mathbb{R}^n)$ if $0 < j < k$ by (5.4.6) and since $\mu > \mu j/k$); so it suffices to show that, for $0 \leq j < k$:

$$\sup_{\tau \in \mathrm{supp}\chi} \left[\int_{|x| \leq 1} \big\| |x|^{-\mu(k-j)/k} [\mathcal{W}((1-\tau)x) - I]^{k-j} S \big\|^r \frac{dx}{|x|^n} \right]^{1/r} < \infty.$$

This is easily obtained by the change of variables $x \mapsto y = (1-\tau)x$ and by using the fact that $S \in \mathscr{C}^{\mu(k-j)/k, r}(\mathbf{F}', \mathbf{F}'')$ by (5.2.7), because $\mu < s$.

To check (5.3.18), we take $k > \mu$ and $\ell > t$ in (5.5.16) and observe that, for some constant $c_1 < \infty$:

$$\left[\int_{|x| \leq 1} \big\| |x|^{-s} [W''(x) - I]^k [\mathcal{W}(x) - I]^\ell \mathscr{I}_\psi[S] \big\|^p \frac{dx}{|x|^n} \right]^{1/p} \leq$$

$$\leq c_1 \sum_{j=0}^{k} \left[\int_{|x| \leq 1} \big\| |x|^{-\mu j/k} [e^{iQ \cdot x} - I]^j \psi \big\|^p_{\mathcal{M}^u} \big\| |x|^{-t-\mu(k-j)/k} [\mathcal{W}(x) - I]^\ell S \big\|^p \frac{dx}{|x|^n} \right]^{1/p}$$

The term with $j = 0$ is finite because $S \in \mathscr{C}^{s,p}(\mathbf{F}', \mathbf{F}'')$. For the other terms we use the Hölder inequality and obtain a bound of the form

$$c_2 \left[\int_{|x| \leq 1} \big\| |x|^{-\mu j/k} [e^{iQ \cdot x} - I]^j \psi \big\|^r_{\mathcal{M}^u} \frac{dx}{|x|^n} \right]^{1/r} \cdot$$

$$\cdot \left[\int_{|x| \leq 1} \big\| |x|^{-t-\mu(k-j)/k} [\mathcal{W}(x) - I]^\ell S \big\|^q \frac{dx}{|x|^n} \right]^{1/q}.$$

The first integral is finite because $\psi \in \mathcal{M}^u_{\mu,r}(\mathbb{R}^n)$ and $\mu \geq \mu j/k$, and the second one is finite because $S \in \mathscr{C}^{s,p}(\mathbf{F}', \mathbf{F}'') \subset \mathscr{C}^{t+\mu(k-j)/k, q}(\mathbf{F}', \mathbf{F}'')$ if $j \neq 0$. $\square$

In the next theorem we isolate a consequence of Proposition 5.5.1 which is especially interesting when $\mathbf{F}'$ and $\mathbf{F}''$ are Hilbert spaces. If $\mathbf{F}$ is a Hilbert space, then $\mathbf{F}_{s,2} = \mathbf{F}_s$ for $s \in \mathbb{Z}$ (see Theorem 3.7.8), and if one defines $\mathbf{F}_s = \mathbf{F}_{s,2}$ for $s \in \mathbb{R}$, then the *continuous* Sobolev scale $\{\mathbf{F}_s\}_{s \in \mathbb{R}}$ may also by obtained by complex (or quadratic) interpolation (see Section 2.8).

From Section 5.4 we recall the notation $\Sigma^{-\mu}(\mathbb{R}^n) = \mathcal{M}^\infty_{\mu,\infty}(\mathbb{R}^n)$ for $\mu > 0$ and the embedding $S^{-\mu}(\mathbb{R}^n) \subset \Sigma^{-\mu}(\mathbb{R}^n)$. For uniformity of notation we also set $\Sigma^0(\mathbb{R}^n) = \mathcal{M}^\infty(\mathbb{R}^n)$ but observe that $S^0(\mathbb{R}^n) \not\subset \Sigma^0(\mathbb{R}^n)$. The following theorem implies that, if $\psi \in S^{-\mu}(\mathbb{R}^n)$ with $\mu > 0$, then $\mathscr{I}_\psi[S]$ is regularity improving exactly of order μ. However, if $\mu = 0$ and $S \in B(\mathbf{F}', \mathbf{F}'')$, then in general $\mathscr{I}_\psi[S]$ is not a bounded operator from $\mathbf{F}'$ to $\mathbf{F}''$ (even in the case where $\mathbf{F}'$ and $\mathbf{F}''$ are Hilbert spaces and W', W'' are unitary groups; see the Appendix to Chapter 6).

We mention that the most difficult situation covered by the following theorem is that where μ is an integer ≥ 1; this case requires real interpolation, Littlewood-Paley theory and Zygmund type conditions of the form (5.4.14).

THEOREM 5.5.2. *Let $(\mathbf{F}', W')$ and $(\mathbf{F}'', W'')$ be two reflexive Banach spaces equipped with polynomially bounded C_0-representations of $\mathbb{R}^n$. Let χ be a measure of compact support on $\mathbb{R}$, $\psi \in \Sigma^{-\mu}(\mathbb{R}^n)$ for some $\mu \geq 0$ and $s > \mu$. Then*

(a) *For each* $S \in \mathscr{C}^{s,2}(A', A''; \mathbf{F}', \mathbf{F}'')$, *the operator* $\mathscr{I}_\psi[S]$ *is a continuous map from* $\mathbf{F}'_{t,q}$ *to* $\mathbf{F}''_{t+\mu,q}$ *for any real* $t \in (-s, s-\mu)$ *and any* $q \in [1, \infty]$, *and also for* $t \in [-s, s-\mu]$ *if* $q = 2$. *In particular, if* $\mathbf{F}'$ *and* $\mathbf{F}''$ *are Hilbert spaces, then* $\mathscr{I}_\psi[S] \in B(\mathbf{F}'_t, \mathbf{F}''_{t+\mu})$ *for all real* $t \in [-s, s-\mu]$.

(b) *For each* t *and* q *as in* (a), *there is a constant* c *such that*

$$(5.5.18) \quad ||\mathscr{I}_\psi[S]||_{\mathbf{F}'_{t,q} \to \mathbf{F}''_{t+\mu,q}} \leq c||S||_{\mathscr{C}^{s,2}} \quad \forall S \in \mathscr{C}^{s,2}(A', A''; \mathbf{F}', \mathbf{F}'').$$

PROOF. (i) Assume that $\mu > 0$. If $S \in \mathscr{C}^{s,2}(\mathbf{F}', \mathbf{F}'')$, then $S^* \in \mathscr{C}^{s,2}(\mathbf{F}''^*, \mathbf{F}'^*)$, and Proposition 5.5.1 (with $p = q = 2$, $r = \infty$) implies that

$$\mathscr{I}_\psi[S] \in B(\mathbf{F}'_{s-\mu,2}, \mathbf{F}''_{s,2}) \text{ and } \mathscr{I}_\psi[S]^* \in B(\mathbf{F}''^*_{s-\mu,2}, \mathbf{F}'^*_{s,2}).$$

From the last relation we get that $\mathscr{I}_\psi[S] \in B(\mathbf{F}'_{-s,2}, \mathbf{F}''_{-s+\mu,2})$. The assertions of (a) follow by interpolating between the preceding two inclusions for $\mathscr{I}_\psi[S]$ (use (2.6.1) and take into account Theorem 3.4.3 (b)). For (b) observe that $\mathscr{I}_\psi$ is a continuous map of the Banach space $B(\mathbf{F}', \mathbf{F}'')$ into itself which maps the B-subspace $\mathscr{C}^{s,2}(\mathbf{F}', \mathbf{F}'')$ into the B-subspace $B(\mathbf{F}', \mathbf{F}'') \cap B(\mathbf{F}'_{t,q}, \mathbf{F}''_{t+\mu,q})$. Hence $\mathscr{I}_\psi : \mathscr{C}^{s,2}(\mathbf{F}', \mathbf{F}'') \to B(\mathbf{F}', \mathbf{F}'') \cap B(\mathbf{F}'_{t,q}, \mathbf{F}''_{t+\mu,q})$ is continuous by the closed graph theorem.

(ii) If $\mu = 0$, the results of the theorem are obtained by using (5.5.15) and by (5.3.14). $\square$

The following result is an immediate consequence of the above theorem; it describes commutator expansions with rather precise estimates on the remainder.

THEOREM 5.5.3. *Let* $(\mathbf{F}', W')$ *and* $(\mathbf{F}'', W'')$ *be two reflexive Banach spaces equipped with polynomially bounded* C_0-*representations of* $\mathbb{R}^n$. *Then for each* $\varphi \in C^\infty_{\text{pol}}(\mathbb{R}^n)$, *each* $S \in B(\mathbf{F}'_\infty, \mathbf{F}''_{-\infty})$ *and each integer* $m \geq 1$ *there are left and right commutator expansion of order* m:

$$(5.5.19) \qquad S\varphi(A') = \sum_{|\alpha| < m} \frac{(-1)^{|\alpha|}}{\alpha!} \varphi^{(\alpha)}(A'') \operatorname{ad}^\alpha_A(S) + \mathscr{R}^L_{m,\varphi}[S],$$

$$(5.5.20) \qquad \varphi(A'')S = \sum_{|\alpha| < m} \frac{1}{\alpha!} \operatorname{ad}^\alpha_A(S) \varphi^{(\alpha)}(A') + \mathscr{R}^R_{m,\varphi}[S].$$

The remainders $\mathscr{R}^L_{m,\varphi}[S]$, $\mathscr{R}^R_{m,\varphi}[S]$ are given by:

$$(5.5.21) \qquad \mathscr{R}^L_{m,\varphi}[S] = \sum_{|\alpha|=m} \frac{(-1)^{|\alpha|}}{\alpha!} \mathscr{I}^L_\alpha[\operatorname{ad}^\alpha_A(S)],$$

$$(5.5.22) \qquad \mathscr{R}^R_{m,\varphi}[S] = \sum_{|\alpha|=m} \frac{1}{\alpha!} \mathscr{I}^R_\alpha[\operatorname{ad}^\alpha_A(S)],$$

where the operators $\mathscr{I}_\alpha^L$ and $\mathscr{I}_\alpha^R$ act in $B(\mathbf{F}'_\infty, \mathbf{F}''_{-\infty})$ according to the formulas

(5.5.23)
$$\mathscr{I}_\alpha^L \equiv \varphi_\alpha(\mathscr{A}';\mathscr{A}'') = |\alpha| \int_0^1 \varphi^{(\alpha)}(\mathscr{A}_\tau)\tau^{|\alpha|-1}d\tau,$$

(5.5.24)
$$\mathscr{I}_\alpha^R \equiv \varphi_\alpha(\mathscr{A}'';\mathscr{A}') = |\alpha| \int_0^1 \varphi^{(\alpha)}(\mathscr{A}_\tau)(1-\tau)^{|\alpha|-1}d\tau.$$

In particular, assume that $\varphi^{(\alpha)} \in \Sigma^{-\mu}(\mathbb{R}^n)$ for some $\mu \geq 0$ and all multi-indices α with $|\alpha| = m$ (for example let $\varphi \in S^{m-\mu}(\mathbb{R}^n)$ for some $\mu > 0$). Then, if $s > \mu$ and $S \in \mathscr{C}^{m+s,2}(A', A''; \mathbf{F}', \mathbf{F}'')$, the operators $\mathscr{R}_{m,\varphi}^L[S]$ and $\mathscr{R}_{m,\varphi}^R[S]$ belong to $B(\mathbf{F}'_{t,q}, \mathbf{F}''_{t+\mu,q})$ for any real $t \in (-s, s - \mu)$ and any $q \in [1,\infty]$ and also for $t = -s$ and $t = s - \mu$ if $q = 2$. If $\mathbf{F}'$ and $\mathbf{F}''$ are Hilbert spaces, then $\mathscr{R}_{m,\varphi}^L[S]$ and $\mathscr{R}_{m,\varphi}^R[S]$ belong to $B(\mathbf{F}'_t, \mathbf{F}''_{t+\mu})$ for any real $t \in [-s, s - \mu]$, with norms bounded by $C\|S\|_{\mathscr{C}^{m+s,2}}$ for some constant C independent of S.

The next result is an example concerning the way in which commutator expansions may be used. We restrict ourselves to the Hilbert space situation only for simplicity of the statement.

PROPOSITION 5.5.4. *Let $(\mathbf{F}', W')$ and $(\mathbf{F}'', W'')$ be two Hilbert spaces equipped with polynomially bounded C_0-representations of $\mathbb{R}^n$. Consider an operator $S \in \mathscr{C}^{m+s,\infty}(A', A''; \mathbf{F}', \mathbf{F}'')$, where $m \geq 1$ is an integer and $s > 0$ is real, and a function $\varphi \in C_{\mathrm{pol}}^\infty(\mathbb{R}^n)$ such that $\varphi^{(\alpha)} \in \Sigma^{-\mu}(\mathbb{R}^n)$ for some real $\mu \in [0, s)$ and all multi-indices α with $|\alpha| = m$.*

(a) *Let $\mathbf{E}$ be a Banach space and $T : \mathbf{E} \to \mathbf{F}'_{-\mu}$ a bounded operator such that $\varphi^{(\alpha)}(A')T \in B(\mathbf{E}, \mathbf{F}')$ if $|\alpha| < m$. Then $\varphi(A'')ST$ is a well defined bounded operator from $\mathbf{E}$ into $\mathbf{F}''$.*

(b) *Let $\mathbf{E}$ be a Banach space and $T : \mathbf{F}''_\mu \to \mathbf{E}$ a bounded operator such that $T\varphi^{(\alpha)}(A'') \in B(\mathbf{F}'', \mathbf{E})$ if $|\alpha| < m$. Then $TS\varphi(A')$ is a well defined bounded operator from $\mathbf{F}'$ into $\mathbf{E}$.*

PROOF. We indicate the proof of (a) which is based on the commutator expansion (5.5.20). Choose s' such that $\mu < s' < s$. Then $S \in \mathscr{C}^{m+s',2}(\mathbf{F}', \mathbf{F}'')$ by (5.2.7). We have $\mathscr{R}_{m,\varphi}^R[S] \in B(\mathbf{F}'_{-\mu}, \mathbf{F}'')$ by Theorem 5.5.3 and $\varphi^{(\alpha)}(A') \in B(\mathbf{F}'_{-\mu}, \mathbf{F}'_{-\mu-m+|\alpha|})$ by Proposition 3.6.5 (b). Finally we observe that $\mathrm{ad}_A^\alpha(S) \in B(\mathbf{F}'_{-\mu-m+|\alpha|}, \mathbf{F}''_{-\mu-m+|\alpha|})$ if $|\alpha| \leq m$ by Proposition 5.2.2 (a) and by (5.3.16). So (5.5.20) holds as an identity in $B(\mathbf{F}'_{-\mu}, \mathbf{F}''_{-\mu-m})$. Upon multiplying it on the right by T and taking into account the hypotheses made on T, one finds that $\varphi(A'')ST \in B(\mathbf{E}, \mathbf{F}'')$. $\square$

As an example (in the situation (a)), let us take $\mu > 0$, $\varphi \in S^{m-\mu}(\mathbb{R}^n)$ and $T = \psi(A')$ with ψ a symbol. We have $\psi(A') \in B(\mathbf{F}', \mathbf{F}'_{-\mu})$ if $\psi \in S^\mu(\mathbb{R}^n)$ and $\varphi^{(\alpha)}(A')\psi(A') \in B(\mathbf{F}')$ if $\varphi^{(\alpha)}\psi \in BC^\infty(\mathbb{R}^n)$ (see Theorems 3.6.9 and 3.7.10). So, if φ and ψ satisfy these conditions, we have $\varphi(A'')S\psi(A') \in B(\mathbf{F}', \mathbf{F}'')$ if S belongs to $\mathscr{C}^{m+s,\infty}(\mathbf{F}', \mathbf{F}'')$ for some $s > \mu$.

Now observe that, if $S \in \mathscr{C}^{m+s,2}(\mathbf{F}', \mathbf{F}'')$, then $S \in \mathscr{C}^{m+s-|t|,2}(\mathbf{F}'_t, \mathbf{F}''_t)$ for each real t with $|t| < m + s$ (see Corollary 5.3.4). So, if $0 < \mu < s - |t|$, we obtain from the preceding result (under the same conditions) that $\varphi(A'')S\psi(A') \in B(\mathbf{F}'_t, \mathbf{F}''_t)$. We state this result in a somewhat more symmetric form:

COROLLARY 5.5.5. *Let* $m \geq 1$ *be an integer, a, b and s real numbers such that* $a \geq 0$, $b \geq 0$, $a + b \leq m$ *and* $s > \max(a, b)$. *Assume that* $\varphi \in S^a(\mathbb{R}^n)$ *and* $\psi \in S^b(\mathbb{R}^n)$ *are such that* $\varphi^{(\alpha)}\psi^{(\beta)}$ *is a bounded function for all multi-indices* α *and* β. *Then* $\varphi(A'')S\psi(A') \in B(\mathbf{F}'_t, \mathbf{F}''_t)$ *for each* $S \in \mathscr{C}^{m+s,\infty}(A', A''; \mathbf{F}', \mathbf{F}'')$ *and each real t satisfying* $|t| < s - \max(a, b)$.

PROOF. Let s' be such that $\max(a, b) < s' < s$. Then $S \in \mathscr{C}^{m+s',2}(\mathbf{F}', \mathbf{F}'')$. If $b > 0$, take $\mu = b$ (then $\varphi \in S^{m-\mu}(\mathbb{R}^n)$) and apply the remarks made before the statement of the corollary. If $b = 0$ and $a > 0$, the argument is similar (use Proposition 5.5.4 (b) instead of Proposition 5.4.4 (a)). If $a = b = 0$, observe that $\psi(A') \in B(\mathbf{F}'_t)$, $S \in B(\mathbf{F}'_t, \mathbf{F}''_t)$ and $\varphi(A'') \in B(\mathbf{F}''_t)$ if $|t| < s'$ by Theorems 3.6.9 and 5.3.3. $\square$

If $S \in C^\infty(A', A''; \mathbf{F}', \mathbf{F}'')$, then one may take $\varphi, \psi \in S^\infty(\mathbb{R}^n)$ and the conclusion in Corollary 5.5.5 holds for each real t. The hypothesis that $\varphi^{(\alpha)}\psi^{(\beta)}$ be bounded for all multi-indices α and β holds for example if $\text{supp}\varphi \cap \text{supp}\psi$ is a compact subset of $\mathbb{R}^n$.

In Section 6.4 we shall use a generalized version of the preceding corollary involving four Hilbert spaces equipped with C_0-groups. We denote these spaces by $\mathbf{E}$, $\mathbf{F}$, $\mathbf{G}$ and $\mathbf{H}$, the associated C_0-groups by $W_\mathbf{E}$, $W_\mathbf{F}$, $W_\mathbf{G}$ and $W_\mathbf{H}$ and their generators by $A_\mathbf{E}$, $A_\mathbf{F}$, $A_\mathbf{G}$ and $A_\mathbf{H}$:

PROPOSITION 5.5.6. *Let* $(\mathbf{E}, W_\mathbf{E})$, $(\mathbf{F}, W_\mathbf{F})$, $(\mathbf{G}, W_\mathbf{G})$ *and* $(\mathbf{H}, W_\mathbf{H})$ *be four Hilbert spaces equipped with polynomially bounded C_0-representations of* $\mathbb{R}^n$. *Consider the following objects:*

(1) *Two symbols* $\varphi, \psi \in S^\infty(\mathbb{R}^n)$ *and a bounded operator* $T : \mathbf{E} \to \mathbf{F}$ *such that* $\varphi^{(\alpha)}(A_\mathbf{F})T\psi^{(\beta)}(A_\mathbf{E}) \in B(\mathbf{E}, \mathbf{F})$ *for all multi-indices* α *and* β,

(2) *Two operators* $S_1 : \mathbf{F} \to \mathbf{H}$ *and* $S_2 : \mathbf{G} \to \mathbf{E}$ *of class* $\mathscr{C}^{m+s,\infty}$ *for some integer* $m \geq 1$ *and some real* $s > 0$.

Furthermore, assume that there are strictly positive real numbers a, b *with* $a + b \leq m$ *and* $\max(a, b) < s$ *such that* $\varphi \in S^a(\mathbb{R}^n)$, $\psi \in S^b(\mathbb{R}^n)$ *and* $T \in B(\mathbf{E}_t, \mathbf{F}_t)$ *for all* $t \in [-b, a]$. *Then* $\varphi(A_\mathbf{H})S_1TS_2\psi(A_\mathbf{G}) \in B(\mathbf{G}, \mathbf{H})$.

PROOF. By using the expressions (5.5.19), (5.5.20) and the abbreviations

$\mathscr{R}_1 = \mathscr{R}^R_{m,\varphi}[S_1]$, $\mathscr{R}_2 = \mathscr{R}^L_{m,\psi}[S_2]$, we obtain

(5.5.25)

$$\varphi(A_{\mathbf{H}})S_1 T S_2 \psi(A_{\mathbf{G}}) = \sum_{\substack{|\alpha|<m \\ |\beta|<m}} \frac{(-1)^{|\beta|}}{\alpha!\beta!} \operatorname{ad}^\alpha_A(S_1)\varphi^{(\alpha)}(A_{\mathbf{F}})T\psi^{(\beta)}(A_{\mathbf{E}})\operatorname{ad}^\beta_A(S_2) +$$

$$+ \sum_{|\beta|<m} \frac{(-1)^{|\beta|}}{\beta!} \mathscr{R}_1 T \psi^{(\beta)}(A_{\mathbf{E}})\operatorname{ad}^\beta_A(S_2)$$

$$+ \sum_{|\alpha|<m} \frac{1}{\alpha!} \operatorname{ad}^\alpha_A(S_1)\varphi^{(\alpha)}(A_{\mathbf{F}})T\mathscr{R}_2 + \mathscr{R}_1 T \mathscr{R}_2,$$

where $\operatorname{ad}^\alpha_A(S_1)$ is given by (5.3.10) with $A' = A_{\mathbf{F}}$, $A'' = A_{\mathbf{H}}$ and $\operatorname{ad}^\alpha_A(S_2)$ by (5.3.10) with $A' = A_{\mathbf{G}}$, $A'' = A_{\mathbf{E}}$. We shall show that each term on the r.h.s. of (5.5.25) belongs to $B(\mathbf{G}, \mathbf{H})$. For the terms in the first sum this is a consequence of the hypotheses (1) and (2); for the remaining terms we have to know the boundedness properties of $\mathscr{R}_1$ and $\mathscr{R}_2$.

If $|\alpha| = m$, we have $\varphi^{(\alpha)} \in S^{a-m}(\mathbb{R}^n) \subset S^{-b}(\mathbb{R}^n) \subset \Sigma^{-b}(\mathbb{R}^n)$ (see Proposition 5.4.5), so Theorem 5.5.3 gives $\mathscr{R}_1 \in B(\mathbf{F}_t, \mathbf{H}_{t+\tau})$ for $t \in (-s, s-b)$ and $\tau \le b$. In particular $\mathscr{R}_1 \in B(\mathbf{F}, \mathbf{H}) \cap B(\mathbf{F}_{-b}, \mathbf{H})$. Similarly one obtains from Theorem 5.5.3 that $\mathscr{R}_2 \in B(\mathbf{G}, \mathbf{E}) \cap B(\mathbf{G}, \mathbf{E}_a)$.

The preceding relations imply that $\mathscr{R}_1 T \mathscr{R}_2 \in B(\mathbf{G}, \mathbf{H})$. The terms in the second sum in (5.5.25) belong to $B(\mathbf{G}, \mathbf{H})$ because $\operatorname{ad}^\beta_A(S_2) \in B(\mathbf{G}, \mathbf{E})$, $\psi^{(\beta)}(A_{\mathbf{E}}) \in B(\mathbf{E}, \mathbf{E}_{-b})$ by Theorem 3.6.9, $T \in B(\mathbf{E}_{-b}, \mathbf{F}_{-b})$ by hypothesis and because $\mathscr{R}_1 \in B(\mathbf{F}_{-b}, \mathbf{H})$. The terms in the third sum are treated similarly by using the relations $\mathscr{R}_2 \in B(\mathbf{G}, \mathbf{E}_a)$, $T \in B(\mathbf{E}_a, \mathbf{F}_a)$, $\varphi^{(\alpha)}(A_{\mathbf{F}}) \in B(\mathbf{F}_a, \mathbf{F})$ and $\operatorname{ad}^\alpha_A(S_1) \in B(\mathbf{F}, \mathbf{H})$. $\square$

5.A. Appendix: Differentiability Properties of Operator-Valued Functions

Let $\mathscr{U}$ be an open subset of $\mathbb{R}^n$, let $\mathbf{E}$ and $\mathbf{F}$ be two Banach spaces, and let Ψ be a function defined on $\mathscr{U}$ with values in $B(\mathbf{E}, \mathbf{F})$. Ψ is said to be *weakly differentiable* at a point x if there are n bounded linear operators $L_j \equiv L_j(x)$: $\mathbf{E} \to \mathbf{F}$ $(j = 1, \ldots, n)$ such that

(5.A.1) $$w - \lim_{\substack{|\varepsilon|\to 0 \\ \varepsilon \in \mathbb{R}^n}} |\varepsilon|^{-1}\left[\Psi(x+\varepsilon) - \Psi(x) - \sum_{i=1}^n \varepsilon_j L_j\right] = 0.$$

Equivalently, for each $f \in \mathbf{E}$ and each $g \in \mathbf{F}^*$, the function $\langle \Psi(\cdot)f, g\rangle : \mathscr{U} \to \mathbb{C}$ has to be differentiable at x.

If the weak limit in (5.A.1) is replaced by strong limit or uniform limit, we get the notion of *strong differentiability* or *norm differentiability* respectively.

Ψ is *weakly C^1* on $\mathscr{U}$ if it is weakly differentiable at each point of $\mathscr{U}$ and if the associated operator-valued functions $L_j : \mathscr{U} \to B(\mathbf{E}, \mathbf{F})$ $(j = 1, \ldots, n)$ are weakly continuous. More generally, if $k \ge 0$ is an integer, Ψ is weakly (or strongly, or in norm) *of class C^k* if all its partial derivatives of order less than or equal to k exist at each $x \in \mathcal{U}$ (in the weak or strong or norm topology on $B(\mathbf{E}, \mathbf{F})$ respectively) and are continuous functions of x in the respective topology. These three notions are distinct, as is easily shown by examples; however, if $k \geq 1$, *Newton's formula* holds in each case, viz.

$$(5.A.2) \qquad \Psi(y) - \Psi(x) = \sum_{j=1}^{n}(y_j - x_j) \int_0^1 (\partial_j \Psi)((1 - \varrho)x + \varrho y)d\varrho$$

if x, $y \in \mathcal{U}$ and the segment $\{(1 - \varrho)x + \varrho y \mid 0 < \varrho < 1\}$ lies entirely in $\mathcal{U}$. The integral in (5.A.2) exists in the Riemann sense in any of the topologies in which Ψ is of class C^1.

Assume now that $\mathbf{F}$ is a reflexive and separable Banach space. If Ψ is weakly C^1, then the operator-valued functions $\partial_j \Psi$ $(j = 1, \ldots, n)$ are weakly continuous, hence locally bounded and strongly measurable; hence the integral in (5.A.2) exists in the Bochner sense in the strong topology. In particular, by using the Lebesgue differentiation theorem, one may then show that Ψ has partial derivatives in the strong topology almost everywhere (Lebesgue's theorem is valid for reflexive Banach space-valued functions). The same conclusion is true if Ψ is a Lipschitz function (Rademacher's theorem), see Lemma 5.A.1 below.

Let θ be a real number such that $0 \leq \theta \leq 1$. We say that Ψ is *Hölder continuous of order θ* if there is a real constant c such that

$$(5.A.3) \qquad \|\Psi(x) - \Psi(y)\|_{B(\mathbf{E}, \mathbf{F})} \leq c|x - y|^\theta \qquad \forall x, y \in \mathcal{U}.$$

If $\theta = 1$, we shall say that Ψ is *Lipschitz*, in agreement with the usual terminology. Let us observe that the notion of Hölder continuity could also be considered in the weak or strong topology. For example, we could say that Ψ is weakly Hölder continuous of order θ if for each $f \in \mathbf{E}$ and each $g \in \mathbf{F}^*$, there is a real constant $c_{f,g}$ such that $|\langle \Psi(x)f, g\rangle - \langle \Psi(y)f, g\rangle| \leq c_{f,g}|x - y|^\theta$ for all x, $y \in \mathcal{U}$. But then the uniform boundedness principle applied to the family of operators $\{|x - y|^{-\theta}[\Psi(x) - \Psi(y)] \mid x, y \in \mathcal{U}, x \neq y\}$ implies that (5.A.3) holds for some finite constant c. Thus there is no difference between weak, strong and norm Hölder continuity of order θ.

We mention the following rather deep property of Lipschitz functions:

LEMMA 5.A.1. *Let $\mathbf{E}$ and $\mathbf{F}$ be reflexive separable Banach spaces and let $\Psi : \mathcal{U} \to B(\mathbf{E}, \mathbf{F})$ be Lipschitz. Then Ψ is strongly differentiable at almost all points of $\mathcal{U}$.*

The following example shows that strong differentiability cannot be replaced by norm differentiability. Let $\mathcal{U} = \mathbb{R}$, $\mathbf{E} = \mathbf{F}$ an infinite-dimensional Hilbert space, H a self-adjoint operator in $\mathbf{F}$ and $\Psi(t) = \int_0^t e^{iH\tau}d\tau$. Clearly the function Ψ is Lipschitz and

$$\frac{\Psi(t) - \Psi(s)}{t - s} = e^{iHs}\frac{1}{t - s}\int_0^{t-s} e^{iH\tau}d\tau.$$

Thus Ψ is norm differentiable at *some* point if and only if it is norm differentiable at $t = 0$, which is the case if and only if it is norm differentiable at *all* points.

Now the strong derivative of Ψ at $t = 0$ is I. So Ψ is norm differentiable if and only if the function

$$\varphi(t) := \left\| \frac{1}{t}\Psi(t) - I \right\|_{B(\mathbf{F})} = \sup_{\lambda \in \sigma(H)} \left| \frac{1}{t}\int_0^t e^{i\lambda\tau}d\tau - 1 \right| = \sup_{\lambda \in \sigma(H)} \left| \frac{e^{i\lambda t} - 1}{i\lambda t} - 1 \right|$$

converges to zero as $t \to 0$. This is the case if and only if H is a bounded operator: $\varphi(t)$ is $O(t)$ as $t \to 0$ if $\sigma(H)$ is a bounded set, whereas $\varphi(t) \geq 1$ for each $t \neq 0$ if $\sigma(H)$ is unbounded.

LEMMA 5.A.2. *Let* $\mathbf{E}$ *and* $\mathbf{F}$ *be Banach spaces and let* $\Phi : \mathscr{U} \to B(\mathbf{E},\mathbf{F})$ *be weakly of class* C^k *for some integer* $k \geq 1$. *Then*
   (a) *for each multi-index* α *with* $|\alpha| \leq k - 1$, *the function* $\partial^\alpha\Phi : \mathscr{U} \to B(\mathbf{E},\mathbf{F})$ *is locally Lipschitz, in particular it is norm continuous,*
   (b) *if for each* α *with* $|\alpha| = k$, *the function* $\partial^\alpha\Phi$ *is strongly continuous, then* Φ *is strongly of class* C^k,
   (c) *if for each* α *with* $|\alpha| = k$, *the function* $\partial^\alpha\Phi$ *is Hölder continuous of order* θ *for some* $0 < \theta \leq 1$, *then* Φ *is of class* C^k *in norm.*

PROOF. The proof is based on Newton' formula (5.A.2).
   (a) We apply (5.A.2) with $\Psi = \partial^\alpha\Phi$. Since $|\alpha| \leq k - 1$, $\partial^\alpha\Phi$ is weakly C^1; hence, by the uniform boundedness principle, $\|\partial_j\partial^\alpha\Phi\|_{B(\mathbf{E},\mathbf{F})}$ is locally bounded for $j = 1, \ldots, n$. Thus, for each compact subset K of $\mathscr{U}$, there is a constant c_K such that

$$\|\partial^\alpha\Phi(y) - \partial^\alpha\Phi(x)\| \leq c_K|y - x| \qquad \text{for all } x, y \in K.$$

   (b) Assume in (5.A.2) that $\partial_j\Psi$ are strongly continuous operator-valued functions. Then the integral exists as a strong integral, and (5.A.2) implies that Ψ is strongly C^1 (take $x_j = y_j$ except for $j = j_0$, and use the Lebesgue dominated convergence theorem for the limit $y_{j_0} \to x_{j_0}$). If we apply this result to $\Psi = \partial^\alpha\Phi$ for $|\alpha| = k - 1$, we find that $\partial^\alpha\Phi$ are strongly C^1; then we use (a).
   (c) In this case the functions $\partial^\alpha\Phi$ are continuous in norm for $|\alpha| = k$. One may repeat the proof of (b) with strong continuity replaced by continuity in norm and strong differentiability replaced by differentiability in norm.   □

COROLLARY 5.A.3. *Let* $\mathbf{F}$ *be a Banach space and* $f : \mathscr{U} \to \mathbf{F}$ *a vector-valued function. If, for some integer* $k \geq 1$, f *is weakly of class* C^k *and* $\partial^\alpha f$ *are strongly continuous functions for each multi-index* α *with* $|\alpha| = k$, *then* f *is strongly of class* C^k.

CHAPTER 6

Unitary Representations and Regularity for Self-adjoint Operators

In this chapter we specialize some of the considerations of Chapter 5 to the case of unitary C_0-groups in a Hilbert space $\mathcal{H}$. The theory of unitary representations $W(x) = e^{iA \cdot x}$ of $\mathbb{R}^n$ is a very well understood classical subject and will not be presented here. However we mention that a n-dimensional version of Stone's theorem states that there is a unique spectral measure E on $\mathbb{R}^n$ such that $W(x) = \int_{\mathbb{R}^n} e^{ix \cdot y} E(dy)$, and this allows one to extend the functional calculus which we already have for functions in $C_{\mathrm{pol}}^\infty(\mathbb{R}^n)$ to all Borel functions $\varphi : \mathbb{R}^n \to \mathbb{C}$. The natural definition of $\varphi(A)$ for a Borel function φ is $\varphi(A) = \int_{\mathbb{R}^n} \varphi(y) E(dy)$, and one may check that for $\varphi \in C_{\mathrm{pol}}^\infty(\mathbb{R}^n)$ the two definitions lead to the same operator $\varphi(A)$.

The main developments of this chapter are contained in Section 6.2 where we present a definition of regularity (of integer or fractional order) with respect to W for an arbitrary self-adjoint operator H in $\mathcal{H}$. One of the main points of this definition is that there is no assumption concerning the invariance of the domain or form domain of H under the group W. Nevertheless, in the case $n = 1$ we are able to describe the class C^1 in terms of the commutator of H with the generator A of W (see Theorem 6.2.10). The principal result of this chapter is Theorem 6.2.5 which describes a large class of functions $\varphi : \mathbb{R} \to \mathbb{C}$ such that $\varphi(H)$ belongs to the same regularity class as H. The proof is based on an integral representation of $\varphi(H)$ in terms of the resolvent of H which we present in Section 6.1. In Section 6.3 we introduce the concept of unitary representations in a Friedrichs couple and study several of its aspects. Finally Section 6.4 contains an application of the commutator expansion formalism of Section 5.5 which has important implications in scattering theory. In an appendix we collect some remarks on the functional calculus associated to the group of automorphisms $\mathcal{W}$ in $B(\mathcal{H})$ induced by W.

We end this introduction with a collection of notations concerning Hilbert spaces and self adjoint operators that will be used throughout a large part of the remainder of the text. In the present and in subsequent chapters we shall deal with a distinguished Hilbert space which will be denoted by the symbol $\mathcal{H}$. The

W. O. Amrein et al., *C$_0$-Groups, Commutator Methods and Spectral Theory of*
N-Body Hamiltonians, Modern Birkhäuser Classics,
DOI: 10.1007/978-3-0348-0733-3_6, © Springer Basel 1996

scalar product and the norm in this distinguished Hilbert space $\mathcal{H}$ will usually be denoted simply by $\langle \cdot, \cdot \rangle$ and $|| \cdot ||$ respectively, whereas the norm $||T||_{B(\mathcal{H})}$ of an operator in $B(\mathcal{H})$ will be written simply $||T||$. We usually identify $\mathcal{H}^*$ with $\mathcal{H}$ through the Riesz isomorphism.

Then, if H is a *self-adjoint* operator in $\mathcal{H}$, we denote by $E(\cdot)$ its spectral measure. We denote by $\sigma(H)$, $\sigma_{\mathrm{p}}(H)$ and $\sigma_{\mathrm{ess}}(H)$ the *spectrum*, the *point spectrum* and the *essential spectrum* of H respectively: the point spectrum is the set of all eigenvalues of H, whereas the essential spectrum is the complement in $\sigma(H)$ of the set of all isolated points in $\sigma(H)$ which are eigenvalues of finite multiplicity. We recall that the *resolvent set* of H is the complement of $\sigma(H)$ in $\mathbb{C}$ and that $\sigma(H) \subseteq \mathbb{R}$. If z is a complex number in the resolvent set of H, we write $R(z)$ for the resolvent $(H - z)^{-1}$ of H at z. We have

$$(6.0.1) \qquad ||R(z)|| = \left[\mathrm{dist}(z, \sigma(H)) \right]^{-1} \leq \frac{1}{|\Im z|}.$$

6.1. Remarks on the Functional Calculus for Self-adjoint Operators

In this section we consider a fixed self-adjoint operator H in a Hilbert space $\mathcal{H}$ and prove an integral representation for functions $\varphi(H)$ of H in terms of its resolvent $R(z) = (H - z)^{-1}$. A formula of a somewhat similar type has been given in [HS]. We shall denote by $E(\cdot)$ the spectral measure of H, so that for any Borel function $\varphi : \mathbb{R} \to \mathbb{C}$ a normal (in general unbounded) operator is associated to H through the formula

$$\varphi(H) = \int_{\mathbb{R}} \varphi(t) E(dt).$$

For a more restricted class of functions φ, it is well known that one may write $\varphi(H)$ explicitly in terms of the resolvent of H; this fact is the content of Lemma 6.1.1 below. Unfortunately, this representation is not a simple integral representation because of the appearance of a weak limit in (6.1.1). However, we shall see that this inconvenience may be easily remedied by an integration by parts.

In what follows we let $\Re T = 2^{-1}(T + T^*)$ and $\Im T = (2i)^{-1}(T - T^*)$ be the real and imaginary part respectively of an operator $T \in B(\mathcal{H})$.

LEMMA 6.1.1. *If $\varphi : \mathbb{R} \to \mathbb{C}$ is a bounded continuous function, then*

$$(6.1.1) \qquad \varphi(H) = \underset{\varepsilon \to +0}{\mathrm{w\text{-}lim}} \, \frac{1}{\pi} \int_{\mathbb{R}} \varphi(\lambda) \Im R(\lambda + i\varepsilon) d\lambda,$$

where the integral exists in the weak sense.

PROOF. Let $f, g \in \mathcal{H}$ and let ν be the complex measure on $\mathbb{R}$ defined by $\nu(dt) = \langle f, E(d\tau)g \rangle$. Furthermore, set $P(x) = [\pi(1 + x^2)]^{-1}$ and, for $\varepsilon > 0$: $P_\varepsilon(x) = \varepsilon^{-1} P(\varepsilon^{-1} x)$. Clearly $P_\varepsilon(x) = P_\varepsilon(-x)$ and

$$(6.1.2) \qquad \int_{\mathbb{R}} P_\varepsilon(x) dx = 1.$$

Now, for $\varepsilon > 0$:

$$(6.1.3) \qquad \frac{1}{\pi}\langle f, \Im R(\lambda + i\varepsilon)g\rangle = \int_{\mathbb{R}} P_\varepsilon(\tau - \lambda)\nu(d\tau) \equiv (P_\varepsilon * \nu)(\lambda)$$

and

$$\frac{1}{\pi}\int_{\mathbb{R}}\langle f, \Im R(\lambda + i\varepsilon)g\rangle\varphi(\lambda)d\lambda = \int_{\mathbb{R}}\nu(d\tau)(P_\varepsilon * \varphi)(\tau),$$

since the double integral is absolutely convergent by (6.1.2) and the boundedness of φ. Furthermore

$$(6.1.4) \qquad |(P_\varepsilon * \varphi)(\tau)| \le ||\varphi||_{L^\infty(\mathbb{R})}\int_{\mathbb{R}} P_\varepsilon(\tau - \lambda)d\lambda = ||\varphi||_{L^\infty(\mathbb{R})} < \infty$$

and

$$(6.1.5) \qquad |\varphi(\tau) - (P_\varepsilon * \varphi)(\tau)| \le \int_{\mathbb{R}} P(x)|\varphi(\tau) - \varphi(\tau + \varepsilon x)|dx.$$

Since φ is bounded and continuous, the r.h.s. of (6.1.5) converges to zero as $\varepsilon \to 0$ by the Lebesgue dominated convergence theorem. This fact and (6.1.4) allow one to conclude, by another application of the Lebesgue dominated convergence theorem, that

$$\left|\langle f, \varphi(H)g\rangle - \frac{1}{\pi}\langle f, \int_{\mathbb{R}} d\lambda\varphi(\lambda)\Im R(\lambda + i\varepsilon)g\rangle\right| = \left|\int_{\mathbb{R}}[\varphi(\tau) - (P_\varepsilon * \varphi)(\tau)]\nu(d\tau)\right|$$

$$\le \int_{\mathbb{R}}|\varphi(\tau) - (P_\varepsilon * \varphi)(\tau)| \cdot |\nu|(d\tau) \to 0 \text{ as } \varepsilon \to +0. \quad \square$$

The basic observation allowing to transform (6.1.1) into an integral representation is the obvious identity

$$(6.1.6) \qquad \frac{d}{d\mu}R(\lambda + i\mu) = i\frac{d}{d\lambda}R(\lambda + i\mu) \qquad (\lambda, \mu \in \mathbb{R}, \mu \ne 0).$$

This expresses the holomorphy of $R(z)$ in the set $\Im z \ne 0$. It implies that for each $\varepsilon > 0$:

$$(6.1.7) \qquad R(\lambda + i\varepsilon) = R(\lambda + i) - i\int_\varepsilon^1 \frac{d}{d\lambda}R(\lambda + i\mu)d\mu.$$

One may insert the last expression into (6.1.1), interchange the order of integration and then integrate by parts. A repetition of this procedure will easily give the formula (6.1.18) below. If the function φ is sufficiently smooth and decays rapidly enough at infinity, it is easy to make the indicated steps rigorous. Our proof of the theorem below is slightly complicated because we wish to get the result for a large class of functions φ.

In order to formulate an essentially optimal result, we use the following gauge on Borel functions $\psi : \mathbb{R} \to \mathbb{C}$:

$$(6.1.8) \qquad |||\psi||| := \sup_{x \in \mathbb{R}}\int_{\mathbb{R}}|\psi(y)|\frac{dy}{\langle x - y\rangle}.$$

This gauge is translation invariant and $|||\psi||| < \infty$ if $\psi \in L^1(\mathbb{R}) + L^p(\mathbb{R})$ for some $p < \infty$. Furthermore

$$(6.1.9) \qquad ||\psi||_{L^1_{loc,unif}} \equiv \sup_{x \in \mathbb{R}} \int_x^{x+1} |\psi(y)| dy \leq \sqrt{2} \, |||\psi|||.$$

In fact the gauge $|||\cdot|||$ is only slightly stronger than $||\cdot||_{L^1_{loc,unif}}$: the assumption $|||\psi||| < \infty$ implies some logarithmic decay of $\int_x^{x+1} |\psi(y)| dy$ at infinity.

In a preliminary lemma we use the gauge $|||\cdot|||$ to estimate certain integrals related to that appearing in (6.1.1).

LEMMA 6.1.2. (a) *For each* $\mu \in (0,1]$ *one has*

$$(6.1.10) \qquad \left\| \int_{\mathbb{R}} \psi(\lambda) \Re R(\lambda + i\mu) d\lambda \right\| \leq \frac{1}{\mu^2} |||\psi|||,$$

$$(6.1.11) \qquad \left\| \int_{\mathbb{R}} \psi(\lambda) \Im R(\lambda + i\mu) d\lambda \right\| \leq \frac{3\pi}{\mu} ||\psi||_{L^1_{loc,unif}}.$$

(b) *Let* $k \in \mathbb{N}$. *Then*

$$(6.1.12) \qquad \left\| \int_0^1 \mu^k d\mu \int_{\mathbb{R}} \psi(\lambda) \Re R(\lambda + i\mu) d\lambda \right\| \leq \pi |||\psi|||,$$

$$(6.1.13) \qquad \left\| \int_0^1 \mu^k d\mu \int_{\mathbb{R}} \psi(\lambda) \Im R(\lambda + i\mu) d\lambda \right\| \leq 3\pi ||\psi||_{L^1_{loc,unif}} \qquad if \ k \geq 1.$$

The integrals appearing in (6.1.10) − (6.1.13) *exist weakly in* $B(\mathcal{H})$ *provided that the norm on the r.h.s. of the respective inequality is finite.*

REMARK 6.1.3. If $\psi \in L^1(\mathbb{R})$ and $k \geq 1$, then the integrals in (6.1.10)-(6.1.13) exist in the norm of $B(\mathcal{H})$, because $||R(\lambda \pm i\mu)|| \leq |\mu|^{-1}$. For more general functions ψ one cannot expect norm convergence of these integrals, because $||\Im R(\lambda + i\mu)|| = |\mu|^{-1}$ if λ belongs to the spectrum of H and $||\Re R(\lambda + i\mu)|| = (2|\mu|)^{-1}$ if λ and $\lambda + \mu$ (or $\lambda - \mu$) belong to the spectrum of H.

PROOF. Let f, $g \in \mathcal{H}$ and let P_ε and ν be as in the proof of Lemma 6.1.1. The following inequalities will be used:

$$(6.1.14) \qquad \frac{1}{\varrho^2 + \mu^2} \leq \frac{1}{\mu^2} \frac{1}{\varrho^2 + 1} \qquad if \ \varrho \in \mathbb{R}, \ 0 < \mu \leq 1,$$

$$(6.1.15) \qquad \arctan y^{-1} \leq \pi \langle y \rangle^{-1} \qquad if \ y > 0.$$

(6.1.15) follows from the fact that $0 \leq \arctan x \leq \min(x, \pi/2)$ if $x \geq 0$.

(i) By using (6.1.3) we get that

$$\int_{\mathbb{R}} |\psi(\lambda)\langle f, \Im R(\lambda + i\mu)g\rangle| d\lambda = \pi \int_{\mathbb{R}} |\psi(\lambda)(P_\mu * \nu)(\lambda)| d\lambda$$

$$\leq \pi \int_{\mathbb{R}} (P_\mu * |\psi|)(\tau) |\nu|(d\tau) \leq \pi ||f|| \, ||g|| \cdot \sup_{\tau \in \mathbb{R}} (P_\mu * |\psi|)(\tau).$$

Now (use (6.1.14)):

$$\pi(P_\mu * |\psi|)(\tau) = \pi \int_{\mathbb{R}} P_\mu(\lambda)|\psi(\tau + \lambda)|d\lambda = \sum_{n\in\mathbb{Z}} \int_n^{n+1} \frac{\mu}{\lambda^2 + \mu^2}|\psi(\tau + \lambda)|d\lambda$$

$$\leq \frac{2}{\mu} \sum_{n\in\mathbb{N}} \frac{1}{n^2 + 1} \int_n^{n+1} |\psi(\tau + \lambda)|d\lambda \leq \frac{2}{\mu}||\psi||_{L^1_{\text{loc,unif}}} \sum_{n\in\mathbb{N}} \frac{1}{n^2 + 1}.$$

This implies the estimate (6.1.11) and the existence of the integral in (6.1.11) in the weak sense if $\psi \in L^1_{\text{loc,unif}}$. Then (6.1.13) follows from (6.1.11).

   (ii) The proof of (6.1.10) is similar. We set

$$Q_\mu(x) = \mu^{-1}xP_\mu(x) \equiv x[\pi(\mu^2 + x^2)]^{-1}$$

and have as in (6.1.3) that

(6.1.16) $$\frac{1}{\pi}\langle f, \Re R(\lambda + i\mu)g\rangle = -(Q_\mu * \nu)(\lambda).$$

Hence

$$\int_{\mathbb{R}} |\psi(\lambda)\langle f, \Re R(\lambda + i\mu)g\rangle|d\lambda \leq \pi||f|| \cdot ||g|| \cdot \sup_{\tau\in\mathbb{R}}(|Q_\mu| * |\psi|)(\tau),$$

and (6.1.10) follows after observing that (use again (6.1.14)):

$$\pi(|Q_\mu| * |\psi|)(\tau) = \int_{\mathbb{R}} \frac{|\lambda - \tau|}{|\lambda - \tau|^2 + \mu^2}|\psi(\lambda)|d\lambda \leq \frac{1}{\mu^2}|||\psi|||.$$

For $k \geq 2$, (6.1.12) is an immediate consequence of (6.1.10). A proof of (6.1.12) valid for all $k \in \mathbb{N}$ is as follows:

$$\int_0^1 \mu^k d\mu \int_{\mathbb{R}} |\psi(\lambda)\langle f, \Re R(\lambda + i\mu)g\rangle|d\lambda$$

$$\leq \int_{\mathbb{R}} |\nu|(d\tau) \int_{\mathbb{R}} d\lambda|\psi(\lambda)| \int_0^1 d\mu \frac{|\lambda - \tau|}{|\lambda - \tau|^2 + \mu^2}$$

$$= \int_{\mathbb{R}} |\nu|(d\tau) \int_{\mathbb{R}} d\lambda|\psi(\lambda)| \arctan\frac{1}{|\lambda - \tau|} \leq ||f|| \cdot ||g|| \cdot \pi|||\psi|||,$$

where the last inequality is obtained by using (6.1.15). $\square$

THEOREM 6.1.4. *Let H be a self-adjoint operator in a Hilbert space $\mathscr{H}$ and $R(z) = (H - z)^{-1}$ for $z \in \mathbb{C} \setminus \sigma(H)$.*

   (a) *If $\varphi : \mathbb{R} \to \mathbb{C}$ is a bounded, absolutely continuous function the derivative of which satisfies $|||\varphi'||| < \infty$, then*

(6.1.17)

$$\varphi(H) = \frac{1}{\pi} \int_{\mathbb{R}} \varphi(\lambda)\Im R(\lambda + i)d\lambda + \frac{1}{\pi} \int_0^1 d\mu \int_{\mathbb{R}} d\lambda\varphi'(\lambda)\Re R(\lambda + i\mu),$$

where the integrals exist in the weak operator topology.

(b) *More generally, let $r \geq 1$ be an integer and let $\varphi : \mathbb{R} \to \mathbb{C}$ be a bounded function of class C^{r-1} such that $\varphi^{(r-1)}$ is absolutely continuous and $|||\varphi^{(k)}||| < \infty$ for $1 \leq k \leq r$. Then*

$$(6.1.18) \qquad \varphi(H) = \sum_{k=0}^{r-1} \frac{1}{\pi k!} \int_{\mathbb{R}} \varphi^{(k)}(\lambda) \Im[i^k R(\lambda + i)] d\lambda$$

$$+ \frac{1}{\pi(r-1)!} \int_0^1 \mu^{r-1} d\mu \int_{\mathbb{R}} d\lambda \varphi^{(r)}(\lambda) \Im[i^r R(\lambda + i\mu)],$$

where all the integrals exist in the weak sense. If $r \geq 2$ and $\varphi^{(k)} \in L^1(\mathbb{R})$ for $k = 0, 1, \ldots, r$, then all the integrals exist in the norm of $B(\mathcal{H})$.

PROOF. (i) For $k = 0, 1, 2, \ldots$ and $z \in \mathbb{C} \setminus \mathbb{R}$ we set $S_k(z) = \Im(i^k R(z))$. Then S_k is a function of class C^∞, and (6.1.7) implies that for $\varepsilon > 0$:

$$(6.1.19) \qquad S_k(\lambda + i\varepsilon) = S_k(\lambda + i) - \int_\varepsilon^1 \frac{d}{d\lambda} S_{k+1}(\lambda + i\mu) d\mu.$$

We claim the following: let $\psi \in BC(\mathbb{R})$ be an absolutely continuous function with $|||\psi'||| < \infty$ and let $k \in \mathbb{N}$; if k is odd, also assume that $|||\psi||| < \infty$. Then

$$(6.1.20)$$
$$\int_{\mathbb{R}} \psi(\lambda) S_k(\lambda + i\varepsilon) d\lambda = \int_{\mathbb{R}} \psi(\lambda) S_k(\lambda + i) d\lambda + \int_\varepsilon^1 d\mu \int_{\mathbb{R}} d\lambda \psi'(\lambda) S_{k+1}(\lambda + i\mu).$$

If ψ is a function of compact support (which is the most important case for applications), (6.1.20) is an immediate consequence of (6.1.19). For the general case, we have to check the convergence of the integrals and then to justify the integration by parts.

We observe that $\psi \in L^1_{\text{loc,unif}}$. By taking into account (6.1.9), one sees from Lemma 6.1.2 that all integrals in (6.1.20) exist weakly. Next, let $0 < N < \infty$. Then (6.1.19) implies that

$$(6.1.21) \qquad \int_{-N}^N \psi(\lambda) S_k(\lambda + i\varepsilon) d\lambda = \int_{-N}^N \psi(\lambda) S_k(\lambda + i) d\lambda$$

$$- \int_{-N}^N d\lambda \int_\varepsilon^1 d\mu \psi(\lambda) \frac{d}{d\lambda} S_{k+1}(\lambda + i\mu),$$

and (6.1.20) will follow if we can show that the second term on the r.h.s. converges weakly to the last term in (6.1.20) as $N \to \infty$. For this we change the order of integration and then integrate by parts in the second term on the r.h.s. of (6.1.21). So this term is equal to

$$(6.1.22) \qquad \int_\varepsilon^1 d\mu \Big[\psi(-N) S_{k+1}(-N + i\mu) - \psi(N) S_{k+1}(N + i\mu) +$$

$$+ \int_{-N}^N d\lambda \psi'(\lambda) S_{k+1}(\lambda + i\mu) \Big].$$

But $|\psi(\pm N)| \leq \text{const.}$, $||S_{k+1}(\pm N + i\mu)|| \leq \mu^{-1} \leq \varepsilon^{-1}$ and $S_{k+1}(\pm N + i\mu) \to 0$ strongly as $N \to \infty$. Moreover, by Lemma 6.1.2 (a):

$$\int_{\mathbb{R}} |\psi'(\lambda)\langle f, S_{k+1}(\lambda + i\mu)g\rangle|d\lambda \leq \frac{6\pi}{\varepsilon^2}||f|| \cdot ||g|| \cdot |||\psi'||| \qquad \forall \mu \in [\varepsilon, 1].$$

By the Lebesgue dominated convergence theorem, the expression (6.1.22) is weakly convergent to the last term in (6.1.20) as $N \to \infty$.

(ii) We now prove (6.1.17). By (6.1.1) and (6.1.20) with $k = 0$ we have

$$\pi\varphi(H) = \underset{\varepsilon \to +0}{\text{w-lim}} \int_{\mathbb{R}} \varphi(\lambda)S_0(\lambda + i\varepsilon)d\lambda$$

$$= \int_{\mathbb{R}} \varphi(\lambda)S_0(\lambda + i)d\lambda + \underset{\varepsilon \to +0}{\text{w-lim}} \int_{\varepsilon}^{1} d\mu \int_{\mathbb{R}} d\lambda \varphi'(\lambda)S_1(\lambda + i\mu).$$

By Lemma 6.1.2, the function $(\lambda, \mu) \mapsto \varphi'(\lambda)S_1(\lambda + i\mu) \in B(\mathscr{H})$ is weakly integrable on $\mathbb{R} \times [0, 1]$, hence we may set $\varepsilon = 0$ in the last term.

(iii) We finally prove (6.1.18) by induction. We may assume that $r \geq 2$.

We first show that the assumptions made on φ imply that $\varphi \in BC^{r-1}(\mathbb{R})$. For this we use the identity

$$(6.1.23) \qquad \psi'(x) = \psi(x + 1) - \psi(x) + \int_{x}^{x+1} (y - x - 1)\psi''(y)dy,$$

which shows that

$$(6.1.24) \qquad ||\psi'||_{L^\infty(\mathbb{R})} \leq 2||\psi||_{L^\infty(\mathbb{R})} + ||\psi''||_{L^1_{\text{loc,unif}}}.$$

By taking successively $\psi = \varphi, \psi = \varphi', \ldots, \psi = \varphi^{(r-2)}$, one obtains from (6.1.23), (6.1.24) and (6.1.9) that $\varphi \in BC^{r-1}(\mathbb{R})$.

Now assume that the following equation holds for some $n \in \{1, 2, \ldots, r - 1\}$:

$$(6.1.25) \qquad \pi\varphi(H) = \sum_{k=0}^{n-1} \frac{1}{k!} \int_{\mathbb{R}} \varphi^{(k)}(\lambda)S_k(\lambda + i)d\lambda +$$

$$+ \frac{1}{(n-1)!} \int_0^1 \mu^{n-1}d\mu \int_{\mathbb{R}} d\lambda \varphi^{(n)}(\lambda)S_n(\lambda + i\mu).$$

The validity of (6.1.25) for $n = 1$ has been shown in (ii); we now prove that (6.1.25) holds with n replaced by $n + 1$. For this we rewrite the last term in (6.1.25) as follows (use (6.1.20) with $\psi = \varphi^{(n)}$ and $k = n$):

$$\frac{1}{(n-1)!} \int_0^1 \mu^{n-1}d\mu \int_{\mathbb{R}} d\lambda \varphi^{(n)}(\lambda)S_n(\lambda + i) +$$

$$+ \frac{1}{(n-1)!} \int_0^1 \mu^{n-1}d\mu \int_\mu^1 d\tau \int_{\mathbb{R}} d\lambda \varphi^{(n+1)}(\lambda)S_{n+1}(\lambda + i\tau).$$

After interchanging the first two integrals in the last term (which is justified because the total integral exists in the weak sense), one obtains the validity of (6.1.25) with n replaced by $n + 1$. $\square$

If φ is a real function, the equation (6.1.18) may be written as

(6.1.26)
$$\pi\varphi(H) = \Im \int_{\mathbb{R}} d\lambda \left\{ \left[\sum_{k=0}^{r-1} \frac{i^k}{k!} \varphi^{(k)}(\lambda) \right] R(\lambda + i) + \frac{i^r}{r!} \varphi^{(r)}(\lambda) \int_0^1 R(\lambda + i\mu) d\mu^r \right\}$$

$$= \Im \int_{\mathbb{R}} d\lambda \int_0^1 d\mu^r \left\{ \sum_{k=0}^{r-1} \frac{i^k}{k!} \varphi^{(k)}(\lambda) R(\lambda + i) + \frac{i^r}{r!} \varphi^{(r)}(\lambda) R(\lambda + i\mu) \right\}.$$

In the square bracket one may recognize a sort of formal Taylor expansion of the function φ at the point λ (if φ were holomorphic in a disc of center λ and radius larger than 1, the expression in that bracket would indeed be the beginning of the Taylor series for $\varphi(\lambda + i)$ at the point λ). This observation establishes the connection with the representation of $\varphi(H)$ given in Proposition 7.2 of [HS], where a quasi-analytic extension of φ into the complex plane is used (for φ in $C_0^\infty(\mathbb{R})$).

6.2. Regularity of Self-adjoint Operators with respect to Unitary C_0-Groups

6.2.1. Let $\mathscr{H}$ be a Hilbert space equipped with a strongly continuous unitary representation $W(x) = e^{iA \cdot x}$ of $\mathbb{R}^n$. Here $A = (A_1, \ldots, A_n)$ is a n-tuple of commuting self-adjoint operators in $\mathscr{H}$. The notion of regularity of a self-adjoint operator H in $\mathscr{H}$ with respect to W that we shall introduce below is justified by the following observation:

LEMMA 6.2.1. *Let $k \in \mathbb{N}$, $0 \le s < \infty$, $1 \le p \le \infty$ and let H be a self-adjoint operator in $\mathscr{H}$. Assume that there is some number z_0 in the resolvent set of H such that $(H - z_0)^{-1}$ belongs to $C_u^k(A; \mathscr{H})$ or to $C^k(A; \mathscr{H})$ or to $\mathscr{C}^{s,p}(A; \mathscr{H})$. Then $(H - z)^{-1}$ belongs to the same class as $(H - z_0)^{-1}$ for each $z \in \mathbb{C} \setminus \sigma(H)$. If H is bounded, then it also belongs to the same class.*

PROOF. If $|z - z_0| < \mathrm{dist}(z_0, \sigma(H))$, then $I - (z - z_0)(H - z_0)^{-1}$ is a bijection of $\mathscr{H}$ onto $\mathscr{H}$, hence its inverse belongs to the same class as $(H - z_0)^{-1}$ by Proposition 5.1.6 or Proposition 5.2.4. It follows that the operator

(6.2.1) $(H - z)^{-1} \equiv (H - z_0)^{-1}[I - (z - z_0)(H - z_0)^{-1}]^{-1}$

belongs to the same class as $(H - z_0)^{-1}$ for all $z \in \mathbb{C} \setminus \sigma(H)$ with $|z - z_0| < \mathrm{dist}(z_0, \sigma(H))$ (use Proposition 5.1.5 or Proposition 5.2.3). By repeating this reasoning as in the usual analytic continuation argument, one obtains that $(H - z)^{-1}$ belongs to the same class as $(H - z_0)^{-1}$ for all z in the connected component of the resolvent set of H containing the point z_0. For the remaining points we use $(H - \bar{z})^{-1} = ((H - z)^{-1})^*$. $\square$

DEFINITION 6.2.2. Let $\mathscr{H}$ be a Hilbert space, $W(x) = e^{iA \cdot x}$ a unitary representation of $\mathbb{R}^n$ in $\mathscr{H}$ and H a self-adjoint operator in $\mathscr{H}$. Let $k \in \mathbb{N}$, $0 \le s < \infty$ and $1 \le p \le \infty$. We say that H *is of class* $C^k(A)$ (or of class $C_u^k(A)$ or $\mathscr{C}^{s,p}(A)$)

if $(H - z)^{-1}$ is of class $C^k(A; \mathcal{H})$ (or of class $C_u^k(A; \mathcal{H})$ or $\mathscr{C}^{s,p}(A; \mathcal{H})$ respectively) for some (and hence for all) $z \in \mathbb{C} \setminus \sigma(H)$. If this is the case, we write $H \in C^k(A)$ (or $H \in C_u^k(A)$ or $H \in \mathscr{C}^{s,p}(A)$ respectively).

If H is of class $C^k(A)$ (or of class $C_u^k(A)$ or $\mathscr{C}^{s,p}(A)$), then Propositions 5.1.5 and 5.2.3 imply that all operators belonging to the algebra generated by the resolvent family $\{(H - z)^{-1}\}_{z \notin \sigma(H)}$ are of the same class. The operators of this algebra are functions of H. By using the representation theorem from Section 6.1 we shall now prove that $\varphi(H)$ is of the same class as H for a considerably larger family of functions φ.

For the proof of the next theorem it is convenient to use a specification of the classes $C^k(A; \mathcal{H})$ and $\mathscr{C}^{s,p}(A; \mathcal{H})$ in terms of commutators. For this we introduce the following notation:

$$(6.2.2) \qquad A_x = \frac{e^{iA \cdot x} - I}{i|x|} \qquad \text{if } x \in \mathbb{R}^n \setminus \{0\}.$$

If $\omega \in \mathbb{R}^n$ is a unit vector, then $A_{\varepsilon\omega}$ converges strongly to $\omega \cdot A = \sum_{j=1}^n \omega_j A_j$ on the domain of $\omega \cdot A$ as $\varepsilon \to 0$. Let us now consider the following linear operators acting in $B(\mathcal{H})$ (the notations are consistent with those of Chapter 5, see (5.3.1), (5.0.1) and (5.1.10)):

$$(6.2.3) \qquad \mathscr{W}''(x)[T] = e^{iA \cdot x}T, \qquad \mathscr{W}'(x)[T] = Te^{iA \cdot x},$$

$$(6.2.4) \qquad \mathscr{A}_x[T] = [T, A_x] = TA_x - A_xT = -\operatorname{ad}_{A_x}(T).$$

Clearly $\mathscr{W}'(x)$, $\mathscr{W}''(x)$ and $\mathscr{A}_x$ are pairwise commuting, $\mathscr{W}'(x)$ and $\mathscr{W}''(x)$ are invertible, and one has

$$(6.2.5)$$
$$\mathscr{W}(x) = \mathscr{W}'(x)\mathscr{W}''(x)^{-1}, \qquad \mathscr{A}_x = \frac{1}{i|x|}[\mathscr{W}'(x) - \mathscr{W}''(x)] = \mathscr{W}''(x)\frac{\mathscr{W}(x) - I}{i|x|}.$$

Thus, if $m \in \mathbb{N}$, then

$$(6.2.6) \qquad \left[\frac{\mathscr{W}(x) - I}{i|x|}\right]^m = \mathscr{W}''(-mx)\mathscr{A}_x^m.$$

Since $\mathscr{W}''(x)$ is an isometric operator in $B(\mathcal{H})$, the following results are obvious (take into account Theorem 5.1.3 and the identity (5.1.14) for (a) and the definition (5.2.1) for (b)):

LEMMA 6.2.3. *Let $k \geq 1$ be an integer, $0 \leq s < \infty$, $1 \leq p \leq \infty$ and $T \in B(\mathcal{H})$. Then*

(a) *T is of class $C^k(A; \mathcal{H})$ if and only if $\liminf_{\varepsilon \to +0} \| \operatorname{ad}_{A_{\varepsilon\omega}}^k(T)\| < \infty$ for each $\omega \in \mathbb{R}^n$ with $|\omega| = 1$. If this is the case, then $\operatorname{ad}_{A_{\varepsilon\omega}}^k(T)$ converges strongly in $B(\mathcal{H})$ to $\operatorname{ad}_{\omega \cdot A}^k(T)$ as $\varepsilon \to +0$, and $\| \operatorname{ad}_{A_x}^k(T)\| \leq c$ for a constant $c < \infty$ and all $x \in \mathbb{R}^n$.*

(b) *T is of class $\mathscr{C}^{s,p}(A; \mathcal{H})$ if and only if for some integer $\ell > s$ one has*

$$\left[\int_{|x| \leq 1} \left\| |x|^{\ell-s} \operatorname{ad}_{A_x}^\ell(T)\right\|^p \frac{dx}{|x|^n}\right]^{1/p} < \infty.$$

REMARK 6.2.4. The property that $T \in C_u^k(A; \mathcal{H})$ is *not* equivalent to the existence of $\lim_{\varepsilon \to +0} \text{ad}_{A_{\varepsilon\omega}}^k(T)$ in the norm of $B(\mathcal{H})$, as can be seen from the following example. Let $\mathcal{H} = L^2(\mathbb{R})$, let $A = P = -i\frac{d}{dx}$ be the generator of the translation group and $T = \varphi(Q)$ the operator of multiplication by a function $\varphi \in C_0^1(\mathbb{R})$. Then $T \in C_u^1(P; \mathcal{H})$ and $i[P, T] = \varphi'(Q)$. Furthermore

$$\left[\frac{e^{iPx} - I}{x}, T\right] - i[P, T] = -e^{iPx}\left\{\frac{e^{-iPx}Te^{iPx} - T}{x} + [iP, T]\right\} - (I - e^{iPx})[iP, T].$$

Since the first term on the r.h.s. is norm convergent to zero, the l.h.s. converges in norm to zero if and only if $||[e^{iPx} - I]\varphi'(Q)|| \to 0$ as $x \to 0$. This cannot hold if $\varphi' \neq 0$ because it would imply that $\varphi'(Q)$ is a compact operator (by the Fréchet-Kolmogorov compactness criterion, cf. §X.1 in [Y]).

THEOREM 6.2.5. *Let $\mathcal{H}$ be a Hilbert space equipped with a unitary representation $\{e^{iA \cdot x}\}_{x \in \mathbb{R}^n}$ of $\mathbb{R}^n$. Let H be a self-adjoint operator in $\mathcal{H}$ of class $C^k(A)$ for some integer $k \geq 1$ or of class $\mathscr{C}^{s,p}(A)$ for some $s \in [0, \infty)$ and $p \in [1, \infty]$. Let $\varphi : \mathbb{R}^n \to \mathbb{C}$ be a function of class $C^{\ell+1}(\mathbb{R})$, where $\ell = k$ in the first case and ℓ is the smallest integer in (s, ∞) in the second case, and assume that $|\varphi^{(j)}(\lambda)| \leq c\langle\lambda\rangle^{-2\ell}$ for $0 \leq j \leq \ell + 1$. Then $\varphi(H)$ is of class $C^k(A; \mathcal{H})$ in the first case and of class $\mathscr{C}^{s,p}(A; \mathcal{H})$ in the second case.*

PROOF. (i) We first estimate $\text{ad}_{A_x}^\ell(R(z))$, where ℓ is the integer mentioned in the statement of the theorem and $R(z) = (H - z)^{-1}$. We fix a number $z_0 \in \mathbb{C}$ with $\Im z_0 \neq 0$. The basic identity is as follows (the calculation may be interpreted in the form sense):

$$(6.2.7) \quad [A_x, R(z)] = R(z)[H - z, A_x]R(z)$$
$$= R(z)(H - z_0)R(z_0)[H - z_0, A_x]R(z_0)(H - z_0)R(z)$$
$$= \{I + (z - z_0)R(z)\}[A_x, R(z_0)]\{I + (z - z_0)R(z)\}.$$

It is now easy to show by induction that for each decomposition $\ell = \ell_1 + \ell_2 + \cdots + \ell_m$ of ℓ into a sum of integers ℓ_j with $\ell_j \geq 1$, there is an integer $c(\ell_1, \ldots, \ell_m)$ such that

$$(6.2.8)$$
$$\text{ad}_{A_x}^\ell(R(z)) = \sum_{\substack{\ell = \ell_1 + \cdots + \ell_m \\ \ell_j \geq 1}} c(\ell_1, \ldots, \ell_m)(z - z_0)^{m-1}\{I + (z - z_0)R(z)\} \cdot$$
$$\cdot \text{ad}_{A_x}^{\ell_1}(R(z_0))\{I + (z - z_0)R(z)\} \cdot \ldots \cdot \text{ad}_{A_x}^{\ell_m}(R(z_0))\{I + (z - z_0)R(z)\}.$$

So we have for any $f \in \mathcal{H}$:

$$(6.2.9) \quad |\langle f, \text{ad}_{A_x}^\ell(R(z))f\rangle| \leq \sum_{\substack{\ell = \ell_1 + \cdots + \ell_m \\ \ell_j \geq 1}} |c(\ell_1, \ldots, \ell_m)| \, |z - z_0|^{m-1} \cdot$$

$$\cdot ||I + (z - z_0)R(z)||^{m-1} \cdot ||[I + (z - z_0)R(z)]f||^2 \cdot \prod_{j=1}^m || \, \text{ad}_{A_x}^{\ell_j}(R(z_0))||.$$

Only the last factor in (6.2.9) is dependent on x. If $\ell = 1$, this factor is just $\|\operatorname{ad}_{A_x}(R(z_0))\|$. If $\ell \geq 2$ and $H \in C^\ell(A)$, then there is a constant c_ℓ such that $\|\operatorname{ad}_{A_x}^{\ell_j}(R(z_0))\| \leq c_\ell$ for each $x \in \mathbb{R}^n$ and $\ell_j \leq \ell$ (see Lemma 6.2.3 (a).). If $\ell \geq 2$ and $H \in \mathscr{C}^{s,p}(A)$ for some $s \in [\ell-1, \ell)$, then $H \in C^{\ell-2}(A)$ (in fact $H \in C^{\ell-1}(A)$ if $s \neq \ell - 1$; see (5.2.7) and (5.2.10)); so $\|\operatorname{ad}_{A_x}^{\ell_j}(R(z_0))\| \leq c_{\ell-2}$ if $\ell_j \leq \ell - 2$. In all cases the last factor in (6.2.9) is bounded by a constant multiple of the quantity $\mathscr{J}_x^{(\ell)}$ defined as

$$(6.2.10) \quad \mathscr{J}_x^{(\ell)} = 1 + \|\operatorname{ad}_{A_x}^\ell(R(z_0))\| + \|\operatorname{ad}_{A_x}^{\ell-1}(R(z_0))\| \cdot \|\operatorname{ad}_{A_x}(R(z_0))\|.$$

Thus for each $\ell \geq 1$ there is a constant C_ℓ independent of z, x and f such that

$$(6.2.11) \quad |\langle f, \operatorname{ad}_{A_x}^\ell(R(z))f\rangle| \leq C_\ell \frac{\langle z\rangle^{2(\ell-1)}}{|\Im z|^{\ell-1}} \mathscr{J}_x^{(\ell)}[\|f\|^2 + \langle z\rangle^2\|R(z)f\|^2].$$

(ii) We now use (6.2.11) to estimate $\|\operatorname{ad}_{A_x}^\ell(\varphi(H))\|$. Without loss of generality we may assume φ to be real. We set

$$S_x^{(j)} = \int \varphi^{(j)}(\lambda)\operatorname{ad}_{A_x}^\ell(R(\lambda+i))d\lambda \qquad \text{if } 0 \leq j \leq \ell$$

and

$$T_x(\mu) = \int \varphi^{(\ell+1)}(\lambda)\operatorname{ad}_{A_x}^\ell(R(\lambda+i\mu))d\lambda \qquad \text{for } \mu > 0.$$

Then, by (6.1.18), $\operatorname{ad}_{A_x}^\ell(\varphi(H))$ is a finite linear combination (with coefficients independent of x) of $S_x^{(0)}, \ldots, S_x^{(\ell)}$ and $\int_0^1 T_x(\mu)\mu^\ell d\mu$ and of their adjoints. To estimate these operators, we use the identity

$$(6.2.12) \quad \int_{-\infty}^\infty \|R(\lambda+i\mu)f\|^2 d\lambda = \frac{\pi}{|\mu|}\|f\|^2$$

and obtain that, for some constant κ_ℓ (independent of x and f):

$$|\langle f, S_x^{(j)})f\rangle| = |\langle f, S_x^{(j)*}f\rangle| \leq \kappa_\ell \mathscr{J}_x^{(\ell)} \sup_{\lambda\in\mathbb{R}}\langle\lambda\rangle^{2\ell}|\varphi^{(j)}(\lambda)| \cdot \|f\|^2$$

and

$$\left|\langle f, \int_0^1 T_x(\mu)\mu^\ell d\mu f\rangle\right| \leq \kappa_\ell \mathscr{J}_x^{(\ell)} \sup_{\lambda\in\mathbb{R}}\langle\lambda\rangle^{2\ell}|\varphi^{(\ell+1)}(\lambda)| \cdot \|f\|^2.$$

Consequently (by the polarization identity) there is a constant d_ℓ such that for all $x \in \mathbb{R}^n$:

$$(6.2.13) \quad \|\operatorname{ad}_{A_x}^\ell(\varphi(H))\| \leq d_\ell \mathscr{J}_x^{(\ell)} \sup_{\lambda\in\mathbb{R}}\langle\lambda\rangle^{2\ell} \sum_{j=0}^{\ell+1}|\varphi^{(j)}(\lambda)|.$$

If $H \in C^\ell(A)$, then (as already said) the r.h.s. is bounded by a constant that is independent of x, hence $\varphi(H) \in C^\ell(A; \mathscr{H})$ by Lemma 6.2.3 (a). If $H \in \mathscr{C}^{s,p}(A)$, then $\varphi(A) \in \mathscr{C}^{s,p}(A; \mathscr{H})$ follows if we can show that $x \mapsto |x|^{\ell-s}\mathscr{J}_x^{(\ell)}$ belongs to L^p on the unit ball of $\mathbb{R}^n$ with respect to the measure $|x|^{-n}dx$ (see Lemma 6.2.3

(b)). For the first two terms in $\mathscr{J}_x^{(\ell)}$ this is evident (take into account Lemma 6.2.3 (b)). So it remains to check that

$$(6.2.14) \quad \left[\int_{|x|\le 1}\Big[|x|^{\ell-s}||\,\mathrm{ad}_{A_x}^{\ell-1}(R(z_0))||\cdot||\,\mathrm{ad}_{A_x}(R(z_0))||\Big]^p\frac{dx}{|x|^n}\right]^{1/p} < \infty.$$

We distinguish three cases. If $s < 1$, then $\ell = 1$ and $\mathrm{ad}_{A_x}^{\ell-1}(R(z_0)) = R(z_0)$; so (6.2.14) holds by Lemma 6.2.3 (b). If $s = 1$, hence $\ell = 2$, we write (6.2.14) as

$$\left[\int_{|x|\le 1}\Big|\Big|\,|x|^{1/2}\,\mathrm{ad}_{A_x}(R(z_0))\Big|\Big|^{2p}\frac{dx}{|x|^n}\right]^{2/2p} < \infty;$$

this is equivalent to $R(z_0) \in \mathscr{C}^{1/2,2p}(A;\mathscr{H})$, and this inclusion holds by (5.2.7). Finally, if $s > 1$, then $R(z_0) \in C^1(A;\mathscr{H})$, so that $x \mapsto ||\,\mathrm{ad}_{A_x}(R(z_0))||$ is bounded; by writing $|x|^{\ell-s} = |x|^{\ell-1-(s-1)}$ and observing that $\ell-1 > s-1$, one obtains (6.2.14) from Lemma 6.2.3 (b) because $R(z_0) \in \mathscr{C}^{s-1,p}(A;\mathscr{H})$. □

For later reference we point out the following inequality which follows from (5.2.1), (6.2.11) and the last paragraph of the preceding proof: if $H \in \mathscr{C}^{s,p}(A)$ and $\ell = [s] + 1$, then there is a constant $c < \infty$ such that for all $z \in \mathbb{C} \setminus \mathbb{R}$:

$$(6.2.15) \quad\quad ||R(z)||_{\mathscr{C}^{s,p}} \le c\langle z\rangle^{2\ell}|\Im z|^{-\ell-1}.$$

COROLLARY 6.2.6. (a) *Let H be a self-adjoint operator of class $C^k(A)$ for some integer $k \ge 1$ and let $\varphi \in C^{k+1}(\mathbb{R})$ be such that $\sum_{j=0}^{k+1}|\varphi^{(j)}(\lambda)| \le c\langle\lambda\rangle^{-2k}$. Then $\varphi(H) \in C^k(A;\mathscr{H})$ and for each multi-index α with $|\alpha| \le k$ one has*

$$(6.2.16)\ \ i^{|\alpha|}\,\mathrm{ad}_A^\alpha(\varphi(H)) = \sum_{\ell=0}^k \frac{1}{\pi\ell!}\int_{\mathbb{R}}\varphi^{(\ell)}(\lambda)\Im[i^{\ell+|\alpha|}\,\mathrm{ad}_A^\alpha(R(\lambda+i))]d\lambda$$

$$+\frac{1}{\pi k!}\int_0^1\mu^k d\mu\int_{\mathbb{R}}d\lambda\varphi^{(k+1)}(\lambda)\Im[i^{k+|\alpha|+1}\,\mathrm{ad}_A^\alpha(R(\lambda+i\mu))].$$

The integrands satisfy the following bound: there is a constant $c < \infty$ such that for all $z \in \mathbb{C} \setminus \mathbb{R}$, all $f \in \mathscr{H}$ and all $|\alpha| \le k$, $\alpha \ne 0$:

$$(6.2.17)\quad\quad |\langle f,\mathrm{ad}_A^\alpha(R(z))f\rangle| \le c\frac{\langle z\rangle^{2(|\alpha|-1)}}{|\Im z|^{|\alpha|-1}}(||f||^2 + \langle z\rangle^2||R(z)f||^2).$$

(b) *If $\varphi \in C^{k+2}(\mathbb{R})$ with $\int_{\mathbb{R}}\langle\lambda\rangle^{2k}|\varphi^{(j)}(\lambda)|d\lambda < \infty$ for $0 \le j \le k+2$ and $H \in C_u^k(A)$, then $\varphi(H) \in C_u^k(A;\mathscr{H})$.*

PROOF. (a) We first show that there is a finite set Ω of unit vectors in $\mathbb{R}^n$ and, for each multi-index α with $|\alpha| \le k$, a set of constants $\{c(\alpha,\omega) \mid \omega \in \Omega\}$ such that for all $T \in C^k(A)$:

$$(6.2.18)\quad \mathrm{ad}_A^\alpha(T) = \sum_{\omega\in\Omega}c(\alpha,\omega)\,\mathrm{ad}_{\omega\cdot A}^{|\alpha|}(T) = \sum_{\omega\in\Omega}c(\alpha,\omega)\lim_{\varepsilon\to+0}\mathrm{ad}_{A_{\varepsilon\omega}}^{|\alpha|}(T),$$

where the limits are strong limits in $B(\mathscr{H})$. The second identity follows from Lemma 6.2.3 (a). For the first one we observe that for $m \in \mathbb{N}$ and $x \in \mathbb{R}^n$:

$$\operatorname{ad}_{x \cdot A}^m = \left(\sum_{j=1}^{n} x_j \operatorname{ad}_{A_j}\right)^m = \sum_{|\alpha|=m} \frac{m!}{\alpha!} x^\alpha \operatorname{ad}_A^\alpha,$$

and it then suffices to apply Proposition 1.1.2 and to observe that $\operatorname{ad}_{y \cdot A} = |y| \operatorname{ad}_{\omega \cdot A}$ if $\omega = |y|^{-1} y$.

Now (6.2.17) follows easily from (6.2.10)-(6.2.11) by using (6.2.18) and Lemma 6.2.3 (a). Similarly (6.2.16) can be obtained from (6.1.18) and (6.2.18) (apply $\operatorname{ad}_{A_{\varepsilon\omega}}^{|\alpha|}$ to both sides of (6.1.18) and then use the dominated convergence theorem in the weak integrals by taking into account (6.2.12)).

(b) Since $\varphi(H) \in C^k(A; \mathscr{H})$ by (a), it is enough to verify that the function $x \mapsto e^{-iA \cdot x} \operatorname{ad}_A^\alpha(\varphi(H)) e^{iA \cdot x}$ is norm continuous in $B(\mathscr{H})$ for $|\alpha| = k$. This can be obtained from the expression (6.2.16) for $\operatorname{ad}_A^\alpha(\varphi(H))$, with k replaced by $k+1$: the function $x \mapsto e^{-iA \cdot x} \operatorname{ad}_A^\alpha(R(z)) e^{iA \cdot x}$ is norm continuous by hypothesis and $\|e^{-iA \cdot x} \operatorname{ad}_A^\alpha(R(z)) e^{iA \cdot x}\| \leq c \langle z \rangle^{|2\alpha|} \cdot |\Im z|^{-|\alpha|-1}$ by (6.2.17), which allows the application of the dominated convergence theorem to the integrals on the r.h.s. of (6.2.16) (with k replaced by $k+1$). $\square$

We now comment the regularity assumptions made on φ in Theorem 6.2.5. The decay assumptions on φ can be relaxed by imposing stronger regularity of H with respect to A. If for example the domain $D(H)$ of H is invariant under $e^{iA \cdot x}$ and $H \in C^k(A; D(H), \mathscr{H})$, where $D(H)$ is equipped with the graph norm, then by using $[A_x, R(z)] = R(z)[H, A_x]R(z)$ in place of (6.2.7) and, more generally, (5.1.19) in place of (6.2.8), one can show that H is of class $C^k(A)$ and $\varphi(H) \in C^k(A; \mathscr{H})$ if $\varphi \in C^k(\mathbb{R})$, $\varphi^{(k)}$ is absolutely continuous and $\int_{\mathbb{R}} \langle \lambda \rangle^k |\varphi^{(j)}(\lambda)| d\lambda < \infty$ for $0 \leq j \leq k+1$.

As regards the local smoothness conditions on φ, let us first consider an example. We denote by $BC_u^k(\mathbb{R}^n)$ the set of all functions $\varphi : \mathbb{R}^n \to \mathbb{C}$ such that φ and its derivatives of order $\leq k$ are bounded and uniformly continuous.

EXAMPLE 6.2.7. Let $\mathscr{H} = L^2(\mathbb{R})$, $A \equiv P = -i\frac{d}{dx}$ and $H \equiv Q$ (the operator of multiplication by the variable x). We have $\|\varphi(Q)\|_{B(\mathscr{H})} = \|\varphi\|_{L^\infty(\mathbb{R})}$ and $e^{-iA \cdot x} \varphi(Q) e^{iA \cdot x} = \varphi(Q - x)$. Thus $\varphi(Q) \in \mathscr{C}^{s,p}(P; \mathscr{H})$ if and only if φ belongs to the Besov space $B_\infty^{s,p}(\mathbb{R})$. Moreover it is easy to show that $\varphi(Q) \in C_u^k(P; \mathscr{H})$ means $\varphi \in BC_u^k(\mathbb{R})$, whereas $\varphi(Q) \in C^k(P; \mathscr{H})$ is equivalent to the conditions that $\varphi \in BC^{k-1}(\mathbb{R})$ and that $\varphi^{(k-1)}$ is Lipschitz. In particular one has $Q \in C_u^k(P)$ for each $k \in \mathbb{N}$.

Let us denote by Φ^k the set of functions φ such that $H \in C^k(A)$ implies $\varphi(H) \in C^k(A; \mathscr{H})$ for all couples of self-adjoint operators A, H and by Φ_u^k the set of all φ such that $H \in C_u^k(A)$ implies $\varphi(H) \in C_u^k(A; \mathscr{H})$ for all couples A, H. The preceding example shows that $\Phi_u^1 \subset BC_u^1(\mathbb{R})$. On the other hand Theorem 6.2.5 implies that $C_0^2(\mathbb{R}) \subset \Phi^1$. It is rather easy to prove that $C_0^{1+\varepsilon}(\mathbb{R}) \subset \Phi^1$ for any $\varepsilon > 0$. For $\varepsilon = 0$ this inclusion is not true, since McIntosh in [Mc] constructed self-adjoint operators A, H and a function φ of class C^1 such that H and $[A, H]$ are bounded but $[A, \varphi(H)]$ is not.

An interesting class is the intersection of all Φ^k, $k \in \mathbb{N}$. A function φ in this set has the property that, for any couple A, H and any $k \in \mathbb{N}$, $\varphi(H)$ belongs to $C^k(A; \mathscr{H})$ if $H \in C^k(A)$. We do not know of an explicit characterization of this class; however one has $\cap_{k \in \mathbb{N}} \Phi^k \subset S^0(\mathbb{R})$. In fact Example 6.2.7 shows that $\cap_{k \in \mathbb{N}} \Phi^k \subset BC^\infty(\mathbb{R})$, and Example 6.2.8 settles the decay properties of φ. We have been unable to prove that a function $\theta \in C^\infty(\mathbb{R})$ such that $\theta(x) = 0$ near $-\infty$ and $\theta(x) = 1$ near $+\infty$ belongs to $\cap \Phi^k$. This fact is quite disturbing for the theory which we shall develop in Chapter 7 (it will force us in certain circumstances to assume that the hamiltonian has a spectral gap). A simple and rich class of functions contained in $\cap_{k \in \mathbb{N}} \Phi^k$ is given by the set of all functions on $\mathbb{R}$ that have a C^∞ extension to the one-point compactification of $\mathbb{R}$. A detailed study of this class is contained in Section 3.2 of [ABG1].

EXAMPLE 6.2.8. Let $\mathscr{H}$ be a Hilbert space equipped with a unitary one-parameter group $\{e^{iAx}\}_{x \in \mathbb{R}}$. We say that a self-adjoint operator H in $\mathscr{H}$ is *A-homogeneous of degree* 1 if $e^{-iAx} H e^{iAx} = e^{-x} H$ for all $x \in \mathbb{R}$; this is equivalent to $[iA, H] = H$. Explicit examples in $\mathscr{H} = L^2(\mathbb{R})$ are:
  (i) $A = -\frac{1}{2}(PQ + QP)$ and $H = P$,
  (ii) $A = -\frac{1}{4}(PQ + QP)$ and $H = P^2$.
If H is A-homogeneous of degree 1, then $H \in C^\infty(A)$. If $\varphi : \mathbb{R} \to \mathbb{C}$ and $k \in \mathbb{N}$, we set $\varphi_k(\lambda) = (\lambda d/d\lambda)^k \varphi(\lambda)$. Then the following is true:
  (a) if $\varphi \in C^k(\mathbb{R})$ and $\varphi, \varphi_1, \ldots, \varphi_k$ are bounded, then $\varphi(H) \in C^k(A; \mathscr{H})$ and $i^k \operatorname{ad}_A^k(\varphi(H)) = \varphi_k(H)$,
  (b) if $\sigma(H) = \mathbb{R}$ and $\varphi \in C^\infty(\mathbb{R})$, then $\varphi(H) \in C^\infty(A; \mathscr{H})$ if and only if $\varphi \in S^0(\mathbb{R})$.

PROOF. We prove the assertion (b). We use the fact that $e^{-iAx} \varphi(H) e^{iAx} = \varphi(e^{-x} H)$ for any Borel function φ (which follows from the A-homogeneity of H). First assume that φ belongs to $C^1(\mathbb{R})$ and is bounded (which is necessary in order to have $\varphi(H) \in B(\mathscr{H})$) and that $\varphi(H) \in C^1(A; \mathscr{H})$. Since $\frac{d}{dx} \varphi(e^{-x}\lambda) = -\varphi_1(e^{-x}\lambda)$, we get by using Fatou's lemma (E denotes the spectral measure of H):

$$\int_{\mathbb{R}} |\varphi_1(\lambda)|^2 \|E(d\lambda)f\|^2 \leq \liminf_{x \to 0} \int_{\mathbb{R}} \left| \frac{\varphi(e^{-x}\lambda) - \varphi(\lambda)}{x} \right|^2 \|E(d\lambda)f\|^2$$
$$= \liminf_{x \to 0} \|x^{-1}[e^{-iA \cdot x} \varphi(H) e^{iA \cdot x} - \varphi(H)]f\|^2 = \|[A, \varphi(H)]f\|^2$$

for any $f \in \mathscr{H}$. Thus $\varphi_1(H) \in B(\mathscr{H})$. Since $\sigma(H) = \mathbb{R}$ and φ_1 is continuous, this means $\sup_{\lambda \in \mathbb{R}} |\varphi_1(\lambda)| < \infty$, i.e. $\varphi_1 \in BC(\mathbb{R})$. Now it is easy to show that $i[A, \varphi(H)] = \varphi_1(H)$. By induction one gets that, if $\varphi \in C^\infty(\mathbb{R})$ and $\varphi(H) \in C^\infty(A; \mathscr{H})$, then $\varphi_k \in BC(\mathbb{R})$ for each $k \in \mathbb{N}$ and $\varphi_k(H) = i^k \operatorname{ad}_A^k(\varphi(H))$. But clearly there are numbers $a_{k\ell}$ with $a_{kk} = 1$ such that $\varphi_k(\lambda) = (\lambda d/d\lambda)^k \varphi = \sum_{\ell=1}^k a_{k\ell} \lambda^\ell \varphi^{(\ell)}(\lambda)$. By recursion this leads to the existence of constants c_ℓ such that $|\lambda^\ell \varphi^{(\ell)}(\lambda)| \leq c_\ell$ for all $\lambda \in \mathbb{R}$. So $\varphi \in S^0(\mathbb{R})$. $\square$

6.2.2. We now turn to the characterization of the regularity classes $C^k(A)$ in terms of commutators. If H is a bounded operator, then one may use the theory of Section 5.1: for an arbitrary multi-index α, one may give a sense to the operator $\mathrm{ad}_A^\alpha(H)$ in the Sobolev scale associated to the group $\{e^{iA\cdot x}\}$ and one has $H \in C^k(A)$ if and only if $\mathrm{ad}_A^\alpha(H) \in B(\mathcal{H})$ for all $|\alpha| \le k$ (Proposition 5.1.2 (b)). If one tries to extend this description of C^k-regularity to unbounded operators, the first problem that one meets is that the meaning of the symbol $\mathrm{ad}_A^\alpha(H)$ is not at all clear in general. We shall give it a sense, for arbitrary α, in Section 6.3, under the assumption that the form domain (or the domain) of H is invariant under the operators $e^{iA\cdot x}$; indeed, this situation will turn out to be a particular case of that considered in Section 5.1. If the form domain of H is not invariant under the group $e^{iA\cdot x}$, one could try to give a meaning to $\mathrm{ad}_A^\alpha(H)$ by using a formula like (5.3.10). Since the case $k \ge 2$ is rather delicate, we shall concentrate here on the situation where $k = 1$ and $n = 1$, which is especially interesting in view of our treatment of the conjugate operator method in Chapter 7. We shall begin by giving a sense to the commutator [A,H] for arbitrary self-adjoint operators A and H and then we shall prove a rather simple characterization of the property $H \in C^1(A)$ in terms of $[A, H]$ (Theorem 6.2.10). Some results in the case $k = 2$ may be found in Proposition 6.3.4.

Let us denote by $\{\mathcal{H}_m\}_{m\in\mathbb{Z}}$ the Sobolev scale associated to the unitary representation $\{e^{iA\cdot x}\}$ of $\mathbb{R}^n$ in the Hilbert space $\mathcal{H}$. In the remainder of this section we shall identify $\mathcal{H}^*$ with $\mathcal{H}$ through the Riesz isomorphism. Then, by Theorem 3.3.28, we have a canonical identification $(\mathcal{H}_m)^* = \mathcal{H}_{-m}$ for any $m \in \mathbb{Z}$. In §5.1.1 we have defined the multiple commutator $\mathrm{ad}_A^\alpha(T)$ for $T \in B(\mathcal{H})$ and each multi-index α as a bounded operator from $\mathcal{H}_{|\alpha|}$ to $\mathcal{H}_{-|\alpha|}$ (see (5.1.10) and (5.3.10)). Moreover we saw that for each integer $m \ge 1$ the following identity strongly in $B(\mathcal{H}_m, \mathcal{H}_{-m})$:

(6.2.19)

$$e^{-iA\cdot x}Te^{iA\cdot x} = \sum_{|\alpha|\le m-1} \frac{(-ix)^\alpha}{\alpha!}\,\mathrm{ad}_A^\alpha(T) +$$

$$+ \sum_{|\alpha|=m} \frac{m(-ix)^\alpha}{\alpha!} \int_0^1 e^{-i\tau A\cdot x}\,\mathrm{ad}_A^\alpha(T)e^{i\tau A\cdot x}(1-\tau)^{m-1}d\tau.$$

From this it was easy to deduce that T is of class $C^k(A; \mathcal{H})$ if and only if $\mathrm{ad}_A^\alpha(T) \in B(\mathcal{H})$ for all $|\alpha| \le k$ (Proposition 5.1.2 (b)); if $T \in C^m(A; \mathcal{H})$, then (6.2.19) holds strongly in $B(\mathcal{H})$.

We now specialize to the case $n = 1$. If A and T are two unbounded operators in $\mathcal{H}$, their commutator is usually defined as a sesquilinear form rather than as an operator in $\mathcal{H}$: $[A, T]$ is the sesquilinear form on $D(A)\cap D(A^*)\cap D(T)\cap D(T^*)$ given as follows:

(6.2.20) $$\langle f, [A,T]g \rangle := \langle A^* f, Tg \rangle - \langle T^* f, Ag \rangle.$$

There is a very simple connection between this sesquilinear form and the operator $\mathrm{ad}_A(T) : \mathcal{H}_1 \to \mathcal{H}_{-1}$ in the case where A is self-adjoint and $T \in B(\mathcal{H})$. Then $\mathcal{H}_1$ is just $D(A)$ equipped with the graph topology (see Remark 3.3.24) and $\mathcal{H}_{-1} = (\mathcal{H}_1)^*$. Clearly $[A, T]$ is a continuous sesquilinear form on $\mathcal{H}_1$, hence there is a unique continuous linear operator $\tilde{T} : \mathcal{H}_1 \to (\mathcal{H}_1)^*$ such that $\langle Af, Tg \rangle_{\mathcal{H}} - \langle T^* f, Ag \rangle_{\mathcal{H}} = \langle f, \tilde{T}g \rangle$ for all $f, g \in \mathcal{H}_1$ ($\langle \cdot, \cdot \rangle_{\mathcal{H}}$ denotes the scalar product in $\mathscr{H}$, while $\langle\cdot,\cdot\rangle$ means anti-duality on $\mathscr{H}_1 \times (\mathscr{H}_1)^*)$. By taking into account the identifications $\mathscr{H}^* = \mathscr{H}$ and $\mathscr{H}_{-1} = (\mathscr{H}_1)^*$, it is straightforward to check that $\widetilde{T} = \mathrm{ad}_A(T)$. Hence $\mathrm{ad}_A(T) : \mathscr{H}_1 \to \mathscr{H}_{-1} \equiv (\mathscr{H}_1)^*$ *is the continuous operator canonically associated to the continuous sesquilinear form* $[A,T]$ *on* $\mathscr{H}_1$ $(= D(A)$ *with the graph topology).* From now on we shall not distinguish between $[A,T]$ and $\mathrm{ad}_A(T)$.

If A and T are closed symmetric operators and $D(A) \cap D(T)$ is dense in $\mathscr{H}$, then $D(A) \cap D(T)$ is a H-space when equipped with the intersection topology (an admissible hilbertian norm is $[\|f\|^2 + \|Af\|^2 + \|Tf\|^2]^{1/2}$) and one has dense continuous embeddings $D(A) \cap D(T) \subset \mathscr{H} \subset [D(A) \cap D(T)]^*$. Then one can interpret the continuous sesquilinear form $i[A,T]$ on $D(A) \cap D(T)$ as a continuous symmetric linear operator from $D(A) \cap D(T)$ into its adjoint space. For unbounded T, however, this point of view is rarely useful, because the space $D(A) \cap D(T)$ is too complicated in general.

The next lemma is a particular case of the statement following Equation (6.2.19). We add a simple self-contained proof. Our aim is to express the property that $T \in C^1(A;\mathscr{H})$ for bounded operators T solely in terms of the sesquilinear form $[A,T]$.

LEMMA 6.2.9. *Let A be a self-adjoint operator and T a bounded operator in the Hilbert space $\mathscr{H}$. Then T is of class $C^1(A;\mathscr{H})$ if and only if the sesquilinear form $[A,T]$ on $D(A)$ is continuous for the topology induced by $\mathscr{H}$, i.e. if and only if there is a constant $c < \infty$ such that*

$$(6.2.21) \qquad |\langle Af, Tf\rangle - \langle T^*f, Af\rangle| \le c\|f\|^2 \qquad \forall f \in D(A).$$

PROOF. If $f, g \in D(A)$, then $x \mapsto \langle e^{iAx}f, Te^{iAx}g\rangle$ is a function of class C^1. Hence

$$(6.2.22) \qquad \langle f, \frac{e^{-iAx}Te^{iAx} - T}{-ix}g\rangle =$$

$$= \frac{1}{x}\int_0^x dy[\langle Ae^{iAy}f, Te^{iAy}g\rangle - \langle e^{iAy}f, TAe^{iAy}g\rangle]$$

$$= \frac{1}{x}\int_0^x dy\langle e^{iAy}f, [A,T]e^{iAy}g\rangle.$$

(i) Assume that $T \in C^1(A;\mathscr{H})$. Then there is a constant c such that

$$\||x|^{-1}(e^{-iAx}Te^{iAx} - T)\| \le c \quad \text{for all } |x| \le 1.$$

Since the integrand in (6.2.22) is continuous if $f, g \in D(A)$, one obtains from (6.2.22) by letting $x \to 0$ that $|\langle f, [A,T]g\rangle| \le c\|f\| \cdot \|g\|$ for $f, g \in D(A)$.

(ii) Assume that (6.2.21) is satisfied. Then there is a unique continuous linear operator $\mathrm{ad}_A(T) : \mathscr{H} \to \mathscr{H}$ such that $\langle g, [A,T]h\rangle = \langle g, \mathrm{ad}_A(T)h\rangle$ for all g, $h \in D(A)$. By using (6.2.22) and the Lebesgue dominated convergence theorem, one finds that

$$\langle f, \frac{e^{-iAx}Te^{iAx} - T}{-ix}g\rangle = \frac{1}{x}\int_0^x dy\langle e^{iAy}f, \mathrm{ad}_A(T)e^{iAy}g\rangle$$

for all $f, g \in \mathcal{H}$. This implies that the weak derivative of $ie^{-iAx}Te^{iAx}$ at $x = 0$ exists and is equal to $\mathrm{ad}_A(T)$. By the group property one then gets that $i\frac{d}{dx}e^{-iAx}Te^{iAx} = e^{-iAx}\,\mathrm{ad}_A(T)e^{iAx}$ weakly. But a weakly differentiable function with strongly continuous derivative is strongly C^1 (by Newton's formula, see Lemma 5.A.2 (b)). Hence $T \in C^1(A; \mathcal{H})$. $\square$

The next result is an extension of Lemma 6.2.9 to the case of unbounded operators T. Although the assumptions below are stronger than those suggested by the lemma, the result is much deeper. Observe that there is no assumption of domain invariance under the group (under such a hypothesis, the theorem would be an easy consequence of Proposition 5.1.2 (b)).

THEOREM 6.2.10. *Let A and H be self-adjoint operators in the Hilbert space $\mathcal{H}$.*

(a) *H is of class $C^1(A)$ if and only if the following two conditions are satisfied:*

(1) *there is a constant $c < \infty$ such that for all $f \in D(A) \cap D(H)$:*

$$(6.2.23) \qquad |\langle Af, Hf \rangle - \langle Hf, Af \rangle| \leq c(\|Hf\|^2 + \|f\|^2),$$

(2) *for some $z \in \mathbb{C} \setminus \sigma(H)$, the set $\{f \in D(A) \mid R(z)f \in D(A) \text{ and } R(\bar{z})f \in D(A)\}$ is a core for A.*

(b) *If H is of class $C^1(A)$, then the following is true:*

(α) *The space $R(z)D(A)$ is independent of $z \in \mathbb{C} \setminus \sigma(H)$ and contained in $D(A)$, it is a core for H and a dense subspace of $D(A) \cap D(H)$ for the intersection topology (i.e. the topology associated to the norm $\|f\| + \|Af\| + \|Hf\|)$;*

(β) *The space $D(A) \cap D(H)$ is a core for H and the form $[A, H]$ has a unique extension to a continuous sesquilinear form on $D(H)$ (equipped with the graph topology); if this extension is denoted by $[A, H]$, the following identity holds on $\mathcal{H}$ (in the form sense):*

$$(6.2.24) \qquad [A, R(z)] = -R(z)[A, H]R(z), \qquad z \in \mathbb{C} \setminus \sigma(H).$$

Before turning to the proof, let us briefly discuss (β). Since $R(z) \in C^1(A; \mathcal{H})$, $[A, R(z)]$ is a bounded operator in $\mathcal{H}$ (given for example as the strong derivative of $x \mapsto ie^{-iAx}R(z)e^{iAx}$ at $x = 0$). (6.2.24) means that $\langle f, [A, R(z)]g \rangle = -\langle R(\bar{z})f, [A, H]R(z)g \rangle$ for all $f, g \in \mathcal{H}$, where the r.h.s. is interpreted as the evaluation of the form $[A, H]$ (extended to $D(H)$) at $R(\bar{z})f, R(z)g \in D(H)$.

A different way of looking at (6.2.24) is as follows: if A and H are self-adjoint operators in $\mathcal{H}$, denote by $\mathcal{E}$ the closure of $D(A) \cap D(H)$ in $D(H)$ (equipped with the graph topology). Assume that (6.2.23) is satisfied. Then the form $t(f, g) = \langle Af, Hg \rangle - \langle Hf, Ag \rangle$ on $D(A) \cap D(H)$ has a bounded extension to $\mathcal{E}$, in other words there is a bounded operator $T : \mathcal{E} \to \mathcal{E}^*$ such that $\langle f, Tg \rangle = \langle Af, Hg \rangle - \langle Hf, Ag \rangle$ for all $f, g \in D(A) \cap D(H)$. If the condition (2) is also satisfied, then (by (β)) $\mathcal{E}$ is identical with $D(H)$, so $[A, H]$ can be identified with a continuous linear operator $D(H) \to D(H)^*$. Since $D(H) \subset \mathcal{H}$ continuously and densely and $D(H)$ is reflexive, we get (after identification of $\mathcal{H}$ with $\mathcal{H}^*$ through the Riesz isomorphism) that $D(H) \subset \mathcal{H} \subset D(H)^*$, with continuous and dense embeddings. Now $R(z) \in B(\mathcal{H}, D(H))$ and (by a simple duality argument) $R(z)$ extends to a continuous linear operator from $D(H)^*$ to $\mathcal{H}$. So the r.h.s. of (6.2.24) may also be interpreted as the product of three bounded operators, viz. $R(z) : \mathcal{H} \to D(H)$, $[A, H] : D(H) \to D(H)^*$ and $R(z) : D(H)^* \to \mathcal{H}$.

PROOF. (i) For $z \in \mathbb{C} \setminus \sigma(H)$, we set $\mathscr{D}(z) = R(z)D(A)$. We first observe that $\mathscr{D}(z)$ is a core for H. Indeed, $R(z)$ is a homeomorphism of $\mathcal{H}$ onto $D(H)$ and $D(A)$ is dense in $\mathcal{H}$; so $R(z)D(A)$ is dense in $D(H)$ (which is always equipped with the graph topology defined by H).

(ii) We now prove the "if" part in (a). We let z be the number occurring in (2) and set $\mathscr{D}_0 = \{f \in D(A) | R(z)f \in D(A) \text{ and } R(\bar{z})f \in D(A)\}$. Each of the operators $AR(z)$, $AR(\bar{z})$ and $R(z)A$ is defined on $\mathscr{D}_0$, and for $f \in \mathscr{D}_0$ we have

$$\langle f, [A, R(z)]f \rangle = \langle f, AR(z)f \rangle - \langle AR(\bar{z})f, f \rangle$$
$$= \langle (H - \bar{z})R(\bar{z})f, AR(z)f \rangle - \langle AR(\bar{z})f, (H - z)R(z)f \rangle$$
$$= \langle HR(\bar{z})f, AR(z)f \rangle - \langle AR(\bar{z})f, HR(z)f \rangle.$$

Together with the assumption (1), this implies that

(6.2.25)
$$|\langle f, [A, R(z)]f \rangle| \leq c(\|HR(\bar{z})f\| + \|R(\bar{z})f\|)(\|HR(z)f\| + \|R(z)f\|)$$
$$\leq c_1 \|f\|^2 \qquad \forall f \in \mathscr{D}_0.$$

Since $\mathscr{D}_0$ is a core for A, this inequality extends to all $f \in D(A)$. So $R(z) \in C^1(A; \mathcal{H})$ by Lemma 6.2.9.

From now on we assume that $H \in C^1(A)$ and prove all consequences of this assumption stated in the proposition.

(iii) Let $z \in \mathbb{C} \setminus \sigma(H)$ and $f \in \mathcal{H}$. Then

$$\frac{e^{iAx} - I}{ix}R(z)f = R(z)\frac{e^{iAx} - I}{ix}f + \frac{e^{iAx}R(z)e^{-iAx} - R(z)}{ix}e^{iAx}f.$$

If $f \in D(A)$, the r.h.s. is strongly convergent to $R(z)Af + [A, R(z)]f$ as $x \to 0$, hence $R(z)f \in D(A)$ and

(6.2.26)
$$AR(z)f = R(z)Af + [A, R(z)]f$$

(remember that $[A, R(z)] \in B(\mathcal{H})$). This shows that $R(z)D(A) \subset D(A)$ (in fact this is a particular case of (5.3.13)) and proves (2). Together with the first resolvent equation $R(z_1) = R(z_2)[I + (z_1 - z_2)R(z_1)]$ it also implies that $R(z_1)D(A) \subset R(z_2)D(A)$ for any $z_1, z_2 \in \mathbb{C} \setminus \sigma(H)$, so that the set $\mathscr{D} := R(z)D(A)$ is independent of z.

(iv) We next complete the proof of (α). Let $f \in D(A) \cap D(H)$ and define $f_m \in \mathscr{D}$ by $f_m = imR(-im)f$ ($m \in \mathbb{N}$). Since $imR(-im) \to I$ strongly in $B(\mathcal{H})$ as $m \to \infty$, we have $\|f_m - f\| + \|H(f_m - f)\| \to 0$ as $m \to \infty$. So the denseness of $\mathscr{D}$ in $D(A) \cap D(H)$ in the intersection topology follows if we can show that $\|A(f_m - f)\| \to 0$ as $m \to \infty$. For this we observe that, by (6.2.26)

(6.2.27) $A(f_m - f) = [imR(-im) - I]Af + im[A, R(-im)]f.$

The first term on the r.h.s. is strongly convergent to zero as $m \to \infty$. To treat the second term on the r.h.s., we fix $z_0 \in \mathbb{C} \setminus \sigma(H)$ and set $g = (H - z_0)f \in \mathcal{H}$. By using (6.2.7), we then get that

$$im[A, R(-im)]f = im(H - z_0)R(-im)[A, R(z_0)]R(-im)(H - z_0)f$$
$$= (H - z_0)R(-im) \cdot [A, R(z_0)] \cdot imR(-im)g.$$

Now $imR(-im) \to I$ and $(H - z_0)R(-im) \to 0$ strongly in $B(\mathcal{H})$, and $[A, R(z_0)]$ belongs to $B(\mathcal{H})$. So the second term on the r.h.s. of (6.2.27) also converges strongly to zero.

(v) We now prove (1). We have for $f, g \in D(A) \cap D(H)$ and $z \in \mathbb{C} \setminus \sigma(H)$:

$$\langle Af, Hg \rangle - \langle Hf, Ag \rangle = \langle Af, (H - z)g \rangle - \langle (H - \bar{z})f, Ag \rangle.$$

Assume that $f, g \in \mathscr{D}$. Then $f = R(\bar{z})f_0$ and $g = R(z)g_0$ for some $f_0, g_0 \in D(A)$, and we get

$$(6.2.28) \qquad \langle Af, Hg \rangle - \langle Hf, Ag \rangle = \langle AR(\bar{z})f_0, g_0 \rangle - \langle f_0, AR(z)g_0 \rangle$$
$$= \langle R(\bar{z})f_0, Ag_0 \rangle - \langle Af_0, R(z)g_0 \rangle = -\langle f_0, [A, R(z)]g_0 \rangle,$$

where we used the fact that $R(z)$ and $R(\bar{z})$ leave $D(A)$ invariant. Hence

$$(6.2.29)$$
$$|\langle Af, Hg \rangle - \langle Hf, Ag \rangle| \le \|[A, R(z)]\| \cdot \|f_0\| \cdot \|g_0\|$$
$$= \|[A, R(z)]\| \cdot \|(H - z)f\| \cdot \|(H - z)g\|$$

for $f, g \in \mathscr{D}$. Since $\mathscr{D}$ is dense in $D(A) \cap D(H)$ in the intersection topology, (6.2.29) remains true for $f, g \in D(A) \cap D(H)$. In particular, (6.2.23) is satisfied.

(vi) Finally we prove (β). By the result of (i), $D(A) \cap D(H)$ is a core for H (because $R(z)D(A) \subset D(A) \cap D(H)$). Then (1) implies that the form $[A, H]$ has a unique continuous extension to $D(H)$. To obtain (6.2.24) we rewrite (6.2.28) as

$$-\langle f_0, [A, R(z)]g_0 \rangle = \langle f, [A, H]g \rangle = \langle R(\bar{z})f_0, [A, H]R(z)g_0 \rangle.$$

This establishes (6.2.24) as an identity in the form sense on $D(A)$, and its validity on $\mathcal{H}$ follows because $D(A)$ is dense in $\mathcal{H}$ and $[A, H]$ extends to a continuous sesquilinear form on $D(H)$. $\square$

6.3. Unitary Groups in Friedrichs Couples

In many applications it is convenient to use commutator methods in the framework consisting of a triplet $(\mathscr{G}, \mathscr{H}; W)$ such that the ordered pair $(\mathscr{G}, \mathscr{H})$ is a Friedrichs couple (see Section 2.8 for the terminology) and $W = \{W(x) \mid x \in \mathbb{R}^n\}$ is a strongly continuous unitary representation of $\mathbb{R}^n$ in $\mathscr{H}$ which leaves $\mathscr{G}$ invariant:

$$(6.3.1) \qquad W(x)\mathscr{G} \subset \mathscr{G} \qquad \forall x \in \mathbb{R}^n.$$

W is called *a unitary representation* (or *group*) *in the Friedrichs couple* $(\mathscr{G}, \mathscr{H})$.

The purpose of this section is to describe some constructions and objects naturally associated to the structure $(\mathscr{G}, \mathscr{H}; W)$. We mention that, in our applications of the conjugate operator method, $\mathscr{H}$ is the Hilbert space of physical states and $\mathscr{G}$ could be either $\mathscr{H}$ (e.g. in the N-body case with highly singular potentials) or the domain of the Hamiltonian H (e.g. in the N-body case if $D(H) = \mathscr{H}^2(X)$, a Sobolev space), or the form domain of H (e.g. in the two-body case we take $\mathscr{G} = \mathscr{H}^1(X)$). In all these cases W will be the unitary group generated by the conjugate operator.

Throughout this section we assume that the adjoint space $\mathscr{H}^*$ has been identified with $\mathscr{H}$ through the Riesz isomorphism. As a consequence, we get a canonical dense embedding $\mathscr{H} \subset \mathscr{G}^*$, hence the scale $\mathscr{G} \subset \mathscr{H} \subset \mathscr{G}^*$ of Hilbert spaces canonically associated to the Friedrichs couple $(\mathscr{G}, \mathscr{H})$. These objects have been studied in Section 2.8 and we shall freely use the notations and results established there; however, we recall the meaning of some symbols:

$$\mathscr{G}^{\theta,p} := (\mathscr{G}, \mathscr{G}^*)_{\frac{1-\theta}{2},p}, \quad \mathscr{G}^\theta \equiv \mathscr{G}^{\theta,2} \equiv [\mathscr{G}, \mathscr{G}^*]_{\frac{1-\theta}{2}} \text{ for } -1 < \theta < 1, \ 1 \le p \le \infty;$$

$$\overset{\circ}{\mathscr{G}}{}^{\theta,\infty} := (\mathscr{G}, \mathscr{G}^*)^{\circ}_{\frac{1-\theta}{2},\infty} = \text{ closure of } \mathscr{G} \text{ in } \mathscr{G}^{\theta,\infty};$$

$$\mathscr{G}^1 \equiv \mathscr{G}, \qquad \mathscr{G}^0 = \mathscr{H}, \qquad \mathscr{G}^{-1} \equiv \mathscr{G}^*.$$

The first fact that one has to notice in connection with the triplet $(\mathscr{G}, \mathscr{H}; W)$ is that W has a natural extension to a C_0-group in $\mathscr{G}^*$:

PROPOSITION 6.3.1. (a) *For each $x \in \mathbb{R}^n$, the map $W(x) : \mathscr{H} \to \mathscr{H}$ has a unique extension to a continuous linear operator $W_{-1}(x) : \mathscr{G}^* \to \mathscr{G}^*$. The family $\{W_{-1}(x)\}_{x \in \mathbb{R}^n}$ is a C_0-representation of $\mathbb{R}^n$ in $\mathscr{G}^*$.*

(b) *Let $W_1(x) : \mathscr{G} \to \mathscr{G}$ be defined by $W_1(x) = W(x)|_{\mathscr{G}} \equiv W_{-1}(x)|_{\mathscr{G}}$. Then $W_1(x) \in B(\mathscr{G})$ for each $x \in \mathbb{R}^n$, and $\{W_1(x)\}_{x \in \mathbb{R}^n}$ is a C_0-representation of $\mathbb{R}^n$ in $\mathscr{G}$.*

(c) *For each $x \in \mathbb{R}^n$ one has $W_1(x)^* = W_{-1}(-x)$.*

PROOF. We first observe that $W_1(x) \equiv W(x)|_{\mathscr{G}}$ belongs to $B(\mathscr{G})$ by the closed graph theorem. The fact that $\{W_1(x)\}$ is a C_0-group in $\mathscr{G}$ follows from Proposition 3.2.5 (b). Now *define* $W_{-1}(x) \in B(\mathscr{G}^*)$ by $W_{-1}(x) := W_1(-x)^*$. By Proposition 3.2.5 (a) the family $\{W_{-1}(x)\}$ is a C_0-group in $\mathscr{G}^*$. For $g \in \mathscr{G}$ and $f \in \mathscr{G}^*$ we have $\langle g, W_{-1}(x)f \rangle = \langle W_1(-x)g, f \rangle = \langle W(-x)g, f \rangle$. If $f \in \mathscr{H}$, the last expression is equal to $\langle g, W(x)f \rangle$ because $W(x)$ is unitary. Since $\mathscr{H}$ is dense in $\mathscr{G}^*$, this proves the first assertion in (a). $\square$

COROLLARY 6.3.2. *Let $\mathscr{E}$ be an interpolation space between $\mathscr{G}$ and $\mathscr{G}^*$ such that $\mathscr{G}$ is dense in $\mathscr{E}$. Then $W_{-1}(x)\mathscr{E} \subset \mathscr{E}$ for each $x \in \mathbb{R}^n$. If $W_{\mathscr{E}}(x) : \mathscr{E} \to \mathscr{E}$ is defined as $W_{\mathscr{E}}(x) = W_{-1}(x)|_{\mathscr{E}}$, then $\{W_{\mathscr{E}}(x)\}_{x \in \mathbb{R}^n}$ is a C_0-representation of $\mathbb{R}^n$ in $\mathscr{E}$.*

PROOF. We have seen at the end of Section 2.1 that there is a constant $C = C(\mathscr{G}, \mathscr{G}^*, \mathscr{E}) < \infty$ such that $\|T\|_{B(\mathscr{E})} \le C \max(\|T\|_{B(\mathscr{G})}, \|T\|_{B(\mathscr{G}^*)})$ for each $T \in B(\mathscr{G}^*)$ satisfying $T\mathscr{G} \subset \mathscr{G}$. Hence $W_{-1}(x)|_{\mathscr{E}}$ belongs to $B(\mathscr{E})$ and $\|W_{\mathscr{E}}(x)\| \le c < \infty$ for some constant c and all $|x| \le 1$. Then the strong continuity of $W_{\mathscr{E}} : \mathbb{R}^n \to B(\mathscr{E})$ follows from the strong continuity of $x \mapsto W_1(x)g$ in $\mathscr{G}$ for each $g \in \mathscr{G}$ and the denseness of $\mathscr{G}$ in $\mathscr{E}$. $\square$

By using Proposition 3.2.5 (b) one can prove the conclusion of the above corollary under a different assumption on $\mathscr{E}$, viz. that $\mathscr{E}$ is a reflexive Banach space, $\mathscr{E} \subset \mathscr{G}^*$ and $\mathscr{E}$ is invariant under W_{-1}.

Corollary 6.3.2 can be applied to $\mathscr{E} = \mathscr{G}^{\theta,p}$ with $-1 < \theta < 1$ and $p < \infty$ and to $\mathscr{E} = \overset{\circ}{\mathscr{G}}{}^{\theta,\infty}$. If there is no risk of confusion, we shall use the same notation W for all representations $W_{\mathscr{E}}$ and write A for the generator of $W_{\mathscr{E}}$ for any $\mathscr{E}$.

Let $\mathscr{E}$ be a Banach space embedded in $\mathscr{G}^*$, invariant under W and such that $W_{\mathscr{E}}$ is a C_0-group in $\mathscr{E}$. Then one can consider the Sobolev scale $\{\mathscr{E}_k\}_{k \in \mathbb{Z}}$ and the Besov scale $\{\mathscr{E}_{s,p} \mid s \in \mathbb{R}, 1 \le p \le \infty\}$ associated to $W_{\mathscr{E}}$. It is easily proven that, if $\mathscr{F}$ is a second space with the same properties and such that $\mathscr{E} \subset \mathscr{F}$, then $\mathscr{E}_k \subset \mathscr{F}_k$ and $\mathscr{E}_{s,p} \subset \mathscr{F}_{s,p}$; moreover these embeddings are dense if $\mathscr{E} \subset \mathscr{F}$ densely and $p < \infty$ (see Theorem 3.3.4). The largest spaces obtained in this way are those associated to $\mathscr{G}^*$. By using the Taylor expansion formula (3.3.16) one finds that for any $k \in \mathbb{N}$:

$$(6.3.2) \qquad \mathscr{E}_k = \{f \in \mathscr{G}_k^* \mid A^{\alpha}f \in \mathscr{E} \text{ for all } |\alpha| \le k\}.$$

In the case $n = 1$, it will be convenient to use the notation $\mathscr{E}_1 = D(A; \mathscr{E})$ (the domain of A in $\mathscr{E}$, i.e. the domain of the generator of the C_0-group $W_{\mathscr{E}}$ in $\mathscr{E}$). Thus

$$(6.3.3) \qquad D(A; \mathscr{E}) = \{f \in \mathscr{E} \mid f \in D(A; \mathscr{G}^*) \text{ and } Af \in \mathscr{E}\}.$$

We consider now the group of automorphisms induced by W. Let us denote by $\mathscr{X}$ the Banach space $B(\mathscr{G}, \mathscr{G}^*)$ and recall that it has a rather rich structure: due to the identification $\mathscr{G}^{**} = \mathscr{G}$, it is equipped with an involution $T \mapsto T^*$ and with a notion of positivity (see Section 2.1). For $x \in \mathbb{R}^n$, let $\mathscr{W}(x) : \mathscr{X} \to \mathscr{X}$ be the linear isomorphism given by:

$$(6.3.4) \qquad \mathscr{W}(x)[T] = W(-x)TW(x) \equiv W_{-1}(-x)TW_1(x).$$

Clearly $(\mathscr{W}(x)[T])^* = \mathscr{W}(x)[T^*]$ and $T \ge 0 \Rightarrow \mathscr{W}(x)[T] \ge 0$, hence $\{\mathscr{W}(x)\}_{x \in \mathbb{R}^n}$ is a n-parameter group of automorphisms of $\mathscr{X}$; it is not of class C_0, but is a C_w-group (Definition 3.2.6).

If $\mathscr{E}$, $\mathscr{F}$ are Banach spaces such that $\mathscr{G} \subset \mathscr{E}$ densely and $\mathscr{F} \subset \mathscr{G}^*$, then there is a canonical embedding $B(\mathscr{E}, \mathscr{F}) \subset B(\mathscr{G}, \mathscr{G}^*)$ (see Section 2.1). For example, $B(\mathscr{G})$, $B(\mathscr{H})$, $B(\mathscr{G}^*)$, $B(\mathscr{G}^*, \mathscr{G})$ and the spaces of type $B(\mathscr{G}^{\theta_1, p_1}, \mathscr{G}^{\theta_2, p_2})$ $(p_1 < \infty)$ are subspaces of $\mathscr{X}$. Clearly, all of them are $\mathscr{W}$-invariant. More generally, if $\mathscr{E}$, $\mathscr{F}$ are interpolation spaces between $\mathscr{G}$ and $\mathscr{G}^*$ and $\mathscr{G}$ is dense in each of them, then $B(\mathscr{E}, \mathscr{F})$ is a $\mathscr{W}$-invariant subspace of $\mathscr{X}$. In such a situation, the regularity classes $C^k(A; \mathscr{E}, \mathscr{F})$, $C_u^k(A; \mathscr{E}, \mathscr{F})$, $\mathscr{C}^{s,p}(A; \mathscr{E}, \mathscr{F})$ are well defined. The next result shows that these classes are related in a natural way.

PROPOSITION 6.3.3. *Let $\mathscr{E}$, $\mathscr{F}'$, $\mathscr{F}''$ be interpolation spaces between $\mathscr{G}$ and $\mathscr{G}^*$ such that $\mathscr{G}$ is dense in each of them. Then, for each $0 < \theta < 1$, the real* *interpolation space* $(\mathscr{F}', \mathscr{F}'')_{\theta,1}$ *is an interpolation space between* $\mathscr{G}$ *and* $\mathscr{G}^*$, *it contains* $\mathscr{G}$ *densely, and:*

$$(6.3.5) \qquad \mathscr{C}^{s,p}(A;\mathscr{E},\mathscr{F}') \cap \mathscr{C}^{t,q}(A;\mathscr{E},\mathscr{F}'') \subset \mathscr{C}^{u,r}(A;\mathscr{E},(\mathscr{F}',\mathscr{F}'')_{\theta,1}).$$

Here $0 \le s,t < \infty$ *and* $1 \le p,q \le \infty$ *are arbitrary, while* u, r *are given by* $u = (1-\theta)s + \theta t$, $\frac{1}{r} = \frac{1-\theta}{p} + \frac{\theta}{q}$.

REMARK. Since we did not require that $\mathscr{F}' \subset \mathscr{F}''$ or the reverse inclusion (this would not be natural in the context of the proposition), the space $(\mathscr{F}', \mathscr{F}'')_{\theta,1}$ is not defined by the rules of Chapter 2. In the applications in which we shall need the preceding result, we shall have $\mathscr{F}' \subset \mathscr{F}''$ in fact. One may observe that in the proof below, the only properties of $(\mathscr{F}', \mathscr{F}'')_{\theta,1}$ which will be used are: 1) it is an interpolation space between $\mathscr{G}$ and $\mathscr{G}^*$; 2) it contains $\mathscr{G}$ densely; 3) we have $|||f||| \le c||f||_{\mathscr{F}'}^{1-\theta}||f||_{\mathscr{F}''}^{\theta}$ for f in $\mathscr{G}$, where $||| \cdot |||$ is an admissible norm on $(\mathscr{F}', \mathscr{F}'')_{\theta,1}$ (see (2.5.4)).

(6.3.5) has to be used in conjunction with the obvious embedding:

$$(6.3.6) \qquad\qquad \mathscr{C}^{u,r}(A;\mathscr{E},\mathscr{F}) \subset \mathscr{C}^{u,r}(A;\mathscr{E},\mathscr{K})$$

if $\mathscr{E}$, $\mathscr{F}$, $\mathscr{K}$ are interpolation spaces between $\mathscr{G}$ and $\mathscr{G}^*$ containing $\mathscr{G}$ densely and such that $\mathscr{F} \subset \mathscr{K}$.

PROOF OF PROPOSITION 6.3.3. Let $\mathscr{F} = (\mathscr{F}', \mathscr{F}'')_{\theta,1}$. Since $\mathscr{F}$ is an interpolation space between $\mathscr{F}'$, $\mathscr{F}''$, and since $\mathscr{F}'$ and $\mathscr{F}''$ are interpolation spaces between $\mathscr{G}$ and $\mathscr{G}^*$, $\mathscr{F}$ will also be an interpolation space between $\mathscr{G}$ and $\mathscr{G}^*$ (this is an immediate application of the definition, see Section 2.1). To prove that $\mathscr{G}$ is dense in $\mathscr{F}$, we use the fact that $\mathscr{F}' \cap \mathscr{F}''$ is a dense subspace of $\mathscr{F}$ (according to the general theory of real interpolation) and that $\mathscr{F}' \cap \mathscr{F}''$ is continuously embedded in $\mathscr{F}$ if we consider on it the intersection topology (for which an admissible norm is $||f||_{\mathscr{F}'} + ||f||_{\mathscr{F}''}$). So it is enough to show that for each $f \in \mathscr{F}' \cap \mathscr{F}''$ there is a sequence $\{f_n\}_{n\in\mathbb{N}}$ in $\mathscr{G}$ such that $f_n \to f$ in $\mathscr{F}'$ and in $\mathscr{F}''$. We construct $\{f_n\}$ with the help of a mollifying sequence of operators ϱ_n associated to the Friedrichs couple $(\mathscr{G},\mathscr{H})$ (see §2.8.2): clearly it is enough to take $f_n = \varrho_n f$. In conclusion, $\mathscr{F}$ is an interpolation space between $\mathscr{G}$ and $\mathscr{G}^*$ and $\mathscr{G} \subset \mathscr{F}$ densely, so the class $\mathscr{C}^{u,r}(A;\mathscr{E},\mathscr{F})$ is well defined.

Now let us show that for each operator $T \in B(\mathscr{G},\mathscr{G}^*)$ we have (the norms are interpreted as gauges):

$$(6.3.7) \qquad\qquad ||T||_{\mathscr{E}\to\mathscr{F}} \le C||T||_{\mathscr{E}\to\mathscr{F}'}^{1-\theta}||T||_{\mathscr{E}\to\mathscr{F}''}^{\theta},$$

where C is a constant independent of T. If the right hand side is infinite, this is trivial, so we may assume $T \in B(\mathscr{E},\mathscr{F}') \cap B(\mathscr{E},\mathscr{F}'')$. Then the interpolation property implies $T \in B(\mathscr{E},\mathscr{F})$, and (6.3.7) is a consequence of (take $f = Te$, $e \in \mathscr{E}$):

$$(6.3.8) \qquad\qquad ||f||_{\mathscr{F}} \le C||f||_{\mathscr{F}'}^{1-\theta}||f||_{\mathscr{F}''}^{\theta}, \qquad f \in \mathscr{F}' \cap \mathscr{F}''.$$

But this is just one of the main properties of $(\mathscr{F}', \mathscr{F}'')_{\theta,1}$ ((2.5.4) states this in the case $\mathscr{F}' \subset \mathscr{F}''$ for example).

Let m be an integer strictly larger than s and t. By using (6.3.7) we get for any $S \in B(\mathscr{G}, \mathscr{G}^*)$ and $x \neq 0$:

$$(6.3.9) \qquad \left\| \frac{(\mathscr{W}(x) - I)^m}{|x|^u} S \right\|_{\mathscr{E} \to \mathscr{F}} \leq$$

$$\leq C \left\| \frac{(\mathscr{W}(x) - I)^m}{|x|^s} S \right\|_{\mathscr{E} \to \mathscr{F}'}^{1-\theta} \left\| \frac{(\mathscr{W}(x) - I)^m}{|x|^t} S \right\|_{\mathscr{E} \to \mathscr{F}''}^{\theta}.$$

Denote by $L_*^r(\mathscr{E}, \mathscr{F})$ the space of strongly Borel $B(\mathscr{E}, \mathscr{F})$-valued functions on the unit ball $|x| < 1$, with norm of power r integrable with respect to the measure $|x|^{-n} dx$. By using the Hölder inequality and (6.3.9) we get:

$$(6.3.10) \qquad \left\| \frac{(\mathscr{W}(\cdot) - I)^m}{|\cdot|^u} S \right\|_{L_*^r(\mathscr{E}, \mathscr{F})} \leq C \left\| \frac{(\mathscr{W}(\cdot) - I)^m}{|\cdot|^s} S \right\|_{L_*^p(\mathscr{E}, \mathscr{F}')}^{1-\theta}$$

$$\cdot \left\| \frac{(\mathscr{W}(\cdot) - I)^m}{|\cdot|^t} S \right\|_{L_*^q(\mathscr{E}, \mathscr{F}'')}^{\theta}. \qquad \square$$

We recall (cf. Section 5.2) that $\mathscr{C}^{0,\infty}(A; \mathscr{E}, \mathscr{F}') = B(\mathscr{E}, \mathscr{F}')$. Hence a particular case of (6.3.5) is:

$$(6.3.11) \qquad B(\mathscr{E}, \mathscr{F}') \cap \mathscr{C}^{t,q}(A; \mathscr{E}, \mathscr{F}'') \subset \mathscr{C}^{\theta t, q/\theta}(A; \mathscr{E}, (\mathscr{F}', \mathscr{F}'')_{\theta,1}).$$

Take $\mathscr{F}' = \mathscr{G}^{\sigma'}$, $\mathscr{F}'' = \mathscr{G}^{\sigma''}$, for some numbers $-1 \leq \sigma', \sigma'' \leq +1$, and set $\sigma = (1 - \theta)\sigma' + \theta\sigma''$. Then

$$(6.3.12) \qquad B(\mathscr{E}, \mathscr{G}^{\sigma'}) \cap \mathscr{C}^{t,q}(A; \mathscr{E}, \mathscr{G}^{\sigma''}) \subset \mathscr{C}^{\theta t, q/\theta}(A; \mathscr{E}, \mathscr{G}^{\sigma,1}).$$

In Proposition 6.3.3 and in the preceding remarks, we considered an operator T defined on a fixed domain $\mathscr{E}$ but with values in two different spaces $\mathscr{F}', \mathscr{F}''$. Sometimes a different situation occurs, namely T is considered with values in a fixed space, but with two different domains. However, in such a case one may apply Proposition 6.3.3 to the adjoint T^* of T and then get the needed result with the help of (5.2.17).

Let us explicitly point out several particular cases of (6.3.12) which will be needed later on. First:

$$(6.3.13) \qquad B(\mathscr{G}, \mathscr{H}) \cap \mathscr{C}^{2,q}(A; \mathscr{G}, \mathscr{G}^*) \subset \mathscr{C}^{1,2q}(A; \mathscr{G}, \mathscr{G}^{-1/2,1})$$

for arbitrary $1 \leq q \leq \infty$, hence (since $C^2 \subset \mathscr{C}^{2,\infty}$):

$$(6.3.14)$$
$$B(\mathscr{G}, \mathscr{H}) \cap C^2(A; \mathscr{G}, \mathscr{G}^*) \subset \mathscr{C}^{1,\infty}(A; \mathscr{G}, \mathscr{G}^{-1/2,1}) \subset \mathscr{C}^{1,\infty}(A; \mathscr{G}, \mathscr{G}^{-1/2}).$$

Then:

$$(6.3.15) \qquad B(\mathscr{G}, \mathscr{H}) \cap \mathscr{C}^{1,1}(A; \mathscr{G}, \mathscr{G}^*) \subset \mathscr{C}^{1/2,2}(A; \mathscr{G}, \mathscr{G}^{-1/2,1}).$$

We shall now discuss several criteria for a self-adjoint operator H in $\mathscr{H}$ to belong to certain regularity classes with respect to a one-parameter unitary group $W(x) = e^{iAx}$. For this we shall assume that the domain of H is invariant under the action of W, but let us first make some remarks in the more general case when only the form domain of H is left invariant by W. Then we may take $\mathscr{G} = D(|H|^{1/2})$ (this is the form domain of H, cf. Section 2.8) equipped with the natural graph topology. The operator H has a unique continuous extension to a map $H : \mathscr{G} \to \mathscr{G}^*$, and for each $z \in \mathbb{C} \setminus \sigma(H)$ the operator $H - z : \mathscr{G} \to \mathscr{G}^*$ is a homeomorphism. Then, for each $j \in \mathbb{N}$, the operator $\mathrm{ad}_A^j(H)$ is well defined as an element of $B(\mathscr{G}_j, \mathscr{G}^*_{-j})$ and we have $H \in C^k(A; \mathscr{G}, \mathscr{G}^*)$ if and only if $\mathrm{ad}_A^j(H) \in B(\mathscr{G}, \mathscr{G}^*)$ for $j = 0, 1, \ldots, k$ (all this holds for n-parameter unitary groups too). By Proposition 5.1.6 we have $H \in C^k(A; \mathscr{G}, \mathscr{G}^*)$ if and only if $R(z) \equiv (H - z)^{-1} \in C^k(A; \mathscr{G}^*, \mathscr{G})$ for some (and hence for all) $z \in \mathbb{C} \setminus \sigma(H)$, but clearly this property is much stronger then $R(z) \in C^k(A; \mathscr{H})$ (i.e. $H \in C^k(A)$).

If the domain of H is invariant under W, then its form domain is invariant too (because $D(|H|^{1/2}) = [D(H), \mathscr{H}]_{1/2}$) so the preceding remarks hold. By taking now $\mathscr{G} = D(H)$ with its natural graph topology, the operators $\mathrm{ad}_A^k(H)$ are well defined elements of $B(\mathscr{G}_k, \mathscr{H}_{-k})$ for each $k \in \mathbb{N}$. Note that, since $H \in B(\mathscr{G}, \mathscr{H}) \subset B(\mathscr{G}, \mathscr{G}^*)$, we also have $\mathrm{ad}_A^k(H) \in B(\mathscr{G}_k, \mathscr{G}^*_{-k})$, and $H \in C^k(A; \mathscr{G}, \mathscr{G}^*)$ is equivalent to $\mathrm{ad}_A^j(H) \in B(\mathscr{G}, \mathscr{G}^*)$ for $j = 0, 1, \ldots, k$.

The following theorem is interesting especially because of the assertion (b).

THEOREM 6.3.4. *Let A and H be self-adjoint operators in a Hilbert space $\mathscr{H}$. Assume that the unitary one-parameter group $\{\exp(iA\tau)\}_{\tau \in \mathbb{R}}$ leaves the domain $D(H)$ of H invariant. Set $\mathscr{G} = D(H)$ equipped with the graph topology. Then*
  (a) *H is of class $C^1(A)$ if and only if $H \in C^1(A; \mathscr{G}, \mathscr{G}^*)$.*
  (b) *H is of class $\mathscr{C}^{1,1}(A)$ if and only if $H \in \mathscr{C}^{1,1}(A; \mathscr{G}, \mathscr{G}^*)$.*
  (c) *If $H \in C^2(A; \mathscr{G}, \mathscr{G}^*)$, then $H \in \mathscr{C}^{1,\infty}(A; \mathscr{G}, \mathscr{G}^{-1/2,1})$; if in addition $H \in C^1(A; \mathscr{G}, \mathscr{G}^{-1/2})$, then H is of class $C^2(A)$.*

PROOF. (a) We set $R = (H - i)^{-1}$ and $A_\tau = (i\tau)^{-1}(e^{iA\tau} - I)$. By Lemma 6.2.3 and Theorem 5.1.3 we have:

$$(6.3.16) \qquad H \in C^1(A) \Longleftrightarrow \liminf_{\tau \to +0} ||[A_\tau, R]||_{B(\mathscr{H})} < \infty,$$

$$(6.3.17) \qquad H \in C^1(A; \mathscr{G}, \mathscr{G}^*) \Longleftrightarrow \liminf_{\tau \to +0} ||[A_\tau, H]||_{\mathscr{G} \to \mathscr{G}^*} < \infty.$$

Now observe that $R \in B(\mathscr{H}, \mathscr{G}) \cap B(\mathscr{G}^*, \mathscr{H})$ and $H \in B(\mathscr{G}, \mathscr{H}) \cap B(\mathscr{H}, \mathscr{G}^*)$ (see Section 2.8; in fact R is an isomorphism $\mathscr{H} \to \mathscr{G}$ and $\mathscr{G}^* \to \mathscr{H}$ with inverse $H - i : \mathscr{G} \to \mathscr{H}$ and $\mathscr{H} \to \mathscr{G}^*$ respectively). Also $A_\tau \in B(\mathscr{G}^\theta)$ for each $\theta \in [-1, 1]$. So we have the following identities:

$$(6.3.18) \quad [A_\tau, R] = -R[A_\tau, H]R, \qquad [A_\tau, H] = -(H - i)[A_\tau, R](H - i).$$

They imply that

$$||[A_\tau, R]||_{B(\mathscr{H})} \le ||R||_{\mathscr{G}^* \to \mathscr{H}} ||[A_\tau, H]||_{\mathscr{G} \to \mathscr{G}^*} ||R||_{\mathscr{H} \to \mathscr{G}},$$

$$||[A_\tau, H]||_{\mathscr{G} \to \mathscr{G}^*} \le ||H - i||_{\mathscr{H} \to \mathscr{G}^*} ||[A_\tau, R]||_{B(\mathscr{H})} ||H - i||_{\mathscr{G} \to \mathscr{H}}.$$

In view of (6.3.16) and (6.3.17), the result of (a) follows.

(b) We denote the norm in $B(\mathscr{G}^\theta, \mathscr{G}^\sigma)$ by $|| \cdot ||_{\theta \to \sigma}$ and, as usual, set $|| \cdot || = || \cdot ||_{0 \to 0} \equiv || \cdot ||_{B(\mathscr{H})}$. By interpolation one obtains that $R \in B(\mathscr{G}^\theta, \mathscr{G}^{\theta+1})$ and $H \in B(\mathscr{G}^{\theta+1}, \mathscr{G}^\theta)$ for each $\theta \in [-1, 0]$. Next we observe that $[A_{-\tau}, H] =$

$-e^{-iA\tau}[A_\tau, H]e^{-iA\tau}$. Since $||e^{-iA\tau}||_{\sigma \to \sigma} \leq c$ for all $|\tau| \leq 1$ and $|\sigma| \leq 1$ and since H is symmetric, we get that for $\tau \leq 1$

$$(6.3.19) \qquad ||[A_\tau, H]||_{1/2 \to -1} = ||[A_\tau, H]^*||_{1 \to -1/2}$$
$$= ||[A_{-\tau}, H]||_{1 \to -1/2} \leq c^2 ||[A_\tau, H]||_{1 \to -1/2}.$$

We need the following two identities which can be deduced from (6.3.18) (note that A_τ leaves $\mathscr{G}$ and $\mathscr{H}$ invariant):

$$(6.3.20) \qquad [A_\tau, [A_\tau, R]] = 2R[A_\tau, H]R[A_\tau, H]R - R[A_\tau, [A_\tau, H]]R,$$
$$(6.3.21) \qquad [A_\tau, [A_\tau, H]] = 2(H-i)[A_\tau, R](H-i)[A_\tau, R](H-i)$$
$$-(H-i)[A_\tau, [A_\tau, R]](H-i).$$

(6.3.20) and (6.3.19) imply that

$$(6.3.22)$$
$$|||[A_\tau, [A_\tau, R]]||| \leq$$
$$\leq 2||R||_{-1 \to 0}||[A_\tau, H]||_{1/2 \to -1}||R||_{-1/2 \to 1/2}||[A_\tau, H]||_{1 \to -1/2}||R||_{0 \to 1} +$$
$$+ ||R||_{-1 \to 0}||[A_\tau, [A_\tau, H]]||_{1 \to -1}||R||_{0 \to 1}$$
$$\leq c_1||[A_\tau, H]||_{1 \to -1/2}^2 + c_1||[A_\tau, [A_\tau, H]]||_{1 \to -1}.$$

Now assume that $H \in \mathscr{C}^{1,1}(A; \mathscr{G}, \mathscr{G}^*)$. Then the second term on the r.h.s. of (6.3.22) is integrable on $[-1, 1]$ with respect to the measure $\tau^{-2}d\tau$. The same is true for the first term because $H \in \mathscr{C}^{1/2,2}(A; \mathscr{G}, \mathscr{G}^{-1/2})$ by (6.3.15). Thus $R \in \mathscr{C}^{1,1}(A; \mathscr{H})$, i.e. H is of class $\mathscr{C}^{1,1}(A)$.

The proof of the reverse implication is similar. From (6.3.21) one finds that

$$||[A_\tau, [A_\tau, H]]||_{1 \to -1} \leq c_2|||[A_\tau, R]||_{0 \to 1/2}||[A_\tau, R^*]||_{0 \to 1/2} + c_2|||[A_\tau, [A_\tau, R]]|||,$$

and it suffices to take into account the fact that R and $R^* \equiv (H+i)^{-1}$ belong to $\mathscr{C}^{1/2,2}(A; \mathscr{H}, \mathscr{G}^{1/2})$ (this follows from (6.3.12) with $\mathscr{E} = \mathscr{H}$, $\sigma' = 1$, $\sigma'' = 0$, $\theta = 1/2$, $t = q = 1$ and the inclusion $\mathscr{G}^{1/2,1} \subset \mathscr{G}^{1/2}$).

(c) The first assertion follows from (6.3.14). If $H \in \mathscr{C}^1(A; \mathscr{G}, \mathscr{G}^{-1/2}) \cap C^2(A; \mathscr{G}, \mathscr{G}^*)$, then $R \in C^2(A; \mathscr{H})$ by (6.3.22), i.e. H is of class $C^2(A)$. $\square$

In the remainder of this section we stay in the framework of Theorem 6.3.4 and present some simple results which are helpful in certain applications. Notice first that, if $T \in B(\mathscr{G}, \mathscr{G}^*)$ is of class $C^1(A; \mathscr{G}, \mathscr{G}^*)$, then for any real $\tau \neq 0$:

$$(6.3.23) \qquad \tau^{-1}[e^{iA\tau}Te^{-iA\tau} - T] = \frac{1}{\tau}\int_0^\tau e^{iA\sigma}[iA, T]e^{-iA\sigma}d\sigma.$$

Let $\mathscr{E}, \mathscr{F}$ be Banach spaces such that $\mathscr{G} \subset \mathscr{E} \subset \mathscr{G}^*$ continuously and densely and $\mathscr{F} \subset \mathscr{G}^*$ continuously; assume moreover that $\mathscr{E}$ and $\mathscr{F}$ are invariant under the group $\{e^{iA\tau}\}$ (induced by A in $\mathscr{G}^*$) and that the groups induced in $\mathscr{E}, \mathscr{F}$ are of class C_0 (see e.g. Corollary 6.3.2). Recall the canonical embedding $B(\mathscr{E}, \mathscr{F}) \subset B(\mathscr{G}, \mathscr{G}^*)$. It follows immediately from (6.3.23) that $[A, T] \in B(\mathscr{E}, \mathscr{F})$ if and only if $e^{iA\tau}Te^{-iA\tau} - T \in B(\mathscr{E}, \mathscr{F})$ $\forall \tau \in \mathbb{R}$ and s-$\lim_{\tau \to 0} \tau^{-1}(e^{iA\tau}Te^{-iA\tau} - T)$ exists in $B(\mathscr{E}, \mathscr{F})$. In particular, if $T \in B(\mathscr{E}, \mathscr{F})$, then $T \in C^1(A; \mathscr{E}, \mathscr{F})$ if and only if $T \in C^1(A; \mathscr{G}, \mathscr{G}^*)$ and $[A, T] \in B(\mathscr{E}, \mathscr{F})$.

Now let H be as in Theorem 6.3.4 and $R = (H - z)^{-1}$ for some fixed $z \in \mathbb{C} \setminus \sigma(H)$. Then $H \in B(\mathscr{G}^\theta, \mathscr{G}^{\theta-1})$ and $R \in B(\mathscr{G}^{\theta-1}, \mathscr{G}^\theta)$ for all $\theta \in [0,1]$. Part (a) of Theorem 6.3.4 states that $R \in C^1(A; \mathscr{H})$ if and only if $H \in C^1(A; \mathscr{G}, \mathscr{G}^*)$. In this case we have (cf. (6.3.18)):

$$(6.3.24) \quad [A, R] = -R[A, H]R, \qquad [A, H] = -(H - z)[A, R](H - z).$$

PROPOSITION 6.3.5. *Let A, H and $\mathscr{G}$ be as in Theorem 6.3.4 and $0 \le \alpha \le 1$. Then the following conditions are equivalent:*
  (a) *there is $\theta \in [\alpha, 1]$ such that $H \in C^1(A; \mathscr{G}^\theta, \mathscr{G}^{\theta-1-\alpha})$;*
  (b) *$H \in C^1(A; \mathscr{G}, \mathscr{G}^*)$ and $[A, H] \in B(\mathscr{G}^\theta, \mathscr{G}^{\theta-1-\alpha})$ for some $\theta \in [\alpha, 1]$;*
  (a') *one has (a) for all $\theta \in [\alpha, 1]$;*
  (b') *one has (b) for all $\theta \in [\alpha, 1]$;*
  (c) *$(H - z)^{-1} \in C^1(A; \mathscr{G}^{\theta-1}, \mathscr{G}^{\theta-\alpha})$ for some $z \in \mathbb{C} \setminus \sigma(H)$ and $\theta \in [\alpha, 1]$;*
  (d) *$(H - z)^{-1} \in C^1(A; \mathscr{G}, \mathscr{G}^*)$ and $[A, (H - z)^{-1}] \in B(\mathscr{G}^{\theta-1}, \mathscr{G}^{\theta-\alpha})$ for some $z \in \mathbb{C} \setminus \sigma(H)$ and $\theta \in [\alpha, 1]$;*
  (c') *one has (c) for all $z \in \mathbb{C} \setminus \sigma(H)$ and all $\theta \in [\alpha, 1]$;*
  (d') *one has (d) for all $z \in \mathbb{C} \setminus \sigma(H)$ and all $\theta \in [\alpha, 1]$.*
  If these conditions are satisfied, then $\varphi(H) \in C^1(A; \mathscr{G}^{-(1-\alpha)}, \mathscr{G}^{1-\alpha})$ for each $\varphi \in C_0^\infty(\mathbb{R})$.

PROOF. The equivalence of the statements (a)-(d') is an easy consequence of (6.3.24) and of the discussion made above. Alternatively, a more direct proof can be based on the identities (6.3.18). If $\varphi \in C_0^\infty(\mathbb{R})$, then Corollary 6.2.6 implies $[A, \varphi(H)] \in B(\mathscr{G}^{\alpha-1}, \mathscr{H}) \cap B(\mathscr{H}, \mathscr{G}^{1-\alpha})$. Since $[A, \varphi_1(H)\varphi_2(H)] = [A, \varphi_1(H)]\varphi_2(H) + \varphi_1(H)[A, \varphi_2(H)]$ and any C_0^∞-function is the product of two such functions, the proposition is proved. $\square$

6.4. Estimates on $\varphi(H_1) - \varphi(H_2)$

Let $(\mathscr{H}, W)$ be a Hilbert space equipped with a strongly continuous unitary representation of $\mathbb{R}^n$, and let H_1, H_2 be self-adjoint operators in $\mathscr{H}$. The purpose of this section is to show that, if $H_1 - H_2$ decays (in some generalized sense) in the spectral representation of the generator A of W and if H_1, H_2 are regular with respect to W, then $\varphi(H_1) - \varphi(H_2)$ decays (at the same rate as $H_1 - H_2$) for a certain class of functions φ. The main tool will be the commutator expansions developed in Section 5.5.

In order to avoid conditions on the domains or the form domains of H_1 and H_2, we shall express the decay assumption in terms of the difference of resolvents $(H_1 - z)^{-1} - (H_2 - z)^{-1}$ rather than in terms of the difference of H_1 and H_2 which could make no sense at all. It will be clear from our arguments below that, if H_1 and H_2 have the same domain or form domain and if this domain (or form domain) is invariant under W and the group induced in it by W is of polynomial growth, then one can easily transfer estimates from $H_1 - H_2$ to the difference of the resolvents by using the identity :

$$(6.4.1) \quad (H_1 - z)^{-1} - (H_2 - z)^{-1} = (H_1 - z)^{-1}(H_2 - H_1)(H_2 - z)^{-1}.$$

We set $R_j(z) = (H_j - z)^{-1}$ for $j = 1, 2$. From the first resolvent equation $R_j(z) = [I + (z - z_0)R_j(z)]R_j(z_0)$ it is straightforward to deduce that, for z and z_0 in the resolvent set of H_1 and H_2:

(6.4.2)
$$R_1(z) - R_2(z) = \{I + (z - z_0)R_1(z)\}[R_1(z_0) - R_2(z_0)]\{I + (z - z_0)R_2(z)\}.$$

Now let us assume that H_1 and H_2 are of class $\mathscr{C}^{m+s,\infty}(A)$ for some integer $m \geq 1$ and some real $s > 0$. Consider two symbols $\psi_1 \in S^{a_1}(\mathbb{R}^n)$, $\psi_2 \in S^{a_2}(\mathbb{R}^n)$, where a_1, a_2 are strictly positive real numbers such that $m \geq a_1 + a_2$ and $s > \max(a_1, a_2)$. Set $S_1 = I + (z - z_0)R_1(z)$, $S_2 = I + (z - z_0)R_2(z)$ and $T = R_1(z_0) - R_2(z_0)$. Then, by a direct application of Proposition 5.5.6 (with all four occurring Hilbert spaces identified with $\mathscr{H}$), one obtains the following result: *if there is a complex number z_0 such that $\psi_1^{(\alpha)}(A)[R_1(z_0) - R_2(z_0)]\psi_2^{(\beta)}(A) \in B(\mathscr{H})$ for all multi-indices α and β, then this inclusion holds with z_0 replaced by any complex number z outside the spectrum of H_1 and H_2.* Our purpose below is to show that one will also have

$$\psi_1^{(\alpha)}(A)[\varphi(H_1) - \varphi(H_2)]\psi_2^{(\beta)}(A) \in B(\mathscr{H})$$

for a much more general class of functions φ than just $\varphi(\lambda) = (\lambda - z)^{-1}$.

We first observe that the preceding inclusion holds for φ of the form $\varphi(\lambda) = (\lambda - z)^{-k}$ if $k \geq 1$ is an integer and $z \notin \sigma(H_1) \cup \sigma(H_2)$; this follows easily from the identity

(6.4.3) $$R_1(z)^k - R_2(z)^k = \sum_{j=0}^{k-1} R_1(z)^j[R_1(z) - R_2(z)]R_2(z)^{k-j-1}$$

and Proposition 5.5.6. Consequently *the mentioned inclusion will hold for any φ in the complex algebra generated by the set of functions of the form $\varphi_z(\lambda) = (\lambda - z)^{-1}$, where $z \in \mathbb{C} \setminus [\sigma(H_1) \cup \sigma(H_2)]$* (this is exactly the algebra of complex rational functions on $\mathbb{R}$ that tend to zero at infinity and have no poles in $\sigma(H_1) \cup \sigma(H_2)$).

The argument for non-rational φ is not so easy; it requires an estimate on the divergence of $\psi_1^{(\alpha)}(A)[R_1(z) - R_2(z)]\psi_2^{(\beta)}(A)$ as $\Im z \to 0$ which we describe in the next lemma:

LEMMA 6.4.1. *Let H_1, H_2 be two self-adjoint operators of class $\mathscr{C}^{m+s,\infty}(A)$ in $\mathscr{H}$, where $m \geq 1$ is an integer and $s > 0$ is a real number. Consider two functions $\psi_1 \in S^{a_1}(\mathbb{R}^n), \psi_2 \in S^{a_2}(\mathbb{R}^n)$, where a_1, a_2 are strictly positive real numbers such that $m \geq a_1 + a_2$ and $s > \max(a_1, a_2)$. Assume that there is a number $z_0 \in \mathbb{C} \setminus [\sigma(H_1) \cup \sigma(H_2)]$ such that for all multi-indices α, β the operator $\psi_1^{(\alpha)}(A)[R_1(z_0) - R_2(z_0)]\psi_2^{(\beta)}(A)$ is bounded. Then there is a finite constant C such that for $0 < |\Im z| \leq 1$:*

(6.4.4) $$\|\psi_1(A)[R_1(z) - R_2(z)]\psi_2(A)\| \leq C \langle z \rangle^{2k} |\Im z|^{-k},$$

with $k = 2m + [s] + 2$.

PROOF. We shall use the following inequality which holds by (6.2.17) and the polarization identity: if H is a self-adjoint operator of class $C^\kappa(A)$, and $R(z) = (H - z)^{-1}$, then there is a finite constant c such that for $|\alpha| \le \kappa$ and $0 < |\Im z| \le 1$:

$$(6.4.5) \qquad \|\operatorname{ad}_A^\alpha(R(z))\| \le c\langle z\rangle^{2|\alpha|}|\Im z|^{-|\alpha|-1}.$$

We now repeat the proof of Proposition 5.5.6 (with all four occurring Hilbert spaces identified with $\mathcal{H}$ and all for C_0-groups identified with $\exp(iA \cdot x)$) by taking into account the dependence on z of the operators in the commutator expansion (5.5.25). More precisely, let

$$T = R_1(z_0) - R_2(z_0), \quad S_j = I + (z - z_0)R_j(z) \text{ with } j = 1, 2$$

and $\mathscr{R}_1 = \mathscr{R}_{m,\psi_1}^R[S_1]$, $\mathscr{R}_2 = \mathscr{R}_{m,\psi_2}^L[S_2]$. Then, by (6.4.2) and (5.5.25):

$$(6.4.6) \qquad \psi_1(A)[R_1(z) - R_2(z)]\psi_2(A) =$$

$$= \sum_{\substack{|\alpha|<m \\ |\beta|<m}} \frac{(-1)^{|\beta|}}{\alpha!\beta!} \operatorname{ad}_A^\alpha(S_1)\psi_1^{(\alpha)}(A)T\psi_2^{(\beta)}(A)\operatorname{ad}_A^\beta(S_2) +$$

$$+ \sum_{|\alpha|<m} \frac{1}{\alpha!}\{(-1)^{|\alpha|}\mathscr{R}_1 T\psi_2^{(\alpha)}(A)\operatorname{ad}_A^\alpha(S_2) +$$

$$+ \operatorname{ad}_A^\alpha(S_1)\psi_1^{(\alpha)}(A)T\mathscr{R}_2\} + \mathscr{R}_1 T\mathscr{R}_2.$$

From (6.4.5) we get a bound of the form $C\langle z\rangle^{4m-2}|\Im z|^{-2m}$ for the first sum. The last term is bounded by $\|T\| \cdot \|\mathscr{R}_1\| \cdot \|\mathscr{R}_2\|$ and Theorem 5.5.3 implies that

$$\|\mathscr{R}_1\| \le \sum_{|\alpha|=m} \frac{1}{\alpha!}\|\mathscr{J}_\alpha^R[\operatorname{ad}_A^\alpha(S_1)]\|$$

$$\le c \sum_{|\alpha|=m} \|\operatorname{ad}_A^\alpha(S_1)\| \le c_1\langle z\rangle^{2m+1}|\Im z|^{-m-1}$$

(for the second inequality take into account (5.5.12) and the fact that $\mathscr{F}\psi_1^{(\alpha)} \in L^1(\mathbb{R}^n)$ if $|\alpha| = m$ by Proposition 1.3.6). The same estimate holds for $\|\mathscr{R}_2\|$. Hence the norm of the last term in (6.4.6) is bounded by $c_2\langle z\rangle^{4m+2}|\Im z|^{-2m-2}$.

By (6.4.5) and the arguments given at the end of the proof of Proposition 5.5.6, one has the following bound for the second sum in (6.4.6):

$$c_3\langle z\rangle^{2m-1}|\Im z|^{-m}[\|\mathscr{R}_1\|_{\mathscr{H}_{-a_2}\to\mathscr{H}} + \|\mathscr{R}_2\|_{\mathscr{H}\to\mathscr{H}_{a_1}}].$$

We show how to obtain a bound for $\|\mathscr{R}_1\|_{\mathscr{H}_{-a_2}\to\mathscr{H}}$; a similar argument leads to the same bound for $\|\mathscr{R}_2\|_{\mathscr{H}\to\mathscr{H}_{-a_1}}$. We shall apply Theorem 5.3.3 with

$$S = S_1 = I + (z - z_0)R_1(z) \in \mathscr{C}^{m+s,\infty}(A;\mathscr{H}) \subset \mathscr{C}^{m+s',2}(A;\mathscr{H})$$

for any $s' < s$ and with $\mu = a_2$ (so that $\psi_1 \in S^{m-\mu}(\mathbb{R}^n)$). So Theorem 5.5.3 and (6.2.15) lead to

$$\|\mathscr{R}_1\|_{\mathscr{H}_{-a_2}\to\mathscr{H}} \le c_4\|S_1\|_{\mathscr{C}^{m+s',2}} \le c_5\langle z\rangle^{2\ell+1}|\Im z|^{-\ell-1},$$

with $\ell = m + [s] + 1$. In view of the choice of k in the lemma, the preceding estimates imply (6.4.4). $\square$

THEOREM 6.4.2. *Let $(\mathscr{H}, \{e^{iA \cdot x}\})$ be a Hilbert space equipped with a unitary representation of $\mathbb{R}^n$. Let H_1 and H_2 be self-adjoint operators in $\mathscr{H}$ of class $\mathscr{C}^{m+s,\infty}(A)$, where $m \geq 1$ is an integer and $s > 0$ is real, and let ψ_1, ψ_2 be functions on $\mathbb{R}^n$ with $\psi_1 \in S^{a_1}(\mathbb{R}^n)$, $\psi_2 \in S^{a_2}(\mathbb{R}^n)$, where a_1, a_2 are strictly positive real numbers such that $m \geq a_1 + a_2$ and $s > \max(a_1, a_2)$. Assume that there is a number $z_0 \in \mathbb{C} \setminus [\sigma(H_1) \cup \sigma(H_2)]$ such that for all multi-indices α, β the operator $\psi_1^{(\alpha)}(A)[(H_1 - z_0)^{-1} - (H_2 - z_0)^{-1}]\psi_2^{(\beta)}(A)$ is bounded.*

Set $k = 2m + [s] + 2$ and let $\varphi : \mathbb{R} \to \mathbb{C}$ be of class $C^{k+1}(\mathbb{R})$ and such that $\int_{\mathbb{R}} \langle \lambda \rangle^{2k} |\varphi^{(j)}(\lambda)| d\lambda < \infty$ for $0 \leq j \leq k+1$. Then $\psi_1(A)[\varphi(H_1) - \varphi(H_2)]\psi_2(A)$ is a bounded operator in $\mathscr{H}$.

PROOF. Without loss of generality we may assume φ to be real. Let $r = k+1$. By using (6.1.26) one gets (e.g. as sesquilinear forms on $\mathscr{H}_\infty$)

$$(6.4.7) \qquad \pi \psi_1(A)[\varphi(H_1) - \varphi(H_2)]\psi_2(A) =$$

$$= \Im \sum_{\ell=0}^{r-1} \frac{i^\ell}{\ell!} \int_{\mathbb{R}} d\lambda \varphi^{(\ell)}(\lambda) \psi_1(A)[R_1(\lambda + i) - R_2(\lambda + i)]\psi_2(A)$$

$$+ \Im \frac{i^r}{(r-1)!} \int_{\mathbb{R}} d\lambda \int_0^1 \mu^{r-1} d\mu \cdot \varphi^{(r)}(\lambda) \psi_1(A)[R_1(\lambda + i\mu) - R_2(\lambda + i\mu)]\psi_2(A).$$

By taking into account the estimate (6.4.4), one sees that each term on the r.h.s. belongs to $B(\mathscr{H})$. $\square$

Let us assume that H_1 and H_2 have the same domain $\mathscr{G}$, that $\mathscr{G}$ is invariant under $e^{iA \cdot x}$ for each $x \in \mathbb{R}^n$ and that the group W induced in $\mathscr{G}$ is polynomially bounded (where $\mathscr{G}$ is equipped with the unique topology for which it becomes a B-subspace of $\mathscr{H}$). Let $\mathscr{G}^*$ be equipped with the adjoint representation of $\mathbb{R}^n$. Then one can get the stronger result that

$$(6.4.8) \qquad \psi_1(A)[\varphi(H_1) - \varphi(H_2)]\psi_2(A) \in B(\mathscr{G}^*, \mathscr{G})$$

if for example $\varphi \in \mathscr{S}(\mathbb{R}^n)$ and if H_1 and H_2 belong to an appropriate regularity class. The proof uses arguments similar to those of the proof of Theorem 6.4.2 and the identity

$$\varphi(H_1) - \varphi(H_2) = R_1(i)[\varphi_2(H_1) - \varphi_2(H_2)]R_2(i) +$$
$$+ \varphi_1(H_1)[R_1(i) - R_2(i)] + [R_1(i) - R_2(i)]\varphi_1(H_2),$$

where $\varphi_j(\lambda) = (\lambda - i)^j \varphi(\lambda)$. Details can be found in Section 3.8 of [ABG1].

6.A. Appendix: Remarks on the Functional Calculus Associated to $\mathscr{W}$ in $B(\mathscr{H})$

Let A be a self-adjoint operator in a Hilbert space $\mathscr{H}$ and $\mathscr{W}$ the associated C_{w}-group of automorphisms of $B(\mathscr{H})$, given by $\mathscr{W}(x)[S] = e^{-iAx}Se^{iAx}$ ($x \in \mathbb{R}$). We have $\mathscr{W}(x) = e^{i\mathscr{A}x}$, where $\mathscr{A}$ is a closed (but non-densely defined) operator acting in $B(\mathscr{H})$ as follows : $\mathscr{A}[S] = [S, A] \equiv SA - AS$.

If $\varphi \in BC^\infty(\mathbb{R})$, then $\varphi(\mathscr{A})[S]$ has a well defined meaning for each $S \in B(\mathscr{H})$ as a sesquilinear form on $D(A^2)$. Indeed, by adapting Corollary 3.6.4 and the argument leading to (3.6.12) to the present situation, one obtains that

$$\left| \int_{\mathbb{R}} \langle f, e^{-iAx}Se^{iAx}g \rangle \widehat{\varphi}(x)\underline{dx} \right| \leq c\|(I + A^2)f\| \cdot \|(I + A^2)g\| \cdot \|S\| \cdot \|\varphi\|_{C^2_{(-1)}}.$$

If the Fourier transform of φ is an integrable measure, this sesquilinear form determines a bounded linear operator $\varphi(\mathscr{A})[S]$ in $\mathscr{H}$, viz.

$$\varphi(\mathscr{A})[S] = \int_{\mathbb{R}} e^{-iAx}Se^{iAx}\widehat{\varphi}(x)\underline{dx} \equiv \int_{\mathbb{R}} \mathscr{W}(x)[S]\widehat{\varphi}(x)\underline{dx}.$$

In this case $\varphi(\mathscr{A})$ is a bounded linear operator acting in $B(\mathscr{H})$. It is important to realize that the set of Fourier transforms of integrable measures is the largest class of functions φ such that $\varphi(\mathscr{A})$ is bounded as an operator in $B(\mathscr{H})$ for arbitrary $\mathscr{H}$ and A (this follows from the example given below). In particular $\varphi(\mathscr{A})$ is in general unbounded in $B(\mathscr{H})$ if $\varphi \in S^0(\mathbb{R})$.

EXAMPLE 6.A.1. Let $\mathscr{H} = L^2(\mathbb{R})$ and $A = P = -i\frac{d}{dx}$. The multiplication operators of the form $S = \eta(Q)$, with $\eta \in L^\infty(\mathbb{R})$, form a subalgebra of $B(\mathscr{H})$ (with $\|S\| = \|\eta\|_{L^\infty(\mathbb{R})}$). For fixed $y \in \mathbb{R}$ one has $\mathscr{W}(y)[S] = e^{-iPy}Se^{iPy} = \eta(Q - y) = (e^{-iPy}\eta)(Q)$ (in the last expression e^{-iPy} denotes the natural action of the translation group in $L^\infty(\mathbb{R})$). So $\varphi(\mathscr{A})[S] = [\varphi(P)\eta](Q)$, and in order for $\varphi(\mathscr{A})$ to be bounded in $B(\mathscr{H})$ it is necessary that $\|\varphi(P)\eta\|_{L^\infty(\mathbb{R})} \leq c\|\eta\|_{L^\infty(\mathbb{R})}$ for all $\eta \in L^\infty(\mathbb{R})$ and some finite constant c. This inequality is satisfied if and only if φ is the Fourier transform of an integrable measure (see the end of Section 3.2). One may treat in a similar way the translation group on the one-dimensional torus to obtain an example in which the spectrum of A is discrete.

It is also interesting to consider the action of $\varphi(\mathscr{A})$ in a certain subalgebras of $\mathscr{B} \equiv B(\mathscr{H})$. We consider in particular the following self-adjoint ideals in $\mathscr{B}$: the set $\mathscr{B}^{(0)}$ of finite rank operators, the set $\mathscr{B}^{(1)}$ of trace class operators, the set $\mathscr{B}^{(2)}$ of Hilbert-Schmidt operators and the set $\mathscr{B}^{(\infty)}$ of compact operators. We cite from §1.15 and §1.19 of [Sa] some details on these algebras: $\mathscr{B}^{(\infty)}$ is naturally equipped with the norm induced by $\mathscr{B}$ for which it is an involutive Banach algebra (in fact a C^*-algebra), but $\mathscr{B}^{(1)}$ and $\mathscr{B}^{(2)}$ have other natural norms, namely $\|S\|_{\mathscr{B}^{(1)}} = \mathrm{Tr}\,|S|$ and $\|S\|_{\mathscr{B}^{(2)}} = [\mathrm{Tr}\,S^*S]^{1/2}$, where Tr denotes the trace function. Equipped with these norms, $\mathscr{B}^{(1)}$ and $\mathscr{B}^{(2)}$ are also involutive Banach algebras, and the ideal $\mathscr{B}^{(0)}$ is dense in $\mathscr{B}^{(1)}$, $\mathscr{B}^{(2)}$ and $\mathscr{B}^{(\infty)}$ but not in $\mathscr{B}$. One may introduce a natural antiduality on $\mathscr{B}^{(1)} \times \mathscr{B}$, namely $\langle S, T \rangle = \mathrm{Tr}\,S^*T$ for $S \in \mathscr{B}^{(1)}$, $T \in \mathscr{B}$, and one can then establish the identifications $(\mathscr{B}^{(1)})^* = \mathscr{B}$ and $(\mathscr{B}^{(\infty)})^* = \mathscr{B}^{(1)}$. Moreover $\langle S, T \rangle$ makes sense for $S, T \in \mathscr{B}^{(2)}$ and defines a scalar product on $\mathcal{B}^{(2)}$ with associated norm equal to $||\cdot||_{\mathcal{B}^{(2)}}$, so that $\mathcal{B}^{(2)}$ is a Hilbert space.

Each of the subalgebras $\mathcal{B}^{(k)}$ ($k = 0, 1, 2, \infty$) of $\mathcal{B}$ is invariant under the group $\mathcal{W}$, and the groups induced by $\mathcal{W}$ in $\mathcal{B}^{(1)}$, $\mathcal{B}^{(2)}$ and $\mathcal{B}^{(\infty)}$ are C_0-groups (although $\mathcal{W}$ is only a C_w-group in $\mathcal{B}$). Let us denote by $\mathcal{W}_k$ the group induced by $\mathcal{W}$ in $\mathcal{B}^{(k)}$. Then the dual group of $\mathcal{W}_1$ is the initial $\mathcal{W}$ in $\mathcal{B}$, the dual group of $\mathcal{W}_\infty$ is just $\mathcal{W}_1$, whereas $\mathcal{W}_2^* = \mathcal{W}_2$, i.e. $\mathcal{W}_2$ is a unitary group in $\mathcal{B}^{(2)}$ (cf. (3.2.7)).

The operator induced by $\mathcal{A}$ in $\mathcal{B}^{(2)}$ (the generator of $\mathcal{W}_2$) is a self-adjoint operator in the Hilbert space $\mathcal{B}^{(2)}$. For any $\varphi \in BC^\infty(\mathbb{R})$, $\varphi(\mathcal{A})$ leaves $\mathcal{B}^{(2)}$ invariant and defines a bounded operator in $\mathcal{B}^{(2)}$, with

$$||\varphi(\mathcal{A})[S]||_{\mathcal{B}^{(2)}} \leq \sup_{x \in \mathbb{R}} |\varphi(x)| \cdot ||S||_{\mathcal{B}^{(2)}}.$$

On the other hand, due to the fact that $\mathcal{W}_1^* = \mathcal{W}$, it is clear that $\varphi(\mathcal{A})$ is in general unbounded as an operator in the Banach space $\mathcal{B}^{(1)}$ if $\widehat{\varphi}$ is not an integrable measure.

It is interesting to observe that $\varphi(\mathcal{A})$ *is bounded in $\mathcal{B}^{(1)}$ if and only if $\varphi(\mathcal{A})$ maps each operator of rank 1 into $\mathcal{B}^{(1)}$.* For the proof of the "if" part of this statement, let $\Phi : \mathcal{H} \times \mathcal{H} \to \mathcal{B}^{(2)}$ be the continuous bilinear mapping $\Phi(f, g) = \varphi(\mathcal{A})[f \otimes g]$, where $f \otimes g$ denotes the rank 1 operator defined by $f \otimes g(h) = \langle g, h \rangle f$. Assume that $\Phi(f, g) \in \mathcal{B}^{(1)}$ for all $f, g \in \mathcal{H}$. Then, by the closed graph theorem $\Phi : \mathcal{H} \times \mathcal{H} \to \mathcal{B}^{(1)}$ will be separately continuous, hence continuous by the principle of uniform boundedness (cf. the Corollary to Theorem III.9 in [RS]). So there will be a constant $c < \infty$ such that $||\phi(f, g)||_{\mathcal{B}^{(1)}} \leq c||f|| \cdot ||g||$ for all f, $g \in \mathcal{H}$. Since each trace class operator S is of the form $S = \sum_{k=1}^\infty \lambda_k f_k \otimes g_k$ with $||f_k|| = ||g_k|| = 1$ and $\sum_{k=1}^\infty |\lambda_k| \leq ||S||_{\mathcal{B}^{(1)}}$, it follows that $||\varphi(\mathcal{A})[S]||_{\mathcal{B}^{(1)}} \leq c||S||_{\mathcal{B}^{(1)}}$ for all $S \in \mathcal{B}^{(1)}$.

So, although $\varphi(\mathcal{A})[f \otimes g]$ is a Hilbert-Schmidt operator for each $\varphi \in BC^\infty(\mathbb{R})$ and $f, g \in \mathcal{H}$, if φ is not the Fourier transform of an integrable measure, then there is $f \in \mathcal{H}$ such that $\varphi(\mathcal{A})[f \otimes f]$ is not a trace class operator. Observe also that $\mathrm{Tr}(\varphi(\mathcal{A})[S]) = \varphi(0) \mathrm{Tr}\, S$ if $\varphi \in \mathcal{M}^0(\mathbb{R})$ and $S \in \mathcal{B}^{(1)}$.

CHAPTER 7

The Conjugate Operator Method

Let H be a self-adjoint operator in a Hilbert space $\mathcal{H}$, $R(z) = (H - z)^{-1}$ its resolvent and λ a real number in the spectrum of H. Since $\|R(\lambda + i\mu)\| = |\mu|^{-1}$, $R(\lambda + i\mu)$ cannot have limits in $B(\mathcal{H})$ as $\mu \to \pm 0$. However, for certain vectors $f \in \mathcal{H}$, the function $F(z) = \langle f, R(z)f \rangle$, which is defined and holomorphic for z outside the spectrum of H, could have a limit as z converges to λ from the upper or lower half-plane (these two limits will be different in general). If this happens for sufficiently many f, one can infer results on the spectral properties of H which are useful for example in scattering theory. This will be elaborated in Section 7.1. The remainder of the chapter is devoted to a detailed description of a method, called the "conjugate operator method", for proving the existence of such limits.

In order to motivate the rather technical developments of the following sections, we here explain the basic ideas of the method in some very simple situations. One way of checking that $\lim_{\mu \to +0} F(\lambda + i\mu)$ exists is to prove that $\int_0^1 |\frac{d}{d\mu} F(\lambda + i\mu)| d\mu < \infty$. A standard technique in the theory of ordinary differential equations for obtaining the finiteness of such integrals is to establish first an estimate for $\frac{d}{d\mu} F(\lambda + i\mu)$ in terms of $F(\lambda + i\mu)$, i.e. to prove a differential inequality of the form $|\frac{d}{d\mu} F(\lambda + i\mu)| \leq U_\lambda(\mu, F(\lambda + i\mu))$ for some function U_λ, and then to use some version of the Gronwall lemma (see e.g. Chapter III in [Hm]). In the following simple examples it is not necessary to invoke the Gronwall lemma, but later on this lemma will be essential.

As a first example we observe that $\frac{d}{dz} R(z) = R(z)^2$ and, formally, $[A, R(z)] = R(z)[H, A]R(z)$ for an arbitrary operator A. Assume that we may choose a self-adjoint A such that $[H, iA] = H$ [1]. Then one has $z \frac{d}{dz} R(z) = [iA, R(z)] - R(z)$; hence, if we assume that $f \in D(A)$, we obtain $z \frac{d}{dz} F(z) = -F(z) - \langle iAf, R(z)f \rangle - \langle R(\bar{z})f, iAf \rangle$. But, if $z = \lambda + i\mu$ and $\mu > 0$, then $\|R(z)f\| = \|R(\bar{z})f\| = $

---

[1] Rigorously this means that $e^{-iA\tau} H e^{iA\tau} = e^\tau H$ for all $\tau \in \mathbb{R}$. We have chosen this special form of the commutation relation because it allows us to do explicitly the calculations that follow, and also because it covers some interesting Schrödinger Hamiltonians, namely $H = \Delta + c|Q|^{-2}$ with $c > -\frac{1}{4}$, by taking for A the generator of the dilation group, i.e. $A = (P \cdot Q + Q \cdot P)/4$.

W. O. Amrein et al., C_0-Groups, Commutator Methods and Spectral Theory of
N-Body Hamiltonians, Modern Birkhäuser Classics,
DOI: 10.1007/978-3-0348-0733-3_7, © Springer Basel 1996

$\mu^{-1/2}|\Im F(z)|^{1/2}$. In conclusion, for $\lambda \neq 0$ and $\mu > 0$, we get the following differential inequality:

$$(7.0.1) \qquad \left|\frac{d}{d\mu}F(\lambda + i\mu)\right| \leq |\lambda|^{-1}(\|f\| + 2\|Af\|)\mu^{-1/2}|F(\lambda + i\mu)|^{1/2}.$$

Since $F(z) \neq 0$ if $\Im z \neq 0$, one may divide both sides by $|F(\lambda + i\mu)|^{1/2}$; upon integrating the resulting inequality for $\left|\frac{d}{d\mu}F(\lambda + i\mu)^{1/2}\right|$, one arrives at the validity of the following estimate for $0 < \mu < 1$:

$$(7.0.2) \qquad |F(\lambda + i\mu)|^{1/2} \leq |F(\lambda + i)|^{1/2} + \frac{1}{|\lambda|}(\|f\| + 2\|Af\|).$$

Finally one may estimate the r.h.s. of (7.0.1) by using (7.0.2). One sees that for each $\delta > 0$ there is a constant c such that

$$(7.0.3) \qquad \left|\frac{d}{d\mu}F(\lambda + i\mu)\right| \leq \frac{c}{\sqrt{\mu}}(\|f\|^2 + \|Af\|^2)$$

for $|\lambda| \geq \delta$ and $0 < \mu < 1$. This obviously implies the existence of $\lim F(\lambda + i\mu)$ as $\mu \to +0$, uniformly in $|\lambda| \geq \delta$.

As a second example, let us point out an important refinement of the preceding arguments. Above, λ was considered fixed and μ was a parameter converging to zero. Clearly we may also get (7.0.2) (with some different constants on the r.h.s.) if we consider $z = \lambda + i\mu$ fixed (with $\mu > 0$), introduce a new parameter $\varepsilon > 0$ and apply the same arguments to the function $\varepsilon \mapsto F(\lambda + i\mu + i\varepsilon)$. Of course, at this level, this is a trivial modification, but it allows us to improve the final result by the following trick. We no more assume $f \in D(A)$ but take a family $\{f_\varepsilon\}_{\varepsilon>0}$ of elements of $D(A)$ depending smoothly on ε and converging to some f as $\varepsilon \to +0$, and we define $F_\varepsilon(z) = \langle f_\varepsilon, R(z + i\varepsilon)f_\varepsilon\rangle$. A computation as above will give (we set $F'_\varepsilon = \frac{d}{d\varepsilon}F_\varepsilon$, etc. and assume again that $[H, iA] = H$):

$$(7.0.4)$$

$$F'_\varepsilon(z) = -\frac{i}{z + i\varepsilon}\Big(F_\varepsilon(z) + \langle iAf_\varepsilon, R(z + i\varepsilon)f_\varepsilon\rangle + \langle R(z + i\varepsilon)^*f_\varepsilon, iAf_\varepsilon\rangle\Big)$$
$$+ \langle f'_\varepsilon, R(z + i\varepsilon)f_\varepsilon\rangle + \langle R(z + i\varepsilon)^*f_\varepsilon, f'_\varepsilon\rangle.$$

By proceeding as in the derivation of (7.0.2), one finds that

$$(7.0.5) \quad |F_\varepsilon(z)|^{1/2} \leq |F_1(z)|^{1/2} + \frac{1}{|\lambda|}\int_0^1 (\|f_\tau\| + 2\|Af_\tau\| + 2|\lambda| \cdot \|f'_\tau\|)\frac{d\tau}{\sqrt{\tau}}$$

for all $\varepsilon \in (0, 1]$ and all $z = \lambda + i\mu$ with $\lambda \in \mathbb{R}$ and $0 < \mu < 1$. The integral on the r.h.s. of (7.0.5) can be made finite by a suitable choice of the family $\{f_\varepsilon\}$ if and only if f belongs to the (real) interpolation space $(D(A), \mathscr{H})_{1/2,1}$ (see Proposition 2.3.3). So, as a consequence of (7.0.4) and (7.0.5), for each $\delta > 0$ there is a constant $c < \infty$ such that for all $\varepsilon \in (0, 1]$, $|\lambda| \geq \delta$ and $\mu \in (0, 1]$:

$$(7.0.6) \quad \left|\frac{d}{d\varepsilon}F_\varepsilon(\lambda + i\mu)\right| \leq c(\|f_1\| + \|f\|_{1/2,1})\frac{1}{\sqrt{\varepsilon}}(\|f_\varepsilon\| + \|Af_\varepsilon\| + \|f'_\varepsilon\|).$$

Since $\langle f, R(\lambda + i\mu)f\rangle = \langle f_1, R(\lambda + i\mu + i)f_1\rangle - \int_0^1 \frac{d}{d\varepsilon}F_\varepsilon(\lambda + i\mu)d\varepsilon$, one obtains the convergence of $\langle f, R(\lambda + i\mu)f\rangle$ as $\mu \to +0$ for each $f \in (D(A), \mathscr{H})_{1/2,1}$ by applying the dominated convergence theorem.

One of the ideas in the preceding argument was to consider a modified resolvent $R(z, \varepsilon) = [(H - i\varepsilon) - z]^{-1}$ depending on a small parameter ε and converging to $R(z)$ as $\varepsilon \to 0$. As a last example we now consider another ε-dependent modification of the resolvent $R(z)$ suggested by some developments in the spectral theory of Schrödinger operators involving the dilation group (more precisely, we have in mind the theory of dilation analytic hamiltonians, for which we refer to [RS]). Let $G_\varepsilon(z) = (e^{-i\varepsilon}H - z)^{-1} = e^{i\varepsilon}R(ze^{i\varepsilon})$ be the new modified resolvent and define $F_\varepsilon(z) = \langle f, G_\varepsilon(z)f\rangle$ for $f \in D(A)$ (we do not consider the improvement which could be obtained by letting f also be ε-dependent). A simple computation based on the relation $[H, iA] = H$ shows that $G'_\varepsilon \equiv \frac{d}{d\varepsilon}G_\varepsilon = [G_\varepsilon, A]$. So, if $z = \lambda + i\mu$:

$$|F'_\varepsilon(z)| \le \|Af\|(\|G_\varepsilon(z)f\| + \|G_\varepsilon(z)^*f\|)$$

$$\le 2\|Af\| \cdot |\lambda\sin\varepsilon + \mu\cos\varepsilon|^{-1/2}|F_\varepsilon(z)|^{1/2}.$$

The estimate we need on $|F_\varepsilon(z)|$ has to be independent of μ. If $\lambda > 0$, then for $\mu \ge 0$ and $\varepsilon \in (0,1)$ we have $\lambda\sin\varepsilon + \mu\cos\varepsilon \ge \lambda\sin\varepsilon \ge 2\pi^{-1}\lambda\varepsilon$, hence $|F'_\varepsilon(z)| \le (2\pi\lambda^{-1}\varepsilon^{-1})^{1/2}\|Af\| \cdot |F_\varepsilon(z)|^{1/2}$. By proceeding as before (division by $|F_\varepsilon(z)|^{1/2}$ and integration), one obtains

$$|F_\varepsilon(z)|^{1/2} \le |F_1(z)|^{1/2} + \sqrt{\frac{2\pi}{\lambda}}\|Af\| \le \sqrt{\frac{2\pi}{\lambda}}(\|f\| + \|Af\|),$$

because $|F_1(z)| \le \|R(e^i z)\| \cdot \|f\|^2 \le \sqrt{2}\lambda^{-1}\|f\|^2$ if $\lambda > 0$ and $\mu \ge 0$. Consequently, for $\lambda, \mu > 0$:

$$\left|\frac{d}{d\varepsilon}F_\varepsilon(\lambda + i\mu)\right| \le \frac{2\pi}{\lambda}\varepsilon^{-1/2}\|Af\|(\|f\| + \|Af\|).$$

The convergence of $\langle f, R(\lambda + i\mu)f\rangle$ as $\mu \to +0$ for $f \in D(A)$ can now be obtained from the relation $\langle f, R(\lambda + i\mu)f\rangle = F_1(\lambda + i\mu) - \int_0^1 \frac{d}{d\varepsilon}F_\varepsilon(\lambda + i\mu)d\varepsilon$ by using the dominated convergence theorem. Notice also that we obtained the estimate [2] $|\langle f, R(z)f\rangle| \le 4\pi\lambda^{-1}(\|f\|^2 + \|Af\|^2)$.

Observe that the chosen form of the modified resolvent $G_\varepsilon(z)$ (we always take $0 < \varepsilon \le 1$) works very well if $\lambda > 0$, $\mu > 0$ and also if $\lambda < 0$, $\mu < 0$. In order to treat the cases $\lambda < 0$, $\mu > 0$ and $\lambda > 0$, $\mu < 0$ we have to take $G_\varepsilon(z) = (e^{i\varepsilon}H - z)^{-1}$. The final estimate then has the same form.

In order to go beyond the special case of A-homogeneous operators H considered so far, we recall that the somewhat formal relation $[H, iA] = H$ should be interpreted as $e^{-iA\tau}He^{iA\tau} = e^\tau H$ for all $\tau \in \mathbb{R}$. This implies that there is a holomorphic map $\xi \mapsto H(\xi)$ defined on the entire complex plane such that $H(i\tau) = e^{-iA\tau}He^{iA\tau}$ for $\tau \in \mathbb{R}$; indeed, we just take $H(\xi) = e^{-i\xi}H$. Formally

---

[2]Remark that the present choice of the modified resolvent gives a somewhat better estimate than that obtained from (7.0.5) for the dependence in λ of the norm of $R(\lambda + i\mu)$ as an operator $\mathscr{K} \to \mathscr{K}^*$, where $\mathscr{K} = D(A)$; one may get the same result if $\mathscr{K} = (D(A), \mathscr{H})_{1/2,1}$. This was remarked and developed in [Mt].

one then has $e^{-\varepsilon A}He^{\varepsilon A} = e^{-i\varepsilon}H$ [3]. By making a first order expansion of $H(\varepsilon)$, one gets $H(\varepsilon) \approx H - i\varepsilon H = H - i\varepsilon[H, iA]$, hence (if $\operatorname{sgn}\lambda = \operatorname{sgn}\mu$)

$$(7.0.7) \qquad G_\varepsilon(z) \approx (H - i\varepsilon[H, iA] - \lambda - i\mu)^{-1}.$$

It was the insight of E. Mourre [M1], [M2] that the arguments of the preceding examples can be extended to situations where there is no simple expression for the commutator $[H, iA]$, provided that this commutator has a definite sign when localized in a neighbourhood of λ (in the expression (7.0.7), $[H, iA]$ should have the same sign as μ). In more precise terms, his condition was as follows: there are real numbers $a > 0$ and $\delta > 0$ such that $E(\lambda; \delta)[H, iA]E(\lambda; \delta) \geq aE(\lambda; \delta)$, where $E(\lambda; \delta)$ is the spectral projection of H associated to the interval $(\lambda - \delta, \lambda + \delta)$. Under some regularity assumptions on H with respect to A, Mourre was able to carry through the ideas outlined in the above examples by using as a modified resolvent the expression on the r.h.s. of (7.0.7) (in fact, in his original papers he worked with $\varphi(H)[H, iA]\varphi(H)$ for some $\varphi \in C_0^\infty(\mathbb{R})$ instead of $[H, iA]$, in order to have a bounded operator; however, this is not necessary under his conditions, see [JMP]).

The abstract Mourre theorem came as a breakthrough in spectral theory because of the ease with which it could be applied in rather complicated situations, for example to N-body Schrödinger hamiltonians. In fact, this application was the principal motivation of Mourre, who treated the case $N = 3$. Soon afterwards Perry, Sigal and Simon [PSS] extended his analysis to an arbitrary N. Moreover, their paper contains an extension of the abstract Mourre theorem which allowed them to eliminate some slightly unnatural conditions that Mourre had to impose on the potentials. More precisely, if $H = \Delta + V(x)$ in $\mathscr{H} = L^2(\mathbb{R}^n)$ and $A = D$ is the generator of the dilation group (cf. (1.2.19)), then Mourre had to require that $V \in B(\mathscr{H}^2, \mathscr{H})$, $[D, V] \in B(\mathscr{H}^2, \mathscr{H})$ and $[D, [D, V]] \in B(\mathscr{H}^2, \mathscr{H}^{-2})$, where $\mathscr{H}^s \equiv \mathscr{H}^s(\mathbb{R}^n)$ are Sobolev spaces (see Section 4.1). The second condition implies that the radial derivative of V has to be square-integrable away from the origin. In [PSS] this hypothesis is replaced by $[D, V] \in B(\mathscr{H}^2, \mathscr{H}^{-1})$, which locally follows from $V \in B(\mathscr{H}^2, \mathscr{H})$. On the other hand, the assumption on the second commutator, although harmless locally, restricts the admissible decay of $V(x)$ at infinity; namely there are classes of short range potentials such that $|V(x)| \leq c\langle x\rangle^{-1-\nu}$, with $0 < \nu < 1$, for which this assumption is not satisfied (e.g. $V(x) = \sin|x| \cdot \langle x\rangle^{-3/2}$). This problem was remarked by several people working on the subject and, as far as we know, the first published solution is due to Yafaev [Ya].

Independently, one of us (W.A.) had the idea of considering a modified resolvent involving an approximate hamiltonian H_ε depending in a new way on the parameter ε, viz. $H_\varepsilon = \Delta + \theta(\varepsilon x)V(x)$, where $\theta \in C_0^\infty(\mathbb{R}^n)$ and $\theta(x) = 1$ if $|x| \leq 1$. Then the arguments of Mourre were easily extended to the modified resolvent $G_\varepsilon = (H - iM_\varepsilon - z)^{-1}$, where $M_\varepsilon = \varepsilon\varphi(H)[H_\varepsilon, iA]\varphi(H)$ for suitable $\varphi \in C_0^\infty(\mathbb{R})$. An abstract version of these result and applications to N-body

---

[3]It is interesting to observe that, for $\xi = \varepsilon \in \mathbb{R} \setminus \{0\}$, the spectrum of the operator $H(\varepsilon)$ is non-real, contained in $e^{-i\xi}\mathbb{R}$. On the other hand, if we set $M = e^{\varepsilon A}$, then we could think of $H(\varepsilon)$ as being equal to $M^{-1}HM$; if M is bounded, such an operator has the same spectrum as H. This shows that A has to be unbounded if $H \neq 0$.

hamiltonians appeared in [ABG2] and, in a more detailed form, in the first part of the unpublished notes [ABG1]. Similar results based on the use of the above operators H_ε were obtained independently by Tamura [Tm1]. The connection between the conditions imposed on the approximating family $\{H_\varepsilon\}$ and real interpolation theory has been observed in [BGM1], [BGM2]; this allowed them to extend the preceding theory to the optimal (in the Besov scale) class of short range potentials (the so-called Enss class). As a by-product of the method, they also got the limiting absorption principle in the optimal Besov space introduced by Agmon and Hörmander [AH] in their analysis of the simply characteristic operators. The fact that this Besov space appears quite naturally in Mourre's original theory was pointed out before by Jensen and Perry [JP] who used estimates proved by Mourre in [M2] (and which show, in fact, that the limiting absorption principle holds in a better space, but not of Besov type; see [BGM2]).

A natural question then was the extension of the theory to the case where the potential V is only form-bounded relative to Δ. A modification of the abstract results of [ABG1] which allows one to cover such a situation appeared in [BMP1]. Unfortunately their results, although quite good in two-body situations, are not very interesting in the N-body case. Finally, the regularity condition $H \in \mathscr{C}^{1,1}(A)$, which is the best possible in the Besov scale, was discovered in [BG5] and shown to give optimal results for large classes of pseudo-differential operators in [BG6].

In the preceding description of some developments of the Mourre theory we concentrated mainly on the abstract aspects of the theory and on the efforts made in order to eliminate the hypothesis on the second commutator $[A, [A, H]]$. Various authors developed the theory and the applications in other directions, but it is not our purpose here to present a review of this work.

The main result of this chapter is Theorem 7.5.4 which gives a version of the conjugate operator method valid in the framework determined by a Friedrichs couple; this theorem is an improvement of that stated without proof in [BG5]. The version of the theory presented in Section 7.3 is an easy consequence of Theorem 7.5.4. We preferred however to give a separate proof of Theorem 7.3.1 because this theorem is sufficient in the case of N-body hamiltonians and its proof is less technical than that of Theorem 7.5.4. We mention that Lemma 7.3.2 is due to M. Mantoiu: he observed that in the context of Section 7.3 it is easy to estimate the modified resolvent G_ε in which the full approximate commutator $B_\varepsilon = [H_\varepsilon, iA]$ appears (so it is not necessary to consider the more complicated object $\varphi(H)B_\varepsilon\varphi(H)$, as in the proof of Theorem 7.5.4; see also [JMP]).

The fact that Theorem 7.3.1 allows one to treat very singular hamiltonians with a spectral gap (e.g. hard-core N-body hamiltonians) has been observed in [BG7] and is developed here in Section 7.4. Neither the domain nor the form domain of the hamiltonian is assumed to be invariant under the group generated by A, and the hypothesis $[A, H] \in B(\mathscr{H}^2, \mathscr{H}^{-1})$ made in [PSS] is replaced by $[A, H] \in B(\mathscr{H}^2, \mathscr{H}^{-2})$. But the main point here is the replacement of the condition on the second commutator by the regularity assumption $H \in \mathscr{C}^{1,1}(A)$, which is optimal in the Besov scale as explained further on in this chapter.

Let us briefly describe the organization of the chapter. Section 7.1 is de-

voted to a description of the limiting absorption principle and its consequences
in spectral and scattering theory. In Section 7.2 we consider two self-adjoint operators A, H such that H is of class $C^1(A)$ and associate to them two functions
$\varrho, \tilde{\varrho} : \mathbb{R} \to (-\infty, +\infty]$ allowing us to express Mourre's operator inequalities in
a way that is convenient for our later developments. Sections 7.3–7.5 contain
several versions of the conjugate operator method, and in the final Section 7.6
we treat the first applications of the method in two-body like situations. There
are three appendices; Appendix 7.B, devoted to the study of the optimality of
our results, is particularly important.

7.1. Locally Smooth Operators and Boundary Values of the Resolvent

7.1.1. If H is a self-adjoint operator in a Hilbert space $\mathscr{H}$ and λ a real number in its spectrum, then the limits $\lim_{\mu \to 0} R(\lambda \pm i\mu)$ do not exist in $B(\mathscr{H})$. They
could exist, however, in a larger space, and this fact has important consequences
in spectral and scattering theory as we are going to explain in this section.

We recall some of the standard terminology in the spectral theory of self-
adjoint operators ([BW], [K1], [RS] or [W]). To each $f \in \mathscr{H}$ one may associate
a positive Borel measure m on $\mathbb{R}$ by $m(B) = ||E(B)f||^2$. The (topological)
support of this measure will be called the H-*support of* f and will be denoted by
$\operatorname{supp}_H f$. We say that f is H-*absolutely continuous* on a set $B \subset \mathbb{R}$ if $m(N) = 0$
for each Borel set $N \subset B$ of Lebesgue measure zero. If each $f \in \mathscr{H}$ is H-
absolutely continuous on the set B, then H is said to have *purely absolutely
continuous spectrum* in B (equivalently, this means that $E(N) = 0$ for each N
as above). Let $m = m_{\text{ac}} + m_{\text{sc}} + m_{\text{p}}$ be the Lebesgue decomposition of m into
an absolutely continuous component m_{ac}, a singularly continuous part m_{sc} and
an atomic (or pure point) part m_{p} (the decomposition is made with respect to
Lebesgue measure). We shall say that H has *no singularly continuous spectrum
in a Borel set* $B \subset \mathbb{R}$ if $m_{\text{sc}}(B) = 0$ for each $f \in \mathscr{H}$.

For real $\mu > 0$, let $\delta_{(\mu)}$ be the function on $\mathbb{R}$ given by:

$$(7.1.1) \qquad \delta_{(\mu)}(x) = \frac{1}{\pi} \frac{\mu}{x^2 + \mu^2} = \frac{1}{\pi} \Im \frac{1}{x - i\mu}.$$

This is just the Poisson kernel, denoted by P_μ in Section 6.1, but it seems more
natural in the present context to use the notation $\delta_{(\mu)}$ for it. Clearly $\delta_{(\mu)} \geq 0$,
$\int_{-\infty}^{\infty} \delta_{(\mu)}(x)dx = 1$ and $\lim_{\mu \to 0} \delta_{(\mu)}(x) = \delta(x)$ (Dirac measure at zero) in the
sense of distributions. Then, for $\lambda \in \mathbb{R}$:

$$(7.1.2) \qquad \delta_{(\mu)}(H - \lambda) = \frac{1}{\pi} \Im R(\lambda + i\mu)$$

$$= \frac{1}{2\pi i}[R(\lambda + i\mu) - R(\lambda - i\mu)] = \frac{\mu}{\pi} R(\lambda \pm i\mu)^* R(\lambda \pm i\mu).$$

The relation (6.1.1) expresses functions of H in terms of $\delta_{(\mu)}(H - \lambda)$. One may
deduce from it a formula giving directly the spectral measure E of H in terms of
its resolvent $R(z)$. This is called *Stone's formula*, a proof of which can be found

in [RS] (Theorem VII.13):

$$(7.1.3) \qquad E((a,b)) + \frac{1}{2}E(\{a\}) + \frac{1}{2}E(\{b\}) = \underset{\mu \to +0}{\text{w-lim}} \int_a^b \delta_{(\mu)}(H - \lambda)d\lambda.$$

This holds for arbitrary real numbers $a < b$. Hence, for any $f \in \mathcal{H}$:

$$(7.1.4) \qquad \frac{1}{b-a}||E((a,b))f||^2 \leq \underset{\substack{a<\lambda<b \\ \mu>0}}{\sup} \langle f, \delta_{(\mu)}(H - \lambda)f \rangle$$

$$= \underset{\substack{a<\lambda<b \\ \mu>0}}{\sup} \frac{\mu}{\pi}||R(\lambda + i\mu)f||^2.$$

This gives a criterion of H-absolute continuity: *If $J \subset \mathbb{R}$ is an open set and $|\langle f, [\Im R(\lambda + i\mu)]f \rangle| \leq C(f) < \infty$ for all $\lambda \in J$ and $\mu > 0$, then f is H-absolutely continuous on J and $\frac{d}{d\lambda}||E_\lambda f||^2 \leq C(f)\pi^{-1}$ a.e. on J (we use the standard notation $E_\lambda = E((-\infty, \lambda]))$; if the preceding condition holds for each f in some dense subset of $\mathcal{H}$, then the spectrum of H in J is purely absolutely continuous.*

It is easy to see that the preceding implication can be partially reversed: if f is H-absolutely continuous on J and K is a subset of J at distance $r > 0$ from the boundary of J, then

$$(7.1.5) \qquad \langle f, \delta_{(\mu)}(H - \lambda)f \rangle \leq \underset{x \in J}{\text{ess sup}} \frac{d}{dx}||E_x f||^2 + \frac{1}{2\pi r}||f||^2$$

$$\text{for } \lambda \in K, \mu > 0.$$

We recall now some simple but useful identities of the Fourier transform type. For $\mu > 0$:

$$(7.1.6) \qquad R(\lambda \pm i\mu) = i \int_0^{\pm\infty} e^{i\lambda t}e^{-iHt-\mu|t|}dt.$$

Together with (7.1.2) this gives:

$$(7.1.7) \qquad \delta_{(\mu)}(H - \lambda) = \frac{1}{2\pi} \int_{-\infty}^\infty e^{i\lambda t}e^{-iHt-\mu|t|}dt.$$

These formulas hold in the strong topology of $\mathcal{H}$ and also in that of $D(H)$. Here and below the domain $D(H)$ of H is equipped with the graph topology associated to H (this is the unique B-space topology on the vector space $D(H)$ such that $D(H) \subset \mathcal{H}$ continuously). The fact that Plancherel's theorem is valid for Hilbert space valued functions leads to the following result: *If $\mathscr{F}$ is any Hilbert space and $T \in B(D(H), \mathscr{F})$, then one has for $f \in D(H)$ and $\mu > 0$:*

$$(7.1.8) \qquad \int_0^\infty e^{-2\mu t}||Te^{\mp iHt}f||^2 dt = \frac{1}{2\pi} \int_{-\infty}^\infty ||TR(\lambda \pm i\mu)f||^2 d\lambda,$$

$$(7.1.9) \qquad \int_{-\infty}^\infty e^{-2\mu|t|}||Te^{-iHt}f||^2 dt = 2\pi \int_{-\infty}^\infty ||T\delta_{(\mu)}(H - \lambda)f||^2 d\lambda.$$

We use the same notation for the norm in $\mathcal{H}$ and in $\mathscr{F}$. It is clear (consider the left-hand sides of these two relations) that the expressions in (7.1.8) and

(7.1.9) are decreasing functions of $\mu > 0$. By applying the monotone convergence theorem on the l.h.s. one finds that

$$(7.1.10) \qquad \int_0^\infty ||Te^{\mp iHt}f||^2 dt = \frac{1}{2\pi} \lim_{\mu \to +0} \int_{-\infty}^\infty ||TR(\lambda \pm i\mu)f||^2 d\lambda,$$

$$(7.1.11) \qquad \int_{-\infty}^\infty ||Te^{-iHt}f||^2 dt = 2\pi \lim_{\mu \to +0} \int_{-\infty}^\infty ||T\delta_{(\mu)}(H - \lambda)f||^2 d\lambda.$$

Of course, one may replace $\lim_{\mu \to +0}$ by $\sup_{\mu > 0}$. If $T \in B(\mathcal{H}, \mathcal{F})$, the preceding identities are valid for each $f \in \mathcal{H}$.

The integrals on the l.h.s. of (7.1.10) and (7.1.11) will be infinite in general. The class of H-smooth operators is defined as the set of those operators T for which the l.h.s. of (7.1.11) is finite *for all* $f \in \mathcal{H}$. This is a very strong requirement on T (whether T is bounded or not; see Remark 14 in §17.1.2 of [BW] for a comment on this question). We shall use the following concept of local H-smoothness. Let $J \subset \mathbb{R}$ be an open set, $\mathcal{F}$ a Hilbert space and $T : D(H) \to \mathcal{F}$ a linear, continuous operator [4]. We shall say that T is *locally H-smooth on J* if for each compact $K \subset J$ there is a constant $C_K < \infty$ such that

$$(7.1.12) \qquad \int_{-\infty}^\infty ||Te^{-iHt}f||^2 dt \leq C_K ||f||^2 \quad \text{if } \mathrm{supp}_H f \subset K.$$

In connection with (7.1.10) we mention that the same class of operators T is obtained if in the preceding estimate the integral over $\mathbb{R}$ is replaced by the integral over $(0, \infty)$ or $(-\infty, 0)$; this follows easily from

$$\int_s^\infty ||Te^{-iHt}f||^2 dt = \int_0^\infty ||Te^{-iHt}e^{-iHs}f||^2 dt.$$

From now on we identify $\mathcal{H} = \mathcal{H}^*$. Then we have dense continuous embeddings $D(H) \subset \mathcal{H} \subset D(H)^*$ and a canonical extension of $R(z)$, $z \notin \sigma(H)$, to an operator in $B(D(H)^*, \mathcal{H}) \cap B(\mathcal{H}, D(H))$. In particular, $\delta_{(\mu)}(H - \lambda) \in B(D(H)^*, D(H))$ and, if $T \in B(D(H), \mathcal{F})$, then $TR(z) \in B(\mathcal{H}, \mathcal{F})$, $R(z)T^* = (TR(\bar{z}))^* \in B(\mathcal{F}^*, \mathcal{H})$ and $T\delta_{(\mu)}(H - \lambda)T^* \in B(\mathcal{F}^*, \mathcal{F})$. Since $||S^*S|| = ||S||^2 = ||S^*||^2$, we clearly have for $\mu > 0$ and $\lambda \in \mathbb{R}$:

$$(7.1.13) \quad ||T\delta_{(\mu)}(H - \lambda)T^*|| = \frac{\mu}{\pi}||TR(\lambda \pm i\mu)||^2 = \frac{\mu}{\pi}||R(\lambda \pm i\mu)T^*||^2.$$

We shall now describe the local smoothness property in time independent terms. All the ideas of the next proof are due to T. Kato.

PROPOSITION 7.1.1. *T is locally H-smooth on J if and only if for each compact $K \subset J$ there is a finite constant C'_K such that*

$$(7.1.14) \qquad ||T[\Im R(z)]T^*|| \leq C'_K \quad \text{if } \Re z \in K \text{ and } 0 < \Im z < 1.$$

---

[4]Unlike other texts we do not require T to be closable as an operator from $\mathcal{H}$ to $\mathcal{F}$. Closability is not necessary in the proofs of the results that we shall need, and the little gain of generality will be useful at one point (Theorem 7.1.5).

PROOF. If $K \subset \mathbb{R}$ is an arbitrary closed set, we define

$$(7.1.15) \qquad C_K^0 = 2\pi \sup_{\substack{\lambda \in \mathbb{R} \\ \mu > 0}} ||TE(K)\delta_{(\mu)}(H - \lambda)T^*||$$

$$= 2\pi \sup_{\substack{\lambda \in \mathbb{R} \\ \mu > 0}} \frac{\mu}{\pi}||TE(K)R(\lambda + i\mu)||^2,$$

where the last equality follows from (7.1.13) with T replaced by $TE(K)$.

(i) We first show that C_K^0 gives the best possible constant C_K in (7.1.12). For any $f \in \mathscr{H}$, we get from (7.1.6):

$$||TR(\lambda + i\mu)f||^2 \leq \left[\int_0^\infty e^{-\mu t}||Te^{-iHt}f||dt\right]^2 \leq \frac{1}{2\mu}\int_0^\infty ||Te^{-iHt}f||^2 dt.$$

Comparison with (7.1.12) shows that $C_K \geq C_K^0$. To see that (7.1.12) holds with $C_K = C_K^0$, observe that, if $E(K)f = f$, then

$$||T\delta_{(\mu)}(H - \lambda)f||^2 \leq \frac{\mu^2}{\pi^2}||TE(K)R(\lambda + i\mu)||^2||R(\lambda - i\mu)f||^2$$

$$= \frac{\mu}{\pi}||TE(K)R(\lambda + i\mu)||^2 \langle f, \delta_{(\mu)}(H - \lambda)f\rangle.$$

Since $\int_{\mathbb{R}} \langle f, \delta_{(\mu)}(H - \lambda)f\rangle d\lambda = ||f||^2$, we get by using (7.1.11) that

$$\int_{-\infty}^\infty ||Te^{-iHt}f||^2 dt \leq 2\pi \sup_{\substack{\lambda \in \mathbb{R} \\ \mu > 0}} \frac{\mu}{\pi}||TE(K)R(\lambda + i\mu)||^2||f||^2.$$

(ii) We claim that

$$(7.1.16) \qquad \sup_{\substack{\lambda \in K \\ \mu > 0}} \frac{\mu}{\pi}||TE(K)R(\lambda + i\mu)||^2 \leq \frac{1}{2\pi}C_K^0 \leq 4 \sup_{\substack{\lambda \in K \\ \mu > 0}} \frac{\mu}{\pi}||TE(K)R(\lambda + i\mu)||^2.$$

The first inequality is evident, and for the second one it suffices to consider the contributions from the points $\lambda \notin K$ to the supremum in (7.1.15). If $\lambda \in \mathbb{R}\backslash K$, let $\lambda_0 \in K$ be such that $\text{dist}(\lambda, K) = |\lambda - \lambda_0|$. Then

$$||TE(K)R(\lambda + i\mu)|| \leq ||TE(K)R(\lambda_0 + i\mu)|| \cdot ||[I + (\lambda - \lambda_0)R(\lambda + i\mu)]E(K)||$$

$$\leq 2||TE(K)R(\lambda_0 + i\mu)||.$$

(iii) Finally, we prove the statement of the proposition. If (7.1.14) holds, then $C_K^0 < \infty$ by the second inequality in (7.1.16) (take into account (7.1.13) and use $||TE(K)R(\lambda + i\mu)|| \leq ||TR(\lambda + i\mu)||$). Conversely, assume that T is locally H-smooth on J. Let K_1 be a compact subset of J such that $K \subset K_1$ and set $\varepsilon = \text{dist}(K, \mathbb{R}\backslash K_1) > 0$. We have $C_{K_1}^0 < \infty$ by assumption, and

$||TE(\mathbb{R}\backslash K_1)R(\lambda + i\mu)|| \leq ||T(H - \lambda)^{-1}E(\mathbb{R}\backslash K_1)|| \leq C(\varepsilon)$ if $\lambda \in K$. So, for any $\lambda \in K$ and $0 < \mu < 1$:

$$||T[\Im R(\lambda + i\mu)]T^*|| = \mu||TE(K_1)R(\lambda + i\mu)||^2 + \mu||TE(\mathbb{R}\backslash K_1)R(\lambda + i\mu)||^2$$

$$\leq \frac{1}{2}C^0_{K_1} + C(\varepsilon)^2. \quad \square$$

COROLLARY 7.1.2. *If $T \in B(\mathcal{H}, \mathcal{F})$, then T is locally H-smooth on J if and only if its adjoint $T^* : \mathcal{F}^* \to \mathcal{H}$ has the following property: for each $g \in \mathcal{F}^*$, the vector T^*g is H-absolutely continuous on J and $\frac{d}{d\lambda}||E_\lambda T^*g||^2$ is locally (essentially) bounded on J.*

PROOF. By the uniform boundedness principle and the polarization identity, (7.1.14) holds if and only if $|\langle T^*g, [\Im R(\lambda + i\mu)]T^*g\rangle| \leq C(K, g) < \infty$ for each $g \in \mathcal{F}^*$ and all $\lambda \in K$, $\mu > 0$. So the assertion of the corollary is true by (7.1.4) and (7.1.5). $\square$

7.1.2. The usefulness of the concept of smooth operators will become clear from the results given in the next subsection. But now let us take advantage of the preceding corollary in order to introduce in a natural way the so-called "limiting absorption principle". We begin by expressing the property of local H-smoothness of a bounded operator T in a different form. If $T \in B(\mathcal{H}, \mathcal{F})$, let $\mathcal{K} = T^*\mathcal{F}^* \subset \mathcal{H}$ be the range of T^* equipped with the norm $||f||_{\mathcal{K}} = \inf\{||g||_{\mathcal{F}^*} \mid T^*g = f\}$. T^* induces an isometric isomorphism from the Hilbert space $\mathcal{F}^*/\ker T^*$ onto $\mathcal{K}$, so $\mathcal{K}$ is a Hilbert space continuously embedded in $\mathcal{H}$. And the fact that T is locally H-smooth on some open set $J \subset \mathbb{R}$ is equivalent to the following assertion: for each compact subset K of J there is a constant $C_K < \infty$ such that $|\langle f, [\Im R(\lambda + i\mu)]f\rangle| \leq C_K||f||^2_{\mathcal{K}}$ if $\lambda \in K$, $\mu > 0$, $f \in \mathcal{K}$. Now forget about T; let $\mathcal{K}$ be any Hilbert space continuously embedded in $\mathcal{H}$ and such that $|\langle f, [\Im R(\lambda + i\mu)]f\rangle| \leq C_K||f||^2_{\mathcal{K}}$ if K, λ, μ, C_K and f are as above. If $\overline{\mathcal{K}}$ denotes the closure of $\mathcal{K}$ in $\mathcal{H}$, then there is a unique positive operator $S \in B(\mathcal{H})$ with $S|\mathcal{H} \ominus \overline{\mathcal{K}} = 0$, $S\mathcal{H} = \mathcal{K}$ and $||Sh||_{\mathcal{K}} = ||h||_{\mathcal{H}}$ for all $h \in \overline{\mathcal{K}}$ (apply Friedrichs theorem for $\mathcal{K}$ in $\overline{\mathcal{K}}$). Clearly, any operator of the form $T = US$, $U \in B(\mathcal{H}, \mathcal{F})$, will be locally H-smooth on J.

The preceding considerations show that the crucial fact is an estimate of the form $|\langle f, [\Im R(\lambda + i\mu)]f\rangle| \leq C_K||f||^2_{\mathcal{K}}$. If this holds with $\Im R(\lambda + i\mu)$ replaced by $R(\lambda + i\mu)$ (we shall see later on that this is a considerably stronger requirement), one usually says that the limiting absorption principle holds in $\mathcal{K}$ locally on J. Of course, this is really useful only if $\mathcal{K}$ is large enough, e.g. if it is dense in $\mathcal{H}$ (then H will have purely absolutely continuous spectrum in J). It will be quite useful to consider spaces $\mathcal{K}$ which are not embedded in $\mathcal{H}$ (this allows the treatment of unbounded locally H-smooth operators T) and which are not Hilbert spaces (this will give a very precise criterion for the existence and the completeness of local wave operators).

We pass now to more formal definitions. Let $\mathcal{K}$ be a Banach space such that $\mathcal{K} \subset D(H)^*$ continuously and densely. This implies a continuous embedding $D(H) \subset \mathcal{K}^*$, but this embedding is not dense in general (and this is the most interesting situation in our applications). Let $\mathcal{K}^{*\circ}$ be the closure of $D(H)$ in $\mathcal{K}^*$, equipped with the Banach space structure induced by $\mathscr{K}^*$. Then the B-space $\mathscr{K}^{*\circ}$ is a closed subspace of $\mathscr{K}^*$ and $D(H) \subset \mathscr{K}^{*\circ}$ continuously and densely. In particular we have canonical embeddings $B(D(H)^*, D(H)) \subset B(\mathscr{K}, \mathscr{K}^{*\circ}) \subset B(\mathscr{K}, \mathscr{K}^*)$. The second embedding here is isometric, so $B(\mathscr{K}, \mathscr{K}^{*\circ})$ is a norm-closed, weak* dense [5] subspace of $B(\mathscr{K}, \mathscr{K}^*)$.

Now recall that, for non-real z, we have $R(z) \in B(D(H)^*, \mathscr{H}) \cap B(\mathscr{H}, D(H))$ and $\Im R(z) \in B(D(H)^*, D(H))$. So we may consider $\Im R(z)$ as an element of $B(\mathscr{K}, \mathscr{K}^{*\circ})$, which in turn is a subspace of $B(\mathscr{K}, \mathscr{K}^*)$. We shall say that the *generalized limiting absorption principle*(G.L.A.P.) *holds for* H *in* $\mathscr{K}$, *locally on an open set* $J \subset \mathbb{R}$, if for each compact $K \subset J$ there is a finite constant C such that $|\langle f, [\Im R(\lambda + i\mu)]f \rangle| \leq C\|f\|_{\mathscr{K}}^2$, for all $f \in \mathscr{K}$, $\lambda \in K$, $\mu > 0$; or, equivalently, if $\sup_{\lambda \in K, \mu > 0} \|\Im R(\lambda + i\mu)\|_{B(\mathscr{K}, \mathscr{K}^*)} < \infty$ for each compact $K \subset J$. We say that the *strong* G.L.A.P. *holds in* $\mathscr{K}$ *locally on* J if $\lim_{\mu \to +0} \Im R(\lambda + i\mu) \equiv \Im R(\lambda + i0)$ exists in the weak* topology of $B(\mathscr{K}, \mathscr{K}^*)$, for each $\lambda \in J$, uniformly in λ on each compact subset of J. Notice that, by virtue of the uniform boundedness principle, the G.L.A.P. follows from the strong G.L.A.P.

PROPOSITION 7.1.3. *Let* $\mathscr{K}$ *be a Banach space with* $\mathscr{K} \subset D(H)^*$ *continuously and densely, and* $J \subset \mathbb{R}$ *open.*

(a) *If the* G.L.A.P. *for* H *holds in* $\mathscr{K}$ *locally on* J, *then* H *has purely absolutely continuous spectrum in* J. *If the strong* G.L.A.P. *holds in* $\mathscr{K}$ *locally on* J, *then for each fixed* $\lambda_0 \in \mathbb{R}$, *the function* $\lambda \mapsto E_\lambda - E_{\lambda_0} \in B(\mathscr{K}, \mathscr{K}^*)$ *is weak*-continuously differentiable on* J, *and its derivative is equal to*

$$(7.1.17) \qquad \frac{d}{d\lambda}E_\lambda = \frac{1}{\pi}\Im R(\lambda + i0).$$

(b) *Assume that* $(\mathscr{K}^{*\circ})^* = \mathscr{K}$ *and that the* G.L.A.P. *holds in* $\mathscr{K}$, *locally on* J. *Let* $\mathscr{F}$ *be a Hilbert space,* $T : D(H) \to \mathscr{F}$ *a linear operator which is continuous when* $D(H)$ *is equipped with the topology induced by* $\mathscr{K}^*$; *in other terms, let* $T \in B(\mathscr{K}^{*\circ}, \mathscr{F})$. *Then* T *is locally* H-*smooth on* J.

PROOF. (a) Notice that, for any $f \in D(H)^*$, $\|E(\cdot)f\|^2$ is a well defined positive Radon measure on $\mathbb{R}$, *unbounded in general*. This is due to the fact that $E(B) \in B(D(H)^*, D(H))$ if B is a bounded Borel set. Clearly (7.1.3) will hold in $B(D(H)^*, D(H))$, hence (7.1.4) holds for any $f \in D(H)^*$. So for $f \in \mathscr{K}$, the measure $\|E(\cdot)f\|^2$ is absolutely continuous on J with locally bounded derivative (and reciprocally, if this holds for all $f \in \mathscr{K}$, then the G.L.A.P. holds in $\mathscr{K}$ locally on J, cf. (7.1.5)). In particular, if $N \subset J$ is a bounded set of Lebesgue measure zero, then $\|E(N)f\|^2 = 0$ for $f \in \mathscr{K}$. But $E(N) : D(H)^* \to \mathscr{H}$ is continuous and $\mathscr{K}$ is dense in $D(H)^*$; so $E(N) = 0$. (7.1.17) follows from (7.1.3).

(b) We have $T^* \in B(\mathscr{F}^*, \mathscr{K})$. Since $\Im R(z)$ maps $\mathscr{K}$ into $D(H) \subset \mathscr{K}^{*\circ}$, we get

$$\|T[\Im R(z)]T^*\| \leq \|T\|_{\mathscr{K}^* \to \mathscr{F}}\|\Im R(z)\|_{\mathscr{K} \to \mathscr{K}^*}\|T^*\|_{\mathscr{F}^* \to \mathscr{K}},$$

and it suffices to apply Proposition 7.1.1. $\square$

---

[5] We recall that, if $\mathscr{F}, \mathscr{G}$ are Banach spaces, then the weak* topology on $B(\mathscr{G}, \mathscr{F}^*)$ is the topology defined by the family of seminorms $S \mapsto |\langle f, Sg \rangle|$ with $f \in \mathscr{F}$ and $g \in \mathscr{G}$. If $\mathscr{F} = \mathscr{G}$, it is enough to consider $f = g$ (by the polarization identity).

The preceding proof shows that, even if $\mathscr{K}$ does not satisfy the condition $(\mathscr{K}^{*\circ})^* = \mathscr{K}$, the conclusion of Proposition 7.1.3 (b) remains true for an arbitrary $T \in B(D(H), \mathscr{F})$ satisfying $T^* \mathscr{F}^* \subset \mathscr{K}$. If T^* has this property, then $T \in B(\mathscr{K}^{*\circ}, \mathscr{F})$ (indeed, let S be the operator T^* considered with values in $\mathscr{K}$; then by the closed graph theorem we have $S \in B(\mathscr{F}^*, \mathscr{K})$, hence $S^* \in B(\mathscr{K}^*, \mathscr{F})$; but clearly $S^*|_{D(H)} = T$). On the other hand, from $T \in B(\mathscr{K}^{*\circ}, \mathscr{F})$ it follows that $T^* \in B(\mathscr{F}^*, (\mathscr{K}^{*\circ})^*)$, and in general we only have $\mathscr{K} \subset (\mathscr{K}^{*\circ})^* \subset D(H)^*$. This comment explains the exact role of the assumption $(\mathscr{K}^{*\circ})^* = \mathscr{K}$.

The usual form of the limiting absorption principle is *formally* obtained by replacing $\Im R(\lambda + i\mu)$ by $R(\lambda + i\mu)$ in the definition of the G.L.A.P. Unfortunately, one cannot work in the preceding framework (without further conditions on $\mathscr{K}$) because $R(z)$ does not belong to $B(D(H)^*, D(H))$ if H is unbounded, hence the expression $\langle f, R(z)f \rangle$ is not a priori well defined for $f \in \mathscr{K}$. In order to bypass this difficulty without introducing too involved conditions, and also because the next condition appears quite naturally in our applications, we shall proceed as follows. Observe first that the consideration of the form domain $D(|H|^{1/2}) \equiv (D(H), \mathscr{H})_{1/2,2} = [D(H), \mathscr{H}]_{1/2}$ of H naturally leads to the following structure:

$$(7.1.18) \qquad D(H) \subset D(|H|^{1/2}) \subset \mathscr{H} = \mathscr{H}^* \subset D(|H|^{1/2})^* \subset D(H)^*.$$

All the spaces here are Hilbert spaces and the embeddings are continuous and dense. The main point is that $R(z) \in B(D(|H|^{1/2})^*, D(|H|^{1/2}))$ for $z \notin \sigma(H)$, with norm $\leq 1 + |z + i| \cdot [\mathrm{dist}(z, \sigma(H))]^{-1}$. So from now on we shall assume that $\mathscr{K}$ is a Banach space such that $\mathscr{K} \subset D(|H|^{1/2})^*$ continuously and densely (in fact, for the main applications one may assume that $\mathscr{K} \subset \mathscr{H}$, cf. Sections 7.3 and 7.4; then it is enough to consider the much simpler structure $\mathscr{K} \subset \mathscr{H} \subset \mathscr{K}^*$, and so $B(\mathscr{H}) \subset B(\mathscr{K}, \mathscr{K}^*)$). This implies that $D(|H|^{1/2}) \subset \mathscr{K}^*$ continuously; hence, if $\mathscr{K}^{*\circ}$ denotes the closure of $D(|H|^{1/2})$ in $\mathscr{K}^*$, we get canonical embeddings

$$(7.1.19) \qquad B(D(|H|^{1/2})^*, D(|H|^{1/2})) \subset B(\mathscr{K}, \mathscr{K}^{*\circ}) \subset B(\mathscr{K}, \mathscr{K}^*).$$

Now, the definition of *the (strong) limiting absorption principle*(L.A.P.) *for H in $\mathscr{K}$, locally on J*, is obtained by replacing $\Im R(\lambda + i\mu)$ by $R(\lambda + i\mu)$ in the definition of the (strong) G.L.A.P. We shall see, in the example we shall discuss in the last part of this section, that the L.A.P. is a much stronger condition on $\mathscr{K}$ than the G.L.A.P. In other terms, the real part of the resolvent $R(\lambda + i\mu)$ is a much more singular object (in the limit $\mu = +0$) than its imaginary part. Observe that the strong L.A.P. holds in $\mathscr{K}$ locally on J if and only if the limits $\lim_{\mu \to +0} R(\lambda \pm i\mu) \equiv R(\lambda \pm i0)$ exist in the weak* topology of $B(\mathscr{K}, \mathscr{K}^*)$ for each $\lambda \in J$, uniformly in λ on each compact subset of J. One can reformulate this in slightly different terms as follows. Let $\mathbb{C}_\pm = \{z \in \mathbb{C} \mid \pm \Im z > 0\}$ and observe that $\mathbb{C}_\pm \ni z \mapsto R(z) \in B(\mathscr{K}, \mathscr{K}^*)$ is a holomorphic function (with values in $B(D(|H|^{1/2})^*, D(|H|^{1/2}))$ in fact). The strong L.A.P. is equivalent to the fact that this function has a weak*-continuous extension to the set $\mathbb{C}_\pm \cup J$. The boundary values $R(\lambda \pm i0)$ of the resolvent on the real axis allow us to express the derivative of the spectral measure on J (see (7.1.17)):

$$(7.1.20) \qquad \frac{d}{d\lambda} E_\lambda = \frac{1}{2\pi i}[R(\lambda + i0) - R(\lambda - i0)].$$

7.1.3. As we explained before, the G.L.A.P. is helpful in finding classes of locally smooth operators. The next theorem is, in fact, an example of what such operators are useful for. We shall give simultaneously a criterion for the *existence and completeness of the so-called local wave operators* and an *invariance principle*. The theorem is due to Kato, Lavine and Schechter, and we give a rather detailed sketch of the proof because we find it instructive. Observe that the class of functions φ allowed by our treatment is considerably larger than in most other formulations of the invariance principle (see, however the work of M. Wollenberg [Wol] or its presentation in [Ko]). In fact, a result which we present in Appendix C to this chapter shows that our functions φ are admissible in the sense of Schechter and reciprocally, if φ has a non-zero derivative almost everywhere, then the condition of admissibility implies $\varphi'(x) > 0$ a.e.

THEOREM 7.1.4. *Let H_1 and H_2 be self-adjoint operators in a Hilbert space $\mathcal{H}$, denote by E_1 and E_2 their spectral measures, and assume that there are a Hilbert space $\mathcal{F}$ and operators $T_j \in B(D(H_j), \mathcal{F})$ such that $H_1 - H_2 = T_1^* T_2$ as forms on $D(H_1) \times D(H_2)$; more explicitly, this means $\langle H_1 f_1, f_2 \rangle - \langle f_1, H_2 f_2 \rangle = \langle T_1 f_1, T_2 f_2 \rangle$ for all $f_j \in D(H_j)$. Assume that $J \subset \mathbb{R}$ is an open set and that T_j is locally H_j-smooth on J $(j = 1, 2)$. Then*

$$(7.1.21) \qquad W_{\pm}(H_1, H_2; J) := \operatorname*{s-lim}_{t \to \pm\infty} e^{iH_1 t} e^{-iH_2 t} E_2(J)$$

exist, are bijective isometries of $E_2(J)\mathcal{H}$ onto $E_1(J)\mathcal{H}$ and satisfy

$$W_{\pm}(H_1, H_2; J)^* = W_{\pm}(H_2, H_1; J),$$
$$W_{\pm}(H_1, H_2; J)\theta(H_2) = \theta(H_1)W_{\pm}(H_1, H_2; J)$$

for each bounded Borel function $\theta : \mathbb{R} \to \mathbb{C}$. Moreover, if $\varphi : \mathbb{R} \to \mathbb{R}$ is Borel and has a strictly positive, finite derivative almost everywhere, then:

$$(7.1.22) \qquad W_{\pm}(H_1, H_2; J) = \operatorname*{s-lim}_{t \to \pm\infty} e^{it\varphi(H_1)} e^{-it\varphi(H_2)} E_2(J).$$

In particular, if $J = \varphi^{-1}(\widetilde{J})$ for some open set $\widetilde{J} \subset \mathbb{R}$ and if $\widetilde{H}_j = \varphi(H_j)$, then $W_{\pm}(\widetilde{H}_1, \widetilde{H}_2; \widetilde{J})$ exist and are equal to $W_{\pm}(H_1, H_2; J)$.

PROOF. (i) The existence of the limits (7.1.21) is a simple consequence of the following assertion: for each $f_2 \in \mathcal{H}$ such that $\operatorname{supp}_{H_2} f_2 \equiv K_2$ is a compact subset of J, and for each function $\theta_1 \in C_0^{\infty}(J)$ with $\theta_1(\lambda) = 1$ on a neighbourhood of K_2, we have:

$$(7.1.23) \qquad \operatorname*{s-lim}_{t \to \pm\infty} \theta_1(H_1) e^{iH_1 t} e^{-iH_2 t} f_2 \text{ exist,}$$

$$(7.1.24) \qquad \operatorname*{s-lim}_{t \to \pm\infty} [I - \theta_1(H_1)] e^{iH_1 t} e^{-iH_2 t} f_2 = 0.$$

(ii) We prove (7.1.23). We set $W(t) = \theta_1(H_1)e^{iH_1 t}e^{-iH_2 t}$ and observe that for $f_1 \in \mathcal{H}$ and $s < t$:

$$
\begin{aligned}
|\langle f_1, [W(t) - W(s)]f_2 \rangle| &= \left| \int_s^t \langle T_1 e^{-iH_1 \sigma} \theta_1(H_1)f_1, T_2 e^{-iH_2 \sigma} f_2 \rangle d\sigma \right| \\
&\leq \left[\int_s^t \|T_1 e^{-iH_1 \sigma} \theta_1(H_1)f_1\|^2 d\sigma \int_s^t \|T_2 e^{-iH_2 \sigma} f_2\|^2 d\sigma \right]^{1/2} \\
&\leq C_{K_1} \|f_1\| \left[\int_s^t \|T_2 e^{-iH_2 \sigma} f_2\|^2 d\sigma \right]^{1/2}.
\end{aligned}
$$

Here $K_1 = \operatorname{supp} \theta_1$ and C_{K_1} is a finite constant furnished by (7.1.12). We obtain $\|[W(t) - W(s)]f_2\| \to 0$ as $s \to +\infty$ or $t \to -\infty$, which proves (7.1.23).

(iii) Now we prove (7.1.24). Let $\theta_2 \in C_0^\infty(J)$ with $\theta_2(\lambda) = 1$ if $\lambda \in K_2$ and $\theta_1 \theta_2 = \theta_2$. Then $f_2 = \theta_2(H_2)f_2$ and $[I - \theta_1(H_1)]\theta_2(H_2) = [I - \theta_1(H_1)][\theta_2(H_2) - \theta_2(H_1)]$. Hence (7.1.24) follows from

$$
\text{(7.1.25)} \qquad \lim_{|t| \to \infty} \|[\theta_2(H_2) - \theta_2(H_1)]e^{-iH_2 t} f_2\| = 0.
$$

We prove this for any function $\theta_2 \in C_\infty(\mathbb{R})$. Since the vector space generated by the family of functions $\{r_z\}_{z \in \mathbb{C} \setminus \mathbb{R}}$, $r_z(x) = (x - z)^{-1}$, is a dense subspace of $C_\infty(\mathbb{R})$, it is enough to show (7.1.25) with θ_2 replaced by r_z. Set $R_j = (H_j - z)^{-1}$ and observe that for any $g_j \in \mathcal{H}$:

$$
\begin{aligned}
|\langle g_1, (R_1 - R_2)g_2 \rangle| &= |\langle R_1^* g_1, H_2 R_2 g_2 \rangle - \langle H_1 R_1^* g_1, R_2 g_2 \rangle| \\
&= |-\langle T_1 R_1^* g_1, T_2 R_2 g_2 \rangle| \leq \|T_1 R_1^*\| \cdot \|g_1\| \cdot \|T_2 R_2 g_2\|.
\end{aligned}
$$

Taking $g_2 = e^{-iH_2 t} f_2$ we see that it is enough to prove that $\|T_2 R_2 e^{-iH_2 t} f_2\| \to 0$ as $|t| \to \infty$. But this is an easy consequence of the fact that both the function $F(t) = T_2 R_2 e^{-iH_2 t} f_2$ and its derivative are square-integrable on $\mathbb{R}$.

(iv) At this moment, the existence of the limits in (7.1.21) is proved. The same arguments show that $W_\pm(H_2, H_1; J)$ exist too, and this implies the assertions made in the sentence following (7.1.21) by standard and easy arguments. It remains to prove (7.1.22), and for this we shall use a method essentially due to Schechter [Sche2].

(v) We begin with an observation of a general order. Let H be a self-adjoint operator in $\mathcal{H}$, f a vector in the domain of H, and T a linear continuous operator from $D(H)$ (equipped with the graph topology) into a Hilbert space $\mathcal{F}$. Assume that the function $F : \mathbb{R} \to \mathcal{F}$ defined by $F(t) = (2\pi)^{-1/2} T e^{-iHt} f$, is square-integrable. We recall that for Hilbert space-valued functions on $\mathbb{R}$ the usual L^2-version of the Fourier transform theory remains valid. In particular, the inverse Fourier transform $\check{F}(\lambda) = (2\pi)^{-1/2} \int_{\mathbb{R}} e^{i\lambda t} F(t)dt$ (which, a priori, is a $\mathcal{F}$-valued tempered distribution) is a square-integrable $\mathcal{F}$-valued function on $\mathbb{R}$. Moreover, Plancherel's theorem gives for any $\alpha \in \mathscr{S}(\mathbb{R})$ the identity $\int_{\mathbb{R}} \alpha(\lambda)\check{F}(\lambda)d\lambda = \int_{\mathbb{R}} \check{\alpha}(t)F(t)dt \equiv T\alpha(H)f$ (the last equality is obvious). From this we first deduce that the support of the function $\check{F}$ is included in $\operatorname{supp}_H f$ (because $\alpha(H)f = 0$ if $\alpha \in C_0^\infty(\mathbb{R} \setminus \operatorname{supp}_H f)$; in fact one may show by using (7.1.7) that $\check{F}(\lambda) = \frac{d}{d\lambda} T E_\lambda f$, but we do not need this). Now assume that $\operatorname{supp}_H f$ is compact. Then, by using a standard limiting procedure, we obtain $T\alpha(H)f = \int \alpha(\lambda)\check{F}(\lambda)d\lambda$ for any bounded Borel function $\alpha : \mathbb{R} \to \mathbb{C}$. In particular, if $\varphi : \mathbb{R} \to \mathbb{R}$ is Borel, we shall have for all $s, t \in \mathbb{R}$:

$$Te^{-isH-it\varphi(H)}f = \int_{-\infty}^{\infty} e^{-is\lambda-it\varphi(\lambda)}\check{F}(\lambda)d\lambda = [e^{-it\varphi(-P)}F](s).$$

The last equality may be interpreted as a definition. Hence:

$$\int_{0}^{\infty} ||Te^{-isH}e^{-it\varphi(H)}f||^2 ds = \int_{0}^{\infty} ||[e^{-it\varphi(-P)}F](s)||^2 ds.$$

Now Corollary 7.C.2 is clearly valid for Hilbert space-valued functions u. In conclusion, if φ is as in the statement of the theorem, we shall have

(7.1.26) $$\lim_{t\to\infty} \int_{0}^{\infty} ||Te^{-isH}e^{-it\varphi(H)}f||^2 ds = 0.$$

(vi) We now prove (7.1.22) for $t \to +\infty$. Let $W = W_+(H_1, H_2; J)$; we first show that:

(7.1.27) $$\lim_{t\to\infty} \langle e^{-it\varphi(H_1)}f_1, (W - I)e^{-it\varphi(H_2)}f_2 \rangle = 0$$

for all $f_j \in E_j(J)\mathcal{H}$. We may suppose that $\mathrm{supp}_{H_j} f_j$ is a compact subset of J. Denoting $f_j^t = e^{-it\varphi(H_j)}f_j$, we have:

$$|\langle f_1^t, (W - I)f_2^t \rangle| = \left| \int_{0}^{\infty} \langle T_1 e^{-isH_1} f_1^t, T_2 e^{-isH_2} f_2^t \rangle ds \right|$$

$$\leq \prod_{j=1,2} \left[\int_{0}^{\infty} ||T_j e^{-isH_j-it\varphi(H_j)} f_j||^2 ds \right]^{1/2},$$

which tends to zero as $t \to +\infty$, by (7.1.26). So (7.1.27) is true. Finally, let $f_2 \in E(J)\mathcal{H}$ and $f_1 = Wf_2$, which belongs to $E_1(J)\mathcal{H}$ because $WE_2(J) = E_1(J)W$. Then:

$$||e^{it\varphi(H_1)}e^{-it\varphi(H_2)}f_2 - Wf_2||^2 = ||f_2^t - f_1^t||^2 = 2\Re\langle f_1^t, (W - I)f_2^t \rangle$$

because $W : E_2(J)\mathcal{H} \to E_1(J)\mathcal{H}$ is an isometry. (7.1.27) shows that the preceding expression tends to zero as $t \to \infty$, which finishes the proof of (7.1.22).

(vii) For the proof of the last assertion of the theorem, observe that the spectral measure $\tilde{E}_j$ of $\tilde{H}_j$ is given by $\tilde{E}_j(B) = E_j(\varphi^{-1}(B))$ for any Borel set $D \subset \mathbb{R}$. $\square$

The factorization assumption $H_1 - H_2 = T_1^* T_2$ made in the preceding theorem is inconvenient in applications and, aesthetically speaking, not very satisfactory. Fortunately, the spaces in which we shall prove the L.A.P. will not only provide us with explicit classes of locally smooth operators, but will also make trivial the verification of this factorization property. In the next theorem we use notions introduced in Section 2.8 (cotype and approximation property). However, if the reader accepts that spaces of the form $(\mathcal{E}_1, \mathcal{E}_0)_{1/2,1}$, where $(\mathcal{E}_1, \mathcal{E}_0)$ is a Friedrichs couple, have cotype 2 and the bounded approximation property, then he will not need to know the precise meaning of these notions, since in our applications only such spaces will appear.

THEOREM 7.1.5. *Let H_1 and H_2 be self-adjoint operators in a Hilbert space $\mathscr{H}$. For each $j = 1, 2$ let $\mathscr{K}_j$ be a Banach space such that $\mathscr{K}_j \subset D(H_j)^*$ continuously and densely; notice the embedding $D(H_j) \subset \mathscr{K}_j^*$, denote by $\mathscr{K}_j^{*\circ}$ the closure of $D(H_j)$ in $\mathscr{K}_j^*$ and suppose that $(\mathscr{K}_j^{*\circ})^* = \mathscr{K}_j$. Furthermore, assume that the following conditions are satisfied:*

(i) *the G.L.A.P. for H_j holds in $\mathscr{K}_j$ locally on some open set $J \subset \mathbb{R}$;*

(ii) *$H_1 - H_2$, considered as a sesquilinear form on $D(H_1) \times D(H_2)$, is continuous for the topology induced by $\mathscr{K}_1^* \times \mathscr{K}_2^*$; in other terms, there is a continuous operator $V : \mathscr{K}_1^{*\circ} \to \mathscr{K}_2 = (\mathscr{K}_2^{*\circ})^*$ such that $\langle H_1 f_1, f_2 \rangle - \langle f_1, H_2 f_2 \rangle = \langle V f_1, f_2 \rangle$ for all $f_j \in D(H_j)$;*

(iii) *the Banach space $\mathscr{K}_j$ ($j = 1, 2$) is of cotype 2 and has the bounded approximation property (these conditions are fulfilled if $\mathscr{K}_j$ is obtained from a Friedrichs couple by real interpolation of order (θ, p) with $0 < \theta < 1$ and $1 \le p \le 2$).*

Then the hypotheses of Theorem 7.1.4 are fulfilled, hence its conclusions are true.

PROOF. By the theorem of Pisier quoted in §2.8.4, there is a Hilbert space $\mathscr{F}$ and there are bounded operators $S_1 : \mathscr{K}_1^{*\circ} \to \mathscr{F}$ and $S_2 : \mathscr{F} \to \mathscr{K}_2$ such that $V = S_2 S_1$. Identify $\mathscr{F}^* = \mathscr{F}$ and let $T_1 = S_1|_{D(H_1)}$ and $T_2 = S_2^*|_{D(H_2)}$. By Proposition 7.1.3, T_j is locally H_j-smooth on J, hence the assumptions of Theorem 7.1.4 are satisfied. $\square$

7.1.4. In this subsection we shall give a rather detailed description of the spectral properties of a certain class of multiplication operators. This will allow us to consider the notions introduced so far in the simplest non-trivial case, and it will provide us with interesting examples for the theory that will be developed later on in this chapter. One should notice the fundamental role played by the translation group (through the notion of derivative) for obtaining non-trivial spectral properties for the operators of multiplication by functions.

Let $\Omega \subset \mathbb{R}^n$ be an open set and let H be the operator of multiplication by a Borel function $h : \Omega \to \mathbb{R}$ in the Hilbert space $\mathscr{H} = L^2(\Omega)$. Then the spectral measure E of H is easily described: for any Borel set $B \subset \mathbb{R}$, $E(B)$ is the operator of multiplication by the characteristic function of the set $h^{-1}(B)$. Hence one may describe several properties of H in terms of h according to the following rules (when we speak about the measure of a subset of a space $\mathbb{R}^k$ without further specification, we always mean Lebesgue measure):

(a) the spectrum of H is equal to the essential range of h (i.e. the set of $\lambda \in \mathbb{R}$ such that $h^{-1}(U)$ is of non-zero measure for any neighbourhood U of λ); if h is continuous, this is just the closure of the range $h(\Omega)$ of h;

(b) $\lambda \in \mathbb{R}$ is an eigenvalue of h if and only if $h^{-1}(\lambda)$ has non-zero measure;

(c) H has purely absolutely continuous spectrum in a set $J \subset \mathbb{R}$ if and only if for any Borel set $N \subset J$ of measure zero, $h^{-1}(N)$ is also of measure zero;

(d) H has a non-trivial singularly continuous component if and only if there is a Borel set $N \subset \mathbb{R}$ of measure zero, such that $h^{-1}(N)$ has non-zero measure but $h^{-1}(\lambda)$ is of zero measure for each $\lambda \in N$.

We leave to a later section of this chapter the study of the operators $H = h(Q)$ in the case $n > 1$; we shall see that this is not a trivial matter, even if h is a smooth function. Here we specialize to the case $n = 1$, where more explicit results can be obtained by classical methods. We shall consider rather singular functions h in order to explain exactly what happens. This will also provide us with a framework in which the abstract tools developed in Sections 7.2-7.5 are easily tested and shown to give optimal results, in a precise sense.

We first describe some known facts about the differentiability of increasing functions (proofs can be found in [S], especially Ch. IV, §9 and Ch. VI, §7; the theorem of de la Vallée Poussin which we use is Theorem 9.6 in Ch. IV). Let $U \subset \mathbb{R}$ be an open interval and $\varphi : U \to \mathbb{R}$ a continuous increasing function. It will be convenient here to make a slight change of notation and to denote its distributional derivative by $\varphi^{(1)}$. It is easily proven (see [Sch]) that $\varphi^{(1)}$ is just the Lebesgue-Stieltjes measure defined by φ (so $\int f(x)\varphi^{(1)}(x)dx = \int f(x)d\varphi(x)$ for $f \in C_0^\infty(U)$, where the l.h.s. is interpreted in the sense of distributions, while the r.h.s. is a Stieltjes integral). Below we shall say L-a.e. or φ-a.e. if we mean that a property holds almost everywhere with respect to Lebesgue measure or $\varphi^{(1)}$-measure; similarly we shall use the terms "L-measure zero" and "φ-measure zero". If the classical derivative of φ at a point $x \in U$ exists, it will be denoted by $\varphi'(x)$ (the value $+\infty$ for $\varphi'(x)$ is permitted); so $\varphi'(x)$ could be an arbitrary number in $[0, +\infty]$. Then $\varphi'(x)$ exists *and is finite* for L-a.e. $x \in U$. The points where $\varphi'(x)$ does not exist are, in fact, rather irrelevant, since they are of L-measure zero and of φ-measure zero. So $\varphi'(x)$ exists L-a.e. *and φ-a.e. on* U. Really important are the two disjoint Borel sets $A_\varphi = \{x \in U \mid \varphi'(x)$ exists and $0 < \varphi'(x) < \infty\}$ and $S_\varphi = \{x \in U \mid \varphi'(x)$ exists and $\varphi'(x) = +\infty\}$. In fact, A_φ and S_φ are Borel-supports [6] for the absolutely continuous and for the singular part of $\varphi^{(1)}$ respectively (this result is due to de la Vallée Poussin). More precisely, if $B \subset U$ is a Borel set, then

(7.1.28)

$$\varphi^{(1)}(B) \equiv \int_B d\varphi(x) = \int_B \varphi'(x)dx + \varphi^{(1)}(B \cap S_\varphi)$$

$$= \int_{B \cap A_\varphi} \varphi'(x)dx + \varphi^{(1)}(B \cap S_\varphi).$$

Observe that the set where $\varphi'(x) = 0$ is of φ-measure zero. However, it could be of L-measure equal to the length of U; this happens if and only if the measure $\varphi^{(1)}$ is purely singular with respect to Lebesgue measure, and then one usually says that the *function* φ is purely singular.

---

[6]We stress the fact that for a measure μ on an open subset $U \subset \mathbb{R}^n$ one can introduce, besides the usual (topological) notion of support, a more subtle one as follows. A Borel set $M \subset U$ is a *Borel-support* for μ if $\mu(U \setminus M) = 0$. Clearly, there are many such supports. But there is only one Borel-support which is closed and is contained in any other closed Borel-support: this is the topological support, denoted $\text{supp}\,\mu$. In measure theory one is usually interested in finding Borel-supports M much smaller than $\text{supp}\,\mu$; notice that if $M \subset \text{supp}\,\mu$, then $\overline{M} = \text{supp}\,\mu$.

PROPOSITION 7.1.6. *Let $\Omega \subset \mathbb{R}$ an open interval and $h : \Omega \to \mathbb{R}$ a continuous, strictly increasing function. Let H be the self-adjoint operator in $\mathscr{H} = L^2(\Omega)$ defined as multiplication by the function h. Then H has purely continuous spectrum equal to the closure of the interval $J = h(\Omega)$. Let S_H be the Borel subset of Ω consisting of the points x where the (classical) derivative $h'(x)$ exists and is equal to zero. Then*

(a) *H has purely absolutely continuous spectrum if and only if S_H is of Lebesgue measure zero.*

(b) *$h(S_H) \equiv \kappa(h)$ is a Borel subset of J of Lebesgue measure zero, and outside $\kappa(H)$ the spectrum of H is purely absolutely continuous.*

PROOF. The function h is a homeomorphism of Ω onto the open interval J; we shall denote by $g \equiv h^{-1} : J \to \Omega$ its inverse, so g is also continuous and strictly increasing. We shall use below the following fact: let $x \in \Omega$ and $y = h(x)$; then h is differentiable at x (with derivative $h'(x)$ finite or not) if and only if g is differentiable at y; and in this case we have $g'(y) = [h'(x)]^{-1}$. In terms of slightly formal notations like $\{h' = 0\} \equiv \{x \in \Omega \mid h'(x)$ exists and is equal to zero$\}$, we get $\{g' = \infty\} = h\{h' = 0\}$ and $\{0 < g' < \infty\} = h\{0 < h' < \infty\}$.

Let Ω be equipped with Lebesgue measure and let the interval J be equipped with the measure $g^{(1)}$. In particular, $\mathscr{H} = L^2(\Omega)$ is constructed with the help of the Lebesgue measure, while $\mathscr{H}_1 = L^2(J) \equiv L^2(J; g^{(1)})$ is constructed in terms of the measure $g^{(1)}$. Observe that the Borel isomorphism $h : \Omega \to J$ is measure preserving: in fact, for any interval $[a, b] \subset \Omega$, we have $b - a = g(h(b)) - g(h(a)) = g^{(1)}([h(a), h(b)]) = g^{(1)}(h([a, b]))$, which implies the assertion. It follows that the map $f \mapsto f \circ g$ induces a unitary operator $\mathscr{H} \to \mathscr{H}_1$, which clearly transforms H into the operator H_1 of multiplication by the free variable $y \in J$ in $\mathscr{H}_1$. This shows that the spectral properties of H are completely described by the measure $g^{(1)}$. For example, H has purely absolutely continuous spectrum if and only if $g^{(1)}$ is absolutely continuous. According to the discussion preceding the statement of the proposition, this happens if and only if $\{g' = \infty\}$ is of $g^{(1)}$-measure zero (see (7.1.28)). Since $h : \Omega \to J$ is measure-preserving and $h\{h' = 0\} = \{g' = \infty\}$, the assertion (a) of the proposition is proved. Moreover, $h(S_H) \equiv \{g' = \infty\}$ and g has a finite derivative L-a.e., so $h(S_H)$ has L-measure zero, and outside this set $g^{(1)}$ is absolutely continuous, which proves (b). $\square$

We consider now some examples in the context of the above proposition. If h is a purely singular function (i.e. $h'(x) = 0$ L-a.e.), then H will have purely singularly continuous spectrum, equal to the interval $\overline{J}$. In order to see this, it is enough to use the formalism introduced in the preceding proof and to observe that $h^{-1} = g$ is also purely singular (indeed, $h\{h' = 0\} = \{g' = \infty\}$ and h transforms Lebesgue measure into $g^{(1)}$, so $g' = \infty$ g-a.e. and this is equivalent to $g'(y) = 0$ L-a.e. by (7.1.28)). In view of the regularity assumptions which will be imposed in the conjugate operator method (cf. Sections 7.3-7.5), it is interesting to note that one can construct strictly increasing, purely singular functions h having any modulus of continuity that does not imply the Lipschitz property (e.g. one may construct such functions h that are Hölder continuous of all orders $\theta < 1$, i.e. such that $|h(x_1) - h(x_2)| \le c(\theta)|x_1 - x_2|^\theta$ for all $\theta < 1$; see [HK] for a much better result). If h is an absolutely continuous function (e.g. if it is locally

Lipschitz), then h^{-1} cannot be purely singular (otherwise $h = (h^{-1})^{-1}$ would be purely singular too), so H will have a non-zero absolutely continuous part. However, H could have a non-trivial singularly continuous component even if h is of class C^∞: according to Proposition 7.1.6 (a) this happens if the derivative of h is zero on a large set (i.e. a set of non-zero Lebesgue measure). More precisely, let $h : \Omega \to \mathbb{R}$ of class C^1 and strictly increasing and set $F = \{x \in \Omega \mid h'(x) = 0\}$. Then F is a closed subset of Ω with empty interior. Reciprocally, for each such F there is a function h as before such that $F = \{h' = 0\}$; for example take h equal to a primitive of the distance function $x \mapsto \text{dist}(x, F)$. If F is also closed as a subset of $\mathbb{R}$, then one may even choose $h \in C^\infty$ (for this, let $\varphi \in C^\infty(\mathbb{R})$, $\varphi \geq 0$, with $F = \{\varphi = 0\}$, which exists by Lemma 1.4.13 in [N], and take for h a primitive of φ; observe that the choice $\varphi(x) = \text{dist}(x, F)$ gives $h \in C^1$ with h' Lipschitz). Since there are plenty of closed sets F with empty interior and non-zero Lebesgue measure, it is easy to construct $h \in C^\infty(\Omega)$, strictly increasing and such that H has a nontrivial singularly continuous component. This fact shows the importance of the requirement of *strict* positivity imposed on the constant a in the Mourre estimate (cf. Section 7.3). By Proposition 7.1.6, the spectrum of the singular component of H is always contained in the critical set $\kappa(h)$ of h (a closed set of measure zero). It is clear that, if each neighbourhood in F of each point of F has non-zero measure, then the spectrum of the singular component of H *is equal* to $\kappa(h)$ (see the proof of Proposition 7.1.6; such F can be constructed).

Now that we have a rather good understanding of the absolutely continuous part of a multiplication operator H, it is instructive to consider the meaning of the limiting absorption principle in this context. If z is a complex number away from $\overline{h(\Omega)}$ (assuming h continuous), then the resolvent $R(z)$ is the operator of multiplication by the function $\Omega \ni x \mapsto (h(x) - z)^{-1}$. Hence $R(\lambda \pm i0)$ should be multiplication by $(h(x) - \lambda \mp i0)^{-1}$, *in some sense*. But if λ belongs to the range of h, it is clear that this is a quite singular object, and we shall see that it is not easy to give a meaning to it if h is not smooth enough ($h \in C^1$ is not sufficient !). We shall continue to consider the one-dimensional case here, our main purpose being to construct a counter-example. Positive, essentially optimal results, will be obtained later on in the n-dimensional case (the division theorem).

Let $\Omega \subset \mathbb{R}$ be an open interval and $h : \Omega \to \mathbb{R}$ a Borel function; as usual, H is the operator of multiplication by h in $\mathscr{H} = L^2(\Omega)$ and $R(\lambda + i\mu) = (H - \lambda - i\mu)^{-1}$ for λ, $\mu \in \mathbb{R}$, $\mu > 0$. Then for $u, v \in L^2(\Omega)$:

$$(7.1.29) \qquad \langle u, R(\lambda + i\mu)v \rangle = \int_\Omega \frac{\overline{u(x)}v(x)}{h(x) - \lambda - i\mu} dx.$$

It is natural to try to get a limiting absorption principle in the framework of distribution theory, i.e. to see whether the above quantity is bounded as $\mu \to 0$ for all $u, v \in C_0^\infty(\Omega)$. Since $u\overline{v}$ is then of class $C_0^\infty(\Omega)$ and each $\varphi \in C_0^\infty(\Omega)$ is such a product, the question may be stated in purely distributional terms as follows. For a fixed $\lambda \in \mathbb{R}$, consider the family $\{(h(\cdot) - \lambda - i\mu)^{-1}\}_{\mu > 0}$ of bounded Borel functions on Ω as a subset of the space of distributions $\mathscr{D}^*(\Omega)$ on Ω; is this family bounded in $\mathscr{D}^*(\Omega)$? If it is, then one can find a sequence of positive numbers such that $\mu_j \to 0$ and $\lim_{j \to \infty}(h(x) - \lambda - i\mu_j)^{-1}$ exists in $\mathscr{D}^*(\Omega)$ (by the compactness of bounded sets in $\mathscr{D}^*(\Omega)$). So it is natural to inquire whether the limit $\lim_{\mu \to +0}(h(x) - \lambda - i\mu)^{-1}$ exists in the sense of distributions, i.e. whether $\lim_{\mu \to +0} \int \varphi(x)(h(x) - \lambda - i\mu)^{-1}dx$ exists for all $\varphi \in C_0^\infty(\Omega)$. If it does, and if we denote the limit by $(h(x) - \lambda - i0)^{-1}$ (a distribution on Ω), then we shall have for $u, v \in C_0^\infty(\Omega)$:

$$(7.1.30) \qquad \lim_{\mu \to +0}\langle u, R(\lambda + i\mu)v \rangle = \int_\Omega \frac{\overline{u(x)}v(x)}{h(x) - \lambda - i0}dx,$$

where the integral has to be understood as the action of the distribution $(h(x) - \lambda - i0)^{-1}$ on the test function $u(x)\overline{v(x)}$.

In order to be able to do explicit calculations, we shall now study the preceding question under a rather strong assumption on h. *We assume that h is an increasing homeomorphism of Ω onto an interval $J \subset \mathbb{R}$ and that its inverse $g \equiv h^{-1} : J \to \mathbb{R}$ is of class C^1* (recall that H has a purely absolutely continuous spectrum if and only if g is an absolutely continuous function). Then h is everywhere differentiable on Ω (the value $+\infty$ for the derivative being allowed) and we have $h'(x) = [g'(y)]^{-1}$ if $y = h(x)$.

Under these assumptions it is quite easy to treat the imaginary part of $(h(x) - \lambda - i\mu)^{-1}$, because we can make the change of variable $x = g(y)$ and get

$$(7.1.31)$$

$$\int_\Omega \varphi(x)\Im[h(x) - \lambda - i\mu]^{-1}dx = \int_J \varphi(g(y))g'(y)\frac{\mu}{(y - \lambda)^2 + \mu^2}dy.$$

Since $\pi^{-1}\mu[(y - \lambda)^2 + \mu^2]^{-1}$ converges to the Dirac distribution with support $\{\lambda\}$ as $\mu \to +0$, we get for any $\varphi \in C_0^\infty(\Omega)$:

$$(7.1.32)$$

$$\lim_{\mu \to +0} \frac{1}{\pi} \int_\Omega \varphi(x)\Im[h(x) - \lambda - i\mu]^{-1}dx = \varphi(g(\lambda))g'(\lambda) = \frac{\varphi(h^{-1}(\lambda))}{h'(h^{-1}(\lambda))}.$$

It follows (cf. (7.1.17)) that for all $\lambda \in J$, the operator $dE_\lambda/d\lambda$ is just multiplication by the distribution $[h'(h^{-1}(\lambda))]^{-1}\delta(x - h^{-1}(\lambda))$ (usually written as $\delta(h(x) - \lambda)$). If Ω is bounded, and if we take for $\mathscr{K}$ the Banach space of bounded continuous functions on $\overline{\Omega}$ that are equal to zero at the end points of Ω, then $\mathscr{K} \subset \mathscr{H} \subset \mathscr{K}^*$ continuously (the first embedding being dense), and we have proved that $\lim_{\mu \to +0}\Im R(\lambda + i\mu)$ exists in the weak* topology of $B(\mathscr{K}, \mathscr{K}^*)$ for each $\lambda \in J$. Notice that, if g is only Lipschitz, then we shall have a uniform bound $\|\Im R(\lambda + i\mu)\|_{\mathscr{K} \to \mathscr{K}^*} \leq M < \infty$ for all λ, $\mu \neq 0$, but the preceding limit will exist only for almost every $\lambda \in J$. This shows the difference between the G.L.A.P. and the strong G.L.A.P.

We shall now treat the real part of $(h(x) - \lambda - i\mu)^{-1}$, a considerably more difficult task. The same change of variable as before gives:

$$(7.1.33)$$

$$\int_\Omega \varphi(x)\Re[h(x) - \lambda - i\mu]^{-1}dx = \int_J \varphi(g(y))g'(y)\frac{y - \lambda}{(y - \lambda)^2 + \mu^2}dy.$$

We shall assume φ at least continuous with compact support, hence $\psi := \varphi \circ g \cdot g'$ belongs to $C_0^0(J)$. The estimate (7.B.1) from the Appendix B to this chapter shows that $\lim_{\mu \to +0} \int \varphi(x) \Re[h(x) - \lambda - i\mu]^{-1} dx$ exists *if and only if the Hilbert transform* $\widetilde{\psi}(\lambda) := \lim_{\mu \to +0} \int_{|y-\lambda| > \mu} \psi(y)(y - \lambda)^{-1} dy$ *of* ψ *at* λ *exists, and the two limits are equal if they exist.* So we can use facts from the theory of the Hilbert transformation in our context (see [Ga], especially Theorem 5.2 page 252 and also pages 110, 105, and Ch. 8 in [Ba]). For each continuous ψ, $\widetilde{\psi}(\lambda)$ exists for almost every $\lambda \in \mathbb{R}$, but there is a function $\psi \in C_0^0(J)$ such that $\widetilde{\psi}(\lambda)$ does not exist for an uncountable set of λ and such that the (almost everywhere defined) function $\widetilde{\psi}$ is essentially unbounded on each subinterval of a given compact subset of J (notice, however, that $\widetilde{\psi} \in L_{\text{loc}}^p$ for all $p < \infty$). In fact, such a function is constructed, as a preliminary step, in Appendix 7.B: the function $f(x) = \sum_{r \in D} a_r \xi(x - r)$ constructed there is not only continuous, but also absolutely continuous. It follows that, even if h is a C^∞-diffeomorphism (e.g. $h(x) \equiv x$), the limiting absorption principle for H cannot hold in a Banach space $\mathscr{K}$ consisting of functions that are not more than absolutely continuous locally. In particular, the real part of the resolvent is effectively more singular than its imaginary part, as observed in §7.1.2.

Now assume that $\varphi \in C_0^\infty(\Omega)$. If g is not more than C^1, then ψ will not be more than C_0^0 in general, so we shall have the same problems again. Appendix B of this chapter is devoted to the construction of a C^1-diffeomorphism $h : \mathbb{R} \to \mathbb{R}$ such that $0 < c \le h'(x) \le c^{-1}$ for some constant c and all x; moreover, h' is absolutely continuous and h is locally of Besov class $B_\infty^{1,q}(\mathbb{R})$ for all $q > 1$; but for any $\varphi \in C_0^1(\mathbb{R})$ with $\varphi(h^{-1}(\lambda)) \ne 0$ the quantity (7.1.33) is unbounded as $\mu \to +0$, and this for all rational numbers λ in a given bounded interval. So the L.A.P. for H cannot hold in a space containing $C_0^\infty(\mathbb{R})$. It is remarkable that the conjugate operator method, as developed in Sections 7.2-7.5, will allow us to prove the strong L.A.P. for H in very simple spaces if $h \in B_\infty^{1,1}(\mathbb{R})$ locally.

One should notice that, if h is a C^1-diffeomorphism, then the spectrum of $H = h(Q)$ is quite nice, e.g. it is purely absolutely continuous. So the validity of the L.A.P. in some space $\mathscr{K}$ is a much stronger assertion than the pure absolute continuity of the spectrum: it involves a certain relation between H and $\mathscr{K}$ which implies *propagation estimates*; these estimates play a fundamental role in the proof of asymptotic completeness (see [M2]).

On the positive side, let us mention that the explanations given in the first part of Appendix 7.B immediately imply that, if h is a C^1-diffeomorphism with Dini-continuous derivative, then the L.A.P. for H will hold in any space $\mathscr{K}$ consisting of functions that are Dini-continuous. We do not insist on this, since a far better result will be obtained further on.

7.2. The Mourre Estimate

In this section we consider a pair of self-adjoint operators A, H in a Hilbert space $\mathscr{H}$. The domain $D(H)$ of H is equipped with the graph topology; hence $D(H)$ is a H-space continuously and densely embedded in $\mathscr{H}$. We shall always identify $\mathscr{H}$ with its adjoint space $\mathscr{H}^*$, so that $D(H) \subset \mathscr{H} \subset D(H)^*$ with continuous and dense embeddings. We define $[A, H] = -[H, A]$ as the sesquilinear form on $D(A) \cap D(H)$ given by the correspondence $(f, g) \mapsto \langle Af, Hg \rangle - \langle Hf, Ag \rangle$.

We shall assume throughout the remainder of this section that H is of class $C^1(A)$. A characterization of such operators in terms of the commutator $[A, H]$ has been specified in Theorem 6.2.10 and explained in the comments following that theorem. In particular $D(A) \cap D(H)$ is dense in the H-space $D(H)$ (i.e. it is a core for H) and $[A, H]$ extends in a unique way to a continuous sesquilinear form on $D(H)$. We keep the notation $[A, H]$ for this extension and also denote by $[A, H]$ the operator in $B(D(H), D(H)^*)$ associated to it by the Riesz lemma. So $[A, H]$ is a continuous linear operator from $D(H)$ into $D(H)^*$ such that $\langle f, [A, H]g \rangle = \langle Af, Hg \rangle - \langle Hf, Ag \rangle$ if $f, g \in D(A) \cap D(H)$ (the bracket on the l.h.s. means anti-duality between $D(H)$ and $D(H)^*$, whereas the brackets on the r.h.s. are scalar products in $\mathcal{H}$). Notice that $[A, H]^* = [H, A] = -[A, H]$, so that $[H, iA] \equiv i[H, A]$ is a symmetric continuous operator from $D(H)$ to $D(H)^*$.

If $z \in \mathbb{C} \setminus \sigma(H)$, then the resolvent $R(z) \equiv (H - z)^{-1}$ extends to a linear homeomorphism $D(H)^* \to \mathcal{H}$ such that $R(z)\mathcal{H} = D(H)$, and one has (cf. (6.2.24)):

$$(7.2.1) \qquad [A, R(z)] = R(z)[H, A]R(z).$$

The r.h.s. is the product of three bounded operators $R(z) : \mathcal{H} \to D(H)$, $[H, A] : D(H) \to D(H)^*$ and $R(z) : D(H)^* \to \mathcal{H}$ (read from right to left), whereas the l.h.s. is independently defined as the following bounded operator in $\mathcal{H}$:

$$(7.2.2)$$

$$[A, R(z)] = \operatorname{s-lim}_{\tau \to 0} \frac{1}{i\tau}\{e^{iA\tau}R(z)e^{-iA\tau} - R(z)\}$$

$$= \operatorname{s-lim}_{\tau \to 0} \frac{1}{i\tau}[e^{iA\tau}, R(z)]e^{-iA\tau} = \operatorname{s-lim}_{\tau \to 0}[A_\tau, R(z)],$$

where $A_\tau = (i\tau)^{-1}(e^{iA\tau} - I)$ for $\tau \neq 0$. We observe that (7.2.1) may also be written as

$$(7.2.3) \qquad\qquad [H, A] = (H - z)[A, R(z)](H - z),$$

where the r.h.s. is a product of three bounded operators $(H - z) : D(H) \to \mathcal{H}$, $[A, R(z)] : \mathcal{H} \to \mathcal{H}$ and $(H - z) : \mathcal{H} \to D(H)^*$.

7.2.1. Let $\{E(\cdot)\}$ be the spectral measure of H. If J is a bounded interval in $\mathbb{R}$, then $E(J)$ belongs to $B(\mathcal{H}, D(H))$ and, by duality, extends to a bounded operator from $D(H)^*$ into $\mathcal{H}$. Hence $E(J)[H, iA]E(J)$ is a bounded symmetric operator in $\mathcal{H}$ which is zero on the orthogonal complement of the subspace $E(J)\mathcal{H}$. Hence there are real constants a', a'' such that $a'E(J) \leq E(J)[iH, A]E(J) \leq a''E(J)$. For various purposes in spectral and scattering theory it is useful to know the optimal values of a' and a'' for which these inequalities hold when J is a small neighbourhood of a point $\lambda \in \mathbb{R}$. A precise expression for these optimal values is given by the quantities $\varrho_H^A(\lambda)$ and $\underline{\varrho}_H^A(\lambda)$ defined as follows: for $\lambda \in \mathbb{R}$ and $\varepsilon > 0$, set $E(\lambda; \varepsilon) = E((\lambda - \varepsilon, \lambda + \varepsilon))$; then

$(7.2.4)\, \varrho_H^A(\lambda) = \sup\{a \in \mathbb{R} \mid \exists \varepsilon > 0 \text{ s.t. } aE(\lambda; \varepsilon) \leq E(\lambda; \varepsilon)[H, iA]E(\lambda; \varepsilon)\}$,

$(7.2.5)\ \underline{\varrho}_H^A(\lambda) = \inf\{a \in \mathbb{R} \mid \exists \varepsilon > 0 \text{ s.t. } E(\lambda; \varepsilon)[H, iA]E(\lambda; \varepsilon) \leq aE(\lambda; \varepsilon)\}$.

Clearly $-\infty < \varrho_H^A(\lambda) \le +\infty$ and $-\infty \le \underline{\varrho}_H^A(\lambda) < +\infty$. In what follows we shall study properties of the function ϱ_H^A; those of $\underline{\varrho}_H^A$ then follow by using the obvious relation

$$(7.2.6) \qquad \underline{\varrho}_H^A(\lambda) = -\varrho_H^{-A}(\lambda).$$

When there is no ambiguity concerning the operators A and H, we shall write simply ϱ_H or ϱ for ϱ_H^A.

The following expression for ϱ_H^A is often convenient:

$$(7.2.7) \qquad \varrho_H^A(\lambda) = \sup\{a \in \mathbb{R} \mid \exists \text{ a real } \varphi \in C_0^\infty(\mathbb{R}) \text{ s.t. } \varphi(\lambda) \neq 0$$
$$\text{and } a\varphi(H)^2 \le \varphi(H)[H, iA]\varphi(H)\}.$$

We omit the simple proof of the equivalence of (7.2.4) and (7.2.7) and point out yet another possible definition of ϱ_H^A which shows in particular that the supremum in (7.2.4) is realized when $\varepsilon \to 0$:

LEMMA 7.2.1. If $\lambda \notin \sigma(H)$, then $\varrho(\lambda) = +\infty$. If $\lambda \in \sigma(H)$, then $\varrho(\lambda)$ is finite and given by

$$(7.2.8) \quad \varrho(\lambda) = \lim_{\varepsilon \to +0} \left(\inf\{\langle f, [H, iA]f\rangle \mid \|f\| = 1 \text{ and } E(\lambda; \varepsilon)f = f\}\right),$$

and there is a sequence $\{f_k\}$ of vectors such that $\|f_k\| = 1$, $E(\lambda; 1/k)f_k = f_k$ and $\lim_{k\to\infty}\langle f_k, [H, iA]f_k\rangle = \varrho(\lambda)$.

The proof is easy. If one adopts the usual convention that $\inf \varnothing = +\infty$, then there is no need to distinguish between the cases $\lambda \notin \sigma(H)$ and $\lambda \in \sigma(H)$ in the above lemma. We mention the following consequence of (7.2.8): If A is a bounded operator, then $\varrho_H^A(\lambda) = 0$ if $\lambda \in \sigma(H)$ (indeed, $|\langle f, [H, iA]f\rangle| = |\langle f, [H - \lambda, iA]f\rangle| \le 2\|(H-\lambda)f\| \cdot \|Af\| \le 2\varepsilon\|A\|$ if $\|f\| = 1$ and $E(\lambda; \varepsilon)f = f$). By using (7.2.8) one can easily get the function ϱ in the following simple situation which will be important further on:

EXAMPLE 7.2.2. Let $\mathscr{H} = L^2(X)$, where X is an euclidean space of non-zero dimension, and let $H = \Delta$ be the usual self-adjoint realization of the Laplace-Beltrami operator (see Chapter 1). Let $D = \sum_{j=1}^n (P_j Q_j + Q_j P_j)/4$, so that $2D$ is the generator of the dilation group in $\mathscr{H}$ (see Section 1.2). Since $[i\Delta, D] = \Delta$, Δ is of class $C^1(D)$ and we have $\varrho_\Delta^D(\lambda) = +\infty$ if $\lambda < 0$ and $\varrho_\Delta^D(\lambda) = \lambda$ if $\lambda \ge 0$. More generally, let k be any vector in X and $D_k = D + (k, Q)/2 = e^{-i(k,Q)}De^{i(k,Q)}$. Then Δ is of class $C^1(D_k)$ and $[i\Delta, D_k] = \Delta + (k, P)$, hence $\varrho_\Delta^{D_k}(\lambda) = +\infty$ if $\lambda < 0$ and $\varrho_\Delta^{D_k}(\lambda) = \lambda^{1/2}(\lambda^{1/2} - |k|)$ if $\lambda \ge 0$.

Part (b) of the next proposition is a technical result that is sometimes useful. In order to clarify its meaning, observe that if $\lambda \in \mathbb{R}$ and θ is a real number such that $\theta < \varrho(\lambda)$, then there is $\varepsilon > 0$ such that $\theta E(\lambda; \varepsilon) \le E(\lambda; \varepsilon)[iH, A]E(\lambda; \varepsilon)$. We shall have to work with such inequalities when λ varies over an interval and θ is a function of λ, and we shall need conditions ensuring that one may choose a number ε that does not depend on λ.

PROPOSITION 7.2.3. (a) *The function* $\varrho : \mathbb{R} \to (-\infty, +\infty]$ *is lower semicontinuous, and* $\varrho(\lambda) < +\infty$ *if and only if* $\lambda \in \sigma(H)$.

(b) *Let* Λ *be a compact subset of* $\mathbb{R}$ *and* $\theta : \Lambda \to \mathbb{R}$ *an upper semicontinuous function such that* $\theta(\lambda) < \varrho(\lambda)$ *for all* $\lambda \in \Lambda$. *Then there is* $\varepsilon > 0$ *such that for all* $\lambda \in \Lambda$:

$$(7.2.9) \qquad \theta(\lambda)E(\lambda; \varepsilon) \le E(\lambda; \varepsilon)[H, iA]E(\lambda; \varepsilon).$$

PROOF. (a) In view of Lemma 7.2.1 it suffices to prove the lower semicontinuity of ϱ. We set $B = [H, iA]$. Let $\lambda_0 \in \mathbb{R}$ and $r \in \mathbb{R}$ be such that $\varrho(\lambda_0) > r$. We must show that there is a neighbourhood of λ_0 on which $\varrho(\lambda) > r$. Since $\varrho(\lambda_0) > r$, there exist $a > r$ and $\varepsilon_0 > 0$ such that $aE(\lambda_0; \varepsilon_0) \le E(\lambda_0; \varepsilon_0)BE(\lambda_0; \varepsilon_0)$. Let $\varepsilon = \varepsilon_0/2$ and $\lambda \in (\lambda_0 - \varepsilon, \lambda_0 + \varepsilon)$. Upon pre- and post-multiplying the preceding inequality by $E(\lambda; \varepsilon)$ and by using the fact that $E(\lambda; \varepsilon)E(\lambda_0; \varepsilon_0) = E(\lambda; \varepsilon)$, one gets that $aE(\lambda; \varepsilon) \le E(\lambda; \varepsilon)BE(\lambda; \varepsilon)$. This implies that $\varrho(\lambda) \ge a > r$ for all $\lambda \in (\lambda_0 - \varepsilon, \lambda_0 + \varepsilon)$.

(b) We first show that each point $\lambda_0 \in \Lambda$ has a neighbourhood $\mathscr{U}(\lambda_0)$ in Λ such that there is $\varepsilon_0 > 0$ with the property

$$(7.2.10) \qquad \lambda \in \mathscr{U}(\lambda_0) \implies \theta(\lambda)E(\lambda; \varepsilon_0) \le E(\lambda; \varepsilon_0)BE(\lambda; \varepsilon_0).$$

To see this, choose $r \in \mathbb{R}$ such that $\theta(\lambda_0) < r < \varrho(\lambda_0)$. By the upper semicontinuity of θ at λ_0 there is a neighbourhood $\mathscr{V}$ of λ_0 in Λ such that $\theta(\lambda) < r$ for all $\lambda \in \mathscr{V}$. Also, there is $\varepsilon > 0$ such that $rE(\lambda_0; \varepsilon) \le E(\lambda_0; \varepsilon)BE(\lambda_0; \varepsilon)$. Now choose a neighbourhood $\mathscr{U}(\lambda_0)$ of λ_0 in Λ and $\varepsilon_0 > 0$ sufficiently small such that $\mathscr{U}(\lambda_0) \subset \mathscr{V}$ and $(\lambda - \varepsilon_0, \lambda + \varepsilon_0) \subset (\lambda_0 - \varepsilon, \lambda_0 + \varepsilon)$ for all $\lambda \in \mathscr{U}(\lambda_0)$ (e.g. take $\varepsilon_0 = \varepsilon/2$ and $\mathscr{U}(\lambda_0) = \mathscr{V} \cap (\lambda_0 - \varepsilon_0, \lambda_0 + \varepsilon_0)$). Then $E(\lambda; \varepsilon_0) \le E(\lambda_0; \varepsilon)$, hence $rE(\lambda; \varepsilon_0) \le E(\lambda; \varepsilon_0)BE(\lambda; \varepsilon_0)$ for all $\lambda \in \mathscr{U}(\lambda_0)$. Since $\theta(\lambda) < r$ in $\mathscr{U}(\lambda_0)$, we get (7.2.10).

Now the proof can be completed as follows. Since Λ is compact, one may choose a finite number of neighbourhoods $\mathscr{U}(\lambda_1), \ldots, \mathscr{U}(\lambda_n)$ covering Λ. Let $\varepsilon_1, \ldots, \varepsilon_n$ be the corresponding numbers for which (7.2.10) holds, and set $\varepsilon = \min\{\varepsilon_1, \ldots, \varepsilon_n\}$. Then for each $\lambda \in \Lambda$ there is $k \in \{1, \ldots, n\}$ such that $\lambda \in \mathscr{U}(\lambda_k)$, hence $\theta(\lambda)E(\lambda; \varepsilon_k) \le E(\lambda; \varepsilon_k)BE(\lambda; \varepsilon_k)$. Since $E(\lambda; \varepsilon) \le E(\lambda; \varepsilon_k)$, one obtains (7.2.9). $\square$

Proposition 7.2.3 has an obvious analogue for the function $\underline{\varrho}$ introduced in (7.2.5); $\underline{\varrho}$ is upper semicontinuous by (7.2.6), and the function θ with $\theta(\lambda) > \underline{\varrho}(\lambda)$ will have to be lower semicontinuous. In the following example we show how to construct a function θ that satisfies the conditions of Proposition 7.2.3 (take $F = \varrho$ below):

EXAMPLE 7.2.4. Let J be a closed interval in $\mathbb{R}$ and $F : J \to (-\infty, +\infty]$ a lower semicontinuous function. For $\nu > 0$ define $\theta_\nu : J \to (-\infty, +\infty]$ by

$$(7.2.11) \qquad \theta_\nu(\lambda) = \inf_{\substack{|\mu - \lambda| < \nu \\ \mu \in J}} F(\mu) - \nu.$$

Then θ_ν is upper semicontinuous and satisfies
   (a) $\lim_{\nu \to 0} \theta_\nu(\lambda) = F(\lambda)$;

(b) $\theta_\nu(\lambda) < F(\lambda)$ if $\theta_\nu(\lambda) \neq +\infty$;

(c) $\theta_\nu(\lambda) < \theta_\mu(\lambda)$ if $\nu > \mu > 0$ and $\theta_\nu(\lambda) \neq +\infty$.

In particular, if $J = \mathbb{R}$ and $F = \varrho$, then $\theta_\nu(\lambda) < \infty$ if $\text{dist}(\lambda, \sigma(H)) < \nu$.

PROOF. Since a lower semicontinuous function on a closed finite interval $[a, b]$ is bounded from below, the function θ_ν does not assume the value $-\infty$. (a) is a consequence of the lower semicontinuity of F. (b) and (c) are evident. The last statement holds by Proposition 7.2.3 (a). For the upper semicontinuity of θ_ν, let $\lambda_0 \in J$ and $r > \theta_\nu(\lambda_0)$. We must show that $r > \theta_\nu(\lambda)$ for all λ in some neighbourhood of λ_0 in J. We choose $\mu_0 \in (\lambda_0 - \nu, \lambda_0 + \nu) \cap J$ such that $F(\mu_0) - \nu < r$ and $\delta > 0$ such that δ is smaller than the distance from μ_0 to the set $\{\lambda_0 - \nu, \lambda_0 + \nu\}$ consisting of the two end points of the interval $(\lambda_0 - \nu, \lambda_0 + \nu)$. If $|\lambda - \lambda_0| < \delta$, then $\mu_0 \in (\lambda - \nu, \lambda + \nu)$, hence (for $\lambda \in J$)

$$\theta_\nu(\lambda) = \inf_{\substack{|\mu - \lambda| < \nu \\ \mu \in J}} F(\mu) - \nu \leq F(\mu_0) - \nu < r.$$

We close this subsection by proving a result which relates the ϱ-function of a self-adjoint operator H to that of $(\lambda_0 - H)^{-1}$ if λ_0 is a real number in the resolvent set of H. This is interesting because $(\lambda_0 - H)^{-1}$ is a *bounded* self-adjoint operator. We may remark that, if ψ is defined as $\psi(\lambda) = (\lambda_0 - \lambda)^{-1}$ for λ outside a neighbourhood of λ_0 that is contained in the resolvent set of H, then we prove that $\varrho^A_{\psi(H)}(\psi(\lambda)) = \psi'(\lambda)\varrho^A_H(\lambda)$; this suggests generalizations which we shall not discuss.

PROPOSITION 7.2.5. *Assume that λ_0 is a real number which belongs to the resolvent set of H (in particular, H must have a spectral gap) and let $R = (\lambda_0 - H)^{-1}$. Then for $\lambda \in \mathbb{R} \setminus \{\lambda_0\}$:*

$$(7.2.12) \qquad \varrho^A_H(\lambda) = (\lambda_0 - \lambda)^2 \varrho^A_R((\lambda_0 - \lambda)^{-1}).$$

PROOF. Since $\lambda \in \sigma(H)$ if and only if $(\lambda_0 - \lambda)^{-1} \in \sigma(R)$, both sides of (7.2.12) are infinite if $\lambda \notin \sigma(H)$. So assume that $\lambda \in \sigma(H)$. Let $\{f_k\}$ be a sequence of vectors with $\|f_k\| = 1$ and such that the support of f_k in the spectral representation of H shrinks to λ as $k \to \infty$ and $\langle f_k, [iH, A]f_k \rangle \to \varrho_H(\lambda)$ as $k \to \infty$ (see Lemma 7.2.1). Then, by using (7.2.1) with $z = \lambda_0$:

$$(\lambda_0 - \lambda)^{-2}\varrho_H(\lambda) = (\lambda_0 - \lambda)^{-2}\lim_{k \to \infty}\langle f_k, [iH, A]f_k \rangle$$
$$= \lim_{k \to \infty}\langle(\lambda_0 - H)^{-1}f_k, [iH, A](\lambda_0 - H)^{-1}f_k \rangle$$
$$= \lim_{k \to \infty}\langle f_k, [iR, A]f_k \rangle \geq \varrho_R((\lambda_0 - \lambda)^{-1}),$$

where the last inequality follows from Lemma 7.2.1 by observing that the support of f_k in the spectral representation of R shrinks to $(\lambda_0 - \lambda)^{-1}$ as $k \to \infty$. The opposite inequality can be obtained similarly from Lemma 7.2.1 and (7.2.3) by choosing a sequence $\{g_k\}$ of vectors with $\|g_k\| = 1$ and such that the support of g_k in the spectral representation of R shrinks to $(\lambda_0 - \lambda)^{-1}$ and $\langle g_k, [iR, A]g_k \rangle \to \varrho_R((\lambda_0 - \lambda)^{-1})$ as $k \to \infty$. $\square$

7.2.2. We now introduce a modification $\widetilde{\varrho}_H^A$ of the function ϱ_H^A and establish its relation to ϱ_H^A. Although ϱ_H^A is the important object, $\widetilde{\varrho}_H^A$ is a useful quantity because it is quite easy to calculate in many practical situations (which is due to the fact that it is invariant under the addition of a compact perturbation to H, see Theorem 7.2.9 for a precise statement).

It will be convenient to use the following notations: if S and T are bounded operators in some Hilbert space $\mathscr{H}$, we write

$$(7.2.13) \qquad S \approx T \text{ if } S - T \text{ is a compact operator,}$$

$$(7.2.14)$$
$$S \lesssim T \text{ if there is a compact operator } K \text{ such that } S \leq T + K.$$

Notice that the compact operators form a closed self-adjoint ideal $K(\mathscr{H})$ in the C^*-algebra $B(\mathscr{H})$, so the quotient $B(\mathscr{H})/K(\mathscr{H})$ is a C^*-algebra too, usually called the *Calkin algebra*. Hence, if we denote by $\overset{\circ}{S}$ the class in the quotient associated to $S \in B(\mathscr{H})$, then $S \approx T$ means that $\overset{\circ}{S} = \overset{\circ}{T}$ and $S \lesssim T$ means that $\overset{\circ}{S} \leq \overset{\circ}{T}$.

Now we define $\widetilde{\varrho}_H^A : \mathbb{R} \to (-\infty, +\infty]$ in one of the following equivalent ways:

$$(7.2.15)$$
$$\widetilde{\varrho}_H^A(\lambda) = \sup\{a \in \mathbb{R} \mid \exists \varepsilon > 0 \text{ s.t. } aE(\lambda; \varepsilon) \lesssim E(\lambda; \varepsilon)[H, iA]E(\lambda; \varepsilon)\}$$
$$= \sup\{a \in \mathbb{R} \mid \exists \varphi \in C_0^\infty(\mathbb{R}) \text{ real s.t. } \varphi(\lambda) \neq 0$$
$$\text{and } a\varphi(H)^2 \lesssim \varphi(H)[H, iA]\varphi(H)\}.$$

The proof of the equality is an exercise. One may also define $\widetilde{\varrho}_H^A$ just by changing $\leq$ into $\lesssim$ in (7.2.5), but since we clearly will have $\underline{\widetilde{\varrho}}_H^A = -\underline{\widetilde{\varrho}}_H^{-A}$, we may restrict our study to $\widetilde{\varrho}_H^A$. As in the case of ϱ_H^A, we shall use the simplified notations $\widetilde{\varrho}_H$ or $\widetilde{\varrho}$ when there is no danger of confusion.

We shall introduce some more terminology, in order to make the connection with several notions which appear in the literature. Let A and H be self-adjoint operators in a Hilbert space $\mathscr{H}$ such that H is of class $C^1(A)$. We say that A is *conjugate to H at the point* $\lambda \in \mathbb{R}$ if $\widetilde{\varrho}_H^A(\lambda) > 0$, and that A is *strictly conjugate to H at λ* if $\varrho_H^A(\lambda) > 0$. If $U \in \mathbb{R}$ is open and $\widetilde{\varrho}_H^A(\lambda) > 0$ for all $\lambda \in U$, we say that A is *locally conjugate to H on U*; if $\varrho_H^A(\lambda) > 0$ for all $\lambda \in U$, A is *locally strictly conjugate to H on U*. Note that bounded operators A are not useful as locally (strictly) conjugate operators. Indeed, we have seen before that $\varrho_H^A(\lambda) = 0$ if $\lambda \in \sigma(H)$ and A is bounded; one may similarly show that $\widetilde{\varrho}_H^A(\lambda) = 0$ if $\lambda \in \sigma_{\mathrm{ess}}(H)$ and A is bounded.

So A is conjugate to H at λ if and only if there are an open interval J containing λ, a strictly positive number a and a compact operator K in $\mathscr{H}$ such that

$$(7.2.16) \qquad E(J)[H, iA]E(J) \geq aE(J) + K.$$

If this inequality holds with $K = 0$ (and $a > 0$), then A is strictly conjugate to H at λ. Now, one can consider an arbitrary Borel set $J \subset \mathbb{R}$: if (7.2.16) holds, in the sense of quadratic forms on $D(H)$, with a number $a > 0$ and a compact operator $K : \mathcal{H} \to \mathcal{H}$, one says that *the Mourre estimate holds on J*, or that *A is conjugate to H on J*. If, moreover, one has $K = 0$, we say that a *strict Mourre estimate holds on J*, or that *A is strictly conjugate to H on J*.

Some simple properties of $\widetilde{\varrho}$ are collected in the following proposition:

PROPOSITION 7.2.6. *The function $\widetilde{\varrho} : \mathbb{R} \to (-\infty, +\infty]$ is lower semicontinuous and satisfies $\widetilde{\varrho} \geq \varrho$. Furthermore $\widetilde{\varrho}(\lambda) < \infty$ if and only if $\lambda \in \sigma_{ess}(\lambda)$.*

PROOF. The lower semicontinuity of $\widetilde{\varrho}$ is obtained similarly to that of ϱ (Proposition 7.2.3 (a)), and the inequality $\widetilde{\varrho} \geq \varrho$ is immediate from the definitions. For the last statement we use the fact that $\lambda \notin \sigma_{ess}(H)$ if and only if $E(\lambda; \varepsilon)$ is compact for some $\varepsilon > 0$ (i.e. $E(\lambda; \varepsilon) \approx 0$). So $\lambda \notin \sigma_{ess}(H)$ implies that $\widetilde{\varrho}(\lambda) = \infty$. Conversely, if $\widetilde{\varrho}(\lambda) = \infty$, let $m = \|E(\lambda; 1)[H, iA]E(\lambda; 1)\|$ and $a > m$. Then there is $\varepsilon \in (0, 1)$ such that $aE(\lambda; \varepsilon) \lesssim E(\lambda; \varepsilon)[H, iA]E(\lambda; \varepsilon)$. On the other hand, from $E(\lambda; 1)[H, iA]E(\lambda; 1) \leq mI$ we get $E(\lambda; \varepsilon)[H, iA]E(\lambda; \varepsilon) \leq mE(\lambda; \varepsilon)$. This implies that $aE(\lambda; \varepsilon) \lesssim mE(\lambda; \varepsilon)$, hence $E(\lambda; \varepsilon) \approx 0$ since $a > m$. $\square$

Proposition 7.2.5 has an analogue for the function $\widetilde{\varrho}$. We state the result without giving the proof, which is a straightforward modification of that of Proposition 7.2.5 (see Proposition 8.3.4 for a more general result).

PROPOSITION 7.2.7. *Let λ_0 be a real number in the resolvent set of H and let $R = (\lambda_0 - H)^{-1}$. Then $\widetilde{\varrho}_H^A(\lambda) = (\lambda_0 - \lambda)^2 \widetilde{\varrho}_R^A((\lambda_0 - \lambda)^{-1})$ for each $\lambda \in \mathbb{R} \setminus \{\lambda_0\}$. In particular A is conjugate to H at some $\lambda \neq \lambda_0$ if and only if it is conjugate to R at $(\lambda_0 - \lambda)^{-1}$.*

An important property of the function $\widetilde{\varrho}$ which is not shared by ϱ is its invariance under a large class of perturbations of H. In particular, this gives a perturbative method for checking the validity of the Mourre estimate (it can be applied for example to two-body hamiltonians with locally very singular potentials). We first recall an elementary fact:

LEMMA 7.2.8. *Let H and H_0 be self-adjoint operators in a Hilbert space $\mathcal{H}$ and assume that $(H - z)^{-1} - (H_0 - z)^{-1}$ is compact for some $z \in \mathbb{C} \setminus [\sigma(H) \cup \sigma(H_0)]$. Then H and H_0 have the same essential spectrum and $\varphi(H) - \varphi(H_0)$ is compact for each $\varphi \in C_\infty(\mathbb{R})$ (i.e. for each continuous $\varphi : \mathbb{R} \to \mathbb{C}$ converging to zero at infinity).*

PROOF. We shall use the following three well-known facts: (i) the set of compact operators in $\mathcal{H}$ is a norm-closed subspace $K(\mathcal{H})$ of $B(\mathcal{H})$; (ii) the map $\varphi \mapsto \varphi(H)$ is a linear norm-continuous map from $C_\infty(\mathbb{R})$ into $B(\mathcal{H})$; (iii) the vector subspace of $C_\infty(\mathbb{R})$ generated by the functions of the form $\lambda \mapsto (\lambda - \xi)^{-1}$, with $\xi \in \mathbb{C} \setminus \mathbb{R}$, is dense in $C_\infty(\mathbb{R})$.

Hence, in order to show that $\varphi(H) - \varphi(H_0) \in K(\mathcal{H})$ for all $\varphi \in C_\infty(\mathbb{R})$, it is enough to prove that $R(\xi) - R_0(\xi) \in K(\mathcal{H})$ if $\xi \notin \mathbb{R}$, where we have denoted

$R(\xi) = (H - \xi)^{-1}$, $R_0(\xi) = (H_0 - \xi)^{-1}$. By hypothesis, $R(z) - R_0(z)$ is compact. Then, using the first resolvent equation, we obtain:

(7.2.17)
$$R(\xi) - R_0(\xi) = [I + (\xi - z)R(\xi)][R(z) - R_0(z)][I + (\xi - z)R_0(\xi)].$$

Since $K(\mathcal{H})$ is a bilateral ideal in $B(\mathcal{H})$, this implies $R(\xi) - R_0(\xi) \in K(\mathcal{H})$.

If $\varphi \in C_0^0(\mathbb{R} \setminus \sigma_{\mathrm{ess}}(H))$, then $\varphi(H) \in K(\mathcal{H})$, hence $\varphi(H_0) \in K(\mathcal{H})$ by the preceding result. If $\lambda \in \mathbb{R} \setminus \sigma_{\mathrm{ess}}(H)$, we may choose $\varphi \in C_0^0(\mathbb{R} \setminus \sigma_{\mathrm{ess}}(H))$ such that $\varphi(x) = 1$ in some neighbourhood of λ. The compactness of $\varphi(H_0)$ then implies that $\lambda \notin \sigma_{\mathrm{ess}}(H_0)$. So $\sigma_{\mathrm{ess}}(H_0) \subset \sigma_{\mathrm{ess}}(H)$. A similar argument leads to the opposite inclusion. $\square$

THEOREM 7.2.9. *Let A, H and H_0 be self-adjoint operators in a Hilbert space $\mathcal{H}$ such that both H and H_0 are of class $C_u^1(A)$. If $(H + i)^{-1} - (H_0 + i)^{-1}$ is compact, then $\widetilde{\varrho}_H^A = \widetilde{\varrho}_{H_0}^A$. In particular, A is conjugate to H at a point $\lambda \in \mathbb{R}$ if and only if it is conjugate to H_0 at λ.*

PROOF. (i) We begin with a general remark concerning expressions of the form $\varphi(H)[A, H]\varphi(H)$ for H of class $C^1(A)$ and $\varphi \in C_0^\infty(\mathbb{R})$. For $k = 1$, 2 we define φ_k by $\varphi_1(x) = x\varphi(x)$ and $\varphi_2(x) = x\varphi^2(x)$. We shall show below that

(7.2.18) $\varphi(H)[iA, H]\varphi(H) = [iA, \varphi_2(H)] - 2\Re\{[iA, \varphi(H)]\varphi_1(H)\}.$

The r.h.s. is easier to handle than the l.h.s. because $\varphi(H)$ and $\varphi_k(H)$ are bounded operators. We recall from Theorem 6.2.5 that each of these operators is of class $C^1(A; \mathcal{H})$ (and of class $C_u^1(A; \mathcal{H})$ if $H \in C_u^1(A)$, see Corollary 6.2.6 (b)). In order to prove (7.2.18), we start from (7.2.3) and use (7.2.2) :

$$\varphi(H)[A, H]\varphi(H) = \underset{\tau \to 0}{\text{s-lim}}(H - z)\varphi(H) \cdot [R(z), A_\tau] \cdot (H - z)\varphi(H)$$
$$= \underset{\tau \to 0}{\text{s-lim}}\{\varphi(H)A_\tau \cdot (H - z)\varphi(H) - (H - z)\varphi(H) \cdot A_\tau\varphi(H)\}$$
$$= \underset{\tau \to 0}{\text{s-lim}}\{A_\tau \cdot H\varphi^2(H) + [\varphi(H), A_\tau] \cdot H\varphi(H)$$
$$- H\varphi^2(H) \cdot A_\tau - H\varphi(H) \cdot [A_\tau, \varphi(H)]\}.$$

(ii) We make a second general remark: if S is a compact operator of class $C_u^1(A; \mathcal{H})$, then $[A, S]$ is also compact. This is obvious, because $[A, S]$ is the norm limit as $\varepsilon \to 0$ of the family of compact operators $(i\varepsilon)^{-1}(e^{iA\varepsilon}Se^{-iA\varepsilon} - S)$.

(iii) By Lemma 7.2.8 we have $\varphi_2(H) \approx \varphi_2(H_0)$ for any $\varphi \in C_0^\infty(\mathbb{R})$. Since $\varphi_2(H) - \varphi_2(H_0) \in C_u^1(A; \mathcal{H})$, we then get by using (ii) that $[A, \varphi_2(H)] \approx [A, \varphi_2(H_0)]$. By treating similarly the second term on the r.h.s. of (7.2.18), one obtains that $\varphi(H)[iH, A]\varphi(H) \approx \varphi(H_0)[iH_0, A]\varphi(H_0)$. Since $\varphi(H)^2 \approx \varphi(H_0)^2$, one sees that, if $a \in \mathbb{R}$, then the relation $\varphi(H)[iH, A]\varphi(H) \gtrsim a\varphi(H)^2$ is satisfied if and only if $\varphi(H_0)[iH_0, A]\varphi(H_0) \gtrsim a\varphi(H_0)^2$. So the assertion of the theorem is an immediate consequence of the second equality in (7.2.15). $\square$

Let us remark that, if the operator H_0 (and hence also the operator H) in Theorem 7.2.9 has a spectral gap, then the conclusion is an easy consequence of Proposition 7.2.7 and of the Remark (ii) of the preceding proof.

We now turn to the main point of this section, the description of the relation between ϱ and $\widetilde{\varrho}$. We begin with a technical result which is useful also in areas of spectral analysis not treated in this text (e.g. in the proof of absence of eigenvalues embedded in the continuous spectrum of self-adjoint operators).

PROPOSITION 7.2.10 (VIRIAL THEOREM). *Let A and H be self-adjoint operators such that H is of class $C^1(A)$. Then $E(\{\lambda\})[A, H]E(\{\lambda\}) = 0$ for each $\lambda \in \mathbb{R}$. In particular, if f is an eigenvector of H, then $\langle f, [A, H]f \rangle = 0$.*

PROOF. We must show that, if $\lambda \in \mathbb{R}$ and $f_1, f_2 \in D(H)$ satisfy $Hf_k = \lambda f_k$, then $\langle f_1, [A, H]f_2 \rangle = 0$. Since $f_1 = (\lambda - i)(H - i)^{-1}f_1$, $f_2 = (\lambda + i)(H + i)^{-1}f_2$, we get by using (7.2.1) and (7.2.2) that

$$\langle f_1, [A, H]f_2 \rangle = -(\lambda + i)^2 \langle f_1, [A, (H + i)^{-1}]f_2 \rangle$$
$$= -(\lambda + i)^2 \lim_{\tau \to 0}[\langle f_1, A_\tau (H + i)^{-1}f_2 \rangle - \langle (H - i)^{-1}f_1, A_\tau f_2 \rangle].$$

For $\tau \neq 0$, the square bracket in the last expression is zero. $\square$

COROLLARY 7.2.11. *Let A and H be self-adjoint operators such that $H \in C^1(A)$. If the Mourre estimate holds on some real Borel set J, then H has at most a finite number of eigenvalues in J, and each of these eigenvalues is of finite multiplicity. In particular, if $\widetilde{\varrho}_H^A(\lambda) > 0$ for some $\lambda \in \mathbb{R}$, then λ has a neighbourhood in which there is at most a finite number of eigenvalues of H, each of finite multiplicity.*

PROOF. Let $a > 0$ and $K \in K(\mathscr{H})$ be such that (7.2.16) is true. If g is an eigenvector of H associated to an eigenvalue in J and $\|g\| = 1$, then (7.2.16) and the virial theorem imply that $\langle g, Kg \rangle \leq -a$. Now assume that the conclusion of the corollary is false. Then there exists an infinite orthonormal sequence $\{g_j\}$ of eigenvectors of H in $E(J)\mathscr{H}$. In particular, $g_j \to 0$ weakly in $\mathscr{H}$ as $j \to \infty$. Since K is compact, one then has $\langle g_j, Kg_j \rangle \to 0$ as $j \to \infty$, which contradicts the inequality $\langle g_j, Kg_j \rangle \leq -a < 0$. $\square$

The next result is also a consequence of the virial theorem:

LEMMA 7.2.12. *Let J be an open bounded real set and let $a \in \mathbb{R}$ be such that $aE(J) \lesssim E(J)[H, iA]E(J)$. Then for each $\lambda \in J$ and each $\eta > 0$ there are a number $\varepsilon > 0$ and a finite rank orthogonal projection F with $F \leq E(\{\lambda\})$ such that*

$$(7.2.19) \qquad (a - \eta)[E(\lambda; \varepsilon) - F] - \eta F \leq E(\lambda; \varepsilon)[H, iA]E(\lambda; \varepsilon).$$

In particular, if λ is not an eigenvalue of H, then

$$(a - \eta)E(\lambda; \varepsilon) \leq E(\lambda; \varepsilon)[H, iA]E(\lambda; \varepsilon),$$

while if λ is an eigenvalue of H, one has only

$$[\min(a - \eta, -\eta)]E(\lambda; \varepsilon) \leq E(\lambda; \varepsilon)[H, iA]E(\lambda; \varepsilon).$$

PROOF. (i) We shall use the following fact: let $\{T_k\}$ be a decreasing sequence of orthogonal projections and $T = \text{s-lim}_{k\to\infty} T_k$, let C be a compact self-adjoint operator and $\nu > 0$; then there is an integer n such that $T_n C T_n \geq TCT - \nu T_n$. In fact we shall have $\|T_k C T_k - TCT\| \to 0$ as $k \to \infty$, so $T_n C T_n - TCT \geq -\nu I$ holds for some n. The assertion follows upon pre- and post-multiplying this last inequality by T_n and observing that $TT_n = T$.

(ii) To simplify the notations, we set $B = [iH, A], E = E(J)$ and $P = E(\{\lambda\})$. If G is any orthogonal projection such that $G \leq E$, we denote by G' the projection $E - G$. The assumption $aE \lesssim EBE$ means that there is a compact self-adjoint operator K such that $aE + K \leq EBE$.

(iii) By using the result of (i) with $T_k = F_k'$, where $\{F_k\}$ is an increasing sequence of finite rank projections such that $F_k \to P$ strongly as $k \to \infty$, one obtains the existence of a finite rank projection F such that $F \leq P$ and $F'KF' \geq P'KP' - (\eta/2)F'$. Then, upon pre- and post-multiplying the inequality $aE+K \leq EBE$ by F' and using the relation $F' \leq E$, one gets that $F'BF' \geq P'KP' + (a - \eta/2)F'$.

(iv) The virial theorem states that $PBP = 0$. Consequently one has $FBF = (P - F)BF = FB(P - F) = 0$. Since $P - F = F' - P'$, we now get that $F'BF = P'BF$ and $FBF' = FBP'$. So, by writing $E = F + F'$ and by taking into account the result of (iii), one obtains

$$(7.2.20) \qquad EBE = P'BF + FBP' + F'BF' \geq (a - \eta/2)F' + K_0,$$

where $K_0 \equiv P'KP' + P'BF + FBP'$ is compact and satisfies $PK_0P = 0$.

(v) Let $E_k = E(\lambda; 1/k)$ for $k \geq k_1$, where k_1 is such that $(\lambda-1/k_1, \lambda+1/k_1) \subset J$. Then $E_k \leq E$ and $E_k F' = E_k - F$. So, after left and right multiplication of (7.2.20) by E_k, we get $E_k B E_k \geq (a - \eta/2)(E_k - F) + E_k K_0 E_k$. Now observe that $\{E_k\}$ is a decreasing sequence of projections converging strongly to P, and $PK_0P = 0$. So (i) implies that one may choose k such that $E_k K_0 E_k \geq -(\eta/2)E_k \geq -(\eta/2)(E_k - F) - \eta F$. Hence we get (7.2.19) with $\varepsilon = 1/k$.

(vi) For the proof of the last assertion, set $b = \min(a - \eta, -\eta)$. Then $(a - \eta)[E(\lambda; \varepsilon) - F] - \eta F \geq b[E(\lambda; \varepsilon) - F] + bF = bE(\lambda; \varepsilon)$. □

We are now ready to prove the main result of this section:

THEOREM 7.2.13. *Let A and H be self-adjoint operators such that H is of class $C^1(A)$, and let $\lambda \in \mathbb{R}$. If λ is an eigenvalue of H and $\widetilde{\varrho}_H^A(\lambda) > 0$, then $\varrho_H^A(\lambda) = 0$. Otherwise $\varrho_H^A(\lambda) = \widetilde{\varrho}_H^A(\lambda)$.*

PROOF. We recall that $\widetilde{\varrho} \geq \varrho$. If λ is not an eigenvalue of H, then Lemma 7.2.12 implies that $\widetilde{\varrho}(\lambda) \leq \varrho(\lambda)$, so these two numbers must be equal. Now assume that λ is an eigenvalue of H. If $\widetilde{\varrho}(\lambda) \leq 0$, then $a \leq 0$ in Lemma 7.2.12, hence $\min(a - \eta, -\eta) = a - \eta$ and we have the same result as before. If $\widetilde{\varrho}(\lambda) > 0$, we may take $a > 0$ in Lemma 7.2.12, which leads to the inequality $\varrho(\lambda) \geq 0$; the opposite inequality $\varrho(\lambda) \leq 0$ follows by using the virial theorem: if $a < \varrho(\lambda)$, there is $\varepsilon > 0$ such that $aE(\lambda; \varepsilon) \leq E(\lambda; \varepsilon)[iH, A]E(\lambda; \varepsilon)$; hence $aE(\{\lambda\}) \leq E(\{\lambda\})[iH, A]E(\{\lambda\}) = 0$. Since $E(\{\lambda\}) \neq 0$, we must have $a \leq 0$. □

It should be clear that the above theorem states that the functions ϱ and $\widetilde{\varrho}$ differ only on a very small set. In fact, if λ is an eigenvalue of H and $\widetilde{\varrho}_H^A(\lambda) > 0$, then there is a neighbourhood of λ which does not contain other eigenvalues of H (see Corollary 7.2.11). So *the set* $\{\lambda \in \mathbb{R} \mid \varrho(\lambda) \neq \widetilde{\varrho}(\lambda)\}$ *is a discrete subset of* $\mathbb{R}$ *consisting of eigenvalues of* H *of finite multiplicity; at these points we have* $\varrho(\lambda) = 0$ *and* $\widetilde{\varrho}(\lambda) > 0$.

7.2.3. The next proposition contains a result concerning the existence of operators A which are locally conjugate to a self-adjoint operator H on some open set $J \subset \mathbb{R}$. We shall show that, if the spectrum of H in J is nice, then there are (many) such A. The conjugate operator method leads to some kind of converse of this result: if there is an operator A that is conjugate to H on J, then the operator H has good spectral properties on J.

PROPOSITION 7.2.14. *Let H be a self-adjoint operator and assume that there is an open interval $J \subset \mathbb{R}$ such that the spectrum of H on J is purely absolutely continuous and of constant multiplicity. Then there exists a self-adjoint operator A such that H is of class $C^\infty(A)$ and which is strictly conjugate to H on each compact subset of J.*

PROOF. The assumption made on H means that there is a Hilbert space $\mathcal{K}$ such that $HE(J)$ is unitarily equivalent to the operator Q of multiplication by the variable x in the Hilbert space $\mathcal{H}_0 = L^2(J; \mathcal{K}; dx)$ of square-integrable $\mathcal{K}$-valued functions on J. We denote by U the unitary operator $E(J)\mathcal{H} \to \mathcal{H}_0$ which realizes the unitary equivalence. Let $\varphi \in BC^\infty(J)$ be such that $\varphi(x) > 0$ for all $x \in J$ and $\int_a^c [\varphi(x)]^{-1} dx = \int_c^b [\varphi(x)]^{-1} dx = \infty$ (where $J = (a, b)$ and $a < c < b$). Then $A_0 = -[\varphi(Q)P + P\varphi(Q)]/2$, with $P = -id/dx$, is a self-adjoint operator in $\mathcal{H}_0$ such that $[iQ, A_0] = \varphi(Q)$ is strictly positive on each compact subset of J. We take $A = U^{-1}A_0 U$ on $E(J)\mathcal{H}$ and $A = 0$ on $E(\mathbb{R} \setminus J)\mathcal{H}$. Observe that, if we set $\varphi(x) = 0$ for $x \notin J$, then $[iH, A] = \varphi(H)$. $\square$

The construction made in the preceding proof explains the terminology "locally conjugate operator". In fact, the operators P and Q satisfy $[iP, Q] = I$ and in quantum mechanics they are called (canonically) conjugate operators.

We end this section with a remark concerning the possibility of localizing in some sense a conjugate operator A in a spectral representation of H. One may find in [BP2] an example which shows the usefulness of such a procedure; in the proof of the essential self-adjointness of S^*AS (see Lemma 7.2.15) we use ideas from Lemma 3.5 of [BP2].

We recall the following simple facts: (a) If A is an arbitrary closed operator and $T \in B(\mathcal{H})$, then the operator AT (with domain equal to the set of $f \in \mathcal{H}$ such that $Tf \in D(A)$) is closed; (b) If A is self-adjoint and $T \in B(\mathcal{H})$, then the adjoint of the operator T^*A (the domain of which is $D(A)$) is equal to AT.

Now assume that A is self-adjoint and $T \in C^1(A; \mathcal{H})$. We have seen that this implies $TD(A) \subset D(A)$, so $D(A) \subset D(AT)$, and $ATg = [A, T]g + TAg$ for all $g \in D(A)$ (cf. Proposition 5.3.1 for a general assertion; or use $A_\tau T = [A_\tau, T] + TA_\tau$ where A_τ is as in (7.2.2)). Let us prove that $D(A)$ *is a core of* AT. For each $\varepsilon > 0$ and $g \in D(A)$ we have $(I + i\varepsilon A)Tg = [\varepsilon A, T]g + T(I + i\varepsilon A)g$.

We take $g = (I + i\varepsilon A)^{-1} f$ with $f \in \mathcal{H}$ and apply the operator $A(I + i\varepsilon A)^{-1}$ to the preceding identity:

$$(7.2.21) \quad AT(I + \varepsilon A)^{-1} f = i\varepsilon A(I + i\varepsilon A)^{-1} \cdot [A, T] \cdot (I + i\varepsilon A)^{-1} f$$
$$+ A(I + i\varepsilon A)^{-1} T f.$$

Fix $f \in D(AT)$ and set $f_\varepsilon = (I + i\varepsilon A)^{-1} f$. Then $f_\varepsilon \in D(A)$ and $\lim_{\varepsilon \to 0} \|f_\varepsilon - f\| = 0$. Moreover, (7.2.21) may be written as $ATf_\varepsilon = i\varepsilon A(I + i\varepsilon A)^{-1}[A, T]f_\varepsilon + (I + i\varepsilon A)^{-1} ATf$, which clearly implies $\lim_{\varepsilon \to 0} \|ATf_\varepsilon - ATf\| = 0$. This finishes the proof.

LEMMA 7.2.15. *Let A be a self-adjoint operator in $\mathcal{H}$ and let $S : \mathcal{H} \to \mathcal{H}$ be of class $C^1(A; \mathcal{H})$. Then S^* and S^*S belong to $C^1(A; \mathcal{H})$, the operators $[A, S]$ and $[A, S^*]$ are bounded, and $[A, S]^* = -[A, S^*]$, $[A, S^*S] = [A, S^*]S + S^*[A, S]$.*

Denote by A_S the closure of the symmetric, densely defined operator S^*AS ($D(S^*AS) = D(AS) \supset D(A)$). Then A_S is self-adjoint and $A_S = [S^*, A]S + AS^*S$. Moreover, $D(A)$ is a core for A_S and we have $A_S|_{D(A)} = S^*[A, S] + S^*SA$.

PROOF. The first assertion of the lemma has been proved in a more general setting in Section 5.1 (one may easily prove it directly starting with the identities $[A_\tau, S]^* = [S^*, A_{-\tau}]$ and $[A_\tau, S^*S] = [A_\tau, S^*]S + S^*[A_\tau, S]$, where A_τ is as in (7.2.2)). By the discussion preceding the lemma (with $T = S$), the set $D(A)$ is a core for AS and $AS|_{D(A)} = [A, S] + SA$. Since S^* is bounded, $D(A)$ is also a core for S^*AS and $S^*AS|_{D(A)} = S^*[A, S] + S^*SA$, hence

$$(S^*AS)^* = (S^*[A, S])^* + (S^*S \cdot A)^* = [S^*, A]S + AS^*S$$

(for the second equality use property (b) mentioned before the lemma with $T = S^*S$). By using once again the remarks made before the lemma (this time with $T = S^*S$), we see that $D(A)$ is a core for AS^*S, hence for $(S^*AS)^*$. In particular, the domain of the symmetric operator S^*AS is a core for its adjoint; this is equivalent with the self-adjointness of A_S. $\square$

PROPOSITION 7.2.16. *Let A, H be self-adjoint operators in $\mathcal{H}$ such that $H \in C^1(A)$ and let $S \in C^1(A; \mathcal{H})$ with $[S, H] = 0$. Denote by A_S the self-adjoint operator given by the closure of S^*AS and define the function $s : \mathbb{R} \to \mathbb{R}$ by $s(\lambda) = \lim_{\varepsilon \to 0} [\inf S^*E(\lambda; \varepsilon)S]$. Then H is of class $C^1(A_S)$, we have $[H, A_S] = S^*[H, A]S$ (the equality holds in $B(D(H), D(H)^*)$; note that $SD(H) \subset D(H)$) and*

$$(7.2.22) \qquad \varrho_H^{A_S}(\lambda) \geq s(\lambda) \varrho_H^A(\lambda) \text{ if } \lambda \in \sigma(H) \text{ and } \varrho_H^A(\lambda) > 0.$$

PROOF. Set $R = (H + i)^{-1}$ and observe that $[A_S, R] = S^*[A, R]S$ as sesquilinear forms on $D(A)$ (recall that $SD(A) \subset D(A)$). Since $D(A)$ is a core for A_S and $S \in B(D(A))$, and since $[A,R]$ is a bounded operator in $\mathcal{H}$, it follows that the sesquilinear form $[A_S, R]$ (with domain $D(A_S)$) is in fact a bounded operator in $\mathcal{H}$. So Lemma 6.2.9 implies that $H \in C^1(A_S)$. Moreover, we shall also have $[A_S, R] = S^*[A, R]S$ as bounded operators on $\mathcal{H}$. Notice that S and S^* belong to $B(D(H))$ and have extensions to operators in $B(D(H)^*)$ that we denote by the same symbols S, S^*. Then, by taking into account (7.2.3), we get:

$$[H, A_S] = (H + i)[A_S, R](H + i) = (H + i)S^*[A, R]S(H + i)$$
$$= S^*(H + i)[A, R](H + i)S = S^*[H, A]S.$$

In order to prove (7.2.22), let $0 < a < \varrho_H^A(\lambda)$ and $b < s(\lambda)$. Then there is $\varepsilon > 0$ such that $E(\lambda; \varepsilon)[H, iA]E(\lambda; \varepsilon) \geq aE(\lambda; \varepsilon)$ and $S^*E(\lambda; \varepsilon)S \geq bE(\lambda; \varepsilon)$. Since

$$E(\lambda; \varepsilon)[H, iA_S]E(\lambda; \varepsilon) = S^*E(\lambda; \varepsilon)[H, iA]E(\lambda; \varepsilon)S \geq aS^*E(\lambda; \varepsilon)S \geq abE(\lambda; \varepsilon),$$

we shall have $\varrho_H^{A_S}(\lambda) \geq ab$. $\square$

If the operator S in Proposition 7.2.16 is of the form $S = \phi(H)$ with $\phi : \mathbb{R} \to \mathbb{C}$ bounded and continuous, then clearly $s(\lambda) = |\phi(\lambda)|^2$. One may easily show in this case that the inequality (7.2.22) becomes an equality and that it holds for all $\lambda \in \sigma(H)$. Recall that $\phi(H)$ is of class $C^1(A; \mathscr{H})$ if ϕ is a bounded rational function without poles in $\sigma(H)$ or if $\phi \in C_0^2(\mathbb{R})$.

For some applications it is important to observe that H could be of class $C^1(A_S)$ for a conveniently chosen S even if H is not of class $C^1(A)$. For example, if $[A, H]$ is a continuous sesquilinear form on $D(H^k)$ for some $k \in \mathbb{N}$, one could take $S = \phi(H)$ with a function ϕ decaying rapidly enough at infinity. We shall meet such a situation in §7.6.3; see also [BP2]. One can give a meaning to the function ϱ_H^A even if H is only locally of class $C^1(A)$, in a sense that will be explained (in a more general setting) in Section 8.3; then (7.2.22) will remain true.

7.3. The Method of Differential Inequalities

Let A and H be self-adjoint operators in a Hilbert space $\mathscr{H}$ and assume that H is of class $C^1(A)$. In the preceding section we associated to such a couple a lower semicontinuous function $\varrho_H^A : \mathbb{R} \to (-\infty, +\infty]$, defined in (7.2.4). Our purpose in this and the next section is to show that H has particularly nice spectral properties in the open set

$$(7.3.1) \quad \mu^A(H) = \{\lambda \in \mathbb{R} \mid \varrho_H^A(\lambda) > 0\}$$
$$\equiv \{\lambda \in \mathbb{R} \mid \exists \varepsilon > 0, a > 0 \text{ s.t. } aE(\lambda; \varepsilon) \leq E(\lambda; \varepsilon)[iH, A]E(\lambda; \varepsilon)\}.$$

This could be called *the Mourre set of H with respect to A*, because it was E. Mourre who understood its relevance for the spectral and propagation properties of H (cf. [M1,2]). It is an easy consequence of the virial theorem that H has no eigenvalues in $\mu^A(H)$. It turns out that it is essential to require some additional regularity of H with respect to A in order get the results stated below and in Section 7.4. We shall impose the condition that $H \in \mathscr{C}^{1,1}(A)$. As explained in Appendix 7.B, *this assumption is optimal in the Besov scale*: even if H is of class $C_u^1(A) \cap \mathscr{C}^{1,q}(A)$ for *all* $q > 1$, the limiting absorption principle presented below can break down in a rather radical way. We recall that $\mathscr{C}^{1,1}(A) \subset C_u^1(A)$, so that the function ϱ_H^A is automatically defined if H is of class $\mathscr{C}^{1,1}(A)$.

This section contains the most technical results. We shall consider the situation where H is a bounded self-adjoint operator, which we denote by S, and prove a very precise form of the limiting absorption principle. The case of un-bounded operators having a spectral gap is easily reduced to this situation, as explained (together with applications in spectral theory) in Section 7.4. For the case of operators without spectral gap, we refer to Section 7.5.

From now on in this section, we shall consider a fixed Hilbert space $\mathscr{H}$ equipped with a strongly continuous unitary group $W(\tau) = e^{iA\tau}$, where A is a self-adjoint operator in $\mathscr{H}$. Then, one can construct the Besov scale $\{\mathscr{H}_{t,p} \mid t \in \mathbb{R}, \; p \in [1,\infty]\}$ associated to this structure. In fact, we shall be interested only in the space $\mathscr{K} \equiv \mathscr{H}_{1/2,1}$ and its dual $\mathscr{K}^* = \mathscr{H}_{-1/2,\infty}$. Let $D(A)$ be the domain of A equipped with the graph topology (this is $\mathscr{H}_1 \equiv \mathscr{H}_{1,2}$ considered as a B-space). We recall that $\mathscr{K}$ is the B-space obtained by real interpolation as : $\mathscr{K} = (D(A), \mathscr{H})_{1/2,1}$. A more intuitive description of $\mathscr{K}$ is in terms of the Littlewood-Paley theory (cf. Sections 2.8 or 3.6), but we shall not need it here. We shall always identify $\mathscr{H}$ and $\mathscr{H}^*$. Since $\mathscr{K} \subset \mathscr{H}$ continuously and densely, we get $\mathscr{K} \subset \mathscr{H} \subset \mathscr{K}^*$ continuously (the second embedding is not dense if A is unbounded). This also implies a continuous embedding $B(\mathscr{H}) \subset B(\mathscr{K}, \mathscr{K}^*)$.

Now let $S \in B(\mathscr{H})$ be a self-adjoint operator. Then for $z \in \mathbb{C} \setminus \sigma(S)$ we have a holomorphic function $z \mapsto (S - z)^{-1} \in B(\mathscr{H})$. Since $||(S - z)^{-1}|| = [\operatorname{dist}(z, \sigma(S))]^{-1}$, there is no chance of extending it to points z of the spectrum of S without leaving the space $B(\mathscr{H})$. But in the preceding framework we can look at $z \mapsto (S - z)^{-1}$ as a $B(\mathscr{K}, \mathscr{K}^*)$-valued function. It turns out that, if S is of class $\mathscr{C}^{1,1}(A; \mathscr{H})$, this function has natural extensions from the upper and lower half-planes to the points of the open subset $\mu^A(S)$ of the real line (see (7.3.1)). We recall that $S \in \mathscr{C}^{1,1}(A; \mathscr{H})$ means

$$(7.3.2) \qquad \int_0^1 ||e^{-iA\tau} S e^{iA\tau} + e^{iA\tau} S e^{-iA\tau} - 2S|| \frac{d\tau}{\tau^2} < \infty.$$

We denote by $||\cdot||$ the norm in $\mathscr{H}$ and in $B(\mathscr{H})$ and set $\mathbb{C}_\pm = \{z \in \mathbb{C} \mid \pm\Im z > 0\}$. We may now state the main result of this chapter.

THEOREM 7.3.1. *Let S be a bounded self-adjoint operator in $\mathscr{H}$ which is of class $\mathscr{C}^{1,1}(A)$. Then the holomorphic function $\mathbb{C}_\pm \ni z \mapsto (S-z)^{-1} \in B(\mathscr{K}, \mathscr{K}^*)$ extends to a weak* continuous function on $\mathbb{C}_\pm \cup \mu^A(S)$.*

In other terms, if $f, g \in \mathscr{K}$, then the function $z \mapsto \langle f, (S - z)^{-1} g \rangle$, which is holomorphic in $\mathbb{C}_\pm$, has a continuous extension to $\mathbb{C}_\pm \cup \mu^A(S)$. The reader may easily convince himself that this is equivalent with the property that the limits $\lim_{\mu \to \pm 0} \langle f, (S - \lambda - i\mu)^{-1} g \rangle$ exist uniformly in λ when λ runs over compact subsets of $\mu^A(S)$; in particular, the limit functions $\langle f, (S - \lambda \mp i0)^{-1} g \rangle$ are continuous in λ. Observe that $\langle f, (S - z)^{-1} g \rangle \to \langle f, (S - \lambda - i0)^{-1} g \rangle$ when z is in the upper half-plane and tends to λ in an arbitrary way (not necessarily on the vertical passing through λ). This fact will be important for the proof of the general limiting absorption principle in Section 7.4. Finally, notice that the expressions $(S - \lambda \mp i0)^{-1}$ are well defined linear continuous operator $\mathscr{K} \to \mathscr{K}^*$.

The remainder of this section is devoted to the proof of the Theorem 7.3.1. The basic ideas have been explained in the introduction of this chapter. We begin with a series of technical lemmas in which the operator $S \in B(\mathcal{H})$ is assumed to be at least of class $C^1(A; \mathcal{H})$ and self-adjoint. We set $B = [iS, A]$.

LEMMA 7.3.2. *Let $\{S_\varepsilon\}_{0<\varepsilon<1}$, $\{B_\varepsilon\}_{0<\varepsilon<1}$ be two families of bounded symmetric operators such that $||S_\varepsilon - S|| + \varepsilon||B_\varepsilon|| \le c\varepsilon$ for some constant c and $||B_\varepsilon - B|| \to 0$ as $\varepsilon \to 0$. Let λ_0 and a be real numbers such that $0 < a < \varrho_S^A(\lambda_0)$. Then there are strictly positive numbers δ, ε_0, b such that for $|\lambda - \lambda_0| \le \delta$, $0 < \varepsilon \le \varepsilon_0$ and $\mu \ge 0$ the following estimate holds for all $g \in \mathcal{H}$:*

$$(7.3.3) \qquad a||g||^2 \le \langle g, B_\varepsilon g \rangle + \frac{b}{\mu^2 + \delta^2} ||[S_\varepsilon - \lambda \mp i(\varepsilon B_\varepsilon + \mu)]g||^2.$$

PROOF. Let E be the spectral measure of S. Choose numbers $a < a_0 < a_1 < \varrho_S^A(\lambda_0)$ and $\delta > 0$ such that $a_1 E \le EBE$ for $E = E((\lambda_0 - 2\delta, \lambda_0 + 2\delta))$. There is $\varepsilon_1 > 0$ such that $||B_\varepsilon - B|| \le a_1 - a_0$ for $0 < \varepsilon \le \varepsilon_1$; this implies $EB_\varepsilon E \ge EBE - (a_1 - a_0)E \ge a_0 E$. Set $E^\perp = I - E$ and consider from now on λ, μ real with $|\lambda - \lambda_0| \le \delta$, $\mu \ge 0$. Observe that $||(S - \lambda \mp i\mu)^{-1}E^\perp|| \le (\mu^2 + \delta^2)^{-1/2}$, hence:

$$||E^\perp g||^2 = ||(S - \lambda \mp i\mu)^{-1}E^\perp[S_\varepsilon - \lambda \mp i(\varepsilon B_\varepsilon + \mu) + S - S_\varepsilon \pm i\varepsilon B_\varepsilon]g||^2$$
$$\le \frac{2}{\mu^2 + \delta^2} ||[S_\varepsilon - \lambda \mp i(\varepsilon B_\varepsilon + \mu)]g||^2 + \frac{2c^2\varepsilon^2}{\mu^2 + \delta^2} ||g||^2.$$

By using the preceding remarks, we get for $\varepsilon \le \varepsilon_1$ and for any $\nu > 0$:

$$a_0||g||^2 = a_0\langle g, Eg \rangle + a_0||E^\perp g||^2 \le \langle g, EB_\varepsilon Eg \rangle + a_0||E^\perp g||^2$$
$$= \langle g, B_\varepsilon g \rangle - 2\Re\langle g, B_\varepsilon E^\perp g \rangle + \langle E^\perp g, B_\varepsilon E^\perp g \rangle + a_0||E^\perp g||^2$$
$$\le \langle g, B_\varepsilon g \rangle + \nu||g||^2 + \nu^{-1}||B_\varepsilon||^2||E^\perp g||^2 + ||B_\varepsilon|| \cdot ||E^\perp g||^2 + a_0||E^\perp g||^2$$
$$\le \langle g, B_\varepsilon g \rangle + \nu||g||^2 + [\nu^{-1}||B_\varepsilon||^2 + ||B_\varepsilon|| + a_0]\frac{2c^2\varepsilon^2}{\mu^2 + \delta^2}||g||^2$$
$$+ [\nu^{-1}||B_\varepsilon||^2 + ||B_\varepsilon|| + a_0]\frac{2}{\mu^2 + \delta^2}||[S_\varepsilon - \lambda \mp i(\varepsilon B_\varepsilon + \mu)]g||^2.$$

We have $||B_\varepsilon|| < c$ and we may assume $c \ge 1$ and $\nu \le 1$. This implies:

$$\left[a_0 - \nu - \frac{2c^2\varepsilon^2(a_0 + 2c^2)}{\nu\delta^2}\right]||g||^2 \le$$
$$\le \langle g, B_\varepsilon g \rangle + \frac{2a_0 + 4c^2}{\nu(\mu^2 + \delta^2)}||[S_\varepsilon - \lambda \mp i(\varepsilon B_\varepsilon + \mu)]g||^2.$$

We can choose $\nu > 0$ and $\varepsilon_0 \in (0, \varepsilon_1)$ such that the expression in the bracket on the l.h.s. is $\ge a$ for $0 < \varepsilon \le \varepsilon_0$. We get (7.3.3) with $b = \nu^{-1}(2a_0 + 4c^2)$. $\square$

LEMMA 7.3.3. *Under the hypotheses of Lemma 7.3.2, the operators $S_\varepsilon - \lambda \mp i(\varepsilon B_\varepsilon + \mu)$ are invertible in $B(\mathcal{H})$ whenever $|\lambda - \lambda_0| \le \delta$, $0 < \varepsilon \le \varepsilon_0$ and $\mu \ge 0$. For fixed λ and μ satisfying these conditions, set $G_\varepsilon^\pm = [S_\varepsilon - \lambda \mp i(\varepsilon B_\varepsilon + \mu)]^{-1}$. Then one has $(G_\varepsilon^\pm)^* = G_\varepsilon^\mp$ and*

$$(7.3.4) \qquad ||G_\varepsilon^\pm|| \le \frac{1}{a\varepsilon + \mu}\left[1 + b\varepsilon\frac{c\varepsilon + ||S|| + |\lambda + i\mu|}{\mu^2 + \delta^2}\right].$$

Moreover, for each $h \in \mathcal{H}$:

$$(7.3.5) \qquad \|G_\varepsilon^\pm h\| \leq \frac{1}{\sqrt{a\varepsilon}} |\Im\langle h, G_\varepsilon^+ h\rangle|^{1/2} + \frac{1}{\delta}\left(\frac{b}{a}\right)^{1/2} \|h\|.$$

PROOF. We use (7.3.3) and denote $T_\varepsilon^\pm = S_\varepsilon - \lambda \mp i(\varepsilon B_\varepsilon + \mu)$:

$$(a\varepsilon + \mu)\|g\|^2 \leq \langle g, (\varepsilon B_\varepsilon + \mu)g\rangle + \frac{b\varepsilon}{\mu^2 + \delta^2}\|T_\varepsilon^\pm g\|^2$$

$$= \mp\Im\langle g, T_\varepsilon^\pm g\rangle + \frac{b\varepsilon}{\mu^2 + \delta^2}\|T_\varepsilon^\pm g\|^2$$

$$\leq \|g\| \cdot \|T_\varepsilon^\pm g\|\left\{1 + \frac{b\varepsilon}{\mu^2 + \delta^2}\|T_\varepsilon^\pm\|\right\}$$

$$\leq \|g\| \cdot \|T_\varepsilon^\pm g\|\left[1 + b\varepsilon\frac{\|S\| + c\varepsilon + |\lambda + i\mu|}{\mu^2 + \delta^2}\right].$$

This shows that $T_\varepsilon^\pm$ are injective operators with closed ranges. Since $(T_\varepsilon^\pm)^* = T_\varepsilon^\mp$, the range of $T_\varepsilon^\pm$ (which is equal to the orthogonal complement of the null space of $(T_\varepsilon^\pm)^*$) must be $\mathcal{H}$. This proves all assertions of the lemma except (7.3.5). To prove it, we take $g = G_\varepsilon^\pm h$ in (7.3.3) to get that

$$a\varepsilon\|G_\varepsilon^\pm h\|^2 \leq \langle G_\varepsilon^\pm h, (\varepsilon B_\varepsilon + \mu)G_\varepsilon^\pm h\rangle + \frac{b\varepsilon}{\mu^2 + \delta^2}\|h\|^2.$$

This implies the inequality (7.3.5) by taking into account the following identity:

$$\langle G_\varepsilon^\pm h, (\varepsilon B_\varepsilon + \mu)G_\varepsilon^\pm h\rangle = \pm(2i)^{-1}\langle h, G_\varepsilon^\mp(T_\varepsilon^\mp - T_\varepsilon^\pm)G_\varepsilon^\pm h\rangle$$

$$= \pm(2i)^{-1}\langle h, (G_\varepsilon^\pm - G_\varepsilon^\mp)h\rangle = \Im\langle h, G_\varepsilon^+ h\rangle. \qquad \square$$

We keep the preceding notations and assumptions and set $G_\varepsilon \equiv G_\varepsilon^+$, $T_\varepsilon \equiv T_\varepsilon^+$. We shall denote derivatives with respect to ε by a prime, e.g. $G_\varepsilon' = \frac{d}{d\varepsilon}G_\varepsilon$.

LEMMA 7.3.4. *Assume that $\varepsilon \mapsto S_\varepsilon$ and $\varepsilon \mapsto B_\varepsilon$ are norm C^1 and that S_ε, $B_\varepsilon \in C^1(A;\mathcal{H})$. Then $\varepsilon \mapsto G_\varepsilon$ is norm C^1, one has $G_\varepsilon \in C^1(A;\mathcal{H})$ and*

$$(7.3.6) \quad G_\varepsilon' + [A, G_\varepsilon] = iG_\varepsilon\{B_\varepsilon - [iS_\varepsilon, A] + \varepsilon(i\varepsilon^{-1}S_\varepsilon' + B_\varepsilon' + [A, B_\varepsilon])\}G_\varepsilon.$$

PROOF. The differentiability of G_ε is easily obtained from the identity $G_\varrho - G_\varepsilon = G_\varrho(T_\varepsilon - T_\varrho)G_\varepsilon$ by using the differentiability of T_ε and the inequality $\|G_\varrho\| \leq c_1\varrho^{-1}$ (see (7.3.4)). We observe that $G_\varepsilon' = -G_\varepsilon T_\varepsilon' G_\varepsilon$. The fact that $G_\varepsilon \in C^1(A;\mathcal{H})$ follows from Proposition 5.1.6 which also gives $[A, G_\varepsilon] = -G_\varepsilon[A, T_\varepsilon]G_\varepsilon$. A short direct proof is obtained by imitating the argument (7.2.2), taking into account that the preceding equality is obvious when A is replaced by the bounded operator A_τ ($\tau \neq 0$). $\square$

LEMMA 7.3.5. *Keep the notations and assumptions of Lemmas 7.3.2-7.3.4, and let $\{f_\varepsilon\}_{0<\varepsilon<1}$ be a family of vectors in $\mathcal{H}$ such that $\varepsilon \mapsto f_\varepsilon \in \mathcal{H}$ is strongly C^1 and $f_\varepsilon \in D(A)$ for each ε. Set $F_\varepsilon = \langle f_\varepsilon, G_\varepsilon f_\varepsilon\rangle$. Then $\varepsilon \mapsto F_\varepsilon$ is of class C^1 and its derivative F_ε' satisfies :*

$$(7.3.7) \quad F_\varepsilon' = \langle f_\varepsilon, (G_\varepsilon' + [A, G_\varepsilon])f_\varepsilon\rangle + \langle G_\varepsilon^* f_\varepsilon, f_\varepsilon' + Af_\varepsilon\rangle + \langle f_\varepsilon' - Af_\varepsilon, G_\varepsilon f_\varepsilon\rangle.$$

Let $\ell(\varepsilon) = ||f'_\varepsilon|| + ||Af_\varepsilon||$ and $q(\varepsilon) = ||\varepsilon^{-1}(B_\varepsilon - [iS_\varepsilon, A]) + i\varepsilon^{-1}S'_\varepsilon + B'_\varepsilon + [A, B_\varepsilon]||$.
Then F_ε satisfies the following differential inequality, where $\omega = a^{-1/2}b^{1/2}\delta^{-1}$
and $0 < \varepsilon \leq \varepsilon_0$:

$$(7.3.8) \quad \frac{1}{2}|F'_\varepsilon| \leq \omega||f_\varepsilon||[\ell(\varepsilon) + \omega\varepsilon q(\varepsilon)||f_\varepsilon||] + \frac{\ell(\varepsilon)}{\sqrt{a\varepsilon}}|F_\varepsilon|^{1/2} + \frac{q(\varepsilon)}{a}|F_\varepsilon|.$$

PROOF. (7.3.7) is obvious and may be rewritten as follows by using (7.3.6):

$$(7.3.9) \quad F'_\varepsilon = i\langle G^*_\varepsilon f_\varepsilon, \{B_\varepsilon - [iS_\varepsilon, A] + \varepsilon(i\varepsilon^{-1}S'_\varepsilon + B'_\varepsilon + [A, B_\varepsilon])\}G_\varepsilon f_\varepsilon\rangle$$
$$+ \langle G^*_\varepsilon f_\varepsilon, f'_\varepsilon + Af_\varepsilon\rangle + \langle f'_\varepsilon - Af_\varepsilon, G_\varepsilon f_\varepsilon\rangle.$$

Hence

$$|F'_\varepsilon| \leq \varepsilon q(\varepsilon)||G_\varepsilon f_\varepsilon|| \cdot ||G^*_\varepsilon f_\varepsilon|| + \ell(\varepsilon)(||G_\varepsilon f_\varepsilon|| + ||G^*_\varepsilon f_\varepsilon||).$$

This implies (7.3.8) after observing that, by (7.3.5):

$$||G^\pm_\varepsilon f_\varepsilon|| \leq \frac{1}{(a\varepsilon)^{1/2}}|F_\varepsilon|^{1/2} + \omega||f_\varepsilon||. \quad \square$$

The differential inequality (7.3.8) (from which the method takes its name)
is quite remarkable in that the spectral variable $z = \lambda + i\mu$ does not appear
explicitly in the coefficients. Of course, we have the conditions $|\lambda - \lambda_0| \leq \delta$ and
$\mu \geq 0$. Let us write (7.3.8) in the form $|F'_\varepsilon| \leq \eta(\varepsilon) + \varphi(\varepsilon)|F_\varepsilon|^{1/2} + \psi(\varepsilon)|F_\varepsilon|$ and
then use $|F_s| \leq |F_{\varepsilon_0}| + \int_s^{\varepsilon_0} |F'_\tau|d\tau$. So, for $0 < \varepsilon < s < \varepsilon_0$ we have

$$|F_s| \leq |F_{\varepsilon_0}| + \int_\varepsilon^{\varepsilon_0} \eta(\tau)d\tau + \int_s^{\varepsilon_0} [\varphi(\tau)|F_\tau|^{1/2} + \psi(\tau)|F_\tau|]d\tau.$$

We may apply the modified Gronwall Lemma presented in Appendix A of this
chapter, with $\theta = 1/2$, and get

$$(7.3.10)$$
$$|F_\varepsilon| \leq \left[\left(|F_{\varepsilon_0}| + \int_\varepsilon^{\varepsilon_0} \eta(\tau)d\tau\right)^{1/2} + \frac{1}{2}\int_\varepsilon^{\varepsilon_0} \varphi(\tau)\exp\left(-\frac{1}{2}\int_\tau^{\varepsilon_0} \psi(s)ds\right)d\tau\right]^2 \cdot$$
$$\cdot \exp\int_\varepsilon^{\varepsilon_0} \psi(\tau)d\tau$$

for all $0 < \varepsilon < \varepsilon_0$. This estimate is optimal (in a sense explained by the last
remark of Appendix A) and we shall use only the obvious consequence:

$$(7.3.11) \quad |F_\varepsilon| \leq 2\left[|F_{\varepsilon_0}| + \int_\varepsilon^{\varepsilon_0} \eta(\tau)d\tau + \left(\int_\varepsilon^{\varepsilon_0} \varphi(\tau)d\tau\right)^2\right]\exp\int_\varepsilon^{\varepsilon_0} \psi(\tau)d\tau.$$

Our purpose is to get a bound $|F_\varepsilon| < \text{const.} < \infty$, independent of $z = \lambda + i\mu$,
as $\varepsilon \to 0$. From the above estimate we see that this is satisfied if $\int_0^{\varepsilon_0}[\eta(\tau) + \varphi(\tau) + \psi(\tau)]d\tau < \infty$. Taking into account the explicit form the functions η, φ,
ψ we see that this means

$$\int_0^{\varepsilon_0}\left[\ell(\varepsilon)||f_\varepsilon|| + \varepsilon q(\varepsilon)||f_\varepsilon||^2 + \frac{\ell(\varepsilon)}{\sqrt{\varepsilon}} + q(\varepsilon)\right]d\varepsilon < \infty.$$

But from the integrability of $\varepsilon^{-1/2}\ell(\varepsilon)$ follows that of $\ell(\varepsilon)$, which in turn implies the boundedness of f_ε (because $\|f_\varepsilon\| \leq \|f_{\varepsilon_0}\| + \int_\varepsilon^{\varepsilon_0} \|f_\tau'\|d\tau$), hence the integrability of $\ell(\varepsilon)\|f_\varepsilon\|$. Then clearly the integrability of $q(\varepsilon)$ will imply that of $\varepsilon q(\varepsilon)\|f_\varepsilon\|^2$. In conclusion, we are left with the condition:

$$(7.3.12) \qquad \int_0^1 [\varepsilon^{-1/2}\ell(\varepsilon) + q(\varepsilon)]d\varepsilon < \infty.$$

The condition $\int_0^1 q(\varepsilon)d\varepsilon < \infty$ will be discussed further on. We shall show in Lemma 7.3.6 that, under the hypothesis of Theorem 7.3.1, one may choose families $\{S_\varepsilon\}$ and $\{B_\varepsilon\}$ such that the preceding integral is finite. Now let us consider the condition $\int_0^1 \varepsilon^{-1/2}\ell(\varepsilon)d\varepsilon < \infty$. Since this implies the boundedness of f_ε as $\varepsilon \to 0$, the condition is equivalent to $\int_0^1 \varepsilon^{1/2}[\|f_\varepsilon'\|+\|f_\varepsilon\|+\|Af_\varepsilon\|]\varepsilon^{-1}d\varepsilon < \infty$. But then, Proposition 2.3.3 states that $f = \lim_{\varepsilon\to 0} f_\varepsilon$ exists in $\mathcal{H}$ and belongs to the interpolation space $(D(A),\mathcal{H})_{1/2,1} \equiv \mathcal{K}$ in our notations; and reciprocally, for any $f \in \mathcal{K}$ there is a family $\{f_\varepsilon\}$ with the preceding properties. In fact, Proposition 2.7.2 provides us with an explicit construction, namely:

$$(7.3.13) \qquad f_\varepsilon = (I + i\varepsilon A)^{-1}f, \qquad \varepsilon > 0.$$

Then $f_\varepsilon \in D(A)$, $f_\varepsilon' = -i(I + i\varepsilon A)^{-1}Af_\varepsilon$, so $\|f_\varepsilon\| \leq \|f\|$, $\|f_\varepsilon - f\| \to 0$ as $\varepsilon \to 0$, $\ell(\varepsilon) \leq 2\|Af_\varepsilon\|$ and $\int_0^1 \ell(\varepsilon)\varepsilon^{-1/2}d\varepsilon \leq c_1\|f\|_\mathcal{K}$ by Proposition 2.7.2, for some constant c_1 independent of f. Furthermore, by taking into account the explicit form of the functions η and φ, one obtains the existence of a constant c_2 such that $\int_0^1 \eta(\tau)d\tau \leq c_2\|f\|_\mathcal{K}^2$ and $\int_0^1 \varphi(\tau)d\tau \leq c_2\|f\|_\mathcal{K}$.

We summarize the preceding considerations: under the assumptions of Theorem 7.3.1, there are two families of operators $\{S_\varepsilon\}_{0<\varepsilon\leq\varepsilon_0}$ and $\{B_\varepsilon\}_{0<\varepsilon\leq\varepsilon_0}$ (constructed explicitly in Lemma 7.3.6), and for each $f \in \mathcal{K}$ there is a family of vectors $\{f_\varepsilon\}_{0<\varepsilon\leq\varepsilon_0}$ (given by (7.3.13) for example), such that $|F_\varepsilon| \leq c\|f\|_\mathcal{K}^2$ for $0 < \varepsilon \leq \varepsilon_0$, $|\lambda - \lambda_0| \leq \delta$ and $\mu \geq 0$, where the constant c is independent of f, ε, λ, μ. From this it is easy to deduce the result of Theorem 7.3.1:

PROOF OF THEOREM 7.3.1. By (7.3.8), (7.3.12) and the preceding estimate on F_ε, there is an integrable function $\kappa : (0,\varepsilon_0) \to \mathbb{R}$ such that $|F_\varepsilon'| \leq \kappa(\varepsilon)$ for all ε, λ, μ as above. Now fix $\mu > 0$. Since $S_\varepsilon - \lambda - i(\varepsilon B_\varepsilon + \mu) \to S - \lambda - i\mu \equiv S - z$ in norm as $\varepsilon \to 0$, we shall have $G_\varepsilon \to (S - z)^{-1}$ in norm too, and:

$$(7.3.14)$$

$$\langle f, (S - z)^{-1}f\rangle = \lim_{\varepsilon\to 0}\langle f_\varepsilon, G_\varepsilon f_\varepsilon\rangle = \langle f_{\varepsilon_0}, G_{\varepsilon_0}(z)f_{\varepsilon_0}\rangle - \int_0^{\varepsilon_0} F_\varepsilon'(z)d\varepsilon.$$

Here we have explicitly indicated the dependence on $z = \lambda + i\mu$ of G_{ε_0} and F_ε'. Let $\Omega = \{\lambda + i\mu \mid |\lambda - \lambda_0| \leq \delta, \mu \geq 0\}$. From (7.3.4) it follows that $\|G_{\varepsilon_0}(z)\| \leq \text{const.} < \infty$ independently of $z \in \Omega$, which clearly implies that $G_{\varepsilon_0}(z)$ is continuous (in norm) as a function of $z \in \Omega$. For each $\varepsilon > 0$, the continuity in $z \in \Omega$ of $F_\varepsilon'(z)$ follows from (7.3.9). By the dominated convergence theorem (use $|F_\varepsilon'| \leq \kappa(\varepsilon)$), (7.3.14) gives the existence of a continuous extension of the function $\langle f, (S - z)^{-1}f\rangle$ from the domain $\{z \in \Omega \mid \Im z = \mu > 0\}$ to all

Ω. The polarization principle shows that this holds for $\langle f, (S - z)^{-1}g \rangle$ with f, $g \in \mathcal{K}$. $\square$

We finally discuss the condition $\int_0^1 q(\varepsilon)d\varepsilon < \infty$. It is natural to try the choice $B_\varepsilon = [iS_\varepsilon, A]$ (since we require that $B_\varepsilon \to B = [iS, A]$). Then $q(\varepsilon) = \||i\varepsilon^{-1}S_\varepsilon' + \frac{d}{d\varepsilon}[iS_\varepsilon, A] - i[[S_\varepsilon, A], A]\||$. Now it is better to think in terms of the group of automorphism $\mathcal{W}(\tau) = e^{i\mathcal{A}\tau}$ associated to W in $B(\mathcal{H})$, namely $\mathcal{W}(\tau)[T] = e^{-iA\tau}Te^{iA\tau}$ and $\mathcal{A}[T] = [T, A]$ (see Chapter 5). So, we require S_ε to belong to the strong domain of $\mathcal{A}^2$, i.e. $S_\varepsilon \in C^2(A; \mathcal{H})$, and the functions $\varepsilon \mapsto S_\varepsilon$, $\varepsilon \mapsto \mathcal{A}[S_\varepsilon]$ should be norm differentiable. If one does not expect compensations inside the norm in $q(\varepsilon)$, we should require that

$$(7.3.15) \qquad \int_0^1 \left[\left\|\frac{d}{d\varepsilon}S_\varepsilon\right\| + \varepsilon\left\|\frac{d}{d\varepsilon}\mathcal{A}[S_\varepsilon]\right\| + \varepsilon\|\mathcal{A}^2[S_\varepsilon]\| \right]\frac{d\varepsilon}{\varepsilon} < \infty.$$

If we let $\widetilde{S}_\tau = S_{\sqrt{\tau}}$, we get

$$\int_0^1 \left[\sqrt{\tau}\left\|\frac{d}{d\tau}\widetilde{S}_\tau\right\| + \tau\left\|\frac{d}{d\tau}\mathcal{A}[\widetilde{S}_\tau]\right\| + \sqrt{\tau}\|\mathcal{A}^2[\widetilde{S}_\tau]\| \right]\frac{d\tau}{\tau} < \infty$$

and $\|\widetilde{S}_\tau - S\| \to 0$. The first part of Proposition 2.3.3 assures us that the existence of such a family $\{S_\varepsilon\}$ implies $S \in (D(\mathcal{A}^2), B(\mathcal{H}))_{1/2,1} = \mathcal{C}^{1,1}(A; \mathcal{H})$ (cf. (5.2.6) and (5.2.22)). But, due to the middle term in (7.3.15), it is not so obvious that the reciprocal assertion holds. We now prove this by a direct construction, suggested by the functional calculus approach that we introduced in Chapter 3.

LEMMA 7.3.6. *Let $S \in B(\mathcal{H})$ be a self-adjoint operator of class $\mathcal{C}^{1,1}(A; \mathcal{H})$, i.e. such that (7.3.2) holds. Then there is a family $\{S_\varepsilon\}_{0<\varepsilon<1}$ of bounded, self-adjoint operators, with the following properties :*

(a) *The function $\varepsilon \mapsto S_\varepsilon \in B(\mathcal{H})$ is of class C^∞, $\|S_\varepsilon - S\| \leq c\varepsilon$ for some constant $c < \infty$, and $\int_0^1 \|S_\varepsilon'\|\varepsilon^{-1}d\varepsilon < \infty$.*

(b) *$S_\varepsilon \in C^\infty(A; \mathcal{H})$ and $\int_0^1 \|[A, [A, S_\varepsilon]]\|d\varepsilon < \infty$.*

(c) *S is of class $C^1(A; \mathcal{H})$ and, if $B = [iS, A]$ and $B_\varepsilon = [iS_\varepsilon, A]$, then one has $\|B_\varepsilon - B\| \to 0$ as $\varepsilon \to 0$, $\varepsilon \mapsto B_\varepsilon \in B(\mathcal{H})$ is of class C^∞ and $\int_0^1 \|B_\varepsilon'\|d\varepsilon < \infty$.*

PROOF. (i) One may exhibit S_ε quite explicitly, for example:

$$(7.3.16) \qquad S_\varepsilon = \int_{-\infty}^{\infty} e^{-iA\varepsilon\tau}Se^{iA\varepsilon\tau}e^{-\tau^2/4}\frac{d\tau}{(4\pi)^{1/2}}.$$

It is straightforward, but not very illuminating, to check that (a), (b), (c) are satisfied. In order to explain what we are really doing, we prefer to present a more theoretical approach, based on the functional calculus associated to the generator $\mathcal{A}$ of the C_w-group $\mathcal{W}(\tau) = e^{i\mathcal{A}\tau}$ in the Banach space $B(\mathcal{H})$. This group has been extensively studied in Chapter 6, but we make an effort to be self-contained here.

(ii) Let $\mathscr{M} = \mathscr{M}(\mathbb{R})$ be the algebra of functions on $\mathbb{R}$ which are Fourier transforms of integrable measures (see Definition 3.1.11). Then for $\varphi \in \mathscr{M}$, we define $\varphi(\mathscr{A}) : B(\mathscr{H}) \to B(\mathscr{H})$ by $\varphi(\mathscr{A}) = \int_{-\infty}^{\infty} \mathscr{W}(\tau)\widehat{\varphi}(\tau)d\tau$ or, more explicitly, $\varphi(\mathscr{A})[T]) = \int_{-\infty}^{\infty} e^{-iA\tau}Te^{iA\tau}\widehat{\varphi}(\tau)d\tau$. Then $\mathscr{M} \ni \varphi \mapsto \varphi(\mathscr{A}) \in B(B(\mathscr{H}))$ is a unital homomorphism, $(\varphi(\mathscr{A})[T])^* = \varphi^+(\mathscr{A})[T^*]$ with $\varphi^+(x) = \overline{\varphi}(-x)$ and $\|\varphi(\mathscr{A})\|_{B(B(\mathscr{H}))} \le \|\varphi\|_{\mathscr{M}}$. A fundamental observation is that, if φ is an even function (i.e. $\varphi(-t) = \varphi(t)$) and $\varphi(0) = 0$, then (observe the connection with (7.3.2)):

$$(7.3.17) \qquad \varphi(\mathscr{A}) = \int_0^{\infty} [\mathscr{W}(\tau) + \mathscr{W}(-\tau) - 2]\widehat{\varphi}(\tau)d\tau.$$

For $0 \le \varepsilon \le 1$, we define $\varphi(\varepsilon\mathscr{A}) = \varphi^{\varepsilon}(\mathscr{A})$ where $\varphi^{\varepsilon}(t) = \varphi(\varepsilon t)$. Then $\varphi^{\varepsilon} \in \mathscr{M}$, $\|\varphi^{\varepsilon}\|_{\mathscr{M}} \le \|\varphi\|_{\mathscr{M}}$, the mapping $\varepsilon \mapsto \varphi(\varepsilon\mathscr{A})[T] \in B(\mathscr{H})$ is strongly continuous and s-$\lim_{\varepsilon \to 0} \varphi(\varepsilon\mathscr{A})[T] = \varphi(0)T$. The point of the transformations $\varphi(\mathscr{A}) : B(\mathscr{H}) \to B(\mathscr{H})$ is that they are regularizing with respect to the group $\mathscr{W}$. Notice that the operator S_{ε} given in (7.3.16) is just $\theta(\varepsilon\mathscr{A})[S]$ with $\theta(x) = e^{-x^2}$, i.e. $S_{\varepsilon} = e^{-\varepsilon^2\mathscr{A}^2}[S]$.

(iii) If φ is differentiable we set $\widetilde{\varphi}(\lambda) = \lambda\varphi'(\lambda)$. The formal calculation

$$(7.3.18) \qquad \frac{d}{d\varepsilon}\varphi(\varepsilon\mathscr{A}) = \mathscr{A}\varphi'(\varepsilon\mathscr{A}) = \varepsilon^{-1} \cdot \varepsilon\mathscr{A}\varphi'(\varepsilon\mathscr{A}) = \varepsilon^{-1}\widetilde{\varphi}(\varepsilon\mathscr{A})$$

can be made rigorous if $\varphi \in \mathscr{M}$, φ is C^1 and $\widetilde{\varphi} \in \mathscr{M}$ by starting from

$$\varphi(\varepsilon\mathscr{A}) = \int_{-\infty}^{\infty} \mathscr{W}(\varepsilon\tau)\widehat{\varphi}(\tau)d\tau = \int_{-\infty}^{\infty} \mathscr{W}(\sigma)\widehat{\varphi}(\sigma/\varepsilon)\frac{d\sigma}{\varepsilon}$$

and justifying the differentiation under the integral. Since for $\varphi \in \mathscr{S}(\mathbb{R})$ this is evident, we do not insist.

(iv) For $k \in \mathbb{N}$ let $\varphi_k(\lambda) = \lambda^k\varphi(\lambda)$. Then the formal identity

$$(7.3.19) \qquad \mathscr{A}\varphi(\varepsilon\mathscr{A}) = \varepsilon^{-1} \cdot \varepsilon\mathscr{A} \cdot \varphi(\varepsilon\mathscr{A}) = \varepsilon^{-1}\varphi_1(\varepsilon\mathscr{A})$$

can be rigorously proved without much difficulty if φ and φ_1 belong to $\mathscr{M}$, cf. Proposition 3.3.3. The simple case we need here can be treated by using the sesquilinear form version of $\mathscr{A}\varphi(\varepsilon\mathscr{A})[T] = [\varphi(\varepsilon\mathscr{A})[T], A]$, by writing for $f \in D(A)$:

$$\langle f, (\mathscr{A}\varphi(\varepsilon\mathscr{A})[T])f \rangle = \langle f, \varphi(\varepsilon\mathscr{A})[T]Af \rangle - \langle Af, \varphi(\varepsilon\mathscr{A})[T]f \rangle$$

$$= \int_{-\infty}^{\infty} \left\{ \langle e^{iA\varepsilon\tau}f, Te^{iA\varepsilon\tau}Af \rangle - \langle e^{iA\varepsilon\tau}Af, Te^{iA\varepsilon\tau}f \rangle \right\}\widehat{\varphi}(\tau)d\tau$$

$$= \int_{-\infty}^{\infty} \left\{ \frac{1}{i\varepsilon}\frac{d}{d\tau}\langle e^{iA\varepsilon\tau}f, Te^{iA\varepsilon\tau}f \rangle \right\}\widehat{\varphi}(\tau)d\tau.$$

If $\varphi \in \mathscr{S}(\mathbb{R})$, one may integrate by parts and get (7.3.19).

(v) Fix a real, even function $\theta \in \mathscr{S}(\mathbb{R})$, with $\theta(0) = 1$, and define $S_{\varepsilon} = \theta(\varepsilon\mathscr{A})[S]$ for $0 < \varepsilon < 1$. Clearly $S_{\varepsilon}^* = S_{\varepsilon}$ and $S_{\varepsilon} \to S$ strongly as $\varepsilon \to 0$. An iteration of (7.3.18) shows that $\varepsilon \mapsto S_{\varepsilon} \in B(\mathscr{H})$ is of class C^{∞}, with $S_{\varepsilon}' = \varepsilon^{-1}\widetilde{\theta}(\varepsilon\mathscr{A})[S]$, $S_{\varepsilon}'' = \varepsilon^{-2}\widetilde{\widetilde{\theta}}(\varepsilon\mathscr{A})[S]$, etc. An iteration of (7.3.19) shows that $S_\varepsilon \in C^\infty(A; \mathcal{H})$ and $\mathcal{A}^k[S_\varepsilon] = \varepsilon^{-k}\theta_k(\varepsilon\mathcal{A})[S]$. In particular $B_\varepsilon = i\mathcal{A}[S_\varepsilon] = i\varepsilon^{-1}\theta_1(\varepsilon\mathcal{A})[S]$ is such that $\varepsilon \mapsto B_\varepsilon \in B(\mathcal{H})$ is also of class C^∞ and $B'_\varepsilon = i\varepsilon^{-2}\theta'_2(\varepsilon\mathcal{A})[S]$ (here $\theta'_2(\lambda) = \lambda^2\theta'(\lambda)$).

(vi) Assume that the integral estimates in (a), (b), (c) are proven, i.e. that

(7.3.20)
$$\int_0^1 \{\varepsilon^{-1}||S'_\varepsilon|| + ||\mathcal{A}^2[S_\varepsilon]|| + ||B'_\varepsilon||\}d\varepsilon \equiv$$
$$\equiv \int_0^1 \{||\widetilde\theta(\varepsilon\mathcal{A})[S]|| + ||\theta_2(\varepsilon\mathcal{A})[S]|| + ||\theta'_2(\varepsilon\mathcal{A})[S]||\}\varepsilon^{-2}d\varepsilon < \infty.$$

Then we have
$$||S - S_\varepsilon|| \le \int_0^\varepsilon ||S'_\tau||d\tau \le \varepsilon \int_0^\varepsilon ||S'_\tau||\frac{d\tau}{\tau} \le c\varepsilon,$$

which proves (a) completely. From $\int_0^1 ||B'_\varepsilon||d\varepsilon < \infty$ we deduce that $B_0 = \lim_{\varepsilon\to 0} B_\varepsilon$ exists in norm. But $S_\varepsilon \to S$ in norm, hence for $f, g \in D(A)$:

$$\langle f, B_0 g\rangle = \lim_{\varepsilon\to 0}\langle f, B_\varepsilon g\rangle = \lim_{\varepsilon\to 0}(\langle S_\varepsilon f, iAg\rangle - \langle Af, iS_\varepsilon g\rangle) = i\langle Sf, Ag\rangle - i\langle Af, Sg\rangle.$$

So $S \in C^1(A; \mathcal{H})$ and $[iS, A] = B_0$; hence $||B_\varepsilon - [iS, A]|| \to 0$, which proves (c) completely.

(vii) It remains to prove (7.3.20). Observe that $\widetilde\theta$ and θ_2 are even functions which vanish at zero. Unfortunately $\theta'_2(\lambda) = \lambda^2\theta'(\lambda)$ is odd (and the corresponding term in (7.3.20), which is $||B'_\varepsilon||$, is the most important one, cf. the remarks preceding the lemma). Now we make a precise choice for θ: we assume that we can factorize $\lambda^2\theta'(\lambda) = \varphi(\lambda)\psi(\lambda)$, where $\varphi, \psi \in \mathcal{M}$ and φ is even and vanishes at zero (this is clearly possible, e.g. if $\theta(\lambda) = e^{-\lambda^2}$, then $\lambda^2\theta'(\lambda) = -2\lambda^3 e^{-\lambda^2}$ and we take $\varphi(\lambda) = \lambda^2 e^{-\lambda^2/2}$, $\psi(\lambda) = -2\lambda e^{-\lambda^2/2}$). Then we use :

$$||\theta'_2(\varepsilon\mathcal{A})[S]|| = ||(\varphi\psi)(\varepsilon\mathcal{A})[S]|| = ||\psi(\varepsilon\mathcal{A})[\varphi(\varepsilon\mathcal{A})[S]]|| \le ||\psi||_\mathcal{M}||\varphi(\varepsilon\mathcal{A})[S]||.$$

(viii) It is now clear that the proof of (7.3.20) (and of the lemma) will be finished once we have shown that, for each even $\varphi \in \mathcal{M}$ with $\varphi(0) = 0$ and such that $\widehat\varphi$ is rapidly vanishing at infinity, one has

(7.3.21)
$$\int_0^1 ||\varphi(\varepsilon\mathcal{A})[S]||\varepsilon^{-2}d\varepsilon < \infty.$$

For this, we use (7.3.17) :

$$\int_0^1 ||\varphi(\varepsilon\mathcal{A})[S]||\varepsilon^{-2}d\varepsilon \le \int_0^1 \varepsilon^{-2}d\varepsilon \int_0^\infty ||[\mathcal{W}(\varepsilon\tau) + \mathcal{W}(-\varepsilon\tau) - 2]S|| \cdot |\widehat\varphi(\tau)|d\tau$$
$$= \int_0^\infty \sigma^{-2}d\sigma||[\mathcal{W}(\sigma) + \mathcal{W}(-\sigma) - 2]S|| \cdot \int_\sigma^\infty \tau|\widehat\varphi(\tau)|d\tau. \quad \square$$

We close this section with some comments concerning the method of proof. The main point is the identity (7.3.6) which, when combined with the *quadratic estimate* (7.3.5), shows that the singularity of G'_ε as $\varepsilon \to 0$ is partly cancelled by $[A, G_\varepsilon]$ (independently of z).

Consider for example the case $S \in C^2(A; \mathcal{H})$; then one may take $S_\varepsilon = S$ and $B_\varepsilon = B$, so that (7.3.6) is $G'_\varepsilon + [A, G_\varepsilon] = i\varepsilon G_\varepsilon[A, B]G_\varepsilon$ (this is essentially the situation studied in the papers of Mourre [M1] and Perry-Sigal-Simon [PSS]). When $\varepsilon \to +0$, we have $G_\varepsilon = O(\varepsilon^{-1})$, so one would expect that $G'_\varepsilon = O(\varepsilon^{-2})$. But for $f \in D(A)$, we shall have $\langle f, G'_\varepsilon f \rangle = -\langle Af, G_\varepsilon f \rangle + \langle G^*_\varepsilon f, Af \rangle + i\varepsilon \langle G^*_\varepsilon f, [A, B]G_\varepsilon f \rangle = O(\varepsilon^{-1})$. After integration we get that $\langle f, G_\varepsilon f \rangle = O(\ln \varepsilon)$, and a repetition of the argument will give $\langle f, G_\varepsilon f \rangle = O(1)$. If S is of higher order regularity with respect to A, the estimates may be very much improved. In fact, the formal solution of the equation $G'_\varepsilon + [A, G_\varepsilon] = 0$, or $\frac{d}{d\varepsilon} G_\varepsilon = -[A, G_\varepsilon]$, with $G_0 = (S - z)^{-1}$, is

$$G_\varepsilon(z) = e^{-A\varepsilon}(S - z)^{-1}e^{A\varepsilon} = \left[e^{-A\varepsilon}Se^{A\varepsilon} - z\right]^{-1}.$$

This argument is rigorous if S is an analytic element for the group $\mathcal{W}$, i.e. if the function $\tau \mapsto e^{-iA\tau}Se^{iA\tau}$ has a holomorphic extension to a complex neighbourhood of $\tau = 0$. The example considered in the introduction of this chapter is of this type.

7.4. Self-adjoint Operators with a Spectral Gap

We say that a self-adjoint operator H *has a spectral gap* if $\sigma(H) \neq \mathbb{R}$. The extension of Theorem 7.3.1 to this class of operators is quite simple and will be stated as Theorem 7.4.1 below. The framework of this section is the same as that of the preceding one: a Hilbert space $\mathcal{H}$ (identified with its adjoint $\mathcal{H}^*$) equipped with a continuous unitary group $W(\tau) = e^{iA\tau}$. Then the Banach space $\mathcal{K} = (D(A), \mathcal{H})_{1/2,1}$ is densely and continuously embedded in $\mathcal{H}$, so $\mathcal{K} \subset \mathcal{H} \subset \mathcal{K}^*$ and $B(\mathcal{H}) \subset B(\mathcal{K}, \mathcal{K}^*)$ continuously. We also recall the definition (7.3.1) of the open set $\mu^A(H) \subset \mathbb{R}$ associated to any self-adjoint operator H in $\mathcal{H}$ of class $C^1(A)$. We prove now *a strong form of the limiting absorption principle* which, according to the theory presented in Section 7.1, implies local smoothness of certain operators.

THEOREM 7.4.1. *Let H be a self-adjoint operator in $\mathcal{H}$ and assume that H has a spectral gap and is of class $\mathcal{C}^{1,1}(A)$. Then, for each $\lambda \in \mu^A(H)$, the limits $\lim_{\mu \to \pm 0}(H - \lambda - i\mu)^{-1} \equiv (H - \lambda \mp i0)^{-1}$ exist in the weak* topology of $B(\mathcal{K}, \mathcal{K}^*)$, uniformly in λ on each compact subset of $\mu^A(H)$. In particular, if T is a linear operator from $\mathcal{H}$ to some Hilbert space, and if T is continuous when $\mathcal{H}$ is equipped with the topology induced by $\mathcal{K}^*$, then T is locally H-smooth on the open set $\mu^A(H) \subset \mathbb{R}$.*

PROOF. Let $\lambda_0 \in \mathbb{R} \setminus \sigma(H)$ and $R = (\lambda_0 - H)^{-1}$. Then R is a bounded self-adjoint operator, and the resolvents of H and R are related by the following identity:

$$(7.4.1) \qquad (H - z)^{-1} = (\lambda_0 - z)^{-1}[R - (\lambda_0 - z)^{-1}]^{-1}R, \qquad \Im z \neq 0.$$

Let $J \subset \mu^A(H)$ be a compact set with $\lambda_0 \notin J$ and $\tilde{J} = \{(\lambda_0 - \lambda)^{-1} \mid \lambda \in J\}$ (there is no restriction on the generality if we assume that λ does not belong to a neighbourhood of λ_0, since $(H - z)^{-1}$ is holomorphic in such a neighbourhood). Then $R \in \mathscr{C}^{1,1}(A; \mathscr{H})$ and $\tilde{J}$ is a compact subset of $\mu^A(R)$, by Proposition 7.2.5. Theorem 7.3.1 says that $\zeta \mapsto (R - \zeta)^{-1} \in B(\mathscr{K}, \mathscr{K}^*)$ extends to a weak*-continuous function on $\mathbb{C}_\pm \cup \tilde{J}$. Since $z \mapsto \zeta = (\lambda_0 - z)^{-1}$ is a homeomorphism of $\mathbb{C}_\pm \cup J$ onto $\mathbb{C}_\pm \cup \tilde{J}$, we see that $z \mapsto [R - (\lambda_0 - z)^{-1}]^{-1} \in B(\mathscr{K}, \mathscr{K}^*)$ extends to a weak*-continuous function on $\mathbb{C}_\pm \cup J$. The result of the theorem now follows from (7.4.1) and the fact that $R\mathscr{K} \subset \mathscr{K}$ (a much more general fact is stated in Theorem 5.3.3; to avoid this reference, observe that $(A + i)R(A + i)^{-1} = R + [A, R](A + i)^{-1} \in B(\mathscr{H})$, so $R \in B(D(A))$, and then interpolate to get $R \in B(\mathscr{K})$). $\square$

Some consequences concerning the spectral properties of H are summarized in the next statement.

THEOREM 7.4.2. *Let H be a self-adjoint operator in $\mathscr{H}$, of class $\mathscr{C}^{1,1}(A)$ and having a spectral gap. Assume that $J \subset \mathbb{R}$ is open and A is conjugate to H on J, i.e. there are a number $a > 0$ and a compact operator $K : \mathscr{H} \to \mathscr{H}$ such that $E(J)[iH, A]E(J) \geq aE(J) + K$ (where E is the spectral measure of H). Then H has at most a finite number of eigenvalues in J (multiplicities counted), and it has no singularly continuous spectrum in J.*

PROOF. The first assertion is a consequence of Corollary 7.2.11. For the second one, observe that $J \setminus \sigma_p(H) \subset \mu^A(H)$, cf. Theorem 7.2.13, and then use the consequences of the limiting absorption principle described in §7.1.2. $\square$

Let us make a comment in connection with the optimality of the preceding theorems. In the context of Theorem 7.4.1, both the regularity assumption $H \in \mathscr{C}^{1,1}(A)$ and the Banach space $\mathscr{K} = (D(A), \mathscr{H})_{1/2,1}$ are optimal in the Besov scales $\mathscr{C}^{s,p}(A)$ and $\mathscr{H}_{t,q}$ associated to A (the first fact has already been mentioned at the beginning of Section 7.3, while the second one will be clear in the simplest examples which will be given later on). But it is quite possible that the assumption $H \in \mathscr{C}^{1,1}(A)$ is stronger than needed in the context of Theorem 7.4.2 (see Section 7.1).

As explained in Section 7.1, one can combine local smoothness techniques with the invariance principle in order to get quite precise criteria for the existence and completeness of (local) wave operators. As an example, we mention the following result.

THEOREM 7.4.3. *Let H_1, H_2 be two self-adjoint operators in the Hilbert space $\mathscr{H}$, denote by E_1, E_2 their spectral measures and assume that they have a common spectral gap, i.e. that $\sigma(H_1) \cup \sigma(H_2) \neq \mathbb{R}$. Let A_1, A_2 be a second pair of self-adjoint operators in $\mathscr{H}$, set $\mathscr{K}_j = (D(A_j), \mathscr{H})_{1/2,1}$, identify $\mathscr{K}_j \subset \mathscr{H} = \mathscr{H}^* \subset \mathscr{K}_j^*$, let $\mathscr{K}_j^{*\circ}$ be the closure of $\mathscr{H}$ in $\mathscr{K}_j^*$ and notice the embedding $B(\mathscr{K}_2^{*\circ}, \mathscr{K}_1) \subset B(\mathscr{H})$. Finally, assume that H_j is of class $\mathscr{C}^{1,1}(A_j)$ and that there is $\lambda_0 \in \mathbb{R} \setminus [\sigma(H_1) \cup \sigma(H_2)]$ such that $(H_1 - \lambda_0)^{-1} - (H_2 - \lambda_0)^{-1} \in B(\mathscr{K}_2^{*\circ}, \mathscr{K}_1)$. Then, if $J = \mu^{A_1}(H_1) \cap \mu^{A_2}(H_2)$, the local wave operators s-$\lim_{t \to \pm\infty} e^{iH_1 t} e^{-iH_2 t} E_2(J)$ exist and their ranges are equal to $E_1(J)\mathscr{H}$.*

PROOF. We assume, without loss of generality, that $\lambda_0 = 0$. Let $\widetilde{H}_j = -H_j^{-1}$; then $\widetilde{H}_j$ is a bounded self-adjoint operator, and $\lambda = 0$ is not an eigenvalue of $\widetilde{H}_j$ $(j = 1, 2)$. If $\widetilde{J} = \mu^{A_1}(\widetilde{H}_1) \cap \mu^{A_2}(\widetilde{H}_2)$, then the strong limiting absorption principle holds for $\widetilde{H}_j$ in $\mathscr{K}_j$, locally on $\widetilde{J}$. Then, according to Theorem 7.1.5, we can apply Theorem 7.1.4 to the pair $\widetilde{H}_1$, $\widetilde{H}_2$ on $\widetilde{J}$ (note that the roles of H_j and $\widetilde{H}_j$ in Theorem 7.1.4 are interchanged). We choose $\varphi : \mathbb{R} \to \mathbb{R}$ in Theorem 7.1.4 such that $\varphi(0) = 0$ and $\varphi(x) = -x^{-1}$ if $x \neq 0$. Then φ is bijective and of class C^1 outside zero with $\varphi'(x) > 0$. Since $\widetilde{J} = \varphi^{-1}(J)$ (by Proposition 7.2.5) and $H_j = \varphi(\widetilde{H}_j)$, we get the conclusion of the theorem. $\square$

The space $\mathscr{K}$ in which the limiting absorption principle has been proved is too small for several applications (e.g. the locally smooth operators furnished by Theorem 7.4.1 are bounded operators on $\mathscr{H}$). Furthermore $\mathscr{K} = (D(A), \mathscr{H})_{1/2,1}$ is not a standard space of distributions in general, since A could be a complicated operator the domain of which is not easily described in the framework of weighted Besov spaces. In order to circumvent these difficulties, one may follow [PSS] and use the formula

$$(7.4.2) \quad R(z) = R(\lambda_0) + (z - \lambda_0)R(\lambda_0)^2 + (z - \lambda_0)^2 R(\lambda_0)R(z)R(\lambda_0),$$

where $R(z) = (H - z)^{-1}$. (7.4.2) is obtained after an iteration of the first resolvent formula $R(z) = R(\lambda_0) + (z - \lambda_0)R(\lambda_0)R(z)$. As an example of this technique, we have the following result. We recall that $\mathbb{C}_{\pm} = \{z \in \mathbb{C} \mid \pm\Im z > 0\}$.

PROPOSITION 7.4.4. *Let H be a self-adjoint operator in $\mathscr{H}$ of class $\mathscr{C}^{1,1}(A)$ and having a spectral gap. Let $(\mathscr{E}_1, \mathscr{E})$ be a Friedrichs couple such that $\mathscr{H} \subset \mathscr{E}$ continuously and densely, identify $\mathscr{E}^* \subset \mathscr{H}^* = \mathscr{H} \subset \mathscr{E}$, and assume that, for some $\lambda_0 \in \mathbb{R} \setminus \sigma(H)$, the operator $(H - \lambda_0)^{-1}$ extends to a continuous operator from $\mathscr{E}$ to $\mathscr{E}^*$ which maps $\mathscr{E}_1$ into $D(A)$. Then*
   (a) *For each $z \in \mathbb{C} \setminus \sigma(H)$, $(H - z)^{-1}$ extends to a continuous operator from $\mathscr{E}$ into $\mathscr{E}^*$ and the map $z \mapsto (H - z)^{-1} \in B(\mathscr{E}, \mathscr{E}^*)$ is holomorphic.*
   (b) *Set $\mathscr{E}_{1/2,1} = (\mathscr{E}_1, \mathscr{E})_{1/2,1}$ and notice the embeddings*

$$\mathscr{E}_{1/2,1} \subset \mathscr{E}, \quad \mathscr{E}^* \subset \mathscr{E}_{1/2,1}^* \quad and \quad B(\mathscr{E}, \mathscr{E}^*) \subset B(\mathscr{E}_{1/2,1}, \mathscr{E}_{1/2,1}^*).$$

Then the holomorphic function $\mathbb{C}_{\pm} \ni z \mapsto (H - z)^{-1} \in B(\mathscr{E}_{1/2,1}, \mathscr{E}_{1/2,1}^*)$ extends to a weak*-continuous function on $\mathbb{C}_{\pm} \cup \mu^A(H)$.

PROOF. The assertion (a) is an immediate consequence of (7.4.2). To prove (b), observe that $R(\lambda_0) \in B(\mathscr{E}, \mathscr{H}) \cap B(\mathscr{E}_1, D(A))$ by hypothesis, which implies $R(\lambda_0) \in B(\mathscr{E}_{1/2,1}, \mathscr{K})$ by interpolation. Since $R(\lambda_0)$ is symmetric, we shall also have $R(\lambda_0) \in B(\mathscr{K}^*, \mathscr{E}_{1/2,1}^*)$. Theorem 7.4.1 and formula (7.4.2) now imply the result of (b). $\square$

For the next corollary, notice that $\mathscr{E}_{1/2,1} \subset \mathscr{E} \subset D(H)^*$ continuously and densely (because $\|f\|_{D(H)^*} \leq C\|(H - \lambda_0)^{-1}f\|_{\mathscr{H}} \leq C'\|f\|_{\mathscr{E}}$ for $f \in \mathscr{H}$) and then apply Proposition 7.1.3.

COROLLARY 7.4.5. *Let T be a Hilbert space-valued linear operator on $\mathscr{E}^*$ which is continuous when $\mathscr{E}^*$ is equipped with the topology induced by $\mathscr{E}_{1/2,1}^* = (\mathscr{E}^*, \mathscr{E}_1^*)_{1/2,\infty}$. Then T is locally H-smooth on the open real set $\mu^A(H)$.*

The statement of Proposition 7.4.4 is unfortunately rather heavy; it is, however, convenient for applications. This will become clear in the later sections of this chapter, but we shall treat right now one of the most important examples in order to clarify the meaning of the various spaces introduced above. We choose a formulation adapted to Schrödinger hamiltonians (even very singular and with many-body potentials) and leave aside some obvious generalizations to higher order elliptic operators. Recall that, if X is an euclidean space, then $\mathscr{H}^s(X)$ is the usual Sobolev space of order $s \in \mathbb{R}$, while $\mathscr{H}_{t,q}^s(X)$ is a weighted Sobolev space ($t \in \mathbb{R}$, $1 \leq q \leq \infty$) and $\mathscr{H}_t^s(X) \equiv \mathscr{H}_{t,2}^s(X)$. If $\varphi : X \to \mathbb{R}$, then $\varphi(Q)$ is the operator of multiplication by φ.

PROPOSITION 7.4.6. *Let X be an euclidean space, $\mathscr{H} = L^2(X)$, $A = D = \frac{1}{4}(P \cdot Q + Q \cdot P)$, so that $2A$ is the generator of the dilation group, and H a self-adjoint operator in $\mathscr{H}$ of class $\mathscr{C}^{1,1}(D)$. Assume that there is $\lambda_0 \in \mathbb{R} \setminus \sigma(H)$ such that $(H - \lambda_0)^{-1}$ and $[\varphi(Q), (H - \lambda_0)^{-1}]$ extend to continuous operators $\mathscr{H}^{-1}(X) \to \mathscr{H}^1(X)$, for each linear function $\varphi : X \to \mathbb{R}$. Then*

$$(H - z)^{-1} \in B(\mathscr{H}^{-1}, \mathscr{H}^1) \subset B(\mathscr{H}_{1/2,1}^{-1}, \mathscr{H}_{-1/2,\infty}^1)$$

for each $z \notin \sigma(H)$ and the limits $\lim_{\mu \to \pm 0}(H - \lambda - i\mu)^{-1}$ exist in the weak* topology of $B(\mathscr{H}_{1/2,1}^{-1}, \mathscr{H}_{-1/2,\infty}^1)$, uniformly in λ on each compact subset of $\mu^D(H)$. If $(H - \lambda_0)^{-1}$ and $[\varphi(Q), (H - \lambda_0)^{-1}]$ belong to $B(\mathscr{H}, \mathscr{H}^2)$ for each φ as before, then the preceding limits will exist in the weak* topology of each space $B(\mathscr{H}_{1/2,1}^s(X), \mathscr{H}_{-1/2,\infty}^{s+2})$, $-2 \leq s \leq 0$, uniformly in λ on compact sets in $\mu^D(H)$.

PROOF. We use Proposition 7.4.4 with $\mathscr{E} = \mathscr{H}^{-1}(X)$, $\mathscr{E}_1 = \mathscr{H}_1^{-1}(X)$. Then $\mathscr{E}^* = \mathscr{H}^1(X)$, $\mathscr{E}_{1/2,1} = \mathscr{H}_{1/2,1}^{-1}(X)$ and $\mathscr{E}_{1/2,1}^* = \mathscr{H}_{-1/2,\infty}^1(X)$. $(H - \lambda_0)^{-1}$ maps $\mathscr{E}_1$ into $D(A)$ if $P \cdot Q(H - \lambda_0)^{-1}\mathscr{H}_1^{-1} \subset \mathscr{H}$, and this follows from the assumption $[Q_j, (H - \lambda_0)^{-1}]\mathscr{H}^{-1} \subset \mathscr{H}^1$ (here Q_j is the operator of multiplication by the coordinate x_j relative to a basis in X). To prove the last part of the proposition observe that the operators $(H - \lambda_0)^{-1}$ and $i[\varphi(Q), (H - \lambda_0)^{-1}]$ will belong to $B(\mathscr{H}^s, \mathscr{H}^{s+2})$ if $-2 \leq s \leq 0$ (because the operators are symmetric, so one can use a duality-interpolation argument). This clearly implies that $(H - \lambda_0)^{-1} \in B(\mathscr{H}_1^s, \mathscr{H}_1^{s+2})$, then, by interpolation, that $(H - \lambda_0)^{-1} \in B(\mathscr{H}_{1/2,1}^s, \mathscr{H}_{1/2,1}^{s+2})$, and finally, by duality, that $(H - \lambda_0)^{-1} \subset B(\mathscr{H}_{-1/2,\infty}^s, \mathscr{H}_{-1/2,\infty}^{s+2})$. By using (7.4.2) and the fact that the strong limiting absorption principle holds in $B(\mathscr{H}_{1/2,1}, \mathscr{H}_{-1/2,\infty})$, we get the result. $\square$

Notice that the assumption $(H - \lambda_0)^{-1} \in B(\mathscr{H}^{-1}, \mathscr{H}^1)$ is fulfilled if the form domain of H is included densely in $\mathscr{H}^1$, because then $D(|H|^{1/2}) \subset \mathscr{H}^1 \subset \mathscr{H} \subset \mathscr{H}^{-1} \subset D(|H|^{1/2})^*$ and $(H - \lambda_0)^{-1} \in B(D(|H|^{1/2})^*, D(|H|^{1/2}))$. Then the hypothesis on $[\varphi(Q), (H - \lambda_0)^{-1}]$ may be checked by justifying its equality with $(H - \lambda_0)^{-1}[H, \varphi(Q)](H - \lambda_0)^{-1}$ and adding some assumptions on $[H, \varphi(Q)]$. In the simplest Schrödinger type situation we have $H = \Delta + V(Q)$, where $V : X \to$

$\mathbb{R}$ is Borel. If $V = V_+ - V_-$, where $V_\pm \geq 0$, $V_+ \in L^1_{\text{loc}}(X)$ and $V_-(Q)$ is Δ-form bounded with relative bound < 1, then the form domain of H is $\mathcal{H}^1(X) \cap D(V_+^{1/2})$, so it is densely included in $\mathcal{H}^1(X)$ (because it contains $C_0^\infty(X)$). In this case $[\varphi(Q), H] = [\varphi(Q), \Delta] \in B(\mathcal{H}^s, \mathcal{H}^{s-1})$ $\forall s \in \mathbb{R}$, so both conditions we put on $(H - \lambda_0)^{-1}$ in the first part of Proposition 7.4.6 are fulfilled. Since H is bounded from below, the only hypothesis that has to be imposed is $H \in \mathscr{C}^{1,1}(D)$.

A straightforward consequence of Proposition 7.4.6 and of the discussion made in §7.1.2 is that *any linear operator* $T : \mathcal{H}^1(X) \to \mathscr{F}$ *($\mathscr{F}$ any Hilbert space) which is continuous for the topology induced by* $\mathcal{H}^1_{-1/2,\infty}(X)$ *on* $\mathcal{H}^1(X)$, *is locally H-smooth on the set* $\mu^D(H)$. Let us denote by $\overset{\circ}{\mathcal{H}}{}^1_{-1/2,\infty}(X)$ the closure of $\mathcal{H}^1(X)$ in $\mathcal{H}^1_{-1/2,\infty}(X)$ and observe that $B(\overset{\circ}{\mathcal{H}}{}^1_{-1/2,\infty}, \mathcal{H}^{-1}_{1/2,1}) \subset B(\mathcal{H}^1, \mathcal{H}^{-1})$. Referring again to the general theory presented in Section 7.1 we obtain the following consequence of Proposition 7.4.6:

COROLLARY 7.4.7. *Let X be an euclidean space, $\mathcal{H} = L^2(X)$, $D = \frac{1}{4}(P \cdot Q + Q \cdot P)$ and H_1, H_2 two self-adjoint operators of class $\mathscr{C}^{1,1}(D)$ in $\mathcal{H}$.*

Assume that each H_j has a spectral gap, that its form domain is $\mathcal{H}^1(X)$ (i.e. $D(|H_j|^{1/2}) = \mathcal{H}^1(X)$, which implies $H_j \in B(\mathcal{H}^1(X), \mathcal{H}^{-1}(X))$ and that $[\varphi(Q), H_j] \in B(\mathcal{H}^1(X), \mathcal{H}^{-1}(X))$ for each linear function φ on X.

Finally, assume that $H_1 - H_2 \in B(\overset{\circ}{\mathcal{H}}{}^1_{-1/2,\infty}(X), \mathcal{H}^{-1}_{1/2,1}(X))$. Set $J = \mu^D(H_1) \cap \mu^D(H_2)$, so that $J \subset \mathbb{R}$ is an open set, and let E_j be the spectral measure of H_j. Then the strong limits

$$(7.4.3) \qquad \text{s-}\lim_{t \to \pm\infty} e^{iH_2 t} e^{-iH_1 t} E_1(J)$$

exist and have ranges equal to $E_2(J)\mathcal{H}$.

From the examples treated in Sections 7.6 and 9.4 it will be seen that the criterion of existence and completeness of wave operators contained in the preceding statement is very precise, both locally and at infinity. Notice that the operators H_j could be quite complicated (of the N-body type, for example).

7.5. Hamiltonians Associated to Symmetric Operators in Friedrichs Couples

The framework of this section coincides with that of Section 6.3, but we make the supplementary assumption $n = 1$. Hence *a triplet* $(\mathcal{G}, \mathcal{H}; A)$ *is given such that $\mathcal{G}$ and $\mathcal{H}$ are Hilbert spaces, $\mathcal{G} \subset \mathcal{H}$ continuously and densely, A is a self-adjoint operator in $\mathcal{H}$, and $W(\tau) = e^{iA\tau}$ leaves $\mathcal{G}$ invariant.* We shall freely use the notations and results of Section 6.3; we recall the identification $\mathcal{G} \subset \mathcal{H} = \mathcal{H}^* \subset \mathcal{G}^*$ and the fact that W naturally induces C_0-groups in $\mathcal{G}$ and $\mathcal{G}^*$, which are denoted by the same symbol W. Moreover, we keep the notation A for the generator of W in each of these spaces and we use the suggestive notations $D(A; \mathcal{G})$, $D(A) \equiv D(A; \mathcal{H})$ and $D(A; \mathcal{G}^*)$ to denote its domain in $\mathcal{G}$, $\mathcal{H}$ and $\mathcal{G}^*$ respectively. Each of these spaces is a H-space for its natural graph topology and we have $D(A; \mathcal{G}) \subset D(A; \mathcal{H}) \subset D(A; \mathcal{G}^*) \subset \mathcal{G}^*$ continuously and densely. For example, $D(A; \mathcal{G}^*)$ is the set of $f \in \mathcal{G}^*$ such that $Af := \lim_{\tau \to 0} (i\tau)^{-1}[W(\tau) - I]f$ exists in $\mathscr{G}^*$ and is equipped with the topology given by the norm $||f||_{\mathscr{G}^*} + ||Af||_{\mathscr{G}^*}$. We have similar definitions for $D(A; \mathscr{G})$ and $D(A; \mathscr{H})$, but also $D(A; \mathscr{G}) = \{f \in \mathscr{G} \mid f \in D(A; \mathscr{G}^*)$ and $Af \in \mathscr{G}\}$, and analogously for $D(A; \mathscr{H})$. Of course, $D(A; \mathscr{H}) = D(A) \equiv$ domain (in $\mathscr{H}$) of the initial self-adjoint operator A. One may also define $D(A; \mathscr{G})$ and $D(A; \mathscr{G}^*)$ directly in terms of the initial self-adjoint operator A in $\mathscr{H}$, namely: (i) $D(A; \mathscr{G}) = \{f \in \mathscr{G} \cap D(A) \mid Af \in \mathscr{G}\}$ with norm $||f||_{\mathscr{G}} + ||Af||_{\mathscr{G}}$; (ii) the operator $A : D(A) \subset \mathscr{G}^* \to \mathscr{G}^*$ is closable when considered as an operator in $\mathscr{G}^*$ and $D(A; \mathscr{G}^*)$ is just the domain of its closure. All these facts have been discussed in Section 6.3, see for example the text following Corollary 6.3.2 where we used the notations $\mathscr{G}_1 = D(A; \mathscr{G})$, $\mathscr{H}_1 = D(A) = D(A; \mathscr{H})$ and $\mathscr{G}_1^* = D(A; \mathscr{G}^*)$, which are consistent with the theory developed in Chapter 3.

The first result that we point out in this framework is a variant of the *virial theorem*. The proof is a simple adaptation of that of Proposition 7.2.10.

PROPOSITION 7.5.1. *Let* $H : \mathscr{G} \to \mathscr{G}^*$ *be a symmetric operator of class* $C^1(A; \mathscr{G}, \mathscr{G}^*)$. *Then* $\langle f, [A, H]f \rangle = 0$ *if* $f \in \mathscr{G}$ *and* $Hf = \lambda f$ *for some* $\lambda \in \mathbb{R}$.

7.5.1. If $H : \mathscr{G} \to \mathscr{G}^*$ is a symmetric operator, then one may naturally associate to it a symmetric operator $\widehat{H}$ in $\mathscr{H}$ by the rule $D(\widehat{H}) = \{f \in \mathscr{G} \mid Hf \in \mathscr{H}\}$ and $\widehat{H} = H|_{D(\widehat{H})}$. If there is no risk of ambiguity, we shall keep the notation H for $\widehat{H}$ and then write $D(H; \mathscr{H})$ for $D(\widehat{H})$ ("domain of H in $\mathscr{H}$"). $\widehat{H}$ will be called *the operator in* $\mathscr{H}$ *associated to* $H : \mathscr{G} \to \mathscr{G}^*$. In many cases, $\widehat{H}$ is a (densely defined) self-adjoint operator in $\mathscr{H}$, and the purpose of this section is to show that the L.A.P. holds for $\widehat{H}$ under simple conditions on H and in spaces quite large and easily accessible in applications. We begin with the case where $D(\widehat{H}) = \mathscr{G}$. The next theorem is a straightforward consequence of the results obtained in Section 7.4. Recall the following notation, introduced and used already in Sections 2.8 and 6.3 : for $-1 < \theta < 1$, $\mathscr{G}^\theta = [\mathscr{G}, \mathscr{G}^*]_{(1-\theta)/2} = (\mathscr{G}, \mathscr{G}^*)_{(1-\theta)/2,2}$ and $\mathscr{G}^1 = \mathscr{G}$, $\mathscr{G}^{-1} = \mathscr{G}^*$; note that $\mathscr{G}^0 = \mathscr{H}$.

THEOREM 7.5.2. *Let* H *be a self-adjoint operator in* $\mathscr{H}$, *with domain* $\mathscr{G}$, *of class* $\mathscr{C}^{1,1}(A; \mathscr{G}, \mathscr{G}^*)$ *and having a spectral gap. Then* H *is of class* $\mathscr{C}^{1,1}(A)$, *hence the open real set* $\mu^A(H)$ *is well-defined, and the strong limiting absorption principle holds locally on* $\mu^A(H)$ *in the Banach space* $(\mathscr{H} \cap D(A; \mathscr{G}^*), \mathscr{G}^*)_{1/2,1}$. *If* $[A, H]\mathscr{G} \subset \mathscr{G}^{-1/2}$ *(i.e. if* $H \in C^1(A; \mathscr{G}, \mathscr{G}^{-1/2})$), *then the strong limiting absorption principle holds in the larger space* $(\mathscr{G}^{-1/2} \cap D(A; \mathscr{G}^*), \mathscr{G}^{-1/2})_{1/2,1}$.

Before starting the proof, we make some comments. We recall that, if $\mathscr{E}, \mathscr{F}$ is a pair of compatible B-spaces (i.e. they are continuously embedded in a separated topological vector space), then $\mathscr{E} \cap \mathscr{F}$ has a canonical B-space structure (if $||\cdot||_{\mathscr{E}}$, $||\cdot||_{\mathscr{F}}$ are admissible norms on $\mathscr{E}$, $\mathscr{F}$, then $||f||_{\mathscr{E}} + ||f||_{\mathscr{F}}$ is an admissible norm on $\mathscr{E} \cap \mathscr{F}$). Then by using $\mathscr{G}^{\theta,p} := (\mathscr{G}, \mathscr{G}^*)_{(1-\theta)/2,p}$ (here $-1 < \theta < 1$, $1 \le p \le \infty$) we get (cf. the reiteration property):

$$(\mathscr{H} \cap D(A; \mathscr{G}^*), \mathscr{G}^*)_{1/2,1} \subset (\mathscr{H}, \mathscr{G}^*)_{1/2,1} = \mathscr{G}^{-1/2,1} \subset \mathscr{G}^{-1/2}.$$

Notice that, since $\mathscr{G}$ is the domain of H, $\mathscr{G}^{1/2}$ *is the form domain of* H (i.e. $\mathscr{G}^{1/2} = D(|H|^{1/2})$) and $(\mathscr{G}^{1/2})^* = \mathscr{G}^{-1/2}$. Hence it makes sense to speak about the L.A.P. for H in the indicated space (see (7.1.19)). A result of P. Grisvard (see Proposition 2.7.4) allows us to give a more explicit description of the main interpolation space involved in the theorem:

$$(7.5.1) \quad (\mathscr{H} \cap D(A; \mathscr{G}^*), \mathscr{G}^*)_{1/2,1} = (\mathscr{H}, \mathscr{G}^*)_{1/2,1} \cap (D(A; \mathscr{G}^*), \mathscr{G}^*)_{1/2,1}$$
$$= \mathscr{G}^{-1/2,1} \cap (D(A; \mathscr{G}^*), \mathscr{G}^*)_{1/2,1}.$$

As we explained in Section 7.1, and as we have seen in Section 7.4, Theorem 7.5.2 implies precise criteria of local H-smoothness and of existence and completeness of local wave operators; we shall, however, not state them explicitly here. Now we make a comment concerning the assumption that H have a spectral gap: it is certainly not needed, but we have not been able to avoid it because of the strategy of our proofs (we reduce everything to the case of bounded operators treated in Section 7.3). Other proofs avoid the spectral gap hypothesis but give results weaker than those of Theorem 7.5.2 (such a proof will be presented below, cf. Theorem 7.5.4). The strongest results of this type known to us are contained in [BGM2], from which we quote the following particular case: If H is self-adjoint on $\mathscr{G}$ in $\mathscr{H}$ and of class $C^1(A; \mathscr{G}, \mathscr{G}^*)$, and if $[A, H]$ is of class $\mathscr{C}^{0,1}(A; \mathscr{G}, \mathscr{G}^*)$ and $[A, H]\mathscr{G} \subset \mathscr{G}^{-1/2}$, then the strong L.A.P. holds on $\mu^A(H)$ in the space (7.5.1). Moreover, the condition $[A, H]\mathscr{G} \subset \mathscr{G}^{-1/2}$ can be replaced by $[A, H]\mathscr{G} \subset \mathscr{G}^{-1+\varepsilon}$ for some $\varepsilon > 0$ (in fact one can take $\varepsilon = +0$ in a precise sense) if the triplet $(\mathscr{G}, \mathscr{H}; A)$ has a certain regularity property (namely: there is a self-adjoint operator S in $\mathscr{H}$ with domain $\mathscr{G}$ such that $\tau \mapsto e^{iA\tau}Se^{-iA\tau} \in B(\mathscr{G}, \mathscr{H})$ is norm-continuous; for this, it is enough to have $[A, S] \in B(\mathscr{G}, \mathscr{H})$). We mention another point which has been discussed in detail in [ABG1], [BGM2] and [BG10], but which is not touched upon here: find natural and optimal spaces $\mathscr{K}$ in which the strong L.A.P. holds *in norm*, i.e. such that the limits $\lim_{\mu \to +0}(H - \lambda \mp i\mu)^{-1}$ exist in norm in $B(\mathscr{K}, \mathscr{K}^*)$, locally uniformly in $\lambda \in \mu^A(H)$, and, moreover, find the precise order of Hölder continuity of the boundary value functions $\lambda \mapsto (H - \lambda \mp i0)^{-1} \in B(\mathscr{K}, \mathscr{K}^*)$.

Finally, we discuss the hypothesis $[A, H]\mathscr{G} \subset \mathscr{G}^{-1/2}$ which appears in the statement of Theorem 7.5.2. In the initial version of the conjugate operator method, due to Mourre, the main assumptions were $[A, H] \in B(\mathscr{G}, \mathscr{H})$ and $[A, [A, H]] \in B(\mathscr{G}, \mathscr{G}^*)$. In our terminology this means $H \in C^1(A; \mathscr{G}, \mathscr{H}) \cap C^2(A; \mathscr{G}, \mathscr{G}^*)$. The first condition was especially restrictive in applications and Perry, Sigal and Simon in [PSS] succeeded in replacing it by $[A, H] \in B(\mathscr{G}, \mathscr{G}^{-1/2})$; in our notations, their main assumption was $H \in C^1(A; \mathscr{G}, \mathscr{G}^{-1/2}) \cap C^2(A; \mathscr{G}, \mathscr{G}^*)$. Now notice that, under the assumption $H \in C^2(A; \mathscr{G}, \mathscr{G}^*)$ (i.e. $[A, [A, H]] \in B(\mathscr{G}, \mathscr{G}^*)$), they could not do essentially better, because our Proposition 6.3.4 (c) shows that this implies $H \in \mathscr{C}^{1,\infty}(A; \mathscr{G}, \mathscr{G}^{-1/2})$. Our main result consists in the replacement of the condition $H \in C^2(A; \mathscr{G}, \mathscr{G}^*)$ by the much weaker one $H \in \mathscr{C}^{1,1}(A; \mathscr{G}, \mathscr{G}^*)$. Then a supplementary hypothesis of the type $[A, H]\mathscr{G} \subset \mathscr{G}^{-1/2}$ is rather irrelevant: its only purpose is to replace the space (7.5.1), in which the L.A.P. holds, by the slightly larger space $(\mathscr{G}^{-1/2} \cap D(A; \mathscr{G}^*), \mathscr{G}^{-1/2})_{1/2,1}$. In all the concrete cases that we shall meet, the difference between these two spaces is not significant; however, the second one is easier to describe (in the N-body case it contains the space $\mathscr{H}^{-1}_{1/2,1}(X)$ which already occurred in Proposition 7.4.6).

PROOF OF THEOREM 7.5.2. Proposition 6.3.4 shows that H is of class $\mathscr{C}^{1,1}(A)$, in particular we may apply Theorem 7.4.1 and get the strong L.A.P. in the space $\mathscr{K} = (D(A;\mathscr{H}),\mathscr{H})_{1/2,1}$. We now show that the strong L.A.P. holds in the space $\widetilde{\mathscr{K}}$ given by (7.5.1). Observe that $\mathscr{K} \subset \widetilde{\mathscr{K}} \subset \mathscr{G}^{-1/2}$ continuously and densely, hence $\mathscr{G}^{1/2} \subset \widetilde{\mathscr{K}}^* \subset \mathscr{K}^*$ and $B(\mathscr{G}^{-1/2}, \mathscr{G}^{1/2}) \subset B(\widetilde{\mathscr{K}},\widetilde{\mathscr{K}}^*) \subset B(\mathscr{K},\mathscr{K}^*)$ continuously. So, by taking into account the identity (7.4.2), it is enough to prove that $R(\lambda_0)\widetilde{\mathscr{K}} \subset \mathscr{K}$ (because then, by using the closed graph theorem and by taking the adjoint, we shall also get $R(\lambda_0) \in B(\widetilde{\mathscr{K}},\mathscr{K}) \cap B(\mathscr{K}^*,\widetilde{\mathscr{K}}^*)$). Since $\widetilde{\mathscr{K}}$ and $\mathscr{K}$ are interpolation spaces and $R(\lambda_0)\mathscr{G}^* = \mathscr{H}$, it is enough to show that $R(\lambda_0)[\mathscr{H} \cap D(A;\mathscr{G}^*)] = D(A;\mathscr{H})$. Let $f \in \mathscr{H}$ such that $f \in D(A;\mathscr{G}^*)$ and set $f_\tau = W(\tau)f$, $R_\tau = W(\tau)R(\lambda_0)W(-\tau)$. Then $R(\lambda_0)f \in \mathscr{H}$ and

$$(7.5.2) \qquad \frac{W(\tau) - I}{\tau} R(\lambda_0)f = R_\tau \cdot \tau^{-1}(f_\tau - f) + \tau^{-1}(R_\tau - R_0)f.$$

Since $\tau \mapsto R_\tau \in B(\mathscr{G}^*,\mathscr{H})$ is strongly continuous, $\tau \mapsto f_\tau \in \mathscr{G}^*$ is (strongly) differentiable and $\tau \mapsto R_\tau \in B(\mathscr{H})$ is strongly differentiable too, we see that the strong limit as $\tau \to 0$ of the above expression exists in $\mathscr{H}$, hence $R(\lambda_0)f \in D(A;\mathscr{H})$. This finishes the proof of the first part of the theorem. Now let us assume that $[A,H]\mathscr{G} \subset \mathscr{G}^{-1/2}$. Then we want to apply directly Proposition 7.4.4 with $\mathscr{E} = \mathscr{G}^{-1/2}$ and $\mathscr{E}_1 = \mathscr{G}^{-1/2} \cap D(A;\mathscr{G}^*)$, and for this we just have to show that $R(\lambda_0)\mathscr{E}_1 \subset D(A;\mathscr{H})$. We argue exactly as above, using the identity (7.5.2). The first term in the right-hand side of (7.5.2) is treated in the same way. For the second one, notice that $R(\lambda_0) \in C^1(A;\mathscr{G}^{-1/2},\mathscr{H})$ by Proposition 6.3.5. $\square$

7.5.2. If $H : \mathscr{G} \to \mathscr{G}^*$ is a symmetric operator, the hypothesis $D(\widehat{H}) = \mathscr{G}$ of Theorem 7.5.2 is equivalent to the existence of a complex number z_0 such that $H - z_0$ is an isomorphism of $\mathscr{G}$ onto $\mathscr{H}$. In the rest of this section, we consider another extreme situation, namely that where $H - z_0 : \mathscr{G} \to \mathscr{G}^*$ is an isomorphism for some $z_0 \in \mathbb{C}$. If H is semibounded, then this is equivalent with saying that $\widehat{H}$ is self-adjoint and $\mathscr{G} = D(|\widehat{H}|^{1/2})$, i.e. in this case the form domain (and not the domain) of the hamiltonian $\widehat{H}$ is invariant under the group $\{e^{iA\tau}\}_{\tau \in \mathbb{R}}$. We mention that the assumptions imposed on H below *are not convenient in the N-body case* (i.e. the results are essentially weaker than those obtained from Theorem 7.5.2). However, this version is quite efficient in two-body type problems. Moreover, Theorem 7.5.4 is in fact a generalization of Theorem 7.3.1, hence it implies indirectly all the results we got until now.

The next lemma describes precisely the class of operators that we have in mind.

LEMMA 7.5.3. *Let $H : \mathscr{G} \to \mathscr{G}^*$ be a symmetric operator such that $H - z_0$ is a bijection of $\mathscr{G}$ onto $\mathscr{G}^*$ for some $z_0 \in \mathbb{C}$. Then $H - z : \mathscr{G} \to \mathscr{G}^*$ is an isomorphism for each $z \in \mathbb{C} \setminus \mathbb{R}$, the function $z \mapsto (H - z)^{-1} \in B(\mathscr{G}^*, \mathscr{G})$ is holomorphic on $\mathbb{C} \setminus \mathbb{R}$, and $\|(H - z)^{-1}\|_{B(\mathscr{G}^*,\mathscr{G})} \leq c\langle z \rangle^2 |\Im z|^{-1}$ for some constant c. The operator $\widehat{H}$ associated to H in $\mathscr{H}$, i.e. the restriction of H to $D(\widehat{H}) = \{f \in \mathscr{G} \mid Hf \in \mathscr{H}\}$,* *is a densely defined self-adjoint operator in $\mathscr{H}$, one has $(\widehat{H}-z)^{-1} = (H-z)^{-1}|_{\mathscr{H}}$ for non-real z and $\varphi(\widehat{H}) \in B(\mathscr{G}^*, \mathscr{G})$ if $\varphi \in C_0^2(\mathbb{R})$. Assume, furthermore, that $H \in C^1(A; \mathscr{G}, \mathscr{G}^*)$. Then $\widehat{H}$ is of class $C^1(A)$ and, if $\varphi \in C_0^3(\mathbb{R})$, then $\varphi(\widehat{H})$ has the stronger regularity property $\varphi(\widehat{H}) \in C^1(A; \mathscr{G}^*, \mathscr{G})$.*

PROOF. A symmetric operator from $\mathscr{G}$ to $\mathscr{G}^*$ is continuous, and a bijective continuous operator $\mathscr{G} \to \mathscr{G}^*$ is an isomorphism. Hence $H-z_0$ and $H-\overline{z}_0 = (H-z_0)^*$ are isomorphisms of $\mathscr{G}$ onto $\mathscr{G}^*$. It is easily seen that $D(\widehat{H}) = (H-z_0)^{-1}\mathscr{H}$; hence, $\mathscr{H}$ being dense in $\mathscr{G}^*$, $D(\widehat{H})$ will be dense in $\mathscr{G}$ and consequently in $\mathscr{H}$. So $\widehat{H}$ is a densely defined symmetric operator in $\mathscr{H}$ and $\widehat{H} - z_0$, $\widehat{H} - \overline{z}_0$ have range $\mathscr{H}$. A well-known criterion for self-adjointness implies then the self-adjointness of $\widehat{H}$. Now let us write (7.4.2) with $\lambda_0 = z_0$ and with H replaced by $\widehat{H}$. We get that $(\widehat{H} - z)^{-1}$ extends to an operator $T \in B(\mathscr{G}^*, \mathscr{G})$ with $\|T\|_{B(\mathscr{G}^*, \mathscr{G})} \leq c\langle z\rangle^2|\Im z|^{-1}$ for a constant c. For $f \in \mathscr{H}$ we have $f = (\widehat{H} - z)Tf = (H - z)Tf$; since $\mathscr{H}$ is dense in $\mathscr{G}^*$ and since $T : \mathscr{G}^* \to \mathscr{G}$ and $H - z : \mathscr{G} \to \mathscr{G}^*$ are continuous, we obtain $f = (H - z)Tf$ for all $f \in \mathscr{G}^*$. Similarly, for $f \in D(H)$ we have $f = T(\widehat{H} - z)f = T(H - z)f$ and so, since $D(\widehat{H})$ is dense in $\mathscr{G}$, we get $f = T(H - z)f$ for all $f \in \mathscr{G}$. In conclusion, $H - z : \mathscr{G} \to \mathscr{G}^*$ is an isomorphism with inverse equal to T if $\Im z \neq 0$ (in fact, if $z \notin \sigma(H)$). The holomorphy of $z \mapsto (H - z)^{-1} \in B(\mathscr{G}^*, \mathscr{G})$ follows from (7.4.2) or from the first resolvent identity which is clearly satisfied in $B(\mathscr{G}^*, \mathscr{G})$. Formula (6.1.18) clearly implies that $\varphi(\widehat{H}) \in B(\mathscr{G}^*, \mathscr{G})$ if $\varphi \in C_0^2(\mathbb{R})$. If $H \in C^1(A; \mathscr{G}, \mathscr{G}^*)$, then the results of Section 5.1 imply that $(H - z)^{-1} \in C^1(A; \mathscr{G}^*, \mathscr{G})$ for non-real z and $[A, (H - z)^{-1}] = (H - z)^{-1}[H, A](H - z)^{-1}$. Of course these facts are stronger than the relation $(\widehat{H} - z)^{-1} \in C^1(A; \mathscr{H})$, so we have $\widehat{H} \in C^1(A)$. Finally, $\varphi(\widehat{H}) \in C^1(A; \mathscr{G}^*, \mathscr{G})$ by virtue of formula of (6.2.16). $\square$

The following comments are meant to clarify the statement of the lemma. First, *if $\widehat{H}$ is an arbitrary self-adjoint operator in $\mathscr{H}$, and if one takes $\mathscr{G} = D(|\widehat{H}|^{1/2})$ (equipped with the graph topology), then there is a unique extension of $\widehat{H}$ to a continuous operator $H : \mathscr{G} \to \mathscr{G}^*$, and this operator fulfills all the conditions of the lemma* (the operator associated to H being the initial $\widehat{H}$). On the other hand, if H is given and $\widehat{H}$ is constructed as in the lemma, then *the equality $\mathscr{G} = D(|\widehat{H}|^{1/2})$ does not hold in general*, as we explained in Section 2.8. There is, however, a very important case when this equality is true, namely, *if H is a semibounded operator* (this assertion is the content of the Friedrichs theorem, see Section 2.8). We could have taken just $\mathscr{G} = D(|\widehat{H}|^{1/2})$ in the theorem which follows, and this would cover most of the two-body problems that appear in the applications. But there are cases in which the conditions of Lemma 7.5.3 are satisfied but there does not seem to be any control on the form domain $D(|\widehat{H}|^{1/2})$. Most notably, this happens for the Dirac operator with Coulomb singularities and with physically natural coupling constants, see [Ne], [TE], [Vo], [MS]. Finally, we mention that the critical assumption of Lemma 7.5.3 is an estimate of the form $\|(H - \overline{z}_0)f\|_{\mathscr{G}^*} \geq c\|f\|_{\mathscr{G}}$ for a constant $c > 0$ and all $f \in \mathscr{G}$. In fact, this implies that $H - \overline{z}_0 : \mathscr{G} \to \mathscr{G}^*$ is injective with closed range, hence its adjoint $H - z_0 : \mathscr{G} \to \mathscr{G}^*$ is surjective. The injectivity of $H - z_0$ follows for example from

$$|\Im z_0| \cdot \|f\|_{\mathscr{H}}^2 = |\Im\langle f, (H - z_0)f\rangle| \leq \|f\|_{\mathscr{G}} \cdot \|(H - z_0)f\|_{\mathscr{G}^*}.$$

Let us return to the context of Lemma 7.5.3 and assume that $H \in C^1(A; \mathscr{G}, \mathscr{G}^*)$. Then $\widehat{H}$ is of class $C^1(A)$, so the function $\varrho_{\widehat{H}}^A$ and the set $\mu^A(\widehat{H})$ are well defined and depend only on H, so we may denote them by ϱ_H^A and $\mu^A(H)$ respectively. However, a direct definition in terms of H is possible. In fact, the commutator $[H, iA]$ is a well defined continuous symmetric operator $\mathscr{G} \to \mathscr{G}^*$ (it is just the strong derivative of $e^{-iA\tau}He^{iA\tau} \in B(\mathscr{G}, \mathscr{G}^*)$ at $\tau = 0$). Furthermore $\varphi(H)$ ($= \varphi(\widehat{H})$) can be defined directly by formula (6.1.18) if $\varphi \in C_0^2(\mathbb{R})$ (due to the estimate $\|(H - z)^{-1}\|_{B(\mathscr{G}^*, \mathscr{G})} \leq c\langle z\rangle^2|\Im z|^{-1}$). Hence the definition (7.2.7) makes sense without reference to the self-adjoint operator $\widehat{H}$ induced by H in $\mathscr{H}$.

If we have the stronger property $H \in \mathscr{C}^{1,1}(A; \mathscr{G}, \mathscr{G}^*)$, then $(H - z)^{-1} \in \mathscr{C}^{1,1}(A; \mathscr{G}^*, \mathscr{G})$ (cf. Section 5.2), which is considerably more than $(\widehat{H} - z)^{-1} \in \mathscr{C}^{1,1}(A; \mathscr{H})$, so $\widehat{H}$ is of class $\mathscr{C}^{1,1}(A)$. If $\widehat{H}$ has a spectral gap, then one may apply Theorem 7.4.1 and Proposition 7.4.4 in order to get the next theorem (take $\mathscr{E} = \mathscr{G}^*, \mathscr{E}_1 = D(A; \mathscr{G}^*)$ and observe that $(H - \lambda_0)^{-1}D(A; \mathscr{G}^*) \subset D(A; \mathscr{G}) \subset D(A)$, because $(H - \lambda_0)^{-1} \in C^1(A; \mathscr{G}^*, \mathscr{G})$). But the result holds without the spectral gap hypothesis.

THEOREM 7.5.4. *Let* $H : \mathscr{G} \to \mathscr{G}^*$ *be a symmetric operator of class* $\mathscr{C}^{1,1}(A; \mathscr{G}, \mathscr{G}^*)$ *and such that* $H - z_0 : \mathscr{G} \to \mathscr{G}^*$ *is bijective for some* $z_0 \in \mathbb{C}$. *Then* $H - z : \mathscr{G} \to \mathscr{G}^*$ *is an isomorphism for each non-real z and the function* $z \mapsto (H-z)^{-1} \in B(\mathscr{G}^*, \mathscr{G})$ *is holomorphic on* $\mathbb{C}\backslash\mathbb{R}$. *Let* $\mathscr{K} = (D(A; \mathscr{G}^*), \mathscr{G}^*)_{1/2,1}$, *so that* $\mathscr{K} \subset \mathscr{G}^*$ *continuously and densely, and identify* $\mathscr{G} \subset \mathscr{K}^*$ *and* $B(\mathscr{G}^*, \mathscr{G}) \subset B(\mathscr{K}, \mathscr{K}^*)$. *Then the limits* $\lim_{\mu \to \pm 0}(H - \lambda - i\mu)^{-1} := (H - \lambda \mp i0)^{-1}$ *exist in the weak* topology in* $B(\mathscr{K}, \mathscr{K}^*)$, *locally uniformly in* $\lambda \in \mu^A(H)$. *Let* $\mathscr{K}^{*\circ}$ *be the closure of* $\mathscr{G}$ *in* $\mathscr{K}^*$ *and let* $\widehat{H}$ *be the self-adjoint operator associated to H in* $\mathscr{H}$, *so that* $D(\widehat{H}) \subset \mathscr{G} \subset \mathscr{K}^{*\circ}$. *Then each bounded linear operator from* $\mathscr{K}^{*\circ}$ *to some Hilbert space is locally* $\widehat{H}$-*smooth on the open set* $\mu^A(H)$.

PROOF. The constructions that we shall make below are more general than needed for the proof of the preceding theorem: we shall, in fact, implicitly prove the main results of [ABG1] and [BGM2]. For example, the reader can easily deduce from the following arguments the proof of the various assertions quoted after the statement of Theorem 7.5.2.

(i) Let $\lambda \in \mathbb{R}$. Let H be an arbitrary self-adjoint operator in $\mathscr{H}$ and φ a function of class $C_0^2(\mathbb{R})$ with $0 \leq \varphi \leq 1$ and $\varphi(\lambda) = 1$. We set $\phi = \varphi(H)$ and $\phi^{\perp} = 1 - \varphi(H)$. Then, let M be a symmetric operator in $\mathscr{H}$ with domain including $D(H)$ and such that $M \geq m\phi^2$ for some number $m \geq 0$. Finally, let μ be a positive number and $H^{\pm} = H - \lambda \mp i(M + \mu) : D(H) \subset \mathscr{H} \to \mathscr{H}$. Our purpose at this stage is to prove the following estimate:

$$(7.5.3) \quad \|H^{\pm}f\| \geq \left[\frac{1}{2} - \|\phi^{\perp}R(\lambda)M\|\right] \cdot [1 + m\|\phi^{\perp}R(\lambda)\|]^{-1} \cdot (m + \mu)\|f\|$$

for all $f \in D(H)$. As usual $R(z) = (H - z)^{-1}$; observe that $\phi^{\perp} R(\lambda)$ is the bounded operator $\mathscr{H} \to D(H)$ associated to the continuous function $x \mapsto (1 - \varphi(x))(x - \lambda)^{-1}$ by the functional calculus. To prove (7.5.3), remark first that

$$\|\phi^{\perp} f\| = \|\phi^{\perp} R(\lambda \pm i\mu)[H^{\pm} \pm iM]f\|$$
$$\leq \|\phi^{\perp} R(\lambda)\| \cdot \|H^{\pm} f\| + \|\phi^{\perp} R(\lambda)M\| \cdot \|f\|.$$

Then:

$$m\|\phi f\|^2 + \mu\|f\|^2 \leq \langle f, (M + \mu)f\rangle = \mp\Im\langle f, H^{\pm} f\rangle \leq \|f\| \cdot \|H^{\pm} f\|,$$

which implies

$$\left(\frac{m}{2} + \mu\right)\|f\|^2 \leq m\|\phi f\|^2 + \mu\|f\|^2 + m\|\phi^{\perp} f\|^2$$
$$\leq \|f\| \cdot \|H^{\pm} f\| + m\|f\| \cdot \|\phi^{\perp} f\|.$$

Upon inserting the estimate obtained above for $\|\phi^{\perp} f\|$, we get (7.5.3).

(ii) Assume that $[\frac{1}{2} - \|\phi^{\perp} R(\lambda)M\|] \cdot [1 + m\|\phi^{\perp} R(\lambda)\|]^{-1} \equiv \frac{1}{\kappa} > 0$ and that one of the numbers m or μ is strictly positive. Then (7.5.3) implies that $H^{\pm} : D(H) \to \mathscr{H}$ are injective operators with closed ranges. If we also *assume M bounded* (in fact, H-bounded with relative bound < 1 would be enough), then $(H^{\pm})^* = H^{\mp}$, hence $H^{\pm} : D(H) \to \mathscr{H}$ are (bijective) isomorphisms. Moreover, if $G^{\pm} \in B(\mathscr{H})$ is the inverse of $H^{\pm}$, then $(G^{\pm})^* = G^{\mp}$ and $\|G^{\pm}\| \leq \kappa(m+\mu)^{-1}$. Observe that $H^{\pm} = (H - \lambda \mp i\mu) \mp iM$, hence $G^{\pm} = R(\lambda \pm i\mu)[I \pm iMG^{\pm}]$. Let us equip $D(H)$ with the norm $\|(H + i)f\| \equiv \|\langle H\rangle f\|$. Since

$$\|(H - \lambda)G^{\pm}\| \leq \|(H - \lambda)R(\lambda \pm i\mu)\| \cdot [1 + \|M\| \cdot \|G^{\pm}\|] \leq 1 + \|M\| \cdot \|G^{\pm}\|,$$

we obtain

$$(7.5.4) \qquad \|G^{\pm}\|_{\mathscr{H} \to D(H)} \leq 1 + \kappa\frac{\|M\| + \langle\lambda\rangle}{m + \mu}.$$

Similarly, we shall have

$$(7.5.5) \qquad \|\phi^{\perp} G^{\pm}\|_{\mathscr{H} \to D(H)} \leq \|\langle H\rangle R(\lambda)\phi^{\perp}\| \cdot \left[1 + \frac{\kappa\|M\|}{m + \mu}\right].$$

(iii) From $H^{\pm} = (H \mp i) - (\lambda \pm i(\mu + M - 1))$ we get the identities:

$$(7.5.6) \qquad G^{\pm} = R(\pm i) + R(\pm i)[\lambda \pm i(\mu + M - 1)]G^{\pm}$$
$$= R(\pm i) + G^{\pm}[\lambda \pm i(\mu + M - 1)]R(\pm i)$$
$$= R(\pm i) + R(\pm i)[\lambda \pm i(\mu + M - 1)]R(\pm i)$$
$$\quad + R(\pm i)[\lambda \pm i(\mu + M - 1)]G^{\pm}[\lambda \pm i(\mu + M - 1)]R(\pm i).$$

Now assume that we are under the conditions of the theorem and that H is the self-adjoint operator in $\mathscr{H}$ associated to the operator $H : \mathscr{G} \to \mathscr{G}^*$ (denoted $\widehat{H}$ in the statement of the theorem). Since $D(H) \equiv D(H, \mathscr{H}) \subset \mathscr{G}$, we have $\|G^{\pm}\|_{\mathscr{H} \to \mathscr{G}} \leq C\|G^{\pm}\|_{\mathscr{H} \to D(H)}$ with a constant C depending only on H and $\mathscr{G}$.

So the second equality in (7.5.6) shows that $G^\pm$ extend to continuous operators $\mathscr{G}^* \to \mathscr{G}$; then (7.5.4) implies

$$(7.5.7) \quad ||G^\pm||_{\mathscr{G}^* \to \mathscr{G}} \le ||R(\pm i)||_{\mathscr{G}^* \to \mathscr{G}}$$

$$+ C\left[1 + \kappa \frac{||M|| + \langle \lambda \rangle}{m + \mu}\right][1 + |\lambda| + |\mu| + ||M||] \cdot ||R(\pm i)||_{\mathscr{G}^* \to \mathscr{H}}.$$

Similarly, (7.5.5) gives

$$(7.5.8) \quad ||\phi^\perp G^\pm||_{\mathscr{G}^* \to \mathscr{G}} \le ||\phi^\perp R(\pm i)||_{\mathscr{G}^* \to \mathscr{G}} +$$

$$+ C||\langle H \rangle R(\lambda)\phi^\perp|| \cdot \left[1 + \frac{\kappa ||M||}{m + \lambda}\right] \cdot [1 + |\lambda| + |\mu| + ||M||] \cdot ||R(\pm i)||_{\mathscr{G}^* \to \mathscr{H}}.$$

(iv) Consider now a new symmetric operator $H_\varepsilon : \mathscr{G} \to \mathscr{G}^*$ such that $||H - H_\varepsilon||_{\mathscr{G} \to \mathscr{G}^*}$ is small enough; more precisely, we require $||G^\pm(H - H_\varepsilon)||_{B(\mathscr{G})} < 1$. Then $H_\varepsilon^\pm := H_\varepsilon - \lambda \mp i(M + \mu) : \mathscr{G} \to \mathscr{G}^*$ is an isomorphism with inverse $G_\varepsilon^\pm$ explicitly given by:

$$G_\varepsilon^\pm = [I - G^\pm(H - H_\varepsilon)]^{-1}G^\pm = G^\pm[I - (H - H_\varepsilon)G^\pm]^{-1}.$$

This leads to the next estimates, which have to be used in tandem with (7.5.7), (7.5.8) :

$$(7.5.9) \quad ||G_\varepsilon^\pm||_{\mathscr{G}^* \to \mathscr{G}} \le ||G^\pm||_{\mathscr{G}^* \to \mathscr{G}} \cdot [1 - ||G^\pm(H - H_\varepsilon)||_{B(\mathscr{G})}]^{-1},$$

$$(7.5.10) \quad ||\phi^\perp G_\varepsilon^\pm||_{\mathscr{G}^* \to \mathscr{G}} \le ||\phi^\perp G^\pm||_{\mathscr{G}^* \to \mathscr{G}} \cdot [1 - ||(H - H_\varepsilon)G^\pm||_{B(\mathscr{G}^*)}]^{-1}.$$

(v) Now we derive the so-called *quadratic estimate*, namely:

$$(7.5.11) \quad ||\phi G_\varepsilon^\pm f||^2 \le \frac{1}{m}|\Im\langle f, G_\varepsilon^\pm f\rangle|, \quad f \in \mathscr{G}^*.$$

From this, using Lemma 7.5.3, we also get for some $c = c(\varphi, H, \mathscr{G}) < \infty$:

$$(7.5.12) \quad ||\phi G_\varepsilon^\pm f||_\mathscr{G} \le \frac{c}{\sqrt{m}}|\Im\langle f, G_\varepsilon^\pm f\rangle|^{1/2}, \quad f \in \mathscr{G}^*.$$

For the proof of (7.5.11), one may argue as follows :

$$m||\phi G_\varepsilon^\pm f||^2 = \langle G_\varepsilon^\pm f, m\phi^2 G_\varepsilon^\pm f\rangle \le \langle G_\varepsilon^\pm f, (M + \mu)G_\varepsilon^\pm f\rangle$$

$$= \langle f, G_\varepsilon^\mp(M + \mu)G_\varepsilon^\pm f\rangle = \mp\frac{1}{2i}\langle f, (G_\varepsilon^- - G_\varepsilon^+)f\rangle = \pm\Im\langle f, G_\varepsilon^\pm f\rangle.$$

(vi) We shall now make a more specific choice for M. Fix $\lambda_0 \in \mathbb{R}$ such that $\varrho_H^A(\lambda_0) > 0$. Denote by B the symmetric operator $[iH, A] \in B(\mathscr{G}, \mathscr{G}^*)$. According to (7.2.7), there are $\delta > 0$, $a_1 > 0$ and $\varphi_1 \in C_0^\infty(\mathbb{R})$, $0 \le \varphi_1 \le 1$, such that $\varphi_1(x) = 1$ if $|x - \lambda_0| \le 2\delta$ and $\varphi_1(H)B\varphi_1(H) \ge a_1\varphi_1^2(H)$. Then, let $\{B_\varepsilon\}_{\varepsilon>0}$ be a family of symmetric operators $\mathscr{G} \to \mathscr{G}^*$ such that $||B_\varepsilon - B||_{\mathscr{G} \to \mathscr{G}^*} \to 0$ as $\varepsilon \to 0$. Fix some $\varphi \in C_0^\infty(\mathbb{R})$, $0 \le \varphi \le 1$, with $\varphi(x) = 1$ if $|x - \lambda_0| \le \delta$ and $\varphi(x) = 0$ if $|x - \lambda_0| \ge 2\delta$, and let $0 < a < a_1$. Then there is $\varepsilon_0 > 0$ such that $\varphi(H)B_\varepsilon\varphi(H) \ge a\varphi^2(H)$ for $0 < \varepsilon < \varepsilon_0$. We shall take $M = \varepsilon\phi B_\varepsilon\phi$ and $m = a\varepsilon$. Moreover, for each $\varepsilon \in (0, \varepsilon_0)$ we assume given a symmetric operator $H_\varepsilon : \mathscr{G} \to \mathscr{G}^*$ such that $\lim_{\varepsilon \to 0} \varepsilon^{-1}||H - H_\varepsilon||_{\mathscr{G} \to \mathscr{G}^*} = 0$. From now on λ is an arbitrary real number with $|\lambda - \lambda_0| \leq \delta$. Then we may assume that the number κ introduced at Step (ii) is finite, strictly positive and independent of ε and λ. Notice that with the preceding choices $H_\varepsilon^\pm = H_\varepsilon - \lambda \mp i(\varepsilon\phi B_\varepsilon\phi + \mu)$. Since $\|G^\pm\|_{\mathscr{G}^*\to\mathscr{G}} \leq \text{const.}(\varepsilon + \mu)^{-1}$ (see (7.5.7)) and $\|\phi^\perp G^\pm\|_{\mathscr{G}^*\to\mathscr{G}} \leq \text{const.}$ (see (7.5.8)), it follows that $\|G^\pm(H - H_\varepsilon)\|_{B(\mathscr{G})} \to 0$ as $\varepsilon \to 0$, so $G_\varepsilon^\pm$ are well defined (ε_0 being assumed small enough) and satisfy the following estimates, cf. (7.5.9), (7.5.10), (7.5.12) :

$$(7.5.13) \qquad \|G_\varepsilon^\pm\|_{\mathscr{G}^*\to\mathscr{G}} \leq \frac{c}{\varepsilon + \mu}, \qquad \|\phi^\perp G_\varepsilon^\pm\|_{\mathscr{G}^*\to\mathscr{G}} \leq c$$

$$(7.5.14) \qquad \|\phi G_\varepsilon^\pm f\|_\mathscr{G} \leq c\varepsilon^{-1/2}|\Im\langle f, G_\varepsilon^\pm f\rangle|^{1/2}, \qquad f \in \mathscr{G}^*.$$

Here $c < \infty$ is independent of $\varepsilon \in (0, \varepsilon_0)$, $\mu \in (0, 1)$, $\lambda \in (\lambda_0 - \delta, \lambda_0 + \delta)$ and f. Notice that $|\Im\langle f, G_\varepsilon^+ f\rangle| = |\Im\langle f, G_\varepsilon^- f\rangle|$ because $(G_\varepsilon^\pm)^* = G_\varepsilon^\mp$. Furthermore, since $\phi + \phi^\perp = I$, we also get:

$$(7.5.15) \qquad \|G_\varepsilon^\pm f\|_\mathscr{G} \leq c\varepsilon^{-1/2}|\Im\langle f, G_\varepsilon^\pm f\rangle|^{1/2} + c\|f\|_{\mathscr{G}^*}.$$

(vii) From now on we assume that the $B(\mathscr{G}, \mathscr{G}^*)$-valued mappings $\varepsilon \mapsto H_\varepsilon$ and $\varepsilon \mapsto B_\varepsilon$ are norm-C^1 on $(0, \varepsilon_0)$. We choose to work with G_ε^+ only and denote it simply by G_ε. It is trivial to prove that $(0, \varepsilon_0) \ni \varepsilon \mapsto G_\varepsilon \in B(\mathscr{G}^*, \mathscr{G})$ is norm-C^1 and that its derivative is

$$(7.5.16) \qquad G_\varepsilon' \equiv \frac{d}{d\varepsilon}G_\varepsilon = G_\varepsilon[-H_\varepsilon' + i\phi B_\varepsilon\phi + i\varepsilon\phi B_\varepsilon'\phi]G_\varepsilon.$$

(viii) So far the operator A has not played any role. Now let us suppose that the operators H, H_ε, B_ε are of class $C^1(A; \mathscr{G}, \mathscr{G}^*)$. By using Lemma 7.5.3 we see that $\phi \in C^1(A; \mathscr{G}^*, \mathscr{G})$, hence $\phi B_\varepsilon\phi \in C^1(A; \mathscr{G}^*, \mathscr{G})$ and $[iA, \phi B_\varepsilon\phi] = 2\Re(\phi B_\varepsilon[iA, \phi]) + \phi[iA, B_\varepsilon]\phi$ (see Proposition 5.1.5). From Proposition 5.1.6 we obtain $G_\varepsilon \in C^1(A; \mathscr{G}^*, \mathscr{G})$ and

$$(7.5.17)$$
$$[A, G_\varepsilon] = G_\varepsilon[H_\varepsilon - i\varepsilon\phi B_\varepsilon\phi, A]G_\varepsilon = G_\varepsilon\{[H_\varepsilon, A] + \varepsilon[iA, \phi B_\varepsilon\phi]\}G_\varepsilon.$$

By adding (7.5.16) and (7.5.17) we obtain the fundamental formula:

$$(7.5.18)$$
$$G_\varepsilon' + [A, G_\varepsilon] = i\varepsilon G_\varepsilon K_\varepsilon G_\varepsilon - iG_\varepsilon\{\phi B_\varepsilon\phi^\perp + \phi^\perp B_\varepsilon\phi + \phi^\perp B_\varepsilon\phi^\perp\}G_\varepsilon,$$

where

$$(7.5.19) \qquad K_\varepsilon = \frac{1}{\varepsilon}(iH_\varepsilon' + i[A, H_\varepsilon] + B_\varepsilon) + \phi B_\varepsilon'\phi + [A, \phi B_\varepsilon\phi].$$

(ix) The last object we have to introduce is a family of vectors $\{f_\varepsilon\}_{0<\varepsilon<\varepsilon_0}$ such that $f_\varepsilon \in D(A; \mathscr{G}^*)$ and the function $\varepsilon \mapsto f_\varepsilon \in \mathscr{G}^*$ is strongly C^1. Then $\langle f_\varepsilon, [A, G_\varepsilon]f_\varepsilon\rangle = \langle Af_\varepsilon, G_\varepsilon f_\varepsilon\rangle - \langle G_\varepsilon^* f_\varepsilon, Af_\varepsilon\rangle$ (notice that here the brackets denote the anti-duality between $\mathscr{G}$ and $\mathscr{G}^*$). To prove this it is enough to observe that we have strongly in $B(\mathscr{G}^*, \mathscr{G})$:

$$\begin{aligned}
[A, G_\varepsilon] &= \lim_{\tau \to 0} \frac{W(\tau) G_\varepsilon W(-\tau) - G_\varepsilon}{i\tau} \\
&= \lim_{\tau \to 0} \frac{W(\tau) G_\varepsilon W(-\tau) - G_\varepsilon}{i\tau} W(\tau) \\
&= \lim_{\tau \to 0} \left\{ \frac{W(\tau) - I}{i\tau} G_\varepsilon - G_\varepsilon \frac{W(\tau) - I}{i\tau} \right\}
\end{aligned}$$

where at the second step we used the fact that W is a C_0-group in $\mathscr{G}^*$. In conclusion, if $F_\varepsilon = \langle f_\varepsilon, G_\varepsilon f_\varepsilon \rangle$, then $\varepsilon \mapsto F_\varepsilon$ is a C^1 function satisfying the identity (7.3.7). By using (7.5.18) we get:

(7.5.20)
$$\begin{aligned}
F_\varepsilon' &= \langle f_\varepsilon' - A f_\varepsilon, G_\varepsilon f_\varepsilon \rangle + \langle G_\varepsilon^* f_\varepsilon, f_\varepsilon' + A f_\varepsilon \rangle \\
&\quad + i\varepsilon \langle G_\varepsilon^* f_\varepsilon, K_\varepsilon G_\varepsilon f_\varepsilon \rangle - i \langle \phi^\perp G_\varepsilon^* f_\varepsilon, B_\varepsilon \phi^\perp G_\varepsilon f_\varepsilon \rangle \\
&\quad - i \langle \phi G_\varepsilon^* f_\varepsilon, B_\varepsilon \phi^\perp G_\varepsilon f_\varepsilon \rangle - i \langle \phi^\perp G_\varepsilon^* f_\varepsilon, B_\varepsilon \phi G_\varepsilon f_\varepsilon \rangle.
\end{aligned}$$

(x) We shall estimate the terms on the right-hand side of (7.5.20) by using (7.5.13)-(7.5.15). So, there is a constant $C > 0$, independent of $\varepsilon \in (0, \varepsilon_0)$, $\mu \in (0, 1)$ and $\lambda \in (\lambda_0 - \delta, \lambda_0 + \delta)$ such that:

$$\begin{aligned}
C^{-1} |F_\varepsilon'| &\leq \|K_\varepsilon\|_{B(\mathscr{G}, \mathscr{G}^*)} |F_\varepsilon| \\
&\quad + \{ \|B_\varepsilon\|_{B(\mathscr{G}, \mathscr{G}^*)} \|f_\varepsilon\|_{\mathscr{G}^*} + \|f_\varepsilon'\|_{\mathscr{G}^*} + \|A f_\varepsilon\|_{\mathscr{G}^*} \} \left| \frac{1}{\varepsilon} F_\varepsilon \right|^{1/2} \\
&\quad + \|f_\varepsilon\|_{\mathscr{G}^*} \{ \varepsilon \|f_\varepsilon\|_{\mathscr{G}^*} \|K_\varepsilon\|_{B(\mathscr{G}, \mathscr{G}^*)} + \|f_\varepsilon\|_{\mathscr{G}^*} \|B_\varepsilon\|_{B(\mathscr{G}, \mathscr{G}^*)} \\
&\quad + \|f_\varepsilon'\|_{\mathscr{G}^*} + \|A f_\varepsilon\|_{\mathscr{G}^*} \}.
\end{aligned}$$

Since $B_\varepsilon \to B$ in $B(\mathscr{G}, \mathscr{G}^*)$ as $\varepsilon \to 0$, we have $\|B_\varepsilon\|_{B(\mathscr{G}, \mathscr{G}^*)} \leq$ const. Moreover, we shall choose $\{f_\varepsilon\}_{\varepsilon > 0}$ such that $f_\varepsilon \to f$ in $\mathscr{G}^*$ as $\varepsilon \to 0$, hence $\|f_\varepsilon\|_{\mathscr{G}^*} \leq$ const. So, by modifying slightly the constant C, we have a simpler estimate:

(7.5.21)
$$\begin{aligned}
C^{-1} |F_\varepsilon'| &\leq \|K_\varepsilon\|_{B(\mathscr{G}, \mathscr{G}^*)} |F_\varepsilon| \\
&\quad + \{ \|f_\varepsilon'\|_{\mathscr{G}^*} + \|f_\varepsilon\|_{D(A; \mathscr{G}^*)} \} \left| \frac{1}{\varepsilon} F_\varepsilon \right|^{1/2} \\
&\quad + \{ \varepsilon \|K_\varepsilon\|_{B(\mathscr{G}, \mathscr{G}^*)} + \|f_\varepsilon'\|_{\mathscr{G}^*} + \|f_\varepsilon\|_{D(A; \mathscr{G}^*)} \}.
\end{aligned}$$

This inequality is of the same nature as (7.3.8).

Now we may finish the proof of the theorem exactly as in Section 7.3. Since A is the generator of a C_0-group in $\mathscr{G}^*$, the operator $1 + i\varepsilon A : D(A; \mathscr{G}^*) \to \mathscr{G}^*$ is bijective for $\varepsilon > 0$ small enough, and if we define $f_\varepsilon = (1 + i\varepsilon A)^{-1} f$ for some $f \in \mathscr{K} = (D(A; \mathscr{G}^*), \mathscr{G}^*)_{1/2, 1}$, we shall have: $f_\varepsilon \in D(A; \mathscr{G}^*)$, $\varepsilon \mapsto f_\varepsilon \in \mathscr{G}^*$ is of class C^1, $f_\varepsilon \to f$ in $\mathscr{G}^*$ as $\varepsilon \to 0$, and finally

$$\int_0^{\varepsilon_0} (\|f_\varepsilon'\|_{\mathscr{G}^*} + \|A f_\varepsilon\|_{\mathscr{G}^*}) \varepsilon^{-1/2} d\varepsilon < \infty$$

(see Proposition 2.7.2). It remains to choose operators H_ε, B_ε in such a way that $\int_0^{\varepsilon_0} \|K_\varepsilon\|_{B(\mathscr{G}, \mathscr{G}^*)} d\varepsilon < \infty$. The method is identical to that used in the proof of Lemma 7.3.6, just replace $B(\mathcal{H})$ by $B(\mathcal{G}, \mathcal{G}^*)$. For example, let H_ε be defined by (7.3.16), where of course $S = H$, and let $B_\varepsilon = i[H_\varepsilon, A]$. Then

$$K_\varepsilon = \frac{i}{\varepsilon} H'_\varepsilon + \phi B'_\varepsilon \phi + [A, \phi] B_\varepsilon \phi + \phi B_\varepsilon [A, \phi] + \phi [A, B_\varepsilon] \phi.$$

Since ϕ and $[A, \phi]$ belong to $B(\mathcal{G}^*, \mathcal{G})$, we shall have:

$$\|K_\varepsilon\|_{B(\mathcal{G}, \mathcal{G}^*)} \leq \frac{1}{\varepsilon} \|H'_\varepsilon\|_{B(\mathcal{G}, \mathcal{G}^*)} + C \|B'_\varepsilon\|_{B(\mathcal{G}, \mathcal{G}^*)}$$
$$+ C \|B_\varepsilon\|_{B(\mathcal{G}, \mathcal{G}^*)} + C \|[A, B_\varepsilon]\|_{B(\mathcal{G}, \mathcal{G}^*)}$$

and the right-hand side is an integrable function. We do not give details, since there is nothing really new with respect to Lemma 7.3.6. The only fact one has to notice is that the function θ from part (v) of the proof of Lemma 7.3.6 has to have a Fourier transform $\hat{\theta}$ decaying at infinity more rapidly than some exponential, since the group $\{e^{iA\tau}\}$ is of exponential growth (in general) in $\mathcal{G}$ and in $\mathcal{G}^*$. Clearly, the choice (7.3.16) is more than sufficient. $\square$

If we combine Theorems 7.5.4 and 7.1.5 we get the following criterion for the existence and completeness of wave operators (this is the analogue of Theorem 7.4.3 in the present context):

THEOREM 7.5.5. *Let $\mathcal{H}$ be a Hilbert space and assume that the following objects are given :*
a) two Hilbert spaces $\mathcal{G}_1$ and $\mathcal{G}_2$ such that $\mathcal{G}_j \subset \mathcal{H}$ continuously and densely (hence $\mathcal{G}_j \subset \mathcal{H} = \mathcal{H}^* \subset \mathcal{G}_j^*$);
b) two symmetric operators $H_j : \mathcal{G}_j \to \mathcal{G}_j^*$, $j = 1, 2$, such that $H_j - z_j : \mathcal{G}_j \to \mathcal{G}_j^*$ are isomorphisms for some $z_j \in \mathbb{C}$; denote by the same symbol H_j the self-adjoint operator in $\mathcal{H}$ associated to H_j;
c) two self-adjoint operators A_1, A_2 in $\mathcal{H}$ such that $e^{iA_j\tau} \mathcal{G}_j \subset \mathcal{G}_j$ ($j = 1, 2$; $\tau \in \mathbb{R}$) and $H_j \in \mathscr{C}^{1,1}(A_j; \mathcal{G}_j, \mathcal{G}_j^*)$; let $D(A_j; \mathcal{G}_j^*)$ be the domain of the closure of the operator A_j in $\mathcal{G}_j^*$ (equipped with the graph topology), define $\mathscr{K}_j = (D(A_j; \mathcal{G}_j^*), \mathcal{G}_j^*)_{1/2,1}$ (hence $\mathcal{G}_j \subset \mathscr{K}_j^*$ continuously but not densely in general) and denote by $\mathscr{K}_j^{*\circ}$ the closure of $\mathcal{G}_j$ in $\mathscr{K}_j^*$.
Finally, assume that there is a continuous operator $V : \mathscr{K}_1^{*\circ} \to \mathscr{K}_2$ such that $H_2 = H_1 + V$ as forms on $D(H_1) \times D(H_2)$, i.e. $\langle f_1, H_2 f_2 \rangle = \langle H_1 f_1, f_2 \rangle + \langle V f_1, f_2 \rangle$ if $f_j \in D(H_j)$.
Let $J = \mu^{A_1}(H_1) \cap \mu^{A_2}(H_2)$, which is an open real set, and let E_j be the spectral measure of H_j. Then the local wave operators

$$W_1^\pm = \operatorname*{s-lim}_{t \to \pm\infty} e^{iH_2 t} e^{-iH_1 t} E_1(J), \qquad W_2^\pm = \operatorname*{s-lim}_{t \to \pm\infty} e^{iH_1 t} e^{-iH_2 t} E_2(J)$$

exist and are complete: $(W_1^\pm)^* = W_2^\pm$, $W_2^\pm W_1^\pm = E_1(J)$, $W_1^\pm W_2^\pm = E_2(J)$.

We stress once again that in most cases the natural choice is $\mathcal{G}_j = D(|H_j|^{1/2})$, i.e. $\mathcal{G}_j$ is just the form domain of H_j; but this is a wrong choice in some singular situations, e.g. in the case of a Dirac operator if one wants to cover Coulomb singularities with optimal (from the point of view of physics) coupling constants.

7.5.3. In the rest of this section we shall describe a *perturbative method* of verifying the hypothesis of Theorem 7.5.4 (so, implicitly, those of Theorem 7.5.5 too). This method is quite efficient in two-body type problems, as will be seen later on.

Let H be a self-adjoint operator in a Hilbert space $\mathscr{H}$ and $J \subset \mathbb{R}$ an open real set; our purpose is to study spectral (and scattering) properties of H in J. The perturbative method consists in assuming that one can decompose H into a sum $H = H_0 + V$, where H_0 is a "simple" operator which can be "explicitly" treated, while V is "small" in some sense. The sum above cannot be interpreted as an operator sum in $\mathscr{H}$ in general. For convenience in applications, we state a version of such a perturbative scheme as a proposition. The result is a straightforward consequence of Theorems 7.5.4 and 7.5.5 combined with Theorems 7.2.9, 7.2.13 and Corollary 7.2.11.

PROPOSITION 7.5.6. *Let $(\mathscr{G}, \mathscr{H})$ be a Friedrichs couple and H_0 a self-adjoint operator in $\mathscr{H}$ such that $D(H_0) \subset \mathscr{G}$ and such that, after the identification $\mathscr{G} \subset \mathscr{H} = \mathscr{H}^* \subset \mathscr{G}^*$, H_0 extends to a symmetric operator $H_0 : \mathscr{G} \to \mathscr{G}^*$ with $(H_0+i)\mathscr{G} = \mathscr{G}^*$. Let J be an open real set and assume that a self-adjoint operator A in $\mathscr{H}$ has been constructed such that:*

(i) $e^{iA\tau}\mathscr{G} \subset \mathscr{G} \quad \forall \tau \in \mathbb{R}$;

(ii) $H_0 \in \mathscr{C}^{1,1}(A; \mathscr{G}, \mathscr{G}^*)$;

(iii) *A is locally conjugate to H_0 on J (i.e. for each $\lambda \in J$ there are a real function $\varphi \in C_0^\infty(\mathbb{R})$ with $\varphi(\lambda) \neq 0$, a number $a > 0$ and a compact operator K in $\mathscr{H}$ such that $\varphi(H_0)[iH_0, A]\varphi(H_0) \geq a\varphi^2(H_0) + K$).*

Let $V : \mathscr{G} \to \mathscr{G}^*$ be a symmetric operator of class $\mathscr{C}^{1,1}(A; \mathscr{G}, \mathscr{G}^*)$ and such that the operator $H = H_0 + V \in B(\mathscr{G}, \mathscr{G}^*)$ has the following properties: $H+i$ is an isomorphism of $\mathscr{G}$ onto $\mathscr{G}^*$ and $(H+i)^{-1} - (H_0+i)^{-1}$ is a compact operator in $\mathscr{H}$. Denote again by H the self-adjoint operator in $\mathscr{H}$ associated to H.

Then H is of class $\mathscr{C}^{1,1}(A; \mathscr{G}, \mathscr{G}^*)$ and A is locally conjugate to H on J, in particular:

1) *H has no singularly continuous spectrum in J.*

2) *The eigenvalues of H in J are of finite multiplicity and have no accumulation points in J.*

3) *$J \setminus \sigma_p(H) \subset \mu^A(H)$, hence the strong limiting absorption principle for H holds on $J \setminus \sigma_p(H)$ in the space $\mathscr{K} = (D(A; \mathscr{G}^*), \mathscr{G}^*)_{1/2,1}$.*

4) *Observe that $\mathscr{G} \subset \mathscr{K}^*$ and let $\mathscr{K}^{*\circ}$ be the closure of $\mathscr{G}$ in $\mathscr{K}^*$; then any continuous Hilbert space-valued operator on $\mathscr{K}^{*\circ}$ is locally H-smooth on $J \setminus \sigma_p(H)$.*

Assume furthermore that $V \in B(\mathscr{K}^{*\circ}, \mathscr{K})$. Then the local wave operators $s\text{-}\lim_{t \to \pm\infty} e^{iHt}e^{-iH_0 t}E_0^c(J)$ exist and have range equal to $E^c(J)$ (where E_0^c and E^c are the continuous components of the spectral measures of H_0 and H respectively).

In many concrete situations the most convenient choice for $\mathscr{G}$ is

$$\mathscr{G} = D(|H_0|^{1/2}) = \text{ form domain of } H_0.$$

Then the simplest way of checking that $H - z$ is an isomorphism of $\mathscr{G}$ onto $\mathscr{G}^*$ for some $z \in \mathbb{C}$ (or equivalently that $(H + i)\mathscr{G} = \mathscr{G}^*$) is to verify an estimate of the form $|\langle g, Vg \rangle| \leq a\langle g, |H_0|g \rangle + b\|g\|_{\mathscr{H}}^2$ for some $a < 1$, $b \in \mathbb{R}$ and all $g \in \mathscr{G}$ (one may write this simply as $\pm V \leq a|H_0| + b$). In this case $\mathscr{G}$ will also be the form domain of the self-adjoint operator associated to H (i.e. $\mathscr{G} = D(|H|^{1/2})$). A more general procedure is to show that there is z such that $\|(H_0 - z)^{-1}V\|_{B(\mathscr{G})} < 1$; see for example [Ne] for the verification of such an estimate in the case of a Dirac operator with Coulomb type singularities.

It is clear that the efficiency of the method described in Proposition 7.5.6 depends on the possibility of verifying the $\mathscr{C}^{1,1}$ property for large classes of operators H_0 and V. We shall describe below criteria which turn out to give quite good results in the situations we met in the applications of the theory.

Let us first recall that an operator $S \in B(\mathscr{G}, \mathscr{G}^*)$ is of class $\mathscr{C}^{1,1}(A; \mathscr{G}, \mathscr{G}^*)$ if

$$(7.5.22) \qquad \int_0^1 \|e^{iA\tau}Se^{-iA\tau} + e^{-iA\tau}Se^{iA\tau} - 2S\|_{B(\mathscr{G}, \mathscr{G}^*)} \frac{d\tau}{\tau^2} < \infty.$$

This implies the existence of the derivative $\frac{d}{d\tau}e^{-iA\tau}Se^{iA\tau}|_{\tau=0} \equiv i[S, A]$ *in norm* in $B(\mathscr{G}, \mathscr{G}^*)$. The justification of the notation $[S, A]$ will be recalled below; then we shall present methods of verifying (7.5.22).

Let $S : \mathscr{G} \to \mathscr{G}^*$ be any symmetric operator. Then $i(SA - AS)$ is a well defined symmetric sesquilinear form on $D(A; \mathscr{G})$, its value at $(f, g) \in D(A; \mathscr{G}) \times D(A; \mathscr{G})$ being $\langle Sf, iAg \rangle + \langle iAf, Sg \rangle$. Recall that $D(A; \mathscr{G})$ is dense in $\mathscr{G}$. We know that S *is of class* $C^1(A; \mathscr{G}, \mathscr{G}^*)$ *if and only if the preceding form is continuous for the topology induced by* $\mathscr{G}$ *on* $D(A; \mathscr{G})$. This is equivalent to the existence of an operator $T \in B(\mathscr{G}, \mathscr{G}^*)$ such that $\langle g, Tg \rangle = 2\Re\langle iAg, Sg \rangle$ for $g \in D(A; \mathscr{G})$. Then T is uniquely defined; it is symmetric, and we denote it by $i[S, A]$.

The most rudimentary way of proving that $S \in \mathscr{C}^{1,1}(A; \mathscr{G}, \mathscr{G}^*)$ is by showing the much stronger property $S \in C^2(A; \mathscr{G}, \mathscr{G}^*)$. Notice that $[A, [A, S]]$ can always be defined as a symmetric sesquilinear form on the domain $D(A^2; \mathscr{G})$ (which is dense in $\mathscr{G}$) by the formula $\langle g, [A, [A, S]]g \rangle = 2\Re\langle A^2 g, Sg \rangle - 2\langle Ag, SAg \rangle$. It is easy to prove that $S \in C^2(A; \mathscr{G}, \mathscr{G}^*)$ if and only if this form is continuous for the topology induced by $\mathscr{G}$ on $D(A^2; \mathscr{G})$; then the continuous (symmetric) operator $[A, [A, S]] : \mathscr{G} \to \mathscr{G}^*$ associated to it is just $-\frac{d^2}{d\tau^2}e^{-iA\tau}Se^{iA\tau}|_{\tau=0}$. In practice it is better to proceed in two steps: one first shows that $S \in C^1(A; \mathscr{G}, \mathscr{G}^*)$ and then that $[iS, A] \in C^1(A; \mathscr{G}, \mathscr{G}^*)$. The second step means that the symmetric sesquilinear form $\langle Af, [A, S]g \rangle + \langle [A, S]f, Ag \rangle$ on $D(A; \mathscr{G})$ is continuous for the topology of $\mathscr{G}$.

In applications one often has $H_0 \in C^2(A; \mathscr{G}, \mathscr{G}^*)$, but the condition $V \in C^2(A, \mathscr{G}, \mathscr{G}^*)$ gives results which are usually considered too rough. Considerably more general situations are covered by the following criterion : S *is of class* $\mathscr{C}^{1,1}(A; \mathscr{G}, \mathscr{G}^*)$ *if the function* $\tau \mapsto e^{-iA\tau}Se^{iA\tau} \in B(\mathscr{G}, \mathscr{G}^*)$ *is differentiable and has a Dini-continuous derivative.* Explicitly, one has to check that $S \in C^1(A; \mathscr{G}, \mathscr{G}^*)$ and $T \equiv i[S, A] \in \mathscr{C}^{0,1}(A; \mathscr{G}, \mathscr{G}^*)$, which means:

$$(7.5.23) \qquad \int_0^1 \|e^{iA\tau}Te^{-iA\tau} - T\|_{B(\mathscr{G}, \mathscr{G}^*)} \frac{d\tau}{\tau} < \infty.$$

But even this condition on S is strictly stronger than $S \in \mathscr{C}^{1,1}(A; \mathscr{G}, \mathscr{G}^*)$, as we explained in Section 5.2. In order to appreciate the quality of this last assumption, we recall that $S \in C^1(A; \mathscr{G}, \mathscr{G}^*)$ or $S \in C^1_u(A; \mathscr{G}, \mathscr{G}^*)$ if and only if

$$\lim_{\varepsilon \to +0} \int_\varepsilon^1 [e^{2iA\tau} S e^{-2iA\tau} - 2e^{iA\tau} S e^{-iA\tau} + S] \frac{d\tau}{\tau^2} < \infty$$

exists strongly or in norm respectively, whereas (7.5.22) is equivalent to the absolute convergence of the preceding integral on $[0,1]$.

We shall now describe methods of a different nature for verifying (7.5.22) and (7.5.23). As explained in Chapter 5, (7.5.22) and (7.5.23) describe certain types of regularity of the operators S, T with respect to the group of automorphisms of $B(\mathscr{G}, \mathscr{G}^*)$ induced by the group $\{e^{iA\tau}\}$. We also saw that an operator having a certain degree of regularity preserves regularity of the vectors in $\mathscr{G}$ if this regularity is defined in terms of the Besov scales $\{\mathscr{G}_{s,p}\}$, $\{\mathscr{G}_{s,p}^*\}$ associated to A in $\mathscr{G}$ and $\mathscr{G}^*$. For example, if $S \in C^1(A; \mathscr{G}, \mathscr{G}^*)$, then $S \in B(\mathscr{G}_{\pm 1}, \mathscr{G}_{\pm 1}^*)$. If S preserves regularity, then S is not regular in general, e.g. in order to have $S \in B(\mathscr{G}_1, \mathscr{G}_1^*)$ it suffices to know that $[A, S] \in B(\mathscr{G}_1, \mathscr{G}^*)$. But if S improves regularity, then one may deduce that S is regular, as illustrated by the fact that the boundedness of $[A, S]$ trivially follows from the boundedness of AS and SA, i.e. from the property $S \in B(\mathscr{G}, \mathscr{G}_1^*) \cap B(\mathscr{G}_{-1}, \mathscr{G}^*)$; moreover, if S is symmetric, then $S\mathscr{G} \subset \mathscr{G}_1^*$ is sufficient (then take the adjoint). We shall develop this observation for the regularity classes $\mathscr{C}^{0,1}$ and $\mathscr{C}^{1,1}$.

We begin with the simplest case, which is $\mathscr{C}^{0,1}$ (see (7.5.23)). For any $T \in B(\mathscr{G}, \mathscr{G}^*)$ we have:

$$e^{iA\tau} T e^{-iA\tau} - T = [e^{iA\tau} - I, T] e^{-iA\tau} = \left[-2 \left(\sin \frac{A\tau}{2} \right)^2 + i \sin A\tau, T \right] e^{-iA\tau}.$$

Hence, if $T : \mathscr{G} \to \mathscr{G}^*$ is symmetric, then (7.5.23), i.e. the property $T \in \mathscr{C}^{0,1}(A; \mathscr{G}, \mathscr{G}^*)$, is a consequence of:

$$(7.5.24) \qquad \int_0^1 \|(\sin A\tau) T\|_{B(\mathscr{G}, \mathscr{G}^*)} \frac{d\tau}{\tau} < \infty.$$

Heuristically, this means that T improves the decay at infinity in the spectral representation of A. It is clear that (7.5.24) holds if $T\mathscr{G} \subset \mathscr{G}_s^*$ for some $s > 0$. But the next result is more precise and easy to use in the examples we have in mind.

PROPOSITION 7.5.7. *Let Λ be a closed densely defined operator in $\mathscr{G}^*$ with domain included in $D(A; \mathscr{G}^*)$ and such that $-ir$ belongs to the resolvent set of Λ and $r\|(\Lambda + ir)^{-1}\|_{B(\mathscr{G}^*)} \le$ const. for all $r > 0$. If $T : \mathscr{G} \to \mathscr{G}^*$ is symmetric and*

$$(7.5.25) \qquad \int_1^\infty \|\Lambda(\Lambda + ir)^{-1} T\|_{B(\mathscr{G}, \mathscr{G}^*)} \frac{dr}{r} < \infty,$$

then T is of class $\mathscr{C}^{0,1}(A; \mathscr{G}, \mathscr{G}^*)$. Assume furthermore that Λ is the generator of a C_0-group $\{e^{i\Lambda\tau}\}_{\tau \in \mathbb{R}}$ of polynomial growth in $\mathscr{G}^*$, or more generally that Λ admits a (LP)-functional calculus (of finite order) in $\mathscr{G}^*$ (see Section 3.5). Let

$\xi \in C^\infty(\mathbb{R})$ such that $\xi(t) = 0$ near zero and $\xi(t) = 1$ near infinity. Then (7.5.25) is a consequence of the inequality

$$(7.5.26) \qquad \int_1^\infty \left\| \xi\left(\frac{\Lambda}{r}\right) T \right\|_{B(\mathscr{G},\mathscr{G}^*)} \frac{dr}{r} < \infty.$$

PROOF. (i) There are finite constants M, ω such that $\|e^{iA\tau}\|_{B(\mathscr{G}^*)} \leq M e^{\omega|\tau|}$ for $\tau \in \mathbb{R}$. Hence any $z \in \mathbb{C}$ with $|\Im z| > \omega$ belongs to the resolvent set of A (considered as an operator in $\mathscr{G}^*$ or $\mathscr{G}$) and $(A + z)^{-1} = -i \int_0^\infty e^{iA\tau + iz\tau} d\tau$ if $\Im z > \omega$. In particular, if $r > \omega$ then $\|(A + ir)^{-1}\|_{B(\mathscr{G}^*)} \leq M(r - \omega)^{-1}$. Since $A(A + ir)^{-1} = I - ir(A + ir)^{-1}$, if we fix $r_0 > \omega$ and consider $r \geq r_0$, we shall have $r\|(A + ir)^{-1}\|_{B(\mathscr{G}^*)} + \|A(A + ir)^{-1}\|_{B(\mathscr{G}^*)} \leq \text{const.} < \infty$. We recall from Chapter 3 that we have a well defined bounded functional calculus for A in $\mathscr{G}^*$: $\varphi(A)$ is defined as an element of $B(\mathscr{G}^*)$ for $\varphi \in \mathscr{M}^\omega(\mathbb{R})$. Moreover, we have $\|\varphi(\varepsilon A)\|_{B(\mathscr{G}^*)} \leq \text{const.} < \infty$ for such a φ and all $\varepsilon \in (0, 1]$. The function $\varphi(x) = \sin x + i\frac{\sin x}{x}$ is admissible, since its Fourier transform is a measure of compact support. If $0 < \varepsilon \leq r_0^{-1}$, we shall have $\sin A\varepsilon = \varepsilon A(\varepsilon A + i)^{-1}\varphi(\varepsilon A)$, so

$$\|(\sin A\varepsilon)T\|_{B(\mathscr{G},\mathscr{G}^*)} \leq \|\varphi(\varepsilon A)\|_{B(\mathscr{G}^*)} \cdot \|\varepsilon A(\varepsilon A + i)^{-1}T\|_{B(\mathscr{G},\mathscr{G}^*)}.$$

Hence, if (7.5.25) is fulfilled with Λ replaced by A and $\int_1^\infty$ replaced by $\int_{r_0}^\infty$, then (7.5.24) is satisfied, so $T \in \mathscr{C}^{0,1}(A; \mathscr{G}, \mathscr{G}^*)$.

(ii) Now let Λ be as in the statement of the proposition. We take $r \geq r_0 > \omega$ and use the identity $I = \Lambda(\Lambda + ir)^{-1} + ir(\Lambda + ir)^{-1}$ and the property $D(\Lambda) \subset D(A; \mathscr{G}^*)$ in order to obtain the next identity on $\mathscr{G}^*$:

$$A(A + ir)^{-1} = A(A + ir)^{-1}\Lambda(\Lambda + ir)^{-1}$$
$$+ ir(A + ir)^{-1}A(\Lambda + ir_0)^{-1}(\Lambda + ir_0)(\Lambda + ir)^{-1}.$$

We have $\|A(\Lambda + ir_0)^{-1}\|_{B(\mathscr{G}^*)} < \infty$ by the closed graph theorem. In view of the estimates obtained before for the resolvent of A, there is $C < \infty$ such that:

$$\|A(A + ir)^{-1}T\|_{B(\mathscr{G},\mathscr{G}^*)} \leq C\|\Lambda(\Lambda + ir)^{-1}T\|_{B(\mathscr{G},\mathscr{G}^*)} + C\|(\Lambda + ir)^{-1}T\|_{B(\mathscr{G},\mathscr{G}^*)}.$$

From the assumption $\|(\Lambda + ir)^{-1}\|_{B(\mathscr{G}^*)} \leq cr^{-1}$, we see that (7.5.25) implies a similar estimate with Λ replaced by A, hence $T \in \mathscr{C}^{0,1}(A; \mathscr{G}, \mathscr{G}^*)$.

(iii) The last part of the proposition is a consequence of the inequality (3.5.25) (with $s = 0$ and A replaced by $-\Lambda$) and of Lemma 3.5.12 (which allows us to reduce ℓ to 1). $\square$

We shall deduce now a criterion of a similar nature for an operator $S \in B(\mathscr{G}, \mathscr{G}^*)$ to be of class $\mathscr{C}^{1,1}(A; \mathscr{G}, \mathscr{G}^*)$. We start with an easily verified identity:

$$e^{iA\tau}Se^{-iA\tau} + e^{-iA\tau}Se^{iA\tau} - 2S =$$
$$= [e^{iA\tau} - 2 + e^{-iA\tau}]Se^{iA\tau} + e^{iA\tau}S[e^{iA\tau} - 2 + e^{-iA\tau}] - 2(e^{iA\tau} - I)S(e^{iA\tau} - I)$$
$$= -4\left(\sin\frac{A\tau}{2}\right)^2 Se^{iA\tau} - 4e^{iA\tau}S\left(\sin\frac{A\tau}{2}\right)^2 + 8e^{iA\tau/2}\left(\sin\frac{A\tau}{2}\right)S\left(\sin\frac{A\tau}{2}\right)e^{iA\tau/2}.$$

Thus the following inequality is *a sufficient condition for a symmetric operator* $S : \mathscr{G} \to \mathscr{G}^*$ *to be of class* $\mathscr{C}^{1,1}(A; \mathscr{G}, \mathscr{G}^*)$:

$$(7.5.27) \qquad \int_0^1 [\|(\sin A\tau)^2 S\|_{B(\mathscr{G}, \mathscr{G}^*)} + \|(\sin A\tau)S(\sin A\tau)\|_{B(\mathscr{G}, \mathscr{G}^*)}] \frac{d\tau}{\tau^2} < \infty.$$

The argument of part (i) of the proof of Proposition 7.5.7 shows that this is a consequence of:

$$(7.5.28) \qquad \int_{r_0}^\infty [\|(A(A + ir)^{-1})^2 S\|_{B(\mathscr{G}, \mathscr{G}^*)}$$

$$+ \|A(A + ir)^{-1} S A(A + ir)^{-1}\|_{B(\mathscr{G}, \mathscr{G}^*)}] dr < \infty.$$

Here r_0 is any sufficiently large positive number. Let $\varphi(x) = (e^{ix} - 1)x^{-1}(x + i)$; then φ is the Fourier transform of a measure of compact support and $(e^{iA\varepsilon} - I)^2 = \varphi(\varepsilon A)^2 [\varepsilon A(\varepsilon A + i)^{-1}]^2$. Hence the finiteness of the first integral in (7.5.28) implies $S \in B(\mathscr{G}, \mathscr{G}_{1,1}^*)$. So S improves decay at infinity in the spectral representation of A by one power at least. On the other hand, if $S \in B(\mathscr{G}, \mathscr{G}_s^*)$ for some $s > 1$, it is easily seen that both integrals in (7.5.28) are finite (this is done by a duality-interpolation argument). The next theorem is a refined version of this observation.

THEOREM 7.5.8. *Let* Λ *be a self-adjoint operator in* $\mathscr{H}$, *bounded from below by a strictly positive constant and such that:*
(1) $e^{i\Lambda\tau}\mathscr{G} \subset \mathscr{G}$ *for all* $\tau \in \mathbb{R}$ *and* $\|e^{i\Lambda\tau}\|_{B(\mathscr{G})} \le c\langle\tau\rangle^m$ *for some finite constants* c, m;
(2) *the operator* $\Lambda^{-2} A^2 : D(A^2; \mathscr{G}) \to \mathscr{G}$ *extends to a continuous operator in* $\mathscr{G}$.

If $S : \mathscr{G} \to \mathscr{G}^*$ *is a symmetric operator such that there is* $\theta \in C_0^\infty((0, \infty))$ *not identically zero with*

$$(7.5.29) \qquad \int_1^\infty \left\|\theta\left(\frac{\Lambda}{r}\right)S\right\|_{B(\mathscr{G}, \mathscr{G}^*)} dr < \infty,$$

then $S \in \mathscr{C}^{1,1}(A; \mathscr{G}, \mathscr{G}^*)$.

PROOF. (i) We first discuss some consequences of the hypothesis (1). As we know from the general theory presented in Section 3.2, the group induced by $\{e^{i\Lambda\tau}\}_{\tau\in\mathbb{R}}$ in $\mathscr{G}$ is of class C_0, and by our hypothesis it is polynomially bounded. We shall use the notation $D(\Lambda^k; \mathscr{G})$ for the spaces of the discrete Sobolev scale associated to this group (only $k = 1, 2, \infty$ will be used). So $D(\Lambda^k; \mathscr{G})$ is the set of $g \in \mathscr{G}$ such that $\tau \mapsto e^{i\Lambda\tau} g \in \mathscr{G}$ is of class C^k. We also have $D(\Lambda; \mathscr{G}) = \{g \in \mathscr{G} \cap D(\Lambda) \mid \Lambda g \in \mathscr{G}\}$ and the generator of the group $e^{i\Lambda\tau}|_{\mathscr{G}}$ is just $\Lambda|_{D(\Lambda;\mathscr{G})}$. For the clarity of the next argument (and also later on when needed), we denote by Λ_1 the operator $\Lambda|_{D(\Lambda;\mathscr{G})}$, considered as an operator in $\mathscr{G}$; hence Λ_1 is just the generator (in $\mathscr{G}$) of the C_0-group $\{e^{i\Lambda\tau}|_{\mathscr{G}}\}_{\tau\in\mathbb{R}}$. We know that $D(\Lambda^\infty; \mathscr{G}) \equiv \cap_{k=1}^\infty D(\Lambda_1^k)$ is a dense subspace of $\mathscr{G}$ (hence of $\mathscr{H}$ too) and that for each $\varphi \in C_{\text{pol}}^\infty(\mathbb{R})$ we may naturally associate to Λ_1 an operator $\varphi(\Lambda_1) : D(\Lambda^\infty; \mathscr{G}) \to \mathscr{G}$ (see §3.6.2). On the other hand, Λ being a self-adjoint operator in $\mathscr{H}$, we can also construct the (unbounded) operator $\varphi(\Lambda)$ in $\mathscr{H}$ (by using the spectral measure of Λ, for example). By using for example (3.6.15), it is easily shown that $\varphi(\Lambda_1)g = \varphi(\Lambda)g$ if $g \in D(\Lambda^\infty; \mathcal{G})$. If φ is a bounded function, $\varphi(\Lambda)$ is a bounded operator in $\mathcal{H}$. Now, from Theorem 3.7.10 and the above remarks it follows that, *if $\varphi \in BC^\infty(\mathbb{R})$, then $\varphi(\Lambda)\mathcal{G} \subset \mathcal{G}$ and there are numbers $C < \infty$, $k \in \mathbb{N}$ such that $\|\varphi(\Lambda)\|_{B(\mathcal{G})} \leq C\|\varphi\|_{BC^k}$.* From now on we shall not distinguish between the operators $\varphi(\Lambda_1)$ and $\varphi(\Lambda)$.

The preceding assertion allows us to prove some facts which are important for the rest of the proof of the theorem. First, we show that *the operators Λ_1 (in $\mathcal{G}$) and Λ (in $\mathcal{H}$) have the same spectrum.* Since $\{e^{i\Lambda_1\tau}\}$ is of polynomial growth, we have $\sigma(\Lambda_1) \subset \mathbb{R}$. Indeed, if $\Im z < 0$ for example, we may take $(\Lambda_1 - z)^{-1} = -i\int_0^\infty e^{i(\Lambda_1-z)\tau}d\tau$. Since Λ is self-adjoint, we also have $\sigma(\Lambda) \subset \mathbb{R}$. Hence it is enough to prove that a real number z belongs to the resolvent set of Λ_1 if and only if it belongs to the resolvent set of Λ. Clearly we may assume $z = 0$. We have $0 \notin \sigma(\Lambda_1)$ if and only if $\Lambda_1 : D(\Lambda_1) \to \mathcal{G}$ is bijective; then $\Lambda_1^{-1} \in B(\mathcal{G})$. Let $\Lambda_{-1} = \Lambda_1^*$, this is a closed densely defined operator in $\mathcal{G}^*$ and zero is outside its spectrum, because $(\Lambda_1^{-1})^* = \Lambda_{-1}^{-1} \in B(\mathcal{G}^*)$. It is trivial to prove that $\Lambda_1 \subset \Lambda \subset \Lambda_{-1}$; in fact, it is clear that Λ_{-1} is the generator of the C_0-group induced in $\mathcal{G}^*$ by $\{e^{i\Lambda\tau}\}_{\tau\in\mathbb{R}}$. The injectivity of $\Lambda : D(\Lambda) \to \mathcal{H}$ is now obvious, but not its surjectivity. Observe that $\Lambda_1 \subset \Lambda_{-1}$ implies that $\Lambda_1^{-1} \subset \Lambda_{-1}^{-1}$ hence, if $S = \Lambda_1^{-1}$, then $S : \mathcal{G} \to \mathcal{G}^*$ is a symmetric operator which belongs to $B(\mathcal{G}) \cap B(\mathcal{G}^*)$. By interpolation one gets $S \in B(\mathcal{H})$. Now let $h \in \mathcal{H}$; then there is $f \in D(\Lambda_{-1})$ such that $h = \Lambda_{-1}f$. Then $f = \Lambda_{-1}^{-1}h = Sh \in \mathcal{H}$. Finally, making a first order Taylor expansion of $e^{i\Lambda\tau}f$ in $\mathcal{G}^*$, one sees that $\frac{d}{d\tau}e^{i\Lambda\tau}f$ exists in $\mathcal{H}$ and is equal to ih, so $f \in D(\Lambda)$ and $\Lambda f = h$. In conclusion, $\Lambda : D(\Lambda) \to \mathcal{H}$ is bijective, i.e. zero does not belong to the spectrum of Λ. Reciprocally, assume that zero is not in $\sigma(\Lambda)$. Let $\varphi \in C^\infty(\mathbb{R})$ such that $\varphi(x) = x^{-1}$ outside a neighbourhood of zero contained in the resolvent set of Λ. Then the restriction of $\Lambda^{-1} = \varphi(\Lambda)$ to $\mathcal{G}$ belongs to $B(\mathcal{G})$ and obviously coincides with Λ_1^{-1}.

Notice that a little variation of the preceding argument shows not only that the spectrum of Λ is the same in each of the spaces $\mathcal{G}$, $\mathcal{H}$, $\mathcal{G}^*$ but also that $\|(\Lambda - z)^{-1}\|_{B(\mathcal{G})} \leq c \max\{\delta^{-1}, \delta^{-1-k}\}$, where $\delta = \text{dist}(z, \sigma(\Lambda))$ (take $\varphi(x) = (x - z)^{-1}$ in a neighbourhood of $\sigma(\Lambda)$ and use $\|\varphi(\Lambda)\|_{B(\mathcal{G})} \leq C\|\varphi\|_{BC^k}$).

By one of the assumptions of the theorem, there is a constant $a > 0$ such that $\Lambda \geq a$ as an operator in $\mathcal{H}$. By what we have just proven, $\sigma(\Lambda_1) \subset [a, \infty)$. Let r be a real positive number and $M = \Lambda(\Lambda + r)^{-1} = I - r(\Lambda + r)^{-1} \in B(\mathcal{G})$. By an elementary version of the spectral mapping theorem, we have $\sigma(M) \subset [a(a+r)^{-1}, 1]$ (independently of the space $\mathcal{G}$, $\mathcal{H}$ or $\mathcal{G}^*$ in which M is considered). So one may define an operator $\log M \in B(\mathcal{G})$ by Cauchy's integral formula such that $M = e^{\log M} = \sum_{n=0}^\infty \frac{1}{n!}(\log M)^n$ (an absolutely convergent series in $B(\mathcal{G})$). One can also proceed in the following way. Let $0 < a_0 < a$ and $\eta \in C^\infty(\mathbb{R})$ with $\eta \geq a_0$, $\eta(x) = a_0$ if $x \leq a_0/2$ and $\eta(x) = x$ if $x \geq a$. Then $\eta(\Lambda) = \Lambda$ as operators in $\mathcal{H}$. Let $\varphi(x) = \log[\eta(x)(\eta(x) + r)^{-1}] = \log\eta(x) - \log(\eta(x) + r)$; then $\varphi(\Lambda) = \log[\Lambda(\Lambda + r)^{-1}]$ as operators in $\mathcal{H}$. Clearly $\varphi \in BC^\infty(\mathbb{R})$ and $|\varphi^{(j)}(x)| \leq C_j$ (with constants independent of r), so $\varphi(\Lambda) \in B(\mathcal{G})$. Since $\Lambda(\Lambda + r)^{-1} = \exp\varphi(\Lambda)$ in $\mathcal{H}$, if we define $\log M = \varphi(\Lambda)$, we shall have $M = \exp\varphi(\Lambda)$ in $B(\mathcal{G})$ too. Then we may define $M^z = \exp[z\log M]$ for all $z \in \mathbb{C}$. We get a holomorphic function $z \mapsto M^z \in B(\mathcal{G})$ on $\mathbb{C}$ and $||M^z||_{B(\mathcal{G})} \le \exp(c_r|z|)$, with $c_r = ||\log M||_{B(\mathcal{G})} < \infty$.

The last fact we shall need is an estimate on $||M^z||_{B(\mathcal{G})}$ independent of r for purely imaginary z. If $-1 \le y \le 1$, we have $M^{iy} = \varphi(\Lambda)^{iy} = \eta(\Lambda)^{iy}(\eta(\Lambda)+r)^{-iy}$. Let $\psi(x) = (\eta(x) + r)^{iy}$; clearly $|\psi^{(j)}(x)| \le c_j$ for all $j \in \mathbb{N}$, with constants independent of $r \ge 0$ and $y \in [-1, +1]$. Hence $(\Lambda + r)^{iy} = (\eta(\Lambda) + r)^{iy} = \psi(\Lambda) \in B(\mathcal{G})$ with norm bounded by a constant independent of r and y. This implies $||M^{iy}||_{B(\mathcal{G})} \le c < \infty$, independently of $r \ge 0$ and $-1 \le y \le 1$. Since $M^{iny} = (M^{iy})^n$, it is clear that we can find another constant c, independent of $r \ge 0$, such that $||M^{iy}||_{B(\mathcal{G})} \le ce^{c|y|}$ for $y \in \mathbb{R}$.

(ii) Let us present now several consequences of the assumption (2) of the theorem. Notice that in part (i) of the proof we have shown that $\Lambda^{-1} \in B(\mathcal{G}) \cap B(\mathcal{G}^*)$ (from now on we do not distinguish between the operators Λ_1, Λ and Λ_{-1}). By using the techniques of the proof of Theorem 3.3.23, it is easy to show that $\Lambda^{-j}\mathcal{G} = D(\Lambda^j; \mathcal{G})$, $\Lambda^{-j}\mathcal{G}^* = D(\Lambda^j; \mathcal{G}^*)$ for each $j \in \mathbb{N}$. By the hypothesis (2), there is $c < \infty$ such that $||\Lambda^{-2}A^2g||_{\mathcal{G}} \le c||g||_{\mathcal{G}}$ for $g \in D(A^2; \mathcal{G})$. Moreover, $||\Lambda^{-2}g||_{\mathcal{G}} \le c||g||_{\mathcal{G}}$ for all $g \in \mathcal{G}$. By making a second order Taylor expansion, we obtain $iA = e^{iA} - I + A^2\varphi(A)$, with $\varphi(A) = \int_0^1 e^{iA\tau}(1 - \tau)d\tau \in B(\mathcal{G})$. This implies that $||\Lambda^{-2}Ag||_{\mathcal{G}} \le c||g||_{\mathcal{G}}$ for $g \in D(A^2; \mathcal{G})$. Recall that $D(A^k; \mathcal{G}) = \mathcal{G}_{-k}$ for $k \in \mathbb{N}$. From Remark 3.3.11, we then see that $||\Lambda^{-2}f||_{\mathcal{G}} \le c||f||_{\mathcal{G}_{-2}}$ for a constant c and all $f \in \mathcal{G}$, i.e. $\Lambda^{-2} \in B(\mathcal{G}_{-2}, \mathcal{G})$. Since Λ^{-2} is symmetric and $(\mathcal{G}_{-2})^* = \mathcal{G}_2^*$ (Theorem 3.3.28), this is equivalent to $\Lambda^{-2} \in B(\mathcal{G}^*, \mathcal{G}_2^*)$. But $\mathcal{G}_2^* = D(A^2; \mathcal{G}^*)$, hence we may write this as $D(\Lambda^2; \mathcal{G}^*) \subset D(A^2; \mathcal{G}^*)$. By interpolating and using Proposition 3.7.7, we get $D(\Lambda; \mathcal{G}^*) \subset D(A; \mathcal{G}^*)$, i.e. $\Lambda^{-1} \in B(\mathcal{G}^*, \mathcal{G}_1^*)$. By taking again the adjoints, we obtain $\Lambda^{-1} \in B(\mathcal{G}_{-1}, \mathcal{G})$. In conclusion, the operators $A\Lambda^{-1}$ and $A^2\Lambda^{-2}$ are bounded in $\mathcal{G}^*$ (we have $\Lambda^{-1}\mathcal{G}^* \subset D(A; \mathcal{G}^*)$ and $\Lambda^{-2}\mathcal{G}^* \subset D(A^2; \mathcal{G}^*)$), while the operators $\Lambda^{-1}A$, $\Lambda^{-2}A^2$ are bounded in $\mathcal{G}$ (a priori, they are defined only on $D(A; \mathcal{G})$ and $D(A^2; \mathcal{G})$ respectively).

(iii) In this step of the proof we show that (7.5.28) is a consequence of

$$(7.5.30) \qquad \int_0^\infty \left\{ \left\|\left(\frac{\Lambda}{\Lambda + r}\right)^2 S\right\|_{B(\mathcal{G}, \mathcal{G}^*)} + \left\|\frac{\Lambda}{\Lambda + r}S\frac{\Lambda}{\Lambda + r}\right\|_{B(\mathcal{G}, \mathcal{G}^*)} \right\}dr < \infty.$$

We use the notations $A_r = A(A + ir)^{-1}$ and $\Lambda_r = \Lambda(\Lambda + r)^{-1}$, where $r \ge r_0$ always. From the identity $I = \Lambda_r + r(\Lambda + r)^{-1}$ we get $I = \Lambda_r^2 + 2r(\Lambda + r)^{-1}\Lambda_r + r^2(\Lambda + r)^{-2}$. This is valid on $\mathcal{G}$, $\mathcal{H}$ and $\mathcal{G}^*$, and it implies

$$A_r^2 = A_r^2\Lambda_r^2 + 2r(A + ir)^{-1}A_rA(\Lambda + r)^{-1}\Lambda_r + [r(A + ir)^{-1}]^2A^2(\Lambda + r)^{-2}$$
$$= \{A_r^2 + 2r(A + ir)^{-1}A_rA\Lambda^{-1} + [r(A + ir)^{-1}]^2A^2\Lambda^{-2}\}\Lambda_r^2.$$

The norm in $B(\mathcal{G}^*)$ of the operator $\{\ldots\}$ is bounded by a constant independent of $r \ge r_0$. Hence the finiteness of the first integral in (7.5.30) implies that of the first integral in (7.5.28). In order to treat the second integral, we start from the following identity valid on $\mathcal{G}^*$:

$$A_r = [A_r + r(A + ir)^{-1}A\Lambda^{-1}]\Lambda_r \equiv B_r\Lambda_r.$$

We have (on $\mathcal{G}^*$) a similar identity : $A_{-r} = B_{-r}\Lambda_r$. By taking adjoints and observing that $(A_{-r})^* = A_r$ (as operators in $\mathcal{G}$), we get $A_r = \Lambda_r B_{-r}^*$ in $\mathcal{G}$. Since $\|B_{\pm r}\|_{B(\mathcal{G}^*)} \leq c$ for $r \geq r_0$, we get also $\|B_{-r}^*\|_{B(\mathcal{G})} \leq c$ and so:

$$\|A_r S A_r\|_{B(\mathcal{G},\mathcal{G}^*)} \leq c^2 \|\Lambda_r S \Lambda_r\|_{B(\mathcal{G},\mathcal{G}^*)}.$$

In conclusion, (7.5.30) implies that $S \in \mathscr{C}^{1,1}(A; \mathcal{G}, \mathcal{G}^*)$.

(iv) Our purpose now is to prove that the second term in (7.5.30) is dominated by the first one. In part (i) of the proof we have introduced a holomorphic family $M^z = [\Lambda(\Lambda + r)^{-1}]^z$ of operators in $B(\mathcal{G})$ such that $\|M^z\|_{B(\mathcal{G})} \leq \exp[c_r|z|]$ and $\|M^{iy}\|_{B(\mathcal{G})} \leq ce^{c|y|}$ for $y \in \mathbb{R}$, with a constant c independent of $r \geq 0$. Let $g \in \mathcal{G}$ and let F be the holomorphic function defined on $\mathbb{C}$ by:

$$F(z) = \langle M^{z^*}g, SM^{2-z}g\rangle e^{(z-1)^2}.$$

We consider only $z = x + iy$ with $0 \leq x \leq 2$. Then F is a continuous function on the closed strip $0 \leq \Re z \leq 2$, it is holomorphic in its interior, and it satisfies a bound of the form $|F(x + iy)| \leq \text{const.}\, e^{-y^2/2}$. By the maximum modulus principle :

$$|F(1)| = |\langle Mg, SMg\rangle| \leq \max\Big\{\sup_{y\in\mathbb{R}}|F(iy)|, \sup_{y\in\mathbb{R}}|F(2+iy)|\Big\}$$

$$\leq \Big[\sup_{y\in\mathbb{R}} e^{1-y^2}\|M^{iy}\|_{B(\mathcal{G})}\Big] \cdot \|M^2 S\|_{B(\mathcal{G},\mathcal{G}^*)} \cdot \|g\|_{\mathcal{G}}^2.$$

The bracket here is bounded by a constant independent of r. In conclusion, there is $c < \infty$ such that for $r \geq 0$ and $S : \mathcal{G} \to \mathcal{G}^*$ symmetric:

$$(7.5.31)\ \ \|\Lambda(\Lambda + r)^{-1}S\Lambda(\Lambda + r)^{-1}\|_{B(\mathcal{G},\mathcal{G}^*)} \leq c\|[\Lambda(\Lambda + r)^{-1}]^2 S\|_{B(\mathcal{G},\mathcal{G}^*)}.$$

(v) It remains to show that (7.5.29) implies the finiteness of the first integral in (7.5.30). For this use (3.5.26) with $s = 1$ and ℓ large enough (and with A replaced by Λ), and then apply Lemma 3.5.12 to get to $\ell = 2$ (note that $\|(\Lambda + r)(\Lambda - ir)^{-1}\|_{B(\mathcal{G})} \leq \text{const.}$ for $r \geq 0$). $\square$

7.6. The Limiting Absorption Principle for Some Classes of Pseudodifferential operators

7.6.1. The main example that one should have in mind when thinking about the conjugate operator method is extremely simple: H is the operator Q of multiplication by the independent variable x in the Hilbert space $\mathscr{H} = L^2(\mathbb{R})$, while the conjugate operator is $A = -P = i\frac{d}{dx}$. Since $(e^{iA\tau}f)(x) = f(x - \tau)$, i.e. $\{e^{-iA\tau}\}$ is just the translation group acting in $L^2(\mathbb{R})$, we have $e^{-iA\tau}Qe^{iA\tau} = Q + \tau$, or $[Q, iA] = I$. In particular, Q is of class $C^\infty(A)$ and $\mu^A(Q) = \mathbb{R}$. Since Q does not have a spectral gap, we apply Theorem 7.5.4 with $\mathcal{G} = \mathscr{H}_{1/2}(\mathbb{R})$ (the form domain of Q). Then $\mathcal{G}^* = \mathscr{H}_{-1/2}(\mathbb{R})$, $D(A; \mathcal{G}^*) = \mathscr{H}_{-1/2}^1(\mathbb{R})$ (a weighted Sobolev space) and $(D(A; \mathcal{G}^*), \mathcal{G}^*)_{1/2,1} = \mathscr{H}_{-1/2}^{1/2,1}(\mathbb{R})$ is a weighted Besov space. It follows that, for each $\lambda \in \mathbb{R}$, the operators $(Q - \lambda \mp i0)^{-1}$: $\mathscr{H}_{-1/2}^{1/2,1}(\mathbb{R}) \to \mathscr{H}_{1/2}^{-1/2,\infty}(\mathbb{R})$ are well defined as weak* limits of $(Q - \lambda \mp i\mu)^{-1}$ as $\mu \to +0$. In particular, the distributions $(x - \lambda \mp i0)^{-1}$ belong to $\mathscr{H}_{\text{loc}}^{-1/2,\infty}(\mathbb{R})$, and this fact is optimal in the Besov scale (even if one considers only its imaginary part, which is $\pi\delta(x - \lambda)$; see page 50 in [P1]), and is *not* quite easy to prove by direct methods.

One may consider natural generalizations of this example by extending the framework in several steps. First, one may replace $\mathbb{R}$ by an open subset $\Omega \subset \mathbb{R}$, but keep the same operator $H = Q$ in $L^2(\Omega)$. Then the preceding choice for A will not work, since Ω is not invariant under translations. We have already explained during the proof of Proposition 7.2.14 how to overcome this problem: it is enough to take $A = -\frac{1}{2}[F(Q)P + PF(Q)]$, where $F : \Omega \to \mathbb{R}$ is of class C^1, $F(x) > 0$ if $x \in \Omega$ and $F(x) \to 0$ rapidly enough when x tends to the boundary of Ω. A second and much less trivial extension is obtained by considering in $L^2(\Omega)$ the operator $H = h(Q)$ with $h : \Omega \to \mathbb{R}$. This situation has been discussed in some detail in §7.1.4: we saw that H can have quite bad spectral properties if h is not locally Lipschitz and that, even if h is of class C^1 with a derivative satisfying $c \leq h'(x) \leq c^{-1}$ for some strictly positive constant c, the strong limiting absorption principle may break down in a very drastic way. In the example constructed in Appendix 7.B, the function h has, moreover, the following supplementary regularity properties: h is locally of Besov class $B_\infty^{1,q}(\mathbb{R})$ for each $q > 1$ and h' is absolutely continuous. Now, if A is as above, then $[h(Q), iA] = F(Q)h'(Q) \geq cF(Q)$. By choosing F conveniently, we get an operator $h(Q)$ of class $C_u^1(A) \cap \mathscr{C}^{1,q}(A)$ for all $q > 1$ and with $\mu^A(h(Q)) = \mathbb{R}$. However, we can arrange things such that $\lim_{\mu \to +0}\langle u, (h(Q) - \lambda \mp i\mu)^{-1}v \rangle$ does not exist for each rational number λ in a given interval included in the spectrum of $h(Q)$ and for all $u, v \in C_0^\infty(\Omega)$ with $u(h^{-1}(\lambda)) \neq 0 \neq v(h^{-1}(\lambda))$. We may assume $F \in C^\infty(\Omega)$, hence $C_0^\infty(\Omega) \subset \cap_{k=1}^\infty D(A^k)$. In conclusion, we see that we need more regularity from h in order to get the strong limiting absorption principle in $C_0^\infty(\Omega)$. We shall prove below that, even if Ω is an open subset of $\mathbb{R}^n$, we can get the best limiting absorption principle (suggested by the case $n = 1$, $h(Q) = Q$ treated above) if h is locally of Besov class $B_\infty^{1,1}(\Omega)$.

7.6.2. Let Ω be an open subset of a n-dimensional euclidean space X and let H be the operator in the Hilbert space $\mathscr{H} = L^2(\Omega)$ of multiplication by a real Borel function h on Ω . We have presented at the beginning of §7.1.4 some elementary facts concerning the spectral properties of H. In order to describe some deeper facts we shall impose stronger regularity conditions on h, namely we assume from now on that *h is at least a locally Lipschitz function* (note that this is a regularity property of H with respect to the translation group, which acts in $L^2(X)$!). Then h is differentiable at (Lebesgue-) almost all points of Ω (this is Rademacher's Theorem, cf. [Sm]), and we shall denote by $h'(x)$ its differential (a vector in X) at $x \in X$, when it exists. We shall use in our arguments below a beautiful formula known in geometric measure theory as the *co-area formula*: for any Borel sets $M \subset \Omega$ and $B \subset \mathbb{R}$ we have

$$(7.6.1) \qquad \int_{M \cap h^{-1}(B)} |h'(x)|dx = \int_B \mathcal{H}^{n-1}(M \cap h^{-1}(y))dy.$$

Here $\mathcal{H}^{n-1}$ denotes $(n-1)$-dimensional Hausdorff measure [7] in X. We point out the following easy consequence of (7.6.1) (see Theorems 3.2.3 (2) and 3.2.12 in [Fe]): for each Borel function $f : \Omega \to [0, \infty]$ and each Borel set $B \subset \mathbb{R}$ one has

$$(7.6.2) \qquad \int_{h^{-1}(B)} f(x)|h'(x)|dx = \int_B dy \int_{h^{-1}(y)} f(x)d\mathcal{H}^{n-1}(x).$$

If we take in (7.6.1) a set B of (Lebesgue) measure zero, the right-hand side of the identity is equal to zero. This implies the assertions (i) and (iii) of the next proposition. To prove (ii) write, for any $N \subset I$ of measure zero, $\{h \in N\} = \{h \in N, h' = 0\} \cup \{h \in N, h' \neq 0\}$ (in slightly formal notations); then the first term on the r.h.s. has measure zero by hypothesis, while the second one has measure zero by the first assertion of the proposition. So we have:

PROPOSITION 7.6.1. *Let $\Omega \subset X$ be an open set and $h : \Omega \to \mathbb{R}$ a locally Lipschitz function. Then, for any set $N \subset \mathbb{R}$ of Lebesgue measure zero, we have $h'(x) = 0$ (Lebesgue-) almost everywhere on the set $h^{-1}(N)$. In particular, if H is the operator of multiplication by h in $L^2(\Omega)$, then the following assertions are true:*

(i) *if $h'(x) \neq 0$ a.e. on Ω, then H has purely absolutely continuous spectrum;*

(ii) *more generally, if $J \subset \mathbb{R}$ is a set such that $h'(x) \neq 0$ for almost every x with $h(x) \in J$, then the spectrum of H in J is purely absolutely continuous;*

(iii) *assume $h \in C^1(\Omega)$ and define the set of critical values of h by:*

$$(7.6.3) \qquad \kappa(h) = \{\lambda \in \mathbb{R} \mid \exists x \in \Omega \text{ such that } h(x) = \lambda \text{ and } h'(x) = 0\}.$$

Then H has purely absolutely continuous spectrum in $\mathbb{R} \setminus \kappa(h)$.

In conclusion, if $h \in C^1(\Omega)$, the singular spectrum of H is contained in the set $\kappa(h) \subset \mathbb{R}$, *the critical set of h.* This set is quite small in general: it has measure zero if $h \in C^n(\Omega)$ and is finite if h is a polynomial (the first assertion is the Morse-Sard theorem, cf. 1.4.6 in [N] or Ch.13 in [Sm], while the second assertion is proved in Section 1.4.3 of [H]). Notice, however, that if h is only of class C^k with $1 \leq k < n$, then $\kappa(h)$ could contain an open non-void interval (cf. [N], loc. cit.). $\kappa(h)$ is also closed under fairly general conditions on h, e.g. if $|h(x)| + |h'(x)| \to \infty$ when $x \to \partial\Omega$.

---

[7]For $n = 1$, this is just the number of points in the set, with the value $+\infty$ admitted. If $n > 1$, the only things one has to know about $\mathcal{H}^{n-1}$ is that it is a Borel measure on X, which when restricted to a C^1-submanifold Y of X of dimension $n - 1$, coincides with the riemannian measure on Y induced by the euclidean measure on X (cf. 3.2.3 in [Fe]). The version (7.6.1) of the co-area formula can be deduced from Theorems 3.2.3 (1) (for $n = 1$) and 3.2.11 (for $n > 1$) of [Fe], by observing first that it is enough to prove it for M with compact closure $\overline{M}$ included in Ω, then taking the function f from [Fe] equal to h on a neighbourhood of $\overline{M}$, and finally replacing M by $M \cap f^{-1}(B) = M \cap h^{-1}(B)$ in the formulas proved by Federer. One may also apply directly the remark at the end of 3.2.1 of [Fe]. Since the treatment of Federer is extremely technical, the reader should consult Section 2.7 of [Zi] for a somewhat simpler discussion of the co-area formula (and also the first chapter of the book for an elementary description of Hausdorff measures). We mention only that the result is non-trivial even if $h \in C^\infty$ and becomes highly non-trivial if $h \in C^k$ with $k < n$ (this is because the Morse-Sard theorem breaks down for such functions h).

The co-area formula allows us to give an explicit description of $dE_\lambda/d\lambda$ (see (7.1.17)) for λ outside the critical set. For simplicity, let h be of class C^1 on Ω and let B be a Borel subset of $\mathbb{R}$, disjoint from $\kappa(h)$; then $h'(x) \neq 0$ on $h^{-1}(B)$, so we may replace $f(x)$ in (7.6.2) by $|g(x)|^2|h'(x)|^{-1}$, where $g : \Omega \to \mathbb{C}$ is a square-integrable function. Clearly, we get

$$(7.6.4) \qquad \langle g, E(B)g \rangle = \int_B d\lambda \int_{h^{-1}(\lambda)} |g(x)|^2 \frac{d\mathcal{H}^{n-1}(x)}{|h'(x)|}.$$

Notice that $h^{-1}(\lambda)$ is a $(n-1)$-dimensional submanifold of class C^1 of Ω (since $h'(x) \neq 0$ if $h(x) = \lambda$), in particular $d\mathcal{H}^{n-1}(x)$ is the usual "surface" measure induced by $\mathbb{R}^n$ on this submanifold. Formally, we shall have:

$$(7.6.5) \qquad \frac{d}{d\lambda}\langle g, E_\lambda g \rangle = \int_{h^{-1}(\lambda)} |g(x)|^2 \frac{d\mathcal{H}^{n-1}(x)}{|h'(x)|}, \quad \lambda \notin \kappa(h).$$

We shall not give a detailed justification of this formula, although this is not difficult. We mentioned it only because it plays a role in some treatments of scattering theory (see Ch. XIV in [H], for example formulas (14.3.4) and (14.6.1)''; cf. also Theorem 6.1.6 in [H]).

In view of the preceding results, it is reasonable to expect that the strong limiting absorption principle for H holds outside $\kappa(h)$ in spaces containing $C_0^\infty(\Omega) \equiv \mathcal{D}(\Omega)$. In particular this would imply, for $\lambda \notin \kappa(h)$, the existence of the limits $\lim_{\mu \to +0}(h(x) - \lambda \mp i\mu)^{-1}$ in the sense of distributions on Ω, i.e. (weakly) in $\mathcal{D}^*(\Omega)$; see the detailed discussion in §7.1.4. But, as explained before, for this we have to require h to be more regular than just of class C^1. We shall prove below that the regularity assumption $h \in B_{\infty,\mathrm{loc}}^{1,1}(\Omega)$ is sufficient; and the example mentioned in §.7.6.1 shows that this condition is optimal in the Besov scale associated to L^∞. We have denoted by $B_{\infty,\mathrm{loc}}^{1,1}(\Omega)$ the space of functions on Ω which are *locally of Besov class* $B_\infty^{1,1}$ on Ω, i.e. this is the space of distributions $h \in \mathcal{D}^*(\Omega)$ such that $\theta h \in B_\infty^{1,1}(X)$ for each $\theta \in C_0^\infty(\Omega)$. We also recall that $B_\infty^{s,p}(X)$ ($s \in \mathbb{R}$, $1 \leq p \leq \infty$) are the spaces of the Besov scale associated to the translation group acting in $L^\infty(X)$. So, $f \in L^\infty(X)$ belongs to $B_\infty^{1,1}(X)$ if

$$\int_{y \in X, |y| < 1} \|e^{iP \cdot y}f - 2f + e^{-iP \cdot y}f\|_{L^\infty(X)} \frac{dy}{|y|^{n+1}} < \infty.$$

It is not difficult to show (using the techniques of our Section 3.4, or those of Section 5.4 in [BS]) that the preceding condition is equivalent to $\int_0^1 \omega_2(f; \varepsilon)\varepsilon^{-2}d\varepsilon < \infty$, where $\omega_2(f; \varepsilon) = \sup\{|f(x+y) - 2f(x) + f(x-y)| \mid x \in X, |y| \leq \varepsilon\}$ is the second order modulus of continuity of f. This allows one to give a more intrinsic definition of $B_{\infty,\mathrm{loc}}^{1,1}(\Omega)$: it is the set of continuous functions h on Ω such that, for each open subset $U \subset \Omega$ with compact closure included in Ω, the second-order modulus of continuity of h on U, i.e.

$$\omega_2(h, U; \varepsilon) = \sup\{|h(x+y) - 2h(x) + h(x-y)| \mid x \in U, |y| \leq \varepsilon, x \pm y \in \Omega\},$$

satisfies $\int_0^1 \omega_2(h, U; \varepsilon)\varepsilon^{-2}d\varepsilon < \infty$. It is an easy consequence of the general theory that *such a function h is necessarily of class $C^1(\Omega)$.*

The limiting absorption principle for $H = h(Q)$ is an easy corollary of our next result, which is a sharp form of a division theorem due to Agmon and Hörmander (see Section 14.2 in [H], where h is assumed to be of class C^2; notice that under this assumption the proof given below becomes quite easy and elementary, since one can take $h_\varepsilon = h$ and $g_\varepsilon = \partial_F h$ independent of ε). Our purpose is to define the distributions $[h(x) - \lambda \mp i0]^{-1}$, but we also wish to describe their local properties (this is essential for the limiting absorption principle). For this, we introduce the local Besov spaces $\mathscr{H}^{s,p}_{\mathrm{loc}}(\Omega) \equiv B^{s,p}_{2,\mathrm{loc}}(\Omega)$ ($s \in \mathbb{R}$, $1 \le p \le \infty$) as the set of distributions f on Ω such that $\theta f \in \mathscr{H}^{s,p}(X) \equiv B^{s,p}_2(X)$ for all $\theta \in C_0^\infty(\Omega)$. We equip $\mathscr{H}^{s,p}_{\mathrm{loc}}(\Omega)$ with the Fréchet space topology defined by the family of seminorms $f \mapsto \|\theta f\|_{\mathscr{H}^{s,p}(X)}$, $\theta \in C_0^\infty(\Omega)$.

THEOREM 7.6.2. *Let Ω be an open subset of an euclidean space and $h \in B^{1,1}_{\infty,loc}(\Omega)$ a real function such that $\nabla h(x) \ne 0$ for all $x \in \Omega$. Then, for each $\lambda \in \mathbb{R}$, the limits $[h(x) - \lambda \mp i0]^{-1} \equiv \lim_{\mu \to +0}[h(x) - \lambda \mp i\mu]^{-1}$ exist (weakly) in $\mathscr{D}^*(\Omega)$, uniformly in λ. The operators of multiplication by the distributions $[h(x) - \lambda \mp i0]^{-1}$ extend to continuous operators $\mathscr{H}^{1/2,1}_{\mathrm{loc}}(\Omega) \to \mathscr{H}^{-1/2,\infty}_{\mathrm{loc}}(\Omega)$.*

PROOF. This theorem is not a consequence of any of the previously proved results, although Theorem 7.3.1 essentially implies it. We prefer to give an independent proof by taking into account the various simplifications that are possible in the present special context.

Let U be an open subset of Ω such that its closure $\overline{U}$ is compact and included in Ω. We fix some numbers $\lambda \in \mathbb{R}$ and $\mu > 0$, and for each $\varepsilon \in (0,1]$ we assume that three functions $h_\varepsilon, g_\varepsilon, \varphi_\varepsilon \in C^\infty(U)$ are given such that:

1) h_ε and g_ε are real;

2) $\varphi_\varepsilon \in C_0^\infty(U)$;

3) the function $(\varepsilon, x) \mapsto h_\varepsilon(x)$ is of class C^∞, and similarly for g_ε and φ_ε.

Further on more precise choices will be made. Finally, let $F : U \to X$ be a C^∞ vector field which, together with its first order derivative, is bounded on U. An explicit construction of F in terms of the given function h will be specified in part (iv) of the proof.

In order to keep the contact with Section 7.3, one should think that the role of the operator A of Theorem 7.3.1 is played by $A = -\frac{1}{2}[F(Q) \cdot P + P \cdot F(Q)] = -F(Q) \cdot P + \frac{i}{2}(\mathrm{div}\,F)(Q)$. In particular, if T is the operator of multiplication by a function $t(x)$, then $[A, T]$ is the operator of multiplication by the function $i\partial_F t(x) = iF(x) \cdot \nabla t(x)$. So ∂_F is the operator of differentiation in the direction of the vector field F.

We shall systematically denote derivatives with respect to ε by a prime and derivatives with respect to x by the symbol ∇ (so, if $s_\varepsilon(x)$ is a function of ε and x, then $s'_\varepsilon(x) = \frac{d}{d\varepsilon}s_\varepsilon(x)$ and $\nabla s_\varepsilon(x) \in X$ is its derivative with respect to x).

(i) We set $G_\varepsilon(x) = [h_\varepsilon(x) - \lambda - i\varepsilon g_\varepsilon(x) - i\mu]^{-1}$; so $G_\varepsilon \in C^\infty(U)$, $|G_\varepsilon(x)| \le \mu^{-1}$ and $(\varepsilon, x) \mapsto G_\varepsilon(x)$ is C^∞. The following identity is easily checked on U:

$$(7.6.6) \qquad G'_\varepsilon + i\partial_F G_\varepsilon = i(g_\varepsilon - \partial_F h_\varepsilon)G_\varepsilon^2 + (i\varepsilon g'_\varepsilon - \varepsilon\partial_F g_\varepsilon - h'_\varepsilon)G_\varepsilon^2.$$

We now make the first choice, namely $g_\varepsilon = \partial_F h_\varepsilon$. Hence $G'_\varepsilon + i\partial_F G_\varepsilon = (i\varepsilon\partial_F h'_\varepsilon - \varepsilon\partial_F^2 h_\varepsilon - h'_\varepsilon)G_\varepsilon^2$. We set $E_\varepsilon = \langle\varphi_\varepsilon, G_\varepsilon\rangle = \int_U \overline{\varphi_\varepsilon}(x)G_\varepsilon(x)dx$ and obtain :

(7.6.7) $\qquad E'_\varepsilon = \langle\varphi'_\varepsilon + i\partial_F^*\varphi_\varepsilon, G_\varepsilon\rangle + \langle\varphi_\varepsilon, [i\varepsilon\partial_F h'_\varepsilon - \varepsilon\partial_F^2 h_\varepsilon - h'_\varepsilon]G_\varepsilon^2\rangle.$

Observe that $\partial_F^* = -\partial_F - \operatorname{div} F$. What is missing now in order to continue the arguments of Section 7.3 is a "quadratic estimate" of the form (7.3.5). To deduce it, we have to assume that $\varphi_\varepsilon = |f_\varepsilon|^2$, where $f_\varepsilon \in C_0^\infty(U)$ and $(\varepsilon, x) \mapsto f_\varepsilon(x)$ is of class C^∞. Then we shall have:

(7.6.8) $\qquad E'_\varepsilon = \langle\overline{G_\varepsilon}f_\varepsilon, f'_\varepsilon + i\partial_F f_\varepsilon\rangle + \langle f'_\varepsilon - i\partial_F f_\varepsilon, G_\varepsilon f_\varepsilon\rangle$

$\qquad\qquad - i\langle f_\varepsilon, (\operatorname{div} F)G_\varepsilon f_\varepsilon\rangle + \langle\overline{G_\varepsilon}f_\varepsilon, [i\varepsilon\partial_F h'_\varepsilon - \varepsilon\partial_F^2 h_\varepsilon - h'_\varepsilon]G_\varepsilon f_\varepsilon\rangle.$

By using the notation $|| \cdot || = || \cdot ||_{L^2(X)}$ and by setting $\ell(\varepsilon) = ||f'_\varepsilon|| + ||\partial_F f_\varepsilon|| + ||\operatorname{div} F||_{L^\infty(U)}||f_\varepsilon||$ and $q(\varepsilon) = ||i\partial_F h'_\varepsilon - \partial_F^2 h_\varepsilon - \varepsilon^{-1}h'_\varepsilon||_{L^\infty(U)}$, we get from (7.6.8) that

(7.6.9) $\qquad |E'_\varepsilon| \le 2\ell(\varepsilon)||G_\varepsilon f_\varepsilon|| + \varepsilon q(\varepsilon)||G_\varepsilon f_\varepsilon||^2.$

Now we assume that $g_\varepsilon(x) \equiv \partial_F h_\varepsilon(x) \ge a > 0$ for $x \in U$ and all ε and observe that the following "quadratic estimate" holds:

(7.6.10) $\qquad ||G_\varepsilon f_\varepsilon||^2 = \langle|f_\varepsilon|^2, |G_\varepsilon|^2\rangle$

$$\le \frac{1}{a\varepsilon}\langle|f_\varepsilon|^2, \frac{\varepsilon g_\varepsilon + \mu}{(h_\varepsilon - \lambda)^2 + (\varepsilon g_\varepsilon + \mu)^2}\rangle$$

$$= \frac{1}{a\varepsilon}\langle|f_\varepsilon|^2, \Im G_\varepsilon\rangle = \frac{1}{a\varepsilon}\Im E_\varepsilon \le \frac{1}{a\varepsilon}|E_\varepsilon|.$$

Using this inequality in (7.6.9) we obtain:

(7.6.11) $\qquad |E'_\varepsilon| \le \frac{2}{\sqrt{a\varepsilon}}\ell(\varepsilon)|E_\varepsilon|^{1/2} + \frac{1}{a}q(\varepsilon)|E_\varepsilon|.$

By applying the modified Gronwall lemma with $\theta = 1/2$ (see Appendix 7.A) we get:

(7.6.12) $\qquad |E_\varepsilon| \le 2\left[|E_1| + \frac{1}{a}\left(\int_\varepsilon^1 \ell(\tau)\frac{d\tau}{\sqrt{\tau}}\right)^2\right]\exp\left[\frac{1}{a}\int_\varepsilon^1 q(\tau)d\tau\right].$

Since F and its first order derivative are bounded on U, the two integrals in (7.6.12) remain finite as $\varepsilon \to +0$ if

(7.6.13) $\qquad \int_0^1 \{\varepsilon^{-1/2}||f'_\varepsilon|| + \varepsilon^{-1/2}||\nabla f_\varepsilon|| + \varepsilon^{-1/2}||f_\varepsilon|| + ||\nabla h'_\varepsilon||_{L^\infty(U)}$

$\qquad\qquad + ||\nabla h_\varepsilon||_{L^\infty(U)} + ||\nabla^2 h_\varepsilon||_{L^\infty(U)} + \varepsilon^{-1}||h'_\varepsilon||_{L^\infty(U)}\}d\varepsilon < \infty.$

(ii) Let K be a fixed compact subset of U and $f \in \mathcal{H}^{1/2,1}(X)$ with $\operatorname{supp} f \subset K$. We shall see that it is possible to choose the families $\{f_\varepsilon\}$ and $\{h_\varepsilon\}$ such that $||f_\varepsilon - f|| \to 0$ and $h_\varepsilon(x) \to h(x)$ uniformly in $x \in U$ as $\varepsilon \to +0$, and such that the r.h.s. of (7.6.12) is bounded by $C||f||^2_{\mathcal{H}^{1/2,1}}$ for some constant C depending only on K and U (i.e. independent of $f, \varepsilon, \lambda, \mu$). One can then complete the proof by an easy argument, cf. the end of the proof of Theorem 7.3.1 (page 304).

(iii) We shall now completely specify the families $\{h_\varepsilon\}$ and $\{f_\varepsilon\}$. Let K and f be as in (ii). Choose a real function $h_0 \in C_0^1(X)$ such that $h_0 \in B_\infty^{1,1}(X)$ and $h_0(x) = h(x)$ on a neighbourhood of the closure of U. More precisely, let $\delta > 0$ such that $\delta < \text{dist}(K, \partial U)$ and $\delta < \text{dist}(U, \partial\Omega)$, and set $U_\delta = \{x \mid \text{dist}(x, U) \le \delta\}$ (so that U_δ is a compact neighbourhood of $\overline{U}$ in Ω), and assume that $h_0|_{U_\delta} = h|_{U_\delta}$. Let $\xi \in \mathscr{S}(X)$ real, with $\xi(0) = 1$, $\xi(-x) = \xi(x)$ and $\text{supp}\,\widehat{\xi} \subset \{x \mid |x| < \delta\}$. For any $\varphi \in \mathscr{S}^*(X)$ and $0 < \varepsilon \le 1$, we shall have $\xi(\varepsilon P)\varphi = \widehat{\xi}_\varepsilon * \varphi$ with $\widehat{\xi}_\varepsilon(x) = \varepsilon^{-n}\widehat{\xi}(\varepsilon^{-1}x)$; in particular, $[\xi(\varepsilon P)\varphi](x)$ depends only on the restriction of φ to the ball of radius δ and center x. Let $f_\varepsilon = \xi(\varepsilon P)f$ and $h_\varepsilon = \xi(\varepsilon P)h_0$ for $0 < \varepsilon \le 1$. Clearly $f_\varepsilon \in C_0^\infty(U)$, $h_\varepsilon \in C_0^\infty(X)$ is real, and $f_\varepsilon(x)$, $h_\varepsilon(x)$ are C^∞ functions of (ε, x). We shall now check the validity of (7.6.13).

Let $\widetilde{\xi}(x) = \sum_{j=1}^n x_j \partial_j \xi(x)$ (coordinates in an orthonormal basis of X), $\xi_j(x) = x_j \xi(x)$, $\xi_{jk}(x) = x_j x_k \xi(x)$ and $\widetilde{\xi}_j(x) = x_j \widetilde{\xi}(x)$. Then $\frac{d}{d\varepsilon}\xi(\varepsilon P)\varphi = P \cdot (\nabla\xi)(\varepsilon P)\varphi = \varepsilon^{-1}\widetilde{\xi}(\varepsilon P)\varphi$ and $P_j\xi(\varepsilon P)\varphi = \varepsilon^{-1}\xi_j(\varepsilon P)\varphi$ for any $\varphi \in \mathscr{S}^*(X)$. Hence we have the following relations:

$$f_\varepsilon' = \varepsilon^{-1}\widetilde{\xi}(\varepsilon P)f, \quad P_j f_\varepsilon = \varepsilon^{-1}\xi_j(\varepsilon P)f, \quad P_j h_\varepsilon' = \varepsilon^{-2}\widetilde{\xi}_j(\varepsilon P)h_0,$$
$$P_j P_k h_\varepsilon = \varepsilon^{-2}\xi_{jk}(\varepsilon P)h_0, \quad \varepsilon^{-1}h_\varepsilon' = \varepsilon^{-2}\widetilde{\xi}(\varepsilon P)h_0.$$

We shall apply the abstract theory of Chapter 3 in two different Banach spaces, namely $\mathbf{F}' = L^2(X)$ and $\mathbf{F}'' = C_\infty(X)$, but with the same C_0-group, namely the group of translations $\{e^{iP \cdot x}\}_{x \in X}$, which act as isometries in both spaces. Then $f \in \mathscr{H}^{1/2,1}(X)$ means $f \in \mathbf{F}'_{1/2,1}$, and $h_0 \in B_\infty^{1,1}(X)$ means $h_0 \in \mathbf{F}''_{1,1}$. According to Theorem 3.6.2 we shall have

$$\int_0^1 [\varepsilon^{-1/2}\|\theta(\varepsilon P)f\| + \varepsilon^{-1}\|\theta(\varepsilon P)h_0\|_{L^\infty(X)}]\frac{d\varepsilon}{\varepsilon} < \infty,$$

where $\theta \in C_0^\infty(X \setminus \{0\})$ is any tauberian function. Consider the framework of Section 3.5 with $\mathscr{M}(X)$ as LP-algebra (of order $r = 0$, see Definition 3.1.11). If $\eta \in \mathscr{S}(X)$ satisfies $\eta(0) = 0$ or $\eta(0) = \eta'(0) = 0$, then η has a zero of order 1 or 2 respectively at the origin, in the $\mathscr{M}(X)$-sense (Example 3.5.7). By using the estimate (3.5.19) with $s = 1/2$, $\ell = 1$ in the first case and $s = 1$, $\ell = 2$ in the second case, and by taking into account Theorem 3.6.2 (a), we get:

$$\int_0^1 \varepsilon^{-3/2}\|\eta(\varepsilon P)f\|d\varepsilon \le c\|f\|_{\mathscr{H}^{1/2,1}} \quad \text{if } \eta \in \mathscr{S}(X) \text{ and } \eta(0) = 0;$$

$$\int_0^1 \varepsilon^{-2}\|\eta(\varepsilon P)h_0\|_{L^\infty(X)}d\varepsilon < \infty \quad \text{if } \eta \in \mathscr{S}(X) \text{ and } \eta(0) = \eta'(0) = 0;$$

here c is a constant depending only on η. We have to take $\eta = \widetilde{\xi}$ or $\eta = \xi_j$ in the case of f and $\eta = \widetilde{\xi}_j$, or $\eta = \xi_{jk}$, or $\eta = \widetilde{\xi}$ in the case of h_0. The only fact which is not obvious is that $\widetilde{\xi}$ has a zero of second order at the origin (this is needed for the last choice of η). But

$$\partial_j \widetilde{\xi}(x) = \sum_{k=1}^{n} [\delta_{jk} \partial_k \xi(x) + x_k \partial_j \partial_k \xi(x)]$$

and $\partial_k \xi(0) = 0$ because $\xi(-x) = \xi(x)$. This finishes the proof of the validity of (7.6.13) for the choices we have made for f_ε, h_ε.

(iv) Let us construct a vector field F of class C^∞, bounded with bounded derivatives on U, such that $\partial_F h_\varepsilon(x) \geq a > 0$ for $x \in U$ and $\varepsilon \leq \varepsilon_0$. Since $h_0 \in C_0^1(X)$, we have $h_\varepsilon(x) \to h_0(x)$ and $\partial_j h_\varepsilon(x) \to \partial_j h_0(x)$ as $\varepsilon \to 0$, uniformly in $x \in X$. So, for any $\nu > 0$, there is $\varepsilon_0 > 0$ such that $|\nabla h_\varepsilon(x) - \nabla h_{\varepsilon_0}(x)| \leq \nu$ for $x \in X$ and $0 \leq \varepsilon \leq \varepsilon_0$. By hypothesis, $\nabla h(x) \neq 0$ for all $x \in \Omega$, and $h(x) = h_0(x)$ if $x \in U_\delta$. Since $\overline{U}$ is compact, we shall have $|\nabla h(x)| \geq \text{const.} > 0$ on $\overline{U}$. By choosing ν small enough, we may find $a > 0$ such that $\langle \nabla h_\varepsilon(x), \nabla h_{\varepsilon_0}(x) \rangle \geq a$ for all $x \in U$ and all $0 \leq \varepsilon \leq \varepsilon_0$. So it suffices to take $F = \nabla h_{\varepsilon_0}$. $\square$

We end this subsection by explicitly pointing out the simplifications which can be made in the preceding proof if $h \in C^2(\Omega)$. If we take $h_\varepsilon = h$, $F = \nabla h$ and $g_\varepsilon = \partial_F h = |\nabla h|^2$, then (7.6.6) becomes $G_\varepsilon' + i\partial_F G_\varepsilon = -\varepsilon(\partial_F^2 h)G_\varepsilon^2$. In order to get the limiting absorption principle in the Besov space $\mathscr{H}^{1/2,1}$, one has however to use an ε-dependent family $\{f_\varepsilon\}$, constructed as in the preceding proof.

7.6.3. Our next purpose is to study operators of the form $H = h(P) + V$ in the Hilbert space $\mathscr{H} = \mathscr{H}(X) = L^2(X)$ by the perturbative technique described for example in Proposition 7.5.6. Here $h : X \to \mathbb{R}$ is a Borel function, $H_0 \equiv h(P) = \mathscr{F}^* h(Q) \mathscr{F}$ is a self-adjoint operator and V is, in some sense, a small symmetric perturbation of H_0. The physical picture behind this formalism is the following: the variable $k \in X$ is interpreted as the momentum of a pseudo-particle, $h(k)$ is the kinetic energy of the particle when its momentum is k, and V is an exterior potential. The free motion of the particle is described by the hamiltonian H_0. From the classical Hamilton equations we see that $h'(k) = \nabla h(k)$ is the free velocity of the system, when the momentum is k. The same interpretation is obtained by considering the quantum mechanical equation of motion:

$$(7.6.14) \qquad e^{iH_0 t} Q e^{-iH_0 t} = Q + h'(P)t,$$

because Q is interpreted as the position observable. It is possible now to understand in physical terms the special role played by the critical values of the function h. These are numbers $\lambda \in \mathbb{R}$ such that $\lambda = h(k)$ for some $k \in X$ with $h'(k) = 0$. In other terms, if the particle has (kinetic) energy λ, it could have momentum k with corresponding velocity zero. So at these energies the particle has bad propagation properties. The phenomenology and terminology developed around the N-body problem then suggest the use of term *threshold energies* for the critical values of the function h. We shall see that we shall have troubles in constructing conjugate operators at threshold energies. This is not surprising, because the existence of a conjugate operator A is related to the fact that the system has good propagation properties in the spectral representation of A (cf. [M2]).

The choice of the locally conjugate operator will be motivated by the following heuristic discussion (in which all operations are assumed to be meaningful; the precise assumptions on h will be stated later on). We fix an orthonormal basis in the euclidean space X and identify $X = \mathbb{R}^n$. Then $[h(P), iQ_j] = (\partial_j h)(P)$ for $1 \leq j \leq n$ (this is equivalent to (7.6.14)). Let $F : X \to X$ be a vector field and

$$(7.6.15) \quad A = \frac{1}{2}[F(P) \cdot Q + Q \cdot F(P)] = \frac{1}{2}\sum_{j=1}^{n}[F_j(P)Q_j + Q_j F_j(P)]$$

$$= F(P) \cdot Q + \frac{i}{2}f(P) = Q \cdot F(p) - \frac{i}{2}f(P),$$

where $f = \operatorname{div} F = \sum_{j=1}^{n} \partial_j F_j$. Then $[h(P), iA] = (Fh')(P)$, where $h' = \nabla h$ and $Fh' = \sum_{j=1}^{n} F_j \partial_j h$. Moreover, $[[h(P), iA], iA] = ((F\nabla)^2 h)(P)$. In order to get local positivity of the first order commutator, the simplest choice is $F = \zeta h'$, where ζ is a positive scalar function. Then $[h(P), iA] = \zeta(P)|h'(P)|^2 \geq 0$. The necessity of the factor ζ comes from the fact that A could have no self-adjoint realizations if the vector field F grows more rapidly than $|x|$ at infinity. If $J \subset \mathbb{R}$ is a Borel set, then the spectral projection $E_0(J)$ of the self-adjoint operator H_0 associated to J is the operator $\chi_J(h(P)) = \chi_\Omega(P)$, where $\Omega = h^{-1}(J)$ and χ_Ω is the characteristic function of Ω. So $E_0(J)[H_0, iA]E_0(J) = \chi_\Omega(P)\zeta(P)|h'(P)|^2$, and this is $\geq mE_0(J)$ for some $m \in \mathbb{R}$ if and only if $\zeta(k)|h'(k)|^2 \geq m$ for all $k \in X$ such that $h(k) \in J$. Since we are interested in having $m > 0$, we see that J must not contain any threshold energies.

To conclude, we see that (7.6.15) is a natural candidate for a locally conjugate operator to H_0 outside the threshold energies (i.e. critical values of h). Suitable conditions on V will make this operator A conjugate to H too. However, in order to be able to treat perturbations V that are very singular (on a compact set), it is convenient to modify the factor Q in A and to consider operators of the form $F(P) \cdot G(Q) + G(Q) \cdot F(P)$, with F and G vector fields on X. The next proposition gives a meaning to a class of such expressions as self-adjoint operators in $\mathscr{H}(X)$.

PROPOSITION 7.6.3. (a) *If $F : X \to X$ is a Lipschitz function, i.e. $|F(x) - F(y)| \leq c|x - y|$ for some constant c and all $x, y \in X$, then the operator A defined by (7.6.15) is essentially self-adjoint in $\mathscr{H}(X)$ on the domain $\mathscr{S}(X)$.*

(b) *Let $F, G : X \to X$ be two functions of class C^2 having bounded derivatives of first and second order. Then the operator $A = F(P) \cdot G(Q) + G(Q) \cdot F(P)$ is essentially self-adjoint in $\mathscr{H}(X)$ on $\mathscr{S}(X)$, the domain of its closure contains $\mathscr{H}^2(X) \cap \mathscr{H}_2(X) = D(P^2 + Q^2)$ and there is a constant c such that $\|Au\| \leq c\|(P^2 + Q^2)u\|$ for all $u \in \mathscr{H}^2(X) \cap \mathscr{H}_2(X)$.*

PROOF. (a) It is easy to describe the unitary group generated by the operator $\hat{A} = \mathscr{F}A\mathscr{F}^* = -\frac{1}{2}[F(Q) \cdot P + P \cdot F(Q)] = -F(Q) \cdot P + \frac{i}{2}f(Q)$. Let $\xi : \mathbb{R} \times X \to X$ be the flow associated to the vector field $-F$, i.e. for each $x \in X$, $\alpha \mapsto \xi(\alpha, x) \equiv \xi_\alpha(x)$ is the unique global solution of the differential equation $\frac{d}{d\alpha}\xi_\alpha(x) = -F(\xi_\alpha(x))$ with initial condition $\xi_0(x) = x$. If we denote by $\eta_\alpha(x) = \det \nabla \xi_\alpha(x)$ the jacobian at x of the mapping $\xi_\alpha : X \to X$, then $(\widehat{W}_\alpha u)(x) = \sqrt{\eta_\alpha(x)} u(\xi_\alpha(x))$ defines a unitary operator $\widehat{W}_\alpha$ in $\mathscr{H}(X)$. It is straightforward to show that $\{\widehat{W}_\alpha\}_{\alpha \in \mathbb{R}}$ is a C_0-group in $\mathscr{H}(X)$ which leaves invariant the dense subspace D consisting of Lipschitz functions of compact support, and that $\frac{d}{d\alpha} \widehat{W}_\alpha = i \widehat{A} \widehat{W}_\alpha$ on this subspace. Nelson's Lemma (or the more general Theorem 3.3.4) will imply the essential self-adjointness of $\widehat{A}$ on D; then an easy argument shows that $\mathscr{S}(X)$ is a core for the closure of $\widehat{A}$ in $\mathscr{H}(X)$. Clearly this proves assertion (a).

(b) We shall deduce the second part of the proposition from Theorem X.37 in [RS] in which we take $N = P^2 + Q^2 + a^2$ with $a \in \mathbb{R}$ large enough, and $D = \mathscr{S}(X)$. We have to show that $A : \mathscr{S}(X) \subset \mathscr{H}(X) \to \mathscr{H}(X)$ is a symmetric operator such that $||Au|| \leq b||Nu||$ and $|\langle Au, Nu \rangle - \langle Nu, Au \rangle| \leq b \langle u, Nu \rangle$ for some constant b and all $u \in \mathscr{S}$. Note that $|F(x)| + |G(x)| \leq c \langle x \rangle$, so we have $D(F(P)) \supset \mathscr{H}^1$ and $D(G(Q)) \supset \mathscr{H}_1$. If $u \in \mathscr{S}$, then $G(Q)u$ is of class C^1 and is rapidly decaying at infinity, and so are its first order derivatives; hence $G(Q)u \in \mathscr{H}^1$. By making a Fourier transformation, one sees that $F(P)u \in \mathscr{H}_1$. So A is well defined and symmetric on $\mathscr{S}$ in $\mathscr{H}$.

Before proving the needed estimates, we observe that

$$N^2 = (P^2 + Q^2 + a^2)^2 = P^4 + 2a^2 P^2 + Q^4 + 2a^2 Q^2 + 2QP^2Q + a^4 - 2n$$
$$\geq \langle P \rangle^4 + \langle Q \rangle^4 + 2QP^2Q + a^4 - 2n - 2,$$

where we have assumed $a \geq 1$. Hence:

(7.6.16)
$$||\langle P \rangle^2 u||^2 + ||\langle Q \rangle^2 u||^2 + 2 \sum_{j,k=1}^{n} ||P_j Q_k u||^2 + (a^4 - 2n - 2)||u||^2 \leq ||Nu||^2$$

for all $u \in \mathscr{S}$. Then, using $||F(P)v|| \leq \text{const.}||v||_{\mathscr{H}^1}$, we obtain:

$$||F(P)G(Q)u|| \leq c||G(Q)u|| + c \sum_{j=1}^{n} ||P_j G(Q)u||$$
$$\leq c'||\langle Q \rangle u|| + c'' \sum_{j,k=1}^{n} ||Q_k P_j u|| \leq c'''||Nu||.$$

Together with a similar estimate for $||G(Q)F(P)u||$, one gets that $||Au|| \leq b||Nu||$ for some constant b and all $u \in \mathscr{S}$. It remains to estimate the commutator $[A, N]$. This is based on the identity

$$[A, N] = F(P)[G(Q), P^2] + [G(Q), P^2]F(P)$$
$$+ [F(P), Q^2]G(Q) + G(Q)[F(P), Q^2].$$

Since $[G(Q), P^2] \in B(\mathscr{H}^1, \mathscr{H})$ and $[F(P), Q^2] \in B(\mathscr{H}_1, \mathscr{H})$, one can easily get the estimate $|\langle u, [A, N]u \rangle \leq b \langle u, Nu \rangle$ by using (7.6.16) once again. $\square$

The next lemma describes in precise terms one of the simplest non-trivial choices of local conjugate operators.

LEMMA 7.6.4. *Let $h : X \to \mathbb{R}$ be a Borel function and $J \subset \mathbb{R}$ an open set such that:*

(i) $h^{-1}(J) \equiv \Omega$ *is an open subset of X and h is of class C^2 on a neighbourhood of the closure of Ω;*

(ii) *there is a constant $m > 0$ such that*

$$(7.6.17) \qquad |h'(x)| \geq m \quad and \quad |h''(x)| \leq m^{-1}|h'(x)|^2 \quad if \ x \in \Omega.$$

Let $\theta \in C_0^\infty(J)$ be real and $F(x) = \theta(h(x))|h'(x)|^{-2}h'(x)$ if $x \in \Omega$, $F(x) = 0$ if $x \notin \Omega$. Then $F \in BC^1(X)$, and the operator A defined in (7.6.15) is essentially self-adjoint on $\mathscr{S}(X)$. The group $\{e^{iA\tau}\}_{\tau \in \mathbb{R}}$ generated by A leaves invariant the domain of $H_0 = h(P)$, and $[H_0, iA] = \theta(H_0)$, $[[H_0, iA], iA] = \theta(H_0)\theta'(H_0)$, $\ldots$, $\operatorname{ad}_{-iA}^k(H_0) = \theta_k(H_0)$ are bounded operators in $\mathscr{H}(X)$ for all $k \geq 1$; here $\theta_k(\lambda) = [\theta(\lambda)\frac{d}{d\lambda}]^k\theta(\lambda)$. In particular, H_0 is of class $C^\infty(A)$. Finally, if $J_\theta = \{\lambda \mid \theta(\lambda) = 1\}$, then $E_0(J_\theta)[H_0, iA]E_0(J_\theta) = E_0(J_\theta)$.

The proof is a straightforward exercise. Notice that, with the notations of the proof of Proposition 7.6.3 (a), we have $e^{iA\alpha}h(P)e^{-iA\alpha} = h(\xi_\alpha(P))$ and $\xi_\alpha(x) = x$ if $x \notin \operatorname{supp} F \subset \overline{\Omega}$. We had to require h to be of class C^2 in order to assure the Lipschitz continuity of F, which is needed for the application of Proposition 7.6.3 (a). In fact, if Ω is a bounded set (e.g. if $h : X \to \mathbb{R}$ is continuous and tends to infinity when $|x| \to \infty$), one can cover the case of $h \in B_{\infty,\mathrm{loc}}^{1,1}(X)$ by using a vector field F of class C^∞ and such that $F(x)h'(x) \geq \mathrm{const.} > 0$ on Ω (see the last part of the proof of Theorem 7.6.2). We leave away such possible generalizations, since they are irrelevant in the present context.

In the preceding lemma the fact that the domain of H_0 is invariant under $e^{iA\tau}$ is obvious because of the explicit form of $e^{iA\alpha}h(P)e^{-iA\alpha}$. We mention now an abstract result which can be used in less explicit situations.

LEMMA 7.6.5. *Let A and H_0 be self-adjoint operators in $\mathscr{H}$, and assume that there is a core D of H_0 with the following properties:*

(i) *for $u \in D$ and $\tau \in \mathbb{R}$, one has $e^{iA\tau}u \in D$ and $\sup_{|\tau| \leq 1}||H_0 e^{iA\tau}u|| < \infty$;*

(ii) *the derivative $\frac{d}{d\tau}e^{-iA\tau}H_0 e^{iA\tau}u|_{\tau=0} \equiv [H_0, iA]u$ exists weakly in $\mathscr{H}$ for each $u \in D$, and $||[H_0, iA]u|| \leq a(||u|| + ||H_0 u||)$ for some finite number a independent of u.*

Then the unitary group $\{e^{iA\tau}\}$ leaves invariant the domain of H_0 (hence its form domain). If there is $0 \leq \theta < 1$ such that $[A, H_0]$ extends to a bounded operator $D(|H_0|^\theta) \to \mathscr{H}$, then $||e^{iA\tau}||_{B(D(H_0))} \leq c\langle\tau\rangle^m$ for all $\tau \in \mathbb{R}$, where $c \in \mathbb{R}$ and $m = (1 - \theta)^{-1}$.

PROOF. Let $u \in D$ and $u_\tau = e^{iA\tau}u$. Then, by using the group property of $e^{iA\tau}$, one sees that the function $g_\tau = e^{-iA\tau}H_0 u_\tau$ is weakly differentiable on $\mathbb{R}$, with derivative $g_\tau' = e^{-iA\tau}[H_0, iA]u_\tau$. By writing $\langle v, g_\tau \rangle = \langle v, H_0 u \rangle + \int_0^\tau \langle v, g_\alpha \rangle d\alpha$, replacing v by $e^{-iA\tau}v$ and using the notation $||f||_{H_0} = ||f|| + ||H_0 f||$, we easily obtain for all $\tau \geq 0$: $||u_\tau||_{H_0} \leq ||u||_{H_0} + a\int_0^\tau ||u_\alpha||_{H_0}d\alpha$. Now the Gronwall Lemma implies that $||u_\tau||_{H_0} \leq e^{a|\tau|}||u||_{H_0}$. This estimate and the density of D in $D(H_0)$ for the norm $|| \cdot ||_{H_0}$ will clearly give $e^{iA\tau}D(H_0) \subset D(H_0)$ and $||e^{iA\tau}||_{B(D(H_0))} \leq e^{a|\tau|}$. To prove the last estimate, observe that

$||[H_0, iA]u_\tau|| \leq c_1||\langle H_0\rangle^\theta u_\tau|| \leq c_2||u_\tau||_{H_0}^\theta||u||^{1-\theta}$, hence $||u_\tau||_{H_0} \leq ||u||_{H_0} +$ $c_2||u||^{1-\theta}\int_0^\tau ||u_\alpha||_{H_0}^\theta d\alpha$. Now, instead of the usual Gronwall Lemma, we use the modified version presented in Appendix 7.A. $\square$

An easy consequence of Lemma 7.6.4 is stated below. For the proof, use Theorem 7.5.4 with $\mathscr{G} = D(|H_0|^{1/2})$ and observe the following embeddings: $(D(A; \mathscr{G}^*), \mathscr{G}^*)_{1/2,1} \supset (D(A; \mathscr{H}), \mathscr{H})_{1/2,1} \supset (\mathscr{H}_1, \mathscr{H})_{1/2,1} = \mathscr{H}_{1/2,1}$. Here A is as in Lemma 7.6.4 with $\theta(\lambda) = 1$ on a given compact subset of J.

PROPOSITION 7.6.6. *Let h and J be as in Lemma 7.6.4. Denote by H_0 the self-adjoint operator $h(P)$ in $\mathscr{H}(X)$. Then the holomorphic function*

$$\mathbb{C}_\pm \ni z \mapsto (H_0 - z)^{-1} \in B(\mathscr{H}(X)) \subset B(\mathscr{H}_{1/2,1}(X), \mathscr{H}_{-1/2,\infty}(X))$$

extends to a weak*-continuous function $\mathbb{C}_\pm \cup J \to B(\mathscr{H}_{1/2,1}(X), \mathscr{H}_{-1/2,\infty}(X))$. In particular, if T is an operator from $\mathscr{H}(X)$ to some Hilbert space, and if T is continuous when $\mathscr{H}(X)$ is equipped with the topology induced by $\mathscr{H}_{-1/2,\infty}(X)$, then T is locally H_0-smooth on J.

Let $h : X \to \mathbb{R}$ be a function of class C^1 such that $|h(x)| + |h'(x)| \to \infty$ as $|x| \to \infty$. This condition can be stated, in physical terms, as follows: if the kinetic energy *and* the velocity are bounded by a constant, then the momentum is bounded by some other constant (note that the set of vectors $u \in \mathscr{H}$ such that $||h(P)u|| + ||h'(P)u|| \leq$ const. will satisfy one of the two assumptions of the Riesz-Kolmogorov compacity criterion). Then, if $J \subset \mathbb{R}$ is open, bounded and with closure disjoint from the set of critical values of h, the first estimate in (7.6.17) is valid (i.e. if the kinetic energy belongs to a compact set disjoint from thresholds, then the velocity is bounded below by a strictly positive constant). Otherwise, we could find a sequence $\{x_j\}_{j\in\mathbb{N}}$ in $\Omega \equiv h^{-1}(J)$ with $|h'(x_j)| \to 0$. This sequence cannot have an accumulation point x in X, because then we would have $x_{j_k} \to x$ for a subsequence, so $h(x_{j_k}) \to h(x) \in \bar{J}$ as $k \to \infty$ and $h'(x) = 0$, which contradicts the hypothesis. Hence $|x_j| \to \infty$, which implies $|h(x_j)| + |h'(x_j)| \to \infty$, which is again a contradiction. In order to satisfy the second estimate in (7.6.17), it is enough to require h to be of class C^2 and, for example, $|h''(x)| \leq c(1 + |h(x)|^2 + |h'(x)|^2)$ on X. The *simply characteristic* polynomials introduced by Agmon and Hörmander (see Section 14.3 in [H]) satisfy a stronger condition, namely $|h^{(\alpha)}(x)| \leq c_\alpha(1 + |h(x)| + |h'(x)|)$ for any α. Notice also that the hypothesis $\Lambda(P_0) = \{0\}$ made in [H] is equivalent, in our notations, with $|h(x)| + |h'(x)| \to \infty$ as $x \to \infty$; cf. Theorem 10.2.9 in [H]. In conclusion, if the function h in Proposition 7.6.6 is a simply characteristic polynomial which effectively depends on all variables (i.e. we have $h(x + ty) = h(x) \forall t \in \mathbb{R}$ only if $y = 0$; this means $\Lambda(h) = \{0\}$), then the proposition holds for any open bounded J, with closure disjoint from the set of critical values of h (which is finite, because h is a polynomial).

Now we study perturbations $H = H_0 + V$ of H_0. The next lemma allows us to isolate a class of functions h for which the perturbative method described in Proposition 7.5.6 and the criteria contained in Proposition 7.5.7 and Theorem 7.5.8 can be used easily and efficiently.

LEMMA 7.6.7. *Let $h : X \to \mathbb{R}$ be a function of class C^m, $m = 1, 2, 3, \ldots$, such that $|h^{(\alpha)}(x)| \leq$ const. for $|\alpha| = m$ (e.g. assume that h is a polynomial of degree m). Let $\mathcal{G} = D(|h(P)|^{1/2})$ be the form domain of the operator $H_0 = h(P)$. Then $\mathcal{G}$ is invariant under the group $\{e^{iQ \cdot x}\}_{x \in X}$ if and only if*

$$\sum_{|\alpha| \leq m} |h^{(\alpha)}(x)| \leq c(1 + |h(x)|)$$

for some constant c. If this condition is satisfied, then the domain of H_0 is also invariant under the group $\{e^{iQ \cdot x}\}$ and the C_0-groups induced in $D(H_0)$ and $D(|H_0|^{1/2})$ are of polynomial growth:

$$(7.6.18) \qquad ||e^{iQ \cdot x}||_{B(D(H_0))} \leq c\langle x \rangle^m, \qquad ||e^{iQ \cdot x}||_{B(\mathcal{G})} \leq c\langle x \rangle^{m/2}.$$

PROOF. Taylor's formula implies the existence of bounded continuous functions $h_\alpha : X \times X \to \mathbb{R}$ such that for all x, $y \in X$ (in this proof we identify $X = \mathbb{R}^n$ with the help of an orthonormal basis in X)

$$h(x + y) = \sum_{|\alpha| \leq m-1} \frac{x^\alpha}{\alpha!} h^{(\alpha)}(y) + \sum_{|\alpha| = m} \frac{x^\alpha}{\alpha!} h_\alpha(y, x).$$

Notice that $e^{-iQ \cdot x} \varphi(P) e^{iQ \cdot x} = \varphi(P + x)$ for each $x \in X$ and each Borel function φ on X. For $u \in \mathscr{S}(X)$ we shall have:

$$|| \, |h(P)|^{1/2} e^{iQ \cdot x} u||^2 = |||h(P + x)|^{1/2} u||^2 = \langle u, |h(P + x)| u \rangle$$
$$\leq \sum_{|\alpha| \leq m-1} \frac{|x^\alpha|}{\alpha!} \langle u, |h^{(\alpha)}(P)| u \rangle + c|x|^m ||u||^2.$$

If the condition $\sum_{|\alpha| \leq m} |h^{(\alpha)}| \leq c\langle h \rangle$ is fulfilled, we clearly get the second estimate in (7.6.18); the first one is obtained similarly. Reciprocally, assume $e^{iQ \cdot x} D(|H_0|^{1/2}) \subset D(|H_0|^{1/2})$ for all x. From the general theory (Section 3.2) we get $|| \, |h(P + x)|^{1/2} u|| = |||h(P)|^{1/2} e^{iQ \cdot x} u|| \leq c(x) ||u||_{\mathcal{G}}$ with $c(x)$ a locally bounded function of x. So

$$\left|\left|\left| \sum_{|\alpha| \leq m-1} \frac{x^\alpha}{\alpha!} h^{(\alpha)}(P) \right|^{1/2} u \right|\right| \leq || \, |h(P + x)|^{1/2} u|| + \left|\left|\left| \sum_{|\alpha| = m} \frac{x^\alpha}{\alpha!} h_\alpha(P, x) \right|^{1/2} u \right|\right|$$

$$\leq c(x) ||(1 + |h(P)|)^{1/2} u|| + c|x|^{m/2} ||u||.$$

By Proposition 1.1.2 there is a finite set $B \subset X$ and there are functions χ_α on B such that for any polynomial L of degree $\leq m$ and any α: $L^{(\alpha)}(x) = \sum_{b \in B} \chi_\alpha(b) L(x + b)$. If $L(x) = \sum_{|\alpha| \leq m-1} \frac{x^\alpha}{\alpha!} \xi_\alpha$, then $L^{(\alpha)}(0) = \xi_\alpha$. By taking $\xi_\alpha = h^{(\alpha)}(P) u$ we obtain from the preceding estimate that $|| \, |h^{(\alpha)}(P)|^{1/2} u|| \leq$ const.$||(1 + |h(P)|)^{1/2} u||$ for $u \in \mathscr{S}(X)$. This easily implies that $|h^{(\alpha)}| \leq$ const. $(1 + |h|)$. $\square$

If h is a hypoelliptic polynomial of degree m, the conditions of the preceding proposition are satisfied. In fact, in this case we have much more: there is $\delta > 0$ such that $|h^{(\alpha)}(x)| \leq c\langle x \rangle^{-|\alpha|\delta}(1 + |h(x)|)$ (Theorem 11.1.3 in [H]), hence $|h(x)| \to \infty$ too.

We have seen after Proposition 7.6.6 that the condition $\lim_{x \to \infty}(|h(x)| + |h'(x)|) = \infty$ is quite natural in our context. Under the assumptions of Lemma 7.6.7, this will imply $\lim_{x \to \infty} |h(x)| = \infty$ (in physical terms, this means: if the kinetic energy is bounded, then the momentum is bounded too). But, if h is a continuous real function on a neighbourhood of infinity in X and $\dim X > 1$, then either $h(x) \to +\infty$ or $h(x) \to -\infty$ (because the region $|x| > R$ in X is connected, so its image through h is a connected subset of $\mathbb{R}$). Hence, when we study such functions, we may assume without loss of generality that $h(x) \to +\infty$ when $|x| \to \infty$.

Let us fix a function $h : X \to \mathbb{R}$ of class C^m, $m = 2, 3, \ldots$, such that :
(a) $\lim_{|x| \to \infty} h(x) = +\infty$;
(b) the derivatives of order m of h are bounded;
(c) $\sum_{|\alpha| \le m} |h^{(\alpha)}(x)| \le c(1 + |h(x)|)$.

We denote by H_0 the self-adjoint operator $h(P)$ in the Hilbert space $\mathscr{H} = L^2(X)$ and by $\mathscr{G}$ its form domain $\mathscr{G} = D(|H_0|^{1/2})$ equipped with the graph topology. Since h is bounded from below, H_0 is bounded from below too and we may consider on $\mathscr{G}$ the admissible norm $\|u\|_{\mathscr{G}} = \langle u, [c + h(P)]u \rangle^{1/2}$, where c is a constant such that $c + h(x) \ge 1$. We identify as usual

$$\mathscr{S}(X) \subset D(H_0) \subset \mathscr{G} \subset \mathscr{H} = \mathscr{H}^* \subset \mathscr{G}^* \subset D(H_0)^* \subset \mathscr{S}^*(X).$$

Notice that the norm in $\mathscr{G}^*$ is $\|u\|_{\mathscr{G}^*} = \langle u, [c + h(P)]^{-1}u \rangle^{1/2}$. The group $\{e^{iQ \cdot x}\}_{x \in X}$ acting in $\mathscr{S}^*(X)$ leaves invariant each of the spaces of the preceding scale and induces in each of them a C_0-group of polynomial growth. Denote by $\mathscr{G}_{s,p}$ and $\mathscr{G}^*_{s,p}$ the spaces of the Besov scales associated to this group in $\mathscr{G}$ and $\mathscr{G}^*$ respectively; here $s \in \mathbb{R}$ and $1 \le p \le \infty$. Notice that $(\mathscr{G}_{s,p})^* = \mathscr{G}^*_{-s,p'}$ if $1 < p < \infty$, $\frac{1}{p} + \frac{1}{p'} = 1$ and, if $\mathscr{G}^\circ_{s,\infty}$ is the closure of $\mathscr{S}(X)$ in $\mathscr{G}_{s,\infty}$, then $(\mathscr{G}^\circ_{s,\infty})^* = \mathscr{G}^*_{-s,1}$. Similarly $(\mathscr{G}^*_{s,p})^* = \mathscr{G}_{-s,p'}$ for $1 \le p < \infty$ and $(\mathscr{G}^{*\circ}_{s,\infty})^* = \mathscr{G}_{-s,1}$. The spaces $\mathscr{G}_s \equiv \mathscr{G}_{s,2}$, $\mathscr{G}^*_s \equiv \mathscr{G}^*_{s,2}$ (for $s \in \mathbb{R}$) constitute the (continuous) Sobolev scales and can be obtained by complex interpolation ($\mathscr{G}_0 = \mathscr{G}$, $\mathscr{G}^*_0 = \mathscr{G}^*$). The most intuitive description of these spaces is offered by the Littlewood-Paley theory (Section 3.6). More precisely, let $\theta \in C_0^\infty(X)$ such that $\theta(x) > 0$ if $a < |x| < b$ and $\theta(x) = 0$ otherwise, and $\widetilde{\theta} \in C_0^\infty(X)$ with $\widetilde{\theta}(x) = 1$ if $|x| < b$; here $0 < a < b$ are fixed numbers. Then the following expression is an admissible norm on $\mathscr{G}_{s,p}$:

$$\|\widetilde{\theta}(Q)u\|_{\mathscr{G}} + \left[\int_1^\infty \|r^s \theta(r^{-1}Q)u\|_{\mathscr{G}}^p \frac{dr}{r} \right]^{1/p}$$

(with the standard modification if $p = \infty$); similarly for $\mathscr{G}^*$. Finally, we recall that, since $\mathscr{G}$ and $\mathscr{G}^*$ are Hilbert spaces, we have a rich functional calculus for Q in all the spaces of these scales: if $\varphi \in BC^\infty(X)$, then $\varphi(Q) : \mathscr{G}_{s,p} \to \mathscr{G}_{s,p}$ and $\|\varphi(Q)\|_{B(\mathscr{G}_{s,p})} \le c_{s,p}\|\varphi\|_{BC^N}$ for a finite, explicit N; similarly for $\mathscr{G}^*$ (see Section 3.7). This estimate will be freely used in the proof of the theorem below.

In the statement of the next theorem we keep the assumptions on h and the notations of the preceding paragraph. Note that the critical set $\kappa(h) = \{h(x) \mid x \in X, h'(x) = 0\}$ is a closed subset of $\mathbb{R}$. Recall that $\mathbb{C}_\pm = \{z \in \mathbb{C} \mid \pm\Im z > 0\}$ and that $\overline{\mathbb{C}_\pm}$ denotes the closure of $\mathbb{C}_\pm$ in $\mathbb{C}$.

THEOREM 7.6.8. *Let H_0 be as above, $\mathcal{G} = D(|H_0|^{1/2})$ and let $V : \mathcal{G} \to \mathcal{G}^*$ be a symmetric operator such that $H_0 + V + i$ is an isomorphism of $\mathcal{G}$ onto $\mathcal{G}^*$ (e.g. assume $\pm V \leq \alpha h(P) + \beta$ for some $\alpha < 1$ and $\beta \in \mathbb{R}$). Denote by H the self-adjoint operator in $\mathcal{H}$ induced by $H_0 + V$ and assume that $(H+i)^{-1} - (H_0+i)^{-1}$ is compact in $\mathcal{H}$. Assume furthermore that V can be decomposed into a sum $V = V_S + V_L$, where $V_S : \mathcal{G} \to \mathcal{G}^*$ (the short-range component) and $V_L : \mathcal{G} \to \mathcal{G}^*$ (the long-range component) are symmetric operators satisfying:*

(S) *there is $\theta \in C_0^\infty(X)$ with $\theta(x) > 0$ in an annulus $0 < a < |x| < b < \infty$ and $\theta(x) = 0$ otherwise, such that:*

$$(7.6.19) \qquad \int_0^\infty \|\theta(r^{-1}Q)V_S\|_{B(\mathcal{G},\mathcal{G}^*)}dr < \infty,$$

(L) *there is $\xi \in C^\infty(X)$ with $\xi(x) = 0$ near zero and $\xi(x) = 1$ near infinity such that:*

$$(7.6.20)$$
$$\sum_{j=1}^n \int_1^\infty \left\{ \|\xi(r^{-1}Q)[Q_j, V_L]\|_{B(\mathcal{G},\mathcal{G}^*)} + \|\xi(r^{-1}Q)|Q|[P_j, V_L]\|_{B(\mathcal{G},\mathcal{G}^*)} \right\} \frac{dr}{r} < \infty.$$

Then

1) *H has no singularly continuous spectrum outside $\kappa(h)$;*

2) *the eigenvalues of H in $\mathbb{R} \setminus \kappa(h)$ are of finite multiplicity and they do not have accumulation points outside $\kappa(h)$;*

3) *the holomorphic map $\mathbb{C}_\pm \ni z \mapsto (H-z)^{-1} \in B(\mathcal{G}^*, \mathcal{G}) \subset B(\mathcal{G}_{1/2,1}^*, \mathcal{G}_{-1/2,\infty}^\circ)$, when considered with values in $B(\mathcal{G}_{1/2,1}^*, \mathcal{G}_{-1/2,\infty})$, extends to a weak *-continuous function on $\overline{\mathbb{C}_\pm} \setminus [\kappa(h) \cup \sigma_p(H)]$;*

4) *if $\mathcal{F}$ is a Hilbert space and $T \in B(\mathcal{G}_{-1/2,\infty}^\circ, \mathcal{F})$, then T is locally H-smooth on $\mathbb{R} \setminus [\kappa(h) \cup \sigma_p(H)]$.*

PROOF. Let J be a bounded open real set with $\overline{J} \cap \kappa(h) = \varnothing$. Then $h^{-1}(\overline{J})$ is a compact subset of X on which $|h'(x)| \geq \text{const.} > 0$. It is easy to construct a vector field F of class $C_0^\infty(X)$ such that $F(x)h'(x) \geq \text{const.} > 0$ on $h^{-1}(J)$. Let A be given by (7.6.15); then $[H_0, iA] = (Fh')(P)$, hence $E_0(J)[H_0, iA]E_0(J) \geq cE_0(J)$ for some constant $c > 0$. Moreover $[[H_0, iA], iA] = ((F\nabla)^2 h)(P)$ is a bounded operator in $\mathcal{H}$ (because h is at least of class C^2, so $H_0 \in \mathcal{C}^{1,1}(A; \mathcal{G}, \mathcal{G}^*)$. By Lemma 7.6.5, the group $\{e^{iA\tau}\}$ induces a (bounded) C_0-group in $\mathcal{G}$. Clearly the conditions (i), (ii), (iii) of Proposition 7.5.6 are satisfied. We shall prove below that $V \in \mathcal{C}^{1,1}(A; \mathcal{G}, \mathcal{G}^*)$. Let us check that $\mathcal{G}_{1/2,1}^* \subset \mathcal{K} = (D(A; \mathcal{G}^*), \mathcal{G}^*)_{1/2,1}$. For this, it is enough to have $D(A; \mathcal{G}^*) \supset \mathcal{G}_1^*$. But this is easy to prove, because $A = \sum_{j=1}^n F_j(P)Q_j + \frac{i}{2}f(P)$, the functions F_j and f are of class C_0^∞ (hence $F_j(P), f(P) \in B(\mathcal{G}^*)$) and $\mathcal{G}_1^* = \{u \in \mathcal{G}^* \mid Q_j u \in \mathcal{G}^*, 1 \leq j \leq n\}$.

In conclusion, everything is a consequence of Proposition 7.5.6 once we have proven that $V_S, V_L \in \mathcal{C}^{1,1}(A; \mathcal{G}, \mathcal{G}^*)$. For this, we shall use Proposition 7.5.7 and Theorem 7.5.8 with $\Lambda = \langle Q \rangle$. Let $\varphi_\tau(x) = e^{i\langle x \rangle \tau}$ for $\tau \in \mathbb{R}$. A straightforward calculation shows that $\varphi_\tau \in BC^\infty(X)$ and $\|\varphi_\tau\|_{BC^N} \leq C_N \langle \tau \rangle^N$ for each $N \in \mathbb{N}$. According to the remarks preceding the theorem, the group $\{e^{iA\tau}\}_{\tau \in \mathbb{R}}$ leaves $\mathcal{G}$ invariant and induces C_0-groups of polynomial growth in $\mathcal{G}$ and $\mathcal{G}^*$. The properties $D(\Lambda; \mathscr{G}^*) \subset D(A; \mathscr{G}^*)$ and $\Lambda^{-2}A^2 \in B(\mathscr{G})$ required in Proposition 7.5.7 and Theorem 7.5.8 are trivial to verify. It is also easy to see that (7.6.19) implies (7.5.29). So, we have $V_S \in \mathscr{C}^{1,1}(A; \mathscr{G}, \mathscr{G}^*)$.

Finally, let us consider V_L. Note that $e^{iA\tau}$ leaves $\mathscr{S}(X)$ invariant (see Section 4.2) and that we have, in $B(\mathscr{S}, \mathscr{S}^*)$:

$$(7.6.21) \quad i[A, V_L] = \sum_{j=1}^{n}\{[iQ_j, V_L]F_j(P) + Q_j[iF_j(P), V_L]\} + \frac{1}{2}[f(P), V_L].$$

It is enough to show that, if T is any of the terms in the right-hand side of this expression, then T belongs to $B(\mathscr{G}, \mathscr{G}^*)$ and satisfies (7.5.26) (assuming (7.6.20)). For $[Q_j, V_L]F_j(P)$ this is obvious, because $F_j(P) \in B(\mathscr{G})$. For the other two types of terms, we need an expression for $[g(P), V_L]$ in terms of $[P_j, V_L]$, when $g \in C_0^\infty(X)$. This is easy by elementary commutator calculus (Chapter 5 gives much better estimates, but they are not useful here):

$$[g(P), V_L] = \sum_{j=1}^{n}\int_0^1 d\tau \int_X e^{iP\cdot x\tau}[P_j, V_L]e^{iP\cdot x(1-\tau)}(\mathcal{F}\partial_j g)(x)dx.$$

Here $g_j \equiv \mathcal{F}\partial_j g \in \mathscr{S}(X)$.

We treat the worst term in (7.6.21), namely $Q_j[F_j(P), V_L]$. Since $Q_j\langle Q\rangle^{-1} \in B(\mathscr{G}^*)$, we see that it is enough to prove that

$$\int_1^\infty \|\xi(r^{-1}Q)\langle Q\rangle[g(P), V_L]\|_{B(\mathscr{G}, \mathscr{G}^*)}\frac{dr}{r} < \infty.$$

Clearly:

$$(7.6.22) \quad \|\xi(r^{-1}Q)\langle Q\rangle[g(P), V_L]\|_{B(\mathscr{G}, \mathscr{G}^*)} \leq$$

$$\leq \sum_{j=1}^{n}\int_0^1 d\tau \int_X \|\xi(r^{-1}(Q - x\tau))\langle Q - x\tau\rangle[P_j, V_L]\|_{B(\mathscr{G}, \mathscr{G}^*)}|g_j(x)|dx$$

$$\leq C\sum_{j=1}^{n}\int_0^1 d\tau \int_X \|\xi(r^{-1}(Q - x\tau))\langle Q\rangle[P_j, V_L]\|_{B(\mathscr{G}, \mathscr{G}^*)}\langle\tau x\rangle^N |g_j(x)|dx d\tau.$$

In fact, if $\varphi(y) = \langle y + z\rangle\langle y\rangle^{-1}$, then $\varphi \in BC^\infty(X)$ and $\|\varphi\|_{BC^N} \leq C_N\langle z\rangle^N$. Now observe that for $r \geq |y| + 1$, assuming $\xi(x) = 0$ for $|x| < 1$ and $\xi(x) = 1$ for $|x| > 2$, we have $\xi(r^{-1}(Q - y)) = \xi(r^{-1}(Q - y))\xi(2(r - |y|)^{-1}Q)$, hence:

$$\int_1^\infty \left\|\xi\left(\frac{Q - y}{r}\right)T\right\|_{B(\mathscr{G}, \mathscr{G}^*)}\frac{dr}{r} = \int_1^{|y|+1}\left\|\xi\left(\frac{Q - y}{r}\right)T\right\|_{B(\mathscr{G}, \mathscr{G}^*)}\frac{dr}{r}$$

$$+ \int_{|y|+1}^\infty \left\|\xi\left(\frac{Q - y}{r}\right)\xi\left(\frac{2Q}{r - |y|}\right)T\right\|_{B(\mathscr{G}, \mathscr{G}^*)}\frac{dr}{r}$$

$$\leq C\|T\|_{B(\mathscr{G}, \mathscr{G}^*)}\ln(|y| + 1) + C\int_{1/2}^\infty \left\|\xi\left(\frac{Q}{r}\right)T\right\|_{B(\mathscr{G}, \mathscr{G}^*)}\frac{dr}{r}.$$

Hence, upon integrating (7.6.22) on $(1, \infty)$ with respect to the measure $r^{-1}dr$, we get a finite quantity. $\square$

REMARK 7.6.9. We remind the reader that the hypotheses (S) and (L) of the last theorem are independent of the choice of θ and ξ (see Section 3.5). Moreover, if we replace θ by ξ in (7.6.19), we get an equivalent condition. In fact, assume (7.6.19) holds for some θ and let ξ be a radial function with the properties stated in (L). We identify $\xi(x) = \xi(|x|)$, so we consider ξ as a function on $(0, \infty)$, equal to zero near zero and to 1 near infinity. Since $\xi(r^{-1}Q) \to 0$ strongly in $B(\mathscr{G}^*)$ as $r \to \infty$, we get:

$$
\int_1^\infty \|\xi(r^{-1}Q)V_S\|_{B(\mathscr{G},\mathscr{G}^*)} dr \leq \int_1^\infty dr \int_r^\infty \left\| \frac{d}{dt}\xi(t^{-1}Q)V_S \right\|_{B(\mathscr{G},\mathscr{G}^*)} dt
$$
$$
= \int_1^\infty dr \int_r^\infty \|\xi_1(t^{-1}Q)V_S\|_{B(\mathscr{G},\mathscr{G}^*)} \frac{dt}{t}
$$
$$
= \int_1^\infty \|\xi_1(t^{-1}Q)V_S\|_{B(\mathscr{G},\mathscr{G}^*)} \frac{t-1}{t} dt < \infty
$$

because $\xi_1(s) = s\xi'(s)$ has compact support in $(0, \infty)$. It is unfortunate that we are not able to deduce condition (L) from a similar condition with ξ replaced by θ. This would give very satisfactory conditions (aesthetically speaking) on the long-range part V_L.

We continue with a comment concerning the short-range assumption. In many cases V_S is a *local operator*, i.e. supp $V_S f \subset$ supp f for any $f \in \mathscr{G}$ (e.g. let V_S be a differential operator). Then hypothesis (S) can be expressed in a form similar to that put forward by Hörmander in Chapter XIV of [H].

THEOREM 7.6.10. *Let $V_S : \mathscr{G} \to \mathscr{G}^*$ be a symmetric, local operator. Then V_S is short-range, in the sense of condition* (S) *of Theorem 7.6.8, if and only if $V_S \in B(\mathscr{G}^\circ_{-1/2,\infty}, \mathscr{G}^*_{1/2,1})$.*

PROOF. 1) Assume first that V_S satisfies (7.6.19). Let a_1, b_1 be numbers such that $0 < a_1 < a < b < b_1 < \infty$ and let $\theta_1 \in C_0^\infty(X)$ such that $\theta_1(x) = 1$ if $|x| \in [a, b]$, $\theta_1(x) > 0$ if $a_1 < |x| < b_1$, and $\theta_1(x) = 0$ otherwise. V_S being local, we shall have $\theta(r^{-1}Q)V_S = \theta(r^{-1}Q)V_S\theta_1(r^{-1}Q)$. So, for $f \in \mathscr{G}$:

$$
\int_1^\infty \|r^{1/2}\theta(r^{-1}Q)V_S f\|_{\mathscr{G}^*} \frac{dr}{r} \leq \int_1^\infty \|\theta(r^{-1}Q)V_S\|_{B(\mathscr{G},\mathscr{G}^*)} dr \cdot
$$
$$
\cdot \sup_{r \geq 1} \|r^{-1/2}\theta_1(r^{-1}Q)f\|_{\mathscr{G}}.
$$

This easily allows one to show that $V_S \in B(\mathscr{G}^\circ_{-1/2,\infty}, \mathscr{G}^*_{1/2,1})$.

2) Reciprocally, assume that this last property holds. Now it is more convenient to use a dyadic version of (7.6.19). Let $\theta_0 \in C_0^\infty(X)$ with $\theta_0(x) > 0$ if $2^{-1} < |x| < 2$ and $\theta_0(x) = 0$ otherwise, and such that $\sum_{j=-\infty}^\infty \theta_0(2^{-j}x) = 1$ if $x \neq 0$ (see the proof of Theorem 3.6.2 (b)). We set $\theta_{-1}(x) = \sum_{j \leq -1} \theta_0(2^{-j}x)$ for $x \neq 0$ and $\theta_{-1}(0) = 1$, and $\theta_j(x) = \theta_0(2^{-j}x)$ if $j \in \mathbb{N}$. Then $\theta_{-1} \in C_0^\infty(X)$, $\theta_{-1}(x) = 1$ for $|x| \leq 2^{-1}$, $\theta_{-1}(x) > 0$ if $|x| < 1$ and $\theta_{-1}(x) = 0$ if $|x| \geq 1$. For $j \geq 0$, we have $\theta_j(x) > 0$ if $x \in X_j = \{x \in X \mid 2^{j-1} < |x| < 2^j\}$ and $\theta_j(x) = 0$ otherwise. Finally $\sum_{j=-1}^\infty \theta_j(x) = 1$ on X, and in this sum only successive functions have supports with non-disjoint interiors. In particular, we shall have

$\widetilde{\theta}_j(x) \equiv \theta_{j-1}(x) + \theta_j(x) + \theta_{j+1}(x) = 1$ on X_j, so $\theta_j \widetilde{\theta}_j = 1$. V_S being local, it is clear that $\theta_j(Q)V_S\widetilde{\theta}_j(Q) = \theta_j(Q)V_S$. It is not difficult to see that V_S satisfies condition (S) of Theorem 7.6.8 if and only if $\sum_{j=0}^{\infty} 2^j ||\theta_j(Q)V_S||_{B(\mathscr{G},\mathscr{G}^*)} < \infty$ (cf. the proof of Theorem 3.6.2 (b)). To estimate the preceding sum, we use an idea from the proof of Theorem 14.2.2 of [H]. Observe first that there are a constant $C < \infty$ and a sequence $f_j \in C_0^{\infty}(\widetilde{X}_j)$, where $\widetilde{X}_j = \overline{X_{j-1}} \cup \overline{X_j} \cup \overline{X_{j+1}}$, with $||f_j||_{\mathscr{G}} = 1$ and $||\theta_j(Q)V_S||_{B(\mathscr{G},\mathscr{G}^*)} \leq C||\theta_j(Q)V_S f_j||_{\mathscr{G}^*}$. In fact, we can find a sequence $g_k \in C_0^{\infty}(X)$ with $||\theta_j(Q)V_S||_{B(\mathscr{G},\mathscr{G}^*)} = \lim_{k\to\infty} ||\theta_j(Q)V_S g_k||_{\mathscr{G}^*} \cdot ||g_k||_{\mathscr{G}}^{-1}$. On the other hand $\theta_j(Q)V_S g_k = \theta_j(Q)V_S\widetilde{\theta}_j(Q)g_k$ and $||\widetilde{\theta}_j(Q)g_k||_{\mathscr{G}} \leq C\, 2^{-1}||g_k||_{\mathscr{G}}$ with a constant independent of j, k. Then we take $f_j = \widetilde{\theta}_j(Q)g_k \cdot ||\widetilde{\theta}_j(Q)g_k||_{\mathscr{G}}^{-1}$ for some large enough k (depending on j). For the next step of the proof, notice the following property: $\theta_j(Q)V_S f_k \neq 0$ only if $|k-j| \leq 1$. Now, let $g = \sum 2^{k/2} f_k$, where the sum is over the *even* integers $k \geq 0$, and let $h = \sum 2^{k/2} f_k$ with sum over *odd* integers $k \geq 1$. For any $j \geq 0$, $\theta_j(Q)g$ and $\theta_j(Q)h$ will be sums of at most two terms, hence we trivially get $2^{-j/2}||\theta_j(Q)g||_{\mathscr{G}} \leq$ const. $< \infty$ and similarly for h. So $g, h \in \mathscr{G}_{-1/2,\infty}$ (see Theorem 3.6.2 (b)). By hypothesis we shall get $||V_S g||_{\mathscr{G}^*_{1/2,1}} < \infty$ and similarly for h. By using Theorem 3.6.2 (b) again, we obtain for example $\sum_{j\geq 0} 2^{j/2}||\theta_j(Q)V_S g||_{\mathscr{G}^*} < \infty$. But, from a remark we made above, it follows for *even* j that $\theta_j(Q)V_S g = \theta_j(Q)2^{j/2}V_S f_j$. Hence $\sum 2^j ||\theta_j(Q)V_S f_j||_{\mathscr{G}^*} < \infty$, where the sum is over even j. Upon replacing g by h we get a similar estimate, but the sum will be over odd j. In conclusion: $\sum_{j=0}^{\infty} 2^j ||\theta_j(Q)V_S||_{B(\mathscr{G},\mathscr{G}^*)} \leq \sum_{j=0}^{\infty} 2^j C ||\theta_j(Q)V_S f_j||_{\mathscr{G}^*} < \infty$. $\square$

We state now a result concerning the existence and completeness of relative wave operators. This is an obvious consequence of Theorem 7.5.5 and of the preceding results.

THEOREM 7.6.11. *Let H be the operator of Theorem 7.6.8 and $U : \mathscr{G} \to \mathscr{G}^*$ a symmetric operator having the following properties:*

(a) $H + U + i$ is an isomorphism of $\mathscr{G}$ onto $\mathscr{G}^*$ and, if $\widetilde{H}$ is the self-adjoint operator in $\mathscr{H}$ associated to $H + U$, then the difference $(\widetilde{H} + i)^{-1} - (H + i)^{-1}$ is compact in $\mathscr{H}$;

(b) U extends to a continuous operator $\mathscr{G}^{\circ}_{-1/2,\infty} \to \mathscr{G}^*_{1/2,1}$;

(c) U is short-range, i.e. there is $\theta \in C_0^{\infty}(X)$ with $\theta(x) > 0$ in a domain $0 < a < |x| < b < \infty$ and $\theta(x) = 0$ otherwise, such that

$$(7.6.23) \qquad \int_1^{\infty} ||\theta(r^{-1}Q)U||_{B(\mathscr{G},\mathscr{G}^*)} dr < \infty.$$

(We recall that conditions (b) and (c) are equivalent if U is local). Let $J = \mathbb{R} \setminus \kappa(h)$ and denote by E_c and $\widetilde{E}_c$ the continuous component of the spectral measure of H and $\widetilde{H}$ respectively. Then the wave operators

$$(7.6.24) \qquad W_{\pm} = \operatorname*{s-lim}_{t\to\pm\infty} e^{i\widetilde{H}t} e^{-iHt} E_c(J)$$

exist and are complete, i.e. their ranges are equal to $\widetilde{E}_c(J)$.

The simplest version of the situation considered above is that in which h is an elliptic symbol of degree $2s > 0$, i.e. $h \in C^\infty(X)$, $|h^{(\alpha)}(x)| \le c_\alpha \langle x \rangle^{2s-|\alpha|}$ for each multi-index α, and we have $|h(x)| \ge c|x|^{2s}$, for some $c > 0$, outside a compact set. Then $\mathscr{G} = \mathscr{H}^s(X)$ and $\mathscr{G}^* = \mathscr{H}^{-s}(X)$ are usual Sobolev spaces, and we get the limiting absorption principle in $B(\mathscr{H}^{-s}_{1/2,1}, \mathscr{H}^s_{-1/2,\infty})$; here $\mathscr{H}^s_{t,q}$ are standard weighted (in the Besov sense) Sobolev spaces. In this context, $U : \mathscr{H}^s \to \mathscr{H}^{-s}$ is short-range if

$$(7.6.25) \qquad \int_1^\infty \|\theta(r^{-1}Q)U\|_{B(\mathscr{H}^s, \mathscr{H}^{-s})} dr < \infty$$

for a function $\theta \in C_0^\infty(X \setminus \{0\})$ such that $\theta(x) > 0$ in a domain $0 < a < |x| < b$; and, for local U, this is equivalent to $U \in B(\overset{\circ}{\mathscr{H}}{}^s_{-1/2,\infty}, \overset{\circ}{\mathscr{H}}{}^s_{1/2,\infty})$ ($\overset{\circ}{\mathscr{H}}{}^s_{t,\infty}$ is the closure of $\mathscr{S}(X)$ in $\mathscr{H}^s_{t,\infty}$). In the Schrödinger case, $s = 1$, we get sharp results in the spectral analysis and scattering theory of operators of the form $\Delta + V_S + V_L$, with both the short-range part V_S and long-range part V_L non-local. Notice that both V_S and V_L could contain second order terms with respect to the derivatives.

There is one implicit condition in Theorem 7.6.8, namely the operator $(H + i)^{-1} - (H_0 + i)^{-1}$ has to be compact in $\mathscr{H}$. Let us describe a general method of checking it. Consider a self-adjoint operator H_0 with form domain $\mathscr{G}$ in a Hilbert space $\mathscr{H}$ and let $V : \mathscr{G} \to \mathscr{G}^*$ be a symmetric operator such that $H_0 + V + i : \mathscr{G} \to \mathscr{G}^*$ is an isomorphism. Denote by H both the operator $H_0 + V : \mathscr{G} \to \mathscr{G}^*$ and the self-adjoint operator in $\mathscr{H}$ associated to it, and set $R = (H + i)^{-1}, R_0 = (H_0 + i)^{-1}$. Then the identity $R - R_0 = -RVR_0$ holds in $B(\mathscr{G}^*, \mathscr{G})$. The domains $D(H_0)$ and $D(H)$ of H_0 and H are equipped with the graph topology and are dense B-subspaces of $\mathscr{G}$, so that we have continuous dense embeddings $D(H_0) \subset \mathscr{G} \subset \mathscr{H} \subset \mathscr{G}^* \subset D(H)^*$. Since R_0 is an isomorphism of $D(H_0)$ onto $\mathscr{H}$ and R extends to an isomorphism of $D(H)^*$ onto $\mathscr{H}$, it follows that $R - R_0$ *is a compact operator in* $\mathscr{H}$ *if and only if* V *is a compact operator from* $D(H_0)$ *to* $D(H)^*$. In the situations where the domain of H is not explicitly known one should rather use the following fact: *if* V *is a compact operator from* $D(H_0)$ *to* $\mathscr{G}^*$, *then* $R - R_0$ *is a compact operator from* $\mathscr{H}$ *to* $\mathscr{G}$ *(hence in* $\mathscr{H}$*).*

If we apply the last criterion in the context of Theorem 7.6.8 with h an elliptic symbol of degree $2s > 0$, and if we take into account that in this case $D(H_0) = \mathscr{H}^{2s}(X)$, we see that the compactness assumption of the theorem is fulfilled provided that the operator $V : \mathscr{H}^s \to \mathscr{H}^{-s}$ induces a compact operator from $\mathscr{H}^{2s}$ to $\mathscr{H}^{-s}$ (then by interpolation we get that $V : \mathscr{H}^t \to \mathscr{H}^{-s}$ is compact for all $t > s$). Now it is not difficult to show that this is equivalent with the condition that the symmetric operator $V : \mathscr{H}^s(X) \to \mathscr{H}^{-s}(X)$ be *small at infinity* in the following sense: there is a function $\xi \in C^\infty(X)$ with $\xi(x) = 0$ if $|x| < 1$ and $\xi(x) = 1$ if $|x| > 2$, and there is a real number $t > s$, such that $\lim_{r\to\infty} \|\xi(Q/r)V\|_{\mathscr{H}^s \to \mathscr{H}^{-t}} = 0$ (see the proof of Lemma 9.4.6). Moreover, one may easily prove that the short-range condition (7.6.25) implies that $U : \mathscr{H}^s(X) \to \mathscr{H}^{-s}(X)$ is small at infinity (see Remark 9.4.14).

It is interesting to apply the preceding considerations to generalized second order elliptic operators of the form

$$(7.6.26) \qquad H = \sum_{j,k=1}^{n} P_j A_{jk} P_k + \sum_{j=1}^{n} (P_j B_j + B_j^* P_j) + C$$

where $P_j = -i\partial_j$ (derivatives with respect to an orthonormal basis in X), $A_{jk} \in B(\mathcal{H}(X))$, $B_j \in B(\mathcal{H}^1(X), \mathcal{H}(X))$ and $C \in B(\mathcal{H}^1(X), \mathcal{H}^{-1}(X))$. We assume that the following uniform ellipticity condition is satisfied: there is $\kappa > 0$ such that $\sum_{j,k=1}^{n} \langle f_j, A_{jk} f_k \rangle \geq \kappa \sum_{j=1}^{n} \|f_j\|^2$ for all $f_1, \ldots, f_n \in \mathcal{H}(X)$. In order to be able to choose $H_0 = \Delta = \sum_{j=1}^{n} P_j^2$ (hence $s = 1$), we have to assume that $A_{jk} = \delta_{jk} I + A_{jk}^0$ where $A_{jk}^0 : \mathcal{H}(X) \to \mathcal{H}(X)$ is small at infinity. For example, if $\int_1^\infty \|\theta(Qr^{-1}) A_{jk}^0\|_{B(\mathcal{H})} dr < \infty$ for some function θ as in (7.6.25), then $\sum_{j,k=1}^{n} P_j A_{jk}^0 P_k$ is a short-range perturbation of Δ. (7.6.26) includes more exotic perturbations of the laplacian like highly oscillating potentials. Indeed, if $B_j = iW_j$ with $W_j : \mathcal{H}^1(X) \to \mathcal{H}(X)$ symmetric and short-range, then the second term in (7.6.26) is of the form $S = \sum_{j=1}^{n} [iP_j, W_j]$ and, as an operator from $\mathcal{H}^1(X)$ to $\mathcal{H}^{-1}(X)$, it is symmetric and short-range (hence small at infinity). If $W_j = w_j(Q)$, then $S = (\operatorname{div} w)(Q)$, hence operators of the form $H = -\frac{d^2}{dx^2} + e^x \langle x \rangle^{-1-\varepsilon} \sin e^x$ in $L^2(\mathbb{R})$ satisfy the conditions of Theorem 7.6.8.

7.A. Appendix: The Gronwall Lemma

We prove a result that is related to the usual Gronwall lemma. For more general results of this type, see Chapter III in [Hm].

LEMMA 7.A.1. *Let* $J = (a, b) \subset \mathbb{R}$ *be an open interval and let* f, φ *and* ψ *be non-negative real functions on* (a, b) *with* f *bounded and* φ, $\psi \in L^1((a, b))$. *Assume that, for some constants* $\omega \geq 0$ *and* $\theta \in [0, 1)$ *and for all* $\lambda \in (a, b)$:

$$(7.A.1) \qquad f(\lambda) \leq \omega + \int_\lambda^b [\varphi(\tau) f(\tau)^\theta + \psi(\tau) f(\tau)] d\tau.$$

Then one has for each $\lambda \in (a, b)$:

$$(7.A.2)$$
$$f(\lambda) \leq \left[\omega^{1-\theta} + (1-\theta) \int_\lambda^b \varphi(\mu) \exp\left\{ (\theta - 1) \int_\mu^b \psi(\tau) d\tau \right\} d\mu \right]^{1/(1-\theta)}$$
$$\cdot \exp\left\{ \int_\lambda^b \psi(\tau) d\tau \right\}.$$

PROOF. We may assume that $\omega > 0$. We set, for $\mu \in (a,b)$:

$$\phi(\mu) = \varphi(\mu) \exp\left\{ (\theta - 1) \int_\mu^b \psi(\tau) d\tau \right\}$$

$$g(\mu) = \omega + \int_\mu^b [\varphi(\tau) f(\tau)^\theta + \psi(\tau) f(\tau)] d\tau$$

$$h(\mu) = g(\mu) \exp\left\{ - \int_\mu^b \psi(\tau) d\tau \right\}.$$

By (7.A.1) we then have for each $\mu \in (a,b)$:

$$(7.A.3) \qquad f(\mu) \le g(\mu) = h(\mu) \exp\left\{ \int_\mu^b \psi(\tau) d\tau \right\}.$$

We use this inequality to deduce that

$$\frac{d}{d\mu} h(\mu) = \left[\frac{d}{d\mu} g(\mu) + g(\mu)\psi(\mu) \right] \exp\left\{ - \int_\mu^b \psi(\tau) d\tau \right\}$$

$$= [-\varphi(\mu) f(\mu)^\theta + \{g(\mu) - f(\mu)\}\psi(\mu)] \exp\left\{ - \int_\mu^b \psi(\tau) d\tau \right\}$$

$$\ge -\varphi(\mu) g(\mu)^\theta \exp\left\{ - \int_\mu^b \psi(\tau) d\tau \right\} = -\phi(\mu) h(\mu)^\theta.$$

We observe that $h(\mu) \ge \omega \exp[-\|\psi\|_{L^1(a,b)}] > 0$, so that

$$\frac{d}{d\mu} [h(\mu)^{1-\theta}] = (1-\theta) h(\mu)^{-\theta} \frac{d}{d\mu} h(\mu) \ge -(1-\theta)\phi(\mu).$$

Since $h(\mu) \to \omega$ as $\mu \to b$, we get upon integrating the preceding inequality on the interval (λ, b) that

$$[h(\lambda)]^{1-\theta} \le \omega^{1-\theta} + (1-\theta) \int_\lambda^b \phi(\mu) d\mu.$$

By inserting the definitions of h and ϕ and by using (7.A.3), one gets (7.A.2). □

The estimate (7.A.2) is optimal: one has equality in (7.A.2) (for all λ) if and only if equality holds in (7.A.1) (for all λ). In other terms, any solution of the inequation (7.A.1) is dominated by the solution of the corresponding equation.

7.B. Appendix: A Counterexample. Optimality of the Results on the Limiting Absorption Principle

This appendix is a natural continuation of §7.1.4; we shall deal with rather subtle local properties of continuous functions of a real variable. Let $\lambda \in \mathbb{R}$ and let f be a continuous function defined in a neighbourhood $|x - \lambda| < 2a$ of λ. We shall say that λ is a *nasty point* for f if the quantity

$$\int_{\varepsilon < |t| < a} f(\lambda + t) \frac{dt}{t} \equiv \int_\varepsilon^a \frac{f(\lambda + t) - f(\lambda - t)}{t} dt$$

is not bounded when $\varepsilon \to +0$. This property depends only on the behaviour of f near λ, in particular the choice of the number a is irrelevant. Observe that

(7.B.1)
$$
\begin{aligned}
J_\lambda &\equiv \left| \int_{|t|<a} f(\lambda+t) \frac{t \, dt}{t^2+\varepsilon^2} - \int_{\varepsilon<|t|<a} f(\lambda+t) \frac{dt}{t} \right| \\
&= \left| \int_{|t|<\varepsilon} [f(\lambda+t)-f(\lambda)] \frac{t \, dt}{t^2+\varepsilon^2} - \int_{\varepsilon<|t|<a} [f(\lambda+t)-f(\lambda)] \frac{\varepsilon^2 \, dt}{t(t^2+\varepsilon^2)} \right| \\
&\leq \ln 2 \cdot \sup_{|t|<\varepsilon} |f(\lambda+t)-f(\lambda)| + \int_{1<|t|<a\varepsilon^{-1}} |f(\lambda+\varepsilon t)-f(\lambda)| \frac{dt}{|t|(1+t^2)}.
\end{aligned}
$$

By using the dominated convergence theorem we see that an equivalent expression for the property of being nasty is as follows: the quantity $\int_{|t|<a} f(\lambda + t)(t^2+\varepsilon^2)^{-1} t \, dt$ is unbounded as $\varepsilon \to +0$. The weaker estimate $J_\lambda \leq \ln 2 \cdot$ ess $\sup_{|t|<a} |f(\lambda+t)|$ (which is obtained as above, but without subtracting $f(\lambda)$), which is valid for an arbitrary bounded Borel function f, is enough for the proof of the equivalence. But the preceding argument gives more: if we replace "unbounded" by "does not have a limit" in the definition of a nasty point (this would give a larger class of nasty points), then we still have two equivalent definitions (for continuous f).

If f_1 is another function defined on a neighbourhood of λ which touches f at λ, i.e. $f(\lambda) = f_1(\lambda)$, and if λ is a nasty point for f, then it could also be a nasty point for f_1, or it could not. However, if the contact of f and f_1 is of order slightly higher than zero, then we are in the first situation. More precisely, let us say that f and f_1 have a *contact of Dini order* at λ if $\int_{|t|<a} |f(\lambda+t)-f_1(\lambda+t)| \cdot |t|^{-1} dt < \infty$ for some $a > 0$. Then clearly λ is nasty for f if and only if it is nasty for f_1. In particular, if θ is defined on a neighbourhood of λ and is *Dini-continuous* at λ, i.e. $\int_{|t|<a} |\theta(\lambda + t) - \theta(\lambda)| \cdot |t|^{-1} dt < \infty$, then a point λ is nasty or non-nasty simultaneously for θf and $\theta(\lambda) f$; if furthermore $\theta(\lambda) \neq 0$, then λ is nasty or non-nasty simultaneously for θf and f. For example, if λ is a nasty point for f and θ is a C^1 function in a neighbourhood of λ with $\theta(\lambda) \neq 0$, then λ is a nasty point for θf. Finally, since $\int_{\varepsilon<|t|<a} ct^{-1} dt = 0$ for any constant c, if f is a Dini-continuous function, then all the points of its domain are non-nasty.

In order to see the interest of the notion introduced above, let us go back to the context of §7.1.4. We have an increasing homeomorphism h of an open interval $\Omega \subset \mathbb{R}$ onto another interval $J \subset \mathbb{R}$, with $g = h^{-1} : J \to \mathbb{R}$ of class C^1. Then for any $\varphi \in C_0^0(J)$ and $\mu \neq 0$:

(7.B.2)
$$
\int_\Omega \varphi(x) \Re[h(x) - \lambda - i\mu]^{-1} dx = \int_J \varphi(g(y)) g'(y) \frac{y-\lambda}{(y-\lambda)^2 + \mu^2} dy.
$$

Set $\psi = \varphi \circ g \cdot g' \in C_0^0(J)$ (if g is not more than C^1, even if φ is C^∞, ψ will not be more than continuous). As a consequence of the preceding discussion we see that $\lim_{\mu \to +0} \pi^{-1} \int_\Omega \varphi(x) \Re[h(x) - \lambda - i\mu]^{-1} dx$ *exists if and only if the Hilbert transform* $\tilde\psi(\lambda) \equiv \lim_{\varepsilon \to +0} \pi^{-1} \int_{|y-\lambda|>\varepsilon} \psi(y)(y-\lambda)^{-1} dy$ *of* ψ *at* λ *exists; in this case the two limits are equal.* Moreover, if $\varphi \in C_0^1(\Omega)$ (so $\varphi \circ g \in C_0^1(J)$), then we can localize the bad points λ in a φ-independent way. More precisely: *if $\lambda \in J$ is a nasty point for the function g' and if $\varphi \in C_0^1(\Omega)$ is such that $\varphi(g(\lambda)) \neq 0$, then*

(7.B.3)
$$\limsup_{\mu \to +0} \left| \int_\Omega \varphi(x) \Re[h(x) - \lambda - i\mu]^{-1} dx \right| = \infty.$$

In particular, the limiting absorption principle breaks down in a very drastic way at nasty points λ of g', namely the family of distributions $\{\Re[h(x) - \lambda - i\mu]^{-1}\}_{\mu>0}$ is not bounded in $\mathscr{D}^*(\Omega)$, so the formal symbol $\Re[h(x) - \lambda - i0]^{-1}$ has no meaning as a distribution.

We shall construct now absolutely continuous functions with many nasty points; our procedure can be traced back to an example due to Lusin in the theory of conjugate trigonometric series (see Section 13, Chapter 8 in [Ba]). Let f be an absolutely continuous function defined in the neighbourhood of a point λ. Then

$$\int_{-a}^{a} f(\lambda + t) \frac{2t \, dt}{t^2 + \varepsilon^2} = [f(\lambda + a) - f(\lambda - a)] \ln(a^2 + \varepsilon^2)$$
$$- \int_{-a}^{a} f'(\lambda + t) \ln(t^2 + \varepsilon^2) dt.$$

Hence λ is nasty for f if and only if $\int_{-a}^{a} f'(\lambda + t) \ln(t^2 + \varepsilon^2) dt$ is unbounded as $\varepsilon \to 0$. Assume, furthermore, that f' is positive in a neighbourhood of λ; then, by using the monotone convergence theorem (observe that $\ln(t^2 + \varepsilon^2)$ decreases to $2 \ln |t|$ as $\varepsilon \to 0$), we obtain the following fact: *if f is an absolutely continuous increasing function on a neighbourhood of λ, then λ is a nasty point for f if and only if $\int_{-a}^{a} f'(\lambda + t) |\ln|t|| \, dt = \infty$ for some $a \in (0,1)$.*

We start now an explicit construction. Let $\alpha > 0$ be a positive real number (only $\alpha \leq 1$ will be of interest). Consider a function $\eta : (0, \infty) \to \mathbb{R}$ with the following properties: (1) $\eta(x) = \alpha x^{-1} |\log x|^{-1-\alpha}$ if $0 < x < e^{-1-\alpha}$; (2) η is of class C^∞; (3) η is decreasing ; (4) $\eta(x) > 0$ if $0 < x < 1$; (5) $\eta(x) = 0$ if $x \geq 1$. Then let $\eta(0) > 0$ be arbitrary and $\eta(x) = \eta(-x)$ if $x < 0$. We get an integrable, positive function $\eta : \mathbb{R} \to \mathbb{R}$ with support equal to $[-1, +1]$. Observe that $\eta(x) |\log |x||$ is *not* integrable if $\alpha \leq 1$. We define $\xi(x) = \int_{-\infty}^{x} \eta(t) dt$, so $\xi : \mathbb{R} \to \mathbb{R}$ is a positive, increasing, absolutely continuous function such that: (1) $\xi(x) = 0$ if $x \leq -1$; (2) $\xi(x) > 0$ if $x > -1$; (3) $\xi(x) = 2 \int_0^1 \eta(t) dt$ if $x \geq +1$; (4) $\xi(x) = \operatorname{sgn} x |\log |x||^{-\alpha} + C$ if $|x| < e^{-1-\alpha}$, where C is a constant. Finally, set $\zeta(x) = \int_{-\infty}^{x} \xi(t) dt = \int_{-\infty}^{x} (x - t)\eta(t) dt$. Then $\zeta : \mathbb{R} \to \mathbb{R}$ is a positive, increasing, convex function of class C^1, with $\zeta(x) = 0$ if $x \leq -1$, $\zeta(x) > 0$ if $x > -1$ and $\zeta(x)$ is linear for $x \geq +1$. So ζ is a convex function of class C^1 which interpolates between the zero function (for $x \leq -1$) and a linear, increasing function (for $x \geq +1$).

Now let D be a countable subset of $\mathbb{R}$ and for each $r \in D$ let a_r be a strictly positive number such that $\sum_{r \in D} a_r < \infty$. Define $f(x) = \sum_{r \in D} a_r \xi(x - r)$. Then $f : \mathbb{R} \to \mathbb{R}$ is a positive, increasing, absolutely continuous function with $f'(x) = \sum_{r \in D} a_r \eta(x - r)$. Moreover, f is bounded, and $f(x) = 0$ if and only if

$x \leq (\inf D) - 1$, $f(x) = $ const. > 0 if $x \geq (\sup D) + 1$. Clearly f is of class C^∞ outside the closure of the set D. For $\lambda \in D$ and $0 < a < 1$ we have:

$$\int_{-a}^{a} f'(\lambda + t)|\ln|t||\,dt = \sum_{r \in D} a_r \int_{-a}^{a} \eta(\lambda + t - r)|\ln|t||\,dt \geq a_\lambda \int_{-\infty}^{\infty} \eta(t)|\ln|t||\,dt.$$

The last integral is infinite if and only if $\alpha \leq 1$. In conclusion, if $0 < \alpha \leq 1$, we get an absolutely continuous, increasing, bounded function $f : \mathbb{R} \to \mathbb{R}$ such that each $\lambda \in D$ is a nasty point for f.

We shall now use f in order to construct a C^1-diffeomorphism h such that (7.B.3) holds for many λ. We shall say that h is of class $HB_\infty^{1,q}(\mathbb{R})$ (*homogeneous Besov class*) if its second modulus of continuity

$$\omega_2(h; \varepsilon) = \sup_{x \in \mathbb{R}} |h(x + \varepsilon) - 2h(x) + h(x - \varepsilon)|$$

has the property $[\int_0^1 [\varepsilon^{-1}\omega_2(h; \varepsilon)]^q \varepsilon^{-1}\,d\varepsilon]^{1/q} < \infty$.

PROPOSITION 7.B.1. *Let $D \subset \mathbb{R}$ be a countable bounded set. There is an increasing concave C^1-diffeomorphism $h : \mathbb{R} \to \mathbb{R}$ which is linear outside a compact set and such that:*

(1) *h' is an absolutely continuous function and there is a constant $c > 0$ such that $c \leq h'(x) \leq c^{-1} \; \forall x \in \mathbb{R}$;*

(2) *h is of class $HB_\infty^{1,q}(\mathbb{R})$ for all $q > 1$;*

(3) *for each $\lambda \in D$ and for each $\varphi \in C_0^1(\mathbb{R})$ with $\varphi(h^{-1}(\lambda)) \neq 0$, (7.B.3) holds.*

PROOF. We shall explicitly define the inverse g of h. Let f be as above and set $F(y) = \int_{-\infty}^{y} f(t)\,dt = \sum_{r \in D} a_r \zeta(y - r)$. Then F is positive, increasing, convex, of class $C^1(\mathbb{R})$ and of class C^∞ outside the closure $\overline{D}$ of D. Since $F'(y) = f(y) \geq $ const. > 0 for $y \geq (\inf D) - 1/2$, we can construct $g(y)$ such as to be linear and strictly increasing for $y < (\inf D) - 1$, $g(y) = F(y)$ for $y \geq (\inf D) - 1/2$, and such that $g : \mathbb{R} \to \mathbb{R}$ is a C^1-diffeomorphism of class C^∞ outside $\overline{D}$. Then g' is absolutely continuous while g is increasing and convex on $\mathbb{R}$ and linear outside a compact set. In particular, there is a finite constant c with $0 < c \leq g'(y) \leq c^{-1}$ for all $y \in \mathbb{R}$. We let $h = g^{-1}$. So $h : \mathbb{R} \to \mathbb{R}$ is an increasing concave C^1-diffeomorphism, linear outside a compact set. From $h'(x) = [g'(h(x))]^{-1}$, we get $c \leq h'(x) \leq c^{-1}$. Then, using the absolute continuity of g', it is straightforward to prove that h' is absolutely continuous and that $h''(x) = -g''(h(x))[g'(h(x))]^{-3}$. The property (3) is clear by the construction of f (with any $\alpha \leq 1$), by the fact that $g' = f$ on a neighbourhood of D, and by the statement (7.B.3). It remains to be shown that (2) holds, which is not so obvious. Observe that, until now, we could have taken any $\alpha \in (0, 1]$. We shall see that, in order to get $h \in HB_\infty^{1,q}$ for all $q > 1$, we must choose $\alpha = 1$. We proceed in two steps. First we show that $F \in HB_\infty^{1,q}$ if $q > \alpha^{-1}$; clearly g will have the same property. Then we prove that the inverse h of g is of the same class.

Observe that $F(x) = \sum_{r \in D} a_r \zeta(y - r)$ implies $\omega_2(F; \varepsilon) \leq \sum_{r \in D} a_r \omega_2(\zeta; \varepsilon)$, so, in order to prove the first assertion, it suffices to show that $\zeta \in HB_\infty^{1,q}(\mathbb{R})$ for

$q > \alpha^{-1}$. For this, we start from the identity:

$$\zeta(x + \varepsilon) - 2\zeta(x) + \zeta(x - \varepsilon) = \int_{|s| < \varepsilon} \zeta''(x + s)(\varepsilon - |s|)ds.$$

Obviously this implies $w_2(\varepsilon)\varepsilon^{-1} \le \sup_{x \in \mathbb{R}} \int_{x-\varepsilon}^{x+\varepsilon} |\zeta''(t)|dt$, where $w_2(\varepsilon) = w_2(\zeta; \varepsilon)$. In our case $\zeta'' = \eta$. We estimate the integral for $x \ge 2\varepsilon$ by taking into account that $\eta(t) \le \eta(\varepsilon)$ for $t \ge \varepsilon$. This will give $\int_{x-\varepsilon}^{x+\varepsilon} \eta(t)dt \le 2\eta(\varepsilon)\varepsilon$ for $x \ge 2\varepsilon$; clearly this holds for $x \le -2\varepsilon$ too. Now assume that $|x| < 2\varepsilon$; then $\int_{x-\varepsilon}^{x+\varepsilon} \eta(t)dt \le \int_{-3\varepsilon}^{3\varepsilon} \eta(t)dt = \xi(3\varepsilon) - \xi(-3\varepsilon)$. For small ε, we may then use the explicit form of η and ξ near zero and we get:

$$\frac{w_2(\varepsilon)}{\varepsilon} \le \max\{2\eta(\varepsilon)\varepsilon, \xi(3\varepsilon) - \xi(-3\varepsilon)\}$$

$$= \max\{2\alpha|\log\varepsilon|^{-1-\alpha}, 2|\log 3\alpha|^{-\alpha}\} \le C|\log\varepsilon|^{-\alpha}.$$

Since $\int_0^{1/2} |\log\varepsilon|^{-\alpha q} \frac{d\varepsilon}{\varepsilon} < \infty$ if $\alpha q > 1$, the assertion concerning g is proved.

In order to estimate the second modulus of continuity of h we use $h'' \le 0$ and the explicit expression of h'' in terms of g:

$$|h(x + \varepsilon) + h(x - \varepsilon) - 2h(x)| = -\int_{x-\varepsilon}^{x+\varepsilon} h''(t)(\varepsilon - |t - x|)dt$$

$$= \int_{x-\varepsilon}^{x+\varepsilon} \frac{g''(h(t))h'(t)}{g'(h(t))^2}(\varepsilon - |t - x|)dt = \int_{h(x-\varepsilon)}^{h(x+\varepsilon)} \frac{g''(s)}{g'(s)^2}(\varepsilon - |g(s) - x|)ds.$$

Let m be a constant such that $0 < h'(x) \le m$. Then $g'(s)^{-1} \le m$ and $h(x+\varepsilon) \le h(x) + m\varepsilon$, $h(x - \varepsilon) \ge h(x) - m\varepsilon$. Moreover, if we set $h(x) = y$, then

$$|g(s) - x| = |g(s) - g(y)| \ge \min_{\tau \in \mathbb{R}} |g'(\tau)| \cdot |s - y| \ge m^{-1}|s - y|.$$

In conclusion:

$$|h(x + \varepsilon) + h(x - \varepsilon) - 2h(x)| \le m^2 \int_{y-\varepsilon m}^{y+\varepsilon m} g''(s)(\varepsilon - m^{-1}|s - y|)ds$$

$$= m \int_{y-\varepsilon m}^{y+\varepsilon m} g''(s)(\varepsilon m - |s - y|)ds = m|g(y + \varepsilon m) + g(y - \varepsilon m) - 2g(y)|.$$

This implies that $w_2(h; \varepsilon) \le m w_2(g; \varepsilon m)$, hence h is of the same homogeneous Besov class as g. $\square$

Let $\mathscr{H} = L^2(\mathbb{R})$ and let H be the self-adjoint operator of multiplication by the function h constructed above. Let $A = -P = i\frac{d}{dx}$, so that $\{e^{-iA\tau}\}_{\tau \in \mathbb{R}}$ is the translation group in $\mathscr{H}$. We have $[iH, A] = h'(Q) \ge cI$, where $c > 0$ is a number. It is clear that H is of class $\mathscr{C}^{1,q}(A)$ for *all* $q > 1$. We may take D dense in a given open bounded interval $J \subset \mathbb{R}$. If $u, v \in C_0^\infty(\mathbb{R})$ and $u(h^{-1}(\lambda))v(h^{-1}(\lambda)) \ne 0$ for some $\lambda \in D \subset J$, then we have

$$(7.\text{B}.4) \qquad \limsup_{\mu \to +0} |\langle u, (H - \lambda - i\mu)^{-1}v\rangle| = \infty.$$

Observe that $\cap_{k=1}^{\infty} D(A^k) = \mathscr{H}^{\infty}(\mathbb{R}) \supset C_0^{\infty}(\mathbb{R})$. So, although A is conjugate to H on J, the limiting absorption principle breaks down in any Banach space containing $D(A)$. This shows that the conjugate operator method does not work if H is only assumed to be of class $C_u^1(A) \cap \cap_{q>1} \mathscr{C}^{1,q}(A)$. The regularity assumption $H \in \mathscr{C}^{1,1}(A)$ is optimal.

Finally, remark that the unboundedness of H plays no role. We could take for h a bounded function and use a local distortion of A of the form $-\frac{1}{2}[F(Q)P + PF(Q)]$, see the proof of Proposition 7.2.14.

7.C. Appendix: Asymptotic Velocity for $H = h(P)$

Let X be an euclidean space and $h : X \to \mathbb{R}$ a Borel function. We recall that, if h is differentiable at some point $x \in X$, then its derivative $h'(x)$ is identified with an element of X (cf. Chapter 1). It is not difficult to prove that the set of points where h is differentiable is a Borel set. The purpose of this appendix is to prove the following:

THEOREM 7.C.1. *Let X be an euclidean space, $h : X \to \mathbb{R}$ an almost everywhere differentiable Borel function and $f : X \to \mathbb{C}$ bounded and Borel. Assume that the set of all $x \in X$ where h is differentiable and $h'(x)$ is a point of discontinuity of f, is a set of measure zero. Then one has on $L^2(X)$:*

(7.C.1) $$\text{s-}\lim_{|t| \to \infty} e^{ith(P)} f(Q/t) e^{-ith(P)} = f(h'(P)).$$

If h is a C^1-function, f is continuous and the strong limit is replaced by a weak limit, then the preceding result is just the Fourier transformed version of Theorem 7.1.29 in [H]. Results of the form (7.C.1) are important in several directions. Firstly, see the way (7.C.1) is used in Ch.14 of [H]. Secondly, see the algebraic approach to scattering theory put forward by Derezinski in [De1]; the title of this appendix is in fact suggested and explained by this paper. Thirdly, there is an obvious connection with the "scattering into cones" theory, cf. [JLN]. It is here that the necessity of considering discontinuous functions f appears, namely f is the characteristic function χ_C of a cone $C \subset X$ with vertex at zero; observe that $\chi_C(x/t) = \chi_C(x)$ if $t > 0$. We shall need Theorem 7.C.1 later on in this text, but already in Section 7.1 the following corollary is used (now the fact that both h and f are allowed to be discontinuous becomes important).

COROLLARY 7.C.2. *Let $\varphi : \mathbb{R} \to \mathbb{R}$ be a Borel function having a finite, strictly positive derivative almost everywhere. Then for any $u \in L^2(\mathbb{R})$ one has:*

(7.C.2) $$\lim_{t \to +\infty} \int_{-\infty}^{0} \left| [e^{-it\varphi(P)} u](x) \right|^2 dx = 0.$$

PROOF. Let $X = \mathbb{R}$, $h = \varphi$ and $f = \chi_-$ the characteristic function of the set $(-\infty, 0]$. Assuming φ differentiable almost everywhere and $\{x \mid \varphi'(x) = 0\}$ of measure zero (observe that 0 is the only point of discontinuity of f), we get

$$\lim_{t \to +\infty} ||\chi_-(Q) e^{-it\varphi(P)} u|| = ||\chi_-(\varphi'(P)) u||$$

for all $u \in L^2(\mathbb{R})$ (because $\chi_-(x/t) = \chi_-(x)$ if $t > 0$). If $\varphi'(x) > 0$ almost everywhere, then $\chi_-(\varphi'(P)) = 0$. $\square$

As a final comment, notice that we have the somewhat formal identity:

$$(7.C.3) \qquad e^{ith(P)} f(Qt^{-1}) e^{-ith(P)} = f(Qt^{-1} + h'(P)),$$

which clarifies (7.C.1).

PROOF OF THE THEOREM. We first consider f such that its Fourier transform $\widehat{f}$ is an integrable measure. Since

$$(7.C.4) \qquad e^{ith(P)} f(Q/t) e^{-ith(P)} = \int_X e^{ith(P)} e^{iQ \cdot x/t} e^{-ith(P)} \widehat{f}(x) \underline{dx},$$

it is enough to consider the case $f(y) = e^{i(x,y)}$ for some $x \in X$. But then:

$$e^{ith(P)} e^{iQ \cdot x/t} e^{-ith(P)} = e^{iQ \cdot x/t} e^{it[h(P+xt^{-1})-h(P)]} \to e^{ixh'(P)}$$

strongly on $L^2(X)$, by the dominated convergence theorem, as $|t| \to \infty$.

So (7.C.1) holds for $f \in \mathscr{S}(X)$. Since $\mathscr{S}(X)$ is dense in $C_\infty(X)$, we easily get that (7.C.1) holds for all $f \in C_\infty(X)$. Now let $u \in L^2(X)$ with $||u|| = 1$ and set $u_t = e^{-ith(P)} u$. From (7.C.1) we obtain for $f \in C_\infty(X)$ ($n = \dim X$):

$$(7.C.5) \qquad \lim_{|t| \to +\infty} \int_X f(x) |t^{n/2} u_t(tx)|^2 \underline{dx} = \int_X f(h'(x)) |\widehat{u}(x)|^2 \underline{dx}.$$

Let μ_t denote the probability measure $|t^{n/2} u_t(tx)|^2 \underline{dx}$ on X. Then the integral on the left-hand side above is $\int_X f(x) \mu_t(dx)$. By the Riesz Theorem, there is a unique Borel probability measure μ on X such that the integral on the right-hand side in (7.C.4) is $\int_X f(x) \mu(dx)$; in fact, μ is just the image of the measure $|\widehat{u}(x)|^2 \underline{dx}$ under the almost everywhere defined Borel mapping $h' : X \to X$. Then the fact that (7.C.4) holds for all $f \in C_\infty(X)$ means that $\mu_t \to \mu$ in the weak* topology of the dual of $C_\infty(X)$ (which is the space of integrable Radon measures on X). Since $\mu_t(X) = \mu(X) = 1$, we can then apply Proposition 22 from §5, no. 12 of Chapter VII in [Bo3] and get $\int_X f(x) \mu_t(dx) \to \int_X f(x) \mu(dx) = \int_X f(h'(x)) |\widehat{u}(x)|^2 \underline{dx}$, for each $f : X \to \mathbb{C}$ which is bounded, Borel and continuous μ-almost everywhere. This continuity requirement means that, if N is the set of points of discontinuity of f, then $\mu(N) = 0$. But $\mu(N) = \int_{\{x|h'(x) \in N\}} |\widehat{u}(x)|^2 \underline{dx}$; hence if $\{x \mid h'(x) \in N\}$ is of zero Lebesgue measure, then we shall have $\mu(N) = 0$ for all $u \in L^2(X)$.

So we have shown that (7.C.1) holds in the weak topology. Now we observe that, if f satisfies the conditions of the theorem, then so does $|f|^2$. Hence

$$||e^{ith(P)} f(Q/t) e^{-ith(P)} u - f(h'(P)) u||^2 = \langle u, e^{ith(P)} |f|^2 (Q/t) e^{-ith(P)} u \rangle -$$

$$- 2\Re \langle e^{ith(P)} f(Q/t) e^{-ith(P)} u, f(h'(P)) u \rangle + ||f(h'(P)) u||^2$$

$$\to \langle u, |f|^2 (h'(P)) u \rangle - 2\Re \langle f(h'(P)) u, f(h'(P)) u \rangle + ||f(h'(P)) u||^2 = 0. \quad \square$$

CHAPTER 8

An Algebraic Framework for the Many-Body Problem

The purpose of this chapter is to develop a framework for the description and study of hamiltonians having a many-channel structure. The term "many-channel" is used here in a rather vague sense: we are thinking of systems consisting of a (large, but finite) number of components which could interact in a complicated way but could also behave independently (i.e. the interaction between some components could be turned off). So, to the "total hamiltonian" H one should be able to associate a collection of "sub-hamiltonians" H_a which, in some sense, should be simpler than H, and then one should construct the spectral and scattering theory of H in terms of the family $\{H_a\}$.

Of course, this is very vague, but our aim is precisely to give a mathematical meaning to the expressions under quotation marks appearing above. For example, we shall define a *many-channel hamiltonian* as a self-adjoint operator H affiliated to a graded C*-algebra, where the grading is given by a finite semilattice. Then to each such hamiltonian we shall be able to associate a family of many-channel hamiltonians H_a which are affiliated to smaller C*-algebras, and in this sense they are simpler. We show that some spectral characteristics of H can easily be determined in terms of those of the H_a (the essential spectrum and the $\widetilde{\varrho}$-function of H are explicitly expressed in terms of the spectra and the ϱ-functions of the H_a).

Our definition of sub-hamiltonians involves an operation which does not have a Hilbert space meaning. Again vaguely speaking, H_a is just the projection of H onto a C-subalgebra of the initial graded C*-algebra.* It is rather remarkable that, in the standard non-relativistic N-body problem, the operators we get in this way coincide with the usual ones (i.e. H_a can also be obtained by replacing by zero the intercluster potentials corresponding to the cluster decomposition a). This fact explains why complicated combinatorial arguments (involving diagrammatic expansions or truncated functions for example) become quite simple or disappear completely in the algebraic approach.

The fact that the general definition of H_a makes sense only at a C*-algebraic level already suggests that the theory should be developed in purely abstract terms. The following phenomena support this point of view. Assume that the graded C*-algebra $\mathscr{C}$, which is the main object of the theory, is naturally re-

W. O. Amrein et al., *C₀-Groups, Commutator Methods and Spectral Theory of N-Body Hamiltonians*, Modern Birkhäuser Classics,
DOI: 10.1007/978-3-0348-0733-3_8, © Springer Basel 1996

alized on a Hilbert space $\mathcal{H}$ (as it happens in the usual N-body case, where $\mathcal{H}$ is the Hilbert space of the "free" system; cf. Chapter 9). Then, even if the total hamiltonian H, affiliated to $\mathcal{C}$, is a densely defined self-adjoint operator in $\mathcal{H}$, its sub-hamiltonians might not be realizable as densely defined operators in $\mathcal{H}$, but only in closed subspaces of $\mathcal{H}$. So the initial Hilbert space $\mathcal{H}$ looses its apparently special position. In other situations one may want to define the hamiltonian by starting from a family of self-adjoint operators defined in a Hilbert space $\mathcal{H}$, by using rather simple and physically natural limiting procedures; in this way one may get many-channel hamiltonians (affiliated to the same algebra $\mathcal{C}$) which do not have an operatorial meaning in $\mathcal{H}$ (so already H could be defined only in a closed subspace of $\mathcal{H}$; this is the case of hard-core hamiltonians). Finally, one obtains important new hamiltonians (the internal hamiltonians H^a, see Section 9.4) by realizing the H_a in new representations of the C*-subalgebra $\mathcal{C}_a$ to which H_a is affiliated (this amounts to fixing the intercluster momentum in the N-body case).

These remarks explain why we present the theory at a Hilbert space independent, purely algebraic level. We believe that this point of view is quite fruitful even for simple many-body systems, i.e. non-relativistic N-body hamiltonians. The formalism allows one to treat very singular hamiltonians (e.g. perturbations of the free hamiltonian which require a change of Hilbert space in order to give an operatorial meaning to the total hamiltonian, as in the hard-core situation). The partition of unity in configuration space introduced by Simon and traditionally used for the decoupling of channels (see [Sim2], [PSS], [ABG1]) is not needed any more, since the relation between the total hamiltonian and the set of sub-hamiltonians is very simple and precise at an algebraic level (although this relation may not have a Hilbert space meaning). Note, however, that one can adopt a mixed (C*-algebraic and hilbertian) point of view which clarifies the role of the partitions of unity (see [BG3,4]).

Two papers played an important role in our early work on this algebraic framework. Our starting point was Polyzou's idea of defining the notion of "a-connected operators" in the standard N-body problem without reference to diagrams [Po]. It seemed to us that, more important than the definition he proposed, was a certain grading of the algebra generated by the a-connected operators. However, we use neither his definition of connectedness nor the associated algebra, which is not a C*-algebra. The grading, at a purely algebraic level, simplifies indeed the combinatorial aspects of the theory, but no deep results can be obtained without a graded C*-algebra structure. The main point is the interplay between the rich functional calculus permitted by the C*-algebra structure and the supplementary operations brought in by the grading. The second work which was crucial for our understanding of the subject was the paper [PSS] of Perry, Sigal and Simon. The C*-algebras $\mathcal{T}(Y)$ which we shall construct in Chapter 9 are modeled after their algebras $\mathrm{Com}(a)$. Also Lemma 9.3.1 (see the next chapter) is clearly suggested by the techniques of the Appendix of [PSS]. Hence the N-body algebra that we shall introduce in Section 9.3 implicitly appears in their paper (it is just $\sum_a \mathrm{Com}(a)$). However, the usefulness and richness of the graded C*-algebra structure seems to have remained unnoticed. For example, the fact that the projections $\mathscr{P}_a$ (see Section 8.4) are C*-morphisms implies that $\mathscr{P}_a[\varphi(H)] = \varphi(H_a)$ if $\varphi : \mathbb{R} \to \mathbb{C}$ is continuous and tends to zero at infinity, which renders unnecessary the use of the truncated function $\varphi_T(H_a)$ introduced in [PSS] (one has $\varphi_T(H_a) = \mathscr{P}(a)[\varphi(H)]$ in the notations of Section 8.4); more importantly, it allows one to eliminate the partition of unity and so to simplify considerably the arguments of [PSS].

Several of the ideas that will be presented below have been introduced in [BG1-4]. Some explanations concerning the connection between this formalism and the more conventional one may found in §10.2.2 or in [BG1,2]. Theorem 8.3.6 in a hilbertian setting (which is much simpler) has been proved in [BG3], where one may also find a detailed presentation of the role that the Simon partition of unity could play in a more hilbertian version the algebraic formalism. Theorem 8.4.1 and its consequences (Proposition 8.4.2 and Theorem 8.4.3) have been proved in [BG4]. Besides this, the point of view and results of this chapter are new. Of course, we do here considerably more than what is necessary for the non-relativistic N-body problem, and this makes the treatment quite technical; however, we find that the ideas are thus developed in their natural setting and generality. The reader will not find examples of graded C*-algebras in this chapter. Those needed in the usual N-body case are constructed in Chapter 9. More interesting examples of graded C*-algebras can be constructed by using the Weyl calculus associated to a symplectic space, see [BG4]. Other classes of such algebras (more in the spirit of Polyzou) are described in [BG2].

The chapter is organized as follows. In Section 8.1 we introduce and study the concept of an observable affiliated to a C*-algebra: this is a purely algebraic version of the notion of a self-adjoint operator. Section 8.2 is devoted to a discussion of tensor products of C*-algebras and observables. In Section 8.3 we show that that the ϱ-functions used in our version of the Mourre theory (cf. Section 7.2) have natural extensions to the C*-algebraic setting, and we describe how they behave under the operations of direct sums and tensor products. Finally, Section 8.4 is the most important one; here we introduce the notion of $\mathscr{L}$-graded C*-algebras and prove the main theorems of this chapter. The first one (Theorem 8.4.2) describes the essential spectrum of a total hamiltonian in terms of the spectra of its sub-hamiltonians. The second one gives an explicit formula for the function $\widetilde{\varrho}$ of the total hamiltonian in terms of the ϱ-functions of the sub-hamiltonians. When combined with Theorem 8.3.6, this gives easily the so-called Mourre estimate for N-body hamiltonians (cf. Theorem 9.4.4).

8.1. Self-adjoint Operators Affiliated to C*-Algebras

8.1.1. A C*-*algebra* is a Banach *-algebra $\mathscr{C}$ (cf. Section 3.1) in which the relation $||T^*T|| = ||T||^2$ holds for all elements T of $\mathscr{C}$. In general, $\mathscr{C}$ need not have a unit element. If it does, then the norm of the unit element will be one. In statements which hold only for algebras with unit, we shall use the term *unital* C*-*algebra*. If $\mathscr{C}_1$, $\mathscr{C}_2$ are C*-algebras, a $\Phi : \mathscr{C}_1 \to \mathscr{C}_2$ is a linear map such that $\Phi(ST) = \Phi(S)\Phi(T)$ and $\Phi(S^*) = \Phi(S)^*$. Such a map Φ is necessarily continuous and $||\Phi|| \leq 1$. Moreover, its range is a C*-subalgebra of $\mathscr{C}_2$ and, if Φ is injective, then Φ is an isometry. If $\mathscr{C}_1$, $\mathscr{C}_2$ are unital C*-algebras, Φ does not necessarily send unit to unit; if it does, we call it a *unital morphism*. All these facts and most of the results we shall use further on can be found in Chapter 1 of [Di2]. A deeper result needed below is the following: for each C*-algebra $\mathscr{C}$, there are a Hilbert space $\mathscr{H}$ and an isometric morphism of $\mathscr{C}$ into $B(\mathscr{H})$ (Theorem 2.6.1 [Di2]); in such a situation we say that $\mathscr{C}$ *is realized on* $\mathscr{H}$. Since we shall often work with such realizations, we denote the elements of C*-algebras by the same symbols as those used for linear operators.

We collect now some facts concerning measures on locally compact second countable (LCSC) topological spaces. To any set Z we associate the unital C*-algebra $M(Z)$ consisting of bounded functions $\varphi : Z \to \mathbb{C}$ with $\|\varphi\| = \sup_{z \in Z} |\varphi(z)|$. If Z is a LCSC space, we denote by $BO(Z)$ the unital C*-subalgebra of $M(Z)$ consisting of Borel functions (the Borel sets of Z are the elements of the σ-algebra generated by the open sets) and by $C_\infty(Z)$ the C*-subalgebra of $BO(Z)$ consisting of all continuous functions that converge to zero at infinity. If Z is not a finite set, $C_\infty(Z)$ is not dense in $BO(Z)$; however, it generates $BO(Z)$ in a rather simple way. Let us say that a sequence $\{\varphi_k\}_{k \in \mathbb{N}}$ of functions from $M(Z)$ *is boundedly convergent* if $\lim_{k \to \infty} \varphi_k(z) \equiv \varphi(z)$ exists for each $z \in Z$ and $|\varphi_k(z)| \le C$ for some constant $C < \infty$ independent of k and z. Then $BO(Z)$ is the smallest vector subspace of $M(Z)$ which contains $C_\infty(Z)$ and is stable under conjugation and under bounded convergence of sequences (i.e. it contains the limit of each boundedly convergent sequence of its elements). This statement is a functional version of the *monotone class theorem* (see [DM]), and its importance lies in the fact that it defines $BO(Z)$ in linear terms starting from $C_\infty(Z)$. As a consequence, each continuous anti-linear form μ on $C_\infty(Z)$ has a canonical extension to $BO(Z)$; this is the content of the Riesz representation theorem. More precisely, each $\mu \in C_\infty(Z)^*$ has a unique extension to a *normal* form $\widetilde{\mu} \in BO(Z)^*$, i.e. a functional having the following property: if $\varphi_k \to \varphi$ boundedly, then $\widetilde{\mu}(\varphi_k) \to \widetilde{\mu}(\varphi)$. It is clear that $\widetilde{\mu}$ can be identified with a complex Borel measure on Z (in the set theoretical sense). Moreover, the norm of $\widetilde{\mu}$ in $BO(Z)^*$ is equal to the norm of μ in $C_\infty(Z)^*$, and if $\mu \ge 0$, then $\widetilde{\mu} \ge 0$. Note that if μ is a positive anti-linear form on $C_\infty(Z)$, then it is automatically continuous. More generally, *if* $\mathscr{C}_1$, $\mathscr{C}_2$ *are* C*-*algebras and* $\Phi : \mathscr{C}_1 \to \mathscr{C}_2$ *is linear and positive* (i.e. $S \ge 0 \Rightarrow \Phi(S) \ge 0$), *then* Φ *is continuous.* (*Proof*: By the uniform boundedness principle, it is enough to show that $\varrho \circ \Phi$ is continuous if ϱ is a state on $\mathscr{C}_2$; this follows from 2.1.8 in [Di2]).

We mention one more general fact which will be needed later on. The map $\mu \mapsto \widetilde{\mu}$ identifies $C_\infty(Z)^*$ with the closed subspace of $BO(Z)^*$ consisting of normal functionals. On the other hand, such a functional can be identified with a complex Borel measure on Z (countably additive; the other elements of $BO(Z)^*$ are identified with finitely additive maps from Borel sets to $\mathbb{C}$). It is then an easy consequence of the Vitali-Hahn-Saks theorem that the subspace of normal functionals is weakly sequentially complete (cf. Corollary 1, §II.2 of [Y]). More precisely, if $\{\nu_k\}_{k \in \mathbb{N}}$ is a sequence of normal forms on $BO(Z)$ and $\lim_{k \to \infty} \nu_k(\varphi) \equiv \nu(\varphi)$ exists for all $\varphi \in BO(Z)$, then ν is normal (it is obvious that $\nu \in BO(Z)^*$, by the uniform boundedness principle; it is also clear that $\nu_k = \widetilde{\mu}_k$ for some $\mu_k \in C_\infty(Z)^*$ and $\mu_k \to \mu$ weakly in $C_\infty(Z)^*$, for some $\mu \in C_\infty(Z)^*$; the non-trivial point is that $\nu = \widetilde{\mu}$, and this is a consequence of the countable additivity of ν).

Consider now a Hilbert space $\mathscr{H}$ and a linear continuous map $\Phi : C_\infty(Z) \to B(\mathscr{H})$. Then Φ has a unique extension to a linear normal continuous map $\widetilde{\Phi} : BO(Z) \to B(\mathscr{H})$, i.e. a map such that $\widetilde{\Phi}(\varphi_k) \to \widetilde{\Phi}(\varphi)$ weakly on $\mathscr{H}$ if $\varphi_k \to \varphi$ boundedly (for the proof, apply the preceding considerations to $\mu_{f,g}(\varphi) = \langle f, \Phi(\varphi)g \rangle$, with $f, g \in \mathscr{H}$, and then apply the Riesz lemma to $(f, g) \mapsto \widetilde{\mu}_{f,g}(\varphi)$). Clearly $\|\widetilde{\Phi}\| = \|\Phi\|$. If Φ is positive, then $\widetilde{\Phi}$ is positive too and $E(\Delta) \equiv \Phi(\chi_\Delta)$ is the so-called *positive operator-valued measure associated to* Φ (χ_Δ being the characteristic function of the Borel set Δ). Finally, if Φ is a morphism, $\widetilde{\Phi}$ is a morphism too, E is a projection-valued measure on Z and we have $\widetilde{\Phi}(\varphi_k) \to \widetilde{\Phi}(\varphi)$ strongly in $\mathscr{H}$ if $\varphi_k \to \varphi$ boundedly. (*Proof*: One shows that $\langle f, \widetilde{\Phi}(\varphi\psi)g \rangle = \langle f, \widetilde{\Phi}(\varphi)\widetilde{\Phi}(\psi)g \rangle$ for all $\varphi, \psi \in BO(Z)$ by first fixing $\varphi \in C_\infty(Z)$ and applying the monotone class theorem in ψ, then fixing $\psi \in BO(Z)$ and applying the monotone class theorem in φ. For the last assertion, we may assume $\varphi = 0$; then $\|\widetilde{\Phi}(\varphi_k)f\|^2 = \langle f, \widetilde{\Phi}(|\varphi_k|^2)f \rangle \to 0$).

We now prove a result which plays an important role in the construction of a functional calculus for commuting observables. We consider two LCSC spaces Z_1, Z_2 and denote by $Z = Z_1 \times Z_2$ their topological product. If $\varphi_j \in C_\infty(Z_j)$, let $\varphi_1 \otimes \varphi_2 \in C_\infty(Z)$ be defined by $(\varphi_1 \otimes \varphi_2)(z_1, z_2) = \varphi_1(z_1)\varphi_2(z_2)$. We recall that the linear subspace $C_\infty(Z_1) \odot C_\infty(Z_2)$ spanned by functions of the form $\varphi_1 \otimes \varphi_2$ is dense in $C_\infty(Z)$ and is canonically identified with the algebraic tensor product of $C_\infty(Z_1)$ and $C_\infty(Z_2)$. In particular, if V is a vector space and $\Phi_0 : C_\infty(Z_1) \times C_\infty(Z_2) \to V$ is bilinear, then there is a unique linear map $\Phi_1 : C_\infty(Z_1) \odot C_\infty(Z_2) \to V$ such that $\Phi_0(\varphi_1, \varphi_2) = \Phi_1(\varphi_1 \otimes \varphi_2)$. In connection with the statement of the next theorem, it is interesting to observe that one cannot replace "positive" by "continuous" (because the largest cross-norm on $C_\infty(Z_1) \odot C_\infty(Z_2)$ is strictly greater then the unique C*-norm on it, cf. [Ta]).

THEOREM 8.1.1. *If $\mathscr{C}$ is a C*-algebra and $\Phi_0 : C_\infty(Z_1) \times C_\infty(Z_2) \to \mathscr{C}$ is bilinear and positive (i.e. $\Phi_0(\varphi_1, \varphi_2) \geq 0$ if $\varphi_1, \varphi_2 \geq 0$), then there is a unique positive linear mapping $\Phi : C_\infty(Z_1 \times Z_2) \to \mathscr{C}$ such that $\Phi(\varphi_1 \otimes \varphi_2) = \Phi_0(\varphi_1, \varphi_2)$.*

PROOF. It is enough to show that the map Φ_1 introduced above (with $V \equiv \mathscr{C}$) is continuous and positive. We may assume that $\mathscr{C} \subset B(\mathscr{H})$ for some Hilbert space $\mathscr{H}$. Let $f \in \mathscr{H}$ and $\varrho(\varphi_1, \varphi_2) = \langle f, \Phi_0(\varphi_1, \varphi_2)f \rangle$. Then ϱ is a positive bilinear form on $C_\infty(Z_1) \times C_\infty(Z_2)$ (positivity means $\varrho(\varphi_1, \varphi_2) \geq 0$ if $\varphi_1, \varphi_2 \geq 0$). Moreover, ϱ is continuous. Indeed, by the uniform boundedness principle, it is enough to see that it is separately continuous, hence to show for example that $\varphi_1 \mapsto \varrho(\varphi_1, \varphi_2)$ is continuous if $\varphi_2 \geq 0$. But this is a consequence of the fact that ϱ is positive, as seen before. So there is $C = C(f) < \infty$ such that $|\varrho(\varphi_1, \varphi_2)| \leq C\|\varphi_1\| \cdot \|\varphi_2\|$. It follows easily from the explanations given above that ϱ has a unique positive bilinear extension $\widetilde{\varrho} : BO(Z_1) \times BO(Z_2) \to \mathbb{C}$ with the following properties: for each $\varphi_1 \in BO(Z_1)$, $\widetilde{\varrho}(\varphi_1, \cdot) : BO(Z_2) \to \mathbb{C}$ is normal, and for each $\varphi_2 \in C_\infty(Z_2)$, $\widetilde{\varrho}(\cdot, \varphi_2) : BO(Z_1) \to \mathbb{C}$ is normal (fix $\varphi_2 \in C_\infty(Z_1)$ and extend $\varrho(\cdot, \varphi_2)$ to $\varrho'(\cdot, \varphi_2) : BO(Z_1) \to \mathbb{C}$; then fix $\varphi_1 \in BO(Z_1)$ and extend $\varrho'(\varphi_1, \cdot)$ to $\widetilde{\varrho}(\varphi_1, \cdot) : BO(Z_2) \to \mathbb{C}$). Clearly $|\widetilde{\varrho}(\varphi_1, \varphi_2)| \leq C\|\varphi_1\| \cdot \|\varphi_2\|$ for all $\varphi_j \in BO(Z_j)$. Let us prove that in fact $\widetilde{\varrho}(\cdot, \varphi_2)$ is normal for each $\varphi_2 \in BO(Z_2)$. By the monotone class theorem, it is enough to show that $\widetilde{\varrho}(\cdot, \varphi_2)$ is normal if

$\varphi_{2,k} \to \varphi_2$ boundedly and $\widetilde{\varrho}(\cdot, \varphi_{2,k})$ is normal for all k. But then $\widetilde{\varrho}(\varphi_1, \varphi_2) = \lim_{k \to \infty} \widetilde{\varrho}(\varphi_1, \varphi_{2,k})$ for all $\varphi_1 \in BO(Z_1)$, so we may apply the Vitali-Hahn-Saks theorem as explained before.

In conclusion, we got a positive bilinear form $\widetilde{\varrho} : BO(Z_1) \times BO(Z_2) \to \mathbb{C}$ which extends ϱ and is separately normal. Clearly $\nu(\Delta_1, \Delta_2) \equiv \widetilde{\varrho}(\chi_{\Delta_1}, \chi_{\Delta_2})$, where Δ_j is a Borel set in Z_j, is a positive bi-measure on $Z_1 \times Z_2$ (i.e. it is a measure in each factor when the other one is fixed) and $0 \leq \nu(\Delta_1, \Delta_2) \leq C$. Since any integrable Borel measure on a LCSC space is regular, we may use a result of Ph.Morando (see page 129 in [DM]) and get a positive Borel measure μ on Z such that $\mu(\Delta_1 \times \Delta_2) \equiv \nu(\Delta_1, \Delta_2)$ for all Δ_1, Δ_2. This is equivalent to $\widetilde{\varrho}(\chi_{\Delta_1}, \chi_{\Delta_2}) = \int_Z \chi_{\Delta_1} \otimes \chi_{\Delta_2} d\mu$, from which it follows by bilinearity that $\widetilde{\varrho}(\varphi_1, \varphi_2) = \int_Z \varphi_1 \otimes \varphi_2 d\mu$ if φ_j are Borel and finitely valued. Since $\widetilde{\varrho}$ is normal in each factor, we see that this formula remains valid for all $\varphi_j \in BO(Z_j)$.

To conclude, we found a positive integrable Borel measure μ on Z such that $\langle f, \Phi_1(\varphi) f \rangle = \int_Z \varphi d\mu$ for $\varphi \in C_\infty(Z_1) \odot C_\infty(Z_2)$. So $\Phi_1(\varphi) \geq 0$ if $\varphi \geq 0$ and $|\langle f, \Phi_1(\varphi) f \rangle| \leq C(f) ||\varphi||$ for $\varphi \in C_\infty(Z_1) \odot C_\infty(Z_2)$ and some constant $C(f) < \infty$. By polarization and the uniform boundedness principle, there is $C < \infty$ such that $||\Phi_1(\varphi)|| \leq C||\varphi||$. Since $C_\infty(Z_1) \odot C_\infty(Z_2)$ is dense in $C_\infty(Z)$ and $\mathscr{C}$ is norm-closed in $B(\mathscr{H})$, the theorem is completely proved. $\square$

8.1.2. We shall introduce now a notion inspired by quantum mechanics and the theory of von Neumann algebras [Di1]. Let Z be a LCSC space and $\mathscr{C}$ a C*-algebra. A *Z-valued observable affiliated to* $\mathscr{C}$ is a morphism $\Phi : C_\infty(Z) \to \mathscr{C}$. Our main interest is in $\mathbb{R}$-valued observables: we call them just *observables* (or *self-adjoint operators*) *affiliated to* $\mathscr{C}$. If $Z = \mathbb{C}$, we shall speak of *normal operators affiliated to* $\mathscr{C}$. The case $Z = \mathbb{R}^n$ will also appear below in some technical arguments (e.g. in the construction of functions of commuting observables).

Let T be a normal element of $\mathscr{C}$, let $\mathscr{C}_I$ be the unital C*-algebra generated by $\mathscr{C}^1$ and $\sigma(T)$ the spectrum of T when T is considered as an element of $\mathscr{C}_I$. Then $\sigma(T)$ is a compact subset of $\mathbb{C}$ and T is a symmetric (also called self-adjoint) element of $\mathscr{C}$ if and only if $\sigma(T) \subset \mathbb{R}$. We shall now canonically associate to T a $\mathbb{C}$-valued observable affiliated to $\mathscr{C}_I$. We recall (cf. 1.5.1 in [Di2]) that there is a unique unital morphism $\Phi_0 : C(\sigma(T)) \to \mathscr{C}_I$ which sends the identity function ($\varphi(z) \equiv z$) into T; the standard notation is $\Phi_0(\varphi) = \varphi(T)$. If $\mathscr{C}$ has no unit, then $0 \in \sigma(T)$; furthermore $\varphi(T) \in \mathscr{C}$ if and only if $\varphi(0) = 0$. Now, for each $\varphi \in C_\infty(\mathbb{C})$ we define $\varphi_T = \varphi|_{\sigma(T)}$ and set $\Phi_T(\varphi) = \varphi_T(T) = \Phi_0(\varphi_T)$. Clearly, Φ_T is a $\mathbb{C}$-valued observable affiliated to $\mathscr{C}_I$. If $\sigma(T) \subset \mathbb{R}$, i.e. if T is a symmetric element of $\mathscr{C}$, we similarly get a ($\mathbb{R}$-valued) observable affiliated to $\mathscr{C}_I$.

We shall describe the observables and the $\mathbb{C}$-valued observables affiliated to $B(\mathscr{H})$ for any Hilbert space $\mathscr{H}$. Several times below and later on, when we refer to an operator in $\mathscr{H}$ which is self-adjoint in the usual sense, we shall say *"densely defined self-adjoint operator in $\mathscr{H}$"*. This convention will be justified by a result

---

[1] If $\mathscr{C}$ is a unital algebra, then $\mathscr{C}_I = \mathscr{C}$. If not, then there is a unique C*-norm on the algebra obtained from $\mathscr{C}$ by adjoining a unit element (see 13.8 in [Di2]), and we denote by $\mathscr{C}_I$ the C*-algebra defined in this way. Then $\mathscr{C}_I$ is unital and $\mathscr{C} \subset \mathscr{C}_I$ is a closed self-adjoint ideal.

that we shall prove in a moment and allows us to distinguish the self-adjoint operators in $\mathcal{H}$ from the self-adjoint operators (or observables) affiliated to $B(\mathcal{H})$. We shall see that each densely defined self-adjoint operator in $\mathcal{H}$ determines an observable affiliated to $B(\mathcal{H})$, but there are observables which can not be constructed in such a way. We recall that there is a one-to-one correspondence between densely defined self-adjoint operators H in $\mathcal{H}$ and $B(\mathcal{H})$-valued *spectral measures* E on $\mathbb{R}$ (i.e. projection-valued measures such that $E(\mathbb{R})$ is the identity operator in $\mathcal{H}$) defined by the relation $H = \int_{\mathbb{R}} \lambda E(d\lambda)$. Then for $\varphi \in C_\infty(\mathbb{R})$ we take $\Phi_H(\varphi) \equiv \varphi(H) = \int_{\mathbb{R}} \varphi(\lambda) E(d\lambda)$. The morphism $\Phi_H : C_\infty(\mathbb{R}) \to B(\mathcal{H})$ is the observable associated to H and the correspondence $H \mapsto \Phi_H$ is easily seen to be injective. From now on we identify the densely defined self-adjoint operator H in $\mathcal{H}$ with the observable Φ_H.

Consider now an arbitrary morphism $\Phi : C_\infty(\mathbb{R}) \to B(\mathcal{H})$. As we explained in §8.1.1, Φ has a canonical extension $\widetilde{\Phi} : BO(\mathbb{R}) \to B(\mathcal{H})$, which in turn determines a projection-valued Borel measure E on $\mathbb{R}$. Let $\mathcal{H}'$ be the closed subspace of $\mathcal{H}$ given by $\mathcal{H}' = E(\mathbb{R})\mathcal{H}$, E' the $B(\mathcal{H}')$-valued spectral measure given by $E'(\Delta) = E(\Delta)|_{\mathcal{H}'}$ and H' the densely defined self-adjoint operator in $\mathcal{H}'$ determined by E', i.e. $H' = \int_{\mathbb{R}} \lambda E'(d\lambda)$. Clearly, for $\varphi \in BO(\mathbb{R})$, we shall have $\widetilde{\Phi}(\varphi)|_{\mathcal{H}'} = \varphi(H') = \int_{\mathbb{R}} \varphi(\lambda) E'(d\lambda)$ and $\widetilde{\Phi}(\varphi)|_{\mathcal{H}'^\perp} = 0$. Reciprocally, if $\mathcal{H}'$ is a closed subspace of $\mathcal{H}$ and H' a densely defined self-adjoint operator in $\mathcal{H}'$, we define a morphism $\Phi : C_\infty(\mathbb{R}) \to B(\mathcal{H})$ as follows: $\Phi(\varphi)|_{\mathcal{H}'} := \varphi(H')$ and $\Phi(\varphi)|_{\mathcal{H}'^\perp} = 0$. To summarize, *one may identify the observables affiliated to $B(\mathcal{H})$ with couples $(\mathcal{H}', H')$, where $\mathcal{H}'$ is a closed subspace of $\mathcal{H}$ and H' is a densely defined self-adjoint operator in $\mathcal{H}'$.* Clearly, if $\mathscr{C}$ is a C*-subalgebra of $B(\mathcal{H})$, Φ will be affiliated to $\mathscr{C}$ if and only if the operator H' has the following property: for each $\varphi \in C_\infty(\mathbb{R})$, the operator on $\mathcal{H}$ equal to $\varphi(H')$ on $\mathcal{H}'$ and equal to zero on $\mathcal{H}'^\perp$, belongs to $\mathscr{C}$. Obviously, the case of $\mathbb{C}$-valued observables is completely similar, except that H' will be a densely defined normal operator in $\mathcal{H}'$.

It will be convenient to use the following notational convention: in the situation just described, we use a symbol H and set $\varphi(H) = \Phi(\varphi)$ if $\varphi \in C_\infty(\mathbb{R})$. In other terms, $\varphi(H)|_{\mathcal{H}'} = \varphi(H')$ (where $\varphi(H')$ is defined by the usual functional calculus for the self-adjoint operator H' in $\mathcal{H}'$) and $\varphi(H)|_{\mathcal{H}'^\perp} = 0$. Formally, one should think of H as a "self-adjoint operator" in $\mathcal{H}$ for which ∞ is an eigenvalue and $\mathcal{H}'^\perp$ is the associated eigenspace (since $\varphi(\infty) = 0$ for $\varphi \in C_\infty(\mathbb{R})$, this is consistent with the preceding convention).

We shall keep the above notational convention even when we work with a C*-algebra $\mathscr{C}$ which is not realized on a Hilbert space. We shall often denote Z-valued observables by the letter H instead of Φ and we shall write $\varphi(H)$ instead of $H(\varphi)$ if $\varphi \in C_\infty(Z)$; we stress that this is justified only if φ tends to zero at infinity. As we explained above, if $\mathscr{C}$ is realized on a Hilbert space, then H gets an operatorial meaning. Let us introduce one more notation which is sometimes convenient. For any Z one may consider the Z-valued observable affiliated to $\mathscr{C}$ defined by the trivial morphism $\Phi(\varphi) = 0$ for each $\varphi \in C_\infty(Z)$; when we use the notation $\varphi(H)$, it is natural to denote this observable by $H = \infty$. Note that, in a Hilbert space realization, $H = \infty$ corresponds to the couple $\mathcal{H}' = \{0\}$, $H' = 0$.

An element S of a C*-algebra $\mathscr{C}$ is called *strictly positive* (written $S > 0$) if $\omega(S) > 0$ for each state ω of $\mathscr{C}$ (see 3.10.5 in [Pd] or Exercise 1.7.3 in [Ta]). A realization of $\mathscr{C}$ on a Hilbert space $\mathscr{H}$ is called *non-degenerate* if $\{Sf \mid S \in \mathscr{C}, f \in \mathscr{H}\}$ spans a dense subspace of $\mathscr{H}$. The following fact can be proved without difficulty (see part (c) of the quoted exercise from [Ta]): *Let H be an observable affiliated to $\mathscr{C}$ such that $\varphi(H) > 0$ for some $\varphi \in C_\infty(\mathbb{R})$; assume that $\mathscr{C}$ is realized on a Hilbert space $\mathscr{H}$ and the realization is non-degenerate. Then the operator in $\mathscr{H}$ associated to the observable H is densely defined* (i.e., with the notations used before, $\mathscr{H}' = \mathscr{H}$).

The observables (i.e. the *real*-valued observables) can be considered from another point of view which is very convenient in many respects. We shall say that $\{R(z) \mid z \in \mathbb{C} \setminus \mathbb{R}\}$ is a *self-adjoint resolvent family in $\mathscr{C}$* (or simply a *$\mathscr{C}$-valued resolvent*) if $R(z) \in \mathscr{C}$ are elements satisfying $R(z)^* = R(\bar{z})$ and $R(z_1) - R(z_2) = (z_1 - z_2)R(z_1)R(z_2)$. If $\Phi : C_\infty(\mathbb{R}) \to \mathscr{C}$ is a morphism and if we define for each non-real z the function $r_z : \mathbb{R} \to \mathbb{C}$ by $r_z(\lambda) = (\lambda - z)^{-1}$, then obviously we get a $\mathscr{C}$-valued resolvent by letting $R(z) = \Phi(r_z)$. In this way we associate to each ($\mathbb{R}$-valued) observable a self-adjoint resolvent family in $\mathscr{C}$. When we use the notation $\Phi(\varphi) = \varphi(H)$, we write $R(z) = (H - z)^{-1}$; but one has to be careful because even if $\mathscr{C}$ is realized on a Hilbert space $\mathscr{H}$, the operator $(H - z)^{-1}$ is not the inverse of an operator in $\mathscr{H}$ (if $\mathscr{H}' \neq \mathscr{H}$, with a notation used above).

Since the subalgebra (and even the vector subspace) generated by the functions $\{r_z \mid z \in \mathbb{C} \setminus \mathbb{R}\}$ is dense in $C_\infty(\mathbb{R})$, *the morphism Φ is uniquely determined by the resolvent associated to it.* We prove now that *any $\mathscr{C}$-valued resolvent is associated to a morphism.* We may assume that $\mathscr{C}$ is realized on a Hilbert space $\mathscr{H}$, so that $\{R(z)\}$ is a family of bounded operators in $\mathscr{H}$ satisfying $R(z)^* = R(\bar{z})$ and $R(z_1) - R(z_2) = (z_1 - z_2)R(z_1)R(z_2)$ for all non-real z, z_1 and z_2. We shall associate to this family an operator H' in $\mathscr{H}$ (non-densely defined in general) by the following well-known argument (see [HP]). Observe that $R(z_1)R(z_2) = R(z_2)R(z_1)$ and $R(z_1)\mathscr{H} = R(z_2)\mathscr{H}$, so that $D(H') \equiv R(z)\mathscr{H}$ is a subspace of $\mathscr{H}$ independent of z. Denote by $\mathscr{H}'$ the closure of $D(H')$ in $\mathscr{H}$ and notice that $R(z)^* = R(\bar{z})$ implies $\ker R(z) = \mathscr{H}'^\perp$. So $R'(z) = R(z)|_{\mathscr{H}'}$ is a family of bounded operators on $\mathscr{H}'$, which satisfies the same identities as $R(z)$, but in addition $R'(z) : \mathscr{H}' \to D(H')$ is bijective. By simple, purely algebraic calculations one may show that the operator $H' = R'(z)^{-1} + z : D(H') \subset \mathscr{H}' \to \mathscr{H}'$ is independent of z, densely defined and symmetric; in fact it is self-adjoint (in $\mathscr{H}'$) because $H' - z = R'(z)^{-1} : D(H') \to \mathscr{H}'$ is surjective for non-real z. For $\varphi \in C_\infty(\mathbb{R})$ we define $\Phi(\varphi) \in B(\mathscr{H})$ as before: $\Phi(\varphi)|_{\mathscr{H}'} = \varphi(H')$, $\Phi(\varphi)|_{\mathscr{H}'^\perp} = 0$. Then $\Phi : C_\infty(\mathbb{R}) \to B(\mathscr{H})$ is a morphism and $\Phi(r_z) = R(z)$. Since Φ is continuous and $\mathscr{C}$ is a norm-closed subalgebra of $B(\mathscr{H})$, we see that Φ is $\mathscr{C}$-valued.

To summarize, we have shown that *there is a bijective correspondence between ($\mathbb{R}$-valued) observables affiliated to $\mathscr{C}$ and self-adjoint resolvent families in $\mathscr{C}$.* It is very easy to define $R(z) \equiv (H - z)^{-1}$ in terms of $\Phi(\varphi) = \varphi(H)$, but the reverse construction that we gave above is not so straightforward. However, for a restricted class of φ, we may use Theorem 6.1.4 in order to get an explicit description of $\varphi(H)$ in terms of $R(z)$. More precisely, if $\varphi \in C^r(\mathbb{R})$ for some

$r \geq 2$ and $\varphi^{(k)} \in L^1(\mathbb{R})$ for $0 \leq k < r$, then

$$(8.1.1) \qquad \varphi(H) = \int_{\mathbb{R}} d\lambda \left\{ \sum_{k=0}^{r-1} \frac{1}{\pi k!} \varphi^{(k)}(\lambda) \Im[i^k R(\lambda + i)] \right.$$

$$\left. + \frac{1}{\pi(r-1)!} \int_0^1 \varphi^{(r)}(\lambda) \Im[i^r R(\lambda + i\mu)] \mu^{r-1} d\mu \right\}.$$

The integrals here exist in the norm topology in $\mathscr{C}$, as follows from the fact that $R(\cdot) : \mathbb{C} \setminus \mathbb{R} \to \mathscr{C}$ is holomorphic and satisfies the estimate $\|R(z)\| \leq |\Im z|^{-1}$. Indeed, we have $R(z) - R(\bar{z}) = 2i\Im z R(z)^* R(z)$ and $\|R(z)\|^2 = \|R(z)^* R(z)\|$; then the holomorphy follows from $R(z) - R(z_0) = (z - z_0)R(z)R(z_0)$, more precisely this identity implies that

$$(8.1.2) \qquad R(z) = R(z_0)[I + (z_0 - z)R(z_0)]^{-1} = \sum_{k=0}^{\infty} (z - z_0)^k R(z_0)^{k+1}$$

if $|z - z_0| < \|R(z_0)\|^{-1}$. Note that, if $\mathscr{C}$ is not unital, the inverse is taken in the unital algebra $\mathscr{C}_I$ generated by $\mathscr{C}$, in which $\mathscr{C}$ is an ideal. For later purposes, let us observe that the first equality in (8.1.2) holds in $\mathscr{C}_I$ for any non-real z and z_0. In slightly different terms, $I + (z_0 - z)R(z_0)$ is always invertible in $\mathscr{C}_I$ and its inverse is equal to $I + (z - z_0)R(z)$ (check that their product is I).

As a first example of the utility of the notion of resolvent family, we shall completely describe the (real) observables affiliated to an abelian C*-algebra. Recall that there is a bijective correspondence between locally compact (LC) spaces and abelian C*-algebras: if X is a LC space, we associate to it the C* algebra $C_\infty(X)$ and reciprocally, if $\mathscr{C}$ is an abelian C*-algebra, then *its spectrum* (or *space of characters*) is a LC space X and there is a canonical identification of $\mathscr{C}$ with $C_\infty(X)$ (Gelfand representation). Then, *there is a bijective correspondence between (real) observables H affiliated to an abelian C*-algebra $\mathscr{C}$ and couples (U, h) consisting of open subsets U of the spectrum X of $\mathscr{C}$ and continuous proper [2] functions $h : U \to \mathbb{R}$* (to $H = \infty$ corresponds $U = \varnothing$). Indeed, let us work in the Gelfand representation: if (U, h) is given we take $\varphi(H)(x) = \varphi(h(x))$ if $x \in U$ and $\varphi(h(x)) = 0$ if $x \in X \setminus U$ (here $\varphi \in C_\infty(\mathbb{R})$); reciprocally, if H is given and $\{R(z) \mid \Im z \neq 0\}$ is its resolvent, then $U = \{x \in X \mid R(z)(x) \neq 0\}$ is an open set independent of z and the function $h(x) = [R(z)(x)]^{-1} + z$ defined for $x \in U$ is independent of z, continuous, proper and real.

We shall explain now a method of defining the *sum* of an observable H affiliated to a C*-algebra $\mathscr{C}$ and of a symmetric element $S \in \mathscr{C}_I$; the result of this operation will be an observable, denoted $H + S$, affiliated to $\mathscr{C}$. We shall define the resolvent family $R_S(z) = (H + S - z)^{-1}$ of $H + S$ in terms of the resolvent family $R(z) = (H - z)^{-1}$ of H. If $\|S\| < |\Im z|$, then $\|SR(z)\| < 1$ and

---

[2] If X_1, X_2 are locally compact spaces, a map $f : X_1 \to X_2$ is called *proper* if $f^{-1}(K)$ is compact for any compact subset K of X_2. In other terms, $f(x_1)$ must tend to infinity in X_2 if x_1 tends to infinity in X_1. Note that an open or closed subspace of a locally compact space is locally compact for the induced topology.

$||R(z)S|| < 1$, hence $I + SR(z)$ and $I + R(z)S$ are invertible in $\mathscr{C}_I$; for such z we define

$$R_S(z) := R(z)[I + SR(z)]^{-1} = [I + R(z)S]^{-1}R(z).$$

Then $\{R_S(z) \mid |\Im z| > ||S||\}$ is a family of elements of $\mathscr{C}$ with the properties $R_S(z)^* = R_S(\bar{z})$ (obvious) and $R_S(z_1) - R_S(z_2) = (z_1 - z_2)R_S(z_1)R_S(z_2)$ (a straightforward calculation). As a consequence, we get $||R_S(z)|| \leq |\Im z|^{-1}$. Since $R_S(z_0) = [I - (z - z_0)R_S(z_0)]R_S(z)$, we have $R_S(z) = \sum_{n=0}^{\infty}(z - z_0)^n R_S(z_0)^{n+1}$ if $|z - z_0| < |\Im z_0|$ and $|\Im z| > ||S||$, $|\Im z_0| > ||S||$. This clearly implies that $\{R_S(z) \mid |\Im z| > ||S||\}$ has a holomorphic extension to the set $\{z \in \mathbb{C} \mid \Im z \neq 0\}$. This extension is a self-adjoint resolvent family in $\mathscr{C}$ and $H + S$ is, by definition, the observable associated to it.

The one-to-one correspondence between observables and resolvent families allows one to construct quite easily observables affiliated to a C*-algebra $\mathscr{C}$ realized on a Hilbert space $\mathscr{H}$. We give below several criteria of a perturbative nature; they will be useful in our applications to the N-body problem. If H is a densely defined self-adjoint operator in $\mathscr{H}$, then H is identified with the observable $\varphi \mapsto \varphi(H)$ affiliated to $B(\mathscr{H})$. If $\varphi(H) \in \mathscr{C}$ for all $\varphi \in C_{\infty}(\mathbb{R})$, then H naturally defines an observable affiliated to $\mathscr{C}$; we shall then say that H itself is affiliated to $\mathscr{C}$. Let us observe that *a densely defined self-adjoint operator H in $\mathscr{H}$ is affiliated to the C*-subalgebra $\mathscr{C}$ of $B(\mathscr{H})$ if and only if there is some $z_0 \in \mathbb{C} \setminus \sigma(H)$ such that $(H - z_0)^{-1} \in \mathscr{C}$.* Indeed, if the last relation holds, one easily obtains from (8.1.2) that $(H - z)^{-1} \in \mathscr{C}$ for all $z \in \mathbb{C} \setminus \sigma(H)$; then the denseness in $C_{\infty}(\mathbb{R})$ of the algebra generated by $\{r_z\}$ in $C_{\infty}(\mathbb{R})$ and the continuity of the morphism $\varphi \mapsto \varphi(H)$ imply that $\varphi(H) \in \mathscr{C}$ for all $\varphi \in C_{\infty}(\mathbb{R})$.

PROPOSITION 8.1.2. *Let H_0 be a densely defined self-adjoint operator in a Hilbert space $\mathscr{H}$ affiliated to a C*-subalgebra $\mathscr{C}$ of $B(\mathscr{H})$.*

(a) *If V is a H_0-bounded symmetric operator in $\mathscr{H}$ with H_0-bound strictly less than 1 and if $V(H_0 - z_0)^{-1} \in \mathscr{C}$ for some $z_0 \in \mathbb{C} \setminus \sigma(H_0)$, then $H = H_0 + V$ is a densely defined self-adjoint operator in $\mathscr{H}$ affiliated to $\mathscr{C}$.*

(b) *Assume that H_0 is bounded from below and let V be a symmetric sesquilinear form on $\mathscr{H}$ which is relatively form-bounded with respect to H_0 with relative bound strictly less than 1. If $(\lambda_0 + H_0)^{-1/2}V(\lambda_0 + H_0)^{-1/2} \in \mathscr{C}$ for some $\lambda_0 > -\inf H_0$, then the operator associated to the form sum $H = H_0 + V$ is a densely defined self-adjoint operator in $\mathscr{H}$ affiliated to $\mathscr{C}$.*

PROOF. (a) From the identity

$$V(H_0 - z)^{-1} = V(H_0 - z_0)^{-1} + (z - z_0)V(H_0 - z_0)^{-1} \cdot (H_0 - z)^{-1}$$

we see that $V(H_0 - z)^{-1} \in \mathscr{C}$ for all $z \in \mathbb{C} \setminus \sigma(H_0)$. Since the H_0-bound of V is less than 1, there is $z \in \mathbb{C} \setminus \sigma(H_0)$ such that $||V(H_0 - z)^{-1}|| < 1$ (see e.g. Theorem X.12 of [RS]); so, for such z:

(8.1.3) $(H - z)^{-1} = (H_0 - z)^{-1}[I + V(H_0 - z)^{-1}]^{-1}$

$$\equiv \sum_{k=0}^{\infty}(H_0 - z)^{-1}[V(z - H_0)^{-1}]^k.$$

Hence (a) is proved.

(b) Let $m = \inf H_0 \in \mathbb{R}$ and $\lambda, \lambda_0 > -m$. If $S = (\lambda_0 - \lambda)(H_0 + \lambda)^{-1}$, then $S \in \mathscr{C}$ and $I + S \geq \min\{1, (\lambda_0 + m)(\lambda + m)^{-1}\}$. So $(I + S)^{1/2}$ exists in $B(\mathscr{H})$ and is of the form $I + T$ for some $T \in \mathscr{C}$. From $(H_0 + \lambda)^{-1} = (H_0 + \lambda_0)^{-1}(I + S)$ we get $(H_0 + \lambda)^{-1/2} = R_0^{1/2}(I + T)$, where $R_0 = (H_0 + \lambda_0)^{-1}$, hence $(H_0 + \lambda)^{-1/2}V(H_0 + \lambda)^{-1/2} = (I + T)R_0^{1/2}VR_0^{1/2}(I + T)$. This shows that $(H_0 + \lambda)^{-1/2}V(H_0 + \lambda)^{-1/2}$ belongs to $\mathscr{C}$ for all $\lambda > -\inf H_0$. Since this operator has norm < 1 if λ is large enough, we have

$$(8.1.4) \quad (H + \lambda)^{-1} =$$
$$= (H_0 + \lambda)^{-1/2}[I + (H_0 + \lambda)^{-1/2}V(H_0 + \lambda)^{-1/2}]^{-1}(H_0 + \lambda)^{-1/2}. \quad \square$$

Norm resolvent limits of densely defined self-adjoint operators provide us with interesting examples of highly singular hamiltonians (non-densely defined) affiliated to quite simple C*-algebras. More precisely, let $\{H_n\}_{n \in \mathbb{N}}$ be a sequence of densely defined self-adjoint operators in $\mathscr{H}$ such that $\lim_{n \to \infty}(H_n - z)^{-1} := R(z)$ exists in norm in $B(\mathscr{H})$ for some fixed z with the property $\inf_n \operatorname{dist}(z, \sigma(H_n)) > 0$. From (8.1.2) and the equality $\|(H_n - z)^{-1}\| = [\operatorname{dist}(z, \sigma(H_n))]^{-1}$ it follows that the preceding limit exists for all non-real z. The family $\{R(z) \mid z \in \mathbb{C} \setminus \mathbb{R}\}$ will be a self-adjoint resolvent family in $B(\mathscr{H})$ and, if each H_n is affiliated to a given C*-algebra $\mathscr{C} \subset B(\mathscr{H})$, $\{R(z)\}$ will be a $\mathscr{C}$-valued resolvent, hence it defines an observable affiliated to $\mathscr{C}$. N-body hamiltonians with hard-core interactions are obtained by such a procedure (see [BGS]).

This method of defining new observables can also be used in the following abstract setting. Let H_0 be an observable affiliated to a C*-algebra $\mathscr{C}$ and $\{S_n\}_{n \in \mathbb{N}}$ a sequence of symmetric elements of $\mathscr{C}_I$. We explained before how to define $H_n := H_0 + S_n$ as an observable affiliated to $\mathscr{C}$. If $\lim_{n \to \infty}(H_n - z)^{-1}$ exists in $\mathscr{C}$ for some $z \in \mathbb{C} \setminus \mathbb{R}$, then it exists for all such z. We thus get a self-adjoint resolvent family in $\mathscr{C}$ which defines an observable H affiliated to $\mathscr{C}$.

We make one more remark in connection with the preceding limiting procedure. Let us denote by H the observable affiliated to $\mathscr{C}$ defined by $\{R(z)\}$, so that (formally) we have $R(z) = (H - z)^{-1}$. Then $\|\varphi(H_n) - \varphi(H)\| \to 0$ as $n \to \infty$ for each $\varphi \in C_\infty(\mathbb{R})$. Indeed, the set of functions φ for which this holds is a C*-subalgebra of $C_\infty(\mathbb{R})$ which contains each of the functions r_z ($\Im z \neq 0$), hence it is equal to $C_\infty(\mathbb{R})$. This suggests the following natural generalization of the notion of convergence for sequences $\{H_n\}_{n \in \mathbb{N}}$ of Z-valued observables affiliated to a C*-algebra $\mathscr{C}$ (where Z may be an arbitrary LCSC space). We shall say that the sequence $\{H_n\}$ is *norm-convergent* to a Z-valued observable H affiliated to $\mathscr{C}$ if $\lim_{n \to \infty}\|\varphi(H_n) - \varphi(H)\| = 0$ for all $\varphi \in C_\infty(Z)$; then we write $H = \lim_{n \to \infty} H_n$.

8.1.3. After this detailed presentation of the notion of observables affiliated to a C*-algebra $\mathscr{C}$, we next describe a rather large bounded and unbounded functional calculus for finite families of *commuting* observables affiliated to $\mathscr{C}$. The bounded functional calculus will produce elements of $\mathscr{C}$; the unbounded one will produce observables affiliated to $\mathscr{C}$. Hence, for non-unital algebras, the distinction between the two kinds of calculi is rather important. We shall begin with the preliminary notions of spectrum and joint spectrum.

Let Z be a LCSC space, $\mathscr{C}$ a C*-algebra and H a Z-valued observable affiliated to $\mathscr{C}$. Then the *spectrum* $\sigma(H)$ of H is the closed subset of Z consisting of the points z with the property: $\varphi \in C_\infty(Z)$ and $\varphi(z) \neq 0 \Rightarrow \varphi(H) \neq 0$. Clearly, the observable $H = \infty$ is characterized by $\sigma(H) = \varnothing$. Let us prove that $z \notin \sigma(H)$ *if and only if there is a neighbourhood V of z such that $\varphi(H) = 0$ for all $\varphi \in C_0(V)$* (which is the set of continuous complex functions with compact support included in V). Indeed, if $z \notin \sigma(H)$, then there is $\varphi_0 \in C_\infty(X)$ with $\varphi_0(z) \neq 0$ and $\varphi_0(H) = 0$. Then we shall have $\varphi(H) = 0$ if $\varphi = \varphi_0 \psi$ for some $\psi \in C_\infty(Z)$; but if V is a neighbourhood of z in which φ_0 does not vanish, then any $\varphi \in C_0(V)$ can be written as such a product; this finishes the proof. We mention that *at a C*-algebra level it is impossible to distinguish between point and continuous spectrum* [3]. Remark that the property $z \in \sigma(H)$ is a local one, in the following precise sense. Let us say that two Z-valued observables H_1, H_2 *are equal on an open set $U \subset Z$* if $\varphi(H_1) = \varphi(H_2)$ for $\varphi \in C_0(U)$; then, if H_1 is equal to H_2 on a neighbourhood of $z \in Z$, we have $z \in \sigma(H_1) \Leftrightarrow z \in \sigma(H_2)$.

An important property of the spectrum of H is the following one: *if $\varphi \in C_\infty(Z)$ and $\varphi|_{\sigma(H)} = 0$, then $\varphi(H) = 0$* (and it is clear that $\sigma(H)$ *is the smallest closed set having this property*; we may say that $\sigma(H)$ *is the support of the morphism $\varphi \mapsto \varphi(H)$*). Indeed, $Y = Z \setminus \sigma(H)$ is an open subset of Z, so it is a LCSC space for the induced topology, and $\varphi|_Y \in C_\infty(Y)$ if φ is as indicated (because for each $\varepsilon > 0$ the set where $|\varphi(z)| \geq \varepsilon$ is compact and included in Y); hence there is a sequence of functions $\varphi_n \in C_0(Y)$ such that $\sup_{z \in Z} |\varphi(z) - \varphi_n(z)| = \sup_{z \in Y} |\varphi(z) - \varphi_n(z)| \to 0$. So we may assume φ to have compact support disjoint from $\sigma(H)$. On the other hand, each $z \notin \sigma(H)$ has a neighbourhood V_z such that $\theta(H) = 0$ if $\theta \in C_0(V_z)$. We can cover the support of φ by a finite number of such neighbourhoods, and then we can construct a partition of unity $\{\theta_z\}$ on supp φ subordinated to the given covering. Since $\varphi = \sum_z \varphi \theta_z$ and $\theta_z(H) = 0$, the assertion is proved.

From the preceding property it follows that, if $\sigma(H) \neq Z$, then there is a natural extension of the morphism $\varphi \mapsto \varphi(H)$ to a class of functions larger than $C_\infty(Z)$. Observe that $\sigma(H)$ is a LCSC space (for the induced topology), hence $C_\infty(\sigma(H))$ is a well defined space. Moreover, $\varphi \mapsto \varphi|_{\sigma(H)}$ is a surjective morphism of $C_\infty(Z)$ onto $C_\infty(\sigma(H))$ (the easiest way to see this is by working with the Alexandroff compactification of Z) and $\varphi(H) = 0$ if and only if φ is in the kernel of this morphism. So, if $\psi \in C_\infty(\sigma(H))$ and φ is any extension of ψ to a function in $C_\infty(Z)$, we may define without ambiguity $\psi(H) \equiv \varphi(H)$. To summarize, *we got an injective morphism $C_\infty(\sigma(H)) \ni \varphi \mapsto \varphi(H) \in \mathscr{C}$; since*

---

[3]See 2.12.12 in [Di2] or consider the following simple situation. Let $\mathscr{H} = L^2([0,1])$ and $\mathscr{C} = C([0,1]) \subset B(\mathscr{H})$. There is a real $h \in \mathscr{C}$ such that $F = h^{-1}(0)$ is a closed set of non-zero measure but with empty interior; so zero is an eigenvalue of the operator of multiplication by h in $\mathscr{H}$. On the other hand, there is a homeomorphism $\theta : [0,1] \to [0,1]$ such that $\theta^{-1}(F)$ has zero measure. Since $f \mapsto f \circ \theta$ is an automorphism of $\mathscr{C}$, h and $h \circ \theta$ have the same spectrum; but clearly zero is not an eigenvalue of the operator $h \circ \theta \in B(\mathscr{H})$. One has to take this phenomenon into account when looking for a natural formulation of the virial theorem in a C*-algebra setting, cf. Proposition 7.2.10 and Section 8.3.

an injective morphism is isometric, we have $\|\varphi(H)\| = \sup\{|\varphi(z)| \mid z \in \sigma(H)\}$ *for each* $\varphi \in C_\infty(\sigma(H))$.

We give now several examples showing the importance of the notion of spectrum and of the extended functional calculus introduced above. First, observe that one can identify the notion of (real-valued) observable with that of complex-valued observable with purely real spectrum. Now let us consider a *lower semi-bounded observable* H, i.e. an observable the spectrum of which is a real set that is bounded from below. Then for any complex λ such that $\Re\lambda > 0$, $e^{-\lambda H}$ is a well defined element of $\mathscr{C}$ and the family $\{e^{-\lambda H} \mid \Re\lambda > 0\}$ is a holomorphic semigroup in $\mathscr{C}$. Finally, let H be a complex observable such that $\sigma(H)$ is a compact set. Then $\varphi(H)$ is a well defined element of $\mathscr{C}$ for any continuous function $\varphi : \mathbb{C} \to \mathbb{C}$. In particular, if Id is the identity function ($\mathrm{Id}(z) \equiv z$), then $T := \mathrm{Id}(H)$ is a normal element of $\mathscr{C}$ and one has $\varphi(T) = \varphi(H)$ for any continuous function on $\mathbb{C}$ such that $\varphi(0) = 0$ [if φ is a polynomial without constant term, this is obvious; then the equality remains true for continuous φ with $\varphi(0) = 0$ because the restrictions of the polynomials in z and $\bar{z}$ to the compact set $\sigma(H)$ form a dense subspace of $C(\sigma(H))$]. If $e(z) = 1$ for all $z \in \mathbb{C}$, then $e(H)$ is a unit element for the C*-subalgebra generated by T and $e(H)$ (this C*-subalgebra is just the set of elements of the form $\varphi(H)$ with $\varphi \in C_\infty(\mathbb{C})$). In particular, if $\mathscr{C} = C_\infty(X)$ with X a locally compact, non-compact, connected space, the only complex observable with compact spectrum affiliated to it is $H = \infty$.

Lower semibounded observables can be defined by using semi-groups in place of resolvents. More precisely, *there is a bijective correspondence between lower semibounded observables H affiliated to $\mathscr{C}$ and families $\{T_\lambda \mid \lambda \in (0,\infty)\}$ of elements of $\mathscr{C}$ having the following properties:* (i) $T_\lambda \geq 0$; (ii) $T_{\lambda+\mu} = T_\lambda T_\mu$; (iii) $\lambda \mapsto T_\lambda$ *is (norm)-continuous.* The correspondence is specified by the condition $T_\lambda = \exp(-\lambda H)$. In order to prove the preceding assertion, note first that H is uniquely defined by this condition (because the linear space generated by the functions of the form $e^{-\lambda x}$, $\lambda > 0$, is dense in $C_\infty([m,\infty))$ for any $m \in \mathbb{R}$). Then let $T := T_1$ and observe that $T^\lambda = T_\lambda$ for all $\lambda > 0$ (if λ is rational, this follows from a simple, purely algebraic argument; by continuity, the equality remains true for all $\lambda > 0$). Now assume that $\mathscr{C}$ is realized on a Hilbert space $\mathscr{H}$, let $\mathscr{H}'$ be the closure of the range of T and $T' = T|_{\mathscr{H}'}$. Then T' is an injective, bounded and positive operator in $\mathscr{H}'$, hence $H' := -\ln T'$ is a lower semibounded densely defined self-adjoint operator in $\mathscr{H}'$ such that $T' = \exp(-H')$. Clearly $T^\lambda|_{\mathscr{H}'} = \exp(-\lambda H')$ and $T^\lambda|_{\mathscr{H}'^\perp} = 0$ if $\lambda > 0$. So the observable $H \equiv (\mathscr{H}', H')$ affiliated to $B(\mathscr{H})$ has the property $\exp(-\lambda H) = T_\lambda \in \mathscr{C}$ for all $\lambda > 0$, hence $\varphi(H) \in \mathscr{C}$ for all $\varphi \in C_\infty(\mathbb{R})$ [by the denseness of the space of exponentials $e^{-\lambda x}$ in $C_\infty([m,\infty))$], i.e. H is affiliated to $\mathscr{C}$.

The semigroup point of view allows one to define, under certain conditions, the sum $H := H_1 + H_2$ of two lower semibounded observables H_1, H_2 affiliated to $\mathscr{C}$. If $\mathscr{C}$ is realized on a Hilbert space, a result of Kato [K2] states that one may define H by the Trotter formula: $\exp(-\lambda H) = \text{s-}\lim_{n\to\infty}[\exp(-\lambda H_1/n)\exp(-\lambda H_2/n)]^n$. If the limit exists in norm (see [Ro] for sufficient conditions), then the observable H defined in this way is affiliated to $\mathscr{C}$. There are modifications of the preceding formula whose norm-convergence is easier to control. The results of [Ro] imply the following fact: if H_1, H_2 are lower semibounded self-adjoint operators in

$\mathscr{H}$ affiliated to $\mathscr{C}$ and if $H = H_1 + H_2$ is densely defined and self-adjoint on $D(H_1) \cap D(H_2)$, then H is affiliated to $\mathscr{C}$.

We make one more remark concerning the spectrum of a (real) observable H. If $R(z) = (H - z)^{-1}$ is its resolvent family, then $||R(z)|| = [\text{dist}(z, \sigma(H))]^{-1}$ as we have seen before in a more general setting. Hence a real number λ does not belong to $\sigma(H)$ if and only if $\liminf_{\mu \to +0} ||R(\lambda + i\mu)|| < \infty$ and also if and only if the map $R(\cdot) : \mathbb{C} \setminus \mathbb{R} \to \mathscr{C}$ has a holomorphic extension to a neighbourhood of λ.

We now introduce the notion of commuting observables and of joint spectrum of a finite family of such observables. Let Z_1, Z_2 be two LCSC spaces, $\mathscr{C}$ a C*-algebra, H_1 a Z_1-valued observable and H_2 a Z_2-valued observable affiliated to $\mathscr{C}$. We say that H_1 and H_2 *commute* if $\varphi_1(H_1)\varphi_2(H_2) = \varphi_2(H_2)\varphi_1(H_1)$ for all $\varphi_j \in C_\infty(Z_j)$. More generally, if we have n LCSC spaces $Z_1, \ldots, Z_n$ and for each j a Z_j-valued observable H_j affiliated to $\mathscr{C}$, we say that $(H_1, \ldots, H_n)$ is a *commuting family of observables* if H_j commutes with H_k for all j, k. We may now state a result which is an immediate consequence of Theorem 8.1.1 (it could also be deduced from Proposition 4.7, Chapter 4 in [Ta]).

THEOREM 8.1.3. *Let $Z_1, \ldots, Z_n$ be LCSC spaces, $Z = Z_1 \times \cdots \times Z_n$ their topological product and $\mathscr{C}$ a C*-algebra. Assume that, for each $j = 1, \ldots, n$, a Z_j-valued observable H_j affiliated to $\mathscr{C}$ is given and that $(H_1, \ldots, H_n)$ is a commuting family. Then there is a unique Z-valued observable $\mathbb{H}$ affiliated to $\mathscr{C}$ such that $\varphi(\mathbb{H}) = \varphi_1(H_1) \cdots \varphi_n(H_n)$ if $\varphi = \varphi_1 \otimes \cdots \otimes \varphi_n$ and $\varphi_j \in C_\infty(Z_j)$.*

We may write $\mathbb{H} = H_1 \times \cdots \times H_n$ and call it *the cartesian product of the observables $H_1, \ldots, H_n$.* The spectrum of $\mathbb{H}$ will be called *joint spectrum* of the commuting family $(H_1, \ldots, H_n)$ and will be denoted by $\sigma(H_1, \ldots, H_n)$; note that it is a closed subset of $Z_1 \times \cdots \times Z_n$. The preceding theorem gives a meaning to $\varphi(H_1, \ldots, H_n) \equiv \varphi(\mathbb{H})$ for any $\varphi \in C_\infty(Z_1 \times \cdots \times Z_n)$. But, since $\varphi(\mathbb{H})$ is well defined for any $\varphi \in C_\infty(\sigma(\mathbb{H}))$, we see that we have obtained a morphism $\varphi \mapsto \varphi(H_1, \ldots, H_n)$ from $C_\infty(\sigma(H_1, \ldots, H_n))$ into $\mathscr{C}$.

This construction is only a preliminary step toward the definition of new observables that can be interpreted as "unbounded functions" of a set of commuting observables (note that $\varphi(H_1, \ldots, H_n)$ is a normal element of $\mathscr{C}$ if $\varphi \in C_\infty(\sigma(H_1, \ldots, H_n))$, hence it is *not* a complex observable affiliated to $\mathscr{C}$ if $\mathscr{C}$ is not unital).

Let us first consider the case of an arbitrary LCSC space Z and of a Z-valued observable H affiliated to $\mathscr{C}$. Let Y be a LCSC space and $f : \sigma(H) \to Y$ a proper continuous function. Then f induces a morphism $f^* : C_\infty(Y) \to C_\infty(\sigma(H))$ by the relation $f^*(\varphi) = \varphi \circ f$, hence we get a new morphism $\varphi \mapsto (f^*(\varphi))(H)$ of $C_\infty(Y)$ into $\mathscr{C}$. To summarize, *for each proper continuous map $f : \sigma(H) \to Y$ there is a unique Y-valued observable H' which is affiliated to $\mathscr{C}$ and such that $\varphi(H') = (\varphi \circ f)(H)$; we shall write $H' \equiv f(H)$, hence $\varphi(f(H)) = (\varphi \circ f)(H)$ for all $\varphi \in C_\infty(Y)$* [4]. Let us notice that *the spectrum of $f(H)$ is equal to the*

---

[4] A more natural notation (from the point of view of our presentation) would be $f^*(H)$ in place of $f(H)$. However, the last notation has the advantage that it preserves its usual meaning in a Hilbert space setting, i.e. if H is a densely defined self-adjoint operator in $\mathscr{H}$ and if $Y = \mathbb{C}$, then $f(H)$ can be obtained by standard functional calculus (the proof of this fact is easy and

image through f of the spectrum of H, i.e. $\sigma(f(H)) = f(\sigma(H))$; in our context this *spectral mapping theorem* is obvious.

Consider for example a real observable H. Since any polynomial function $P : \mathbb{R} \to \mathbb{R}$ is a proper map, the preceding construction gives a meaning to $P(H)$ as a real observable affiliated to $\mathscr{C}$. The simplest case is $P(x) = x + \lambda$ for some $\lambda \in \mathbb{R}$; observe that the resolvent family of $H + \lambda$ is related to that of H in a very simple manner.

If we go back to the setting of Theorem 8.1.3, we see that for each proper continuous map $f : \sigma(H_1, \ldots, H_n) \to Y$, we get a Y-valued observable $f(H_1, \ldots, H_n)$ $\equiv f(\mathbb{H})$ affiliated to $\mathscr{C}$ and with spectrum equal to $f(\sigma(H_1, \ldots, H_n))$. The simplest example of this general construction is the following. Let $(H_1, \ldots, H_n)$ be a commuting family of (real) observables affiliated to $\mathscr{C}$ and let $P : \mathbb{R}^n \to \mathbb{C}$ be a polynomial whose restriction to the joint spectrum $\sigma(H_1, \ldots, H_n) \subset \mathbb{R}^n$ is a proper map. Then $P(H_1, \ldots, H_n)$ is a complex (real if P is real-valued) observable affiliated to $\mathscr{C}$ with $P(\sigma(H_1, \ldots, H_n))$ as spectrum. The following very special case is particularly interesting: if $H_1, \ldots, H_n$ are lower semibounded, then $H_1 + \cdots + H_n$ is a well defined, lower semibounded observable affiliated to $\mathscr{C}$ and $\sigma(H_1 + \cdots + H_n) = \{\sum_{i=1}^n \lambda_i \mid (\lambda_1, \ldots, \lambda_n) \in \sigma(H_1, \ldots, H_n)\}$. Explicitly, we have by definition $\varphi(H_1 + \ldots + H_n) = (\varphi \circ \Sigma)(H_1, \ldots, H_n)$, where $\Sigma(\lambda_1, \ldots, \lambda_n) = \lambda_1 + \cdots + \lambda_n$.

8.1.4. In this subsection we shall describe two new types of operations which may be defined in the set of Z-valued observables for some fixed Z, namely the image of an observable through a morphism and the direct sum of observables. We shall leave to Section 8.2 the definition of tensor products, which is not so straightforward. In contradistinction to the operations considered so far, if one starts with one or several observables affiliated to a C*-algebra $\mathscr{C}$, then the result of the new operations is an observable affiliated to a C*-algebra different from $\mathscr{C}$.

Let H be a Z-valued observable affiliated to a C*-algebra $\mathscr{C}$, and let $\mathscr{P} : \mathscr{C} \to \mathscr{C}'$ be a morphism of $\mathscr{C}$ into a C*-algebra $\mathscr{C}'$. Then $\varphi \mapsto \mathscr{P}[\varphi(H)]$ is a morphism of $C_\infty(Z)$ into $\mathscr{C}'$, hence it is a Z-valued observable H' affiliated to $\mathscr{C}'$ with the property $\varphi(H') = \mathscr{P}[\varphi(H)]$ for each $\varphi \in C_\infty(Z)$. We shall use the notation $\mathscr{P}[H] = H'$ and call this new observable *the image through $\mathscr{P}$ of H*. So $\varphi(\mathscr{P}[H]) := \mathscr{P}[\varphi(H)]$ for each $\varphi \in C_\infty(Z)$ gives a natural extension of the action of $\mathscr{P}$ from elements of the C*-algebra to observables affiliated to the C*-algebra. Clearly $\mathscr{P}[H]$ is a Z-valued observable affiliated to $\mathscr{C}'$ with spectrum $\sigma(\mathscr{P}[H]) \subset \sigma(H)$, and for each proper continuous map $f : \sigma(H) \to Y$ we have $\mathscr{P}[f(H)] = f(\mathscr{P}[H])$. *The operation $\mathscr{P}$ does not have a meaning in purely Hilbert space terms, and the possibility of using it is the main advantage of the C*-algebra setting.* An example (important in our applications to the N-body problem) is the following one. Let H be a densely defined self-adjoint operator in a Hilbert space $\mathscr{H}$, so that H is an observable affiliated to $B(\mathscr{H})$. Let $K(\mathscr{H})$ be the self-adjoint closed ideal in $B(\mathscr{H})$ consisting of all compact

---

s left as an exercise). Later on we shall define the image of a Z-valued observable H affiliated to $\mathscr{C}$ through a morphism $\mathscr{P} : \mathscr{C} \to \mathscr{B}$; this will be a Z-valued observable affiliated to $\mathscr{B}$, and we shall denote it by $\mathscr{P}[H]$ (rather than by $\mathscr{P}(H)$) in order to avoid any confusion.

operators. Denote by $\mathscr{P}$ the natural morphism of $B(\mathscr{H})$ onto the quotient C*-algebra $B(\mathscr{H})/K(\mathscr{H})$ (named *Calkin algebra*). Then $\mathscr{P}[H]$ is an observable affiliated to the Calkin algebra which has no simple (or natural) Hilbert space interpretation. It is easily seen that its spectrum $\sigma(\mathscr{P}[H])$ coincides with the essential spectrum $\sigma_{\mathrm{ess}}(H)$ of H (in the Hilbert space sense; we shall consider this question in §8.1.5 in a more general setting).

We pass to direct sums. Let $H_1, \ldots, H_n$ be Z-valued observables affiliated to C*-algebras $\mathscr{C}_1, \ldots, \mathscr{C}_n$. Recall that the C*-direct sum $\mathscr{C} = \mathscr{C}_1 \oplus \cdots \oplus \mathscr{C}_n$ is defined as follows (1.3.3 in [Di2]): in the usual linear direct sum, the multiplication and the involution are defined componentwise and the norm of $S = (S_1, \ldots, S_n) \equiv \oplus_{i=1}^{n} S_i$ is $\sup_{1 \le i \le n} \|S_i\|$. We shall define the *direct sum* $H \equiv \oplus_{i=1}^{n} H_i$ of the Z-valued observables $H_1, \ldots, H_n$ as the Z-valued observable affiliated to $\mathscr{C}$ given by the formula: $\varphi(H) := \oplus_{i=1}^{n} \varphi(H_i)$. The spectrum of H is given by $\sigma(H) = \cup_{i=1}^{n} \sigma(H_i)$. In fact, $z \notin \sigma(H)$ means: there is $\varphi \in C_{\infty}(Z)$ with $\varphi(z) \neq 0$ and $\varphi(H) = 0$, i.e. $\varphi(H_i) = 0$ for $1 \le i \le n$; hence $z \notin \cup_{i=1}^{n} \sigma(H_i)$. Reciprocally, for such a z and for each i we can find a neighbourhood V_i of z such that $\varphi \in C_0(V_i) \Rightarrow \varphi(H_i) = 0$; then $V = \cap_{i=1}^{n} V_i$ is a neighbourhood of z and $\varphi(H) = 0$ if $\varphi \in C_0(V)$, so $z \notin \sigma(H)$. It is easily shown that $f(H) = \oplus_{i=1}^{n} f(H_i)$ if $f : \sigma(H) \to Y$ is a proper continuous function.

8.1.5. We now develop a remark made in §8.1.4 in connection with the essential spectrum (in the Hilbert space sense) of a densely defined self-adjoint operator H in a Hilbert space $\mathscr{H}$ and then extend this notion to a C*-algebra setting. We recall that a real number λ belongs to $\sigma_{\mathrm{ess}}(H)$ if and only if the projection $E((\lambda - \varepsilon, \lambda + \varepsilon))$ is infinite dimensional for each $\varepsilon > 0$; here E is the spectral measure of H. This is clearly equivalent to: $\varphi \in C_{\infty}(\mathbb{R})$ and $\varphi(\lambda) \neq 0 \Rightarrow \varphi(H)$ is not a compact operator in $\mathscr{H}$, i.e. $\varphi(H)$ does not belong to the closed bilateral ideal $K(\mathscr{H})$ of $B(\mathscr{H})$. This suggests the following general notion.

Let $\mathscr{C}$ be a C*-algebra and $\mathfrak{J}$ a closed bilateral ideal in $\mathscr{C}$. We recall (cf. 1.8.2 in [Di2]) that a closed ideal $\mathfrak{J}$ of $\mathscr{C}$ is bilateral if and only if it is self-adjoint, and in this case the quotient $*$-algebra $\mathscr{C}/\mathfrak{J}$ is a C*-algebra for the quotient norm. We denote by π the canonical morphism of $\mathscr{C}$ onto $\mathscr{C}/\mathfrak{J}$. Let Z be a LCSC space and H a Z-valued observable affiliated to $\mathscr{C}$. Then $\widetilde{H} := \pi[H]$ is a Z-valued observable affiliated to $\widetilde{\mathscr{C}} := \mathscr{C}/\mathfrak{J}$. We define the $\mathfrak{J}$-*essential spectrum* $\mathfrak{J}$-$\sigma_{ess}(H)$ *of* H as the set of points $z \in Z$ such that: $\varphi \in C_{\infty}(Z)$ and $\varphi(z) \neq 0 \Rightarrow \varphi(H) \notin \mathfrak{J}$. Obviously we have $\mathfrak{J}$-$\sigma_{\mathrm{ess}}(H) = \sigma(\widetilde{H})$. In particular, $z \notin \mathfrak{J}$-$\sigma_{\mathrm{ess}}(H)$ if and only if there is a neighbourhood V of z such that $\varphi(H) \in \mathfrak{J}$ for all $\varphi \in C_0(V)$.

It is clear that, if $\mathscr{C} \subset B(\mathscr{H})$ for some Hilbert space $\mathscr{H}$, $\mathfrak{J} = K(\mathscr{H}) \subset \mathscr{C}$ and $Z = \mathbb{R}$, then we shall have $K(\mathscr{H})$-$\sigma_{\mathrm{ess}}(H) = \sigma_{\mathrm{ess}}(H')$; here H' is the densely defined self-adjoint operator in the closed subspace $\mathscr{H}'$ of $\mathscr{H}$ that is canonically associated to the observable H. In Section 8.2 we shall present more interesting examples (relevant in the N-body case).

8.2. Tensor products

8.2.1. If $\mathbf{F}$, $\mathbf{G}$ are vector spaces, we shall use the notation $\mathbf{F} \odot \mathbf{G}$ for their algebraic tensor product and $(f, g) \mapsto f \otimes g$ for the canonical bilinear map

$\mathbf{F} \times \mathbf{G} \to \mathbf{F} \odot \mathbf{G}$. If $\mathbf{F}'$, $\mathbf{G}'$ is a second pair of vector spaces and $S : \mathbf{F} \to \mathbf{F}'$, $T : \mathbf{G} \to \mathbf{G}'$ are linear, we denote by $S \odot T$ their algebraic tensor product; hence $S \odot T$ is a linear map $\mathbf{F} \odot \mathbf{G} \to \mathbf{F}' \odot \mathbf{G}'$. Assume now that $\mathbf{F}$, $\mathbf{G}$ are Hilbert spaces. Then there is a unique hilbertian norm on $\mathbf{F} \odot \mathbf{G}$ such that $||f \otimes g|| = ||f|| \cdot ||g||$ for all $f \in \mathbf{F}$, $g \in \mathbf{G}$; the completion of $\mathbf{F} \odot \mathbf{G}$ under this norm will be denoted by $\mathbf{F} \otimes \mathbf{G}$ and will be called the *Hilbert tensor product* of $\mathbf{F}$ and $\mathbf{G}$. If $\mathbf{F}'$, $\mathbf{G}'$ are also Hilbert spaces and S, T are continuous, then $S \odot T$ has a unique extension to a linear continuous operator from $\mathbf{F} \otimes \mathbf{G}$ into $\mathbf{F}' \otimes \mathbf{G}'$. We denote this extension by $S \otimes T$. Moreover, if S and T are continuous, we shall not distinguish between $S \odot T$ and $S \otimes T$. If $\mathscr{B}$, $\mathscr{C}$ are $*$-subalgebras of $B(\mathbf{F})$, $B(\mathbf{G})$ (here $\mathbf{F}$, $\mathbf{G}$ are Hilbert spaces), we write $\mathscr{B} \odot \mathscr{C}$ for the linear subspace of $B(\mathbf{F} \otimes \mathbf{G})$ generated by operators of the form $B \otimes C$, with $B \in \mathscr{B}$, $C \in \mathscr{C}$. Clearly $\mathscr{B} \odot \mathscr{C}$ is a $*$-subalgebra of $B(\mathbf{F} \otimes \mathbf{G})$. If $\mathscr{B}$, $\mathscr{C}$ are C*-subalgebras of $B(\mathbf{F})$, $B(\mathbf{G})$, we denote by $\mathscr{B} \otimes \mathscr{C}$ the closure $\mathscr{B} \odot \mathscr{C}$ in $B(\mathbf{F} \otimes \mathbf{G})$. Then $\mathscr{B} \otimes \mathscr{C}$ is a C*-algebra and the notation is consistent with that used below.

The definition of tensor products of C*-algebras and of observables affiliated to them is more delicate, and we treat only the case which will be of interest to us. We begin with some remarks concerning the tensor product of two C*-algebras $\mathscr{C}_1$, $\mathscr{C}_2$ (see Ch.4 in [Ta], where the case of algebras without unit is explicitly treated, or Ch.11 in [KR]). We denote by $\mathscr{C}_1 \odot \mathscr{C}_2$ the algebraic tensor product of the vector spaces $\mathscr{C}_1$, $\mathscr{C}_2$, equipped with the structure of $*$-algebra defined by the rules $(S_1 \otimes S_2)(T_1 \otimes T_2) = (S_1 T_1) \otimes (S_2 T_2)$ and $(S \otimes T)^* = S^* \otimes T^*$. A C*-norm on $\mathscr{C}_1 \odot \mathscr{C}_2$ is a norm satisfying $||AB|| \leq ||A|| \cdot ||B||$ and $||A^* A|| = ||A||^2$ for all A, $B \in \mathscr{C}_1 \odot \mathscr{C}_2$. The following facts are true:

(1) any C*-norm is a cross-norm, i.e. satisfies $||S_1 \otimes S_2|| = ||S_1|| \cdot ||S_2||$;

(2) there are a smallest C*-norm $|| \cdot ||_{\min}$ and a greatest C*-norm $|| \cdot ||_{\max}$, hence any C*-norm satisfies $||A||_{\min} \leq ||A|| \leq ||A||_{\max}$ for all $A \in \mathscr{C}_1 \odot \mathscr{C}_2$;

(3) in general there are many C*-norms, but there are some important cases in which $||A||_{\min} = ||A||_{\max}$, for example if *one* of the algebras $\mathscr{C}_1$, $\mathscr{C}_2$ is abelian (or, more generally, of type I), and this will be the case in all our applications.

For reasons of simplicity we shall work from now on with only one C*-norm on $\mathscr{C}_1 \odot \mathscr{C}_2$, namely with $|| \cdot ||_{\min}$, and we shall denote it simply by $|| \cdot ||$; moreover, we denote by $\mathscr{C}_1 \otimes \mathscr{C}_2$ the C*-algebra obtained by completing $\mathscr{C}_1 \odot \mathscr{C}_2$ with respect to this norm (the usual notation is $\mathscr{C}_1 \otimes_{\min} \mathscr{C}_2$). $\mathscr{C}_1 \otimes \mathscr{C}_2$ is sometimes called the *spatial tensor product* (e.g. in [KR]), and it can be explicitly defined as follows. Assume that $\mathscr{C}_j$ is realized on a Hilbert space $\mathscr{H}_j$ (i.e. an injective morphism $\pi_j : \mathscr{C}_j \to B(\mathscr{H}_j)$ is given and we identify $\mathscr{C}_j \equiv \pi_j(\mathscr{C}_j) \subset B(\mathscr{H}_j)$); form the Hilbert tensor product $\mathscr{H}_1 \otimes \mathscr{H}_2$ and realize $\mathscr{C}_1 \odot \mathscr{C}_2 \subset B(\mathscr{H}_1 \otimes \mathscr{H}_2)$ in the standard way; then $||A|| \equiv ||A||_{\min}$ is just the norm of the operator $A \in B(\mathscr{H}_1 \otimes \mathscr{H}_2)$ for any $A \in \mathscr{C}_1 \odot \mathscr{C}_2$ (the main point is that the norm defined in this way is independent of the chosen representations π_1, π_2). This construction shows that the tensor products which will be used in Chapter 9 coincide with those used here. We shall now explicitly describe the tensor product algebra in two situations which will often appear in the next chapter.

Recall that $K(\mathscr{H})$ is the C*-algebra of compact operators in the Hilbert space $\mathscr{H}$. We shall prove now that for *any two Hilbert spaces* $\mathscr{H}_1$, $\mathscr{H}_2$, if $\mathscr{H}_1 \otimes \mathscr{H}_2$ denotes *their Hilbert tensor product, then* $K(\mathscr{H}_1) \otimes K(\mathscr{H}_2) = K(\mathscr{H}_1 \otimes \mathscr{H}_2)$. Since we have a natural embedding $K(\mathcal{H}_1) \odot K(\mathcal{H}_2) \subset K(\mathcal{H}_1 \otimes \mathcal{H}_2)$, it is enough to show that this embedding is dense. For this, it suffices to prove that for each $u \in \mathcal{H}$, $||u|| = 1$, the rank one orthogonal projection P_u with range $\mathbb{C}u$ can be approximated in norm by linear combinations of operators of the form $T_1 \otimes T_2$, with $T_j \in K(\mathcal{H}_j)$. Remark that if $u = u_1 \otimes u_2$ for some $u_j \in \mathcal{H}_j$, then $P_u = P_{u_1} \otimes P_{u_2}$ (indeed, $P_u v = \langle u, v \rangle u$). Hence, if u is a linear combination of decomposable vectors, then $P_u \in K(\mathcal{H}_1) \odot K(\mathcal{H}_2)$. If u is arbitrary, we may find a sequence $\{u_n\}$ of linear combinations of decomposable vectors such that $||u_n - u|| \to 0$. But then $||P_{u_n} - P_u|| \to 0$, which finishes the proof.

For the second example, we need a preliminary observation. Let $\mathbf{E}$ be a Banach space, X a locally compact space and $C_\infty(X; \mathbf{E})$ the Banach space of all functions $f : X \to \mathbf{E}$ that are continuous and converge to zero at infinity $(||f|| = \sup_{x \in X} ||f(x)||_{\mathbf{E}})$. Then the linear subspace of functions of the form $\sum_{j=1}^{n} \varphi_j e_j$, with $\varphi_j \in C_\infty(X)$ and $e_j \in \mathbf{E}$, is dense in $C_\infty(X; \mathbf{E})$. Indeed, if $f \in C_\infty(X; \mathbf{E})$ and $\varepsilon > 0$, then there is a finite open covering $(\Omega_0, \Omega_1, \ldots, \Omega_n)$ of X and there are vectors $e_0 = 0, e_1, \ldots, e_n \in \mathbf{E}$, such that Ω_0 is a neighbourhood of infinity, $\Omega_1, \ldots, \Omega_n$ are relatively compact, and $||f(x) - e_j||_{\mathbf{E}} < \varepsilon$ if $x \in \Omega_j$, $j = 0, 1, \ldots, n$. Let $\varphi_0, \varphi_1, \ldots, \varphi_n$ be positive continuous functions on X, with $\operatorname{supp} \varphi_j \subset \Omega_j$ and $\varphi_0 + \varphi_1 + \cdots + \varphi_n = 1$. Then clearly $||f(x) - \sum \varphi_j(x) e_j||_{\mathbf{E}} < \varepsilon$ for all $x \in X$.

Consider now an arbitrary locally compact space X and a C*-algebra $\mathscr{C}$. Then $C_\infty(X; \mathscr{C})$ has an obvious C*-algebra structure and there is a canonical identification $C_\infty(X) \otimes \mathscr{C} \equiv C_\infty(X; \mathscr{C})$, determined by the requirement that the function $x \mapsto \varphi(x) S$ corresponds to the element $\varphi \otimes S$. For the proof, observe that to an element $\sum_{j=1}^{n} \varphi_j \otimes S_j$ of $C_\infty(X) \odot \mathscr{C}$ will correspond the function $f(x) = \sum_{j=1}^{n} \varphi_j(x) S_j$ hence, by what we just proved, it is enough to show that $|| \sum_{j=1}^{n} \varphi_j \otimes S_j|| = ||f||$ $(\equiv \sup_{x \in X} ||f(x)||)$. As explained above, in order to calculate the norm on the left-hand side of this formula, we have to consider some Hilbert space realizations of $C_\infty(X)$ and $\mathscr{C}$. Choose any Hilbert space realization $\mathscr{C} \subset B(\mathcal{H})$ of $\mathscr{C}$ and let $\varphi \in C_\infty(X)$ be realized as a multiplication operator on the space $\ell^2(X) = \{u : X \to \mathbb{C} \mid \sum_{x \in X} |u(x)|^2 < \infty\}$, so that $C_\infty(X) \subset B(\ell^2(X))$ (in place of the counting measure one may consider any positive Borel measure on X which gives non-zero mass to each open, non-empty set). Then $\ell^2(X) \otimes \mathcal{H} \cong \ell^2(X; \mathcal{H}) = \{u : X \to \mathcal{H} \mid \sum_{x \in X} ||u(x)||^2 < \infty\}$, and in this representation $\sum_{j=1}^{n} \varphi_j \otimes S_j$ becomes the operator of multiplication by $f : X \to \mathscr{C} \subset B(\mathcal{H})$, which is easily seen to have norm equal to $\sup_{x \in X} ||f(x)||$. This finishes the proof.

There is an obvious generalization of the preceding theory from the case of two algebras to the case of n C*-algebras $\mathscr{C}_1, \ldots, \mathscr{C}_n$; for example, one may define $\mathscr{C}_1 \otimes \cdots \otimes \mathscr{C}_n \equiv ((\mathscr{C}_1 \otimes \mathscr{C}_2) \otimes \mathscr{C}_3) \otimes \ldots$. However, one gets a more symmetrical treatment by working from the beginning with all the algebras, see [KR]. We do not insist on this point since in our applications only the case $n = 2$ will appear.

8.2.2. Let us fix n C*-algebras $\mathscr{C}_1, \ldots, \mathscr{C}_n$ and denote by $\mathscr{C} = \mathscr{C}_1 \otimes \cdots \otimes \mathscr{C}_n$ their (spatial) tensor product. We assume that n LCSC spaces $Z_1, \ldots, Z_n$ are given, and for each $j = 1, \ldots, n$ we consider a Z_j-valued observable H_j affiliated to $\mathscr{C}_j$. Let $Z = Z_1 \times \cdots \times Z_n$ be the topological product of $Z_1, \ldots, Z_n$.

THEOREM 8.2.1. *There is a unique Z-valued observable $\mathbb{H}$ affiliated to $\mathscr{C}$ such that $\varphi(\mathbb{H}) = \varphi_1(H_1) \otimes \dots \otimes \varphi_n(H_n)$ if $\varphi = \varphi_1 \otimes \dots \otimes \varphi_n$ and $\varphi_j \in C_\infty(Z_j)$. One has $\sigma(\mathbb{H}) = \sigma(H_1) \times \dots \times \sigma(H_n)$.*

PROOF. We assume $n = 2$; the general case is similar. Then the existence and uniqueness of $\mathbb{H}$ follow immediately from Theorem 8.1.1 applied to the map $\Phi_0(\varphi_1, \varphi_2) := \varphi_1(H_1) \otimes \varphi_2(H_2)$; the map $\Phi : C_\infty(Z) \to \mathscr{C}$ obtained by extension is obviously a morphism (by continuity and linearity it is enough to check that $\Phi((\varphi_1 \otimes \varphi_2) \cdot (\psi_1 \otimes \psi_2)) = \Phi(\varphi_1 \otimes \varphi_2) \cdot \Phi(\psi_1 \otimes \psi_2)$, which is trivial, hence it is a Z-valued observable affiliated to $\mathscr{C}$, and we use the notation $\Phi(\varphi) = \varphi(\mathbb{H})$. Now we prove that $\sigma(\mathbb{H}) \subset \sigma(H_1) \times \sigma(H_2)$. Since the right-hand side is a closed subset of Z, if $z = (z_1, z_2)$ does not belong to it we can find neighbourhoods V_1 and V_2 of z_1 and z_2 respectively such that $V_j \cap \sigma(H_j) = \varnothing$. Let $\varphi_j \in C_0(V_j)$ with $\varphi_j(z_j) \neq 0$; then $\varphi = \varphi_1 \otimes \varphi_2$ has the property $\varphi(z) \neq 0$ and $\Phi(\varphi) = \varphi_1(H_1) \otimes \varphi_2(H_2) = 0$, so $z \notin \sigma(\Phi)$. Reciprocally let $z \in \sigma(H_1) \times \sigma(H_2)$ and $\varphi \in C_\infty(Z)$ with $\varphi(z) \neq 0$. Then there are functions $\varphi_j \in C_0(Z_j)$ with $\varphi_j(z_j) \neq 0$, $\varphi_j \geq 0$ and $|\varphi|^2 \geq \varphi_1 \otimes \varphi_2$, from which we get:

$$\Phi(\varphi)^* \Phi(\varphi) = \Phi(|\varphi|^2) \geq \Phi(\varphi_1 \otimes \varphi_2) = \varphi_1(H_1) \otimes \varphi_2(H_2) \neq 0$$

because $\varphi_j(H_j) \neq 0$; so $\Phi(\varphi) \neq 0$. This proves that $z \in \sigma(\mathbb{H})$; hence the equality $\sigma(\mathbb{H}) = \sigma(H_1) \times \sigma(H_2)$ is established. $\square$

Now we can proceed exactly [5] as in §8.1.3 and define for each proper continuous function $f : \sigma(H_1) \times \dots \times \sigma(H_n) \to Y$ a Y-valued observable $f(\mathbb{H})$ affiliated to $\mathscr{C}$ as the morphism $\varphi \mapsto (\varphi \circ f)(\mathbb{H})$ of $C_\infty(Y)$ into $\mathscr{C}$. The spectrum of $f(\mathbb{H})$ is clearly equal to $f(\sigma(H_1) \times \dots \times \sigma(H_n))$.

8.2.3. We shall need only a very special case of the preceding construction, namely that where $n = 2$, $Z_1 = Z_2 = Y = \mathbb{R}$ and $f(z_1, z_2) = z_1 + z_2$. Observe that this function f is proper when considered as defined on a set of the form $J_1 \times J_2$ where $J_i \subset \mathbb{R}$ are bounded from below. Hence, if H_1 and H_2 are real, *lower semibounded* observables affiliated to C*-algebras $\mathscr{C}_1$ and $\mathscr{C}_2$ respectively, then there is a unique observable H affiliated to $\mathscr{C}_1 \otimes \mathscr{C}_2$ such that $\varphi(H) = \Phi(\varphi \circ \Sigma)$ for $\varphi \in C_\infty(\mathbb{R})$. Here $\Sigma : \mathbb{R}^2 \to \mathbb{R}$ is the map $(\lambda_1, \lambda_2) \mapsto \lambda_1 + \lambda_2$ and $\Phi : C_\infty(\mathbb{R}^2) \to \mathscr{C}_1 \otimes \mathscr{C}_2$ is the morphism determined by the condition $\Phi(\varphi_1 \otimes \varphi_2) = \varphi(H_1) \otimes \varphi_2(H_2)$. We have $\sigma(H) = \sigma(H_1) + \sigma(H_2)$. We shall

---

[5] We mention that if $\mathscr{C}_1, \dots, \mathscr{C}_n$ are unital algebras, then what we do now is in fact a particular case of the situation considered in §8.1.3. Indeed, assuming $n = 2$ for simplicity of notations, let $H_1 \otimes I$ be the Z_1-valued observable affiliated to $\mathscr{C} = \mathscr{C}_1 \otimes \mathscr{C}_2$ given by $\varphi(H_1 \otimes I) := \varphi(H_1) \otimes I$; define similarly $I \otimes H_2$. Clearly $H_1 \otimes I$ commutes with $I \otimes H_2$, so we may apply Theorem 8.1.3 in order to construct $\mathbb{H}$. The only new fact here is that the joint spectrum of $H_1 \otimes I$ and $I \otimes H_2$ is just equal to $\sigma(H_1) \times \sigma(H_2)$. If H_1, H_2 are real-valued and bounded from below, then the mapping $f : \sigma(H_1) \otimes \sigma(H_2) \to \mathbb{R}$ given by $f(\lambda_1, \lambda_2) = \lambda_1 + \lambda_2$ is proper and we may consider the observable $f(\mathbb{H})$; according to the notation introduced at the end of §8.1.3 we denote it by $H_1 \otimes I + I \otimes H_2$. Below we shall keep this notation even if the algebras C_j do not have a unit, but we insist that in such a case the objects $H_1 \otimes I$ and $I \otimes H_2$ are not defined. However, we could define them as observables affiliated to $(\mathscr{C}_1)_I \otimes (\mathscr{C}_2)_I$, which contains $\mathscr{C}_1 \otimes \mathscr{C}_2$ (here $(\mathscr{C}_j)_I$ is the unital algebra generated by $\mathscr{C}_j$). We avoid this procedure because it is rather unnatural in our applications.

keep for H the standard notation $H = H_1 \otimes I + I \otimes H_2$ (see the note $(^5)$). Let us mention now a criterion which can be used in order to decide that a given lower semibounded real observable H affiliated to $\mathscr{C}_1 \otimes \mathscr{C}_2$ is in fact equal to $H_1 \otimes I + I \otimes H_2$: it is enough to check that for any real $\tau > 0$ one has $e^{-\tau H} = e^{-\tau H_1} \otimes e^{-\tau H_2}$. Indeed, if $m \in \mathbb{R}$, then the linear subspace generated by the functions $\{E_\tau \mid \tau > 0\}$, with $E_\tau(x) = e^{-\tau x}$, is dense in $C_\infty([m, \infty))$, and $E_\tau \circ \Sigma = E_\tau \otimes E_\tau$.

Assume now that $\mathscr{C}_1$ and $\mathscr{C}_2$ are realized on Hilbert spaces $\mathscr{H}_1$ and $\mathscr{H}_2$ and that H_1, H_2 are densely defined self-adjoint operators in $\mathscr{H}_1$ and $\mathscr{H}_2$ respectively. Since $\mathscr{C} \equiv \mathscr{C}_1 \otimes \mathscr{C}_2$ is a C*-subalgebra of $B(\mathscr{H})$, where $\mathscr{H} \equiv \mathscr{H}_1 \otimes \mathscr{H}_2$ (recall that we have chosen the spatial tensor product), Theorem 8.2.1 assures the existence of a unique morphism $\Phi : C_\infty(\mathbb{R}^2) \to B(\mathscr{H})$ such that $\Phi(\varphi_1 \otimes \varphi_2) = \varphi_1(H_1) \otimes \varphi_2(H_2)$ if $\varphi_j \in C_\infty(\mathbb{R})$. Let $\widetilde{\Phi}$ be the canonical extension of Φ to a normal morphism $\widetilde{\Phi} : BO(\mathbb{R}^2) \to B(\mathscr{H})$ (cf. §8.1.1). Then we have $\widetilde{\Phi}(\varphi_1 \otimes \varphi_2) = \varphi_1(H_1) \otimes \varphi_2(H_2)$ for $\varphi_j \in BO(\mathbb{R})$ (to prove this, one applies twice the monotone class theorem mentioned in §8.1.1, first with a fixed $\varphi_1 \in C_\infty(\mathbb{R})$ and then with a fixed $\varphi_2 \in BO(\mathbb{R})$). In particular, by setting $e_s(x) = \exp(isx)$, we get $\widetilde{\Phi}(e_s \otimes e_t) = \exp(isH_1) \otimes \exp(itH_2)$ for all $s, t \in \mathbb{R}$. If $\Sigma : \mathbb{R}^2 \to \mathbb{R}$ is the function "sum" introduced above, we have a unital morphism $\varphi \mapsto \varphi \circ \Sigma$ of $BO(\mathbb{R})$ into $BO(\mathbb{R}^2)$, hence we may consider the unital morphism $\varphi \mapsto \widetilde{\Phi}(\varphi \circ \Sigma)$ of $BO(\mathbb{R})$ into $B(\mathscr{H})$. This last morphism is clearly normal (i.e. $\widetilde{\Phi}(\varphi_k \circ \Sigma) \to \widetilde{\Phi}(\varphi \circ \Sigma)$ strongly if $\varphi_k \to \varphi$ boundedly, cf. §8.1.1), hence $E(\Delta) := \widetilde{\Phi}(\chi_\Delta \circ \Sigma)$ is a $B(\mathscr{H})$-valued spectral measure on $\mathbb{R}$ such that $\widetilde{\Phi}(\varphi \circ \Sigma) = \int_{\mathbb{R}} \varphi(\lambda) E(d\lambda)$ for all $\varphi \in BO(\mathbb{R})$. In other terms, if $H = \int_{\mathbb{R}} \lambda E(d\lambda)$ is the densely defined self-adjoint operator in $\mathscr{H}$ determined by E, we have $\widetilde{\Phi}(\varphi \circ \Sigma) = \varphi(H)$. Let $t \in \mathbb{R}$ and observe that $e_t \circ \Sigma = e_t \otimes e_t$. It follows that $\exp(iHt) = \exp(iH_1t) \otimes \exp(iH_2t)$ for all $t \in \mathbb{R}$. By applying both sides of this equality to some vector $f_1 \otimes f_2$ with $f_j \in D(H_j)$ and taking derivatives, one sees that $f_1 \otimes f_2 \in D(H)$ and $H(f_1 \otimes f_2) = (H_1 f_1) \otimes f_2 + f_1 \otimes (H_2 f_2)$. It follows that the algebraic tensor product $D(H_1) \odot D(H_2)$ (identified with a dense subspace of $\mathscr{H} = \mathscr{H}_1 \otimes \mathscr{H}_2$) is included in $D(H)$ and that $H|_{[D(H_1) \odot D(H_2)]} = H_1 \odot I + I \odot H_2$, where the right-hand side has to be interpreted in the algebraic sense. Finally, since $D(H_1) \odot D(H_2)$ is invariant under $\exp(iHt) = \exp(iH_1t) \otimes \exp(iH_2t)$, Nelson's lemma (see Theorem 3.3.4) implies the essential self-adjointness of H on $D(H_1) \odot D(H_2)$, which justifies the notation $H = H_1 \otimes I + I \otimes H_2$ for the closure of $H_1 \odot I + I \odot H_2$. Observe that the operators H_j need not be lower semibounded in this construction; but if H_1 and H_2 are bounded from below, then clearly the operator H constructed here coincides with that constructed above in an abstract setting, in particular it is affiliated to $\mathscr{C}_1 \otimes \mathscr{C}_2$ (it is clear from the following example that this fact does not hold in general: the operator Q of multiplication by the independent variable in $L^2(\mathbb{R})$ is affiliated to $C_\infty(\mathbb{R})$; but $Q \otimes I + I \otimes Q$ is not affiliated to $C_\infty(\mathbb{R}) \otimes C_\infty(\mathbb{R}) = C_\infty(\mathbb{R}^2)$, because the function $(x, y) \mapsto x + y$ is not divergent at infinity on $\mathbb{R}^2$). It is easy to see that in general $\sigma(H)$ is equal to the closure of $\sigma(H_1) + \sigma(H_2)$.

Let us remain in the preceding Hilbert space setting and make some remarks concerning the spectral properties of H in $\mathscr{H}$. It will be useful to know that, *if one of the operators H_1, H_2 has purely absolutely continuous spectrum, then H has purely absolutely continuous spectrum too.* For this it is sufficient to find a set of vectors $f \in \mathcal{H}$ which span a dense linear subspace of $\mathcal{H}$ and such that the function $F(t) \equiv \langle f, \exp(iHt)f \rangle$ is square-integrable on $\mathbb{R}$ (because then the Fourier transform $\widehat{F}$ of F belongs to $L^2(\mathbb{R})$, so the measure $\langle f, E(\cdot)f \rangle$ is absolutely continuous). So assume that H_1 is purely absolutely continuous and take $f = f_1 \otimes f_2$; then $|F(t)| \leq |\langle f_1, \exp(iH_1t)f_1 \rangle| \cdot ||f_2||^2$, and there is a dense subset of vectors $f_1 \in \mathcal{H}_1$ such that the function $t \mapsto \langle f_1, \exp(iH_1t)f_1 \rangle$ is square-integrable on $\mathbb{R}$.

We make one more remark, in the same context, concerning the eigenvalues of H. The number $\lambda \in \mathbb{R}$ is an eigenvalue of H if and only if $E(\{\lambda\}) = \widetilde{\Phi}(\chi_{\{\lambda\}} \circ \Sigma) \neq 0$, and then $E(\{\lambda\})\mathcal{H}$ is the corresponding eigenspace. Let $\Delta_\lambda = \{(\mu, \lambda - \mu) \mid \mu \in \mathbb{R}\} \subset \mathbb{R}^2$, so $E(\{\lambda\}) = \widetilde{\Phi}(\chi_{\Delta_\lambda}) \equiv P(\Delta_\lambda)$, where P is the $B(\mathcal{H})$-valued spectral measure on $\mathbb{R}^2$ associated to $\widetilde{\Phi}$. Clearly $P(A_1 \times A_2) = E_1(A_1) \otimes E_2(A_2)$, where E_j is the spectral measure of H_j and $A_1, A_2 \subset \mathbb{R}$ are Borel sets (the equality is equivalent to $\widetilde{\Phi}(\chi_{A_1} \otimes \chi_{A_2}) = \chi_{A_1}(H_1) \otimes \chi_{A_2}(H_2)$, and we know that this is true). Let $f = f_1 \otimes f_2$, then the measure $||P(\cdot)f||^2$ on $\mathbb{R}^2$ is just the product (in the sense of measure theory) of the measures $||E_1(\cdot)f_1||^2$ and $||E_2(\cdot)f_2||^2$ on $\mathbb{R}$. By Fubini's theorem we get

$$||P(\Delta_\lambda)f||^2 = \int_{\mathbb{R}} ||E_1(\{\lambda - \mu\})f_1||^2 \cdot ||E_2(d\mu)f_2||^2.$$

Since $E_1(\{\lambda - \mu\})$ is a projection orthogonal to $E_1(\{\lambda - \mu'\})$ if $\mu \neq \mu'$, the function $\mu \mapsto ||E_1(\{\lambda - \mu\})f_1||^2$ is non-zero at most for a countable number of values of μ. Hence the integral above is in fact a sum and we have:

$$\langle f, P(\Delta_\lambda)f \rangle = \sum_{\mu \in \mathbb{R}} ||E_1(\{\lambda - \mu\})f_1||^2 ||E_2(\{\mu\})f_2||^2$$

$$= \sum_{\mu \in \mathbb{R}} \langle f_1 \otimes f_2, E_1(\{\lambda - \mu\}) \otimes E_2(\{\mu\}) \cdot f_1 \otimes f_2 \rangle.$$

In other terms:

$$(8.2.1) \qquad E(\{\lambda\}) = \sum_{\lambda_1 + \lambda_2 = \lambda} E_1(\{\lambda_1\}) \otimes E_2(\{\lambda_2\}).$$

Hence λ *is an eigenvalue of H if and only if it is of the form $\lambda = \lambda_1 + \lambda_2$ with λ_j an eigenvalue of H_j*, and we have an explicit description of the eigenspace $E(\{\lambda\})\mathcal{H}$ as an *orthogonal* direct sum of tensor products

$$[E_1(\{\lambda_1\})\mathcal{H}_1] \otimes [E_2(\{\lambda_2\})\mathcal{H}_2].$$

We shall end this subsection with a more explicit description (on an abstract level) of the operator $H := H_1 \otimes I + I \otimes H_2$ under the assumption that one of the C*-algebras, for example $\mathcal{C}_1$, is abelian. Remember that, in order for H to be well defined, one has to assume that H_1 and H_2 are bounded from below. Working in the Gelfand representation, we may assume that $\mathcal{C}_1 = C_\infty(X_1)$ for some locally compact space X_1 (the character space of $\mathcal{C}_1$) and we may identify H_1 with a couple (U_1, h_1), where $U_1 \subset X_1$ is open and $h_1 : U_1 \to \mathbb{R}$

is continuous and proper (see §8.1.2). Note that $h_1(U_1)$ is a closed subset of $\mathbb{R}$ and is equal to $\sigma(H_1)$; hence the lower semiboundedness of H_1 is equivalent to that of h_1. Then, as we explained in §8.2.1, we may identify $\mathscr{C}_1 \otimes \mathscr{C}_2 = C_\infty(X_1; \mathscr{C}_2)$. If $x_1 \in U_1$, then $h_1(x_1)$ is a real number, so $h_1(x_1) + H_2$ is a well defined real observable affiliated to $\mathscr{C}_2$ (cf. §8.1.3). Now for $\varphi \in C_\infty(\mathbb{R})$ define $\varphi(H) \in C_\infty(X_1; \mathscr{C}_2)$ as follows: $\varphi(H)(x_1) = \varphi(h_1(x_1) + H_2)$ if $x \in U_1$ and $\varphi(H)(x_1) = 0$ otherwise (note that, if m_2 is the lower bound of H_2, we have $\|\varphi(h_1(x_1) + H_2)\| \leq \sup\{|\varphi(\lambda)| \mid \lambda \geq h(x_1) + m_2\}$, which tends to zero when x_1 tends to the boundary of U_1 or to infinity (in X_1); moreover, the set of $\varphi \in C_\infty(\mathbb{R})$ for which $x_1 \mapsto \varphi(h_1(x_1) + H_2) \in \mathscr{C}_2$ is continuous is a closed $*$-subalgebra of $C_\infty(\mathbb{R})$ which contains the functions $\varphi(\lambda) = (\lambda - z)^{-1}$, $z \notin \mathbb{R}$, hence it is equal to $C_\infty(\mathbb{R})$; these facts prove that $\varphi(H) \in C_\infty(X_1; \mathscr{C}_2)$). It is straightforward to check that $e^{-\tau H} = e^{-\tau H_1} \otimes e^{-\tau H_2}$ if $\tau > 0$, so the observable H defined above is just $H_1 \otimes I + I \otimes H_2$.

8.2.4. In the rest of this section we shall discuss several questions concerning arbitrary observables affiliated to C*-algebras of the form $\mathscr{B} \otimes \mathscr{C}$ with $\mathscr{B}$ abelian. This is a natural extension of the point of view presented at the end of §8.2.3, and its main purpose is to allow one to study many-body hamiltonians whose kinetic energy is not a quadratic function of momentum. We shall work in the Gelfand representation of $\mathscr{B}$, hence we take $\mathscr{B} = C_\infty(X)$ for some locally compact space X. Then $\mathscr{B} \otimes \mathscr{C} \equiv C_\infty(X; \mathscr{C})$ and our first step will be to give a more explicit representation of the observables affiliated to such an algebra. For this it is necessary to define what is meant by a family of observables depending continuously on a parameter.

Let Z be a LCSC space, $\mathscr{C}$ a C*-algebra, X a topological space and $\{H_x\}_{x \in X}$ a family of Z-valued observables affiliated to $\mathscr{C}$. We say that *the map $x \mapsto H_x$ is continuous* if, for each $\varphi \in C_\infty(Z)$, the function $x \mapsto \varphi(H_x) \in \mathscr{C}$ is (norm)-continuous. Observe that, in any case, the set of $\varphi \in C_\infty(Z)$ for which $x \mapsto \varphi(H_x)$ is continuous is a C*-subalgebra of $C_\infty(Z)$. Hence, in order to establish the continuity of the map $x \mapsto H_x$, it is sufficient to show that $x \mapsto \varphi(H_x)$ is continuous for a set of $\varphi \in C_\infty(Z)$ which generates $C_\infty(Z)$ (as a C*-algebra). Assume now $Z = \mathbb{R}$; then $\{H_x\}_{x \in X}$ *is a continuous family of observables affiliated to $\mathscr{C}$ if and only if there is a non-real number z such that $x \mapsto (H_x - z)^{-1} \in \mathscr{C}$ is continuous* (if the preceding continuity property is true for some $z \in \mathbb{C} \setminus \mathbb{R}$, then it will hold for each non-real z, by (8.1.2) and the continuity of the involution in $\mathscr{C}$). By taking $X = \mathbb{N} \cup \{\infty\}$, we thus give a meaning to the notion of a convergent sequence of Z-valued observables affiliated to $\mathscr{C}$ (note that $\lim_{n \to \infty} H_n = H$ corresponds to the so-called norm resolvent convergence in the Hilbert space setting). More generally, if X is a locally compact space, we can give a meaning to $\lim_{x \to \infty} H_x = H$ by considering the one-point compactification $X \cup \{\infty\}$ of X. We shall be interested especially in the case $H = \infty$, which is the Z-valued observable defined by the morphism $\Phi(\varphi) = 0$ for all $\varphi \in C_\infty(Z)$. It will be convenient (and natural) to say that a family $\{H_x\}_{x \in X}$ of Z-valued observables affiliated to $\mathscr{C}$ is *proper* if $\lim_{x \to \infty} H_x = \infty$; this means $\|\varphi(H_x)\| \to 0$ as $x \to \infty$ (in X) and, if $Z = \mathbb{R}$, this is equivalent to the existence of a non-real number z such that $\|(H_x - z)^{-1}\| \to 0$ as $x \to \infty$.

Let $\{H_x\}_{x \in X}$ be a proper family of Z-valued observables affiliated to $\mathscr{C}$, where X is a locally compact space. Then the spectrum $\sigma(H_x)$ of H_x is localized near infinity for large x in the following sense: for any compact $K \subset Z$, there is a compact $L \subset X$ such that $x \notin L \Rightarrow \sigma(H_x) \cap K = \varnothing$. (For the proof, let $\varphi \in C_\infty(Z)$ with $\varphi(z) = 1$ on K; since $\|\varphi(H_x)\| \to 0$ as $x \to \infty$, we can find a compact set $L \subset X$ such that $\|\varphi(H_x)\| < 1$ if $x \notin L$; but $\|\varphi(H_x)\| = \sup\{|\varphi(z)| \mid z \in \sigma(H_x)\}$, hence $|\varphi(z)| < 1$ if $z \in \sigma(H_x)$ and $x \notin L$). Now let us assume furthermore that the map $x \mapsto H_x$ is continuous; then $\cup_{x \in X} \sigma(H_x)$ is a closed subset of Z.

PROOF. If z belongs to the closure of $\cup \sigma(H_x)$, then there are sequences $\{x_n\}$ in X and $\{z_n\}$ in Z such that $z_n \in \sigma(H_{x_n})$ and $z_n \to z$. Let K be a compact set such that $z_n \in K$ for each n and let L be as above. Then $x_n \in L$ for all n hence, after replacing $\{x_n\}$ by a subsequence, we may assume that $\{x_n\}$ converges to some $x \in L$. Let $\varphi \in C_\infty(Z)$, $\varphi(z) \neq 0$; then $\varphi(z_n) \to \varphi(z)$, hence $|\varphi(z_n)| \geq |\varphi(z)|/2 > 0$ for all sufficiently large n. So $\|\varphi(H_{x_n})\| = \sup\{|\varphi(\zeta)| \mid \zeta \in \sigma(H_{x_n})\} \geq |\varphi(z_n)| \geq |\varphi(z)|/2 > 0$ for n large. But $\|\varphi(H_{x_n})\| \to \|\varphi(H_x)\|$, hence $\varphi(H_x) \neq 0$. This proves that $z \in \sigma(H_x)$. $\square$

Consider now a locally compact space X, a C*-algebra $\mathscr{C}$, and set $\mathscr{C}^X = C_\infty(X) \otimes \mathscr{C} \equiv C_\infty(X; \mathscr{C})$. Note that for each $x \in X$ there is a canonical surjective morphism $\theta_x : \mathscr{C}^X \to \mathscr{C}$ defined by $\theta_x[S] = S_x$; here $S \in \mathscr{C}^X$ is the function $x \mapsto S_x \in \mathscr{C}$. This allows us to define a *bijective correspondence* between Z-valued observables H affiliated to $\mathscr{C}^X$ and proper continuous families $\{H_x\}_{x \in U}$ of Z-valued observables H_x affiliated to $\mathscr{C}$, indexed by open subsets U of X (to $U = \varnothing$ corresponds $H = \infty$). Indeed, if H is given, we let $H_x = \theta_x[H]$ (i.e. $\varphi(H_x) = \theta_x[\varphi(H)]$ for each $\varphi \in C_\infty(Z)$) and $U = \{x \in X \mid H_x \neq \infty\} = \{x \in X \mid \exists \varphi \in C_\infty(Z) \text{ such that } \theta_x[\varphi(H)] \neq 0\}$. Clearly U is an open subset of X, and the map $x \mapsto H_x$ is continuous and proper on U (for the last property, note that for each $\varepsilon > 0$ and $\varphi \in C_\infty(Z)$, the set of $x \in X$ with $\|\varphi(H_x)\| \geq \varepsilon$ is compact and contained in U). Reciprocally, if $\{H_x\}_{x \in U}$ is given, we define H as the morphism $C_\infty(Z) \to \mathscr{C}^X$ which associates to φ the function $X \to \mathscr{C}$ equal to $x \mapsto \varphi(H_x)$ on U and equal to zero on $X \setminus U$.

With the preceding notations, let us prove that $\sigma(H) = \cup_{x \in X} \sigma(H_x)$. If $z \in \sigma(H_x)$ for some x, then: $\varphi \in C_\infty(Z)$, $\varphi(z) \neq 0 \Rightarrow \varphi(H_x) \neq 0$; hence $\theta_x[\varphi(H)] \neq 0$, in particular $\varphi(H) \neq 0$. So $\cup \sigma(H_x) \subset \sigma(H)$. Reciprocally, assume that $z \notin \cup \sigma(H_x)$. We have seen above that this union is a closed set, hence there is a neighbourhood V of z such that $V \cap \sigma(H_x) = \varnothing$ for each $x \in U$. Then for each $\varphi \in C_0(V)$ we have $\varphi(H_x) = 0$ for each $x \in U$, hence $\varphi(H) = 0$. This implies that $z \notin \sigma(H)$, and the proof of the assertion is finished. In particular, we have $\|\varphi(H)\| = \sup_x \|\varphi(H_x)\| = \sup\{|\varphi(z)| \mid z \in \sigma(H_x), x \in U\}$.

The reader may have noticed that the constructions made here are natural generalizations of the direct sum constructions of §8.1.4: there X was the finite set $\{1, 2, \ldots, n\}$ equipped with the discrete topology.

Let $\mathfrak{J}$ be a closed self-adjoint ideal in $\mathscr{C}$ and $\mathfrak{J}^X = C_\infty(X) \otimes \mathfrak{J} \equiv C_\infty(X; \mathfrak{J})$. Clearly $\mathfrak{J}^X$ is a closed self-adjoint ideal in $\mathscr{C}^X$. We observe that there is a canonical isomorphism between the quotient C*-algebra $\mathscr{C}^X/\mathfrak{J}^X$ and the C*-algebra $C_\infty(X) \otimes (\mathscr{C}/\mathfrak{J}) \equiv C_\infty(X; \mathscr{C}/\mathfrak{J}) \equiv (\mathscr{C}/\mathfrak{J})^X$. Indeed, if $\pi : \mathscr{C} \to \mathscr{C}/\mathfrak{J}$

is the canonical surjection, two functions $S, T \in \mathscr{C}^X$ are equal modulo $\mathfrak{J}^X$ if and only if the functions $x \mapsto \pi[S_x] \in \mathscr{C}/\mathfrak{J}$ and $x \mapsto \pi[T_x] \in \mathscr{C}/\mathfrak{J}$ are identical. Hence we have a natural injective morphism of $\mathscr{C}^X/\mathfrak{J}^X$ into $(\mathscr{C}/\mathfrak{J})^X$, and this morphism is surjective because it has closed range (which is true for any morphism between C*-algebras) and its range obviously contains the algebraic tensor product $C_\infty(X) \odot (\mathscr{C}/\mathfrak{J})$. From now on we identify $\mathscr{C}^X/\mathfrak{J}^X \equiv (\mathscr{C}/\mathfrak{J})^X$; then the canonical surjection $\pi^X : \mathscr{C}^X \to \mathscr{C}^X/\mathfrak{J}^X$ acts as follows: $\pi^X[S]_x = \pi[S_x]$ for all $x \in X$.

Now let us consider a Z-valued observable H affiliated to $\mathscr{C}^X$ and the family $\{H_x\}_{x \in U}$ of Z-valued observables affiliated to $\mathscr{C}$ which corresponds to H. Then $\pi^X[H]$ is affiliated to $\mathscr{C}^X/\mathfrak{J}^X = (\mathscr{C}/\mathfrak{J})^X$, and it is clear that $\{\pi[H_x]\}_{x \in U}$ is the corresponding family of Z-valued observables affiliated to $\mathscr{C}/\mathfrak{J}$. Then

$$(8.2.2) \qquad \mathfrak{J}^X\text{-}\sigma_{\mathrm{ess}}(H) = \sigma(\pi^X[H]) = \bigcup_{x \in U} \sigma(\pi[H_x]) = \bigcup_{x \in U} \mathfrak{J}\text{-}\sigma_{\mathrm{ess}}(H_x).$$

Finally, we give an explicit example in which this identity gives a result of interest in the N-body problem. Let $\mathscr{H}$ be a Hilbert space, $\mathscr{C} = B(\mathscr{H})$ and $\mathfrak{J} = K(\mathscr{H})$ the ideal of compact operators on $\mathscr{H}$. Then $K(\mathscr{H})\text{-}\sigma_{\mathrm{ess}}(T)$ is the usual essential spectrum $\sigma_{\mathrm{ess}}(T)$ (see §8.1.5). Assume that the locally compact space X is equipped with a positive Radon measure with support equal to X and let $L^2(X)$ be the corresponding Hilbert space of complex-valued, square-integrable functions on X. Then $C_\infty(X) \subset B(L^2(X))$ canonically and $L^2(X) \otimes \mathscr{H} \equiv L^2(X; \mathscr{H})$. If $\{H_x\}_{x \in X}$ is a proper continuous family of densely defined self-adjoint operators in $\mathscr{H}$ (in particular each H_x is an observable affiliated to $B(\mathscr{H})$), then the observable H affiliated to $C_\infty(X) \otimes B(\mathscr{H}) = B(\mathscr{H})^X$ associated to it can obviously be realized as a densely defined self-adjoint operator in $L^2(X; \mathscr{H})$. We shall then have:

$$(8.2.3) \qquad K(\mathscr{H})^X\text{-}\sigma_{\mathrm{ess}}(H) = \bigcup_{x \in X} \sigma_{\mathrm{ess}}(H_x).$$

For an even more explicit example, assume (as at the end of §8.2.3) that $H_x = h_1(x) + H_2$, where H_2 is a lower semibounded, densely defined self-adjoint operator in $\mathscr{H}$, $h_1 : X \to \mathbb{R}$ is continuous and $\lim h_1(x) = +\infty$ as $x \to \infty$. In other terms, $H = H_1 \otimes I + I \otimes H_2$ with H_1 the operator of multiplication by h_1 in $L^2(X)$. Then $\sigma_{\mathrm{ess}}(H_x) = h_1(x) + \sigma_{\mathrm{ess}}(H_2)$, hence $K(\mathscr{H})^X\text{-}\sigma_{\mathrm{ess}}(H) = \cup_{x \in X}[h_1(x) + \sigma_{\mathrm{ess}}(H_2)]$. If X is a connected set, then the result is particularly simple, because then h_1 maps X onto $[\min h_1, +\infty)$, hence

$$(8.2.4) \qquad K(\mathscr{H})^X\text{-}\sigma_{\mathrm{ess}}(H) = [\min h_1 + \min \sigma_{\mathrm{ess}}(H_2), \infty).$$

8.3. ϱ-Functions in a C*-Algebra Setting

8.3.1. Let $\mathscr{C}$ be a C*-algebra and $\{\mathscr{U}_\tau\}_{\tau \in \mathbb{R}}$ a one-parameter group of automorphisms of $\mathscr{C}$. So for each $\tau \in \mathbb{R}$ a bijective morphism $\mathscr{U}_\tau : \mathscr{C} \to \mathscr{C}$ is given and $\mathscr{U}_\tau \mathscr{U}_\sigma = \mathscr{U}_{\tau+\sigma}$ holds for all real τ and σ. We shall not make any continuity assumption, although in our applications $\{\mathscr{U}_\tau\}$ is in fact a C_0-group. Our rules from Sections 3.3 and 3.4 allow us to define the Sobolev (half) scale $\{\mathscr{C}_k \mid k \geq 0$ integer$\}$ and the Besov (half) scale $\{\mathscr{C}_{s,p} \mid 0 < s < \infty, 1 \leq p \leq \infty\}$. It is easily seen (observe that an automorphism of $\mathscr{C}$ is automatically an isometry) that $\mathscr{C}_0$, which is the space of $T \in \mathscr{C}$ such that $\tau \mapsto \mathscr{W}_\tau[T] \in \mathscr{C}$ is (norm)-continuous, is a C*-subalgebra of $\mathscr{C}$, invariant under the morphisms $\mathscr{W}_\tau$, and that the group of automorphisms induced by $\{\mathscr{W}_\tau\}$ in $\mathscr{C}_0$ is of class C_0. Moreover, all the spaces $\mathscr{C}_k$, $\mathscr{C}_{s,p}$ are *-subalgebras of $\mathscr{C}_0$, invariant under the automorphism group (cf. Sections 5.1 and 5.2). However, the only space which will be of interest for us in this chapter is $\mathscr{C}_1$, the domain of the generator of $\{\mathscr{W}_\tau\}$. In order to avoid ambiguities later on, we shall adopt a special notation for $\mathscr{C}_1$ (suggested by the developments of Chapter 5). Moreover, we shall define below the generator of $\{\mathscr{W}_\tau\}$ in a way which is better suited to our present context; more precisely we normalize the generator $\mathscr{A}$ such that $\mathscr{W}_\tau = \exp(\mathscr{A}\tau)$ (and *not* $\mathscr{W}_\tau = \exp(i\mathscr{A}\tau)$ as was the rule in Chapter 5).

Let $k \geq 0$ be an integer and $S \in \mathscr{C}$. We shall say that S *is of class* $C_u^k(\mathscr{A})$ if the function $\tau \mapsto \mathscr{W}_\tau[S] \in \mathscr{C}$ is of class C^k (in the norm topology of $\mathscr{C}$; since this is the natural topology on a C*-algebra, the subscript u in the notation C_u^k may seem redundant; however, we prefer to keep it in order to stress the similarity with the classes considered in Chapter 5). We denote by $C_u^k(\mathscr{A})$ the set of $S \in \mathscr{C}$ that are of class $C_u^k(\mathscr{A})$ (hence $C_u^k(\mathscr{A}) = \mathscr{C}_k$ in the notation used above). It is trivial to see that $C_u^k(\mathscr{A})$ is a *-subalgebra of $\mathscr{C}$, that $C_u^k(\mathscr{A}) \subset C_u^m(\mathscr{A})$ if $0 \leq m \leq k$, and that $C_u^0(\mathscr{A})$ is a C*-subalgebra of $\mathscr{C}$ (i.e. it is norm-closed). $C_u^1(\mathscr{A})$ is just the domain of the generator $\mathscr{A}$ of $\mathscr{W}_\tau$, where $\mathscr{A}$ is the linear operator in $\mathscr{C}$ defined by $\mathscr{A}[S] = \lim_{\varepsilon \to 0} \varepsilon^{-1}(\mathscr{W}_\varepsilon[S] - S)$ (observe that if this limit exists, then S necessarily belongs to $C_u^1(\mathscr{A})$; indeed, $\lim_{\varepsilon \to 0} \varepsilon^{-1}(\mathscr{W}_{\tau+\varepsilon}[S] - \mathscr{W}_\tau[S])$ will exist uniformly in $\tau \in \mathbb{R}$). Then $C_u^k(\mathscr{A})$ is the domain of the operator $\mathscr{A}^k$. We know from Section 3.3 that $C_u^\infty(\mathscr{A}) = \cap_{k=0}^\infty C_u^k(\mathscr{A})$ is a dense subspace of $C_u^0(\mathscr{A})$. It is easy to see that for $S, T \in C_u^1(\mathscr{A})$ one has $\mathscr{A}[ST] = \mathscr{A}[S]T + S\mathscr{A}[T]$ and $\mathscr{A}[S]^* = \mathscr{A}[S^*]$. Since $\mathscr{A}: C_u^1(\mathscr{A}) \subset \mathscr{C} \to \mathscr{C}$ is a closed operator in $\mathscr{C}$, it is clear that $C_u^1(\mathscr{A})$ equipped with the new norm $\|S\|_{\mathscr{A}} := \|S\| + \|\mathscr{A}[S]\|$ is a Banach *-algebra.

As an example, assume that $\mathscr{C}$ is realized on a Hilbert space $\mathscr{H}$ and $\{\mathscr{W}_\tau\}$ is induced by a unitary group $\{e^{iA\tau}\}$ in $\mathscr{H}$, i.e. $\mathscr{W}_\tau[S] = e^{-iA\tau}Se^{iA\tau}$. Then, with the notations of Section 6.2, S is of class $C_u^k(\mathscr{A})$ if and only if $S \in C_u^k(A; \mathscr{H})$. Moreover, if $k = 1$, we shall have $\mathscr{A}[S] = [S, iA]$.

Now let H be a (real) observable affiliated to $\mathscr{C}$. We shall say that H *is locally of class* $C_u^1(\mathscr{A})$ if $\varphi(H) \in C_u^1(\mathscr{A})$ for all $\varphi \in C_0^\infty(\mathbb{R})$. This definition can be stated in slightly different terms as follows. Denote by $H_\tau = \mathscr{W}_\tau[H]$ the image of H through the morphism $\mathscr{W}_\tau$ (see §8.1.4). Then H is locally of class $C_u^1(\mathscr{A})$ if and only if the map $\tau \mapsto \varphi(H_\tau) \in \mathscr{C}$ is differentiable (hence of class C^1) for each $\varphi \in C_0^\infty(\mathbb{R})$. We could similarly introduce more general notions like "locally of class $C_u^k(\mathscr{A})$ on an open set $J \subset \mathbb{R}$" by requiring that $\varphi(H) \in C_u^k(\mathscr{A})$ if $\varphi \in C_0^\infty(J)$, but we do not need such generalizations. It turns out that it is difficult to verify directly that H is locally of class $C_u^1(\mathscr{A})$, so we shall introduce a more restricted [6] class of regularity as follows: we say that H *is of class* $C_u^1(\mathscr{A})$

---

[6]The difference between the two notions, more precisely the fact that the second one has a global character, is clarified by the following example. Let $\mathscr{H} = L^2(\mathbb{R})$, H the operator of multiplication by a function $h: \mathbb{R} \to \mathbb{R}$ and let $\mathscr{W}_\tau$ be induced by the translation group, i.e. $A = P = -i\frac{d}{dx}$ and $\mathscr{A}[S] = [S, iP]$. If h is a homeomorphism of class C^1 on $\mathbb{R}$, then

if there is $z \in \mathbb{C} \setminus \sigma(H)$ such that $(H - z)^{-1} \in C_u^1(\mathscr{A})$. Then this property will hold for *all* $z \in \mathbb{C} \setminus \sigma(H)$ (*Proof*: Apply $\mathscr{U}_\tau$ to (8.1.2) and observe that the right-hand side is of class C^1 as function of τ if $R(z_0) \in C_u^1(\mathscr{A})$). Now let us prove that, *if H is of class $C_u^1(\mathscr{A})$, then $\varphi(H) \in C_u^1(\mathscr{A})$ for all $\varphi \in C_0^3(\mathbb{R})$, hence H will be locally of class $C_u^1(\mathscr{A})$.* We shall need the formula $(R(z) = (H - z)^{-1})$

$$(8.3.1) \qquad \mathscr{A}[R(z)] = \{I + (z - z_0)R(z)\}\mathscr{A}[R(z_0)]\{I + (z - z_0)R(z)\}$$

which holds for all non-real z and z_0; as usual, I denotes the identity of the unital algebra $\mathscr{C}_I$ generated by $\mathscr{C}$. To prove (8.3.1), we apply $\mathscr{U}_\tau$ to the relation $R(z) = R(z_0) + (z - z_0)R(z)R(z_0)$ and then we take derivatives with respect to τ at $\tau = 0$. This gives:

$$\mathscr{A}[R(z)]\{I + (z_0 - z)R(z_0)\} = \{I + (z - z_0)R(z)\}\mathscr{A}[R(z_0)].$$

This implies (8.3.1), according to a remark made after (8.1.2). We shall only need the following immediate consequence of (8.3.1):

$$(8.3.2) \qquad \|\mathscr{A}[R(z)]\| \le (1 + |z - z_0| \cdot |\Im z|^{-1})^2 \|\mathscr{A}[R(z_0)]\|.$$

Let us apply $\mathscr{U}_\tau$ to (8.1.1) in which we take $r = 3$. The preceding estimate shows that we may take derivatives under the integral (by the dominated convergence theorem), hence $\varphi(H) \in C_u^1(\mathscr{A})$.

We have proved more, in fact. It is clear that for any H of class $C_u^1(\mathscr{A})$ we shall have $\varphi(H) \in C_u^1(\mathscr{A})$ if $\varphi \in C_\infty(\mathbb{R})$ and $\int_{\mathbb{R}} \sum_{k=0}^3 |\varphi^{(k)}(\lambda)|(1 + \lambda^2)d\lambda < \infty$. Moreover, an easy computation gives the following estimate:

$$(8.3.3) \qquad \|\mathscr{A}[\varphi(H)]\| \le \|\mathscr{A}[R(i)]\| \int_{\mathbb{R}} \sum_{k=0}^3 |\varphi^{(k)}(\lambda)|(1 + \lambda^2)d\lambda.$$

If the observable H has a spectral gap, i.e. if there is a number $\lambda_0 \in \mathbb{R} \setminus \sigma(H)$, then one can improve this by the following procedure. For each $\varphi \in C_\infty(\mathbb{R})$ such that $\varphi(x) = 0$ in a neighbourhood of λ_0, let us define $\tilde{\varphi} : \mathbb{R} \to \mathbb{C}$ by $\tilde{\varphi}(y) = \varphi(\lambda_0 - y^{-1})$ if $y \ne 0$ and $\tilde{\varphi}(0) = 0$. Then $\tilde{\varphi} \in C_0(\mathbb{R})$ and $\tilde{\varphi}(R) = \varphi(H)$ if $R = (\lambda_0 - H)^{-1} \in \mathscr{C}$. We shall use Corollary 3.2.33 from [BR] with $\delta = \mathscr{A}$ (in the C*-algebra with unit generated by $\mathscr{C}$), $A = R$ and $f = \tilde{\varphi}$. It follows that $\varphi(H)$ is of class $C_u^1(\mathscr{A})$ if $\varphi \in C_\infty(\mathbb{R}) \cap C^2(\mathbb{R})$ and if the integral below is finite; furthermore, the next estimate holds:

$$(8.3.4) \qquad \|\mathscr{A}[\varphi(H)]\| \le \|\mathscr{A}[R]\| \left[\frac{\pi}{2}\int_{\mathbb{R}} \left|\lambda\frac{d}{d\lambda}(\lambda^2\varphi'(\lambda)) - \lambda\varphi'(\lambda)\right|^2 d\lambda\right]^{1/2}.$$

For example, if $\varphi \in C_\infty^2(\mathbb{R})$ and there is $\alpha \in \mathbb{R}$ such that $\varphi''(\lambda) = \alpha\lambda^{-3} + O(\lambda^{-4})$ as $|\lambda| \to \infty$, then the preceding integral is finite; we have $\tilde{\varphi} \in C_0^2(\mathbb{R})$ if and only if $\varphi''(\lambda) = \alpha\lambda^{-3} + \beta\lambda^{-4} + o(\lambda^{-4})$ for some $\alpha, \beta \in \mathbb{R}$.

H is *locally* of class $C^1(\mathscr{A})$. But since $\mathscr{A}[(H - z)^{-1}]$ is the operator of multiplication by $h'(h - z)^{-2}$, H is of class $C_u^1(\mathscr{A})$ only if $|h'(x)| \le C(1 + h(x)^2)$. See also Proposition 6.2.10.

8.3.2. If $\mathbb{H}$ is a $\mathbb{R}^n$-valued observable affiliated to $\mathscr{C}$, one may extend to it the notion introduced above in a straightforward way: $\mathbb{H}$ is locally of class $C_u^1(\mathscr{A})$ if $\varphi(\mathbb{H}) \in C_u^1(\mathscr{A})$ for any $\varphi \in C_0^\infty(\mathbb{R}^n)$. Clearly, if $f : \sigma(\mathbb{H}) \to \mathbb{R}^m$ is such that there is an open neighbourhood V in $\mathbb{R}^n$ of $\sigma(\mathbb{H})$ and a proper C^∞ extension $f_1 : V \to \mathbb{R}^m$ of f, then the $\mathbb{R}^m$-valued observable $f(\mathbb{H})$ will also be locally of class $C_u^1(\mathscr{A})$. We shall prove now a rather important technical fact:

THEOREM 8.3.1. *Let $\mathbb{H}$ be a $\mathbb{R}^n$-valued observable affiliated to $\mathscr{C}$ such that $\varphi(\mathbb{H}) \in C_u^1(\mathscr{A})$ if $\varphi = \varphi_1 \otimes \cdots \otimes \varphi_n$ and $\varphi_j \in C_0^\infty(\mathbb{R})$ for all j. Then $\mathbb{H}$ is locally of class $C_u^1(\mathscr{A})$.*

We first mention an obvious corollary (see Theorem 8.1.3):

COROLLARY 8.3.2. *Let $(H_1, \ldots, H_n)$ be a commuting family of real observables affiliated to $\mathscr{C}$, $\mathbb{H} = H_1 \times \cdots \times H_n$ their cartesian product, and $\sigma(H_1, \ldots, H_n)$ their joint spectrum. If each H_j is locally of class $C_u^1(\mathscr{A})$, then $\mathbb{H}$ is locally of class $C_u^1(\mathscr{A})$. If the function $f : \sigma(H_1, \ldots, H_n) \to \mathbb{R}^m$ has a proper C^∞ extension to a neighbourhood of $\sigma(H_1, \ldots, H_n)$ in $\mathbb{R}^n$, then $f(H_1, \ldots, H_n)$ is locally of class $C_u^1(\mathscr{A})$.*

In order to simplify the notations we shall prove Theorem 8.3.1 in the case $n = 2$. To make our presentation self-contained, we first give a simple proof of the so-called *kernel theorem*.

LEMMA 8.3.3. *Let $\mathbf{E}$ be a Banach space and $\Phi_0 : C_0^\infty(\mathbb{R}) \times C_0^\infty(\mathbb{R}) \to \mathbf{E}$ a bilinear, separately continuous map. Then there is a unique continuous linear map $\Phi : C_0^\infty(\mathbb{R}^2) \to \mathbf{E}$ such that $\Phi(\varphi_1 \otimes \varphi_2) = \Phi_0(\varphi_1, \varphi_2)$.*

PROOF. (i) The algebraic tensor product $C_0^\infty(\mathbb{R}) \odot C_0^\infty(\mathbb{R})$ is naturally realized as a dense subspace of $C_0^\infty(\mathbb{R}^2)$ and Φ_0 has a unique extension to a linear map $\Phi_1 : C_0^\infty(\mathbb{R}) \odot C_0^\infty(\mathbb{R}) \to \mathbf{E}$ such that $\Phi_1(\varphi_1 \otimes \varphi_2) = \Phi_0(\varphi_1, \varphi_2)$. Hence the uniqueness of Φ is immediate and, for its existence, it is sufficient to prove that Φ_1 is continuous when $C_0^\infty(\mathbb{R}) \odot C_0^\infty(\mathbb{R})$ is equipped with the topology induced by $C_0^\infty(\mathbb{R}^2)$. Let $a > 0$ real, $J = [-a, a]$ and $C_0^\infty(J) = \{\varphi \in C_0^\infty(\mathbb{R}) \mid \operatorname{supp} \varphi \subset J\}$. It is clearly enough to consider the restriction of Φ_0 to $C_0^\infty(J) \times C_0^\infty(J)$.

(ii) For $k \in \mathbb{N}$ let $\mathscr{H}_0^k(J) = \{f \in \mathscr{H}^k(\mathbb{R}) \mid \operatorname{supp} f \subset J\}$; this is a closed subspace of the usual Sobolev space $\mathscr{H}^k(\mathbb{R})$, with norm denoted by $\|f\|_k \equiv \|\langle P \rangle^k f\|$ (see Section 4.1). It is easily shown that $C_0^\infty(J)$ is a dense subspace of $\mathscr{H}_0^k(J)$ and that $\cap_{k=0}^\infty \mathscr{H}_0^k(J) = C_0^\infty(J)$. Hence the natural (Fréchet space) topology of $C_0^\infty(J)$ is defined by the increasing family of norms $\{\|\cdot\|_k\}_{k \in \mathbb{N}}$. We shall need below the following fact: the canonical inclusion $\mathscr{H}_0^{k+1}(J) \to \mathscr{H}_0^k(J)$ is a Hilbert-Schmidt operator (for the proof, observe that it is enough that $P^j \chi(Q) \langle P \rangle^{-k-1}$ be of Hilbert-Schmidt class in $L^2(\mathbb{R})$ for some $\chi \in C_0^\infty(\mathbb{R})$ with $\chi(x) = 1$ on J and each $j = 0, \ldots, k$; commuting P^j with $\chi(Q)$, this last fact is an immediate consequence of the results of Section 4.1). More precisely, we shall use this Hilbert-Schmidt property in the form: there is $M_k < \infty$ such that for any orthonormal family $\{\xi_\alpha\}$ in $\mathscr{H}_0^{k+1}(J)$ the following estimate holds: $\sum_\alpha \|\xi_\alpha\|_k^2 \leq M_k$.

(iii) Since $C_0^\infty(J)$ is a Fréchet space, the separate continuity of the map $\Phi_0 : C_0^\infty(J) \times C_0^\infty(J) \to \mathbf{E}$ implies its continuity (uniform boundedness principle).

So there is a neighbourhood U of zero in $C_0^\infty(J)$ such that $\|\Phi_0(\varphi_1, \varphi_2)\| \leq 1$ if $\varphi_j \in U$. Then there are $k \in \mathbb{N}$ and $C > 0$ such that $\varphi \in U$ if $\|\varphi\|_k \leq C^{-1/2}$. Hence $\|\Phi_0(\varphi_1, \varphi_2)\| \leq C\|\varphi_1\|_k \|\varphi_2\|_k$ for all $\varphi_1, \varphi_2 \in C_0^\infty(J)$.

(iv) Let $\varphi \in C_0^\infty(J) \odot C_0^\infty(J)$. Then one can find a finite family $\{\xi_\alpha\}$ of elements of $C_0^\infty(J)$ such that $\varphi = \sum_{\alpha,\beta} a_{\alpha\beta} \xi_\alpha \otimes \xi_\beta$ with $a_{\alpha\beta} \in \mathbb{C}$. By virtue of the Gram-Schmidt method, we may assume that $\{\xi_\alpha\}$ is an orthonormal family in $\mathscr{H}_0^{k+1}(J)$. Then:

$$\|\Phi_1(\varphi)\| \leq \sum |a_{\alpha\beta}| \cdot \|\Phi_0(\xi_\alpha, \xi_\beta)\| \leq C \sum |a_{\alpha\beta}| \cdot \|\xi_\alpha\|_k \|\xi_\beta\|_k$$

$$\leq C \Big[\sum |a_{\alpha\beta}|^2\Big]^{1/2} \Big[\sum \|\xi_\alpha\|_k^2 \|\xi_\beta\|_k^2\Big]^{1/2}$$

$$\leq C M_k \Big[\sum |a_{\alpha\beta}|^2\Big]^{1/2} = C M_k \|\varphi\|_{\mathscr{H}_0^{k+1} \otimes \mathscr{H}_0^{k+1}}.$$

Here $\mathscr{H}_0^{k+1}(J) \otimes \mathscr{H}_0^{k+1}(J)$ is the Hilbert tensor product. Since $\mathscr{H}_0^{k+1}(J)$ is a closed subspace of $\mathscr{H}^{k+1}(\mathbb{R})$, $\mathscr{H}_0^{k+1}(J) \otimes \mathscr{H}_0^{k+1}(J)$ is a closed subspace of $\mathscr{H}^{k+1}(\mathbb{R}) \otimes \mathscr{H}^{k+1}(\mathbb{R})$ (equipped with the induced norm). By using a Fourier transformation, it is easily seen that $\mathscr{H}^{k+1}(\mathbb{R}) \otimes \mathscr{H}^{k+1}(\mathbb{R}) \supset \mathscr{H}^m(\mathbb{R}^2)$ (as B-spaces) if $m = 2(k+1)$. Finally, since $\operatorname{supp}\varphi \subset J \times J$, which is compact, there is a finite constant C_1 such that $\|\Phi_1(\varphi)\| \leq C_1 \|\varphi\|_{BC^m(J\times J)}$ for all $\varphi \in C_0^\infty(J) \odot C_0^\infty(J)$. This finishes the proof of the lemma. $\square$

PROOF OF THEOREM 8.3.1. Let us denote by $\mathbf{F}$ the Banach space $\mathscr{C}$ and by $\mathbf{E}$ the Banach space obtained by providing $C_u^1(\mathscr{A})$ with the norm $\|S\|_{\mathscr{A}} = \|S\| + \|\mathscr{A}[S]\|$. Observe that $\mathbf{E} \subset \mathbf{F}$ continuously. Then $\mathbb{H}$ is in fact a linear continuous map $\varphi \mapsto \varphi(\mathbb{H})$ of $C_\infty(\mathbb{R}^2)$ into $\mathbf{F}$. By hypothesis, this map sends an element $\varphi = \varphi_1 \otimes \varphi_2$ with $\varphi_j \in C_0^\infty(J)$ (we use the notations of the preceding lemma) into $\mathbf{E}$. Let $\Phi_0 : C_0^\infty(\mathbb{R}) \times C_0^\infty(\mathbb{R}) \to \mathbf{E}$ be given by $\Phi_0(\varphi_1, \varphi_2) = (\varphi_1 \otimes \varphi_2)(\mathbb{H})$. Then Φ_0 is separately continuous: $\varphi_1 \mapsto \Phi_0(\varphi_1, \varphi_2)$ is a continuous map of the Fréchet space $C_0^\infty(J)$ into the Banach space $\mathbf{F}$ and its range is contained in $\mathbf{E}$; since $\mathbf{E}$ is a Banach space continuously embedded in $\mathbf{F}$, the closed graph theorem implies the continuity of $\varphi_1 \mapsto \Phi_0(\varphi_1, \varphi_2)$ as a map $C_0^\infty(J) \to \mathbf{E}$. Then let Φ be as in Lemma 8.3.3. If $\varphi \in C_0^\infty(\mathbb{R}^2)$, then we can find a sequence $\{\varphi_k\}_{k\in\mathbb{N}}$ in $C_0^\infty(J) \odot C_0^\infty(J)$ (for some fixed J) such that $\varphi_k \to \varphi$ in $C_0^\infty(\mathbb{R}^2)$. This will imply $\Phi(\varphi_k) \to \Phi(\varphi)$ in $\mathbf{E}$ and also $\Phi(\varphi_k) \equiv \varphi_k(\mathbb{H}) \to \varphi(\mathbb{H})$ in $\mathbf{F}$. So $\varphi(\mathbb{H}) = \Phi(\varphi) \in \mathbf{E}$. $\square$

8.3.3. Our next purpose is to give a meaning to the symbol $\mathscr{A}[H]$ for an arbitrary observable H affiliated to $\mathscr{C}$ and locally of class $C_u^1(\mathscr{A})$. For this we take up on an abstract level an idea already used in the proof of Theorem 7.2.9. Observe that $S\mathscr{A}[T]S = \mathscr{A}[STS] - 2\Re(\mathscr{A}[S]TS)$ if S, T are symmetric elements of $\mathscr{C}$ of class $C_u^1(\mathscr{A})$. Let us take $S = \varphi(H)$ and $T = \psi(H)$ with real $\varphi, \psi \in C_0^\infty(\mathbb{R})$. Then we get an identity the right-hand side of which depends on ψ only through the product $\varphi\psi$ (in particular, it makes sense for any $\psi \in C^\infty(\mathbb{R})$). Hence, we can *define* $\varphi(H)\mathscr{A}[H]\varphi(H)$ for any real $\varphi \in C_0^\infty(\mathbb{R})$ by the following procedure: we choose an arbitrary real $\psi \in C_0^\infty(\mathbb{R})$ such that $\varphi(x)\psi(x) = \varphi(x)x$ for all $x \in \mathbb{R}$ and take $\varphi(H)\mathscr{A}[H]\varphi(H) := \varphi(H)\mathscr{A}[\psi(H)]\varphi(H)$. So $\varphi(H)\mathscr{A}[H]\varphi(H)$ is a well defined symmetric element of $\mathscr{C}$ for any real $\varphi \in C_0^\infty(\mathbb{R})$. As explained during the proof of Theorem 7.2.9, if we are in the setting of Section 7.2, the two possible definitions of $\varphi(H)\mathscr{A}[H]\varphi(H) \equiv \varphi(H)[H, iA]\varphi(H)$ give the same object.

We can now associate to H a function $\varrho_H \equiv \varrho_H^{\mathscr{A}} : \mathbb{R} \to (-\infty, +\infty]$ by a natural extension to a C*-algebra level of the relation (7.2.7). Namely, $\varrho_H(\lambda)$ is the least upper bound of the numbers $a \in \mathbb{R}$ such that $a\varphi(H)^2 \leq \varphi(H)\mathscr{A}[H]\varphi(H)$ for some real $\varphi \in C_0^\infty(\mathbb{R})$ with $\varphi(\lambda) \neq 0$. It is easy to prove that ϱ_H *is a lower semicontinuous function and* $\varrho_H(\lambda) < \infty$ *if and only if* $\lambda \in \sigma(H)$ (for the proof of the second assertion, observe that $\varphi(H)\mathscr{A}[H]\varphi(H) \leq c\varphi(H)^2$ for some fixed constant $c \in \mathbb{R}$ if supp φ is included in a fixed compact set).

If $\mathscr{C}$ is a unital algebra, each symmetric element S of $\mathscr{C}$ is canonically identified with an observable affiliated to $\mathscr{C}$, hence ϱ_S is well defined if $S \in C_u^1(\mathscr{A})$. If $\mathscr{C}$ is not unital, the observable defined by S is affiliated only to the unital algebra $\mathscr{C}_I$ generated by $\mathscr{C}$. However, the group $\{\mathscr{W}_\tau\}$ has a canonical extension to a group of (unital) automorphisms of $\mathscr{C}_I$, and this clearly allows us to define the function ϱ_S even if $\mathscr{C}$ is not unital. Notice that, for $\lambda \neq 0$, $\varrho_S(\lambda)$ may be defined very simply without leaving the algebra $\mathscr{C}$. Indeed, we have $\mathscr{A}[S] \in \mathscr{C}$ (here $\mathscr{A}[S] = \lim_{\varepsilon \to 0} \varepsilon^{-1}(\mathscr{W}_\varepsilon[S] - S)$) and $\varphi(S)$ is a well defined element of $\mathscr{C}$ for any $\varphi \in C(\mathbb{R})$ with $\varphi(0) = 0$ by standard functional calculus in C*-algebras. Hence $\varrho_S(\lambda)$ is the least upper bound of the numbers $a \in \mathbb{R}$ such that $a\varphi(S)^2 \leq \varphi(S)\mathscr{A}[S]\varphi(S)$ for some real $\varphi \in C_0^\infty(\mathbb{R})$ with $\varphi(\lambda) \neq 0$ and $\varphi(0) = 0$ (note that the supremum is realized as the support of φ shrinks to λ).

It is possible now to extend Proposition 7.2.5 to an algebraic level and so to express the ϱ-function of an observable H having a spectral gap in terms of the ϱ-function of its resolvent.

PROPOSITION 8.3.4. *Let* H *be an observable affiliated to* $\mathscr{C}$, *of class* $C_u^1(\mathscr{A})$ *and such that* $\sigma(H) \neq \mathbb{R}$. *Let* $\lambda_0 \in \mathbb{R} \setminus \sigma(H)$ *and* $R = (\lambda_0 - H)^{-1}$, *so that* $R \in \mathscr{C}$ *is symmetric and of class* $C_u^1(\mathscr{A})$. *Then for each* $\lambda \in \mathbb{R} \setminus \{\lambda_0\}$ *we have* $\varrho_H(\lambda) = (\lambda_0 - \lambda)^2 \varrho_R((\lambda_0 - \lambda)^{-1})$.

PROOF. We assume $\lambda_0 = 0$ and define $\varphi_1(x) = x\varphi(x)$ for each real $\varphi \in C_0^\infty(\mathbb{R})$. Then we remark that $\varphi(H)\mathscr{A}[H]\varphi(H) = \varphi_1(H)\mathscr{A}[R]\varphi_1(H)$. Indeed, if $\psi \in C_0^\infty(\mathbb{R})$ is real and $\psi(x) = x$ on supp φ, we have:

$$\varphi(H)\mathscr{A}[H]\varphi(H) = \varphi(H)\mathscr{A}[\psi(H)]\varphi(H)$$
$$= \mathscr{A}[\varphi(H)^2\psi(H)] - 2\Re(\mathscr{A}[\varphi(H)]\varphi(H)\psi(H))$$
$$= -\mathscr{A}[\varphi_1(H)^2 R] + 2\Re(\mathscr{A}[\varphi_1(H)R]\varphi_1(H))$$
$$= \varphi_1(H)\mathscr{A}[R]\varphi_1(H).$$

It follows that for any $\lambda \in \mathbb{R}$:

$$\varphi(H)\mathscr{A}[H]\varphi(H) = \lambda^2\varphi(H)\mathscr{A}[R]\varphi(H) + 2\Re\{\lambda\varphi(H)\mathscr{A}[R](\varphi_1(H) - \lambda\varphi(H))\}$$
$$+ (\varphi_1(H) - \lambda\varphi(H))\mathscr{A}[R](\varphi_1(H) - \lambda\varphi(H)).$$

Let $\varphi \in C_0^\infty(\mathbb{R})$ real such that $|\varphi(x)| \leq 1$, $\varphi(x) = 0$ in a neighbourhood of zero, and $\varphi(x) = 1$ in a neighbourhood of λ (which is assumed $\neq 0$). We set $\tilde{\varphi}(x) = \varphi(-x^{-1})$, so that $\tilde{\varphi} \in C_0^\infty(\mathbb{R})$, $|\tilde{\varphi}(x)| \leq 1$, $\tilde{\varphi}(x) = 1$ near $x = -\lambda^{-1}$,

$\widetilde{\varphi}(x) = 0$ near $x = 0$ and $\varphi(H) = \widetilde{\varphi}(R)$. If δ is the diameter of $\operatorname{supp}\varphi$, then $\|\varphi_1(H) - \lambda\varphi(H)\| \leq \delta$, hence:

$$\|\varphi(H)\mathscr{A}[H]\varphi(H) - \lambda^2\widetilde{\varphi}(R)\mathscr{A}[R]\widetilde{\varphi}(R)\| \leq \|\mathscr{A}[R]\|(2|\lambda| + \delta)\delta.$$

The assertion of the proposition follows easily from this estimate. □

8.3.4. We may define an abstract version of the function $\widetilde{\varrho}$ of §7.2.2 if a closed bilateral ideal $\mathfrak{J}$ of $\mathscr{C}$ is given such that $\mathscr{W}_\tau\mathfrak{J} \subset \mathfrak{J}$ for all $\tau \in \mathbb{R}$. In this case, $\{\mathscr{W}_\tau\}$ induces a group $\{\widetilde{\mathscr{W}}_\tau\}$ of automorphisms of the quotient C*-algebra $\widetilde{\mathscr{C}} := \mathscr{C}/\mathfrak{J}$ whose generator will be denoted by $\widetilde{\mathscr{A}}$, so $\widetilde{\mathscr{W}}_\tau = \exp(\widetilde{\mathscr{A}}\tau)$. Let $\pi : \mathscr{C} \to \widetilde{\mathscr{C}}$ be the canonical morphism. It is obvious that for each $S \in C^1_u(\mathscr{A})$ we have $\pi[S] \in C^1_u(\widetilde{\mathscr{A}})$ and $\widetilde{\mathscr{A}}[\pi[S]] = \pi[\mathscr{A}[S]]$. In particular, if H is an observable affiliated to $\mathscr{C}$, locally of class $C^1_u(\mathscr{A})$, then its image $\widetilde{H} = \pi[H]$ through π is locally of class $C^1_u(\widetilde{\mathscr{A}})$. Moreover, for each $\varphi \in C^\infty_0(\mathbb{R})$ we shall have $\widetilde{\mathscr{A}}[\varphi(\widetilde{H})] = \widetilde{\mathscr{A}}[\pi[\varphi(H)]] = \pi[\mathscr{A}[\varphi(H)]]$. Let us define $\widetilde{\varrho}_H := \varrho_{\widetilde{H}}$. The preceding remarks clearly imply the following alternative definition of the function $\widetilde{\varrho}_H$: for each $\lambda \in \mathbb{R}$, $\widetilde{\varrho}_H(\lambda)$ is the least upper bound of the numbers $a \in \mathbb{R}$ such that there are a real function $\varphi \in C^\infty_0(\mathbb{R})$ with $\varphi(\lambda) \neq 0$ and an element $K \in \mathfrak{J}$ such that $\varphi(H)\mathscr{A}[H]\varphi(H) \geq a\varphi(H)^2 + K$. The function $\widetilde{\varrho}_H : \mathbb{R} \to (-\infty, +\infty]$ is lower semicontinuous and we have $\widetilde{\varrho}_H(\lambda) < \infty$ if and only if $\lambda \in \mathfrak{J}\text{-}\sigma_{\mathrm{ess}}(H) \equiv \sigma(\widetilde{H})$. The function $\widetilde{\varrho}_H$ plays an important but auxiliary role in the theory developed here (cf. Sections 8.4 and 7.2).

8.3.5. In the rest of this section we shall show how one may calculate the ϱ-function of a direct sum or a tensor product of observables. We begin with the case of direct sums, which is rather trivial. Let $\mathscr{C}_1, \ldots, \mathscr{C}_n$ be C*-algebras and $\mathscr{C}$ their C*-direct sum (see §8.1.4). We assume that $\mathscr{C}_k$ is equipped with a group $\{\mathscr{W}_{k,\tau}\}_{\tau \in \mathbb{R}}$ of automorphisms and denote by $\mathscr{W}_\tau = \oplus^n_{k=1}\mathscr{W}_{k,\tau}$ the automorphism of $\mathscr{C}$ defined by $\mathscr{W}_\tau[\oplus^n_{k=1}S_k] = \oplus^n_{k=1}\mathscr{W}_{k,\tau}[S_k]$. Then $\{\mathscr{W}_\tau\}_{\tau \in \mathbb{R}}$ is a one-parameter group of automorphisms of $\mathscr{C}$ and its generator $\mathscr{A}$ is the direct sum $\mathscr{A} = \oplus^n_{k=1}\mathscr{A}_k$ of the generators $\mathscr{A}_1, \ldots, \mathscr{A}_n$ of the groups $\{\mathscr{W}_{1,\tau}\}, \ldots, \{\mathscr{W}_{n,\tau}\}$ in the following sense: an element $S = \oplus^n_{k=1}S_k \in \mathscr{C}$ is of class $C^1_u(\mathscr{A})$ if and only if $S_k \in C^1_u(\mathscr{A}_k)$ for $k = 1, \ldots, n$, and then $\mathscr{A}[S] = \oplus^n_{k=1}\mathscr{A}_k[S_k]$. The proof of this assertion is straightforward.

PROPOSITION 8.3.5. *Assume that the observable H affiliated to $\mathscr{C}$ is the direct sum of the observables H_k affiliated to $\mathscr{C}_k$ ($k = 1, \ldots, n$).*

(a) *H is locally of class $C^1_u(\mathscr{A})$ if and only if each H_k is locally of class $C^1_u(\mathscr{A}_k)$.*

(b) *Assume that H is locally of class $C^1_u(\mathscr{A})$ and let $\varrho_H = \varrho^\mathscr{A}_H$, $\varrho_{H_k} = \varrho^{\mathscr{A}_k}_{H_k}$ ($k = 1, \ldots, n$). Then $\varrho_H = \min\limits_{1 \leq k \leq n} \varrho_{H_k}$.*

PROOF. (a) This is obvious by what we said above.

(b) Let $a \in \mathbb{R}$ and recall that $a < \varrho_H(\lambda)$ means: there is $\varphi \in C^\infty_0(\mathbb{R})$ real, with $\varphi(\lambda) \neq 0$, such that $\varphi(H)\mathscr{A}[H]\varphi(H) \geq a\varphi(H)^2$. It is clear, according to the preceding remarks, that $\varphi(H)\mathscr{A}[H]\varphi(H) = \oplus^n_{k=1}(\varphi(H_k)\mathscr{A}_k[H_k]\varphi(H_k))$ and $\varphi(H)^2 = \oplus^n_{k=1}\varphi(H_k)^2$. Moreover, for $S = \oplus^n_{k=1}S_k \in \mathscr{C}$ we have $S \geq 0$ if and only if $S_k \geq 0$ for each $k = 1, \ldots, n$. Hence $\varphi(H)\mathcal{A}[H]\varphi(H) \geq a\varphi(H)^2$ is equivalent to $\varphi(H_k)\mathcal{A}_k[H_k]\varphi(H_k) \geq a\varphi(H_k)^2$ for each k. This clearly shows that $a < \varrho_H(\lambda) \Rightarrow a \leq \varrho_{H_k}(\lambda)$ for each k, i.e. $\varrho_H \leq \min \varrho_{H_k}$.

Reciprocally, assume that $a < \varrho_{H_k}(\lambda)$ for $k = 1, \ldots, n$. Then for each k there is $\varphi_k \in C_0^\infty(\mathbb{R})$ real, with $\varphi_k(\lambda) \neq 0$, such that $\varphi_k(H_k)\mathcal{A}_k[H_k]\varphi_k(H_k) \geq a\varphi_k(H_k)^2$. Let $\psi_k \in C_0^\infty(\mathbb{R})$ be real with $\varphi_k(x)\psi_k(x) = 1$ in a neighbourhood V_k of λ. Then $\varphi_k(H_k)\psi_k(H_k)\mathcal{A}_k[H_k]\varphi_k(H_k)\psi_k(H_k) \geq a(\varphi_k(H_k)\psi_k(H_k))^2$ for all k. Finally, let $\varphi \in C_0^\infty(\mathbb{R})$ real, with $\varphi(\lambda) \neq 0$ and $\operatorname{supp}\varphi \subset \cap_k V_k$. Multiplying (left and right) the preceding inequality by $\varphi(H_k)$ we get $\varphi(H_k)\mathcal{A}_k[H_k]\varphi(H_k) \geq a\varphi(H_k)^2$ for each k, which implies that $a \leq \varrho_H(\lambda)$. $\square$

8.3.6. We now consider tensor products of observables and assume, for simplicity of notations, that $n = 2$. If $\mathscr{C}_1$, $\mathscr{C}_2$ are C*-algebras equipped with one-parameter groups of automorphisms $\{\mathscr{W}_{1,\tau}\}_{\tau \in \mathbb{R}}$ and $\{\mathscr{W}_{2,\tau}\}_{\tau \in \mathbb{R}}$ respectively, then the (spatial) tensor product $\mathscr{C} = \mathscr{C}_1 \otimes \mathscr{C}_2$ is naturally equipped with the one-parameter group of automorphisms $\mathscr{W}_\tau = \mathscr{W}_{1,\tau} \otimes \mathscr{W}_{2,\tau}$ whose action is determined by the condition $\mathscr{W}_\tau[S_1 \otimes S_2] = \mathscr{W}_{1,\tau}[S_1] \otimes \mathscr{W}_{2,\tau}[S_2]$. For the generator $\mathscr{A}$ of $\{\mathscr{W}_\tau\}$ we may use the natural notation $\mathscr{A} = \mathscr{A}_1 \otimes I + I \otimes \mathscr{A}_2$, where $\mathscr{A}_k$ is the generator of $\{\mathscr{W}_{k,\tau}\}$. It is clear that $S = S_1 \otimes S_2$ is of class $C_u^1(\mathscr{A})$ if S_k is of class $C_u^1(\mathscr{A}_k)$ $(k = 1, 2)$, and then $\mathscr{A}[S] = \mathscr{A}_1[S_1] \otimes I + I \otimes \mathscr{A}_2[S_2]$.

THEOREM 8.3.6. *Let H_1, H_2 be observables affiliated to the algebras $\mathscr{C}_1$, $\mathscr{C}_2$ and assume that they are bounded from below and locally of class $C_u^1(\mathscr{A}_1)$, $C_u^1(\mathscr{A}_2)$ respectively ; set $\varrho_k \equiv \varrho_{H_k}^{\mathscr{A}_k}$ for $k = 1, 2$. If $H = H_1 \otimes I + I \otimes H_2$, then H is a lower semibounded observable affiliated to $\mathscr{C}$, H is locally of class $C_u^1(\mathscr{A})$ and, with the notation $\varrho \equiv \varrho_H^{\mathscr{A}}$, one has for all $\lambda \in \mathbb{R}$:*

$$(8.3.5) \qquad \varrho(\lambda) = \inf_{\lambda = \lambda_1 + \lambda_2} [\varrho_1(\lambda_1) + \varrho_2(\lambda_2)].$$

PROOF. (i) If $\mathbb{H}$ is the $\mathbb{R}^2$-valued observable affiliated to $\mathscr{C}$ defined in Theorem 8.2.1, then $\mathbb{H}$ is locally of class $C_u^1(\mathscr{A})$ (cf. Theorem 8.3.1 and the remark we made before Theorem 8.3.6). Since $\sigma(\mathbb{H}) = \sigma(H_1) \times \sigma(H_2)$, the restriction of the map $\Sigma(\lambda_1, \lambda_2) = \lambda_1 + \lambda_2$ to some open neighbourhood of $\sigma(\mathbb{H})$ is proper and of class C^∞; hence H, the image of $\mathbb{H}$ through Σ, is locally of class $C_u^1(\mathscr{A})$ (see the beginning of §8.3.2).

The rest of the proof will be devoted to the proof of the formula (8.3.5). We recall that $\sigma(H) = \sigma(H_1) + \sigma(H_2)$, so (8.3.5) is clearly true for $\lambda \notin \sigma(H)$, for then both sides of the equation are equal to ∞. Hence it will be enough to assume $\lambda \in \sigma(H)$ from now on. Moreover, since $\varrho_H(\lambda) = \varrho_{H + \mu}(\lambda + \mu)$, we may assume that $H_k \geq 0$ (i.e. $\sigma(H_k) \subset [0, \infty)$), hence $H \geq 0$ too. From now on we consider a fixed $\lambda \in \sigma(H)$, so $\lambda \geq 0$, and observe that in (8.3.5) it suffices to consider decompositions $\lambda = \lambda_1 + \lambda_2$ with $\lambda_k \in \sigma(H_k)$, so $\lambda_k \geq 0$. We set $\Lambda = [0, \lambda + 1]$.

(ii) We make some preliminary remarks concerning expressions of the form $\varphi(H)\mathscr{A}[H]\varphi(H)$ with $\varphi \in C_0^\infty(\mathbb{R})$ real and such that $\operatorname{supp}\varphi \subset (\lambda - 1, \lambda + 1)$ (note that, for $\lambda < 1$, the behaviour of $\varphi(x)$ for $x \leq 0$ is of no importance). Let us choose two real functions $\psi, \xi \in C_0^\infty(\mathbb{R})$ with $\psi(x) = x$ and $\xi(x) = 1$ on a neighbourhood of Λ. Then for any $x_1, x_2 \geq 0$ we shall have $\psi(x_1 + x_2)\varphi(x_1 + x_2) = [\psi(x_1)\xi(x_2) + \xi(x_1)\psi(x_2)]\varphi(x_1 + x_2)$, because $x_1, x_2 \in \Lambda$ if $\varphi(x_1 + x_2) \neq 0$. In other terms, the restrictions of the functions $\psi \circ \Sigma \cdot \varphi \circ \Sigma$ and $\psi \otimes \xi \cdot \varphi \circ \Sigma + \xi \otimes \psi \cdot \varphi \circ \Sigma$ to $\sigma(\mathbb{H})$ are equal. Since for example $(\psi \circ \Sigma)(\mathbb{H}) = \psi(H)$, we get $\psi(H)\varphi(H) = \psi(H_1) \otimes \xi(H_2) \cdot \varphi(H) + \xi(H_1) \otimes \psi(H_2) \cdot \varphi(H)$. Then we have:

$$
\begin{aligned}
\varphi(H)\mathscr{A}[H]\varphi(H) &= \varphi(H)\mathscr{A}[\psi(H)]\varphi(H) \\
&= \mathscr{A}[\psi(H)\varphi(H)^2] - 2\Re\{\mathscr{A}[\varphi(H)]\psi(H)\varphi(H)\} \\
&= \mathscr{A}[\psi(H_1) \otimes \xi(H_2)\varphi(H)^2] - 2\Re\{\mathscr{A}[\varphi(H)]\psi(H_1) \otimes \xi(H_2)\varphi(H)\} \\
&\quad + \mathscr{A}[\xi(H_1) \otimes \psi(H_2)\varphi(H)^2] - 2\Re\{\mathscr{A}[\varphi(H)]\xi(H_1) \otimes \psi(H_2)\varphi(H)\} \\
&= \varphi(H)\mathscr{A}[\psi(H_1) \otimes \xi(H_2)]\varphi(H) + \varphi(H)\mathscr{A}[\xi(H_1) \otimes \psi(H_2)]\varphi(H).
\end{aligned}
$$

On the other hand, since $\xi(x) = 1$ on a neighbourhood of Λ, there is $\eta \in C_0^\infty(\mathbb{R})$ real with $\eta(x) = 1$ on Λ and $\xi(x) = 1$ on $\operatorname{supp}\eta$, in particular $\xi\eta = \eta$. Moreover, $\varphi(x_1 + x_2) = \eta(x_1)\eta(x_2)\varphi(x_1 + x_2)$ if $x_1, x_2 \geq 0$, so that $\varphi(H) = \eta(H_1) \otimes \eta(H_2)\varphi(H)$. Hence:

$$
\begin{aligned}
\varphi(H)\mathscr{A}[\psi(H_1) \otimes \xi(H_2)]\varphi(H) &= \\
&= \varphi(H)\mathscr{A}_1[\psi(H_1)] \otimes \xi(H_2)\varphi(H) + \varphi(H)\psi(H_1) \otimes \mathscr{A}_2[\xi(H_2)]\varphi(H) \\
&= \varphi(H)\mathscr{A}_1[\psi(H_1)] \otimes \xi(H_2)\varphi(H) \\
&\quad + \varphi(H)\{\psi(H_1)\eta(H_1)^2\} \otimes \{\eta(H_2)\mathscr{A}_2[\xi(H_2)]\eta(H_2)\}\varphi(H).
\end{aligned}
$$

Observe that:

$$
\eta(H_2)\mathscr{A}_2[\xi(H_2)]\eta(H_2) = \mathscr{A}_2[\xi(H_2)\eta(H_2)^2] - 2\Re\{\mathscr{A}_2[\eta(H_2)]\xi(H_2)\eta(H_2)\} = 0.
$$

In conclusion, we shall have:

$$
\begin{aligned}
(8.3.6) \qquad \varphi(H)\mathscr{A}[H]\varphi(H) &= \varphi(H) \cdot \mathscr{A}_1[\psi(H_1)] \otimes \xi(H_2) \cdot \varphi(H) \\
&\quad + \varphi(H) \cdot \xi(H_1) \otimes \mathscr{A}_2[\psi(H_2)] \cdot \varphi(H).
\end{aligned}
$$

From now on we use the abbreviations $B_k = \mathscr{A}_k[\psi(H_k)]$ and $B^1 = B_1 \otimes \xi(H_2)$, $B^2 = \xi(H_1) \otimes B_2$.

(iii) We fix a number $0 < \nu < 1$ and we define $\theta_k^0 : \mathbb{R} \to (-\infty, +\infty]$ by the procedure indicated in Example 7.2.4, namely $\theta_k^0(x) = \inf\{\varrho_k(y) - \nu \mid |y - x| < \nu\}$. Then θ_k^0 is upper semicontinuous and $\theta_k^0(x) < \infty$ if (and only if) $\operatorname{dist}(x, \sigma(H_k)) < \nu$. The set of $x \in \Lambda$ such that $\operatorname{dist}(x, \sigma(H_k)) \leq \nu/2$ is compact and θ_k^0 is bounded from above on it. Let C be a real constant such that C is strictly larger than the supremum of θ_k^0 on $\{x \in \Lambda \mid \operatorname{dist}(x, \sigma(H_k)) \leq \nu/2\}$ for $k = 1, 2$. Now we define $\theta_k : \Lambda \to \mathbb{R}$ as follows: $\theta_k(x) = \theta_k^0(x)$ if $x \in \Lambda$ and $\operatorname{dist}(x, \sigma(H_k)) < \nu/2$, and $\theta_k(x) = C$ if $x \in \Lambda$ and $\operatorname{dist}(x, \sigma(H_k)) \geq \nu/2$. Clearly θ_k is upper semicontinuous and $\theta_k(x) < \varrho_k(x)$ for all $x \in \Lambda$. We assert that there is a real function $\varphi \in C_0^\infty((\lambda - \nu, \lambda + \nu))$, with $\varphi(x) = 1$ near $x = \lambda$, such that for all $\lambda_1, \lambda_2 \geq 0$:

$$
(8.3.7) \qquad \varphi(H_1 + \lambda_2)B_1\varphi(H_1 + \lambda_2) \geq \theta_1(\lambda - \lambda_2)\varphi(H_1 + \lambda_2)^2,
$$

$$
(8.3.8) \qquad \varphi(\lambda_1 + H_2)B_2\varphi(\lambda_1 + H_2) \geq \theta_2(\lambda - \lambda_1)\varphi(\lambda_1 + H_2)^2.
$$

Observe that we have to consider only $\lambda_2 \in \Lambda$ in (8.3.7) and $\lambda_1 \in \Lambda$ in (8.3.8) (if $\lambda_2 > \lambda + 1$, then $H_1 + \lambda_2 \geq \lambda + 1$, so $\varphi(H_1 + \lambda_2) = 0$ because $\operatorname{supp}\varphi \subset$

$(\lambda - 1, \lambda + 1))$. The proof of (8.3.7) and (8.3.8) is a slight modification of the proof of Proposition 7.2.3 (b) and is left as an exercise.

(iv) For $x \geq 0$, $\lambda_2 \geq 0$ we have $\varphi(x + \lambda_2)^2 = \varphi(x + \lambda_2)^2 \xi(x)$, hence $\varphi(H_1 + \lambda_2)^2 = \xi(H_1)\varphi(H_1 + \lambda_2)^2$, and also $\varphi(H_1 + \lambda_2) = \xi(\lambda_2)\varphi(H_1 + \lambda_2)$. By using (8.3.7) we obtain $\varphi(H_1 + \lambda_2)B_1\xi(\lambda_2)\varphi(H_1 + \lambda_2) \geq \xi(H_1)\theta_1(\lambda - \lambda_2)\varphi(H_1 + \lambda_2)^2$ for all $\lambda_2 \geq 0$, and we next show that this inequality implies $\varphi(H)B^1\varphi(H) \geq \xi(H_1) \otimes \theta_1(\lambda - H_2)\varphi(H)^2$. For this, let $\mathscr{B}_2$ be the C*-subalgebra of $\mathscr{C}_2$ generated by H_2; then $\mathscr{B}_2$ is abelian, and it is enough to establish the preceding inequality in the C*-subalgebra $\mathscr{C}_1 \otimes \mathscr{B}_2$ of $\mathscr{C}_1 \otimes \mathscr{C}_2$. We shall work in the Gelfand representation of $\mathscr{B}_2$, so we identify $\mathscr{B}_2 \equiv C_\infty(X_2)$ for some locally compact space X_2, and $\mathscr{C}_1 \otimes \mathscr{B}_2 \equiv C_\infty(X_2; \mathscr{C}_1)$. Then, as explained in §8.1.2, we may identify H_2 with a couple (U_2, h_2), where $U_2 \subset X_2$ is open and $h_2 : U_2 \to \mathbb{R}$ is proper and continuous (notice that $U_2 = X_2$, because H_2 generates $\mathscr{B}_2$; however, this fact is irrelevant here). Moreover, as we saw at the end of §8.2.3, the element $\varphi(H)$ of $C_\infty(X_2; \mathscr{C}_1)$ is given by $\varphi(H)(x_2) = \varphi(H_1 + h_2(x_2))$. So the inequality $\varphi(H)B^1\varphi(H) \geq \xi(H_1) \otimes \theta_1(\lambda - H_2)\varphi(H)^2$ is equivalent to $\varphi(H_1 + h_2(x_2))B_1\xi(h_2(x_2)) \cdot \varphi(H_1 + h_2(x_2)) \geq \xi(H_1)\theta_1(\lambda - h_2(x_2))\varphi(H_1 + h_2(x_2))^2$ for all $x_2 \in U_2$. But the last inequality is certainly true, because we may take above $\lambda_2 = h_2(x_2) \geq 0$. An identical argument gives $\varphi(H)B^2\varphi(H) \geq \theta_2(\lambda - H_1) \otimes \xi(H_2)\varphi(H)^2$.

(v) If we use (8.3.6) and the preceding consequences of (8.3.7) and (8.3.8), we obtain:

(8.3.9)
$$\varphi(H)\mathscr{A}[H]\varphi(H) \geq \{\xi(H_1) \otimes \theta_1(\lambda - H_2) + \theta_2(\lambda - H_1) \otimes \xi(H_2)\}\varphi(H)^2.$$

By working as above in the Gelfand representations of the C*-algebras $\mathscr{B}_1$, $\mathscr{B}_2$ generated by H_1, H_2, it is easily seen that the right-hand side of (8.3.9) is bounded below by

$$\inf\{\xi(h_1(x_1))\theta_1(\lambda - h_2(x_2)) + \theta_2(\lambda - h_1(x_1))\xi(h_2(x_2))|\varphi(h_1(x_1) + h_2(x_2)) \neq 0\} \cdot \varphi(H)^2.$$

Since $\varphi(\tau_1 + \tau_2) = 0$ if $|\tau_1 + \tau_2 - \lambda| \geq \nu$, we clearly obtain:

$$\varrho(\lambda) \geq \inf\{\theta_1(\lambda - \tau_2) + \theta_2(\lambda - \tau_1) \mid \tau_1, \tau_2 \geq 0 \text{ and } \lambda - \nu < \tau_1 + \tau_2 < \lambda + \nu\}.$$

In terms of the new variables $\lambda_1 = \lambda - \tau_2$ and $\lambda_2 = \lambda - \tau_1$ we see that we must estimate $\inf\{\theta_1(\lambda_1) + \theta_2(\lambda_2)\}$ over $\lambda_1, \lambda_2 \leq \lambda$ under the restriction $|\lambda_1 + \lambda_2 - \lambda| < \nu$. By taking into account the definition of the functions θ_1, θ_2 we shall have:

$$\varrho(\lambda) \geq \inf\{\theta_1(\lambda_1) + \theta_2(\lambda_2) \mid \lambda_1, \lambda_2 \leq \lambda, |\lambda_1 + \lambda_2 - \lambda| < \nu\}$$
$$= \inf\{\theta_1(\lambda_1) + \theta_2(\lambda_2) \mid \lambda_1, \lambda_2 \leq \lambda, |\lambda_1 + \lambda_2 - \lambda| < \nu$$
$$\text{and } \text{dist}(\lambda_k, \sigma(H_k)) < \nu/2\}$$

(to obtain the equality notice that $\lambda \in \sigma(H) = \sigma(H_1) + \sigma(H_2)$, so there is a decomposition $\lambda = \lambda_1 + \lambda_2$ with $\lambda_k \in \sigma(H_k)$). Then

$$\varrho(\lambda) \geq \inf\{\varrho_1(\mu_1) + \varrho_2(\mu_2) - 2\nu \mid \mu_1, \mu_2 \leq \lambda + \nu, |\mu_1 + \mu_2 - \lambda| < 3\nu\}$$
$$\geq \inf\{\varrho_1(\mu_1) + \varrho_2(\mu_2) - 2\nu \mid |\mu_1 + \mu_2 - \lambda| < 3\nu\}$$
$$= \inf_{|\mu - \lambda| < 3\nu} \inf_{\mu_1 + \mu_2 = \mu} \{\varrho_1(\mu_1) + \varrho_2(\mu_2)\} - 2\nu.$$

(vi) Since the number $\nu \in (0, 1)$ is arbitrary, the inequality $\varrho(\lambda) \geq f(\lambda) :=$ $\inf\{\varrho_1(\lambda_1) + \varrho_2(\lambda_2) \mid \lambda_1 + \lambda_2 = \lambda\}$ will be obvious once we have shown that the function $f : \mathbb{R} \to (-\infty, +\infty]$ is lower semicontinuous. Since $f(\lambda) = \infty$ if $\lambda < 0$, we have to consider only the restriction of f to $\mathbb{R}_+ = [0, \infty)$. Moreover, since $\varrho_k(\lambda_k) = \infty$ if $\lambda_k < 0$, it is enough to consider decompositions $\lambda = \lambda_1 + \lambda_2$ with $\lambda_k \geq 0$ in the definition of f. Let $f_j = \varrho_j|_{\mathbb{R}_+}$ and $F(\lambda_1, \lambda_2) = f_1(\lambda_1) + f_2(\lambda_2)$. Then $F : \mathbb{R}_+^2 \to (-\infty, \infty]$ is lower semicontinuous. For $\lambda \geq 0$ set $I_\lambda = \{(\lambda_1, \lambda_2) \mid \lambda_j \geq 0 \text{ and } \lambda_1 + \lambda_2 = \lambda\}$. I_λ is a compact subset of $\mathbb{R}_+^2$ and $f(\lambda) = \inf\{F(\lambda_1, \lambda_2) \mid (\lambda_1, \lambda_2) \in I_\lambda\}$. Assume $f(\lambda) > a$; we have to show that $f(\mu) > a$ for μ in some neighbourhood of λ. We have $F(\lambda_1, \lambda_2) > a$ for all $(\lambda_1, \lambda_2) \in I_\lambda$, hence each $(\lambda_1, \lambda_2) \in I_\lambda$ has a neighbourhood $U(\lambda_1, \lambda_2)$ in $\mathbb{R}_+^2$ on which F is strictly greater than a. Since I_λ is compact, it can be covered by a finite set $U_1, \ldots, U_n$ of such neighbourhoods. Then $U = U_1 \cup U_2 \cup \ldots \cup U_n$ is a neighbourhood of I_λ. Since I_λ is compact, U will contain a set of the form $I_\lambda(\varepsilon) = \{(\lambda_1, \lambda_2) \in \mathbb{R}_+^2 \mid \lambda - \varepsilon \leq \lambda_1 + \lambda_2 \leq \lambda + \varepsilon\}$ with $\varepsilon > 0$. So $F(\lambda_1, \lambda_2) > a$ on $I_\lambda(\varepsilon)$. Since F is lower semicontinuous, it attains its lower bound on compact sets, so we have $f(\mu) > a$ for $\lambda - \varepsilon \leq \mu \leq \lambda + \varepsilon$.

(vii) It remains to show the opposite inequality $\varrho(\lambda) \leq f(\lambda)$ for $\lambda \geq 0$, $\lambda \in \sigma(H)$. It is enough to prove that $\varrho(\lambda) \leq \varrho_1(\lambda_1) + \varrho_2(\lambda_2)$ if $\lambda_k \in \sigma(H_k)$ and $\lambda = \lambda_1 + \lambda_2$. Let $a < \varrho(\lambda)$. Then there are $\varepsilon \in (0, 1)$ and $\varphi \in C_0^\infty(\mathbb{R})$ real, with $\operatorname{supp}\varphi \subset (\lambda - 1, \lambda + 1)$ and $\varphi(x) = 1$ if $|x - \lambda| < \varepsilon$, such that $a\varphi(H)^2 \leq \varphi(H)B\varphi(H)$, where $B = \mathcal{A}[\psi(H)]$ and ψ is the same as before. If $\varphi_k \in C_0^\infty((\lambda_k - \varepsilon/2, \lambda_k + \varepsilon/2))$ is real, we get by using $\varphi(x_1 + x_2)\varphi_1(x_1)\varphi_2(x_2) = \varphi_1(x_1)\varphi_2(x_2)$ and (8.3.6) that

(8.3.10)
$$a\varphi_1(H_1)^2 \otimes \varphi_2(H_2)^2 \leq [\varphi_1(H_1)B_1\varphi_1(H_1)] \otimes \varphi_2(H_2)^2$$
$$+ \varphi_1(H_1)^2 \otimes [\varphi_2(H_2)B_2\varphi_2(H_2)].$$

We can assume that $\mathscr{C}_k$ is realized on a Hilbert space $\mathscr{H}_k$. Then we have the preceding inequality in $B(\mathscr{H})$ with $\mathscr{H} = \mathscr{H}_1 \otimes \mathscr{H}_2$. It is easily shown (see Lemma 7.2.1) that there is a sequence of vectors $\{f_{kj}\}_{j \in \mathbb{N}}$ in $\mathscr{H}_k$ such that $\|f_{kj}\| = 1$, $\varphi_k(H_k)f_{kj} = f_{kj}$ and $\langle f_{kj}, B_k f_{kj}\rangle \to \varrho_k(\lambda_k)$ as $j \to \infty$. Let $f_j = f_{1j} \otimes f_{2j}$; then (8.3.10) implies:

$$a = \langle f_j, a\varphi_1(H_1)^2 \otimes \varphi_2(H_2)^2 f_j\rangle \leq \langle f_{1j}, B_1 f_{1j}\rangle + \langle f_{2j}, B_2 f_{2j}\rangle$$

for any $j \in \mathbb{N}$. Hence $a \leq \varrho_1(\lambda_1) + \varrho_2(\lambda_2)$, which completes the proof of the theorem. $\square$

8.4. Graded C*-Algebras

8.4.1. In this section we shall consider an arbitrary *finite semilattice* $\mathscr{L}$, i.e. a finite partially ordered set in which the upper bound of each non-empty subset exists. We use symbols a, b, c, ... to denote the elements of $\mathscr{L}$, $\leq$ is the order relation, $<$ means strict inequality and we write $a \lessdot b$ if b *covers* a, i.e. if $a < b$ and there is no element of $\mathscr{L}$ between a and b. Since $\mathscr{L}$ is finite, it has a largest element, which we denote by $\max \mathscr{L}$. If $\mathscr{L}$ has a least element, we denote it by $\min \mathscr{L}$; this happens if and only if $\mathscr{L}$ is a *lattice*, i.e. if and only if the upper and the lower bound of each non-empty subset of $\mathscr{L}$ exist. For $a, b \in \mathscr{L}$ we denote by $a \vee b$ their upper bound (which exists by hypothesis) and by $a \wedge b$ their lower bound (when it exists).

An alternative point of view is sometimes more convenient. If $\mathscr{L}$ is a finite lattice, then the couple $(\mathscr{L}, \vee)$ is a *finite abelian monoid*, i.e. $\vee$ is a binary operation in $\mathscr{L}$ which is commutative, associative (i.e. $(a \vee b) \vee c = a \vee (b \vee c)$) and has a unit element (namely $\min \mathscr{L}$); moreover, we have $a \vee a = a$ for all $a \in \mathscr{L}$. Reciprocally, if $(\mathscr{L}, \vee)$ is a finite abelian monoid such that $a \vee a = a$ for all a, then we may define an order in $\mathscr{L}$ by saying that $a \leq b$ if $a \vee b = b$; this provides $\mathscr{L}$ with a lattice structure for which $a \vee b$ is just the upper bound of a and b.

One should think of $\mathscr{L}$ as being related to some sort of "generalized N-body problem". In order to explain the meaning of N in terms of $\mathscr{L}$, it is convenient to introduce the *corank function* of $\mathscr{L}$. This is a map $a \mapsto |a|$ of $\mathscr{L}$ into $\{1, 2, 3, \dots\}$ defined inductively as follows: (1) $|\max \mathscr{L}| = 1$; (2) if a is a maximal element of the set $\mathscr{L} \setminus \{\max \mathscr{L}\}$, then $|a| = 2, \dots, (k)$ if a is a maximal element of the set $\mathscr{L} \setminus \{b \in \mathscr{L} \mid |b| \leq k - 1\}$, then $|a| = k$; etc. Then $N := \sup\{|a| \mid a \in \mathscr{L}\}$ is *the corank of $\mathscr{L}$*. Note that for $a < b$ we have $|b| < |a|$, but if $a \lessdot b$ we *do not* have $|a| = |b| + 1$ in general. One may also define $|a|$ as one plus the maximal length of all *chains* (totally ordered subsets) connecting a with $\max \mathscr{L}$. We set $\mathscr{L}_k = [a \in \mathscr{L} \mid |a| \geq k]$, $\mathscr{L}(k) = \{a \in \mathscr{L} \mid |a| = k\}$ and say that $\mathscr{L}(k)$ is the *k-level* of $\mathscr{L}$. Most important for what follows is $\mathscr{L}(2)$, the set of maximal elements of $\mathscr{L} \setminus \{\max \mathscr{L}\} = \mathscr{L}_2$. More generally, $\mathscr{L}(k)$ is the set of maximal elements of $\mathscr{L}_k$. Observe that $\mathscr{L}_k$ is not a semilattice in general. But if, for any $a \in \mathscr{L}$, we define $\mathscr{L}_a = \{b \in \mathscr{L} \mid b \leq a\}$, then $\mathscr{L}_a$ is a semilattice for the order relation induced by $\mathscr{L}$.

Many arguments involving an induction over the semilattice $\mathscr{L}$ can be avoided by using the so-called *Möbius function* associated to $\mathscr{L}$. This is a map $\mu : \mathscr{L} \times \mathscr{L} \to \mathbb{Z}$ uniquely characterized by the following two properties:

(i) $\mu(a, b) = 0$ if $a \not\leq b$;

(ii) if E is a vector space and f, g are functions $\mathscr{L} \to E$, then the next two relations are equivalent:

$$(8.4.1) \qquad g(a) = \sum_{\substack{b \in \mathscr{L} \\ b \leq a}} f(b) \qquad \text{for all } a \in \mathscr{L};$$

$$(8.4.2) \qquad f(a) = \sum_{\substack{b \in \mathscr{L} \\ b \leq a}} g(b) \mu(b, a) \qquad \text{for all } a \in \mathscr{L}.$$

These are called *Möbius inversion formulas*. We refer to books on combinatorics (like [Ai], [Be]) for the theory and the applications of the Möbius function, but we present here a simple proof of the existence and uniqueness of μ.

PROOF. Let $\mathscr{L}$ be an arbitrary finite partially ordered set (the semilattice property is irrelevant here) and assume $E = \mathbb{R}$ (which can be done without loss of generality). Let $M \equiv M(\mathscr{L})$ be the finite-dimensional real vector space of all functions $u : \mathscr{L} \to \mathbb{R}$ and $\{\varepsilon_a\}_{a \in \mathscr{L}}$ its natural basis ($\varepsilon_a(b) = 1$ if $b = a$ and $\varepsilon_a(b) = 0$ otherwise). We shall identify a linear operator $\xi : M \to M$ with its kernel (or matrix) $\xi : \mathscr{L} \times \mathscr{L} \to \mathbb{R}$ through the formula $\xi\varepsilon_a = \sum_{b \in \mathscr{L}} \xi(a,b)\varepsilon_b$, or $(\xi u)(a) = \sum_{b \in \mathscr{L}} u(b)\xi(b,a)$. Let $\mathscr{L}^a = \{b \in \mathscr{L} \mid b \geq a\}$ and $M^a \equiv M(\mathscr{L}^a)$ the subspace generated by $\{\varepsilon_b\}_{b \in \mathscr{L}^a}$; so $M^a = \{u \in M \mid u(b) = 0 \text{ if } a \not\leq b\}$. Then $\{M^a\}_{a \in \mathscr{L}}$ is a decreasing family of subspaces of M and we have $\xi M^a \subset M^a$ for each a if and only if $\xi(a,b) = 0$ if $a \not\leq b$ (if $\mathscr{L} = \{1, 2, \ldots, n\}$ this means that ξ is a lower triangular matrix). Clearly the set of these ξ is a unital subalgebra of $B(M)$, usually called the *incidence algebra* of $\mathscr{L}$. One can show quite easily that a kernel ξ in the incidence algebra with the property $\xi(a,a) = 0$ for all $a \in \mathscr{L}$ is a nilpotent operator in M. Let δ be the kernel of the identity operator and $\zeta(a,b) = 1$ if $a \leq b$ and $= 0$ otherwise. Then δ and ζ belong to the incidence algebra and $\eta := \zeta - \delta$ is nilpotent, so ζ is invertible in the incidence algebra. The Möbius function μ is just the inverse of ζ, hence $\mu = \zeta^{-1} = \sum_{k \geq 0} (-\eta)^k$, where the sum is finite; in particular μ is $\mathbb{Z}$-valued. $\square$

8.4.2. We shall say that a C*-algebra $\mathscr{C}$ is $\mathscr{L}$-*graded* if a family $\{\mathscr{C}(a)\}_{a \in \mathscr{L}}$ of C*-subalgebras of $\mathscr{C}$ is given such that $\mathscr{C} = \sum_{a \in \mathscr{L}} \mathscr{C}(a)$, where the sum is direct (in the vector space sense) and such that $\mathscr{C}(a) \cdot \mathscr{C}(b) \subset \mathscr{C}(a \vee b)$ for all $a, b \in \mathscr{L}$. At a purely algebraic level, this notion is a particular case of that introduced by Bourbaki in Chapter 3, §3 of [Bo1]. The first condition we put on $\mathscr{C}$ means that each $S \in \mathscr{C}$ can be written in a unique way as a sum $S = \sum_{a \in \mathscr{L}} S(a)$ with $S(a) \in \mathscr{C}(a)$; we shall say that $S(a)$ is the a-*homogeneous*, or a-*connected*, *component* of S (the terminology "a-connected" will be justified in §10.2.2). We denote by $\mathscr{P}(a) : \mathscr{C} \to \mathscr{C}(a)$ the *linear projection* determined by this decomposition, i.e. $\mathscr{P}(a)[S] = S(a)$. Then $\mathscr{P}(a)$ is a continuous linear operator in $\mathscr{C}$ such that $\mathscr{P}(a)[S]^* = \mathscr{P}(a)[S^*]$ and $\mathscr{P}(a)^2 \equiv \mathscr{P}(a) \circ \mathscr{P}(a) = \mathscr{P}(a)$ (in order to prove the continuity of these projections, consider the Banach space $\oplus_{a \in \mathscr{L}} \mathscr{C}(a)$ and observe that the map which sends an element $(S(a))_{a \in \mathscr{L}}$ into $\sum_a S(a)$ is a bijective continuous map from $\oplus \mathscr{C}(a)$ onto $\mathscr{C}$, hence its inverse is continuous too). Note that $\mathscr{P}(a)$ *is not a morphism*.

We set $\mathscr{C}_a = \sum_{b \leq a} \mathscr{C}(b)$. It is clear that $\mathscr{C}_a$ is a C*-subalgebra of $\mathscr{C}$ which is $\mathscr{L}_a$-graded in a natural way. The family $\{\mathscr{C}_a\}_{a \in \mathscr{L}}$ is a *filtration* of $\mathscr{C}$, i.e. we have $\mathscr{C}_a \subset \mathscr{C}_b$ if $a \leq b$; moreover $\mathscr{C}_{\max \mathscr{L}} = \mathscr{C}$. We have a natural *linear continuous projection* $\mathscr{P}_a$ of $\mathscr{C}$ onto $\mathscr{C}_a$, namely $\mathscr{P}_a = \sum_{b \leq a} \mathscr{P}(b)$. If $S \in \mathscr{C}$ we shall often write $S_a = \mathscr{P}_a[S] \equiv \sum_{b \leq a} S(b)$. It is quite remarkable that $\mathscr{P}_a$ is also a morphism of $\mathscr{C}$ onto $\mathscr{C}_a$, i.e. we have $\mathscr{P}_a[ST] = \mathscr{P}_a[S]\mathscr{P}_a[T]$ and $\mathscr{P}_a[S]^* = \mathscr{P}_a[S^*]$. The second equality is obvious, while for the first one we notice that

$$ST = \sum_{x,y \in \mathscr{L}} S(x)T(y) = \sum_{a \in \mathscr{L}} \sum_{x \vee y = a} S(x)T(y),$$

hence the component of ST in $\mathscr{C}(a)$ is equal to $\sum_{x \vee y = a} S(x)T(y)$; now observe that $\{(x,y) \in \mathscr{L} \times \mathscr{L} \mid x \vee y \leq a\} = \mathscr{L}_a \times \mathscr{L}_a$, so

$$\mathscr{P}_a[ST] = \sum_{b \leq a} \mathscr{P}(b)[ST] = \sum_{x \vee y \leq a} S(x)T(y)$$

$$= \sum_{x,y \in \mathscr{L}_a} S(x)T(y) = \Big[\sum_{x \leq a} S(x)\Big] \cdot \Big[\sum_{y \leq a} T(y)\Big].$$

The fact that the projections $\mathscr{P}_a$ are morphisms is the fundamental property of $\mathscr{L}$-graded C*-algebras and is extremely useful in applications (note that it is a consequence of the property $a \vee a = a$ for all a in the monoid $(\mathscr{L}, \vee)$). For example, it implies $\mathscr{P}_a[\varphi(S)] = \varphi(\mathscr{P}_a[S])$ for each normal $S \in \mathscr{C}$ and each complex continuous functions φ on the spectrum of S (if $\mathscr{C}$ has no unit, then $0 \in \sigma(S)$ and we have to assume $\varphi(0) = 0$). We mention that $\mathscr{P}_a\mathscr{P}_b = \mathscr{P}_b\mathscr{P}_a = \mathscr{P}_a$ if $a \leq b$ (because then $\mathscr{C}_a \subset \mathscr{C}_b$). On may express the projections $\mathscr{P}(a)$ in terms of the morphisms $\mathscr{P}_a$ by using the Möbius function of $\mathscr{L}$:

$$(8.4.3) \qquad \mathscr{P}(a) = \sum_{b \leq a} \mathscr{P}_b \mu(b, a).$$

The second important observation is that $\mathscr{C}(\max \mathscr{L})$ is a closed bilateral ideal in $\mathscr{C}$. In fact, $\mathscr{C}$ has many natural (i.e. related to the $\mathscr{L}$-graded structure) closed bilateral ideals, for example $\ker \mathscr{P}_a \equiv \sum_{b \not\leq a} \mathscr{C}(b)$, $\sum_{b \geq a} \mathscr{C}(b)$, or $\sum_{b > a} \mathscr{C}(b)$ for any $a \in \mathscr{L}$, but $\mathscr{C}(\max \mathscr{L})$ is the smallest non-trivial one. If H is an observable affiliated to $\mathscr{C}$, then we may speak of the *essential spectrum of H relative to the ideal* $\mathscr{C}(\max \mathscr{L})$; see §8.1.5. Since we view $\mathscr{C}(\max \mathscr{L})$ as *the* ideal canonically associated to $\mathscr{C}$, we shall use in the context of $\mathscr{L}$-graded C*-algebras the notation $\mathscr{C}\text{-}\sigma_{\mathrm{ess}}(H)$ in place of $\mathscr{C}(\max \mathscr{L})\text{-}\sigma_{\mathrm{ess}}(H)$.

We shall now associate to each $\mathscr{L}$-graded C*-algebras $\mathscr{C}$ two (non-graded) C*-algebras, namely

$$(8.4.4) \qquad \widetilde{\mathscr{C}} := \mathscr{C}/\mathscr{C}(\max \mathscr{L}), \qquad \widehat{\mathscr{C}} := \oplus_{a \in \mathscr{L}(2)} \mathscr{C}_a.$$

So $\widetilde{\mathscr{C}}$ is the quotient algebra of $\mathscr{C}$ with respect to the natural "minimal" ideal $\mathscr{C}(\max \mathscr{L})$ and $\widehat{\mathscr{C}}$ is *the* C*-*direct sum* (cf. §8.1.4) of the natural "maximal" C*-subalgebras $\mathscr{C}_a$ of $\mathscr{C}$ (note that for each $b \in \mathscr{L}$, $b \neq \max \mathscr{L}$, there is $a \in \mathscr{L}$ with $|a| = 2$ and $b \leq a$, hence $\mathscr{C}_b \subset \mathscr{C}_a$). The third (and last) important fact is that there is a canonical embedding of $\widetilde{\mathscr{C}}$ into $\widehat{\mathscr{C}}$. Before explaining this, let us observe that we have a canonical identification of the vector space $\widetilde{\mathscr{C}}$ with the vector space $\sum_{|a| \geq 2} \mathscr{C}(a)$; indeed $\mathscr{C}(\max \mathscr{L})$ is just one term of the *linear direct sum* decomposition $\mathscr{C} = \sum_{a \in \mathscr{L}} \mathscr{C}(a)$. But $\sum_{|a| \geq 2} \mathscr{C}(a)$ *is not* an algebra; consequently, in order to understand the multiplication in $\widetilde{\mathscr{C}}$, we have to realize $\widetilde{\mathscr{C}}$ in a different way (in our applications the algebras $\mathscr{C}_a$ are explicitly realized on Hilbert spaces, so the embedding $\widetilde{\mathscr{C}} \subset \widehat{\mathscr{C}}$ will provide us with a natural Hilbert space realization of the abstract algebra $\widetilde{\mathscr{C}}$). Below we shall use the notation $\oplus_{|a|=2} S_a$ for an element $(S_a)_{a \in \mathscr{L}(2)}$ of $\widehat{\mathscr{C}}$.

THEOREM 8.4.1. *The map $S \mapsto \oplus_{|a|=2}\mathscr{P}_a[S]$ is a morphism of $\mathscr{C}$ into $\widehat{\mathscr{C}}$ with kernel equal to $\mathscr{C}(\max \mathscr{L})$. The map $\widetilde{\mathscr{C}} \to \widehat{\mathscr{C}}$ induced by this morphism is an isometric isomorphism of $\widetilde{\mathscr{C}}$ onto the C*-subalgebra of $\widehat{\mathscr{C}}$ consisting of all the elements $\oplus_{|a|=2}S_a \in \widehat{\mathscr{C}}$ having the property: $|a| = |b| = 2$ and $c \leq a$, $c \leq b \Rightarrow \mathscr{P}_c[S_a] = \mathscr{P}_c[S_b]$.*

From now on, we shall identify $\widetilde{\mathscr{C}}$ with its image in $\widehat{\mathscr{C}}$; in particular, if we denote by $\widetilde{S}$ the image of $S \in \mathscr{C}$ in the quotient algebra $\widetilde{\mathscr{C}}$, then $\widetilde{S} \equiv \oplus_{|a|=2}S_a$, where $S_a = \mathscr{P}_a[S]$.

PROOF. Since each $\mathscr{P}_a$ is a morphism and $\widehat{\mathscr{C}}$ is defined as a C*-direct sum, the map $S \mapsto \oplus_{|a|=2}\mathscr{P}_a[S]$ is clearly a morphism. If $\mathscr{P}_a[S] = 0$ for all $a \in \mathscr{L}(2)$, then $\mathscr{P}_b[S] = 0$ for all $b \neq \max \mathscr{L}$ (because for each such b there is a $a \in \mathscr{L}(2)$ with $b \leq a$, and then $\mathscr{P}_b = \mathscr{P}_b\mathscr{P}_a$). Then (8.4.3) implies $\mathscr{P}(a)[S] = 0$ for all $a \neq \max \mathscr{L}$, i.e. $S \in \mathscr{C}(\max \mathscr{L})$. In order to prove the assertion concerning the range of the morphism $S \mapsto \oplus_{|a|=2}\mathscr{P}_a[S]$, observe first that $\mathscr{P}_c[\mathscr{P}_a[S]] = \mathscr{P}_c[S] = \mathscr{P}_c[\mathscr{P}_b[S]]$ if $c \leq a$, $c \leq b$. Reciprocally, assume that $\{S_a\}_{a \in \mathscr{L}(2)}$, with $S_a \in \mathscr{C}_a$, is such that $\mathscr{P}_c[S_a] = \mathscr{P}_c[S_b]$ if $|a| = |b| = 2$ and $c \leq a$, $c \leq b$; from (8.4.3) we then get that $\mathscr{P}(c)[S_a] = \mathscr{P}(c)[S_b]$. Since each $c \neq \max \mathscr{L}$ is less than some $a \in \mathscr{L}(2)$, we can define $S(c) = \mathscr{P}(c)[S_a]$ without ambiguity and the element $S = \sum_{|c| \geq 2} S(c) \in \mathscr{C}$ will have the property $\mathscr{P}_a[S] = S_a$ for all $a \in \mathscr{L}(2)$. $\square$

8.4.3. Let $\mathscr{C}$ be a $\mathscr{L}$-graded C*-algebra and H an observable affiliated to $\mathscr{C}$. Since $\mathscr{P}_a : \mathscr{C} \to \mathscr{C}_a$ are morphisms, we may define for each $a \in \mathscr{L}$ an observable H_a affiliated to $\mathscr{C}_a$ by the relation $H_a = \mathscr{P}_a[H]$. We recall that this means $\varphi(H_a) = \mathscr{P}_a[\varphi(H)]$ for $\varphi \in C_\infty(\mathbb{R})$. If H is interpreted as the hamiltonian of a physical system, then H_a is called a *sub-hamiltonian*; if $\mathscr{L}$ has a least element, $H_{\min \mathscr{L}}$ is said to be the *free hamiltonian* and $H = H_{\max \mathscr{L}}$ the *total hamiltonian*. Our aim is to describe some spectral properties of the total hamiltonian H in terms of sub-hamiltonians H_a with $a \neq \max \mathscr{L}$.

There is another interesting observable which may be associated to the initial H, namely the image of H through the canonical surjection $\mathscr{C} \to \widetilde{\mathscr{C}} = \mathscr{C}/\mathscr{C}(\max \mathscr{L})$; we shall denote by $\widetilde{H}$ the observable affiliated to $\widetilde{\mathscr{C}}$ obtained in this way. We can express $\widetilde{H}$ in other terms by using the embedding $\widetilde{\mathscr{C}} \subset \widehat{\mathscr{C}}$ described in Theorem 8.4.1: by taking into account the considerations of §8.1.4, we have $\widetilde{H} = \oplus_{|a|=2}H_a$. Since $\widetilde{H}$ is affiliated to $\widetilde{\mathscr{C}}$, it is also affiliated to the larger algebra $\widehat{\mathscr{C}}$. Note that the spectrum of an observable affiliated to a C*-subalgebra of a given C*-algebra does not depend on the algebra with respect to which it is calculated. By taking into account the remarks made in §8.1.5 and §8.1.4 we clearly have:

$$\mathscr{C}\text{-}\sigma_{\mathrm{ess}}(H) = \sigma(\widetilde{H}) = \sigma(\oplus_{|a|=2}H_a) = \bigcup_{|a|=2} \sigma(H_a).$$

Observe, furthermore, that $\mathscr{P}_a[H_b] = H_a$ if $a \leq b$ (H_b is affiliated to $\mathscr{C}_b$, so to $\mathscr{C}$ too). Hence $\sigma(H_a) \subset \sigma(H_b)$ for $a \leq b$, in particular we have $\sigma(H_a) \subset \sigma(H)$ for any $a \in \mathscr{L}$. We shall summarize all these remarks in the next proposition, which is the first interesting result of the theory (HVZ stands for Hunziker, Van Winter, Weinberg, Zhislin, cf. [RS]).

PROPOSITION 8.4.2 (ABSTRACT HVZ THEOREM). *Let $\mathscr{C}$ be a $\mathscr{L}$-graded C^*-algebra, H an observable affiliated to it and $H_a = \mathscr{P}_a[H]$ for each $a \in \mathscr{L}$. Then $\sigma(H_a) \subset \sigma(H_b) \subset \sigma(H)$ if $a \leq b$. The essential spectrum of H with respect to the ideal $\mathscr{C}(\max \mathscr{L})$ is given by the formula :*

$$(8.4.5) \qquad \mathscr{C}\text{-}\sigma_{ess}(H) = \bigcup_{a \in \mathscr{L}(2)} \sigma(H_a).$$

For $a \neq \max \mathscr{L}$, we have $\mathscr{C}\text{-}\sigma_{ess}(H_a) = \sigma(H_a)$.

The last assertion of the proposition is obvious and shows that $\mathscr{C}\text{-}\sigma_{ess}(H_a)$ is not an interesting object if $|a| \geq 2$. On the other hand, H_a is also affiliated to $\mathscr{C}_a$, which is a $\mathscr{L}_a$-graded C*-algebra with "minimal" ideal $\mathscr{C}(a)$ (because $a = \max \mathscr{L}_a$). So we may also consider the essential spectrum of H_a with respect to the ideal $\mathscr{C}(a)$, i.e. the set $\mathscr{C}_a\text{-}\sigma_{ess}(H_a)$. It turns out that this is a non-trivial subset of $\sigma(H_a)$ [7].

The next result will play an important role in the calculation of the ϱ-function for a N-body hamiltonian. It is an obvious consequence of the considerations of §8.3.4 and §8.3.5 together with the fact that $\widetilde{H} = \oplus_{|a|=2} H_a$.

THEOREM 8.4.3. *Let $\mathscr{C}$ be an $\mathscr{L}$-graded C^*-algebra equipped with a one-parameter group $\{\mathscr{W}_\tau\}$ of automorphisms which are compatible with the grading, i.e. $\mathscr{W}_\tau \mathscr{C}(a) \subset \mathscr{C}(a)$ for all $\tau \in \mathbb{R}$ and $a \in \mathscr{L}$. Assume that H is an observable affiliated to $\mathscr{C}$, locally of class $C_u^1(\mathscr{A})$. Then each H_a is locally of class $C_u^1(\mathscr{A})$. Set $\varrho_a = \varrho_{H_a}^{\mathscr{A}}$ and let $\widetilde{\varrho} = \widetilde{\varrho}_H$ be defined as in §8.3.4 relative to the ideal $\mathscr{C}(\max \mathscr{L})$. Then $\widetilde{\varrho} = \min_{a \in \mathscr{L}(2)} \varrho_a$.*

Note that the assumption of compatibility of the group $\{\mathscr{W}_\tau\}$ with the grading is equivalent to each of the next two conditions:

(i) $\mathscr{W}_\tau \mathscr{P}(a) = \mathscr{P}(a) \mathscr{W}_\tau$ for all $\tau \in \mathbb{R}$, $a \in \mathscr{L}$;

(ii) $\mathscr{W}_\tau \mathscr{P}_a = \mathscr{P}_a \mathscr{W}_\tau$ for all $\tau \in \mathbb{R}$, $a \in \mathscr{L}$ (use (8.4.3)).

Consequently, an element $S \in \mathscr{C}$ is of class $C_u^k(\mathscr{A})$ (for some $k \in \mathbb{N}$) if and only if its components $S(a) \in \mathscr{C}(a)$ are of class $C_u^k(\mathscr{A})$, and also if and only if S_a is of class $C_u^k(\mathscr{A})$ for each $a \in \mathscr{L}$. If $k \geq 1$ we shall have $\mathscr{P}(a)[\mathscr{A}[S]] = \mathscr{A}[S(a)]$, $\mathscr{P}_a[\mathscr{A}[S]] = \mathscr{A}[S_a]$. Formally this means $\mathscr{P}(a)\mathscr{A} = \mathscr{A}\mathscr{P}(a)$ and $\mathscr{P}_a \mathscr{A} = \mathscr{A}\mathscr{P}_a$. Similar assertions hold for observables affiliated to $\mathscr{C}$.

8.4.4. In our applications, the $\mathscr{L}$-graded C*-algebra $\mathscr{C}$ is realized in a Hilbert space $\mathscr{H}$ and we would like to have more effective criteria for a self-adjoint operator H in $\mathscr{H}$ to be affiliated to $\mathscr{C}$ and, moreover, to have a more explicit description of the operators H_a. Results of this type are contained in the next proposition.

---

[7] If we are in the N-body non-relativistic case, and if we use the operator H^a which can be introduced in this situation, then for $a \neq \max \mathscr{L}$ we have $\mathscr{C}_a\text{-}\sigma_{ess}(H_a) = [\inf \sigma_{ess}(H^a), \infty)$ and $\mathscr{C}\text{-}\sigma_{ess}(H_a) = [\inf \sigma(H^a), \infty)$. See §8.2.4 and the next chapter.

PROPOSITION 8.4.4. *Let $\mathscr{L}$ be a finite semilattice having a least element* $\min \mathscr{L} \equiv 0$, *and let $\mathscr{C}$ be a $\mathscr{L}$-graded C^*-algebra realized on a Hilbert space* $\mathscr{H}$. *Assume that a densely defined self-adjoint operator $H_0 \equiv H(0)$ is given on* $\mathscr{H}$ *such that H_0 is affiliated to $\mathscr{C}_0 \equiv \mathscr{C}(0)$. Furthermore, suppose that one is in one of the following two situations:*

(i) *For each $a \in \mathscr{L} \setminus \{0\}$, a symmetric operator $H(a)$ in $\mathscr{H}$ is given such that* $D(H(a)) \supset D(H_0)$ *and $H(a)(H_0 - \lambda_a)^{-1} \in \mathscr{C}(a)$ for some $\lambda_a \in \mathbb{C} \setminus \sigma(H_0)$.*

(ii) *H_0 is bounded from below and, for each $a \in \mathscr{L} \setminus \{0\}$, a continuous symmetric sesquilinear form $H(a)$ is given on $D(|H_0|^{1/2})$ such that*

$$(H_0 + \lambda_a)^{-1/2} H(a)(H_0 + \lambda_a)^{-1/2} \in \mathscr{C}(a)$$

for some $\lambda_a \in \mathbb{R}$ with $-\lambda_a < \inf H_0$.

Let $\mu(a)$ be the operator bound (in situation (i)) or form bound (in situation (ii)) of $H(a)$ relative to H_0, and assume that $\sum_{a \neq 0} \mu(a) < 1$. Then for each $a \in \mathscr{L}$, the densely defined self-adjoint operator $H_a = \sum_{b \leq a} H(b)$ in $\mathscr{H}$ (operator sum in case (i), form sum in case (ii)) is affiliated to the C^*-subalgebra $\mathscr{C}_a$ of $B(\mathscr{H})$. If $H \equiv H_{\max \mathscr{L}}$, then $\mathscr{P}_a[H] = H_a$ for each $a \in \mathscr{L}$; in other terms: $\mathscr{P}_a[\varphi(H)] = \varphi(H_a)$ *for all $\varphi \in C_\infty(\mathbb{R})$.*

PROOF. We remark first that we shall have $H(a)(H_0 - z)^{-1} \in \mathscr{C}(a)$ for all $z \in \mathbb{C} \setminus \sigma(H_0)$ in case (i) and $(H_0 + \lambda)^{-1/2} H(a)(H_0 + \lambda)^{-1/2} \in \mathscr{C}(a)$ for all $\lambda > -\inf H_0$ in case (ii). This is shown exactly as in the proof of Proposition 8.1.2 by taking into account the property $\mathscr{C}(0) \cdot \mathscr{C}(a) \subset \mathscr{C}(a)$ (note that $0 = \min \mathscr{L} \leq a$, so that $0 \vee a = a$). Since we may now take λ_a independent of a, we may clearly apply Proposition 8.1.2 and obtain that H_a is affiliated to $\mathscr{C}_a$ (this also follows directly from the next arguments). It remains to prove that $\mathscr{P}_a[H] = H_a$, and for this we first consider the case (i). We fix a number $z = i\lambda$ with $\lambda \in \mathbb{R}$ so large that $\sum_{a \neq 0} \|H(a)(H_0 - z)^{-1}\| < 1$ and set $T(a) = H(a)(H_0 - z)^{-1}$, $T_a = \sum_{0 \neq b \leq a} T(b)$, $T = T_{\max \mathscr{L}}$. Then we have

$$(8.4.6) \qquad (H_a - z)^{-1} = (H_0 - z)^{-1}(I + T_a)^{-1}$$
$$= (H_0 - z)^{-1} - (H_0 - z)^{-1} T_a (I + T_a)^{-1}.$$

Clearly $\mathscr{P}_a[T] = T_a$ and $\mathscr{P}_a[T(I + T)^{-1}] = T_a(I + T_a)^{-1}$ because $\mathscr{P}_a$ is a morphism. Let us write (8.4.6) for $a = \max \mathscr{L}$; if we apply $\mathscr{P}_a$ to both sides and use $\mathscr{P}_a[(H_0 - z)^{-1}] = (H_0 - z)^{-1}$ and the fact that $\mathscr{P}_a$ is a morphism, we get:

$$\mathscr{P}_a[(H - z)^{-1}] = \mathscr{P}_a[(H_0 - z)^{-1}] + \mathscr{P}_a[(H_0 - z)^{-1}]\mathscr{P}_a[T(I + T)^{-1}]$$
$$= (H_a - z)^{-1}.$$

This finishes the proof of the proposition in the situation (i). The proof in case (ii) runs along the same lines. We first choose $\lambda \in \mathbb{R}$ so large that the operators $T(a) \equiv (H_0 + \lambda)^{-1/2} H(a)(H_0 + \lambda)^{-1/2} \in \mathscr{C}(a)$ have the property $\sum_{a \neq 0} \|T(a)\| < 1$. Then we use in place of (8.4.6) the identity:

$$(H_a + \lambda)^{-1} = (H_0 + \lambda)^{-1/2}(I + T_a)^{-1}(H_0 + \lambda)^{-1/2}$$
$$= (H_0 + \lambda)^{-1} - (H_0 + \lambda)^{-1/2} T_a (I + T_a)^{-1}(H_0 + \lambda)^{-1/2}. \qquad \square$$

In some applications, one may improve the results of Proposition 8.4.4 by using approximation procedures, as explained at the end of §8.1.2. For example, assume that $H_0 \geq 0$ and that, for each $a \neq 0$, a continuous symmetric sesquilinear form $H(a)$ is given on $D(H_0)$ such that $(H_0 + I)^{-1}H(a)(H_0 + I)^{-1} \in \mathscr{C}(a)$. For each $\varepsilon > 0$ let $H^\varepsilon(a) = (I + \varepsilon H_0)^{-1}H(a)(I + \varepsilon H_0)^{-1}$, which clearly is an element of $\mathscr{C}(a)$. Then $H_a^\varepsilon = H_0 + \sum_{0 \neq b \leq a} H^\varepsilon(b)$ is a densely defined self-adjoint operator in $\mathscr{H}$ affiliated to $\mathscr{C}_a$. If $H^\varepsilon \equiv H^\varepsilon_{\max} \mathscr{L}$, then clearly $\mathscr{P}_a[H^\varepsilon] = H_a^\varepsilon$. If $\lim_{\varepsilon \to +0} H_a^\varepsilon \equiv H_a$ exists in the norm resolvent sense, then H_a is affiliated to $\mathscr{C}_a$ and $\mathscr{P}_a[H] = H_a$. This technique will be used in the proof of Proposition 9.4.9. One may find other versions and applications of this method in [BGS].

8.4.5. We end this chapter with some considerations concerning the meaning of the so-called Weinberg-Van Winter (WVW) equation in our formalism. This is interesting for historical reasons but will not be needed in the rest of this text.

Let $\mathscr{L}$ be a finite lattice with least element, which we denote by 0. We assume that a $\mathscr{L}$-graded C*-algebra $\mathscr{C} = \sum_{a \in \mathscr{L}} \mathscr{C}(a)$ is given and that $H_0 = H(0)$ is a free hamiltonian, i.e. an observable affiliated to $\mathscr{C}_0 = \mathscr{C}(0)$. Our purpose is to study total hamiltonians H affiliated to $\mathscr{C}$ that are, in some sense, perturbations of H_0. We begin with the simplest case where $H = H_0 + \sum_{a \neq 0} H(a)$ for some $H(a) = H(a)^* \in \mathscr{C}(a)$ (the sum is then well defined, cf. §8.1.2). The two-body problem corresponds to the case where $\mathscr{L}$ contains just two elements, and in this situation the so-called second resolvent equation plays an important role in the spectral analysis of H. More precisely, if $\mathscr{L} = \{0, \max \mathscr{L}\}$ and in the notation $H(\max \mathscr{L}) = V$, the second resolvent equation is as follows (we assume $z \in \mathbb{C} \setminus \sigma(H)$ and recall that $\sigma(H_0) \subset \sigma(H)$):

$$(8.4.7) \qquad (z - H)^{-1} = (z - H_0)^{-1} + (z - H_0)^{-1}V(z - H)^{-1}.$$

In order to understand what is the natural generalization of this equation to the N-body case (i.e. corank $\mathscr{L} = N \geq 3$), observe that the decomposition given by the right-hand side of (8.4.7) is just the decomposition of the element $(z - H)^{-1} \in \mathscr{C} = \mathscr{C}(0) + \mathscr{C}(\max \mathscr{L})$ into homogeneous components:

$$(8.4.8) \qquad \mathscr{P}(0)[(z - H)^{-1}] = (z - H_0)^{-1},$$
$$\mathscr{P}(\max \mathscr{L})[(z - H)^{-1}] = (z - H_0)^{-1}V(z - H)^{-1}.$$

This is obvious because $(z - H_0)^{-1} \in \mathscr{C}(0)$ and $V \in \mathscr{C}(\max \mathscr{L})$ which is an ideal in $\mathscr{C}$.

Now consider an arbitrary $\mathscr{L}$. Since $(z - H)^{-1} \in \mathscr{C}$ for $z \in \mathbb{C} \setminus \sigma(H)$, we may write it as a sum of homogeneous (or "connected") components:

$$(8.4.9) \qquad (z - H)^{-1} = \sum_{a \in \mathscr{L}} \mathscr{P}(a)[(z - H)^{-1}].$$

Let us recall that $\sigma(H_a) \subset \sigma(H)$ for any $a \in \mathscr{L}$, where $H_a = H_0 + \sum_{0 \neq b \leq a} H(b)$. Below we shall explicitly calculate the terms of the decomposition (8.4.9) in terms of the family $\{(z - H_a)^{-1}\}$. As a result we shall obtain a generalization of the formulas (8.4.8), and then a comparison with the expressions appearing in the usual treatment of the N-body problem (see [RS], [ABG1] or §10.2.2 here) will show that (8.4.9) *is just the generalization to our context of the standard WVW equation.* Note that *no perturbative expansion is needed* in our argument.

The component of $(z-H)^{-1}$ in $\mathscr{C}(0)$ is trivial to compute because $\mathscr{P}(0) = \mathscr{P}_0$ and $\mathscr{P}_0[H] = H_0$; so we have $\mathscr{P}(0)[(z-H)^{-1}] = (z-H_0)^{-1}$. The case $a \neq 0$ is more complicated and is treated in the next proposition. Recall that a *chain* in $\mathscr{L}$ is any totally ordered subset $\mathcal{M}$ of $\mathscr{L}$; the number $\operatorname{card}\mathcal{M} - 1$ is called the *length* of the chain $\mathcal{M}$. For arbitrary $a, b \in \mathscr{L}$ we shall write $H_{ab} = \sum H(c)$ where we sum over all $c \in \mathscr{L}$ such that $c \not\leq a$ and $a \vee c = b$; by convention, the sum over an empty set is equal to zero. So $H_{ab} \neq 0$ only if $a < b$, and in this case the sum is over all $c \in \mathscr{L}$ such that $a \vee c = b$.

PROPOSITION 8.4.5. *Let $\mathscr{L}$ be a finite lattice with least element denoted by 0, and let $\mathscr{C} = \sum \mathscr{C}(a)$ be a $\mathscr{L}$-graded C^*-algebra. Assume that $H_0 \equiv H(0)$ is an observable affiliated to $\mathscr{C}_0 \equiv \mathscr{C}(0)$ and that for each $a \neq 0$ a symmetric element $H(a) \in \mathscr{C}(a)$ is given. Set $H_a = \sum_{b \leq a} H(b)$, which is an observable affiliated to $\mathscr{C}_a$, and $H = H_{\max \mathscr{L}}$. For a fixed $z \in \mathbb{C} \setminus \sigma(H) = \mathbb{C} \setminus \cup_a \sigma(H_a)$, let $R_a = (z - H_a)^{-1} \in \mathscr{C}_a$ and $R = R_{\max \mathscr{L}} \in \mathscr{C}$. For each chain $\mathcal{M} \subset \mathscr{L}$ of length ≥ 1, written $\mathcal{M} = \{a_1, \ldots, a_m\}$ with uniquely determined $a_1 < a_2 < \cdots < a_m$ and $m \geq 2$, define:*

$$(8.4.10) \qquad R(\mathcal{M}) = R_{a_1} H_{a_1 a_2} R_{a_2} \ldots H_{a_{m-1} a_m} R_{a_m};$$

here $H_{a_k a_{k+1}} = \sum H(c)$, where the sum is over all $c \in \mathscr{L}$ such that $c \vee a_k = a_{k+1}$. Then $\mathscr{P}(0)[R] = R_0$ and, if $a \neq 0$, then $\mathscr{P}(a)[R] = \sum R(\mathcal{M})$, where the sum is over all chains $\mathcal{M} \subset \mathscr{L}$ such that $\min \mathcal{M} = 0$ and $\max \mathcal{M} = a$.

PROOF. If $a < b$ we have

$$H_b = H_a + \sum_{\substack{c \not\leq a \\ c \leq b}} H(c) = H_a + \sum_{c \leq b} \sum_{\substack{d \not\leq a \\ d \vee a = c}} H(d) = H_a + \sum_{c \leq b} H_{ac}.$$

This clearly implies $R_b = R_a + \sum_{c \leq b} R_a H_{ac} R_b$. Upon iterating this identity one obtains:

$$R_b - R_a = \sum_{a_1 \leq b} R_a H_{aa_1} R_{a_1} + \sum_{a_1, a_2 \leq b} R_a H_{aa_1} R_{a_1} H_{aa_2} R_{a_2} + \cdots$$

Since $H_{cd} \neq 0 \Rightarrow c < d$, a term on the right-hand side can be non-zero only if the corresponding sequence $a_1, \ldots, a_n$ is strictly increasing. So we may rewrite the preceding formula as $R_b - R_a = \sum R(\mathcal{M})$, where sum is over all chains $\mathcal{M}$ with $\min \mathcal{M} = a$ and $\max \mathcal{M} \leq b$. In particular:

$$(8.4.11) \qquad R = R_0 + \sum R(\mathcal{M})$$

where the sum is over all chains $\mathcal{M} \subset \mathscr{L}$ of length ≥ 1 and such that $\min \mathcal{M} = 0$. *It is clear that (8.4.11) is just the WVW equation* (see [ABG1] for a more classical treatment).

In order to prove the proposition, it is enough to show that $R(\mathcal{M}) \in \mathscr{C}(\max \mathcal{M})$ if $\mathcal{M}$ is a chain of length ≥ 1 and such that $\min \mathcal{M} = 0$. We do this by induction over the length of $\mathcal{M}$. Let $R(\mathcal{M})$ be given by (8.4.10), but with $a_1 = 0$. If $m = 2$, then $H_{0a_2} = H(a_2)$, hence $R(\mathcal{M}) = R_0 H(a_2) R_{a_2}$; since $H(a_2) \in \mathscr{C}(a_2)$, which is an ideal in $\mathscr{C}_{a_2}$, and since R_0, $R_{a_2} \in \mathscr{C}_{a_2}$, we get $R(\mathscr{M}) \in \mathscr{C}(\max \mathscr{M})$ in this case. Now assume that this holds for all chains of length $\leq m-1$. Then we may write $R(\mathscr{M}) = SH_{a_{m-1}a_m}R_{a_m}$ with $S \in \mathscr{C}(a_{m-1})$. $H_{a_{m-1}a_m}$ is a sum of terms $H(c)$ with $c \vee a_{m-1} = a_m$. For each such term we have $SH(c) \in \mathscr{C}(a_{m-1}) \cdot \mathscr{C}(c) \subset \mathscr{C}(a_{m-1} \vee c) = \mathscr{C}(a_m)$. Hence $SH_{a_{m-1}a_m} \in \mathscr{C}(a_m)$. But $\mathscr{C}(a_m)$ is an ideal in $\mathscr{C}_{a_m}$ and $R_{a_m} \in \mathscr{C}_{a_m}$, so $R(\mathscr{M}) \in \mathscr{C}(a_m)$. $\square$

If we are in a Hilbert space setting, as in Proposition 8.4.4, then it is quite easy to extend the assertion of Proposition 8.4.5 to the case where $H(a)$ are unbounded operators. For example, one may check by going through the preceding proof that the result of Proposition 8.4.5 remains valid under either one of the hypotheses of Proposition 8.4.4. Or else, one may use an approximation procedure, as explained at the end of §8.4.4.

CHAPTER 9

Spectral Theory of N-Body Hamiltonians

In this chapter we shall apply the techniques developed so far in this text to the spectral theory of N-body Hamiltonians. The principal results are contained in Section 9.4. In particular we prove the Mourre estimate for a very large class of short range and long range (local or non-local) many-body interactions; our method of proof is based on the algebraic approach described in Chapter 8 and is quite different from the methods of Perry, Sigal, Simon [PSS] and Froese, Herbst [FH1].

The first three sections are of a preparatory nature. After collecting some conventions in Section 9.1, we define and study in Section 9.2 the concept of semicompact operators. The main conclusions are given in Theorem 9.2.4 which embraces various results that had played a role in the development of the N-body problem (cf. [Cb1], [Cb2], [PSS]). In Section 9.3 we define a class of graded C*-algebras which characterize in some sense the N-body hamiltonians, and we describe a large class of (not necessarily non-relativistic) hamiltonians affiliated to them.

In a preliminary version of this text [ABG1] we adopted a more geometric point of view that originated in the papers [A] and [FH1] of Agmon and Froese, Herbst. Here we have abandoned it in favor of an algebraic approach which seems more powerful to us. A detailed description of the geometric methods can be found in Part I of [ABG1]. The geometric language has been much used in the recent literature on the N-body problem, see e.g. [De1] and references therein.

9.1. Tensorial Factorizations of $\mathcal{H}(X)$

In Sections 1.1 and 1.2 we described several spaces, groups and operators that one can associate in a canonical way to an arbitrary euclidean space X. We shall recall here some of these objects, and then we shall make some supplementary remarks concerning them. Let $\mathcal{H}(X) = L^2(X)$ be the Hilbert space of square-integrable functions $f : X \to \mathbb{C}$ with respect to the Fourier measure $\underline{d}x$. We write $B(X)$ for the C*-algebra $B(\mathcal{H}(X))$ of all bounded linear operators in $\mathcal{H}(X)$ and $\mathbb{K}(X)$ for the C*-subalgebra $K(\mathcal{H}(X))$ consisting of all compact operators. The translation group $\{T(x)\}_{x \in X}$ and the dilation group $\{W_\tau\}_{\tau \in \mathbb{R}}$ act according to

W. O. Amrein et al., *C₀-Groups, Commutator Methods and Spectral Theory of N-Body Hamiltonians*, Modern Birkhäuser Classics,
DOI: 10.1007/978-3-0348-0733-3_9, © Springer Basel 1996

the rules $[T(x)f](y) = f(y-x)$ and $[W(\tau)f](x) = \exp(\tau \dim X/2)f(e^\tau X)$. Then $\{T(x)\}$ is a continuous unitary representation of the additive group X in $\mathcal{H}(X)$ whose generator is $-P$, where P is the momentum observable. And $\{W(\tau)$ is a continuous unitary one-parameter group in $\mathcal{H}(X)$ whose generator is $2D$, with D defined in (1.2.19). Another operator which will play an important role in this chapter is the Laplace-Beltrami operator Δ. Since $\mathcal{F}^*\Delta\mathcal{F} = Q^2$, Δ is essentially selfadjoint on $\mathcal{S}(X)$ in $\mathcal{H}(X)$, and the domain of its closure is the Sobolev space $\mathcal{H}^2(X)$.

In the above we have implicitly assumed that $X \neq \mathbf{O}$, where $\mathbf{O} = \{0\}$ is the vector space consisting only of the zero vector. By convention, if $X = \mathbf{O}$ we take: $\mathcal{H}(\mathbf{O}) = \mathbb{C}$, $\mathbb{B}(\mathbf{O}) = \mathbb{K}(\mathbf{O}) = \mathbb{C}$, $\mathcal{S}(\mathbf{O}) = \mathcal{S}^*(\mathbf{O}) = \mathbb{C}$, $\mathcal{F} = \mathcal{F}^* = I$, $\Delta = D = P = Q = 0$.

Now let Y be a subspace of X, considered with the induced euclidean structure. Then one can associate to it objects like $L^p(Y)$, $\mathcal{H}(Y)$, Δ, ... as above. Below we shall point out the relations between these objects and the corresponding objects associated to X. Before doing this, it is necessary to specify by suitable notations what subspace objects like Δ, $\mathcal{F}$, ... refer to. We shall do this by adding the symbol Y of the subspace as superscript. More precisely, Δ^Y for example denotes the Laplace-Beltrami operator associated to Y, considered as a self-adjoint operator in $\mathcal{H}(Y)$ or as an operator in $\mathcal{S}^*(Y)$. Similarly $\mathcal{F}^Y$, T^Y and W^Y will be the Fourier transformation, the translation group and the dilation group respectively in $\mathcal{H}(Y)$ or in $\mathcal{S}^*(Y)$.

We shall always consider X to be a given fixed space, and hence, if $Y = X$, we shall omit these superscripts (which is consistent with the definitions given before). The reason for choosing superscripts rather than subscripts is twofold:

(i) we shall have to associate to objects like Δ, $\mathcal{F}$ certain operators in $\mathcal{H}(X)$ or in $\mathcal{S}^*(X)$ depending on a subspace Y; in that case we shall use subscripts (examples can be found in Table 1.1 in Section 1.2);

(ii) the above conventions concerning superscripts and subscripts are consistent with the notations used in the literature on the quantum-mechanical many-body problem (cf. Chapter 10).

If Y is a subspace of X, then the factorization property (1.2.1) of the Fourier measure allows us to identify $\mathcal{H}(Y) \otimes \mathcal{H}(Y^\perp)$ with $\mathcal{H}(X) = \mathcal{H}(Y \oplus Y^\perp)$. More precisely, we identify a function $f : Y \times Y^\perp \to \mathbb{C}$ with the function $f_0 : X \to \mathbb{C}$ defined by $f_0(x) = f(\pi_Y(x), \pi_{Y^\perp}(x))$, and this identification defines a Hilbert space isomorphism between $\mathcal{H}(Y) \otimes \mathcal{H}(Y^\perp)$ and $\mathcal{H}(X)$. If S is a linear operator in $\mathcal{H}(Y)$ and T a linear operator in $\mathcal{H}(Y^\perp)$, we shall write $S \otimes_Y T$ for their tensor product (viewed as an operator in $\mathcal{H}(X)$, by making the identification of $\mathcal{H}(Y) \otimes \mathcal{H}(Y^\perp)$ and $\mathcal{H}(X)$). The subscript Y on the symbol for the tensor product is introduced to specify the spaces in which S and T act; in fact in most situations we shall not work with a fixed subspace Y of X but with an entire semilattice of subspaces of X, and we shall need to consider products of the form $S \otimes_Y S' \cdot T \otimes_Z T'$, where $S \in B(\mathcal{H}(Y))$, $S' \in B(\mathcal{H}(Y^\perp))$, $T \in B(\mathcal{H}(Z))$, $T' \in B(\mathcal{H}(Z^\perp))$ and Y and Z are two different subspaces of X. It is seen from this example that the subscripts on the symbol for the tensor product are useful to avoid ambiguities. We remark that $S \otimes_Y T = T \otimes_{Y^\perp} S$.

By using the preceding conventions, it is clear that we shall have $\mathcal{F} = \mathcal{F}^Y \otimes_Y$

$\mathcal{F}^{Y^\perp}$, and also $T(x) = T^Y(\pi_Y(x)) \otimes_Y T^{Y^\perp}(\pi_{Y^\perp}(x))$ $(x \in X)$, $W(\tau) = W^Y(\tau) \otimes_Y$ $W^{Y^\perp}(\tau)$ $(\tau \in \mathbb{R})$. The operator Δ has an extremely important factorization property, namely $\Delta = \Delta^Y \otimes_Y I + I \otimes_Y \Delta^{Y^\perp}$ for any subspace $Y \subset X$. This relation may be interpreted in two different ways:

(i) if we consider Δ as differential operators acting on functions or distributions, then it is an immediate consequence of (1.1.3) (take for $\{v_1, \ldots, v_m\}$ a basis of Y and for $\{v_{m+1}, \ldots, v_n\}$ a basis of $Y^\perp$);

(ii) if we consider Δ as self-adjoint operators in the corresponding L^2-spaces, then the preceding identity is true in the sense of the definition given in §8.2.3 (use the fact that $\mathscr{S}(X)$ is a core for Δ).
Similarly, if we define D^Y to be the self-adjoint operator in $\mathscr{H}(Y)$ such that $W^Y(\tau) = \exp(2i\tau D^Y)$, then we have as a consequence of the factorization property of $W(\tau)$ that $D = D^Y \otimes_Y I + I \otimes_Y D^{Y^\perp}$.

We now make some further notational conventions. If Y is a subspace and if an operator bears the symbol Y as a subscript, we mean that this operator acts in $\mathscr{H}(X)$ but it is related in some way to the subspace Y. Examples of such operators are given in Table 1.1. Or let D_Y be the operator defined in (1.2.22) viewed as an operator in $\mathscr{H}(X)$, in which case we have $D_Y = D^Y \otimes_Y I$. Further operators of this type will be introduced throughout the remainder of this text. In particular we set (as operators in $\mathscr{H}(X)$) : $\Delta_Y = \Delta^Y \otimes_Y I$, $\mathcal{F}_Y = \mathcal{F}^Y \otimes_Y I$. With these notations we have: $\Delta = \Delta_Y + \Delta_{Y^\perp}$, $\mathcal{F} = \mathcal{F}_Y \mathcal{F}_{Y^\perp}$.

The tensor products of various operators introduced so far referred to a factorization of $\mathscr{H}(X)$ into $\mathscr{H}(Y) \otimes \mathscr{H}(Y^\perp)$ for some subspace Y of X. Later we shall use tensor products of a similar type for the case where the role played above by X is assumed by some subspace Z containing Y. If $Y \subset Z$ and $Z \ominus Y \equiv Y^\perp \cap Z$, one can write $Z = Y \oplus (Z \ominus Y)$, so that $\mathscr{H}(Z)$ can be canonically identified with $\mathscr{H}(Y) \otimes \mathscr{H}(Z \ominus Y)$. Now if S is an operator in $\mathscr{H}(Y)$ and T an operator in $\mathscr{H}(Z \ominus Y)$, we denote by $S \otimes_Y^Z T$ the operator in $\mathscr{H}(Z)$ associated by this canonical identification to the operator $S \otimes T$ in $\mathscr{H}(Y) \otimes \mathscr{H}(Z \ominus Y)$. As an example, we shall have $W_Z(\tau) = W^Y(\tau) \otimes_Y^Z W^{Y^\perp \cap Z}(\tau)$.

9.2. Semicompact Operators

In this section we consider a fixed euclidean space X. For each subspace Y of X we write $\mathbb{B}(Y)$ for $B(\mathscr{H}(Y))$ and $\mathbb{K}(Y)$ for $K(\mathscr{H}(Y))$ (according to the conventions made in the preceding section). If $\mathscr{B}$ and $\mathscr{C}$ are *-subalgebras of $\mathbb{B}(Y)$ and $\mathbb{B}(Y^\perp)$ respectively, we write $\mathscr{B} \odot_Y \mathscr{C}$ for the linear subspace of $\mathbb{B}(X)$ generated by operators of the form $B \otimes_Y C$, with $B \in \mathscr{B}$ and $C \in \mathscr{C}$; clearly $\mathscr{B} \odot_Y \mathscr{C}$ is a *-subalgebra of $\mathbb{B}(X)$. If $\mathscr{B}$ and $\mathscr{C}$ are norm-closed (hence are C^*-algebras), then we denote by $\mathscr{B} \otimes_Y \mathscr{C}$ the norm-closure of $\mathscr{B} \odot_Y \mathscr{C}$ in $\mathbb{B}(X)$. According to the explanations we gave in §8.2.1, the C^*-subalgebra $\mathscr{B} \otimes_Y \mathscr{C}$ of $\mathbb{B}(X)$ is canonically identified with the spatial tensor product of $\mathscr{B}$ and $\mathscr{C}$. Observe that we have (see §8.2.1):

(9.2.1) $$\mathbb{K}(Y) \otimes_Y \mathbb{K}(Y^\perp) = \mathbb{K}(X).$$

We now define:

$$(9.2.2) \qquad \mathscr{K}(Y) := \mathbb{K}(Y) \otimes_Y \mathbb{B}(Y^\perp).$$

The elements of this C*-subalgebra of $\mathbb{B}(X)$ will be called Y-*semicompact operators*. The main purpose of this section is to study properties of products of Y-semicompact and Z-semicompact operators when Y and Z are different subspaces of X.

The main technical point for the proof of the principal results of this section is presented in the next lemma. We denote by $\mathbb{B}_2(Y)$ the set of all Hilbert-Schmidt operators in $\mathscr{H}(Y)$ and by $||\cdot||_2$ the Hilbert-Schmidt norm in $\mathbb{B}_2(Y)$.

LEMMA 9.2.1. *Let Y and Z be subspaces of X such that $Y \cap Z = \mathbf{O}$ and $Y + Z = X$. Then, if $S \in \mathbb{B}_2(Y)$ and $T \in \mathbb{B}_2(Z)$, one has $S \otimes_Y I \cdot T \otimes_Z I \in \mathbb{B}_2(X)$. Moreover, there is a finite constant κ, depending only on Y and Z, such that:*

$$(9.2.3) \qquad ||S \otimes_Y I \cdot T \otimes_Z I||_2 = \kappa ||S||_2 ||T||_2.$$

PROOF. (i) We first introduce some notations. If x is a point in X, we can write $x = (y, y') \in Y \times Y^\perp$ and $x = (z, z') \in Z \times Z^\perp$, with $y = \pi_Y(x)$, $y' = \pi_{Y^\perp}(x)$, etc. The correspondence $(y, y') \mapsto (z, z')$ may be interpreted as a change of coordinate system in X and can be written as $z = \Theta y + \Lambda y'$, $z' = \Xi y + \Omega y'$, where $\Theta : Y \to Z$, $\Lambda : Y^\perp \to Z$, $\Xi : Y \to Z^\perp$ and $\Omega : Y^\perp \to Z^\perp$ are defined as follows:

$$\Theta = \pi_Z|_Y, \quad \Lambda = \pi_Z|_{Y^\perp}, \quad \Xi = \pi_{Z^\perp}|_Y, \quad \Omega = \pi_{Z^\perp}|_{Y^\perp}.$$

The hypotheses made on Y and Z imply that $\dim Z = \dim Y^\perp$, $\dim Z^\perp = \dim Y$ and that Λ and Ξ are bijections.

We also recall that, by the conventions made in Section 9.1, we have identified $\mathscr{H}(Y) \otimes \mathscr{H}(Y^\perp)$ and $\mathscr{H}(Z) \otimes \mathscr{H}(Z^\perp)$ with $\mathscr{H}(X)$. In the present context it is useful to take the identification maps explicitly into account; we shall denote them by $U(Y)$ and $U(Z)$ respectively and recall that for example $[U(Y)f](x) = f(\pi_Y(x), \pi_{Y^\perp}(x)) = f(y, y')$ for $f \in \mathscr{H}(Y) \otimes \mathscr{H}(Y^\perp)$. In these notations, we must show that

$$S \otimes_Y I \cdot T \otimes_Z I \equiv U(Y)[S \otimes I]U(Y)^{-1} \cdot U(Z)[T \otimes I]U(Z)^{-1} \in \mathbb{B}_2(X)$$

or equivalently (since $U(Y)$ and $U(Z)^{-1}$ are unitary, and with the definition $U(Z; Y) := U(Y)^{-1}U(Z)$) that $S \otimes I \cdot U(Z; Y) \cdot T \otimes I$ is an integral operator from $\mathscr{H}(Z) \otimes \mathscr{H}(Z^\perp)$ to $\mathscr{H}(Y) \otimes \mathscr{H}(Y^\perp)$ with square-integrable kernel. For later reference we observe that, for $f \in \mathscr{H}(Z) \otimes \mathscr{H}(Z^\perp)$:

$$(9.2.4) \qquad [U(Z; Y)f](y, y') = f(\Theta y + \Lambda y', \Xi y + \Omega y').$$

(ii) By assumption, the operators S and T are Hilbert-Schmidt operators in $\mathscr{H}(Y)$ and $\mathscr{H}(Z)$ respectively. In other words one has:

$$(Sg)(y, y') = \int_Y dy_1 \sigma(y, y_1) g(y_1, y') \quad \text{for } g \in \mathscr{H}(Y) \otimes \mathscr{H}(Y^\perp),$$

$$(Tf)(z, z') = \int_Z dz_1 \tau(z, z_1) f(z_1, z') \quad \text{for } f \in \mathscr{H}(Z) \otimes \mathscr{H}(Z^\perp),$$

with

$$||S||_2^2 = \int_Y dy \int_Y dy_1 |\sigma(y, y_1)|^2, \qquad ||T||_2^2 = \int_Z dz \int_Z dz_1 |\tau(z, z_1)|^2.$$

Thus, by using (9.2.4), one finds that for $f \in \mathscr{H}(Z) \otimes \mathscr{H}(Z^\perp)$:

$$[S \otimes I \cdot U(Z; Y) \cdot T \otimes If](y, y') = \int_Y dy_1 \sigma(y, y_1) \cdot$$

$$\cdot \int_Z dz_1 \tau(\Theta y_1 + \Lambda y', z_1) f(z_1, \Xi y_1 + \Omega y').$$

Now, for each fixed y', the correspondence $y_1 \mapsto z_1' \equiv \Xi y_1 + \Omega y'$ defines a bijection from Y onto $Z^\perp$. Thus, by making in the first integral the change of variables $y_1 = \Xi^{-1}(z_1' - \Omega y')$, one sees that $S \otimes I \cdot U(Z; Y) \cdot T \otimes I$ is indeed an integral operator from $\mathscr{H}(Z) \otimes \mathscr{H}(Z^\perp)$ to $\mathscr{H}(Y) \otimes \mathscr{H}(Y^\perp)$ with kernel $\gamma(y, y'; z_1, z_1')$ given by $\gamma(y, y'; z_1, z_1') = |\det \Xi|^{-1} \sigma(y, \Xi^{-1} z_1' - \Xi^{-1}\Omega y') \tau(\Theta\Xi^{-1} z_1' + \{\Lambda - \Theta\Xi^{-1}\Omega\} y', z_1)$. Then

$$||S \otimes_Y I \cdot T \otimes_Z I||_2^2 = \int_{Y \times Y^\perp} dy dy' \int_{Z \times Z^\perp} dz_1 dz_1' |\gamma(y, y'; z_1, z_1')|^2.$$

In this multiple integral we shall make the change of variables $(y', z_1') \mapsto (y_2, z_2) \in Y \times Z$, where

(9.2.5) $$y_2 = -\Xi^{-1}\Omega y' + \Xi^{-1} z_1',$$

(9.2.6) $$z_2 = (\Lambda - \Theta\Xi^{-1}\Omega)y' + \Theta\Xi^{-1} z_1'.$$

This change of variables is a linear map $V : Y^\perp \times Z^\perp \to Y \times Z$. We have $\dim(Y^\perp \times Z^\perp) = \dim(Y \times Z) - \dim X$, and V is injective: if $V(y', z_1') = 0$, then $z_1' = \Omega y'$ by (9.2.5) (because Ξ is a bijection), so that $\Lambda y' = 0$ by (9.2.6); since Λ is also a bijection, this implies that $y' = 0$, hence $z_1' \equiv \Omega y' = 0$, which proves the injectivity of V. Thus the above change of variables is justified, and we obtain that

$$||S \otimes_Y I \cdot T \otimes_Z I||_2^2 = |\det \Xi|^{-2}|\det V|^{-1} \int_{Y \times Y} dy dy_2 |\sigma(y, y_2)|^2 \cdot$$

$$\cdot \int_{Z \times Z} dz_1 dz_2 |\tau(z_2, z_1)|^2$$

$$= |\det \Xi|^{-2}|\det V|^{-1}||S||_2^2||T||_2^2. \quad \square$$

Observe that the preceding proof shows more than we stated. It is clear that, if $S \in \mathbb{B}(Y)$ and $T \in \mathbb{B}(Z)$ are both non-zero, and if Y and Z are as in Lemma 9.2.1, then the product $S \otimes_Y I \cdot T \otimes_Z I$ is Hilbert-Schmidt if and only if S and T are Hilbert-Schmidt operators. Moreover, if Y and Z are not mutually orthogonal, this proof shows that, even if S and T are rank 1 orthogonal projections, then the product $S \otimes_Y I \cdot T \otimes_Z I$ is not a finite rank operator in $\mathscr{H}$ (in general not even of trace class).

Part (c) of the next proposition contains the principal assertion on products of semicompact operators, namely: if S is Y-semicompact, T is Z-semicompact and $X = Y + Z$, then ST is a compact operator in $\mathscr{H}(X)$.

PROPOSITION 9.2.2. *Let Y and Z be subspaces of X. One has:*
(a) $\mathscr{K}(\mathbf{O}) = \mathbb{B}(X)$ *and* $\mathscr{K}(X) = \mathbb{K}(X)$,
(b) *if* $Y \subset Z$, *then* $\mathscr{K}(Z) \subset \mathscr{K}(Y)$,
(c) *if* $X = Y + Z$, *then* $\mathscr{K}(Y) \cdot \mathscr{K}(Z) \subset \mathbb{K}(X)$ *and* $\mathscr{K}(Y) \cap \mathscr{K}(Z) = \mathbb{K}(X)$.

PROOF. (b) It suffices to show that $S \otimes_Z T \in \mathscr{K}(Y)$ if $S \in \mathbb{K}(Z)$ and $T \in \mathbb{B}(Z^\perp)$. For this, we write $Z = Y \oplus W$ and observe that, by (9.2.1), it suffices to consider the case where S is of the form $A \otimes_Y^Z B$ with $A \in \mathbb{K}(Y)$ and $B \in \mathbb{K}(W)$. But

$$(A \otimes_Y^Z B) \otimes_Z T = A \otimes_Y (B \otimes_W^{Y^\perp} T)$$

(because $Y^\perp = W \oplus Z^\perp$), and the operator on the right-hand side is in $\mathscr{K}(Y)$.

(c) For the first assertion in (c) it suffices to verify that ST is a compact operator in $\mathscr{H}(X)$ if $S = A \otimes_Y A'$ with $A \in \mathbb{K}(Y)$, $A' \in \mathbb{B}(Y^\perp)$ and $T = B \otimes_Z B'$ with $B \in \mathbb{K}(Z)$ and $B' \in \mathbb{B}(Z^\perp)$. We observe that

$$ST = I \otimes_Y A'[A \otimes_Y I \cdot B \otimes_Z I]I \otimes_Z B'.$$

Hence it suffices to show that

(9.2.7) $S \in \mathbb{K}(Y)$ and $T \in \mathbb{K}(Z) \Longrightarrow S \otimes_Y I \cdot T \otimes_Z I \in \mathbb{K}(X)$.

(i) If $Y \cap Z = \mathbf{O}$, the validity of (9.2.7) follows from Lemma 9.2.1 by using the fact that $\mathbb{B}_2(W)$ is dense in $\mathbb{K}(W)$ with respect to the operator norm.

(ii) In the general case, we set $W = Y \cap Z$ and define Y_0, Z_0 to be such that $Y = Y_0 \oplus W$, $Z = Z_0 \oplus W$. By virtue of (9.2.1), it suffices to consider the case where $S = S_0 \otimes_{Y_0}^Y S_1$, $T = T_0 \otimes_{Z_0}^Z T_1$ with $S_0 \in \mathbb{K}(Y_0)$, $T_0 \in \mathbb{K}(Z_0)$ and $S_1, T_1 \in \mathbb{K}(W)$. Now, by using the fact that, if E, F are mutually orthogonal subspaces of X, then $U \otimes_E I \cdot V \otimes_F I = V \otimes_F I \cdot U \otimes_E I$, we get:

$$
\begin{aligned}
S \otimes_Y I \cdot T \otimes_Z I &= S_0 \otimes_{Y_0} I \cdot S_1 \otimes_W I \cdot T_0 \otimes_{Z_0} I \cdot T_1 \otimes_W I \\
&= S_0 \otimes_{Y_0} I \cdot T_0 \otimes_{Z_0} I \cdot S_1 \otimes_W I \cdot T_1 \otimes_W I \\
&\equiv \{S_0 \otimes_{Y_0}^{W^\perp} I \cdot T_0 \otimes_{Z_0}^{W^\perp} I\} \otimes_{W^\perp} I \cdot (S_1 T_1) \otimes_W I \\
&= \{S_0 \otimes_{Y_0}^{W^\perp} I \cdot T_0 \otimes_{Z_0}^{W^\perp} I\} \otimes_{W^\perp} (S_1 T_1).
\end{aligned}
$$

Since $W^\perp = Y_0 + Z_0$ and $Z_0 \cap Y_0 = \mathbf{O}$, the operator in the curly bracket belongs to $\mathbb{K}(W^\perp)$ by the result of (i). Since $S_1, T_1 \in \mathbb{K}(W)$, (9.2.7) follows by using again (9.2.1).

To prove the second assertion in (c), assume that $S \in \mathscr{K}(Y) \cap \mathscr{K}(Z)$. Then S^* is also in $\mathscr{K}(Y) \cap \mathscr{K}(Z)$. Thus, since $S^* \in \mathscr{K}(Y)$ and $S \in \mathscr{K}(Z)$, the result of the first part of (c) shows that $S^* S \in \mathbb{K}(X)$ which in turn implies that $S \in \mathbb{K}(X)$. $\square$

Before turning to some particular classes of semicompact operators, we point out a useful property of such operators in relation to the translation group $\{T(x)\}$.

PROPOSITION 9.2.3. *Let Y be a subspace of X and $R, S \in \mathscr{K}(Y)$. Then*

(9.2.8) $\text{s-}\lim_{|\pi_Y(x)| \to \infty} RT(x) = 0$ *and* $\lim_{|\pi_Y(x)| \to \infty} \|RT(x)S\| = 0.$

PROOF. It suffices to consider the case where $R = R_0 \otimes_Y R_1$, $S = S_0 \otimes_Y S_1$ with $R_0, S_0 \in \mathbb{K}(Y)$ and $R_1, S_1 \in \mathbb{B}(Y^\perp)$. For $x \in X$, set $y_0 = \pi_Y(x)$ and $y_1 = \pi_{Y^\perp}(x)$. Then

$$RT(x) = [R_0 T^Y(y_0)] \otimes_Y [R_1 T^{Y^\perp}(y_1)],$$

$$RT(x)S = [R_0 T^Y(y_0)S_0] \otimes_Y [R_1 T^{Y^\perp}(y_1)S_1].$$

(9.2.8) now follows by using the compactness of R_0 and S_0 and the fact that $T^Y(y_0)$ converges weakly to zero when $|y_0| \to \infty$. $\square$

In Proposition 9.2.2 we saw that $\mathcal{K}(Y) \cdot \mathcal{K}(Z) \subset \mathbb{K}(Y + Z)$ if $Y + Z = X$. We shall now show that for certain C*-subalgebras of $\mathcal{K}(Y)$, such an inclusion remains true even if $Y + Z \neq X$.

THEOREM 9.2.4. *Assume that to each subspace Y of X there is associated a C*-algebra $\mathbb{F}(Y) \subset \mathbb{B}(Y)$ such that, if $Y \perp Z$ and $W = Y \oplus Z$, then $\mathbb{F}(W) = \mathbb{F}(Y) \otimes \mathbb{F}(Z)$ (more precisely, $\mathbb{F}(W)$ is the norm closure in $\mathbb{B}(W)$ of the subspace generated by operators of the form $S \otimes_Y^W T$ with $S \in \mathbb{F}(Y)$, $T \in \mathbb{F}(Z)$). Define a C*-algebra $\mathcal{K}_\mathbb{F}(Y)$ by*

(9.2.9) $$\mathcal{K}_\mathbb{F}(Y) = \mathbb{K}(Y) \otimes_Y \mathbb{F}(Y^\perp).$$

Then one has for all subspaces Y, Z of X:

(9.2.10) $$\mathcal{K}_\mathbb{F}(Y) \cdot \mathcal{K}_\mathbb{F}(Z) \subset \mathcal{K}_\mathbb{F}(Y + Z).$$

PROOF. Let Y, Z be two subspaces of X. Set $X_0 = Y + Z$ and $W = X_0^\perp$. Let U and V be such that $X_0 = Y \oplus U = Z \oplus V$. Then $Y^\perp = U \oplus W$ and $Z^\perp = V \oplus W$. It suffices to show that $R_1 R_2 \in \mathcal{K}_\mathbb{F}(X_0)$ if R_1 and R_2 are operators of the form $R_1 = S \otimes_Y S'$, $R_2 = T \otimes_Z T'$ with $S \in \mathbb{K}(Y)$, $T \in \mathbb{K}(Z)$, $S' \in \mathbb{F}(Y^\perp)$ and $T' \in \mathbb{F}(Z^\perp)$. By the hypothesis made on the algebras $\{\mathbb{F}(Y)\}$, we may assume without loss of generality that S' and T' are of the following form: $S' = S_1 \otimes_U^{Y^\perp} S_2$, $T' = T_1 \otimes_V^{Z^\perp} T_2$ with $S_1 \in \mathbb{F}(U)$, $T_1 \in \mathbb{F}(V)$ and $S_2, T_2 \in \mathbb{F}(W)$. Under these assumptions we have:

$$R_1 R_2 = [S \otimes_Y^{X_0} S_1] \otimes_{X_0} S_2 \cdot [T \otimes_Z^{X_0} T_1] \otimes_{X_0} T_2$$

$$= [S \otimes_Y^{X_0} S_1 \cdot T \otimes_Z^{X_0} T_1] \otimes_{X_0} (S_2 T_2).$$

By Proposition 9.2.2 (c), the operator in the square brackets on the r.h.s. belongs to $\mathbb{K}(X_0)$. Hence $R_1 R_2 \in \mathcal{K}_\mathbb{F}(X_0)$. $\square$

COROLLARY 9.2.5. *If $Y_1, Y_2, \dots, Y_n$ are subspaces of X, then*

(9.2.11) $$\mathcal{K}_\mathbb{F}(Y_1) \cdot \mathcal{K}_\mathbb{F}(Y_2) \cdot \dots \cdot \mathcal{K}_\mathbb{F}(Y_n) \subset \mathcal{K}_\mathbb{F}(Y_1 + \dots + Y_n).$$

In particular, if $Y_1 + \dots + Y_n = X$ and $S_j \in \mathcal{K}_\mathbb{F}(Y_j)$ $(j = 1, \dots, n)$, then $S_1 \cdot \dots \cdot S_n$ is a compact operator in $\mathcal{H}(X)$.

We now give some examples of algebras $\mathbb{F}(Y)$ that appear in our applications. The simplest choice is $\mathbb{F}(Y) = \mathbb{C} \cdot I_{\mathscr{H}(Y)}$. This gives the following results:

(a) if Y, Z are subspaces of X, $S \in \mathbb{K}(Y)$, $T \in \mathbb{K}(Z)$, then there is an operator $R \in \mathbb{K}(Y + Z)$ such that $S \otimes_Y I \cdot T \otimes_Z I = R \otimes_{(Y+Z)} I$;

(b) if $Y_1, \ldots, Y_n$ are subspaces of X such that $Y_1 + \cdots + Y_n = X$ and if $S_j \in \mathbb{K}(Y_j)$ $(j = 1, \ldots, n)$, then $S_1 \otimes_{Y_1} I \cdot \ldots \cdot S_n \otimes_{Y_n} I \in \mathbb{K}(X)$.

In our next example we use the C*-algebra $C_\infty(X)$ of continuous functions $f : X \to \mathbb{C}$ that converge to zero at infinity. We consider it embedded in $\mathbb{B}(X)$ by interpreting its elements as multiplication operators. One easily sees that, if we set $\mathbb{F}(Y) = C_\infty(Y)$, then the hypotheses imposed on the family $\{\mathbb{F}(Y)\}$ in Theorem 9.2.4 are satisfied. Hence, if we define $\mathscr{K}_Q(Y) = \mathbb{K}(Y) \otimes_Y C_\infty(Y^\perp)$, we have the following results:

(a) if Y, Z are subspaces of X, then $\mathscr{K}_Q(Y) \cdot \mathscr{K}_Q(Z) \subset \mathscr{K}_Q(Y + Z)$;

(b) If $Y_1, \ldots, Y_n$ are subspaces of X such that $Y_1 + \cdots + Y_n = X$, then $\mathscr{K}_Q(Y_1) \cdot \ldots \cdot \mathscr{K}_Q(Y_n) \subset \mathbb{K}(X)$.

REMARK 9.2.6. In the applications we shall often work in a representation in which the elements of the algebra $\mathscr{K}_Q(Y)$ have a simple explicit form. Namely, we notice that $\mathscr{H}(X)$ is canonically isomorphic to $L^2(Y^\perp; \mathscr{H}(Y))$ (the Hilbert space of square-integrable $\mathscr{H}(Y)$-valued functions defined on $Y^\perp$). Under this isomorphism the C*-algebra $\mathscr{K}_Q(Y)$ becomes the algebra of multiplication operators by functions $\Phi : Y^\perp \to \mathbb{K}(Y)$ which are norm continuous and tend to zero at infinity:

$$(9.2.12) \qquad \mathscr{K}_Q(Y) \cong C_\infty(Y^\perp; \mathbb{K}(Y)).$$

This fact has been proved in a more general setting in §8.2.1.

In order to define our last and most important class of semicompact operators, we have to introduce a new C*-algebra $\mathbf{T}_\infty(X)$ in $\mathbb{B}(X)$, namely the C*-algebra naturally associated to the representation $\{T(x)\}$ of the translation group introduced in (1.2.12). More precisely, $\mathbf{T}_\infty(X)$ is the norm closure in $\mathbb{B}(X)$ of the set of operators of the form $\int_X T(x) f(x) dx$ with $f \in L^1(X)$. Equivalently, $\mathbf{T}_\infty(X)$ is given by

$$(9.2.13) \qquad \mathbf{T}_\infty(X) = \mathcal{F}^* C_\infty(X) \mathcal{F} = \{\varphi(P) \mid \varphi \in C_\infty(X)\}.$$

It is now clear that, if we take $\mathbb{F}(Y) = \mathbf{T}_\infty(Y)$ in Theorem 9.2.4, then the hypotheses of that theorem are satisfied. We define $\mathscr{T}(Y)$ by

$$(9.2.14) \qquad \mathscr{T}(Y) = \mathbb{K}(Y) \otimes_Y \mathbf{T}_\infty(Y^\perp).$$

We clearly have

$$(9.2.15) \qquad \mathscr{T}(Y) = \mathcal{F}^* \mathscr{K}_Q(Y) \mathcal{F} = \mathcal{F}_{Y^\perp}^* \mathscr{K}_Q(Y) \mathcal{F}_{Y^\perp}.$$

In particular it would be more natural to use the notation $\mathscr{K}_P(Y)$ for $\mathscr{T}(Y)$. But, by taking into account the importance of these algebras for what follows, we prefer to have a special notation for them. As a consequence of Theorem 9.2.4 we get $\mathscr{T}(Y) \cdot \mathscr{T}(Z) \subset \mathscr{T}(Y + Z)$ for any subspaces Y, Z of X.

9.3. The N-Body Algebra

Let X be an euclidean space and let $\Pi(X)$ be the set of all linear subspaces of X equipped with the partial order relation given by inclusion. We shall sometimes write $Y \leq Z$ in place of $Y \subset Z$. Then $\Pi(X)$ is a complete lattice, i.e. the lower bound $\wedge_{i \in \mathscr{I}} Y_i$ and the upper bound $\vee_{i \in \mathscr{I}} Y_i$ of an arbitrary family of subspaces $\{Y_i\}_{i \in \mathscr{I}}$ exists in $\Pi(X)$. More precisely, we have $\wedge_{i \in \mathscr{I}} Y_i = \cap_{i \in \mathscr{I}} Y_i$, and $\vee_{i \in \mathscr{I}} Y_i = \sum_{i \in \mathscr{I}} Y_i$ is the linear subspace generated by $\cup_{i \in \mathscr{I}} Y_i$. $\Pi(X)$ has a least element, namely $\mathbf{O} = \{0\}$, and a greatest element, namely X.

In the preceding section we have associated to each subspace Y of X a C*-subalgebra $\mathscr{T}(Y)$ of $\mathbb{B}(X)$ (see (9.2.14)). We shall now describe several interesting properties of the family of C*-algebras $\{\mathscr{T}(Y) \mid Y \in \Pi(X)\}$. The following facts are either obvious or have already been proved in Section 9.2:

(i) $\mathscr{T}(\mathbf{O}) = \mathbf{T}_\infty(X)$ and $\mathscr{T}(X) = \mathbb{K}(X)$;

(ii) for any $Y, Z \in \Pi(X)$ we have $\mathscr{T}(Y) \cdot \mathscr{T}(Z) \subset \mathscr{T}(Y + Z) \equiv \mathscr{T}(Y \vee Z)$;

(iii) $\mathscr{T}(Y)$ is a separable C*-algebra without unit (if $X \neq \mathbf{O}$) and $\mathscr{T}(Y)$ is abelian if and only if $Y = \mathbf{O}$;

(iv) the groups of automorphisms of $\mathbb{B}(X)$ induced by the unitary groups $\{W(\tau)\}_{\tau \in \mathbb{R}}$ (dilations), $\{T(x)\}_{x \in X}$ (translations) and $\{\exp i(Q, x)\}_{x \in X}$ leave each algebra $\mathscr{T}(Y)$ invariant, and the groups of automorphisms of $\mathscr{T}(Y)$ induced by them are of class C_0.

The last assertion is an immediate consequence of the factorization property of the indicated groups. For example, since $W(\tau) = W^Y(\tau) \otimes_Y W^{Y^\perp}(\tau)$, if $S = K \otimes_Y \varphi(P^{Y^\perp})$ for some $K \in \mathbb{K}(Y)$ and $\varphi \in C_\infty(Y^\perp)$, we have:

$$W(\tau)^* SW(\tau) = [W^Y(-\tau)KW^Y(\tau)] \otimes_Y \varphi(e^\tau P^{Y^\perp}).$$

Note that $W^Y(\tau) \to 0$ weakly as $|\tau| \to \infty$ if $Y \neq \mathbf{O}$; since K is compact, we shall then have $W^Y(-\tau)KW^Y(\tau) \to 0$ strongly if $|\tau| \to \infty$. On the other hand, $\varphi(e^\tau P) \to 0$ strongly if $\tau \to +\infty$ and $\varphi(e^\tau P) \to \varphi(0)$ strongly if $\tau \to -\infty$ (here $\varphi \in C_\infty(X)$ and $X \neq \mathbf{O}$). This proves the next property:

(v) For each $Y \in \Pi(X)$ with $Y \neq \mathbf{O}$ and each $S \in \mathscr{T}(Y)$, we have

$$\underset{|\tau| \to \infty}{\text{s-lim}} \, W(\tau)^* SW(\tau) = 0;$$

if $Y = \mathbf{O}$ and $S = \varphi(P) \in \mathscr{T}(\mathbf{O})$, then s-$\lim_{\tau \to +\infty} W(\tau)^* \varphi(P)W(\tau) = 0$, while s-$\lim_{\tau \to -\infty} W(\tau)^* \varphi(P)W(\tau) = \varphi(0)$.

For an arbitrary subset $\mathscr{L} \subset \Pi(X)$ we denote by $\mathscr{T}(\mathscr{L})$ the linear subspace of $\mathbb{B}(X)$ generated by $\cup_{Y \in \mathscr{L}} \mathscr{T}(Y)$. In other terms, $\mathscr{T}(\mathscr{L}) = \sum_{Y \in \mathscr{L}} \mathscr{T}(Y)$. It is clear that the property (ii) implies: if $\mathscr{L}$ is such that $Y, Z \in \mathscr{L} \Rightarrow Y + Z \in \mathscr{L}$, then $\mathscr{T}(\mathscr{L})$ is a *-subalgebra of $\mathbb{B}(X)$. In particular, $\mathscr{T}(\Pi(X))$ is a *-subalgebra of $\mathbb{B}(X)$ and its closure $\overline{\mathscr{T}}$ is a C*-subalgebra, which is non-trivial in the following sense: for each $S \in \overline{\mathscr{T}}$, we have s-$\lim_{\tau \to +\infty} W(\tau)^* SW(\tau) = 0$, and $\theta(S) := $ s-$\lim_{\tau \to -\infty} W(\tau)^* SW(\tau)$ exists and defines a non-zero character of the C*-algebra $\overline{\mathscr{T}}$ (i.e. θ is a non-zero morphism $\overline{\mathscr{T}} \to \mathbb{C}$). The next result is important for what follows.

LEMMA 9.3.1. *Let $\mathscr{L}$ be a finite subset of $\Pi(X)$. If $Y \in \mathscr{L}$ and $Y \neq X$, let Y^+ be the set of $x \in Y^\perp$ such that $x \notin W^\perp$ if $W \in \mathscr{L}$ and $W \not\subset Y$. Then Y^+ is a dense cone in $Y^\perp$ and for each $\omega \in Y^+$, $Z \in \mathscr{L}$ and $S \in \mathscr{T}(Z)$ we have:*

$$(9.3.1) \qquad \underset{\lambda \to \infty}{\text{s-lim}}\, T(\lambda\omega)^* S T(\lambda\omega) = \begin{cases} S & \text{if } Z \subset Y \\ 0 & \text{if } Z \not\subset Y. \end{cases}$$

PROOF. If $W \not\subset Y$, then $Y^\perp \not\subset W^\perp$, hence $Y^\perp \cap W^\perp$ is a subspace of $Y^\perp$ of dimension strictly smaller than $\dim Y^\perp$ (observe that $Y^\perp \neq \mathbf{O}$). So Y^+ is equal to the complement in $Y^\perp$ of the union of a finite number of strict subspaces; in particular Y^+ is dense in $Y^\perp$ and is a cone. It suffices to prove (9.3.1) for $S = S_1 \otimes_Z S_2$ with $S_1 \in \mathbb{K}(Z)$ and $S_2 \in \mathbf{T}_\infty(Z^\perp)$. If $Z \subset Y$, then $\omega \in Y^\perp \subset Z^\perp$, hence $T(\lambda\omega) = I \otimes_Z T^{Z^\perp}(\lambda\omega)$. Therefore $T(\lambda\omega)^* S T(\lambda\omega) = S$. If $Z \not\subset Y$, then the orthogonal projection ω' of ω onto Z is different from zero. By (9.2.8) we have $S_1 T^Z(\lambda\omega') \to 0$ strongly as $\lambda \to \infty$, which proves (9.3.1) completely. $\square$

The main property of the family of C*-algebras $\{\mathscr{T}(Y)\}$ is described in the following theorem:

THEOREM 9.3.2. *The sum $\sum_{Y \in \Pi(X)} \mathscr{T}(Y)$ is direct, i.e. each element S in the linear subspace of $\mathbb{B}(X)$ generated by $\cup_{Y \in \Pi(X)} \mathscr{T}(Y)$ can be expressed in a unique way as a sum $S = \sum_Y S(Y)$, with $S(Y) \in \mathscr{T}(Y)$ for each $Y \in \Pi(X)$ and $S(Y) \neq 0$ only for a finite number of subspaces Y. Moreover, for each finite subset $\mathscr{L} \subset \Pi(X)$, the subspace $\mathscr{T}(\mathscr{L}) = \sum_{Y \in \mathscr{L}} \mathscr{T}(Y)$ is norm-closed in $\mathbb{B}(X)$ and the sum is direct in the topological sense (i.e. the component $S(Y)$ of $S \in \mathscr{T}(\mathscr{L})$ in $\mathscr{T}(Y)$ is a norm-continuous function of S).*

PROOF. It is sufficient to prove the second part of the theorem. For each $Y \in \mathscr{L}$ with $Y \neq X$, let us choose some $\omega_Y \in Y^+$ (we use the notations of Lemma 9.3.1) and let us define $\mathscr{R}_Y[S] = \text{s-lim}_{\lambda \to \infty} T(\lambda\omega_Y)^* S T(\lambda\omega_Y)$ for each $S \in \mathscr{T}(\mathscr{L})$. From (9.3.1) it follows that this limit exists and defines a map $\mathscr{R}_Y : \mathscr{T}(\mathscr{L}) \to \mathscr{T}(\mathscr{L})$. Indeed, if $S = \sum_{Z \in \mathscr{L}} S(Z)$ with $S(Z) \in \mathscr{T}(Z)$, then we shall have $\mathscr{R}_Y[S] \equiv S_Y = \sum_{Z \in \mathscr{L}, Z \subset Y} S(Z)$. In order to prove the uniqueness of the decomposition of S into components $S(Z) \in \mathscr{T}(Z)$, it suffices to show that $S = 0 \Rightarrow S(Z) = 0$ for all Z. If $S = 0$, we shall clearly have $S_Y = 0$ for all $Y \in \mathscr{L}$ (with the convention that $S_X = S$). Let μ be the Möbius function of the finite partially ordered set $\mathscr{L}$; then (8.4.1) and (8.4.2) imply that $S(Y) = \sum_{Z \in \mathscr{L}} S_Z \mu(Z, Y) = 0$, so the assertion concerning the uniqueness is proved. But the same formulas give more, namely the projection $\mathscr{P}(Y) : \mathscr{T}(\mathscr{L}) \to \mathscr{T}(Y)$ determined by the linear direct sum decomposition $\mathscr{T}(\mathscr{L}) = \sum_{Y \in \mathscr{L}} \mathscr{T}(Y)$ (i.e. $\mathscr{P}(Y)[S] = S(Y)$) will be given by $\mathscr{P}(Y) = \sum_{Z \in \mathscr{L}} \mathscr{P}_Z \mu(Z, Y)$. Here $\mathscr{P}_X[S] = S$. Since each $\mathscr{R}_Y$ is norm-continuous by the explicit formula we gave at the beginning of the proof, each $\mathscr{P}(Y)$ will be continuous too. $\square$

We add some immediate consequences of the theorem and of its proof:

COROLLARY 9.3.3. *Let $\mathscr{L}$ be a finite semilattice of subspaces of X, i.e. $\mathscr{L} \subset \Pi(X)$ is finite and $Y, Z \in \mathscr{L} \Rightarrow Y \vee Z \equiv Y + Z \in \mathscr{L}$. Then $\mathscr{T} \equiv \mathscr{T}(\mathscr{L}) = \sum_{Y \in \mathscr{L}} \mathscr{T}(Y)$ is a C*-subalgebra of $\mathbb{B}(X)$ which is $\mathscr{L}$-graded by the family of*

C^*-algebras $\{\mathcal{T}(Y)\}_{Y\in\mathscr{L}}$. For each $Y \in \mathscr{L}$, $Y \neq X$, the canonical morphism (and projection) $\mathscr{P}_Y$ of $\mathcal{T}$ onto its C^*-subalgebra $\mathcal{T}_Y := \sum_{Z\in\mathscr{L}, Z\subset Y} \mathcal{T}(Z)$ can be expressed as follows: for each $\omega_Y \in Y^+$ (see Lemma 9.3.1) and $S \in \mathcal{T}$:

$$(9.3.2) \qquad \mathscr{P}_Y[S] = \text{s-}\lim_{\lambda\to\infty} T(\lambda\omega_Y)^* ST(\lambda\omega_Y).$$

If μ is the Möbius function of $\mathscr{L}$, then the canonical projection $\mathscr{P}(Y)$ of $\mathcal{T}$ onto $\mathcal{T}(Y)$ is given by:

$$(9.3.3) \qquad \mathscr{P}(Y) = \sum_{\substack{Z\in\mathscr{L} \\ Z\subset Y}} \mathscr{P}_Z \mu(Z, Y),$$

where $\mathscr{P}_X$ is the identity map $\mathcal{T} \to \mathcal{T}$.

We call N-*body algebra* any C^*-subalgebra $\mathcal{T}$ of $\mathbb{B}(X)$ of the form $\mathcal{T} = \mathcal{T}(\mathscr{L})$, where $\mathscr{L}$ is a finite semilattice of subspaces of X such that $\max \mathscr{L} = X$ and corank $\mathscr{L} = N$ (see §8.4.1). It is important to notice that in such a case we have $\mathcal{T}(\max\mathscr{L}) = \mathcal{T}(X) = \mathbb{K}(X)$; i.e. the natural "minimal" ideal of the $\mathscr{L}$-graded C^*-algebra $\mathcal{T}$ is just the ideal of compact operators in $\mathscr{H}(X)$. In particular, if H is a densely defined self-adjoint operator in $\mathscr{H}(X)$ and if H is affiliated to $\mathcal{T}$, then the usual (in the Hilbert space sense) essential spectrum $\sigma_{\text{ess}}(H)$ coincides with $\mathcal{T}\text{-}\sigma_{\text{ess}}(H)$ (cf. §8.4.2).

Let $\mathcal{T} = \sum_{Y\in\mathscr{L}} \mathcal{T}(Y)$ be a N-body algebra and H a (real) observable affiliated to it. We shall then use the notations $\mathcal{T}_Y$, $\mathscr{P}(Y)$ and $\mathscr{P}_Y$ with their natural meaning, cf. Corollary 9.3.3. Furthermore, let $H_Y = \mathscr{P}_Y[H]$, which is an observable affiliated to $\mathcal{T}_Y$ (note that the expression $H(Y) = \mathscr{P}(Y)[H]$ does not make sense in general). Since all the C^*-algebras $\mathcal{T}_Y$ are realized on the Hilbert space $\mathscr{H}(X)$, each H_Y has an operatorial meaning (cf. §8.1.2), i.e. there are a closed subspace $\mathscr{H}'_Y \subset \mathscr{H}(X)$ and a densely defined self-adjoint operator H'_Y in the Hilbert space $\mathscr{H}'_Y$ such that H_Y can be identified with the couple $(\mathscr{H}'_Y, H'_Y)$ in the sense of §8.1.2. The fact that in general $H \equiv H_X$ and H_Y are realized as non-densely defined operators is a real advantage in applications, because this allows one to treat N-body systems with very singular interactions (for the hard-core case, see [BGS]). However, in the rest of this text we shall restrict ourselves to the case of densely defined operators, mainly for the sake of simplicity of exposition.

We make one more remark concerning the essential spectrum (in the Hilbert space sense) of the operators H_Y (if H_Y is not densely defined, when we refer to a spectral property of H_Y of a Hilbert space character, we mean the corresponding property of the densely defined self-adjoint operator H'_Y in $\mathscr{H}'_Y$; recall that $\mathscr{H}(X) \ominus \mathscr{H}'_Y$ should be interpreted as the eigenspace of H_Y associated to the eigenvalue ∞). If we apply the abstract HVZ theorem (Proposition 8.4.2), we get that $\sigma(H_Z) \subset \sigma(H_Y) \subset \sigma_{\text{ess}}(H)$ if $Z, Y \in \mathscr{L}$, $Z \subset Y \neq X$, and

$$9.3.4 \qquad \sigma_{\text{ess}}(H) = \bigcup_{Y\in\mathscr{L}(2)} \sigma(H_Y).$$

Moreover, $\sigma_{\text{ess}}(H_Y) = \sigma(H_Y)$ if $Y \neq X$. But now we can prove more, namely H_Y has no eigenvalues of finite multiplicity if $Y \neq X$. Indeed, since the operator

$(H_Y - z)^{-1}$ belongs to $\mathscr{F}_Y$, we have $(H_Y - z)^{-1}T(y') = T(y')(H_Y - z)^{-1}$ for all $y' \in Y^\perp$ and $z \in \mathbb{C} \setminus \sigma(H_Y)$. Thus, the eigenspace of H_Y corresponding to some eigenvalue $\lambda \in \mathbb{R}$ is invariant under the unitary group $\{T(y') \mid y' \in Y^\perp\}$. But $Y^\perp \neq \mathbf{O}$, hence the generators of this group have purely absolutely continuous spectrum, in particular the group cannot have a finite dimensional non-zero invariant subspace.

We shall now explicitly describe a large class of observables (to be interpreted as hamiltonians) affiliated to a N-body algebra. If $\mathscr{L} = \{\mathbf{O}\}$ (so we are talking about free hamiltonians), then this is very easy because $\mathscr{T}(\mathbf{O}) = \mathbf{T}_\infty(X)$ is an abelian C*-algebra with X as spectrum, and we have described all observables affiliated to such algebras in §8.1.2. But one may check without any difficulty that a densely defined, lower semibounded self-adjoint operator $H_{\mathbf{O}}$ in $\mathscr{H}(X)$ is affiliated to $\mathscr{T}(\mathbf{O})$ if and only if there is a continuous function $h : X \to \mathbb{R}$, with $\lim_{|x| \to \infty} h(x) = +\infty$, such that $H_{\mathbf{O}} = h(P)$ (choose $a \in \mathbb{R}$ such that $H_{\mathbf{O}} + a \geq$ const. > 0 and note that $(H_{\mathbf{O}} + a)^{-1} \in \mathbf{T}_\infty(X)$). In particular, the Laplace-Beltrami operator Δ is affiliated to $\mathscr{T}(\mathbf{O})$. In order to cover more general $\mathscr{L}$, we shall use the perturbative criterion given in Proposition 8.4.4 (ii).

In the next proposition we shall use spaces which will now be introduced. Assume that a continuous function $h : X \to \mathbb{R}$ is given, with the property $h(x) \to +\infty$ as $|x| \to \infty$. If $Y \subset X$ is a subspace, define $h^Y : Y \to \mathbb{R}$ by $h^Y(y) = \inf\{h(y + y') \mid y' \in Y^\perp\}$. Observe that h^Y is continuous and divergent at infinity. Then let $\mathscr{H}^h(Y)$ be the domain in $\mathscr{H}(Y)$ of the operator $|h^Y(P^Y)|^{1/2}$ equipped with the graph norm $\langle f, (1 + |h^Y(P^Y)|)f \rangle^{1/2}$. After identifying $\mathscr{H}(Y)$ and $\mathscr{H}(Y)^*$, we get embeddings $\mathscr{H}^h(Y) \subset \mathscr{H}(Y) \subset \mathscr{H}^h(Y)^*$. Obviously, if $h(x) = |x|^2$, hence $h(P) = \Delta$, we shall have $\mathscr{H}^h(Y) = \mathscr{H}^1(Y)$, $\mathscr{H}^h(Y)^* = \mathscr{H}^{-1}(Y)$, which are usual Sobolev spaces. Recall that $K(\mathscr{H}_1, \mathscr{H}_2)$ is the Banach space of all compact operators $\mathscr{H}_1 \to \mathscr{H}_2$.

PROPOSITION 9.3.4. *Let $\mathscr{L}$ be a semilattice of subspaces of X with $\mathbf{O}, X \in \mathscr{L}$ and $h : X \to \mathbb{R}$ a continuous function such that $h(x) \to +\infty$ as $|x| \to \infty$. Define $H_{\mathbf{O}} = H(\mathbf{O}) = h(P)$. Assume that for each $Y \in \mathscr{L}$, $Y \neq \mathbf{O}$, a norm-continuous function $V^Y : Y^\perp \to K(\mathscr{H}^h(Y), \mathscr{H}^h(Y)^*)$ is given, that $V^Y(y')$ is symmetric for each $y' \in Y^\perp$, and that for each $\varepsilon > 0$ there is a locally bounded function $\delta_\varepsilon : Y^\perp \to \mathbb{R}$ such that $\delta_\varepsilon(y')[h^{Y^\perp}(y')]^{-1} \to 0$ as $|y'| \to \infty$ and $\pm V^Y(y') \leq \varepsilon h^Y(P^Y) + \delta_\varepsilon(y')$ for all $y' \in Y^\perp$. Let $H(Y)$ be the sesquilinear form in $\mathscr{H}(X)$ defined as follows: in the representation $\mathscr{H}(X) \cong L^2(Y^\perp; \mathscr{H}(Y))$ (cf. Remark 9.2.6), $\mathscr{F}_{Y^\perp} H(Y) \mathscr{F}_{Y^\perp}^*$ is multiplication by the operator-valued function V^Y. Then $H(Y)$ is a symmetric form on $\mathscr{H}(X)$, form-bounded with respect to $H_{\mathbf{O}}$ with relative bound zero. For each $Y \in \mathscr{L}$, let $H_Y = \sum_{Z \in \mathscr{L}, Z \subset Y} H(Z)$. Then H_Y is a densely defined self-adjoint operator in $\mathscr{H}(X)$ affiliated to $\mathscr{F}_Y = \sum_{Z \in \mathscr{L}, Z \subset Y} \mathscr{T}(Z)$, and $\mathscr{P}_Y[H] = H_Y$ if $H \equiv H_X$.*

PROOF. We assume, without loss of generality, that $h(x) \geq$ const. > 0. We wish to apply Proposition 8.4.4, and for this we first prove that we have $H_{\mathbf{O}}^{-1/2} H(Y) H_{\mathbf{O}}^{-1/2} \in \mathscr{T}(Y)$ for all $Y \neq \mathbf{O}$. Since the case $Y = X$ is trivial, we may assume that $Y \neq X$. We work in the representation $\mathscr{H}(X) = L^2(Y^\perp; \mathscr{H}(Y))$ and make a Fourier transformation in the variable $y' \in Y^\perp$, see (9.2.15) and Remark 9.2.6. Then $H_{\mathbf{O}}^{-1/2}$ becomes the operator of multiplication by the operator-valued function $F : Y^\perp \to \mathbb{B}(Y)$ given by $F(y') = [h(P^Y + y')]^{-1/2}$. This function is norm-continuous because $\|F(y_1') - F(y_2')\| = \sup_{y \in Y} |h(y+y_1')^{-1/2} - h(y+y_2')^{-1/2}|$ and $h^{-1/2}$ is uniformly continuous, and it is norm-convergent to zero at infinity because $\|F(y')\| = \sup_{y \in Y} |h(y + y')|^{-1/2} = h^{Y^\perp}(y')^{-1/2}$. The operator $\mathcal{F}_{Y^\perp} H_{\mathbf{O}}^{-1/2} H(Y) H_{\mathbf{O}}^{-1/2} \mathcal{F}_{Y^\perp}^*$ belongs to $\mathscr{K}_{\mathbf{Q}}(Y)$ (see (9.2.15)) if and only if it is the operator of multiplication by a function $G \in C_\infty(Y^\perp; \mathbb{K}(Y))$. But clearly

$$G(y') = F(y')V^Y(y')F(y')$$
$$= \left[\frac{h^Y(P^Y)}{h(P^Y + y')}\right]^{1/2} \cdot h^Y(P^Y)^{-1/2} V^Y(y') h^Y(P^Y)^{-1/2} \cdot \left[\frac{h^Y(P^Y)}{h(P^Y + y')}\right]^{1/2}$$
$$\equiv f(y')v(y')f(y').$$

Here $v(y') = h^Y(P^Y)^{-1/2} V^Y(y') h^Y(P^Y)^{-1/2}$ is a compact operator in $\mathscr{H}(Y)$ by hypothesis and $f(y') = [h^Y(P^Y)\{h(P^Y + y')\}^{-1}]^{1/2}$ is bounded because $0 \leq f(y') \leq I$. Hence G is compact-operator valued. We assumed v norm-continuous, and f is easily seen to be strongly continuous. By writing $G(y') = f(y')[v(y') - v(y_0')]f(y') + f(y')v(y_0')f(y')$ and by taking into account the compactness of $v(y_0')$, we see that $G : Y^\perp \to \mathbb{K}(Y)$ is norm-continuous. Since $h(P^Y + y')^{-1} \leq \sup_{y \in Y} h(y + y')^{-1} = h^{Y^\perp}(y')^{-1}$, we have for each $\varepsilon > 0$:

$$\pm G(y') \leq \varepsilon f(y')^2 + \delta_\varepsilon(y')h(P^Y + y')^{-1} \leq \varepsilon + \delta_\varepsilon(y')[h^{Y^\perp}(y')]^{-1},$$

from which we easily get $\|G(y')\| \to 0$ as $|y'| \to \infty$.

In order to apply Proposition 8.4.4, it remains to show that the form bound of $H(Y)$ with respect to $H_{\mathbf{O}}$ is equal to zero. This is equivalent to the property that $\lim_{\lambda \to \infty} \|(H_{\mathbf{O}} + \lambda)^{-1/2} H(Y)(H_{\mathbf{O}} + \lambda)^{-1/2}\| = 0$. A computation similar to the preceding one gives:

$$\pm [h(P^Y + y') + \lambda]^{-1/2} V^Y(y')[h(P^Y + y') + \lambda]^{-1/2} \leq \varepsilon + \delta_\varepsilon(y')[h^{Y^\perp}(y') + \lambda]^{-1}.$$

For large y', say $|y'| \geq r$, the right-hand side above is $\leq 2\varepsilon$. For $|y'| \leq r$, it will be $\leq \varepsilon + c\lambda^{-1}$ for a finite constant $c = c(r)$. This finishes the proof. $\square$

Let us see what class of hamiltonians we obtain if we take $h(x) = |x|^2$ in the preceding proposition. Clearly $h^Y(y) = |y|^2$ for each subspace $Y \subset X$, hence $\mathscr{H}^h(Y) = \mathscr{H}^1(Y)$, $\mathscr{H}^h(Y)^* = \mathscr{H}^{-1}(Y)$. Then $H_{\mathbf{O}} = H(\mathbf{O}) = \Delta$ and the "potentials" V^Y for $Y \neq \mathbf{O}$ are as follows:

(1) If $Y = X$, then V^X is just a symmetric, compact operator $\mathscr{H}^1(X) \to \mathscr{H}^{-1}(X)$.

(2) If $Y \neq \mathbf{O}, X$, then V^Y is a norm-continuous function defined on $Y^\perp$ with values in $K(\mathscr{H}^1(Y), \mathscr{H}^{-1}(Y))$ whose growth at infinity is restricted by the following condition: for each $\varepsilon > 0$ there is a function $\delta_\varepsilon : Y^\perp \to \mathbb{R}$ such that $\delta_\varepsilon(y')|y'|^{-2} \to 0$ as $|y'| \to \infty$ and $\pm V^Y(y') \leq \varepsilon \Delta^Y + \delta_\varepsilon(y')$ as forms on $\mathscr{H}^1(Y)$, for each $y' \in Y^\perp$.

We see that, in the standard non-relativistic N-body problem, the potentials are allowed to depend on the inter-cluster momentum (formally, $H(Y)$ is a function of the observables Q^Y, P^Y and $P^{Y^\perp}$).

It is not our purpose here to study finer spectral properties of the hamiltonians introduced in Proposition 9.3.4. Instead, we shall concentrate on non-relativistic N-body hamiltonians for which the analysis can be continued in a very simple way, due to some special factorization properties.

9.4. Non-Relativistic N-Body Hamiltonians

In this section we consider a fixed euclidean space X and a finite family $\mathscr{L}$ of subspaces of X such that $\mathbf{O}, X \in \mathscr{L}$ and $Y, Z \in \mathscr{L} \Rightarrow Y \vee Z \equiv Y + Z \in \mathscr{L}$. We denote by $\mathscr{T} = \sum_{Y \in \mathscr{L}} \mathscr{T}(Y)$ the N-body algebra determined by $\mathscr{L}$, and we keep the notations introduced in Corollary 9.3.3 and in the comments which follow it.

9.4.1. An important role will be played by a factorization property of the algebras $\mathscr{T}_Y$ which we shall now describe. Let $Z \in \mathscr{L}_Y$ (i.e. $Z \in \mathscr{L}$ and $Z \subset Y$); then $Y^\perp \subset Z^\perp$ and we have an orthogonal decomposition $Z^\perp = (Y \ominus Z) \oplus Y^\perp$ which induces a tensorial factorization $\mathbf{T}_\infty(Z^\perp) = \mathbf{T}_\infty(Y \ominus Z) \otimes^{Z^\perp}_{Y \ominus Z} \mathbf{T}_\infty(Y^\perp)$ (see Section 9.2). Then, from Definition (9.2.14) we obtain:

$$(9.4.1) \qquad \mathscr{T}(Z) = [\mathbb{K}(Z) \otimes^Y_Z \mathbf{T}_\infty(Y \ominus Z)] \otimes_Y \mathbf{T}_\infty(Y^\perp).$$

Observe that, for each $Y \in \mathscr{L}$, $\mathscr{L}_Y$ is a family of subspaces of Y which has exactly the same properties with respect to Y as $\mathscr{L}$ with respect to X; let N_Y be the corank of the lattice $\mathscr{L}_Y$. So, we may define the N_Y-body algebra $\mathscr{T}^Y$ associated to $\mathscr{L}_Y$: $\mathscr{T}^Y$ is a C*-subalgebra of $\mathbb{B}(Y)$, which is $\mathscr{L}_Y$-graded by $\mathscr{T}^Y = \sum_{Z \in \mathscr{L}_Y} \mathscr{T}^Y(Z)$, where the C*-subalgebra $\mathscr{T}^Y(Z)$ of $\mathbb{B}(Y)$ is given by:

$$(9.4.2) \qquad \mathscr{T}^Y(Z) = \mathbb{K}(Z) \otimes^Y_Z \mathbf{T}_\infty(Y \ominus Z).$$

Note that $\mathscr{T}^X(Z) = \mathscr{T}(Z)$, $\mathscr{T}^X = \mathscr{T}$, so our notations are consistent with the conventions made in Section 9.1. We clearly have

$$(9.4.3) \qquad \mathscr{T}_Y = \mathscr{T}^Y \otimes_Y \mathbf{T}_\infty(Y^\perp),$$

which is the factorization property of $\mathscr{T}_Y$ mentioned above. Let us stress the fact that the C*-algebra $\mathscr{T}^Y$ is naturally realized on the Hilbert space $\mathscr{H}(Y)$, while the C*-algebra $\mathscr{T}_Y$ is realized on $\mathscr{H}(X)$; the preceding tensor factorization of $\mathscr{T}_Y$ refers to the factorization $\mathscr{H}(X) \cong \mathscr{H}(Y) \otimes \mathscr{H}(Y^\perp)$ of $\mathscr{H}(X)$.

We can now define in precise terms what we mean by a non-relativistic N-body hamiltonian.

DEFINITION 9.4.1. A densely defined self-adjoint lower semibounded operator H in $\mathscr{H}(X)$ is called a *non-relativistic N-body hamiltonian* (with respect to the family $\mathscr{L}$ of subspaces of X) if H is affiliated to $\mathscr{T}$ and if, for each $Y \in \mathscr{L}$, there is a densely defined self-adjoint operator H^Y in $\mathscr{H}(Y)$ such that $H_Y = H^Y \otimes_Y I + I \otimes_Y \Delta^{Y^\perp}$. Furthermore, we assume $H^{\mathbf{O}} = 0$.

We shall use the abbreviation (NR)-*hamiltonian* or N-*body* (NR)-*hamiltonian* in place of "non-relativistic N-body hamiltonian". The operators H^Y will be called *internal hamiltonians*, while H_Y are the *sub-hamiltonians*; clearly $H^X = H_X = H$. Examples of such operators are obtained by taking $H_O = \Delta \equiv \Delta^X$ (i.e. $h(x) = |x|^2$) and V^Y a constant function in Proposition 9.3.4. So V^Y is just a symmetric compact operator $\mathscr{H}^1(Y) \to \mathscr{H}^{-1}(Y)$, and $H(Y) \equiv V_Y = V^Y \otimes_Y I :$ $\mathscr{H}^1(X) \to \mathscr{H}^{-1}(X)$ is a symmetric sesquilinear form on $\mathscr{H}(X)$, Δ-form bounded with relative bound zero [1]. But the preceding definition covers much more general classes of potentials. We have assumed all H^Y to be densely defined only in order to somewhat simplify the presentation. In fact, most of what follows remains valid for non-densely defined operators, and this is important in applications to highly singular interactions (see [BGS]). Note that $\mathscr{H}(O) = \mathbb{C}$, $\mathscr{T}^O = \mathbb{C}$, so H^O could, in general, be a real constant. The hypothesis $H^O = 0$, which is equivalent to $H_O = \Delta^X$, is in fact a normalization condition which fixes the lower bound of the free hamiltonian H_O.

It is an easy consequence of Definition 9.4.1 that each H^Y will be a N_Y-body (NR)-hamiltonian with respect to the lattice $\mathscr{L}_Y$, of corank N_Y [2]; more precisely, H^Y is a densely defined self-adjoint lower semibounded operator in $\mathscr{H}(Y)$ which is affiliated to $\mathscr{T}^Y$ and has factorization properties similar to those of H with respect to subspaces $Z \in \mathscr{L}_Y$. This explains the simplicity of the results we obtain below and allows one to prove many important facts by a simple induction procedure. As a first example, we have the following rather explicit description of the essential spectrum of H (this is the classical HVZ theorem):

THEOREM 9.4.2. *Let H be a non-relativistic N-body hamiltonian and let $\{H^Y\}_{Y\in\mathscr{L}}$ be the family of internal hamiltonians associated to it. For each $Y \in \mathscr{L}$, let $\tau_Y = \inf \sigma(H^Y)$ and $\tau^Y = \inf \sigma_{\mathrm{ess}}(H^Y)$. Then:*

(i) $\sigma_{\mathrm{ess}}(H) = [\tau^X, \infty)$ *and* $\tau^X = \min_{Y\in\mathscr{L}(2)} \tau_Y$.

(ii) *If $Z \subset Y$ and $Z \neq Y$, then* $-\infty < \tau_Y \leq \tau^Y \leq \tau_Z \leq 0$.

(iii) *If $Y \neq X$, then the spectrum of H_Y is purely absolutely continuous and* $\sigma(H_Y) = [\tau_Y, \infty)$.

---

[1] The tensor product $V^Y \otimes_Y I$ may be interpreted as follows: first take the Hilbert tensor product of the bounded operator $V^Y : \mathscr{H}^1(Y) \to \mathscr{H}^{-1}(Y)$ with the identity map $\mathscr{H}(Y^\perp) \to \mathscr{H}(Y^\perp)$, then observe that $X = Y \oplus Y^\perp$ provides us with canonical embeddings $\mathscr{H}^1(X) \subset \mathscr{H}^1(Y) \otimes \mathscr{H}(Y^\perp) \subset \mathscr{H}(X) \subset \mathscr{H}^{-1}(Y) \otimes \mathscr{H}(Y^\perp) \subset \mathscr{H}^{-1}(X)$, and finally define $V^Y \otimes_Y I$ as the restriction of $V^Y \otimes I$ to $\mathscr{H}^1(X)$, considered with values in $\mathscr{H}^{-1}(X)$. Of course, this definition of $\Pi(Y)$ is equivalent to that of Proposition 9.3.4.

[2] Indeed, H^Y will be bounded from below because $\sigma(H_Y) \subset \sigma(H)$ and $\sigma(H_Y) = \sigma(H^Y) + \mathbb{R}_+$, cf. §8.2.3. Moreover, $H_Y = \mathscr{P}_Y[H]$ is affiliated to $\mathscr{T}_Y$ which is the tensor product of $\mathscr{T}^Y$ with the abelian C*-algebra $\mathbf{T}_\infty(Y^\perp)$. By taking into account the description of the observables affiliated to such C*-algebras given in §8.2.4 and by noticing that to work in the Gelfand representation of $\mathbf{T}_\infty(Y^\perp)$ is equivalent with making a Fourier transformation in the variable of $Y^\perp$, one sees that there is a proper, continuous family $\{H_{y'} \mid y' \in U \subset Y^\perp\}$ of observables $H_{y'}$ affiliated to $\mathscr{T}^Y$ such that $\mathscr{F}_{Y^\perp}(H_{y'} - z)^{-1}\mathscr{F}^*_{Y^\perp}$ is the operator of multiplication by the operator-valued function $y' \mapsto (H_{y'} - z)^{-1}$ (see Remark 9.2.6). But $H_Y = H^Y \otimes_Y I + I \otimes_Y \Delta^{Y^\perp}$ implies $H_{y'} = H^Y + |y'|^2$, hence H^Y is affiliated to $\mathscr{T}^Y$. If $Z \subset Y$, we have $\mathscr{P}_Z[H_Y] = \mathscr{P}_Z[H]$, so H^Y has factorization properties similar to those of $H = H^X$.

PROOF. The abstract HVZ theorem (Proposition 8.4.2) and the relation $\mathscr{T}(X)$ $= \mathbb{K}(X)$ imply that $\sigma_{\mathrm{ess}}(H) = \mathscr{T}\text{-}\sigma_{\mathrm{ess}}(H) = \cup_{Y\in\mathscr{L}(2)}\sigma(H_Y)$. If $Y \neq X$, then $\Delta^{Y^\perp}$ has a purely absolutely continuous spectrum, equal to $\mathbb{R}_+ = [0,\infty)$. Since $H_Y = H^Y \otimes_Y I + I \otimes_Y \Delta^{Y^\perp}$, the results of §8.2.3 imply the absolute continuity of the spectrum of H_Y and $\sigma(H_Y) = \sigma(H^Y) + \mathbb{R}_+ = [\tau_Y,\infty)$. Hence (i) and (iii) are proved. If $Z \subset Y$, we have $\sigma(H_Z) \subset \sigma(H_Y)$, so we shall have $\tau^X \leq \tau_Z$ for all $Z \neq X$. Then (ii) follows by replacing H by H^Y and by taking into account that $\tau^{\mathbf{O}} = \tau_{\mathbf{O}} = 0$. $\square$

We use the notation $\sigma_p(H)$ for the set of eigenvalues of a self-adjoint operator H. If H is a non-relativistic N-body hamiltonian, one associates to it two new remarkable sets: the *set of thresholds* of H, defined as $\tau(H) := \cup_{Y\neq X}\sigma_p(H^Y)$, and the *critical set* of H, defined as $\kappa(H) := \cup_{Y\in\mathscr{L}}\sigma_p(H^Y)$. Observe that these are countable subsets of $\mathbb{R}$. The main point of the spectral analysis of (NR)-hamiltonians is that H has rather simple spectral properties outside $\tau(H)$. Hence it is important to be able to prove that $\tau(H)$ is a closed set (one may think of a countable closed set as being a small set). The closedness of $\tau(H)$ will be shown to be a consequence of the Mourre estimate. We should like to add several comments concerning the set of thresholds of H:

(1) It is easy to show that $\tau^X \in \tau(H)$, so τ^X is simultaneously the bottom of the spectrum of some H^Y with $Y \neq X$ and an eigenvalue of H^Y; clearly $\tau^X = \min\tau(H)$.

(2) By induction one sees that τ_Y, $\tau^Y \in \tau(H)$ for all $Y \neq X$, in particular $0 \in \tau(H)$.

(3) For a large class of non-relativistic N-body hamiltonians of the form $H = \Delta + \sum_{Y\neq\mathbf{O}} V^Y \otimes_Y I$, with V^Y multiplication operators, it has been shown that $\tau(H) \subset [\tau^X,0]$, see [FH2]; this is equivalent with the non-existence of strictly positive eigenvalues for all H^Y; note, however, that H could have negative eigenvalues embedded in its continuous spectrum.

(4) If $Y \neq X$, then H_Y is a (NR)-hamiltonian with purely continuous spectrum; however, its set of thresholds could be very rich, because $\tau(H_Y) = \kappa(H_Y) = \kappa(H^Y) \supset \tau(H^Y)$ (the proof of these relations is a straightforward exercise).

In the spectral analysis of (NR)-hamiltonians the role of the conjugate operator will be played by the generator of the dilation group in X, denoted $D \equiv D^X$ and explicitly defined in (1.2.19). One should always have in mind the relation $D^X = D^Y \otimes_Y I + I \otimes_Y D^{Y^\perp}$ between the generators of the dilation groups in X, Y and $Y^\perp$.

LEMMA 9.4.3. *If the (NR)-hamiltonian H is of class $C_u^1(D)$, then each H_Y is of class $C_u^1(D)$ and each internal hamiltonian H^Y is of class $C_u^1(D^Y)$.*

PROOF. The hypothesis means that the function $\tau \mapsto W(\tau)^*(H+i)^{-1}W(\tau) \in \mathbb{B}(X)$ is of class C^1 in norm, where $W \equiv W^X$ is the dilation group in $\mathscr{H}(X)$. We know (cf. the beginning of Section 9.3) that the group of automorphisms $\mathscr{W}_\tau[S] = W(\tau)^*SW(\tau)$ induced by W leaves $\mathscr{T}$ invariant, its restriction to $\mathscr{T}$ is of class C_0 and is compatible with the grading $\mathscr{T} = \sum_{Y\in\mathscr{L}}\mathscr{T}(Y)$ (i.e. $\mathscr{P}_Y\mathscr{W}_\tau = \mathscr{W}_\tau\mathscr{P}_Y$ on $\mathscr{T}$). Since $(H+i)^{-1} \in \mathscr{T}$ and $\mathscr{P}_Y[(H+i)^{-1}] = (H_Y+i)^{-1}$, it follows immediately that H_Y is of class $C_u^1(D)$. In view of the factorization property of $W(\tau)$, we clearly have:

$$W(\tau)^* H_Y W(\tau) = [W^Y(\tau)^* H^Y W^Y(\tau)] \otimes_Y I + I \otimes_Y [e^{2\tau} \Delta^{Y^\perp}].$$

If we make a Fourier transformation in the variable of $Y^\perp$ and work in the representation $\mathcal{H}(X) \cong L^2(Y^\perp; \mathcal{H}(Y))$, then $W(\tau)^*(H_Y + i)^{-1}W(\tau)$ becomes the operator of multiplication by the operator-valued function $F_\tau : Y^\perp \to \mathbb{B}(Y)$ defined by $F_\tau(y') = [W^Y(\tau)^* H^Y W^Y(\tau) + e^{2\tau}|y'|^2 + i]^{-1}$. So the function $\tau \mapsto F_\tau \in C_\infty(Y^\perp; \mathbb{B}(Y))$ is norm C^1. But $S \mapsto S(0)$ is a morphism of $C_\infty(Y^\perp; \mathbb{B}(Y))$ onto $\mathbb{B}(Y)$. Hence $\tau \mapsto F_\tau(0) = W^Y(\tau)^*(H^Y + i)^{-1}W^Y(\tau)$ is norm C^1. $\square$

Assume that H is a (NR)-hamiltonian of class $C_u^1(D)$. Then we may consider the functions $\varrho = \varrho_H^D$ and $\widetilde{\varrho} = \widetilde{\varrho}_H^D$ defined according to the rules of Section 7.2. We stress the fact that $\widetilde{\varrho}$ is defined with respect to the ideal $\mathbb{K}(X)$ of all compact operators in $\mathcal{H}(X)$, and that $\mathbb{K}(X) = \mathscr{T}(\max \mathscr{L})$, so that we may apply Theorem 8.4.3 (see also §8.3.4). When we want to insist on the fact that we consider the (NR)-hamiltonian $H = H^X$ on the euclidean space X, we write $\varrho \equiv \varrho^X$, $\widetilde{\varrho} \equiv \widetilde{\varrho}^X$. This implicitly defines $\varrho^Y = \varrho_{H^Y}^{D^Y}$ and $\widetilde{\varrho}^Y = \widetilde{\varrho}_{H^Y}^{D^Y}$ for all $Y \in \mathscr{L}$; note that H^Y is an operator in the Hilbert space $\mathcal{H}(Y)$ and $\widetilde{\varrho}^Y$ is defined in terms of the ideal $\mathbb{K}(Y)$ of compact operators in $\mathcal{H}(Y)$. On the other hand, we may also consider the (NR)-hamiltonians H_Y in $\mathcal{H}(X)$, and so we may define $\varrho_Y = \varrho_{H_Y}^D$ and $\widetilde{\varrho}_Y = \widetilde{\varrho}_{H_Y}^D$. But if $Y \neq X$, we have $\varrho_Y = \widetilde{\varrho}_Y$ by Theorem 7.2.13. Now observe that $\widetilde{\varrho}^X = \min_{Y \in \mathscr{L}(2)} \varrho_Y$, as a consequence of Theorem 8.4.3. But we may compute ϱ_Y in terms of ϱ^Y if $Y \neq X$ (if $Y = X$, we have $\varrho_X = \varrho^X$) by using the Theorem 8.3.6 and Example 7.2.2. Obviously, if $Y \neq X$ and $\lambda \in \mathbb{R}$:

$$(9.4.4) \qquad \varrho_Y(\lambda) = \inf_{\nu \geq 0} [\varrho^Y(\lambda - \nu) + \nu] = \inf_{\mu \leq \lambda} [\varrho^Y(\mu) + \lambda - \mu].$$

Finally, one may express ϱ^Y in terms of $\widetilde{\varrho}^Y$ by using again Theorem 7.2.13. The result will be an explicit expression of $\widetilde{\varrho}^X$ in terms of the functions $\widetilde{\varrho}^Y$ with $Y \in \mathscr{L}(2)$. So a question concerning a N-body hamiltonian (the calculation of $\widetilde{\varrho}^X$) has been reduced to a similar question concerning $(N-1)$-body hamiltonians (calculation of $\widetilde{\varrho}^Y$ with $Y \in \mathscr{L}(2)$). By induction, we can now compute $\widetilde{\varrho}^X$ explicitly. We state the precise result in the next theorem.

THEOREM 9.4.4. *Let H be a non-relativistic N-body hamiltonian and let D be the generator of the dilation group, normalized as in (1.2.19). Assume that H is of class $C_u^1(D)$ and let $\widetilde{\varrho} = \widetilde{\varrho}_H^D$. Then $\tau(H)$ and $\kappa(H)$ are closed countable sets, $\kappa(H) \setminus \tau(H)$ is a discrete subset of $\mathbb{R}$ consisting of eigenvalues of H of finite multiplicity, and $\mu^D(H) = \mathbb{R} \setminus \kappa(H)$ (see (7.3.1)). Moreover, for each $\lambda \in \mathbb{R}$ we have*

$$(9.4.5) \qquad \widetilde{\varrho}(\lambda) = \inf\{\lambda - \mu \mid \mu \in \tau(H) \text{ and } \mu \leq \lambda\},$$

with the convention that the infimum over an empty set is $+\infty$.

PROOF. The main ideas of the proof have been presented above, we now give the details. We first consider two very simple cases, which correspond to $N = 1$ and $N = 2$. If $Y = \mathbf{O}$, then obviously $\widetilde{\varrho}^{\mathbf{O}}(\lambda) = +\infty$ for all λ, while $\varrho^{\mathbf{O}}(0) = 0$

and $\varrho^{\mathbf{O}}(\lambda) = +\infty$ if $\lambda \neq 0$. Now let $Y \in \mathscr{L}$ be an element which covers $\mathbf{O}$, i.e. $\mathscr{L}_Y = \{\mathbf{O}, Y\}$. Then H^Y is a 2-body hamiltonian and we may use Example 7.2.2 and Theorem 7.2.13 in order to compute $\widetilde{\varrho}^Y$. We obtain $\widetilde{\varrho}^Y(\lambda) = +\infty$ if $\lambda < 0$ and $\widetilde{\varrho}^Y(\lambda) = \lambda$ if $\lambda \geq 0$. Now Theorem 7.2.13 will give $\varrho^Y(\lambda) = 0$ if $\lambda \in \sigma_p(H^Y) \cup \sigma_p(H^{\mathbf{O}}) = \kappa(H^Y)$ and $\varrho^Y(\lambda) = \widetilde{\varrho}^Y(\lambda)$ if $\lambda \notin \kappa(H^Y)$ (note that $\sigma_p(H^{\mathbf{O}}) = \{0\} = \tau(H^Y)$ in the case under consideration). In view of the last assertion of Corollary 7.2.11, the theorem is proved if $N \leq 2$.

We now prove the theorem by induction over the lattice $\mathscr{L}$. We assume that the assertions of the theorem are proved for all H^Y with $Y < Z$, for some fixed Z, and we prove them for H^Z. There is no loss of generality and it is notationally convenient to assume $Z = X$. Observe first that, for each $Y \neq X$, $\kappa(H^Y) = \tau(H^Y) \cup [\kappa(H^Y) \setminus \tau(H^Y)]$ is a closed set (since $\tau(H^Y)$ is closed and the points of $\kappa(H^Y) \setminus \tau(H^Y)$ may accumulate only at points of $\tau(H^Y)$ or at $\pm\infty$). On the other hand, $\tau(H)$ is clearly equal to $\cup_{Y \neq X} \kappa(H^Y)$, so it is closed too. If we succeed in proving (9.4.5), the other assertions of the theorem will follow from Corollary 7.2.11. We know that $\widetilde{\varrho} = \min_{Y \in \mathscr{L}(2)} \varrho_Y$. We shall prove below that for $Y \neq X$:

$$(9.4.6) \qquad \varrho_Y(\lambda) = \inf\{\lambda - \mu \mid \mu \in \kappa(H^Y), \mu \leq \lambda\}.$$

This allows us to conclude as follows:

$$\widetilde{\varrho}(\lambda) = \min_{Y \in \mathscr{L}(2)} \varrho_Y(\lambda) = \inf\{\lambda - \mu \mid \mu \in \kappa(H^Y), \mu \leq \lambda, Y \in \mathscr{L}(2)\}$$

$$= \inf\{\lambda - \mu \mid \mu \in \cup_{Y \in \mathscr{L}(2)} \kappa(H^Y), \mu \leq \lambda\} = \inf\{\lambda - \mu \mid \mu \in \tau(H), \mu \leq \lambda\}.$$

It remains to determine ϱ_Y for $Y \neq X$. From the induction hypothesis and Theorem 7.2.13 we obtain that $\varrho^Y(\mu) = 0$ if $\mu \in \kappa(H^Y)$ and $\varrho^Y(\mu) = \widetilde{\varrho}^Y(\mu) = \inf\{\mu - \nu \mid \nu \in \tau(H^Y), \nu \leq \mu\}$ if $\mu \notin \kappa(H^Y)$. Note that $\kappa(H^Y)$ is closed, so $\varrho^Y(\mu) > 0$ if (and only if) $\mu \notin \kappa(H^Y)$. We first check (9.4.6) for $\lambda \in \kappa(H^Y)$. Then the r.h.s. of (9.4.6) is zero; by (9.4.4) its l.h.s. is zero too, because $\varrho^Y(\mu) \geq 0$ for all $\mu \in \mathbb{R}$ and $\varrho^Y(\lambda) = 0$. Now assume that $\lambda \notin \kappa(H^Y)$. Then λ belongs to a unique connected component of the open set $\mathbb{R} \setminus \kappa(H^Y)$, so there are uniquely determined numbers $a < b$ with $a = -\infty$ or $a \in \kappa(H^Y)$ and $b = +\infty$ or $b \in \kappa(H^Y)$, such that $\lambda \in (a,b)$ and $(a,b) \cap \kappa(H^Y) = \varnothing$. If $a = -\infty$, then $\lambda < \inf \kappa(H^Y)$, so $\varrho_Y(\lambda) = +\infty$ and (9.4.6) is true. If a is finite, then the r.h.s. of (9.4.6) is $\lambda - a$. To determine its l.h.s., we use again (9.4.4). The infimum over $\mu \leq a$ in (9.4.4) is attained for $\mu = a$ and is equal to $\lambda - a$; for $a < \mu \leq \lambda$ one has

$$\varrho^Y(\mu) + \lambda - \mu = \widetilde{\varrho}^Y(\mu) + \lambda - \mu$$

$$= \inf\{\mu - \nu + \lambda - \mu \mid \nu \in \tau(H^Y), \nu \leq \mu\} \geq \lambda - a,$$

because $\tau(H^Y) \subset \kappa(H^Y)$. So $\varrho_Y(\lambda) = \lambda - a$. $\square$

If H is a hamiltonian as in Theorem 9.4.4 and λ is an eigenvalue of H, then the following three situations for the value of $\widetilde{\varrho} = \widetilde{\varrho}_H^D$ at λ may be considered:

(1) $\widetilde{\varrho}(\lambda) = +\infty$; this happens precisely when λ is an eigenvalue of finite multiplicity which is isolated from the remainder of the spectrum of H;

(2) $0 < \widetilde{\varrho}(\lambda) < +\infty$; this occurs when λ is an eigenvalue of finite multiplicity, belongs to the essential spectrum of H but is isolated from the other eigenvalues of H (i.e. λ is an isolated point of $\sigma_p(H)$);

(3) $\widetilde{\varrho}(\lambda) = 0$; this occurs when λ is of infinite multiplicity or an accumulation point of eigenvalues of H.

The following example illustrates the situation (2). It corresponds in some sense to a system of two electrons interacting with a nucleus but not between themselves. Let X be two-dimensional and denote the coordinates of a point x by $x = (x_1, x_2)$ (with respect to some orthonormal basis of X). We take $H = H_1 + H_2$, with $H_k = -\frac{d^2}{dx_k^2} + V_k(x_k)$, where V_k satisfy $V_k(s) \to 0$ as $|s| \to \infty$ ($k = 1, 2$). We assume that V_1, V_2 are such that H_1 has exactly one eigenvalue $\lambda_1 < 0$, whereas H_2 has exactly two negative eigenvalues which we denote by λ_2 and λ_3; furthermore we assume the following relations between these eigenvalues: $\lambda_2 < \lambda_1 + \lambda_3 < \lambda_1 < \lambda_3$. Then $\tau_X = \lambda_1 + \lambda_2$ is an isolated eigenvalue of H, the bottom of its essential spectrum is $\tau^X = \lambda_2$, the threshold set of H is $\tau(H) = \{\lambda_1, \lambda_2, \lambda_3, 0\}$ and $\lambda_1 + \lambda_3$ is an eigenvalue of H that is embedded in its essential spectrum. We have $0 < \lambda_1 + \lambda_3 - \lambda_2 = \widetilde{\varrho}(\lambda_1 + \lambda_3) < +\infty$.

Let H be a hamiltonian of class $C^1(D)$. Then, besides the functions $\varrho \equiv \varrho_H^D$ and $\widetilde{\varrho} \equiv \widetilde{\varrho}_H^D$, one may associate to it two other functions $\underline{\varrho} \equiv \underline{\varrho}_H^D$ and $\underline{\widetilde{\varrho}} \equiv \underline{\widetilde{\varrho}}_H^D$, defined in (7.2.5) and in the comments after (7.2.15) respectively. We have seen in Section 7.2 that one has $\underline{\varrho} = -\varrho_H^{-D}$ and $\underline{\widetilde{\varrho}} = -\widetilde{\varrho}_H^{-D}$. This allows us to calculate explicitly these new functions for (NR)-hamiltonians. Note that it is sufficient to compute $\underline{\widetilde{\varrho}}$; then $\underline{\varrho}$ is given by Theorem 7.2.13.

PROPOSITION 9.4.5. *Let the assumptions of Theorem 9.4.4 be satisfied and set $\underline{\widetilde{\varrho}} = \underline{\widetilde{\varrho}}_H^D$. Then one has for each $\lambda \in \mathbb{R}$:*

(9.4.7) $\quad \underline{\widetilde{\varrho}}(\lambda) = \sup\{\lambda - \mu \mid \mu \in \tau(H) \text{ and } \mu \leq \lambda\} = \begin{cases} -\infty & \text{if } \lambda < \tau^X, \\ \lambda - \tau^X & \text{if } \lambda \geq \tau^X, \end{cases}$

We recall the convention that the supremum over an empty set is $-\infty$.

PROOF. The proof of (9.4.7) is somewhat simpler than that of (9.4.5), but runs along the same lines. If $X = \mathbf{O}$, then $\tau^X = +\infty$ and the result is obvious. By using Example 7.2.2, Theorem 7.2.13 and the equality $\underline{\varrho}_\Delta^D = -\varrho_\Delta^{-D}$, we obtain for $X \neq \mathbf{O}$: $\underline{\varrho}_\Delta^D(\lambda) = -\infty$ if $\lambda < 0$ and $\underline{\varrho}_\Delta^D(\lambda) = \lambda$ if $\lambda \geq 0$. Hence (9.4.7) is true if H is the free hamiltonian. We adopt notations similar to those of the proof of Theorem 9.4.4, namely $\underline{\varrho}^Y = \underline{\varrho}_{H^Y}^{D^Y}$, $\underline{\widetilde{\varrho}}^Y = \underline{\widetilde{\varrho}}_{H^Y}^{D^Y}$, $\underline{\varrho}_Y = \underline{\varrho}_{H_Y}^D$, $\underline{\varrho}_Y = \underline{\varrho}_{H_Y}^D$. If $Y \neq X$, we have by Theorems 7.2.13 and 8.3.6:

(9.4.8) $\quad \underline{\widetilde{\varrho}}_Y(\lambda) = \underline{\varrho}_Y(\lambda) = \sup_{\mu \leq \lambda}[\underline{\varrho}^Y(\mu) + \lambda - \mu].$

Assume that (9.4.7) is true for all internal hamiltonians H^Y with $Y \neq X$ (this is the induction hypothesis). In other terms, if $Y \neq X$, then $\underline{\widetilde{\varrho}}^Y(\lambda) = -\infty$ if $\lambda < \tau^Y$ and $\underline{\widetilde{\varrho}}^Y(\lambda) = \lambda - \tau^Y$ if $\lambda \geq \tau^Y$. Then Theorem 7.2.13 gives the following expression for the function $\underline{\varrho}^Y$: (1) $\underline{\varrho}^Y(\lambda) = 0$ if $\lambda < \tau^Y$ and $\lambda \in \sigma_p(H^Y)$; (2)

$\varrho^Y(\lambda) = -\infty$ if $\lambda < \tau^Y$ and $\lambda \notin \sigma_p(H^Y)$; (3) $\varrho^Y(\lambda) = \lambda - \tau^Y$ if $\lambda \geq \tau^Y$. By using (9.4.8) it is quite easy now to prove that

$$\underline{\varrho}_Y(\lambda) = \sup\{\lambda - \mu \mid \mu \in \kappa(H^Y) \text{ and } \mu \leq \lambda\} = \begin{cases} -\infty & \text{if } \lambda < \tau_Y, \\ \lambda - \tau_Y & \text{if } \lambda \geq \tau_Y. \end{cases}$$

Finally, we use Theorem 8.4.3 and get

$$\widetilde{\varrho}^X = \sup_{Y \in \mathscr{L}(2)} \underline{\varrho}_Y.$$

Since $\tau^X = \min_{Y \in \mathscr{L}(2)} \tau_Y$ (cf. Theorem 9.4.2), the relation (9.4.7) is proved. $\square$

Let us state more explicitly the content of relations (9.4.5) and (9.4.7). *Assume that H is a non-relativistic N-body hamiltonian of class $C_u^1(D)$, where D is given by (1.2.19). Let $\lambda \in \sigma_{\mathrm{ess}}(H) = [\tau^X, \infty)$ and*

$$a_1 = \inf\{\lambda - \mu \mid \mu \in \tau(H), \mu \leq \lambda\}, \quad a_2 = \lambda - \tau^X.$$

Then for each couple of numbers $\delta_1, \delta_2 > 0$ there is a number $\varepsilon > 0$ and there are compact operators K_1, K_2 in $\mathscr{H}(X)$ such that

$$(a_1 - \delta_1)E(\lambda; \varepsilon) + K_1 \leq E(\lambda; \varepsilon)[H, iD]E(\lambda; \varepsilon) \leq (a_2 + \delta_2)E(\lambda; \varepsilon) + K_2.$$

Here E is the spectral measure of H and $E(\lambda; \varepsilon) = E((\lambda - \varepsilon, \lambda + \varepsilon))$. Moreover, a_1 and a_2 are the best possible constants such that the preceding statement is true.

We add a remark concerning the regularity condition we put on H in Theorem 9.4.4. *Let H be a non-relativistic N-body hamiltonian; then H is of class $C_u^1(D)$ if and only if there is $z \in \mathbb{C} \setminus \sigma(H)$ such that the sesquilinear form $[D, (H-z)^{-1}]$, with domain equal to the domain of D in $\mathscr{H}(X)$, extends to a bounded operator belonging to $\mathscr{T}$* (see also Theorem 6.2.10). Indeed, if H is of class $C_u^1(D)$, then the derivative at $\tau = 0$ of $F_\tau \equiv W(\tau)^*(H-z)^{-1}W(\tau)$ exists in norm and is equal to $[(H-z)^{-1}, iD]$. Since $F_\tau \in \mathscr{T}$ which is norm-closed, we see that $[(H-z)^{-1}, D] \in \mathscr{T}$. Reciprocally, if this holds, then F_τ is a strongly differentiable function of τ (see Section 6.2) and $F_\tau' = W(\tau)^* F_0' W(\tau)$. Since $F_0' \in \mathscr{T}$ and the group $\mathscr{W}_\tau$ restricted to $\mathscr{T}$ is of class C_0, we see that F_τ' is a norm-continuous function of τ, so F_τ is in fact norm-C^1, i.e. H is of class $C_u^1(D)$.

9.4.2. We finally present some general but easily verifiable conditions for a (NR)-hamiltonian to be of class $C_u^1(D)$ or $\mathscr{C}^{1,1}(D)$. We shall restrict ourselves to hamiltonians of the form $H = \Delta + \sum_{Y \in \mathscr{L}} V_Y$ with $V_Y = V^Y \otimes_Y I$ and V^Y a symmetric Δ^Y-bounded operator in $\mathscr{H}(Y)$. However, we shall not require that V^Y be relatively compact with respect to Δ^Y (nor that the operator $\mathscr{H}^1(Y) \to \mathscr{H}^{-1}(Y)$ induced by V^Y be compact) because this condition is locally too strong, e.g. it eliminates second order differential operators (even with coefficients of class $C_0^\infty(Y)$).

We shall replace the compactness assumptions made before (e.g. in Proposition 9.3.4) by a condition of smallness at infinity which will be introduced below (this is partly inspired by developments in [JW]). Then we make some comments concerning the definition of H and give a new criterion for an operator to be affiliated to $\mathscr{T}$ (Proposition 9.3.4 will not apply any more). It is only after these preliminary considerations that we shall turn to the question of the regularity class of H with respect to D.

LEMMA 9.4.6. *If $S : \mathscr{H}^2(X) \to \mathscr{H}(X)$ is a linear continuous operator, then the following conditions are equivalent:*

(a) *There are a number $s > 0$ and a function $\xi \in C^\infty(X)$, with $\xi(x) = 0$ near the origin and $\xi(x) = 1$ in a neighbourhood of infinity such that*

$$\lim_{r \to \infty} \|\xi(Q/r)S\|_{\mathscr{H}^2 \to \mathscr{H}^{-s}} = 0.$$

(a') *For each function $\xi \in BC^\infty(X)$ with $\xi(x) = 0$ near the origin and for each number $s > 0$ one has $\lim_{r \to \infty} \|\xi(Q/r)S\|_{\mathscr{H}^2 \to \mathscr{H}^{-s}} = 0$.*

(b) *There is $s > 0$ such that S is a compact operator from $\mathscr{H}^2(X)$ to $\mathscr{H}^{-s}(X)$.*

(b') *For each $s > 0$, the operator $S : \mathscr{H}^2(X) \to \mathscr{H}^{-s}(X)$ is compact.*

PROOF. We shall show that (a)$\Rightarrow$(a')$\Rightarrow$(b')$\Rightarrow$(b)$\Rightarrow$(a) and immediately observe that the third implication is obvious.

(i) (a)$\Rightarrow$(a'): We assume that (a) holds for some function ξ and some number s. Let $\eta \in BC^\infty(X)$ be such that $\eta(x) = 0$ near the origin, and let $t > 0$. Then clearly there are a number $\lambda > 0$ and a function $\varphi \in BC^\infty(X)$ such that $\eta(x) = \varphi(x)\xi(\lambda x) \; \forall x \in X$. Hence

$$\|\eta(Q/r)S\|_{\mathscr{H}^2 \to \mathscr{H}^{-t}} \leq \|\varphi(Q/r)\|_{\mathscr{H}^{-t} \to \mathscr{H}^{-t}} \|\xi(\lambda Q/r)S\|_{\mathscr{H}^2 \to \mathscr{H}^{-t}}$$
$$\leq C\|\xi(\lambda Q/r)S\|_{\mathscr{H}^2 \to \mathscr{H}^{-t}}$$

for some constant C independent of $r \geq 1$. So it is sufficient to show that

$$\lim_{r \to \infty} \|\xi(Q/r)S\|_{\mathscr{H}^2 \to \mathscr{H}^{-t}} = 0 \text{ for all } t > 0.$$

If $t \geq s$, this is obvious. If $0 < t < s$, then $(\mathscr{H}, \mathscr{H}^{-s})_{\theta,2} = \mathscr{H}^{-t}$ if $\theta = t/s$, and we can use Corollary 2.6.2.

(ii)(a')$\Rightarrow$(b'): The function $1 - \xi$ is of class $C_0^\infty(X)$, hence $I - \xi(Q/r)$ is a compact operator $\mathscr{H}(X) \to \mathscr{H}^{-s}(X)$ for each $s > 0$ (cf. Proposition 4.1.3); so $S_r \equiv [I - \xi(Q/r)]S : \mathscr{H}^2(X) \to \mathscr{H}^{-s}(X)$ is also a compact operator for each $r > 0$. Since $S_r \to S$ in norm in $B(\mathscr{H}^2, \mathscr{H}^{-s})$ as $r \to \infty$, we obtain (b').

(iii)(b)$\Rightarrow$(a): Let s be as in (b) and let ξ be an arbitrary function of class $BC^\infty(X)$ with $\xi(x) = 0$ in a neighbourhood of the origin. Then the operator $\xi(Q/r)$ converges strongly to zero in $\mathscr{H}^{-s}(X)$ as $r \to \infty$. Hence for any compact operator $S : \mathscr{H}^2(X) \to \mathscr{H}^{-s}(X)$ we shall have $\|\xi(Q/r)S\|_{\mathscr{H}^2 \to \mathscr{H}^{-s}} \to 0$ as $r \to \infty$. $\square$

DEFINITION 9.4.7. A linear continuous operator $S : \mathscr{H}^2(X) \to \mathscr{H}(X)$ satisfying the conditions of the preceding lemma is called *small at infinity*.

Each compact operator from $\mathscr{H}^2(X)$ to $\mathscr{H}(X)$ is small at infinity. But, although a second order differential operator is never compact as an operator from $\mathscr{H}^2(X)$ to $\mathscr{H}(X)$ (unless it is of lower order), it could be small at infinity. For example, if $S = \sum a_{jk}(Q)P_jP_k$ (derivatives with respect to a basis of X) with

$a_{jk} \in C_\infty(X)$, then S satisfies (a) with $s = 0$, so is small at infinity when considered as an operator from $\mathscr{H}^2(X)$ to $\mathscr{H}(X)$. We mention that the condition (b') of the lemma does not imply that $\lim_{r\to\infty} ||\xi(Q/r)S||_{\mathscr{H}^2\to\mathscr{H}} = 0$ (indeed, let $X = \mathbb{R}$ and $S = a(Q)(P^2 + I)$ with $a \in L^\infty(\mathbb{R}) \cap L^2(\mathbb{R})$; then $S : \mathscr{H}^2(\mathbb{R}) \to \mathscr{H}^{-2}(\mathbb{R})$ is a Hilbert-Schmidt operator and $||\xi(Q/r)S||_{\mathscr{H}^2\to\mathscr{H}} = \operatorname{ess\,sup}_{x\in\mathbb{R}} |\xi(x/r)a(x)|$ does not tend to zero in general as $r \to \infty$).

Let Y be a subspace of X and $V^Y : \mathscr{H}^2(Y) \to \mathscr{H}(Y)$ a linear continuous operator. Then the operator $V_Y = V^Y \otimes I : \mathscr{H}^2(Y) \otimes \mathscr{H}(Y^\perp) \to \mathscr{H}(Y) \otimes \mathscr{H}(Y^\perp)$ will be interpreted as a densely defined operator in $\mathscr{H}(X)$, which is possible due to the natural dense embedding $\mathscr{H}^2(Y) \otimes \mathscr{H}(Y^\perp) \subset \mathscr{H}(X)$ and to the canonical identification of $\mathscr{H}(Y) \otimes \mathscr{H}(Y^\perp)$ with $\mathscr{H}(X)$ (see Section 9.1). According to the rules of Section 9.1, we have to set $V_Y = V^Y \otimes_Y I$ when V_Y is considered as operator in $\mathscr{H}(X)$. One gets a very convenient expression for the operator V_Y when working in the representation $\mathscr{H}(X) = L^2(Y^\perp; \mathscr{H}(Y))$. Then the domain of V_Y is $L^2(Y^\perp; \mathscr{H}^2(Y))$ and we have $(V_Y f)(y') = V^Y f(y')$ for $f \in L^2(Y^\perp; \mathscr{H}^2(Y))$ and $y' \in Y^\perp$.

LEMMA 9.4.8. *With the preceding hypotheses and notations, one has* $D(V_Y) \supset \mathscr{H}^2(X)$. *Furthermore, if a and b are positive numbers such that* $||V^Y g||^2_{\mathscr{H}(Y)} \le a||\Delta^Y g||^2_{\mathscr{H}(Y)} + b||g||^2_{\mathscr{H}(Y)}$ *for all $g \in \mathscr{H}^2(Y)$, then* $||V_Y f||^2_{\mathscr{H}(X)} \le a||\Delta f||^2_{\mathscr{H}(X)} + b||f||^2_{\mathscr{H}(X)}$ *for all $f \in \mathscr{H}^2(X)$.*

PROOF. As in the proof of Proposition 9.3.4, it is useful to work in the representation $\mathscr{H}(X) = L^2(Y^\perp; \mathscr{H}(Y))$ and to make a Fourier transformation in the variable $y' \in Y^\perp$. Then the operator V_Y will have the same expression as above, while the operator Δ, with domain $\mathscr{H}^2(X)$, becomes the operator of multiplication by the operator valued-function $y' \mapsto \Delta^Y + |y'|^2$ with domain equal to the set of $f \in L^2(Y^\perp; \mathscr{H}(Y))$ such that $\int_{Y^\perp} || |y'|^2 f(y')||^2_{\mathscr{H}(Y)} dy' < \infty$. Now it is easy to check the assertions of the lemma. $\square$

We now choose, once and for all, a family of operators $\{V^Y\}_{Y\in\mathscr{L}}$ with the following properties:

(i) V^Y is a symmetric operator in $\mathscr{H}(Y)$ with domain equal to $\mathscr{H}^2(Y)$;

(ii) if $\alpha(Y)$ denotes the operator bound of V^Y with respect to Δ^Y, then $\sum_{Y\in\mathscr{L}} \alpha(Y) < 1$;

(iii) $V^O = 0$.

Further conditions will be imposed on V^Y later on. If we set $V_Y = V^Y \otimes_Y I$, then by Lemma 9.4.8 each V_Y is a symmetric operator in $\mathscr{H}(X)$ with domain containing $\mathscr{H}^2(X)$, and the operator bound of V_Y with respect to Δ is $\alpha(Y)$. Hence the operator $H = \Delta + \sum_{Y\in\mathscr{L}} V_Y$ is self-adjoint on $\mathscr{H}^2(X)$ in $\mathscr{H}(X)$ (by the Kato-Rellich theorem, since $\sum_{Y\in\mathscr{L}} \alpha(Y) < 1$). This is the operator that will be studied in the remainder of this section. For each $Y \in \mathscr{L}$ we set $H_Y = \Delta + \sum_{Z\in\mathscr{L}, Z\le Y} V_Z$; then H_Y is also a self-adjoint operator in $\mathscr{H}(X)$ with domain $\mathscr{H}^2(X)$.

PROPOSITION 9.4.9. *Assume that for each $Y \in \mathscr{L}$ the operator $V^Y : \mathscr{H}^2(Y) \to \mathscr{H}(Y)$ is small at infinity. Then the self-adjoint operator H is affiliated to $\mathscr{T}$ and $\mathscr{R}_Y[H] = H_Y$ for all $Y \in \mathscr{L}$.*

PROOF. (i) For each $\varepsilon > 0$ we set $V^{\varepsilon Y} = (I + \varepsilon \Delta^Y)^{-1} V^Y (I + \varepsilon \Delta^Y)^{-1}$. Since $V^Y : \mathscr{H}^2(Y) \to \mathscr{H}^{-2}(Y)$ is compact, $V^{\varepsilon Y}$ is a compact symmetric operator in $\mathscr{H}(Y)$ and we have $\lim_{\varepsilon \to 0} V^{\varepsilon Y} = V^Y$ in norm in $B(\mathscr{H}^2(Y), \mathscr{H}^{-2}(Y))$ (note that $(I + \varepsilon \Delta^Y)^{-1}$ converges strongly to I in $B(\mathscr{H}^{\pm 2}(Y))$ as $\varepsilon \to 0$). We set $V_Y^\varepsilon = V^{\varepsilon Y} \otimes_Y I$, $V^\varepsilon = \sum_{Y \in \mathscr{L}} V_Y^\varepsilon$ and $H^\varepsilon = \Delta + V^\varepsilon$, so that H^ε is a self-adjoint operator in $\mathscr{H}(X)$ with domain $\mathscr{H}^2(X)$.

Observe that for each $\lambda > 0$ we have

$$(9.4.9) \qquad \|[(\lambda + \Delta^Y) \otimes_Y I](\lambda + \Delta)^{-1}\|_{\mathbb{B}(X)} \le 1.$$

Since $\|V^Y (\lambda + \Delta^Y)^{-1}\|_{\mathbb{B}(Y)} \to \alpha(Y)$ as $\lambda \to \infty$, one sees that for each $\nu > 0$ there is $\lambda_\nu > 0$ such that, if $\lambda \ge \lambda_\nu$, then

$$\|(\lambda + \Delta)^{-1} V^\varepsilon\|_{\mathbb{B}(X)} \le \sum_{Y \in \mathscr{L}} \|(\lambda + \Delta^Y)^{-1} V^{\varepsilon Y}\|_{\mathbb{B}(Y)} \le \sum_{Y \in \mathscr{L}} (\alpha(Y) + \nu).$$

By choosing ν small enough it follows that there are numbers $\alpha < 1$ and $\lambda > 0$ such that $\|(\lambda + \Delta)^{-1} V^\varepsilon\|_{\mathbb{B}(X)} \le \alpha$ for all $\varepsilon > 0$. Since λ can be choosen as large as we wish we may assume that $-\lambda$ belongs to the resolvent set of H and of each H^ε. So, if we set $V = \sum_{Y \in \mathscr{L}} V_Y$, we obtain:

$$(\lambda + H^\varepsilon)^{-1} - (\lambda + H)^{-1} = (\lambda + H^\varepsilon)^{-1} (V - V^\varepsilon)(\lambda + H)^{-1}$$
$$= [I + (\lambda + \Delta)^{-1} V^\varepsilon]^{-1} \cdot (\lambda + \Delta)^{-1} (V - V^\varepsilon)(\lambda + \Delta)^{-1} \cdot (\lambda + \Delta)(\lambda + H)^{-1}.$$

Now it is easy to prove that $(\lambda + H^\varepsilon)^{-1} \to (\lambda + H)^{-1}$ in norm in $\mathbb{B}(X)$ as $\varepsilon \to 0$. Indeed, by using (9.4.9) again we have for some finite constants C_1, C_2:

$$\|(\lambda + H^\varepsilon)^{-1} - (\lambda + H)^{-1}\|_{\mathbb{B}(X)} \le$$
$$\le (1 - \alpha)^{-1} \|(\lambda + \Delta)^{-1}(V - V^\varepsilon)(\lambda + \Delta)^{-1}\|_{\mathbb{B}(X)} \cdot \|(\lambda + \Delta)(\lambda + H)^{-1}\|_{\mathbb{B}(X)}$$
$$\le C_1 \sum_{Y \in \mathscr{L}} \|(\lambda + \Delta^Y)^{-1}(V^Y - V^{\varepsilon Y})(\lambda + \Delta^Y)^{-1}\|_{\mathbb{B}(Y)}$$
$$\le C_2 \sum_{Y \in \mathscr{L}} \|V^Y - V^{\varepsilon Y}\|_{\mathscr{H}^2(Y) \to \mathscr{H}^{-2}(Y)} \to 0 \text{ as } \varepsilon \to 0.$$

(ii) By an elementary particular case of Proposition 9.3.4, the operator H^ε is affiliated to $\mathscr{T}$ and $\mathscr{R}_Y[H^\varepsilon] = H_Y^\varepsilon \equiv \Delta + \sum_{Z \in \mathscr{L}, Z \le Y} V_Z^\varepsilon$. By the result established in step (i) we have $\lim_{\varepsilon \to 0} H_Y^\varepsilon = H_Y$ in the norm resolvent sense (cf. the end of §8.1.2) for each $Y \in \mathscr{L}$. This implies that H is affiliated to $\mathscr{T}$ and that $\mathscr{R}_Y[H] = H_Y$ for $Y \in \mathscr{L}$. $\square$

By taking ino account the factorization property of the dilation group $\{W(\tau)\}$ and the fact that $W(\tau)$ leaves $\mathscr{H}^2(X)$ invariant, one sees that the operators $W(\tau) V_Y W(-\tau)$ and $[W^Y(\tau) V^Y W^Y(-\tau)] \otimes_Y I$ have the same restriction to $\mathscr{H}^2(X)$. It follows that V_Y is of class $C^1(D; \mathscr{H}^2(X), \mathscr{H}^{-2}(X))$ if V^Y is of class $C^1(D^Y; \mathscr{H}^2(Y), \mathscr{H}^{-2}(Y))$, and then $[D, V_Y] = [D^Y, V^Y] \otimes_Y I$. On the other hand, if V^Y is an arbitrary symmetric operator in $B(\mathscr{H}^2(Y), \mathscr{H}(Y))$, the commutator $[D^Y, V^Y]$ is a well defined element of $B(\mathscr{H}_1^2(Y), \mathscr{H}_{-1}^{-2}(Y))$, and V^Y is of class $C^1(D^Y; \mathscr{H}^2(Y), \mathscr{H}^{-2}(Y))$ if and only if $[D^Y, V^Y] \in B(\mathscr{H}^2(Y), \mathscr{H}^{-2}(Y))$

(this is a straightforward consequence of (6.3.23); the details of the proof are left to the reader).

THEOREM 9.4.10. *Assume that for each $Y \in \mathscr{L}$ the operator $V^Y : \mathscr{H}^2(Y) \to \mathscr{H}(Y)$ is small at infinity and that $[D^Y, V^Y]$ is a compact operator $\mathscr{H}^2(Y) \to \mathscr{H}^{-2}(Y)$. Then H is a* (NR)-*hamiltonian of class $C_u^1(D)$. In particular $\tau(H)$ is a closed countable set, the eigenvalues of H outside $\tau(H)$ are of finite multiplicity and do not have accumulation points outside $\tau(H)$, and D is locally conjugate to H on $\mathbb{R} \setminus \tau(H)$.*

PROOF. The fact that H is a (NR)-hamiltonian is a straightforward consequence of Proposition 9.4.9. It remains to be shown that H is of class $C_u^1(D)$, since the other assertions of the theorem will then follow from Theorem 9.4.4. According to the comments made before the statement of the theorem, and by taking into account the equality $[\Delta, iD] = \Delta$, we have $H \in C^1(D; \mathscr{H}^2(X), \mathscr{H}^{-2}(X))$. Then it follows from Theorem 6.3.4 (a) that H is of class $C^1(D)$ and that for $R = (H - z)^{-1}$, $z \notin \sigma(H)$, we have

$$[iD, R] = R[H, iD]R = R + zR^2 + \sum_{Y \in \mathscr{L}} R([V_Y, iD] - V_Y)R.$$

The property $R \in C_u^1(D; \mathscr{H})$ is equivalent to $[iD, R] \in \mathscr{T}$ (cf. the remarks that follow Proposition 9.4.5). Since $R \in \mathscr{T}$, it suffices to show that $R U_Y R \in \mathscr{T}$ for each Y, where $U_Y = U^Y \otimes_Y I$ and $U^Y = [V^Y, iD^Y] - V^Y$. The operator $U^Y : \mathscr{H}^2(Y) \to \mathscr{H}^{-2}(Y)$ is compact, so $U^Y = \lim_{\varepsilon \to 0} U^{\varepsilon Y}$ in norm in $B(\mathscr{H}^2(Y), \mathscr{H}^{-2}(Y))$, with $U^{\varepsilon Y} = (I + \varepsilon \Delta^Y)^{-1} U^Y (I + \varepsilon \Delta^Y)^{-1}$ a compact symmetric operator in $\mathscr{H}(Y)$. If we set $U_Y^\varepsilon = U^{\varepsilon Y} \otimes_Y I$, then we shall have $\lim_{\varepsilon \to 0} U_Y^\varepsilon = U_Y$ in norm in $B(\mathscr{H}^2(X), \mathscr{H}^{-2}(X))$, hence $\lim_{\varepsilon \to 0} \|R U_Y^\varepsilon R - R U_Y R\|_{\mathbb{B}(X)} = 0$ (because R is a continuous operator $\mathscr{H}(X) \to \mathscr{H}^2(X)$ and $\mathscr{H}^{-2}(X) \to \mathscr{H}(X)$). Since $\mathscr{T}$ is a norm closed subalgebra of $\mathbb{B}(X)$, the proof of the theorem will be finished once we have shown that $U_Y^\varepsilon R \in \mathscr{T}$.

We prove more generally that $K \otimes_Y I \cdot T \in \mathscr{T}(Y + Z)$ if $K \in \mathbb{K}(Y)$ and $T \in \mathscr{T}(Z)$. Now $\mathscr{T}(Y + Z)$ is a closed subspace of $\mathbb{B}(X)$. So, by the definition (9.2.14), it suffices to consider $T = L \otimes_Z S$ with $L \in \mathbb{K}(Z)$ and $S \in \mathbf{T}_\infty(Z^\perp)$. Then, if we set $E = Y + Z$, we have

$$K \otimes_Y I \cdot L \otimes_Z S = [K \otimes_Y I \cdot L \otimes_Z I] \cdot I \otimes_Z S = M \otimes_E I \cdot I \otimes_Z S$$

for some $M \in \mathbb{K}(E)$ (see the example (a) after Corollary 9.2.5). Since $Z \subset E$ we may write $Z^\perp = E^\perp \oplus F$ with F a subspace of E. So S is a norm limit of linear combinations of operators of the form $S_1 \otimes_F^{Z^\perp} S_2$ with $S_1 \in \mathbf{T}_\infty(F)$ and $S_2 \in \mathbf{T}_\infty(E^\perp)$. But

$$M \otimes_E^X I \cdot I \otimes_Z^X (S_1 \otimes_F^{Z^\perp} S_2) = M \otimes_E^X I \cdot [(S_1 \otimes_F^E I) \otimes_E^X S_2]$$
$$= [M \cdot (S_1 \otimes_F^E I)] \otimes_E^X S_2 \in \mathscr{T}(E). \quad \square$$

We shall now point out several sufficient conditions on the potentials V^Y implying that the hamiltonian H is of class $\mathscr{C}^{1,1}(D)$. Since the domain of H is invariant under the dilation group, we see from Theorem 6.3.4 (b) that H is of class $\mathscr{C}^{1,1}(D)$ if and only if $H \in \mathscr{C}^{1,1}(D; \mathscr{H}^2(X), \mathscr{H}^{-2}(X))$. The operator Δ is of class $C^\infty(D; \mathscr{H}^2(X), \mathscr{H}(X))$, so H is of class $\mathscr{C}^{1,1}(D)$ if for each $Y \in \mathscr{L}$ one has $V_Y \in \mathscr{C}^{1,1}(D; \mathscr{H}^2(X), \mathscr{H}^{-2}(X))$. On the other hand we clearly have

$$\|W(\tau)V_Y W(-\tau) + W(-\tau)V_Y W(\tau) - 2V_Y\|_{\mathscr{H}^2(X) \to \mathscr{H}^{-2}(X)} \le$$
$$\le \|W^Y(\tau)V^Y W^Y(-\tau) + W^Y(-\tau)V^Y W^Y(\tau) - 2V^Y\|_{\mathscr{H}^2(Y) \to \mathscr{H}^{-2}(Y)}.$$

In conclusion, if V^Y belongs to $\mathscr{C}^{1,1}(D^Y; \mathscr{H}^2(Y), \mathscr{H}^{-2}(Y))$ for each $Y \in \mathscr{L}$, then H is of class $\mathscr{C}^{1,1}(D)$. So it is sufficient to find some simple but efficient criteria for a symmetric operator in $B(\mathscr{H}^2(Y), \mathscr{H}(Y))$ to be of class $\mathscr{C}^{1,1}(D^Y; \mathscr{H}^2(Y), \mathscr{H}^{-2}(Y))$. Since X is an arbitrary euclidean space it suffices to study this question for the case $Y = X$. In order to simplify the notations, we shall not specify explicitly the space X, e.g. we set $\mathscr{H}^s \equiv \mathscr{H}^s(X)$; moreover, we sometimes denote the norm in $B(\mathscr{H}^s(X), \mathscr{H}^t(X))$ by $\|\cdot\|_{s,t}$. The discussion that follows is parallel to that of §7.5.3.

Consider first an arbitrary operator $S \in B(\mathscr{H}^2, \mathscr{H}^{-2})$. Then the simplest way of checking that $S \in \mathscr{C}^{1,1}(D; \mathscr{H}^2, \mathscr{H}^{-2})$ is by showing that S belongs to $C^2(D; \mathscr{H}^2, \mathscr{H}^{-2})$. Note that $[D, [D, S]]$ is always a well defined continuous operator $\mathscr{S}(X) \to \mathscr{S}^*(X)$ or equivalently a continuous sesquilinear form on $\mathscr{S}(X)$. It is now a standard matter to show that S is of class $C^2(D; \mathscr{H}^2, \mathscr{H}^{-2})$ if and only if $[D, [D, S]] \in B(\mathscr{H}^2, \mathscr{H}^{-2})$ (see our notational conventions in Section 2.1) or equivalently if and only if the sesquilinear form $[D, [D, S]]$ is continuous for the topology induced by $\mathscr{H}^2(X)$ on $\mathscr{S}(X)$.

A more general assumption on S which ensures that $S \in \mathscr{C}^{1,1}(D; \mathscr{H}^2, \mathscr{H}^{-2})$ is $S \in C^{1+0}(D; \mathscr{H}^2, \mathscr{H}^{-2})$, i.e. $[D, S] \in B(\mathscr{H}^2, \mathscr{H}^{-2})$ and

$$\int_0^1 \|W(\tau)[D, S]W(-\tau) - [D, S]\|_{2,-2}\, \tau^{-1} d\tau < \infty.$$

As we explained in §7.5.3, the last integrability condition is satisfied if $S : \mathscr{H}^2 \to \mathscr{H}^{-2}$ is a symmetric operator and

$$(9.4.10) \qquad \int_{r_0}^\infty \|D(D + ir)^{-1}[D, S]\|_{2,-2}\, r^{-1} dr < \infty$$

for some $r_0 < \infty$ (see part (i) of the proof of Proposition 7.5.7). (9.4.10) is clearly satisfied if the range of the operator $[D, S]$ (defined on $\mathscr{H}^2$) is included in an interpolation space $(\mathscr{K}, \mathscr{H}^{-2})_{\theta,p}$ with $0 < \theta < 1$, where $\mathscr{K}$ is the domain of D in $\mathscr{H}^{-2}$. Since $\mathscr{K} \supset \mathscr{H}_1^{-1}$, we see that (9.4.10) follows from $[D, S]\mathscr{H}^2 \subset (\mathscr{H}_1^{-1}, \mathscr{H}^{-2})_{\theta,2}$ for some $\theta < 1$. Then, by using the relation $(0 < \theta < 1)$

$$(\mathscr{H}_1^{-1}, \mathscr{H}^{-2})_{\theta,2} \supset (\mathscr{H}_1^{-1}, \mathscr{H}^{-1})_{\theta,2} = \mathscr{H}_{1-\theta,2}^{-1} = \mathscr{H}_{1-\theta}^{-1}$$

one finds that (9.4.10) follows from $[D, S]\mathscr{H}^2 \subset \mathscr{H}_\varepsilon^{-1}$ for some $\varepsilon > 0$ (in fact a more careful interpolation argument shows that $[D, S]\mathscr{H}^2 \subset \mathscr{H}_\varepsilon^{-2+\varepsilon}$ would be sufficient). The next proposition improves this result.

PROPOSITION 9.4.11. *Let* $S : \mathscr{H}^2(X) \to \mathscr{H}^{-2}(X)$ *be a symmetric operator such that* $[D, S] \in B(\mathscr{H}^2(X), \mathscr{H}^{-1}(X))$. *Assume that there is a function* $\xi \in$

$C^\infty(X)$ *with* $\xi(x) = 0$ *near the origin and* $\xi(x) = 1$ *in a neighbourhood of infinity,*
such that

$$(9.4.11) \qquad \int_1^\infty \|\xi(Q/r)[D,S]\|_{\mathscr{H}^2(X)\to\mathscr{H}^{-1}(X)}\, r^{-1}dr < \infty.$$

Then S is of class $\mathscr{C}^{1,1}(D;\mathscr{H}^2(X),\mathscr{H}^{-2}(X))$.

PROOF. The proof is almost identical with parts (ii) and (iii) of the proof of
Proposition 7.5.7. We take $\Lambda = \langle Q \rangle$ and set $D_r = D(D+ir)^{-1}$, $\Lambda_r = \Lambda(\Lambda+r)^{-1}$,
where r is a large positive number. Then

$$(9.4.12) \\ D_r = [D_r + r(D+ir)^{-1}\cdot D\Lambda^{-1}]\Lambda_r = \Lambda_r[D_r + \Lambda^{-1}D\cdot r(D+ir)^{-1}]$$

and $D\Lambda^{-1} \in B(\mathscr{H}^{-1},\mathscr{H}^{-2})$ (this is the only difference with respect to the con-
ditions of Proposition 7.5.7). Hence $\|D_r[D,S]\|_{2,-2} \le c\|\Lambda_r[D,S]\|_{2,-1}$ for some
constant c and all sufficiently large r. So (9.4.10) follows from

$$\int_1^\infty \|\Lambda_r[D,S]\|_{2,-1}\, r^{-1}dr < \infty$$

which in turn is a consequence of $\int_1^\infty \|\eta(\Lambda/r)[D,S]\|_{2,-1}\, r^{-1}dr < \infty$ if η is a
function in $C^\infty(\mathbb{R})$ such that $\eta(t) = 0$ near the origin and $\eta(t) = 1$ for large $|t|$
(use (3.4.25) and Lemma 3.5.12 for example). To finish the proof note that we
may choose η such that $\xi(x) = \xi(x)\eta(\langle x\rangle)\,\forall x \in X$. $\square$

The last criterion that we shall discuss is the analogue of Theorem 7.5.8. Note
first that, according to (7.5.28), if $S : \mathscr{H}^2 \to \mathscr{H}^{-2}$ is a symmetric operator and
D_r has the same meaning as above and if

$$(9.4.13) \qquad \int_{r_0}^\infty \Big[\|D_r^2 S\|_{2,-2} + \|D_r S D_r\|_{2,-2}\Big]dr < \infty$$

for some finite r_0, then $S \in \mathscr{C}^{1,1}(D;\mathscr{H}^2,\mathscr{H}^{-2})$. Now let $\mathscr{K}$ be the domain of D^2
in $\mathscr{H}^{-2}$. We have $\mathscr{K} \supset \mathscr{H}_2$ and one may show without difficulty that (9.4.13)
is a consequence of $S\mathscr{H}^2 \subset (\mathscr{H}_2,\mathscr{H}^{-2})_{\theta,2}$ for some $\theta < 1/2$. Since $\mathscr{H}^{-2} \supset \mathscr{H}$
and $(\mathscr{H}_2,\mathscr{H})_{\theta,2} = \mathscr{H}_{2(1-\theta),2}$, we see that it is sufficient to have $S\mathscr{H}^2 \subset \mathscr{H}_{1+\varepsilon}$ for
some $\varepsilon > 0$ (it is more difficult to show that $S\mathscr{H}^2 \subset \mathscr{H}_{1+\varepsilon}^{-1+\varepsilon}$ for some $\varepsilon > 0$ is
still sufficient). We now improve this.

PROPOSITION 9.4.12. *Let $S : \mathscr{H}^2(X) \to \mathscr{H}(X)$ be a symmetric operator.*
Assume that there is a function $\theta \in C_0^\infty(X \setminus \{0\})$ *with* $\theta(x) \ne 0$ *for* $1 \le |x| \le 2$
and such that

$$(9.4.14) \qquad \int_1^\infty \|\theta(Q/r)S\|_{\mathscr{H}^2(X)\to\mathscr{H}(X)}dr < \infty.$$

Then $S \in \mathscr{C}^{1,1}(D;\mathscr{H}^2(X),\mathscr{H}^{-2}(X))$.

PROOF. This is essentially a repetition of the proof of Theorem 7.5.8. We use the same notations as in the proof of Proposition 9.4.11, the identities (9.4.12) and

$$D_r^2 = \left\{ D_r^2 + 2r(D+ir)^{-1} \cdot D_r \cdot D\Lambda^{-1} + [r(D+ir)^{-1}]^2 D^2\Lambda^{-2} \right\}\Lambda_r^2.$$

By taking into account the properties $D\Lambda^{-1}$, $\Lambda^{-1}D \in B(\mathscr{H}^s, \mathscr{H}^{s-1})$ and $D^2\Lambda^{-2} \in B(\mathscr{H}^s, \mathscr{H}^{s-2})$ for all real s, we then see that (9.4.13) is a consequence of

$$\int_1^\infty [\|\Lambda_r^2 S\|_{2,0} + \|\Lambda_r S\Lambda_r\|_{1,-1}]dr < \infty.$$

The contribution of the first term above is dominated by the integral (9.4.14), cf. part (v) of the proof of Theorem 7.5.8. So it is sufficient to show that there is a finite constant C such that for all $r \geq 0$

(9.4.15) $\|\Lambda_r S\Lambda_r\|_{1,-1} \leq C\|\Lambda_r^2 S\|_{2,0}.$

Observe that the norm on the l.h.s. is equal to $\|\langle P\rangle^{-1}\Lambda_r S\Lambda_r\langle P\rangle^{-1}\|$ while that on the r.h.s. is $\|\Lambda_r^2 S\langle P\rangle^{-2}\| \equiv \|\langle P\rangle^{-2}S\Lambda_r^2\|$. So it is natural to consider the family of bounded operators in $\mathscr{H}$ defined by $G(z) = \langle P\rangle^{-z}\Lambda_r^{2-z}S\Lambda_r^z\langle P\rangle^{z-2}$ for $0 \leq \Re z \leq 2$. Indeed, we have $G(0) = \Lambda_r^2 S\langle P\rangle^{-2}$, $G(1) = \langle P\rangle^{-1}\Lambda_r S\Lambda_r\langle P\rangle^{-1}$ and $G(2) = \langle P\rangle^{-2}S\Lambda_r^2$. Now (9.4.15) follows by the analytic interpolation argument described in part (iv) of the proof of Theorem 7.5.8. In fact the situation here is much simpler since the operators $\Lambda_r = \langle Q\rangle(\langle Q\rangle + r)^{-1}$ are quite explicit and easy to manipulate. For example, note that if $z = x + iy$ with $0 \leq x \leq 2$:

$$\|G(z)\| = \|\langle P\rangle^{-z}\Lambda_r^{2-z}S\Lambda_r^z\langle P\rangle^{x-2}\|$$

$$\leq \|\Lambda_r^{2-z}\|_{-x,-x}\|S\|_{2-x,-x}\|\Lambda_r^z\|_{2-x,2-x},$$

and for $-2 \leq s \leq 2$ we have $\|u(Q)\|_{s,s} \leq C\|u\|_{BC^2(X)}$. □

REMARK 9.4.13. It is unfortunately impossible to replace the function ξ in (9.4.11) by a function θ with the properties required in Proposition 9.4.12. On the other hand, if (9.4.14) holds, then we shall also have the apparently stronger property: $\int_1^\infty \|\xi(Q/r)S\|_{\mathscr{H}^2(X)\to\mathscr{H}(X)}dr < \infty$ for any bounded Borel function ξ such that $\xi(x) = 0$ in a neighbourhood of zero. Indeed, it is sufficient to prove this under the supplementary assumptions $\xi \in C^\infty(X)$ and $\xi(x) = 1$ for large $|x|$, and then we may use the argument of Remark 7.6.9.

REMARK 9.4.14. If S is as in Proposition 9.4.12, then $S : \mathscr{H}^2(X) \to \mathscr{H}(X)$ is small at infinity. This is an immediate consequence of the preceding remark.

The classes of potentials isolated by the three criteria studied above are described in the next definition.

DEFINITION 9.4.15. Let $U : \mathscr{H}^2(X) \to \mathscr{H}(X)$ be a linear symmetric operator.
   (i) We say that U is a *Mourre potential*, and we write $U \in \mathrm{M}(X)$, if $[D, [D, U]]$ belongs to $B(\mathscr{H}^2(X), \mathscr{H}^{-2}(X))$ (i.e. the symmetric sesquilinear form $[D, [D, U]]$ with domain $\mathscr{S}(X)$ is continuous for the topology induced by $\mathscr{H}^2(X)$ on $\mathscr{S}(X)$).

(ii) We say that U is a *long-range potential*, and we write $U \in \mathrm{LR}(X)$, if $[D, U] \in B(\mathcal{H}^2(X), \mathcal{H}^{-1}(X))$ and there is a function $\xi \in C^\infty(X)$ with $\xi(x) = 0$ if $|x| \leq 1$ and $\xi(x) = 1$ if $|x| \geq 2$ such that

$$\int_1^\infty \|\xi(Q/r)[D, U]\|_{\mathcal{H}^2(X) \to \mathcal{H}^{-1}(X)} \, r^{-1} dr < \infty.$$

(iii) We say that U is a *short-range potential*, and we write $U \in \mathrm{SR}(X)$, if

$$\int_1^\infty \|\chi_r(Q)U\|_{\mathcal{H}^2(X) \to \mathcal{H}(X)} \, dr < \infty,$$

where χ_r is the characteristic function of the annulus $r \leq |x| \leq 2r$.

If U is a differential operator (necessarily of order ≤ 2), then explicit conditions on the coefficients of U ensuring that U is a long-range or short-range potential may be easily deduced from the results presented in the last part of Section 1.3. Observe that, if U is the operator of multiplication by a function u and U belongs to $B(\mathcal{H}^2(X), \mathcal{H}(X))$, then $u \in L^2_{\mathrm{loc}}(X)$ and $2[iD, U]$ is the operator of multiplication by the distribution $r\partial_r u(x)$ (here $r = |x|$ and ∂_r is the radial derivative).

We are now able to describe explicitly a large class of (NR)-hamiltonians for which one can make a detailed spectral analysis by the conjugate operator method.

DEFINITION 9.4.16. Let X be an euclidean space and $\mathscr{L}$ a finite family of subspaces of X such that $\mathbf{O}, X \in \mathscr{L}$ and $Y, Z \in \mathscr{L} \Rightarrow Y + Z \in \mathscr{L}$. An operator $H \in \mathcal{H}(X)$ is called an *admissible hamiltonian* if H has the form $H = \Delta + \sum_{Y \in \mathscr{L}} V_Y$ with $V_{\mathbf{O}} = 0$, $V_Y = V^Y \otimes_Y I$ and the operators V^Y have the following properties

(a) $V^Y : \mathcal{H}^2(Y) \to \mathcal{H}(Y)$ is symmetric and small at infinity;
(b) the operator bound $\alpha(Y)$ of V^Y with respect to Δ^Y is such that

$$\sum_{Y \in \mathscr{L}} \alpha(Y) < 1.$$

(c) One may write $V^Y = V_M^Y + V_L^Y + V_S^Y$ with $V_M^Y \in \mathrm{M}(Y)$, $V_L^Y \in \mathrm{LR}(Y)$ and $V_S^Y \in \mathrm{SR}(Y)$.

Proposition 9.4.9 implies that *an admissible hamiltonian is a (NR)-hamiltonian (with respect to the lattice $\mathscr{L}$) and has domain $\mathcal{H}^2(X)$.*

Moreover, according to the study made before, *an admissible hamiltonian is of class $\mathscr{C}^{1,1}(D; \mathcal{H}^2(X), \mathcal{H}^{-2}(X))$, hence of classes $\mathscr{C}^{1,1}(D)$ and $C_u^1(D)$* (Theorem 9.4.10 is not needed in the present context).

So if H is an admissible hamiltonian, we may apply Theorems 9.4.4 and 7.5.2 with $\mathscr{G} = \mathcal{H}^2(X)$ and $A = D$. Since $\mathscr{G}^* = \mathcal{H}^{-2}$, $\mathscr{G}^{-1/2,1} = \mathcal{H}^{-1,1}$ and $(D(A; \mathscr{G}^*), \mathscr{G}^*)_{1/2,1} \supset (\mathcal{H}_1^{-1}, \mathcal{H}^{-2})_{1/2,1} \supset (\mathcal{H}_1^{-1}, \mathcal{H}^{-1})_{1/2,1} = \mathcal{H}_{1/2,1}^{-1}$, it follows from (7.5.1) that *the strong limiting absorption principle for H holds locally on $\mu^D(H) \equiv \mathbb{R} \setminus \kappa(H)$ in the Banach space $\mathcal{H}^{-1,1}(X) \cap \mathcal{H}_{1/2,1}^{-1}(X)$.* This space is slightly smaller than $\mathcal{H}_{1/2,1}^{-1}(X)$ but it can be replaced by $\mathcal{H}_{1/2,1}^{-1}(X)$ without essential loss of generality, as we shall see below. We first state a simple and general consequence of the preceding results.

THEOREM 9.4.17. *Let H be an admissible hamiltonian. Then $\tau(H)$ is a closed countable set, the eigenvalues of H outside $\tau(H)$ are of finite multiplicity and can accumulate only at points belonging to $\tau(H)$, and H has no singularly continuous spectrum. The limits $\lim_{\mu \to \pm 0}(H - \lambda - i\mu)^{-1}$ exist in the weak* topology of $B(\mathscr{H}_{1/2,1}(X), \mathscr{H}_{-1/2,\infty}(X))$, uniformly in λ on each compact subset of $\mathbb{R} \setminus \kappa(H)$, where $\kappa(H) = \tau(H) \cup \sigma_p(H)$ is a closed countable real set. Finally, if $\mathscr{K}$ is a Hilbert space and $T : \mathscr{H}(X) \to \mathscr{K}$ is a linear operator which is continuous when $\mathscr{H}(X)$ is equipped with the topology induced by $\mathscr{H}_{-1/2,\infty}(X)$, then T is locally H-smooth on $\mathbb{R} \setminus \kappa(H)$.*

COROLLARY 9.4.18. *Let H_1, H_2 be two admissible hamiltonians (with respect to different lattices $\mathscr{L}_1, \mathscr{L}_2$ in general) with spectral measures E_1, E_2. Set $J = \mathbb{R} \setminus (\kappa(H_1) \cup \kappa(H_2))$. If $H_1 - H_2 \in B(\overset{\circ}{\mathscr{H}}_{-1/2,\infty}(X), \mathscr{H}_{1/2,1}(X))$, then the operators s-$\lim_{t \to \pm 0} \exp(iH_2 t) \exp(-iH_1 t) E_1(J)$ exist and have ranges equal to $E_2(J)\mathscr{H}$.*

We now give conditions under which the Banach space in which the strong limiting absorption principle holds can be improved.

THEOREM 9.4.19. *Assume that the admissible hamiltonian H has one of the following two properties:*

(1) $[D^Y, V^Y] \in B(\mathscr{H}^2(Y), \mathscr{H}^{-1}(Y))$ *for each $Y \in \mathscr{L}$;*

(2) *for each $Y \in \mathscr{L}$ and each linear function $\varphi : Y \to \mathbb{R}$ the operator $[\varphi(Q^Y), V^Y]$ belongs to $B(\mathscr{H}^1(Y), \mathscr{H}^{-1}(Y))$.*

Then the two limits $\lim_{\mu \to \pm 0}(H - \lambda - i\mu)^{-1}$ exist in the weak* topology of $B(\mathscr{H}_{1/2,1}^{-1}(X); \mathscr{H}_{-1/2,\infty}^1(X))$, uniformly in λ on each compact subset of $\mathbb{R} \setminus \kappa(H)$. If $\mathscr{K}$ is a Hilbert space and $T : \overset{\circ}{\mathscr{H}}_{-1/2,\infty}^1(X) \to \mathscr{K}$ is a linear continuous operator, then T is locally H-smooth on $\mathbb{R} \setminus \kappa(H)$. In particular, the conclusion of Corollary 9.4.18 remains true if H_1 and H_2 satisfy one of the conditions (1), (2) and if $H_1 - H_2 \in B(\overset{\circ}{\mathscr{H}}_{-1/2,\infty}^1(X), \mathscr{H}_{1/2,1}^{-1}(X))$.

PROOF. If (1) holds, then $[D, H] \in B(\mathscr{H}^2(X), \mathscr{H}^{-1}(X))$, so we may apply the last part of Theorem 7.5.2 and the result follows because $\mathscr{G}^{-1/2} = \mathscr{H}^{-1}$, $D(D; \mathscr{H}^{-2}) \supset \mathscr{H}_1^{-1}$ and $(\mathscr{H}_1^{-1}, \mathscr{H}^{-1})_{1/2,1} = \mathscr{H}_{1/2,1}^{-1}$. If (2) is satisfied, we use Proposition 7.4.6. $\square$

We also mention the following consequence of Proposition 7.4.6:

PROPOSITION 9.4.20. *If H is an admissible hamiltonian and $[\varphi(Q^Y), V^Y] \in B(\mathscr{H}^2(Y), \mathscr{H}(Y))$ for each $Y \in \mathscr{L}$ and each linear function $\varphi : Y \to \mathbb{R}$, then for each $s \in [-2, 0]$ the limits $\lim_{\mu \to \pm 0}(H - \lambda - i\mu)^{-1}$ exist in the weak* topology of $B(\mathscr{H}_{1/2,1}^s(X), \mathscr{H}_{-1/2,\infty}^{s+2}(X))$, uniformly in λ on compact sets in $\mathbb{R} \setminus \kappa(H)$.*

9.A. Appendix: Remarks on the $\mathscr{C}^{1,1}$ Property

In §§7.5.3 and 9.4.2 we have described several methods of checking that an operator is of class $\mathscr{C}^{1,1}(A)$. We shall present here another approach to this question, based on the so-called trace method in interpolation theory. In particular this will explain the connection between the theory developed in Chapter 7 (in which the regularity classes $\mathscr{C}^{s,p}$ play a fundamental role) ad the earlier versions of the theory (in which some approximation properties were essential; see [ABG2], [BGM1,2] and [Tm1]).

We shall first place ourselves in the abstract setting of Section 6.3: $(\mathscr{G}, \mathscr{H})$ is a Friedrichs couple and A is a self-adjoint operator in $\mathscr{H}$ such that $e^{iA\tau}\mathscr{G} \subset \mathscr{G}$ for all $\tau \in \mathbb{R}$. Then, according to Theorem 6.3.4 (b), if H is a self-adjoint operator in $\mathscr{H}$ with domain $\mathscr{G}$, then H is of class $\mathscr{C}^{1,1}(A)$ if and only if $H \in \mathscr{C}^{1,1}(A; \mathscr{G}, \mathscr{G}^*)$. So it is sufficient to give criteria for an operator $S : \mathscr{G} \to \mathscr{G}^*$ to be of class $\mathscr{C}^{1,1}(A; \mathscr{G}, \mathscr{G}^*)$.

We begin by observing that (5.2.22) has the following consequence:

$$(9.A.1) \qquad \mathscr{C}^{1,1}(A; \mathscr{G}, \mathscr{G}^*) = (C^2(A; \mathscr{G}, \mathscr{G}^*), B(\mathscr{G}, \mathscr{G}^*))_{1/2,1}.$$

We recall that $C^2(A; \mathscr{G}, \mathscr{G}^*)$ is a Banach space equipped with the norm (cf. (5.1.11))

$$(9.A.2) \; ||T||_{C^2(A;\mathscr{G},\mathscr{G}^*)} = [||T||^2_{\mathscr{G}\to\mathscr{G}^*} + ||\mathscr{A}[T]||^2_{\mathscr{G}\to\mathscr{G}^*} + ||\mathscr{A}^2[T]||^2_{\mathscr{G}\to\mathscr{G}^*}]^{1/2},$$

where $\mathscr{A}[T] = [T, A]$. By using the trace method of interpolation described in Proposition 2.3.3, we then obtain the following result: a continuous operator $S : \mathscr{G} \to \mathscr{G}^*$ belongs to $\mathscr{C}^{1,1}(A; \mathscr{G}, \mathscr{G}^*)$ if and only if there is a family $\{S_\varepsilon\}_{0<\varepsilon<1}$ of bounded operators $S_\varepsilon : \mathscr{G} \to \mathscr{G}^*$ such that

(i) $\lim_{\varepsilon\to 0} S_\varepsilon = S$ in norm in $B(\mathscr{G}, \mathscr{G}^*)$;

(ii) $S_\varepsilon \in C^2(A; \mathscr{G}, \mathscr{G}^*)$ for each $\varepsilon \in (0, 1)$ and the map $\varepsilon \mapsto S_\varepsilon \in C^2(A; \mathscr{G}, \mathscr{G}^*)$ is of class C^∞;

(iii) $\int_0^1 [\varepsilon^{-1}||S_\varepsilon'||_{\mathscr{G}\to\mathscr{G}^*} + ||S_\varepsilon||_{C^2(A;\mathscr{G},\mathscr{G}^*)}]d\varepsilon < \infty$.

We have normalized the family $\{S_\varepsilon\}$ differently than in Proposition 2.3.3 in order to facilitate comparison with the approach in [ABG2] and [BGM1,2]. More precisely, if $\{S(\tau)\}_{\tau>0}$ is the family of operators furnished by part (b) of that proposition, then we take $S_\varepsilon = S(\varepsilon^2)$. By mimicking the proof of Proposition 2.3.3 (a) we shall now obtain the following sufficient condition for an operator to be of class $\mathscr{C}^{1,1}(A; \mathscr{G}, \mathscr{G}^*)$:

PROPOSITION 9.A.1. *Let $\{S_\varepsilon\}_{0<\varepsilon<1}$ be a family of bounded operators S_ε from $\mathscr{G}$ to $\mathscr{G}^*$ such that $S_\varepsilon \in C^2(A; \mathscr{G}, \mathscr{G}^*)$, $\varepsilon \mapsto S_\varepsilon \in B(\mathscr{G}, \mathscr{G}^*)$ is weakly of class C^1 and*

$$(9.A.3) \qquad \int_0^1 [\varepsilon^{-1}||S_\varepsilon'||_{\mathscr{G}\to\mathscr{G}^*} + ||[A, [A, S_\varepsilon]]||_{\mathscr{G}\to\mathscr{G}^*}]d\varepsilon < \infty.$$

Then $\lim_{\varepsilon\to 0} S_\varepsilon \equiv S$ exists in norm in $B(\mathscr{G}, \mathscr{G}^*)$ and $S \in \mathscr{C}^{1,1}(A; \mathscr{G}, \mathscr{G}^*)$.

PROOF. (i) The fact that S_ε is norm-convergent as $\varepsilon \to 0$ follows from $\int_0^1 \|S'_\tau\|_{\mathscr{G} \to \mathscr{G}^*} d\tau < \infty$ by taking into account the identity $S_\varepsilon = S_{\varepsilon_0} - \int_\varepsilon^{\varepsilon_0} S'_\tau d\tau$. In particular we have $\|S_\varepsilon\|_{\mathscr{G} \to \mathscr{G}^*} \leq C_1$ for some finite constant C_1 and all $\varepsilon \in (0,1)$. Then note that (5.1.8) gives for each $T \in C^2(A; \mathscr{G}, \mathscr{G}^*)$:

$$i\mathscr{A}[T] = e^{-iA}Te^{iA} - T + \int_0^1 e^{-iA\tau}\mathscr{A}^2[T]e^{iA\tau}(1 - \tau)d\tau.$$

Since $\{e^{iA\tau}\}_{\tau \in \mathbb{R}}$ is a C_0-group in $\mathscr{G}$ and in $\mathscr{G}^*$, we see that there is a constant $C_2 < \infty$ such that for all such T:

$$\|\mathscr{A}[T]\|_{\mathscr{G} \to \mathscr{G}^*} \leq C_2\|T\|_{\mathscr{G} \to \mathscr{G}^*} + C_2\|\mathscr{A}^2[T]\|_{\mathscr{G} \to \mathscr{G}^*}.$$

By taking $T = S_\varepsilon$ and by using (9.A.3) we then obtain $\int_0^1 \|S_\varepsilon\|_{C^2(A;\mathscr{G},\mathscr{G}^*)}d\varepsilon < \infty$.

(ii) We denote by $K(\tau, T)$ the K-functional of the pair of Banach spaces $C^2(A; \mathscr{G}, \mathscr{G}^*) \equiv C^2$ and $B(\mathscr{G}, \mathscr{G}^*) \equiv B$. Then (2.2.1) implies that

$$(9.A.4) \quad K(\varepsilon^2, S) \leq \varepsilon^2\|S_\varepsilon\|_{C^2} + \|S - S_\varepsilon\|_B \leq \varepsilon^2\|S_\varepsilon\|_{C^2} + \int_0^\varepsilon \|S'_\tau\|_B d\tau.$$

On the other hand, from (9.A.1) and the Definition 2.3.3, we see that $S \in \mathscr{C}^{1,1}(A; \mathscr{G}, \mathscr{G}^*)$ if and only if

$$\int_0^1 K(\tau, S)\tau^{-3/2}d\tau = 2\int_0^1 K(\varepsilon^2, S)\varepsilon^{-2}d\varepsilon < \infty.$$

This condition is satisfied as a consequence of (9.A.4), (9.A.3) and of the conclusion of step (i) of the proof. $\square$

Another sufficient condition for $S \in \mathscr{C}^{1,1}(A; \mathscr{G}, \mathscr{G}^*)$ is: $S \in C^1(A; \mathscr{G}, \mathscr{G}^*)$ and $T := [S, iA] \in \mathscr{C}^{0,1}(A; \mathscr{G}, \mathscr{G}^*)$. Since $\mathscr{C}^{0,1}$ is not an interpolation space, the property $T \in \mathscr{C}^{0,1}$ cannot be deduced from the trace method. However, the next result is easy to prove (see Lemma 2.3 in [BGM2]):

PROPOSITION 9.A.2. *An operator $T \in B(\mathscr{G}, \mathscr{G}^*)$ is of class $\mathscr{C}^{0,1}(A; \mathscr{G}, \mathscr{G}^*)$ if and only if there is a family of operators $\{T_\varepsilon\}_{0 < \varepsilon < 1}$ such that $T_\varepsilon \in C^1(A; \mathscr{G}, \mathscr{G}^*)$ and*

$$\int_0^1 [\varepsilon^{-1}\|T - T_\varepsilon\|_{\mathscr{G} \to \mathscr{G}^*} + \|[A, T_\varepsilon]\|_{\mathscr{G} \to \mathscr{G}^*}]d\varepsilon < \infty.$$

If this is satisfied, then the family $\{T_\varepsilon\}$ may be chosen to be weakly C^1 as function of ε and such that $\int_0^1 \|T'_\varepsilon\|_{\mathscr{G} \to \mathscr{G}^*} d\varepsilon < \infty$.

As an example, we shall explain now how one may deduce a slightly weaker form of Proposition 9.4.12 from Proposition 9.A.1 (one may similarly deduce Proposition 9.4.11 from Proposition 9.A.2). We shall take $\mathscr{G} = \mathscr{H}^2(X)$, $\mathscr{H} = \mathscr{H}(X)$ and $A = D$. Let $\varphi \in C_0^\infty(X)$ real with $\varphi(x) = 1$ in a neighbourhood of the origin. We set $\varphi_\varepsilon = \varphi(\varepsilon Q)$ and choose $S_\varepsilon = \varphi_\varepsilon S \varphi_\varepsilon$ (cf. [ABG1,2]). Here S is assumed to be a symmetric operator in $B(\mathscr{H}^2, \mathscr{H})$ and we abbreviate $\mathscr{H}^s = \mathscr{H}^s(X)$, $\|\cdot\|_{s,t} = \|\cdot\|_{\mathscr{H}^s \to \mathscr{H}^t}$. Our purpose is to find conditions on S such that the hypotheses of Proposition 9.A.1 are satisfied.

Set $\widetilde{\varphi}(x) = x \cdot \nabla\varphi(x)$. Then $\frac{d}{d\varepsilon}\varphi(\varepsilon x) = \varepsilon^{-1}\widetilde{\varphi}(\varepsilon x)$, hence the map $\varepsilon \mapsto S_\varepsilon \in B(\mathscr{H}^2, \mathscr{H})$ is of class C^∞ on $(0, \infty)$ and we have $\varepsilon S_\varepsilon' = \widetilde{\varphi}_\varepsilon S \varphi_\varepsilon + \varphi_\varepsilon S \widetilde{\varphi}_\varepsilon$. Since $||\varphi(\varepsilon Q)||_{2,2} \leq C < \infty$ for $0 < \varepsilon < 1$, we have

$$\int_0^1 ||S_\varepsilon'||_{2,-2}\frac{d\varepsilon}{\varepsilon} \leq 2C \int_0^1 ||\widetilde{\varphi}_\varepsilon S||_{2,-2}\frac{d\varepsilon}{\varepsilon^2} = 2C \int_1^\infty ||\widetilde{\varphi}(Q/r)S||_{2,-2}dr,$$

and the last integral is finite if (9.4.14) holds. Observe that we have $S_\varepsilon \to S$ strongly (hence in norm too, cf. Proposition 9.A.1) in $B(\mathscr{H}^2, \mathscr{H})$ as $\varepsilon \to 0$.

We have $[D, [D, S_\varepsilon]] = D^2 S_\varepsilon + S_\varepsilon D^2 - 2DS_\varepsilon D$ and each term on the r.h.s. belongs to $B(\mathscr{H}^2, \mathscr{H}^{-2})$. Hence S_ε is of class $C^2(D; \mathscr{H}^2, \mathscr{H}^{-2})$ and

$$||[D, [D, S_\varepsilon]]||_{2,-2} \leq 2||D^2 S_\varepsilon||_{2,-2} + 2||DS_\varepsilon D||_{2,-2}.$$

Now a straightforward computation shows that the contribution of the second term in the integral in (9.A.3) will be finite provided that

$$\int_0^1 [||Q_j Q_k \varphi(\varepsilon Q)S||_{2,0} + ||Q_j \varphi(\varepsilon Q)S\varphi(\varepsilon Q)Q_k||_{1,-1} + ||Q_j \varphi(\varepsilon Q)S||_{2,-1}]d\varepsilon < \infty$$

for all $j, k = 1, \ldots, \dim X$. One may estimate the three terms of the preceding integral by using the identity $\varphi(\varepsilon x) = \varphi(x) - \int_\varepsilon^1 \widetilde{\varphi}(\tau x)\tau^{-1}d\tau$. For example, if we set $\psi_j(x) = x_j\widetilde{\varphi}(x)$, we obtain

$$||Q_j \varphi(\varepsilon Q)S\varphi(\varepsilon Q)Q_k||_{1,-1} \leq ||Q_j \varphi(Q)S\varphi(Q)Q_k||_{1,-1}$$

$$+ \int_\varepsilon^1 ||Q_j \varphi(Q)S\psi_k(\tau Q)||_{1,-1}\tau^{-2}d\tau + \int_\varepsilon^1 ||\psi_j(\tau Q)S\varphi(Q)Q_k||_{1,-1}\tau^{-2}d\tau$$

$$+ \int_\varepsilon^1 \sigma^{-2}d\sigma \int_\varepsilon^1 \tau^{-2}d\tau ||\psi_j(\sigma Q)S\psi_k(\tau Q)||_{1,-1}.$$

In order to ensure the integrability (with respect to $\varepsilon \in (0, 1)$) of each term on the r.h.s. of this expression it suffices to assume that $\int_0^1 (||\theta(\varepsilon Q)S||_{2,0} + ||S\theta(\varepsilon Q)||_{2,0})\varepsilon^{-2}d\varepsilon < \infty$ for all $\theta \in C_0^\infty(X \setminus \{0\})$. For the proof note that if $T \in B(\mathscr{H}^2, \mathscr{H}) \cap B(\mathscr{H}, \mathscr{H}^{-2})$ then

$$||T||_{1,-1} \leq [||T||_{2,0} \cdot ||T||_{0,-2}]^{1/2} = [||T||_{2,0} \cdot ||T^*||_{2,0}]^{1/2} \leq \frac{1}{2}||T||_{2,0} + \frac{1}{2}||T^*||_{2,0},$$

hence

$$2||Q_j \varphi(Q)S\psi_k(\tau Q)||_{1,-1} \leq ||Q_j \varphi(Q)S\psi_k(\tau Q)||_{2,0} + ||\psi_k(\tau Q)S\varphi(Q)Q_j||_{2,0}.$$

Moreover, if we set $F(\sigma, \tau) = ||\psi_j(\sigma Q)S\psi_k(\tau Q)||_{1,-1}$, then

$$\int_0^1 d\varepsilon \int_\varepsilon^1 \frac{d\sigma}{\sigma^2} \int_\varepsilon^1 \frac{d\tau}{\tau^2}F(\sigma, \tau) = \int_0^1 \frac{d\sigma}{\sigma^2} \int_0^1 \frac{d\tau}{\tau^2} \min(\sigma, \tau)F(\sigma, \tau).$$

We mention that, although the choice $S_\varepsilon = \varphi(\varepsilon Q)S\varphi(\varepsilon Q)$ gives (essentially) the result of Proposition 9.4.12, it will not allow us to cover operators S with the property $S\mathscr{H}^2 \subset \mathscr{H}_{1+\varepsilon}^{-1+\varepsilon}$ (see the comments made before Proposition 9.4.12). However, such operators can be treated by using a more complicated regularization of S (in Q and P simultaneously; see Section 5 in [BGM2]).

CHAPTER 10

Quantum-Mechanical N-Body Systems

The purpose of this chapter is to explain how quantum-mechanical N-body systems $(N \geq 2)$ fit into the geometric framework presented in this text. Section 10.1 is concerned with the appropriate semilattice of subspaces and Section 10.2 with the associated N-body Hamiltonians.

10.1. Clustering of Particles

10.1.1. We begin with some purely combinatorial considerations. We consider a set $\mathcal{N}$ of $N \geq 2$ (distinguishable) particles which we label by the numbers $1, 2, \ldots, N$. $\mathcal{N}$ may then be identified with the set $1, 2, \ldots, N$; so we shall have $\mathcal{N} = \{1, 2, \ldots, N\}$. A *cluster* of the system is defined as a non-empty subset A of $\mathcal{N}$. We denote by $\mathcal{C}$ the set of all clusters consisting of at least two particles:

$$(10.1.1) \qquad \mathcal{C} = \{C \subset \mathcal{N} \mid \operatorname{card} C \geq 2\},$$

where $\operatorname{card} C$ is the number of elements of the set C. We observe that $\mathcal{C}$ is a partially ordered set (the order relation is the usual inclusion of subsets). The elements of $\mathcal{C}$ will be called *composite clusters*.

A *cluster decomposition* of the system is a partition $\mathfrak{a}$ of $\mathcal{N}$. More precisely, $\mathfrak{a}$ is a family of subsets of $\mathcal{N}$ such that
  (i) each $A \in \mathfrak{a}$ is a cluster, i.e. a non-empty subset of $\mathcal{N}$,
  (ii) if A, $A' \in \mathfrak{a}$ are distinct, then $A \cap A' = \varnothing$,
  (iii) $\bigcup_{A \in \mathfrak{a}} A = \mathcal{N}$.
We shall denote by $|\mathfrak{a}| \equiv \operatorname{card} \mathfrak{a}$ the number of clusters in $\mathfrak{a}$. If $\mathfrak{a}$, $\mathfrak{b}$ are cluster decompositions, we write $\mathfrak{a} \leq \mathfrak{b}$ if the partition $\mathfrak{a}$ is finer than the partition $\mathfrak{b}$ (this convention is that usually adopted in combinatorics, see e.g. Chapter 1, §2.B of [Ai], and moreover has the advantage that the correspondence $\mathfrak{a} \mapsto X^{\mathfrak{a}}$ defined in §10.1.3 below is an isomorphism). Hence $\mathfrak{a} \leq \mathfrak{b}$ means that one of the following three equivalent conditions is fulfilled:
  (i) $A \in \mathfrak{a} \Rightarrow \exists B \in \mathfrak{b}$ such that $A \subset B$,
  (ii) $B \in \mathfrak{b} \Rightarrow B = \bigcup_{A \in \mathfrak{a},\, A \subset B} A$,

W. O. Amrein et al., *C₀-Groups, Commutator Methods and Spectral Theory of
N-Body Hamiltonians*, Modern Birkhäuser Classics,
DOI: 10.1007/978-3-0348-0733-3_10, © Springer Basel 1996

(iii) $A \in \mathfrak{a},\, B \in \mathfrak{b} \Rightarrow A \subset B$ or $A \cap B = \varnothing$.

Clearly $\mathfrak{a} \leq \mathfrak{b}$ implies that $|\mathfrak{a}| \geq |\mathfrak{b}|$. We write $\mathfrak{a} < \mathfrak{b}$ if $\mathfrak{a} \leq \mathfrak{b}$ and $\mathfrak{a} \neq \mathfrak{b}$ (i.e. if $\mathfrak{a} \leq \mathfrak{b}$ and $|\mathfrak{a}| > |\mathfrak{b}|$).

We denote by $\mathscr{D}$ the set of all cluster decompositions provided with the relation $\leq$, which is obviously a partial order relation. Then $\mathscr{D}$ is a finite lattice, which means that it has the following four properties:

(1) It has a *smallest element*, namely the cluster decomposition

$$\mathfrak{a}_{\min} = \{\{1\}, \{2\}, \ldots, \{N\}\}$$

uniquely characterized by $|\mathfrak{a}_{\min}| = N$. Clearly $\mathfrak{a}_{\min} \leq \mathfrak{a}$ for all $\mathfrak{a} \in \mathscr{D}$.

(2) It has a *largest element*, namely

$$\mathfrak{a}_{\max} = \{\mathcal{N}\},$$

uniquely characterized by $|\mathfrak{a}_{\max}| = 1$. One has $\mathfrak{a} \leq \mathfrak{a}_{\max}$ for all $\mathfrak{a} \in \mathscr{D}$.

(3) If $\mathfrak{a}, \mathfrak{b} \in \mathscr{D}$, then they have a *greatest lower bound* (g.l.b.) denoted by $\mathfrak{a} \wedge \mathfrak{b}$, namely the partition the clusters of which are the sets $A \cap B$ with $A \in \mathfrak{a}$, $B \in \mathfrak{b}$ and $A \cap B \neq \varnothing$. Clearly, for $\mathfrak{c} \in \mathscr{D}$, one has $\mathfrak{c} \leq \mathfrak{a}$ and $\mathfrak{c} \leq \mathfrak{b}$ if and only if $\mathfrak{c} \leq \mathfrak{a} \wedge \mathfrak{b}$.

(4) If $\mathfrak{a}, \mathfrak{b} \in \mathscr{D}$, then they have a *least upper bound* (l.u.b.) denoted by $\mathfrak{a} \vee \mathfrak{b}$; this follows from the facts that $\mathscr{D}$ is finite, has a largest element and from (3): $\mathfrak{a} \vee \mathfrak{b}$ is just the g.l.b. of all $\mathfrak{c} \in \mathscr{D}$ such that $\mathfrak{a} \leq \mathfrak{c}$ and $\mathfrak{b} \leq \mathfrak{c}$.

In order to describe more explicitly $\mathfrak{a} \vee \mathfrak{b}$, and also for later purposes, it is useful to introduce the following notation: if $i, j \in \mathcal{N}$ and $\mathfrak{a} \in \mathscr{D}$, then

$$(10.1.2) \qquad i \sim_{\mathfrak{a}} j \Leftrightarrow \exists A \in \mathfrak{a} \text{ such that } i \in A \text{ and } j \in A.$$

Then $(i \not\sim_{\mathfrak{a}} j)$ means that $i \sim_{\mathfrak{a}} j$ is not satisfied. Now it is easily seen that for i, $j \in \mathcal{N}$ and $\mathfrak{a}, \mathfrak{b} \in \mathscr{D}$:

$$(10.1.3) \qquad i \sim_{\mathfrak{a} \vee \mathfrak{b}} j \Leftrightarrow \exists i_0, i_1, \ldots, i_n \in \mathcal{N} \text{ such that } i_0 = i,\, i_n = j$$

$$\text{and } i_{k-1} \sim_{\mathfrak{a}} i_k \text{ or } i_{k-1} \sim_{\mathfrak{b}} i_k \text{ for each } k = 1, \ldots, n.$$

For another explicit description of $\mathfrak{a} \vee \mathfrak{b}$ the reader may consult [PSS], p. 524.

For $k = 1, \ldots, N$ we define $\mathscr{D}_k = \{\mathfrak{a} \in \mathscr{D} \mid |\mathfrak{a}| \geq k\}$ and $\mathscr{D}(k) = \{\mathfrak{a} \in \mathscr{D} \mid |\mathfrak{a}| = k\}$. Clearly $\mathscr{D}(k) \cap \mathscr{D}(j) = \varnothing$ if $j \neq k$ and $\mathscr{D}_1 = \mathscr{D} = \cup_{k=1}^{N} \mathscr{D}(k)$. $\mathscr{D}(k)$ is the set of maximal elements of $\mathscr{D}_k$, and one has $\mathscr{D}_2 = \mathscr{D} \setminus \{\mathfrak{a}_{\max}\}$ and $\mathscr{D}_N \equiv \{\mathfrak{a}_{\min}\} \subset \mathscr{D}_{N-1} \subset \cdots \subset \mathscr{D}_1 \equiv \mathscr{D}$.

We shall embed $\mathcal{C}$ in $\mathscr{D}$ by identifying a cluster $C \subset \mathcal{N}$ such that $\operatorname{card} C \geq 2$ with the cluster decomposition consisting of the cluster C and of $N - \operatorname{card} C$ clusters of the form $\{i\}$, with $i \in \mathcal{N} \setminus C$. Hence $\mathcal{C}$ becomes the set of cluster decompositions that contain exactly one composite cluster. If $\operatorname{card} C = 2$, we say that C is a *pair*; it is usual to use the symbol ℓ for denoting a pair. Observe that $\operatorname{card} \ell = 2$ but that, when ℓ is considered as a cluster decomposition, then $|\ell| = N - 1$, i.e. $\ell \in \mathscr{D}(N-1)$. Moreover each cluster decomposition in $\mathscr{D}(N-1)$ is of the form ℓ for some pair $\ell \in \mathcal{C}$; $\mathscr{D}(N-1)$ is the set of minimal elements of $\mathscr{D} \setminus \{\mathfrak{a}_{\min}\}$. We also remark that the embedding $\mathcal{C} \subset \mathscr{D}$ respects the order relation: if $A, B \in \mathcal{C}$, then $A \subset B$ as subsets of $\mathcal{N}$ if and only if $A \leq B$ as cluster decompositions. However the least upper bound and the greatest lower bound in

$\mathscr{D}$ do not correspond to those in $\mathcal{C}$ in general. Clearly $\mathcal{N} \equiv \mathfrak{a}_{\max}$ under the above identification of $\mathcal{C}$ with a subset of $\mathscr{D}$. Furthermore observe that, according to our conventions, if $C \in \mathcal{C}$ and $\mathfrak{a} \in \mathscr{D}$, then $C \leq \mathfrak{a}$ means that there exists $A \in \mathfrak{a}$ such that $C \subset A$.

For $\mathfrak{a} \in \mathscr{D}$, define $\mathcal{C}(\mathfrak{a}) = \{A \in \mathfrak{a} \mid \operatorname{card} A \geq 2\} \subset \mathcal{C} \subset \mathscr{D}$. Then one has

$$(10.1.4) \qquad \mathfrak{a} = \vee_{A \in \mathcal{C}(\mathfrak{a})} A,$$

where on the right-hand side the clusters A are considered as elements of $\mathscr{D}$. If $\mathfrak{a} = \mathfrak{a}_{\min}$, then $\mathcal{C}(\mathfrak{a}) = \varnothing$ and we make the usual convention that $\vee_{i \in \varnothing} \mathfrak{a}_i = \mathfrak{a}_{\min}$. This shows that $\mathcal{C}$ generates $\mathscr{D}$, more precisely that each $\mathfrak{a} \in \mathscr{D}$ is the least upper bound of a uniquely defined family in $\mathcal{C}$.

10.1.2. We now consider the configuration space of a system of N particles. We denote by $m_1, \ldots, m_N$ the masses of the particles ($m_k > 0$) and by ν ($\nu = 1, 2, 3 \ldots$) the dimension of the physical space $\mathbb{R}^\nu$ in which the motion of the system takes place. Let $\mathscr{X} = \mathbb{R}^{\nu N}$ and denote the points in $\mathscr{X}$ by $x = (x_1, \ldots, x_N)$, with $x_k \in \mathbb{R}^\nu$. We provide $\mathscr{X}$ with the Euclidean structure given by the following scalar product:

$$(10.1.5) \qquad (x, y)_{\mathscr{X}} = \sum_{k=1}^{N} 2 m_k \, x_k \cdot y_k,$$

where $x_k \cdot y_k$ is the usual scalar product in $\mathbb{R}^\nu$. We set $|x| = [(x, x)_{\mathscr{X}}]^{1/2}$. This Euclidean space $\mathscr{X}$ is called the *configuration space of the system of particles $\mathcal{N}$*.

REMARK. To justify the definition (10.1.5) of the scalar product in $\mathscr{X}$, let us calculate the expression of the Laplace-Beltrami operator $\Delta_{\mathscr{X}}$ in the canonical basis $\{e_{k\alpha} \mid k = 1, \ldots, N; \alpha = 1, \ldots, \nu\}$ of $\mathscr{X}$ coming from the identification $\mathscr{X} = \mathbb{R}^{\nu N}$. It is an orthogonal but not an orthonormal basis of $\mathscr{X}$ (since, for fixed k, the set $\{e_{k\alpha} \mid \alpha = 1, \ldots, \nu\}$ is an orthonormal basis of $\mathbb{R}^\nu$ with respect to the usual scalar product in $\mathbb{R}^\nu$). The dual basis is given by the vectors

$$e^{k\alpha} = (2 m_k)^{-1} e_{k\alpha}.$$

Hence

$$g^{j\alpha, k\beta} \equiv (e^{j\alpha}, e^{k\beta})_{\mathscr{X}} = (2 m_j)^{-1} \delta_{jk} \delta_{\alpha\beta}$$

($j, k = 1, \ldots, N; \alpha, \beta = 1, \ldots, \nu$). Thus, in the usual coordinates of $\mathbb{R}^{\nu N}$ one has (cf. Eq. (1.1.3)):

$$(10.1.6) \qquad \Delta_{\mathscr{X}} = -\sum_{k=1}^{N} \sum_{\alpha=1}^{\nu} \frac{1}{2 m_k} \partial_{x^{k\alpha}}^2 \qquad \left(x_k = \sum_{\alpha=1}^{\nu} x^{k\alpha} e_{k\alpha}\right).$$

Further justifications for (10.1.5) may be found in the nice decomposition properties that arise in relation with the notions of subsystems and of cluster decompositions; we refer to [A], [DHSV], [FH2], [Sim2] for details on this formalism).

We are mainly interested in studying N-body systems interacting by translation invariant forces. For this reason we shall restrict ourselves to a subspace of $\mathscr{X}$, although the general geometric formalism covers easily the case when there are also external fields. The *configuration space of the internal motion of the system* is defined to be

$$(10.1.7) \qquad X = \left\{ x \in \mathscr{X} \ \middle| \ \sum_{k=1}^{N} m_k x_k = 0 \right\},$$

provided with the induced Euclidean structure. A point $x \in X$ represents the configuration of the system in the center-of-mass frame. One may write

$$\mathscr{X} = X \oplus \mathscr{X}_N,$$

where *the configuration space $\mathscr{X}_N$ for the movement of the center of mass of the system* is given by $\mathscr{X}_N = \{ x \in \mathscr{X} \mid x_1 = x_2 = \cdots = x_N \}$. We denote by π_N the orthogonal projection of $\mathscr{X}$ onto $\mathscr{X}_N$ and by $\pi^N = I - \pi_N$ that of $\mathscr{X}$ onto X. The following are explicit expressions for these two projections:

$$(10.1.8) \qquad \pi_N(x) = \left(\frac{1}{M} \sum_{k=1}^{N} m_k x_k, \ldots, \frac{1}{M} \sum_{k=1}^{N} m_k x_k \right),$$

$$(10.1.9) \qquad \pi^N(x) = \left(x_1 - \frac{1}{M} \sum_{k=1}^{N} m_k x_k, \ldots, x_N - \frac{1}{M} \sum_{k=1}^{N} m_k x_k \right),$$

where $M = \sum_{k=1}^{N} m_k$ is the *total mass* of the system and where the quantity $x_j - M^{-1} \sum_{k=1}^{N} m_k x_k$ represents the position of the particle j relative to the center of mass of the system.

10.1.3. We next introduce the appropriate semilattice $\mathscr{L}$ for the study of N-particle systems. We first define two mappings $\mathfrak{a} \mapsto X_\mathfrak{a}$ and $\mathfrak{a} \mapsto X^\mathfrak{a}$ from $\mathscr{D}$ to the set of subspaces of X, by setting

$$(10.1.10) \qquad X_\mathfrak{a} = \{ x \in X \mid x_i = x_j \text{ if } i \sim_\mathfrak{a} j \},$$

$$X^\mathfrak{a} = \{ x \in X \mid \sum_{k \in A} m_k x_k = 0 \text{ for each } A \in \mathfrak{a} \}.$$

Clearly $X_\mathfrak{a}$ and $X^\mathfrak{a}$ are mutually orthogonal, and their dimensions are $\nu(|\mathfrak{a}| - 1)$ and $\nu(N - |\mathfrak{a}|)$ respectively, so that one has

$$(10.1.11) \qquad X = X_\mathfrak{a} \oplus X^\mathfrak{a}.$$

Some additional properties of these subspaces are collected in the following proposition.

PROPOSITION 10.1.1. *One has*
(a) $\mathfrak{a} \leq \mathfrak{b} \Leftrightarrow X^\mathfrak{a} \subset X^\mathfrak{b} \Leftrightarrow X_\mathfrak{a} \supset X_\mathfrak{b}$,
(b) $\mathfrak{a} \neq \mathfrak{b} \Leftrightarrow X^\mathfrak{a} \neq X^\mathfrak{b}$,
(c) $X^{\mathfrak{a}_{\min}} = \mathbf{O}$, $\quad X_{\mathfrak{a}_{\min}} = X$, $\quad X^{\mathfrak{a}_{\max}} \equiv X^N = X$, $\quad X_N = \mathbf{O}$,
(d) $X_{\mathfrak{a} \vee \mathfrak{b}} = X_\mathfrak{a} \cap X_\mathfrak{b}$, $\quad X^{\mathfrak{a} \vee \mathfrak{b}} = X^\mathfrak{a} + X^\mathfrak{b}$,

(e) $X_{a \wedge b} \supset X_a + X_b$, $X^{a \wedge b} \subset X^a \cap X^b$, *with strict inclusions in general.*

PROOF. (a), (b) and (c) are simple consequences of the definitions of the occurring notions. For the first identity in (d) we observe that, since $a \leq a \vee b$ and $b \leq a \vee b$, one has $X_{a \vee b} \subset X_a \cap X_b$ by using (a). To obtain the converse inclusion, we must show that $x \in X_a \cap X_b$ and $i \sim_{a \vee b} j \Rightarrow x_i = x_j$. Now, if $i \sim_{a \vee b} j$, then by (10.1.3) there is a sequence $i_0 = i, i_1, i_2, \ldots, i_n = j$ such that $i_{k-1} \sim_a i_k$ or $i_{k-1} \sim_b i_k$ for each k; for $x \in X_a \cap X_b$, this implies that $x_{i_{k-1}} = x_{i_k}$ for each k, hence $x_i = x_j$.

The second identity in (d) follows from the first one by using (10.1.11) and the fact that $(Y \cap Z)^\perp = Y^\perp + Z^\perp$ if Y, Z are any subspaces of X. Next we combine (a) with the fact that $a \wedge b \leq a$ and $a \wedge b \leq b$ to find that $X_a \subset X_{a \wedge b}$ and $X_b \subset X_{a \wedge b}$, which implies the first inclusion in (e). The second one then follows as above from the first one, (10.1.11) and the relation $(Y \cap Z)^\perp = Y^\perp + Z^\perp$:

$$X^{a \wedge b} = (X_{a \wedge b})^\perp \subset (X_a + X_b)^\perp = X_a^\perp \cap X_b^\perp \equiv X^a \cap X^b.$$

To show that the inclusions in (e) are strict in general, it suffices to give a counterexample for the second inclusion. For this, we let $\mathcal{N} = \{1, 2, 3, 4\}$, $a = \{A, A'\}$, $b = \{B, B'\}$ with $A = \{1, 2\}$, $A' = \{3, 4\}$, $B = \{1, 3\}$, $B' = \{2, 4\}$. Clearly $a \wedge b = a_{\min} = \{\{1\}, \{2\}, \{3\}, \{4\}\}$. We take all masses m_k equal to 1 and $x = (x_1, x_2, x_3, x_4)$ such that $x_1 = x_4 = -x_2 = -x_3 = y \neq 0$. Then $x \notin X^{a \wedge b} = X^{a_{\min}} = \mathbf{O}$, but $x \in X^a$ (since $x_1 + x_2 = x_3 + x_4 = 0$) and $x \in X^b$ (since $x_1 + x_3 = x_2 + x_4 = 0$), so that $x \in X^a \cap X^b$. □

For the case of an N-body system with translation invariant interaction, the euclidean space X used in Chapter 9 is the space defined in (10.1.7) with scalar product (10.1.5), and the semilattice $\mathscr{L}$ is just $\mathscr{L} = \{X^a \mid a \in \mathscr{D}\}$.

The correspondence $\mathscr{D} \ni a \mapsto X^a \in \mathscr{L}$ is an isomorphism of partially ordered sets. Remark that, if $a, b \in \mathscr{D}$, then $a \wedge b \in \mathscr{D}$ but $X^a \cap X^b \notin \mathscr{L}$ in general. However, the greatest lower bound $X^a \wedge X^b$ in $\mathscr{L}$ exists, since one may define, for $Y, Z \in \mathscr{L}$:

$$Y \wedge Z = \sup\{E \in \mathscr{L} \mid E \subset Y \text{ and } E \subset Z\}.$$

Thus, since $a \mapsto X^a$ is an isomorphism, one will have

$$X^{a \wedge b} = X^a \wedge X^b \subset X^a \cap X^b,$$

with *strict* inclusion in general.

Let us consider the particular case of a cluster decomposition $C \in \mathcal{C} \subset \mathscr{D}$ containing exactly one composite cluster $C \subset \mathcal{N}$ (card $C \geq 2$). Then clearly

$$(10.1.12) \qquad X^C = \{x \in X \mid \sum_{k \in C} m_k x_k = 0 \text{ and } x_j = 0 \ \forall j \notin C\}.$$

Thus X^C is just the *configuration space for the internal movement of the cluster* C. Also the relation (10.1.12) implies that

$$(10.1.13) \qquad A, B \in \mathcal{C} \text{ and } A \cap B = \varnothing \Rightarrow X^A \perp X^B.$$

Now let $\mathfrak{a}$ be an arbitrary cluster decomposition. Then, if $A, B \in \mathcal{C}(\mathfrak{a})$ and $A \neq B$, one has $A \cap B = \varnothing$. Hence (10.1.4), Proposition 10.1.1 (d) and (10.1.13) imply that

$$(10.1.14) \qquad X^{\mathfrak{a}} = \oplus_{A \in \mathcal{C}(\mathfrak{a})} X^A \text{ for each } \mathfrak{a} \in \mathcal{D}.$$

This justifies the interpretation of $X^{\mathfrak{a}}$ as the configuration space of the internal movements of the clusters belonging to $\mathfrak{a}$. Remark that it is also natural to define $X^{\varnothing} = O$ and $X^A = O$ if $A \subset \mathcal{N}$ and card $A = 1$, so that $X^{\mathfrak{a}} = \oplus_{A \in \mathfrak{a}} X^A$. On the other hand it is clear from (10.1.10) that *one can interpret $X_{\mathfrak{a}}$ as the configuration space of the relative movement of the (centers of mass of the) clusters belonging to $\mathfrak{a}$.* More precisely, in $X_{\mathfrak{a}}$ the clusters of $\mathfrak{a}$ are considered to be particles and $X_{\mathfrak{a}}$ describes them in their center-of-mass frame (which coincides with the center-of-mass frame of the total system).

It is easy to calculate explicitly the orthogonal projections $\pi_{\mathfrak{a}}$ and $\pi^{\mathfrak{a}}$ of X onto $X_{\mathfrak{a}}$ and $X^{\mathfrak{a}}$ respectively. If $M(A) \equiv \sum_{j \in A} m_j$ denotes the total mass of the cluster A, then, if $i \in A \in \mathfrak{a}$

$$(10.1.15) \qquad (\pi_{\mathfrak{a}}(x))_i = \frac{1}{M(A)} \sum_{j \in A} m_j x_j,$$

$$(10.1.16) \quad (\pi^{\mathfrak{a}}(x))_i = x_i - \frac{1}{M(A)} \sum_{j \in A} m_j x_j \equiv \frac{1}{M(A)} \sum_{j \in A} m_j (x_i - x_j).$$

EXAMPLE 10.1.2. Let $\mathfrak{a} = \ell$, where $\ell \in \mathcal{C}$ is a pair: $\ell = \{i, j\}$ with $i, j \in \mathcal{N}$ and $i < j$. Then clearly:

$$X_\ell = \{x \in X \mid x = (x_1, \ldots, x_{i-1}, y, x_{i+1}, \ldots, x_{j-1}, y, x_{j+1}, \ldots, x_n) \text{ with}$$
$$x_k, y \in \mathbb{R}^\nu\},$$

$$X^\ell = \{x \in X \mid x = (0, \ldots, 0, \frac{1}{m_i} y, 0, \ldots, 0, -\frac{1}{m_j} y, 0, \ldots, 0) \text{ with } y \in \mathbb{R}^\nu\},$$

where the two non-zero entries are at the i-th and j-th position. Furthermore one has

$$\pi^\ell(x) = (0, \ldots, 0, \frac{m_j}{m_i + m_j}(x_i - x_j), 0, \ldots, 0, -\frac{m_i}{m_i + m_j}(x_i - x_j), 0, \ldots, 0).$$

Let us define $J^\ell : \mathbb{R}^\nu \to X$ by

$$J^\ell(q) = (0, \ldots, 0, \frac{m_j}{m_i + m_j} q, 0, \ldots, 0, -\frac{m_i}{m_i + m_j} q, 0, \ldots, 0).$$

J^ℓ is linear and injective, its range is X^ℓ, and one has $|J^\ell(q)|^2_X = 2\mu_\ell |q|^2_{\mathbb{R}^\nu}$, where $\mu_\ell = m_i m_j / (m_i + m_j)$ is the *reduced mass* of the pair ℓ. Hence we may identify functions $V^\ell : X^\ell \to \mathbb{C}$ with functions $v_\ell : \mathbb{R}^\nu \to \mathbb{C}$ by using the formula $V^\ell(J^\ell(q)) \equiv v_\ell(q)$. With this identification, the function $V_\ell \equiv V^\ell \circ \pi^\ell : X \to \mathbb{C}$ is given by $V_\ell(x) = v_\ell(x_i - x_j)$. This formula identifies functions $V_\ell : X \to \mathbb{C}$ satisfying $V_\ell(x) = V_\ell(\pi^\ell(x))$ with functions $v_\ell : \mathbb{R}^\nu \to \mathbb{C}$.

10.2. Quantum-Mechanical N-Body Hamiltonians

10.2.1. Example 10.1.2 illustrates the form of translation invariant two-body interactions in the usual presentation of N-body hamiltonians. One may consider the following more general situation. A function $V : X \to \mathbb{R}$ is called a *k-body potential* $(2 \leq k \leq N)$ if there is a cluster $C \subset \mathcal{N}$ with $\operatorname{card} C = k$ such that $V(x) = V(\pi^C(x))$ for all $x \in X$. This is equivalent to the following: there is a function $V^C : X^C \to \mathbb{R}$ such that $V \equiv V_C = V^C \circ \pi^C$; here X^C is given by (10.1.12) and $(\pi^C(x))_i = 0$ if $i \notin C$, $(\pi^C(x))_i = \sum_{j \in C} m_j(x_i - x_j)/M(C)$ if $i \in C$. V_C depends only on the relative coordinates $x_j - x_k$ of the particles in C $(j, k \in C)$, as in Example 10.1.2. For instance if $C = \{i_1, \ldots, i_n\}$ with $i_1 < i_2 < \cdots < i_n$, then there is a unique function $v_C : \mathbb{R}^{\nu(n-1)} \to \mathbb{R}$ such that

$$V_C(x) = v_C(x_{i_1} - x_{i_2}, x_{i_2} - x_{i_3}, \ldots, x_{i_{n-1}} - x_{i_n}).$$

These considerations justify the following definition:

DEFINITION 10.2.1. For each $C \in \mathcal{C}$, let $V^C : \mathscr{H}^2(X^C) \to \mathscr{H}(X^C)$ be a compact operator which is symmetric as an operator in $\mathscr{H}(X^C)$, and set $V_C = V^C \otimes_{X^C} I \in B(\mathscr{H}^2(X), \mathscr{H}(X))$. Then the associated *N-body hamiltonian* is the operator

$$H = \Delta_X + \sum_{C \in \mathcal{C}} V_C.$$

If $\operatorname{card} C = k$, then V_C is called a *k-body interaction*; it involves only particles from the cluster C. If $V_C = 0$ for all $C \in \mathcal{C}$ with $\operatorname{card} C \neq 2$, we say that H *contains only two-body forces.*

It is clear that a N-body hamiltonian is a particular case of a non-relativistic N-body hamiltonian (see Definition 9.4.1): we just have $V_{X^a} = 0$ if $a \in \mathscr{D}$ but $a \notin \mathcal{C}$ (i.e. if a contains more than one composite cluster). In particular, all the theory developed in Chapter 9 is applicable to N-body hamiltonians with translation invariant forces (to cover interactions including external fields, it would suffice to introduce some simple changes). It is clear that one may also consider N-body hamiltonians defined in the form-sense by requiring that each V^C be a symmetric compact operator from $\mathscr{H}^1(X^C)$ to $\mathscr{H}^{-1}(X^C)$.

We next point out some factorization properties of the N body hamiltonians introduced in Definition 10.2.1 that follow essentially from the decomposition (10.1.14) of X^a. We first observe that Δ_X is the free hamiltonian of the *internal* movement of the N particles in $\mathcal{N}$ (cf. (10.1.6)). If a is a cluster decomposition, then this hamiltonian should be equal to the free hamiltonian of the relative movement of the centers of mass of the clusters in a plus the free hamiltonian of these clusters. This property can be deduced from (10.1.11) and (10.1.14) which imply that

(10.2.1) $$X = X_a \oplus X^a = X_a \oplus \left(\oplus_{A \in \mathcal{C}(a)} X^A \right).$$

This in turn leads to the expected decomposition of the kinetic energy operator Δ^X:

(10.2.2)
$$\Delta^X = \Delta^{X_\mathfrak{a}} \otimes_{X_\mathfrak{a}} I + I \otimes_{X_\mathfrak{a}} \Delta^{X^\mathfrak{a}} \text{ and}$$
$$\Delta^{X^\mathfrak{a}} = \sum_{A \in \mathcal{C}(\mathfrak{a})} \otimes_{A' \in \mathcal{C}(\mathfrak{a})} T^{AA'},$$

where $T^{AA'} = \Delta^{X^A}$ if $A' = A$ and $T^{AA'} = I_{\mathcal{H}(X^{A'})}$ if $A' \neq A$. The tensor product that appears in the second equation in (10.2.2) corresponds to the identification

$$\mathcal{H}(X^\mathfrak{a}) = \otimes_{A' \in \mathcal{C}(\mathfrak{a})} \mathcal{H}(X^{A'})$$

which is determined by (10.1.14).

It is customary in the N-body literature to use the following notations:

$$\mathcal{H} = \mathcal{H}(X), \quad \mathcal{H}_\mathfrak{a} = \mathcal{H}(X_\mathfrak{a}), \quad \mathcal{H}^\mathfrak{a} = \mathcal{H}(X^\mathfrak{a}).$$

This also defines $\mathcal{H}^A \equiv \mathcal{H}(X^A)$ if $A \in \mathcal{C}$ ($\subset \mathcal{D}$). With these notations one has the following identifications

(10.2.3)
$$\mathcal{H} = \mathcal{H}_\mathfrak{a} \otimes \mathcal{H}^\mathfrak{a}, \quad \mathcal{H}^\mathfrak{a} = \otimes_{A \in \mathcal{C}(\mathfrak{a})} \mathcal{H}^A.$$

(Of course, if $A \subset \mathcal{N}$ and card $A = 1$, we take $\mathcal{H}^A \equiv \mathbb{C}$).

Now let H be a N-body hamiltonian as in Definition 10.2.1 and $\mathfrak{a} \in \mathcal{D}$. Then $H_{X^\mathfrak{a}}$ and $H^{X^\mathfrak{a}}$ are defined as in Section 9.4, since $X^\mathfrak{a} \in \mathcal{L}$. It is usual to denote these operators by $H_\mathfrak{a}$ and $H^\mathfrak{a}$ respectively. One then has

$$H_\mathfrak{a} = \Delta_X + \sum_{C \leq \mathfrak{a}} V_C,$$

where $C \leq \mathfrak{a}$ means that there is a cluster $A \in \mathfrak{a}$ such that $C \subset A$ (cf. the conventions made in §10.1.1). Thus $H_\mathfrak{a}$ is the hamiltonian in $\mathcal{H}$ which corresponds to the situation where all interactions between different clusters in $\mathfrak{a}$ have been replaced by zero. If $\mathbb{I}_\mathfrak{a} \equiv \sum_{C \nleq \mathfrak{a}} V_C$ denotes the total interaction between different clusters of $\mathfrak{a}$, then $H = H_\mathfrak{a} + \mathbb{I}_\mathfrak{a}$. On the other hand, $H^\mathfrak{a}$ is the total hamiltonian (in $\mathcal{H}^\mathfrak{a}$) determining the internal movement of the clusters (when there are no interactions between the clusters), and one has

$$H_\mathfrak{a} = \Delta^{X^\mathfrak{a}} \otimes_{X_\mathfrak{a}} I + I \otimes_{X_\mathfrak{a}} H^\mathfrak{a}.$$

An explicit expression for $H^\mathfrak{a}$ is

(10.2.4)
$$H^\mathfrak{a} = \Delta^{X^\mathfrak{a}} + \sum_{C \leq \mathfrak{a}} V^C \otimes_{X^C}^{X^\mathfrak{a}} I.$$

All this is a particularization of the general theory to the present situation and there is nothing new. But now $H^\mathfrak{a}$ has a further decomposition property related to the fact that in the N-body case $V^{X^\mathfrak{b}} \neq 0$ only if the cluster decomposition $\mathfrak{b}$ belongs to $\mathcal{C}$ (i.e. $\mathfrak{b}$ contains only one composite cluster). This decomposition expresses the fact that H^a describes $|a|$ clusters which do not interact between themselves; it is easily deduced from (10.2.4) and (10.2.2):

$$(10.2.5) \qquad H^a = \sum_{A \in \mathcal{C}(a)} \otimes_{A' \in \mathcal{C}(a)} H^{AA'},$$

where the tensor product is relative to the factorization (10.2.3) of $\mathcal{H}^a$ and $H^{AA'} = H^A$ if $A = A'$, $H^{AA'} = I_{\mathcal{H}^{A'}}$ if $A' \neq A$. Here H^A is the internal hamiltonian of the cluster A (acting in $\mathcal{H}^A$); it is well defined as a particular case of an operator of the form H^b, since A determines uniquely a cluster decomposition by the convention $\mathcal{C} \subset \mathcal{D}$ if card $A \geq 2$; if card $A = 1$ it is natural to set $H^A = 0$ in $\mathcal{H}^A = \mathbb{C}$.

The expression (10.2.5) allows one to describe the *spectral properties* of H^a in terms of those of H^A, $A \in \mathcal{C}(a)$, by using the results of Section 8.2. For example one will have:

(a) $\sigma(H^a) = \sum_{A \in \mathcal{C}(a)} \sigma(H^A)$,

(b) λ is an eigenvalue of H^a if and only if for each $A \in \mathcal{C}(a)$ there is an eigenvalue λ^A of H^A such that $\lambda = \sum_{A \in a} \lambda^A$. Moreover, the eigenvectors associated to the eigenvalues λ of H^a are just the vectors in the closed subspace of H^a spanned by the products $\otimes_{A \in a} e^A$, where $e^A \in \mathcal{H}^A$ is an eigenvector of H^A associated to the eigenvalue λ^A (and the eigenvalues $\{\lambda^A\}$ are such that $\sum_{A \in a} \lambda^A = \lambda$). In particular, the multiplicity $n^a(\lambda)$ of the eigenvalue λ of H^a is given by

$$n^a(\lambda) = \sum \prod_{A \in a} n^A(\lambda^A),$$

where the sum is over all $|a|$-tuples of real numbers $\{\lambda^A\}_{A \in a}$ such that $\sum_{A \in a} \lambda^A = \lambda$. The eigenvectors of H^A are often called *bound states* of H^A.

The *thresholds* of a quantum mechanical N-body system are real numbers of the form $\sum_{A \in \mathcal{C}(a)} \lambda^A$, where $a \in \mathcal{D}$, $|a| \geq 2$ and where λ^A denotes an eigenvalue of the operator H^A. It is clear that the correspondence $\mathcal{D} \ni a \mapsto X^a \in \mathcal{L}$ determines a bijection between $\mathcal{D}_k$ and $\mathcal{L}_k$ as well as between $\mathcal{D}(k)$ and $\mathcal{L}(k)$. The k-*body threshold* τ_k is defined to be the minimal energy at which the break-up of the system into k clusters becomes possible; more precisely:

$$(10.2.6) \qquad \tau_k = \inf_{|a|=k} \sigma(H_a) = \inf_{|a|=k} \sigma(H^a).$$

We have $\tau_1 \leq \tau_2 \leq \tau_3 \leq \cdots \leq \tau_{N-1} \leq \tau_N = 0$, where $\tau_1 = \inf \sigma(H)$. Furthermore, Theorem 9.4.2 implies that $[\tau_2, \infty) = \sigma_{\text{ess}}(H)$.

10.2.2. Let us make the connection between our presentation of the *Weinberg-van Winter equation* and the diagrammatic technique used in the N-body literature (see e.g. [Cb1] or [RS]). Since V_C is Δ-bounded with relative bound zero, we have $\|V_C(z - \Delta)^{-1}\| < (\text{card}\,\mathcal{C})^{-1}$ if $z \in \mathbb{C}$ is sufficiently far from $\mathbb{R}_+ = \sigma(\Delta)$. Then the Neumann series for $(z - H)^{-1}$ is absolutely convergent in norm and may be written as

$$(10.2.7) \quad (z - H)^{-1} = (z - \Delta)^{-1} +$$

$$+ (z - \Delta)^{-1} \sum_{n=1}^{\infty} \sum_{C_1, \ldots, C_n \in \mathcal{C}} V_{C_1}(z - \Delta)^{-1} \ldots V_{C_n}(z - \Delta)^{-1}.$$

A term in the double sum is uniquely specified by a sequence $(C_1, C_2, \ldots, C_n)$ with $n = 1, 2, \ldots$ and $C_i \in \mathcal{C}$. Physicists like to represent such sequences in terms of graphs. For this one draws N horizontal lines (representing the particles $1, 2, \ldots, N$) and n vertical lines (representing the clusters $C_1, \ldots, C_n$) and marks with a dot the intersection of the j-th horizontal line with the k-th vertical line for all values of j $(j = 1, \ldots, N)$ and k $(k = 1, \ldots, n)$ for which $j \in C_k$ (meaning that particle j is contained in the cluster C_k). As an example, consider the following graph:

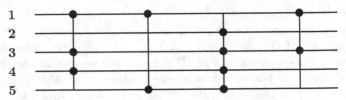

Here $N = 5$, $n = 4$, $C_1 = \{1, 3, 4\}$, $C_2 = \{1, 5\}$, $C_3 = \{2, 3, 4, 5\}$ and $C_4 = \{1, 3\}$.

To each pair of dots on the same horizontal or vertical line of a graph one may associate a *segment*, namely the part of the line between the two dots. Then each graph Γ determines uniquely a cluster decomposition $\mathfrak{a}(\Gamma)$ as follows: i and j $(1 \leq i < j \leq N)$ belong to the same cluster in $\mathfrak{a}(\Gamma)$, i.e. $i \sim_{\mathfrak{a}(\Gamma)} j$, if and only if the corresponding horizontal lines can be joined by a (connected) sequence of such segments.

A *chain* starting from $\mathfrak{a}_{\min}$ will be a family $(\mathfrak{a}_0, \mathfrak{a}_1, \ldots, \mathfrak{a}_k)$ of cluster decompositions such that $\mathfrak{a}_0 = \mathfrak{a}_{\min}$ and $\mathfrak{a}_{i-1} \leq \mathfrak{a}_i$, $\mathfrak{a}_{i-1} \neq \mathfrak{a}_i$ for each $i = 1, \ldots, k$. Each graph Γ determines uniquely such a chain: for $j = 1, \ldots, n-1$, let Γ_j be the subgraph of Γ to the left of a vertical cut on Γ between its j-th and $(j+1)$-st vertical lines, and let $\mathfrak{b}_j = \mathfrak{a}(\Gamma_j)$ be the cluster decomposition associated to the graph Γ_j by the prescription given above. Also set $\mathfrak{b}_0 = \mathfrak{a}_{\min}$ and $\mathfrak{b}_n = \mathfrak{a}(\Gamma)$. Clearly $\mathfrak{b}_{j-1} \leq \mathfrak{b}_j$ for each $j = 1, \ldots, n$, and the chain $(\mathfrak{a}_0, \mathfrak{a}_1, \ldots, \mathfrak{a}_k)$ is obtained from the sequence $(\mathfrak{b}_0, \mathfrak{b}_1, \ldots, \mathfrak{b}_n)$ by leaving out those $\mathfrak{b}_j$ that satisfy $\mathfrak{b}_{j-1} = \mathfrak{b}_j$. In the example given above, one has $\mathfrak{a}_0 = \mathfrak{a}_{\min}$, $\mathfrak{a}_1 = \{\{1, 3, 4\}, \{2\}, \{5\}\}$, $\mathfrak{a}_2 = \{\{1, 3, 4, 5\}, \{2\}\}$ and $\mathfrak{a}_3 = \mathfrak{a}(\Gamma) = \{\{1, 2, 3, 4, 5\}\} = \mathfrak{a}_{\max}$; this particular graph is connected, i.e. $\mathfrak{a}(\Gamma) = \mathfrak{a}_{\max}$.

We set $\mathscr{T}(\mathfrak{a}) = \mathscr{T}(X^{\mathfrak{a}})$, with $\mathscr{T}(X^{\mathfrak{a}})$ defined as in (9.2.14). Then we have $\mathscr{T}(\mathfrak{a})\mathscr{T}(\mathfrak{b}) \subset \mathscr{T}(\mathfrak{a} \vee \mathfrak{b})$. It is clear that $V_C(z - \Delta)^{-1} \in \mathscr{T}(C)$. Hence a term $V_{C_1}(z - \Delta)^{-1} \ldots V_{C_n}(z - \Delta)^{-1}$ in the expansion (10.2.7) belongs to $\mathscr{T}(\mathfrak{a}(\Gamma))$ (in particular, a non-zero term associated to a graph Γ is a compact operator in $\mathscr{H}$ if and only if Γ is connected). Since $\mathscr{T}(\mathfrak{a})$ is norm-closed for each $\mathfrak{a}$ and the series (10.2.7) is convergent in norm, it follows that the sum of all terms associated to graphs Γ such that $\mathfrak{a}(\Gamma) = \mathfrak{a}$ belongs to $\mathscr{T}(\mathfrak{a})$. By virtue of Proposition 8.4.5, the $\mathfrak{a}$-connected component $\mathscr{P}(\mathfrak{a})[R]$ of $R \equiv (z - H)^{-1}$ is the sum over all $\mathfrak{a}$-connected graphs (i.e. those graphs Γ that satisfy $\mathfrak{a}(\Gamma) = \mathfrak{a}$). This establishes the connection between the physicists' presentation of the WVW equation and that of our §8.4.5. Moreover, it explains the term "$\mathfrak{a}$-connected component of R" adopted in Section 8.4 for $\mathscr{P}(\mathfrak{a})[R]$.

Bibliography

[A]   S. Agmon, *Lectures on Exponential Decay of Solutions of Second Order Elliptic Equations*, Princeton Univ. Press, Princeton, 1982.

[ABG1]   W.O. Amrein, A. Boutet de Monvel and V. Georgescu, *Notes on the N-body Problem, I & II*, Preprints UGVA-DPT 1988/11-598a & 1991/04-178, Université de Genève, 1988 & 1991.

[ABG2]   _____, On Mourre's Approach to Spectral Theory, *Helv. Phys. Acta* **62** (1989), 1–20.

[ABG3]   _____, Commutators of Schrödinger Hamiltonians and Applications in Scattering Theory, *Lett. Math. Phys.* **18** (1989), 223–228.

[ABG4]   _____, Semicompactness and Spectral Analysis of N-body Hamiltonians, *C.R. Acad. Sci. Paris, Sér. I* **307** (1988), 813–818.

[AG]   N. Aronszajn and E. Gagliardo, Interpolation Spaces and Interpolation Methods, *Ann. Mat. Pura Appl.* **68** (1965), 51–117.

[AH]   S. Agmon and L. Hörmander, Asymptotic Properties of Solutions of Differential Equations with Simple Characteristics, *J. Analyse Math.* **30** (1976), 1–38.

[Ai]   M. Aigner, *Combinatorial Theory*, Springer, Berlin, 1979.

[AR]   R. Abraham and J. Robbin, *Transversal Mappings and Flows*, Benjamin, New York, 1967.

[As]   G. Arsu, Spectral Analysis for Simply Characteristic Operators by Mourre's Method: I, in *Linear Operators in Function Spaces*, Timisoara 1988, *Operator Theory: Advances and Applications*, vol. 43, Birkhäuser Verlag, Basel, 1990, 89–99; II, Preprint INCREST no. 37 (1990).

[Ba]   N. Bary, *A Treatise on Trigonometric Series*, Pergamon Press, Oxford, 1964.

[BB]   P.L. Butzer and H. Berens, *Semi-Groups of Operators and Approximation*, Springer, Berlin, 1967.

[Bd]   G. Bourdaud, *Analyse fonctionnelle dans l'espace euclidien*, Publ. Mathématiques de l'Université Paris VII, Paris.

[Be]   C. Berge, *Principles of combinatorics*, Academic Press, 1971.

[Ber]   H. Berens, *Interpolationsmethoden zur Behandlung von Approximationsprozessen auf Banachräumen*, Lecture Notes in Math., vol. 64, Springer, Berlin, 1968.

[BG1]   A. Boutet de Monvel and V. Georgescu, Graded C^*-Algebras and Many-Body Perturbation Theory. I: The N-body problem, *C.R. Acad. Sci. Paris, Sér. I* **312** (1991), 477–482.

[BG2]   _____, Graded C^*-Algebras in the N-Body Problem, *J. Math. Phys.* **32** (1991), 3101–3110.

[BG3]   _____, Graded C^*-Algebras and Many-Body Perturbation Theory. II: The Mourre estimate, in "Méthodes semi-classiques", vol. 2, Colloque international, Nantes, juin 1991, *Astérisque*, vol. 210, Société Mathématique de France, 1992, pp. 75–96.

W. O. Amrein et al., *C₀-Groups, Commutator Methods and Spectral Theory of N-Body Hamiltonians*, Modern Birkhäuser Classics,
DOI: 10.1007/978-3-0348-0733-3, © Springer Basel 1996

[BG4]   _____, Graded C^*-Algebras Associated to Symplectic Spaces and Spectral Analysis of Many-Channel Hamiltonians, in *Dynamics of complex and irregular systems*, Bielefeld encounters in Math. & Physics VIII, 1991, World Scientific, 1993, pp. 22–66.

[BG5]   _____, Locally Conjugate Operators, Boundary Values of the Resolvent and Wave Operators, *C.R. Acad. Sci. Paris, Sér. I* **313** (1991), 13–18.

[BG6]   _____, Spectral Theory and Scattering Theory by the Conjugate Operator Method, *Algebra i Analyz* **4** (1992), no. 3, 73–116, [=*St Petersburg Math. J.* **4** (1993), no. 3, 469–501].

[BG7]   _____, Some Developments and Applications of the Abstract Mourre Theory, in "Méthodes semi-classiques", vol. 2, Colloque international, Nantes, juin 1991, *Astérisque*, vol. 210, Société Mathématique de France, 1992, 27–48.

[BG8]   _____, Limiting Absorption Principle for Long-Range Perturbations of Pseudo-Differential Operators, in *Advances in Dynamical Systems and Quantum Physics*, Capri 1993, Conference in honour of G. Dell'Antonio, World Scientific, 1994 [Preprint BiBoS no. 616/2/94].

[BG9]   _____, The Method of Differential Inequalities, in *Recent Developments in Quantum Mechanics, Brasov International Summer School, 1989*, eds. A. Boutet de Monvel & al., Math. Physics Studies 12, Kluwer Academic Publishers (1991), pp. 279–298.

[BG10]   _____, Boundary Values of the Resolvent of a Self-adjoint Operator: Higher Order Estimates, in *Proceedings of the First Ukrainian- French- Romanian School "Algebraic and Geometric Methods in Mathematical Physics"*, Kaciveli, Crimea, September 1993, eds. A. Boutet de Monvel and V.A. Marchenko, Math. Physics Studies, Kluwer Academic Publishers (1995).

[BGM1]   A. Boutet de Monvel, V. Georgescu and M. Mantoiu, Mourre Theory in a Besov Space Setting, *C.R. Acad. Sci. Paris, Sér. I* **310** (1990), 233–237.

[BGM2]   _____, Locally Smooth Operators and the Limiting Absorption Principle for N-Body Hamiltonians, *Reviews in Math. Physics* **5** (1993), no. 1, 105–189.

[BGS]   A. Boutet de Monvel, V. Georgescu and A. Soffer, N-body Hamiltonians with Hard Core Interactions, *Reviews in Math. Physics* **6** (1994), no. 3, 1–82.

[Bi]   P. Billingsley, *Convergence of Probability Measures*, Wiley, New York, 1968.

[BK]   Yu.A. Brudnyi and N.Ya. Krugljak, *Interpolation Functors and Interpolation Spaces*, North-Holland Math. Library, Amsterdam, 1991.

[BL]   J. Bergh and J. Löfström, *Interpolation Spaces*, Springer, Berlin, 1976.

[BM1]   A. Boutet de Monvel and D. Manda, The Conjugate Operator Method for Strongly Singular Operators, *Letters in Math. Phys.* **27** (1993), 1–14.

[BM2]   _____, Spectral and Scattering Theory for Wave Propagation in Perturbed Stratified Media, *J. Math. Analysis and Applications* **191** (1995), 137–167.

[BMP1]   A. Boutet de Monvel, D. Manda and R. Purice, The Commutator Method for Form-Relatively Compact Perturbations, *Letters in Math. Phys.* **22** (1991), 211–223.

[BMP2]   _____, Limiting Absorption Principle for the Dirac operator, *Annales Inst. Henri Poincaré, Physique théorique* **58** (1993), 413–431.

[Bo1]   N. Bourbaki, *Algèbre*, chap. 1–4, Hermann, Paris, 1970.

[Bo2]   _____, *Topological Vector Spaces*, Springer, Berlin, 1987.

[Bo3]   _____, *Intégration*, Hermann, Paris, 1967.

[Bo4]   _____, *Topologie générale*, chap. 1–4 & chap. 5–10, Hermann, Paris, 1971& 1974.

[BP1]   A. Boutet de Monvel and R. Purice, The Conjugate Operator Method for Magnetic Hamiltonians, *C.R. Acad. Sci. Paris, Sér. I* **316** (1990), 239–244.

[BP2]   _____, Limiting Absorption Principle for Schrödinger Hamiltonians with Magnetic Field, *Comm. Partial Differential Equations* **19** (1994), 89–118.

[BP3]   _____, On Regularity with Respect to an Automorphism Group and Applications, Minneapolis Conference, 1994, to appear.

[Br]   O. Bratteli, *Derivations, Dissipations and Group Actions on C^*-Algebras*, Lecture Notes in Math., vol. 1229, Springer, Berlin, 1986.

[BR]   O. Bratteli and D.W. Robinson, *Operator Algebras and Quantum Statistical Mechanics*, Springer, New York, 1979.

[BS]   C. Bennett and R. Sharpley, *Interpolation of Operators*, Academic Press, Boston, 1988.

[BuS]   P.L. Butzer and K. Scherer, *Approximationsprozesse und Interpolationsmethoden*, Bibliographisches Institut Mannheim/Zürich, 1968.

[BW]   H. Baumgaertel and M. Wollenberg, *Mathematical Scattering Theory*, Akademie-Verlag, Berlin, 1983.

[C1]   A.P. Calderón, Lebesgue Spaces of Differentiable Functions and Distributions, *Proc. Symp. Pure Math.*, vol. IV, Amer. Math. Soc., 1961, pp. 33–49.

[C2]   _____, Intermediate Spaces and Interpolation, the Complex Method, *Studia Math.* **24** (1964), 113–190.

[Cb1]   J.-M. Combes, Relatively Compact Interactions in Many-Particle Systems, *Comm. Math. Phys.* **12** (1969), 283–295.

[Cb2]   _____, Properties of Some Connected Kernels in Multiparticle Systems, *J. Math. Phys.* **12** (1971), 1719–1731.

[CFKS]   H.L. Cycon, R.G. Froese, W. Kirsch, and B. Simon, *Schrödinger Operators with Application to Quantum Mechanics and Global Geometry*, Springer, Berlin, 1987.

[Cm]   F. Combes, Représentations des groupes localement compacts et applications aux algèbres d'opérateurs, Séminaire d'analyse fonctionnelle, Orléans 1973/74, *Astérisque*, vol. 55, Société Mathématique de France, Paris, 1978.

[CP]   H.L. Cycon and P. Perry, Local Time Decay of High Energy Scattering States for the Schrödinger Equation, *Math. Zeitschrift* **188** (1984), 125–142.

[Cr]   H.O. Cordes, On Compactness of Commutators of Multiplications and Convolutions, and Boundedness of Pseudodifferential Operators, *J. Funct. Analysis* **18** (1975), 115–131.

[D]   J. Dieudonné, *Foundations of Modern Analysis*, Vols. 1 and 2, Academic Press, New York, 1960.

[Da]   E.B. Davies, *One-Parameter Semigroups*, Academic Press, London, 1980.

[DBHS]   S. De Bièvre, P.D. Hislop and I.M. Sigal, Scattering Theory for the Wave Equation on Non-compact Manifolds, *Rev. Math. Phys.* **4** (1992), 575–619.

[DBP]   S. De Bièvre and D.W. Pravica, Spectral Analysis for Optical Fibres and Stratified Fluids, I, The Limiting Absorption Principle, *J. Funct. Anal.* **98** (1991), 404–436.

[De1]   J. Derezinsky, Algebraic Approach to the *N*-Body Long Range Scattering, *Rev. Math. Phys.* **3** (1991), 1–62.

[De2]   _____, *N*-Body Observables in the Calkin Algebra, *Trans. Amer. Math. Soc.* **332** (1992), 571–582.

[De3]   _____, A New Proof of the Propagation Theorem for *N*-Body Quantum Systems, *Comm. Math. Phys.* **122** (1989), 203–231.

[DF]   A. Defant and K. Floret, *Tensor Norms and Operator Ideals*, Math. Studies, vol. 176, North-Holland, Amsterdam, 1993.

[DHSV]   P. Deift, W. Hunziker, B. Simon, and E. Vock, Pointwise Bounds on Eigenfunctions and Wave Packets in *N*-Body Quantum Systems, *Comm. Math. Phys.* **64** (1978), 1–34.

[Di1]   J. Dixmier, *Les algèbres d'opérateurs dans l'espace Hilbertien*, Gauthier-Villars, Paris, 1969.

[Di2]   _____, *C*-algebras*, North-Holland, Amsterdam, 1977.

[DM]   C. Dellacherie and P.-A. Meyer, *Probabilités et Potentiel*, chap. I–IV, Hermann, Paris, 1975.

[Do]   W.F. Donoghue, The Interpolation of Quadratic Norms, *Acta Math.* **118** (1967), 251–270.

[E]   V. Enss, Quantum Scattering Theory for Two and Three-Body Systems for Potentials of Short and Long Range, in *Schrödinger Operators*, Lecture Notes in Math., vol. 1159, Springer, Berlin, 1985, 39–176.

[Fe]   H. Federer, *Geometric Measure Theory*, Springer, Berlin, 1969.

[FH1]   R.G. Froese and I. Herbst, A New Proof of the Mourre Estimate, *Duke Math. J.* **49** (1982), 1075–1085.

[FH2]   ———, Exponential Bounds and Absence of Positive Eigenvalues for N-Body Schrödinger Operators, *Comm. Math. Phys.* **87** (1982), 429–447.

[FHi]   R.G. Froese and P. Hislop, Spectral Analysis of Second-Order Elliptic Operators on Non-compact Manifolds, *Duke Math. J.* **58** (1989), 103–128.

[FHP]   R.G. Froese, P. Hislop, and P. Perry, A Mourre Estimate and Related Bounds for Hyperbolic Manifolds with Cusps of Nonmaximal Rank, *J. Funct. Anal.* **98** (1991), 292–310.

[FJW]   M. Frazier, B. Jawerth, and G. Weiss, *Littlewood-Paley Theory and the Study of Function Spaces*, Regional Conference Series in Mathematics, vol. 79, Amer. Math. Soc., Providence R.I., 1991.

[Ga]   J.B. Garnett, *Bounded Analytic Functions*, Academic Press, Orlando, 1981.

[Ge]   Ch. Gérard, The Mourre Estimate for Regular Dispersive Systems, *Ann. Inst. H. Poincaré, Physique théorique* **54** (1991), 59–88.

[Gr1]   P. Grisvard, Commutativité de deux foncteurs d'interpolation et applications, *J. Math. Pures et Appliquées* **45** (1966), 143–290.

[Gr2]   ———, Semi-groupes faiblement continus et interpolation, *C.R. Acad. Sci. Paris, Sér. A* **259** (1964), 27–29.

[Gr3]   ———, Interpolation non commutative, *Atti. Acad. Naz. Lincei Rend. Cl. Sci. Fis. Mat. Natur.* **52** (1972), 11–15.

[Gu]   M. Guenin, On the Derivation and Commutation of Operator Functionals, *Helv. Phys. Acta* **41** (1968), 75–76.

[H]   L. Hörmander, *The Analysis of Linear Partial Differential Operators*, Vols. I-IV, Springer, Berlin, 1983, 1985.

[Hi]   P.D. Hislop, The Geometry and Spectra of Hyperbolic Manifolds, *Proc. Indian Acad. Sci., Math. Sci.,* **104** (1994), 715–777.

[HK]   Ph. Hartmann and R. Kershner, The Structure of Monotone Functions, *Amer. J. Math.* **59** (1937), 809–822.

[HLP]   G.H. Hardy, J.E. Littlewood and G. Polya, *Inequalities*, Cambridge Univ. Press, Cambridge, 1952.

[Hm]   Ph. Hartmann, *Ordinary Differential Equations*, Wiley, New York, 1964.

[HP]   E. Hille and R.S. Phillips, *Functional Analysis and Semi-Groups*, Amer. Math. Soc., Providence R.I., 1957.

[HS]   B. Helffer and J. Sjöstrand, Equation de Schrödinger avec champ magnétique et équation de Harper, *Proceedings of a Conference on Operator Theory*, Lecture Notes in Phys., vol. 345, Springer, Berlin, 1989, pp. 118–197.

[If]   V. Iftimie, Le principe d'absorption limitée pour une classe d'opérateurs formellement hypoelliptiques, *Rev. Roumaine Math. Pures et Appl.* **30** (1985), 345–361.

[IP]   V. Iftimie and R. Purice, Hamiltoniens à N corps avec champs magnétiques très singuliers du type courte portée, *Letters Math. Phys.* **33** (1995), 127–138.

[Is]   H. Isozaki, On N-body Schrödinger Operators, *Proc. Indian Acad. Sci., Math. Sci.,* **104** (1994), 667–705.

[Iw]   H. Iwashita, Spectral Theory for Symmetric Systems in an Exterior Domain, I, *Tsukuba J. Math.* **11** (1987), 241–256 & II, *J. Funct. Anal.* **82** (1989), 91–112.

[J1]   A. Jensen, Propagation Estimates for Schrödinger-Type Operators, *Trans. Amer. Math. Soc.* **291** (1985), 129–144.

[J2]   A. Jensen, High Energy Resolvent Estimates for Generalized Many-Body Schrödinger Operators, *Publi. RIMS, Kyoto University* **25** (1989), 155–167.

[J3]   A. Jensen, Commutator Methods and Schrödinger Operators, in *"Rigorous Results in Quantum Dynamics"*, Liblice, Czechoslovakia, June 1990, J. Dittrich and P. Exner (eds.), World Scientific, Singapore, 1991, 3–15.

[J4]   A. Jensen, Scattering Theory for Stark Hamiltonians, *Proc. Indian Acad. Sci., Math. Sci.,* **104** (1994), 599–653.

[JLN]   J.M. Jauch, R. Lavine and R.G. Newton, Scattering into Cones, *Helv. Phys. Acta* **45** (1972), 325–330.

[JM]   P.E.T. Joergensen and R.T. Moore, *Operator Commutation Relations*, Reidel, Dordrecht, 1984.

[JMP]   A. Jensen, E. Mourre and P. Perry, Multiple Commutator Estimates and Resolvent Smoothness in Quantum Scattering Theory, *Ann. Inst. H. Poincaré, Physique théorique* **41** (1984), 207–225.

[JP]   A. Jensen and P. Perry, Commutator Methods and Besov Space Estimates for Schrödinger Operators, *J. Operator Theory* **14** (1985), 181–189.

[JT]   S. Janson and M.H. Taibleson, I Teoremi di Rappresentazione di Calderón, *Rend. Sem. Mat. Univ. Politecn. Torino* **39** (1981), 27–35.

[JW]   K. Jörgens and J. Weidmann, *Spectral Properties of Hamiltonian Operators*, Lecture Notes in Math., vol. 313, Springer, Berlin, 1973.

[K1]   T. Kato, *Perturbation Theory for Linear Operators*, Springer, Berlin, 1966.

[K2]   ———, Trotter's Product Formula for an Arbitrary Pair of Self-adjoint Contraction Semigroups, in *Topics in Functional Analysis*, eds I. Gohberg, M. Kac, *Advances in Math. Suppl. Studies* **3** (1978), 185–195.

[Km]   H. Komatsu, Fractional Powers of Operators, I, *Pacific J. Math.* **19** (1966), 285–346; II, *Pacific J. Math.* **21** (1967), 89–111; III, *J. Math. Soc. Japan* **21** (1969), 205–220; IV, *J. Math. Soc. Japan* **21** (1969), 221–228.

[Ko]   A.Y. Konstantinov, The Invariance Principle for Wave Operators, *Dokl. Akad. Nauk* **281** (1985), 1041–1044.

[KPR]   K.L. Kowalski, W.N. Polyzou and E.F. Redish, Partition Combinatorics and Multiparticle Scattering Theory, *J. Math. Phys.* **9** (1981), 1965–1982.

[KPS]   S.G. Krein, Ju.I. Petunin and E.M. Semenov, *Interpolation of Linear Operators*, Amer. Math. Soc., Translations of Mathematical Monographs, vol. 54, 1982.

[KR]   R.V. Kadison and J.R. Ringrose, *Fundamentals of the Theory of Operator Algebras*, Vols. 1 & 2, Academic Press, New York, 1983 & 1986.

[La]   R. Lavine, Absolute Continuity of Positive Spectrum for Schrödinger Operators with Long-Range Potentials, *J. Funct. Anal.* **12** (1973), 30–54.

[LP]   J.-L. Lions and J. Peetre, Sur une classe d'espaces d'interpolation, *Publ. Math. I.H.E.S.* **19** (1964), 5–68.

[M1]   E. Mourre, Absence of Singular Continuous Spectrum for Certain Self-Adjoint Operators, *Comm. Math. Phys.* **78** (1981), 391–408.

[M2]   ———, Opérateurs conjugués et Propriétés de Propagation, *Comm. Math. Phys.* **91** (1983), 279–300.

[Ma]   N. Mandache, On C_0-groups in Hilbert Spaces, *C.R. Acad. Sci. Paris, Sér. I* **316** (1993), 873–878.

[Mc]   A. McIntosh, Functions and Derivations of C*-Algebras, *J. Funct. Anal.* **30** (1978), 264–275.

[Me]   Y. Meyer, *Ondelettes*, Hermann, Paris, 1990.

[MP]   B. Maurey and G. Pisier, Séries de variables aléatoires vectorielles indépendantes et propriétés géométriques des espaces de Banach, *Studia Math.* **58** (1976), 45–90.

[MS]   E. Mourre and I.M. Sigal, *Phase Space Analysis and Scattering Theory for N-Particle Systems. I. Single Channel Systems*, Preprint, 1983.

[Mt]   M. Mantoiu, Weighted Estimates in the Framework of Mourre Theory, *C.R. Acad. Sci. Paris, Sér. I* **313** (1991), 715–720.

[Mu]   P. Muthuramalingam, A Conjecture for some Partial Differential Operators in $L^2(\mathbb{R}^n)$, *Proc. Indian Acad. Sci., Math. Sci.*, **104** (1994), 705–715.

[MuS]   P. Muthuramalingam and K.B. Sinha, Existence and Completeness of Wave Operators for the Dirac Operator in an Electromagnetic Field with Long Range Potentials, *J. Indian Math. Soc.*, New Ser. **50** (1986), 103–130.

[N]   R. Narasimhan, *Analysis on Real and Complex Manifolds*, North-Holland, Amsterdam, 1968.

[Ne]   G. Nenciu, Distinguished Self-adjoint Extensions for Dirac Operators with Potentials Dominated by Multicenter Coulomb Potentials, *Helv. Phys. Acta* **50** (1977), 1–3.

[P]   D.B. Pearson, A Commutator Approach to the Limiting Absorption Principle, *Helv. Phys. Acta* **62** (1989), 21–41.

[P1]   J. Peetre, *New Thoughts on Besov Spaces*, Mathematics Series, vol. I, Duke Univ. Press, Durham, 1976.

[P2]   _____ , *A Theory of Interpolation of Normed Spaces*, Notas di Matematica, vol. 39, Rio de Janeiro, 1968.

[Pd]   G.K. Pedersen, *C^*-Algebras and their Automorphism Groups*, Academic Press, London, 1979.

[Pi]   G. Pisier, Un théorème sur les opérateurs linéaires entre les espaces de Banach qui se factorisent par un espace de Hilbert, *Ann. Sci. Ec. Norm. Sup.*, 4ème série **13** (1980), 23–43.

[Po]   W.N. Polyzou, Combinatorics, Partitions and Many-Body Physics, *J. Math. Phys.* **21** (1980), 506–513.

[Pr]   P. Perry, Propagation of States in Dilation Analytic Potentials and Asymptotic Completeness, *Comm. Math. Phys.* **81** (1981), 243–259.

[Ps]   R.T. Prosser, A Double Scale of Weighted L^2-Space, *Bull. Amer. Math. Soc.* **81** (1975), 615–618.

[PSS]   P. Perry, I.M. Sigal and B. Simon, Spectral Analysis of N-Body Schrödinger Operators, *Ann. Math.* **114** (1981), 519–567.

[Pu]   C.R. Putnam, *Commutation Properties of Hilbert Space Operators and Related Topics*, Springer, Berlin, 1967.

[RaS]   Ch. Radin and B. Simon, Invariant Domains for the Time-Dependent Schrödinger Equation, *J. Diff. Equa.* **29** (1978), 289–296.

[Ro]   D.L. Rogova, Error Bounds for Trotter-type Formulas for Self-Adjoint Operators, *Funct. Anal. and Its Applications* **27** (1993), 217–219.

[RS]   M. Reed and B. Simon, *Methods of Modern Mathematical Physics*, Vol. I, II, III & IV, Academic Press, London, 1972-1979.

[Ru]   W. Rudin, *Real and Complex Analysis*, Mc Graw-Hill, New York, 1966.

[S]   S. Saks, *Theory of the Integral*, Hafner, New York, 1947.

[Sa]   S. Sakai, *C^*-Algebras and W^*-Algebras*, Springer, Berlin, 1971.

[Sch]   L. Schwartz, *Théorie des distributions*, Hermann, Paris, 1969.

[Sche1]   M. Schechter, *Spectra of Partial Differential Operators, 2nd ed.*, North-Holland, Amsterdam, 1986.

[Sche2]   _____ , The Invariance Principle, *Comm. Math. Helv.* **54** (1979), 111–125.

[Sh1]   H.S. Shapiro, *Smoothing and Approximation of Functions*, Van Nostrand, New York, 1969.

[Sh2]   H.S. Shapiro, A Tauberian Theorem Related to Approximation Theory, *Acta Math.* **120** (1968), 279–292.

[Sh3]   H.S. Shapiro, *Topics in Approximation Theory*, Lecture Notes in Math., vol. 187, Springer, Berlin, 1971.

[Sh4]   H.S. Shapiro, Monotonic Singular Functions of High Smoothness, *Michigan Math. J.* **15** (1968), 265–275.

[Sim1]   B. Simon, Lower Semicontinuity of Positive Quadratic Forms, *Proc. Royal Soc. Edinburgh* **79** (1977), 267–273.

[Sim2]   _____ , Geometric Methods in Multiparticle Quantum Systems, *Comm. Math. Phys.* **55** (1977), 259–274.

[Sm]   K.T. Smith, *Primer of Modern Analysis*, Springer, New York, 1983.

[SS]   I.M. Sigal and A. Soffer, The N-Particle Scattering Problem: Asymptotic Completeness for Short Range Systems, *Ann. Math.* **125** (1987), 35–108.

[St1]   E.M. Stein, *Singular Integrals and Differentiability Properties of Functions*, Princeton Univ. Press, Princeton N.J., 1970.

[St2]   _____ , *Harmonic Analysis*, Princeton Univ. Press, Princeton N.J., 1993.

[SW]   E.M. Stein and G. Weiss, *Introduction to Fourier Analysis on Euclidean Spaces*, Princeton Univ. Press, Princeton N.J., 1971.

[Ta]   M. Takesaki, *Theory of Operator Algebras, I*, Springer, New York, 1979.

[TE]   B. Thaller and V. Enss, Asymptotic Observables and Coulomb Scattering for the Dirac Equation, *Annales Inst. H. Poincaré, Physique théorique* **45** (1986), 147–171.

[Ti]   E.C. Titchmarsh, *Theory of Fourier Integrals*, Clarendon Press, Oxford, 1948.

[Tm1]   H. Tamura, Principle of Limiting Absorption for N-Body Schrödinger Operators, *Lett. Math. Phys.* **17** (1989), 31–36.

[Tm2]   ──────, Spectral and Scattering Theory for Symmetric Systems of Non-Constant Deficit, *J. Funct. Anal.* **67** (1986), 73–104.

[Tr]   H. Triebel, *Interpolation Theory, Function Spaces, Differential Operators*, VEB Deutscher Verlag der Wissenschaften, Berlin, 1978.

[Vo]   V. Vogelsang, Remark on Essential Selfadjointness of Dirac Operators with Coulomb Potentials, *Math. Z.* **196** (1987), 517–521.

[W]   J. Weidmann, *Linear Operators in Hilbert Spaces*, Springer, New York, 1980.

[We1]   R. Weder, Spectral Analysis of Strongly Propagative Systems, *J. für reine angew. Math.* **354** (1984), 95–122.

[We2]   ──────, Spectral and Scattering Theory for Wave Propagation in Perturbed Stratified Media, Applied Math. Sc., vol. 87, Springer, Berlin, 1991.

[Wo]   P. Wojtaszczyk, *Banach Spaces for Analysts*, Cambridge Univ. Press, Cambridge, 1991.

[Wol]   M. Wollenberg, Wave Algebras and Scattering Theory, I & II, *ZIMM-Report* R-07/79 & R-10/79, Akad. Wiss. DDR, Berlin, 1979.

[WZ]   M. Weiss and A. Zygmund, A Note on Smooth Functions, *Nederl. Akad. Wetensch. Proc. Ser. A 62, Indag. Math.* **21** (1959), 52–58.

[Ya]   D.R. Yafaev, Remarks on Spectral Theory for Multiparticle Schrödinger Operators, *Zap. Nauch. Sem. LOMI* **133** (1984), 277–298.

[Y]   K. Yosida, *Functional Analysis*, Springer, Berlin, 1980.

[Z]   A. Zygmund, *Trigonometric Series I, II*, Cambridge Univ. Press, Cambridge, 1959.

[Zi]   W.P. Ziemer, *Weakly Differentiable Functions*, Springer, New York, 1989.

Notations

$(\cdot, \cdot)$, 1
$(\cdot, \cdot)_X$, 1
$\langle \cdot, \cdot \rangle$, 6, 29
$|\cdot|$, 1
$|\cdot|_X$, 1
$\|\cdot\| \equiv \|\cdot\|_{L^2(X)}$, 6
$\|\cdot\|_1 \sim \|\cdot\|_2$, 31
$\|\cdot\|_{BC^k}$, 3
$\|\cdot\|_{BC^\tau}$, 16
$\|\cdot\|_{C^k}$, 195
$\|\cdot\|_{C^k_{(s)}}$, 152
$\|\cdot\|_{\mathbf{E} \to \mathbf{F}}$, 32
$\|\cdot\|_{\theta, p}$, 38
$\|\cdot\|_{\mathbf{F}_m}$, 94
$\|\cdot\|_{\mathbf{F}_{-m}}$, 101
$\|\cdot\|_{L^p(X)}$, 6
$\|\cdot\|_{L^p_*(\mathbf{G})}$, 42
$\|\cdot\|_{\mathscr{M}^r}$, 216
$\|\cdot\|_{\mathscr{M}_w}$, 78
$\|\cdot\|_{\mathscr{N}}$, 137
$\|\cdot\|_{s,p}^{(\ell)}$, 200
$\|\cdot\|_{s,p}^{(m)}$, 127

$a < b$, 391
$a \leq b$, 391
$a \ll b$, 391
$a \vee b$, 391
$a \wedge b$, 391
α, 3
$|\alpha|$, 3
$\alpha!$, 3
$\binom{\alpha}{\beta}$, 3
$\alpha - \beta$, 3
A, 95
A^α, 95
A_j, 95
A^*, 120
A_x, 243
$\langle A \rangle^k$, 157
$A \cdot x$, 90
$\mathscr{A}$, 193, 381
$\mathfrak{a}$, 433
$|\mathfrak{a}|$, 433
$\mathfrak{a}_{\max}$, 434
$\mathfrak{a}_{\min}$, 434
ad_A^α, 194

B, 272
B_ε, 271, 301
$B_r^{s,p}(X)$, 172

$BC^k(X)$, 3
$BC(X)$, 3
$BC_u^k(\mathbb{R}^n)$, 247
$B(\mathbf{E})$, 30
$B(\mathbf{E}, \mathbf{F})$, 30
$B(\mathbf{F}_{-\infty})$, 154
$BH^\omega(\mathbb{R}^n)$, 162
$BO(Z)$, 360
$\mathbb{B}(X)$, 6
$\mathscr{B} = B(\mathbf{F}', \mathbf{F}'')$, 193
$\mathscr{B}_{-\infty}$, 193
$\mathscr{B}_{s,p}$, 214
$\mathscr{B}^k$, 214
$\mathscr{B}^{s,p}$, 214
$\mathscr{B}_u$, 191
$\mathscr{B} \otimes \mathscr{C}$, 373
$\mathscr{B} \odot \mathscr{C}$, 373
$\mathscr{B} \otimes_Y \mathscr{C}$, 403
$\beta \leq \alpha$, 3

$C(X)$, 3
$C_0(X)$, 3
$C_\infty(X)$, 3
$C^k(X)$, 3
$C_0^k(X)$, 3
$C^k(A)$, 242
$C_\infty^k(X)$, 3
$C_\infty(X; \mathbf{E})$, 374
$C_{\mathrm{pol}}^k(X)$, 3
$C_{(s)}^k(\mathbb{R}^n)$, 152
$C_u^k(A)$, 242
$C_u^k(\mathscr{A})$, 381
$C^k(A; \mathbf{F})$, 195
$C^k(\mathbf{F})$, 195
$C^k(A', A''; \mathbf{F}', \mathbf{F}'')$, 192, 195
$C_u^k(A', A''; \mathbf{F}', \mathbf{F}'')$, 192, 195
$\overset{\circ}{C}{}^k(\mathbb{R}^n)$, 152
$\overset{\circ}{C}{}_{(s)}^k$, 152
$\mathscr{C}$, 433
$\mathscr{C}^X$, 379
$\mathscr{C}_I$, 362
$\mathscr{C}(a)$, 392
$\mathscr{C}\text{-}\sigma_{\mathrm{ess}}(H)$, 393
$\widetilde{\mathscr{C}}$, 393
$\widehat{\mathscr{C}}$, 393
$\mathscr{C}^{s,p}(A)$, 242
$\mathscr{C}^{s,p}(A; \mathbf{F})$, 200
$\mathscr{C}^{s,p}(A', A''; \mathbf{F}', \mathbf{F}'')$, 192, 200
$\mathscr{C}^{s,p}(\mathbf{F})$, 200

W. O. Amrein et al., *C_0-Groups, Commutator Methods and Spectral Theory of N-Body Hamiltonians*, Modern Birkhäuser Classics,
DOI: 10.1007/978-3-0348-0733-3, © Springer Basel 1996

Index

W. O. Amrein et al., *C₀-Groups, Commutator Methods and Spectral Theory of N-Body Hamiltonians*, Modern Birkhäuser Classics,
DOI: 10.1007/978-3-0348-0733-3, © Springer Basel 1996